S0-CAS-651

Organic Syntheses via Metal Carbonyls

Volume 2

Organic Syntheses via Metal Carbonyls

Volume 2

Edited by

IRVING WENDER

Pittsburgh Energy Research Center
U.S. Energy Research and
Development Administration
Pittsburgh, Pennsylvania

and

PIERO PINO

Swiss Federal Institute of Technology
Zurich, Switzerland

A Wiley-Interscience Publication
JOHN WILEY & SONS

New York · London · Sydney · Toronto

Library of Congress Cataloging in Publication Data (Revised)

Wender, Irving.

Organic syntheses via metal carbonyls.

Vol. 2 has imprint: New York, Wiley.

"A Wiley-Interscience publication."

Includes bibliographies and indexes.

1. Metal carbonyls. 2. Chemistry, Organic—Synthesis. I. Pino, Piero, joint author. II. Title.

QD411.W4 547'.2 67-13965

ISBN 0-471-93367-8 (v. 2)

Printed in the United States of America

10 9 8 7 6 5 4 3 2 1

Preface

Volume 1 of this series, published in 1968, covered the preparation, structure and properties of the metal carbonyls and their derivatives and discussed their use in three types of organic synthesis. This second volume is devoted entirely to a whole range of organic syntheses in which these transition metal compounds are used chiefly as catalysts and occasionally as stoichiometric reagents.

The metal carbonyls have for years been the most versatile and the most investigated homogeneous catalysts in nonaqueous phases. Studies of the properties of and organic syntheses via metal carbonyls since the late 1940s have led to concepts such as π-complex formation between unsaturated compounds and transition metal compounds, ligand and substrate insertions, associative and dissociative mechanisms, and oxidative addition and reductive eliminations. These reaction steps and mechanisms have usually been presented in reports of specific investigations of the hydroformylation (oxo) reaction but were then designated by different names.

Indeed, work with the metal carbonyls provided the impetus for the very rapid growth of the whole field of homogeneous catalysis that occurred during 1955–1970. In this period, it was shown that the above concepts had wide application in numerous homogeneous reactions related to those catalyzed by metal carbonyls; these included the vital fields of homogeneous hydrogenation, Ziegler–Natta type homogeneous polymerization, and olefin oxidation.

Interestingly, after opening the way to understanding homogeneous catalysis by transition metals, these same metal carbonyls and derivatives have, because of world events, become of paramount importance today. With a high price of oil and a shortage of natural gas, the countries with abundant indigenous or available coal reserves must fall back on this natural resource. This is resulting in the revival of carbon monoxide chemistry in general and carbonylation in particular, connected with attempts to synthesize fuels, petrochemicals (including olefins), and fine

chemicals starting with carbon monoxide or mixtures of carbon monoxide and hydrogen (synthesis gas).

The way to fuels and chemicals in the future will likely start with the conversion of coal to synthesis gas. Synthesis gas can be converted to liquid fuels via the Fischer–Tropsch reaction and other processes, and to methanol and hence to formaldehyde. It also will furnish the hydrogen for ammonia production. We can now go from methanol to acetic acid via an improved homogeneously catalyzed process. It is possible that the ethylene of the future will be produced via ethanol obtained by the homogeneous, metal carbonyl catalyzed reaction of methanol with synthesis gas (a reaction discovered in 1951, incidently). And it may well be that a significant part of our petrochemical and fuel economy in the last fifteen years of this century will rely on synthesis gas from coal. Therefore, the role of metal carbonyls and their derivatives, so well known as catalysts for reactions involving carbon monoxide, synthesis gas, and molecular hydrogen—already of great importance—will grow even larger in the years to come.

There is, of course, much research still to be done in catalysis by metal carbonyls, despite its relatively long history. Chemists must learn to use these and related metal complexes to synthesize organic compounds via homogeneous catalysis directly from carbon monoxide and hydrogen. Until recently, this direct synthesis could only be accomplished by using heterogeneous catalysts. There are many advantages in using homogeneous catalysts: milder conditions, greater selectivity, and less sensitivity to poisons. A breakthrough has been announced in the synthesis of ethylene glycol and similar alcohols from synthesis gas while using soluble, transition metal complexes. Further problems to be solved include the homogeneously catalyzed synthesis of methanol and methane from synthesis gas and the carbonylation of saturated carbon–hydrogen bonds.

We hope this second volume will furnish some of the necessary background to synthetic organic chemists and in general to researchers entering the field of carbon monoxide chemistry. It should also point the way to specific instances and general areas that need further research in the fields of carbon monoxide chemistry and homogeneous catalysis.

The editors contemplated compiling a list of known reactions of organic substrates catalyzed by metal carbonyls and their derivatives. This would obviously be of great value, especially if classified according to substrate, catalyst, other reactants, and references. But the number of these reactions is so large and the almost daily appearance of new homogeneously catalyzed reactions precluded this endeavor. We are thus most aware that these volumes merely scratch the surface by including only the more well-known reactions.

Due to the breadth of the field, the authors were asked for critical rather than for complete reviews of the literature. For this reason and despite occasional unavoidable delays in publication resulting in different periods of coverage of certain topics, the chapters should nevertheless give a view of how the chemistry of the different reactions was developed. No particular emphasis is given to industrial application of the reactions discussed, especially when reviews on these subjects have appeared. The one exception is the description of the synthesis of acrylic acid from acetylene, carbon monoxide, and water.

The various chapters in this volume are presented in what the authors consider a logical order: The first chapter relates the carbonylation of saturated organic compounds. The second and third chapters treat the homogeneously catalyzed reactions of olefins. The fourth and fifth chapters deal with allylic compounds and acetylenic substrates, respectively, which have more possibilities and pathways for reaction. The sixth chapter moves to the carbonylation of halogenated compounds. The seventh chapter summarizes a number of reactions involving iron carbonyls. The eighth chapter on decarbonylation involving transition metal complexes supplements and helps us understand all the chapters that precede it. The ninth and tenth chapters concern homogeneous reactions of unsaturated compounds that do not involve carbon monoxide but in which metal carbonyls are usually the catalyst precursors. The last chapter introduces us to symmetry-restricted reactions catalyzed by transition metal complexes, in which field further interesting developments are expected.

We thank the following persons who in various ways aided in the preparation of this book: R. F. Heck, J. F. Roth, N. v. Kutepow and W. Himmele of BASF, C. Zahn, and G. F. Pregaglia.

IRVING WENDER
PIERO PINO

Pittsburgh, Pennsylvania
Zurich, Switzerland
December 1975

Authors of Volume 2

H. ALPER
University of Ottawa, Ottawa, Ontario, Canada

M. BIANCHI
University of Pisa, Pisa, Italy

G. BRACA
University of Pisa, Pisa, Italy

E. S. BROWN
Union Carbide Corporation, South Charleston, West Virginia

L. CASSAR
Montecatini Edison, S.p.A., Novara, Italy

A. J. CHALK
Givaudan Corporation, Clifton, New Jersey

G. P. CHIUSOLI
Montecatini Edison, S.p.A., Novara, Italy

M. FOÁ
Montecatini Edison, Novara, Italy

J. HALPERN
University of Chicago, Chicago, Illinois

J. F. HARROD
McGill University, Montreal, Quebec, Canada

F. PIACENTI
University of Florence, Florence, Italy

P. PINO
Swiss Federal Institute of Technology, Zurich, Switzerland

J. TSUJI
Tokyo Institute of Technology, Tokyo, Japan

T. A. WEIL
Amoco Chemical Corporation, Naperville, Illinois

Contents of Volume 2

Contents of Volume 1

Organic Syntheses via
Metal Carbonyls

Volume 2

Carbonylation of Saturated Oxygenated Compounds

F. PIACENTI and M. BIANCHI, *Istituto di Chimica Organica dell'Università, Via Gino Capponi 9, 50121 Firenze, Italy*

I. INTRODUCTION

This chapter deals with the carbonylation of saturated aliphatic and aromatic compounds having oxygenated functional groups. No division will be made between aliphatic and aromatic substrates since aromatic unsaturation does not, as a rule, react under conditions chosen for carbonylation reactions.

The term "carbonylation" is used here as a generic term, including reactions such as formylation, hydroformylation, carboxylation, and homologation—all involving the introduction of a carbonyl group into an organic substrate.

TABLE 1
Reactions of Saturated Oxygenated Substrates with CO

Reaction	Catalyst[a]
$ROH + CO \rightarrow RCOOH$	$Co_2(CO)_8 + I_2$; $Rh + HI$
$ROH + CO \rightarrow HCOOR$	$Co_2(CO)_8$
$ROH + CO + H_2 \rightarrow RCH_2OH$	$Co_2(CO)_8$
$ROR + CO \rightarrow RCOOR$	$Co_2(CO)_8 + I_2$
$RCH\overset{O}{—}CHR + CO + H_2 \rightarrow RCH(CHO)CHOHR$ (epoxide ring: O bridging both CH)	$Co_2(CO)_8$
$H_2C(OR)_2 + CO + H_2 \xrightarrow{-ROH} H_2C(OR)(CHO)$	$Co_2(CO)_8$
$R'C(OR)_3 + CO + H_2 \rightarrow R'COOR + ROH + RCHO$	$Co_2(CO)_8$
$RCOOR' + CO \rightarrow RCO—O—OCR'$	$CoBr_2 + Et_4NBr$
$(CH_2)_n(CO)O$ (cyclic, lactone) $+ CO + H_2 \xrightarrow{H_2O} HOOC(CH_2)_nCOOH$	$Ni(CO)_4 + I_2$
$RCH(OH)NHCOCH_3 + CO \rightarrow RCH(COOH)NHCOCH_3$	$Co_2(CO)_8$

[a] Catalyst generally used.

The reactions that will be considered here are listed in Table 1.

These reactions are catalyzed by group VIII metals, especially Fe, Co, Ni, in the form of metal carbonyls or other derivatives which under reaction conditions are transformed into carbonyls. More recently, Ir, Rh, Ru, Os, Pt, and Pd derivatives have been used as catalysts for some of these reactions.

In some cases (especially in carbonylation of alcohols, esters, and ethers) halogens and organic or inorganic halides are used as promoters.

The presence of H_2 or of compounds containing active hydrogen generally increases the rate of these reactions so that they may be carried out at lower temperatures.

All the listed reactions, apart from the synthesis of acetic acid, have received little attention as compared with the hydroformylation of olefins.

This is probably due to the lack of practical importance of these reactions which, in some cases, is related to their low selectivity.

II. CARBONYLATION OF ALCOHOLS

A. Introduction

Different products, depending on the catalyst, reagents, and reaction conditions used, may be obtained by carbonylation of alcohols in the presence of metal carbonyls.

The simple addition of CO to an alcohol may yield either its formic ester or a carboxylic acid having one more carbon atom (reaction 1). Both reactions are thermodynamically possible in the same range of pressures

$$ROH + CO \longrightarrow \begin{cases} RCOOH \\ HCOOR \end{cases} \tag{1}$$

and temperatures (Table 2), the synthesis of acids being favored below 200°C. By using the appropriate catalyst and reaction conditions, the formation of either may be favored.

When H_2 is also present, the alcohol having one more carbon atom in the chain is formed instead of the carboxylic acid (see homologation of alcohols, II-D).

B. Synthesis of Carboxylic Acids

The synthesis of carboxylic acids from alcohols and CO (Eq. 2) in the presence of acid catalysts has been known for a long time (1); although it

TABLE 2

Thermodynamic Data of Some Carbonylation Reactions of Alcohols[a]

Reaction	ΔH°_{298} (kcal/mole)	ΔG°_{298} (kcal/mole)
$CH_3OH_{(g)} + CO_{(g)} = HCOOCH_{3(g)}$	−10.20	+0.06
$CH_3OH_{(g)} + CO_{(g)} = CH_3COOH_{(g)}$	−30.23	−19.81
$CH_3OH_{(g)} + CO_{(g)} + 2H_{2(g)} = CH_3CH_2OH_{(g)} + H_2O_{(g)}$	−35.53	−23.44

[a] Calculated according to O. A. Hougen, K. M. Watson, and R. A. Ragatz, *Chemical Process Principles*, John Wiley, New York, 2nd ed., 1954, using the values of the heats and of the free energies of formation of the reactants and products listed in *Chemical Engineers Handbook*, R. H. Perry, C. H. Chilton, and S. D. Kirkpatrick, Eds., McGraw-Hill, New York, 4th ed., 1963, pp. 3–134.

has been studied by numerous workers, it has never achieved practical importance, apart from the Koch synthesis (2), owing to its low yield.

$$ROH + CO \rightarrow RCOOH \qquad (2)$$

The use of metal carbonyls or of other transition metal complexes as catalysts for this synthesis first received attention only after Reppe's work (3,4).

Because of its practical importance (4–20), most of the activity in the field has been concentrated on the synthesis of acetic acid from methanol and CO. These investigations (4,7,21–25), however, have been extended to the higher homologs of methanol and to secondary (7,21,26), tertiary (21,26), unsaturated (27,28), and aryl-substituted alcohols (21), as well as to diols (4,21,25,26,29).

Table 3 lists a selection of results obtained from the carbonylation of alcohols to acids. The metals generally used as catalysts are Ni and Co; Fe shows only low activity (10). Halogens are generally used as promoters. The temperature and pressure necessary for this reaction (200–300°C; 200–700 atm) are higher than those generally used for the carbonylation of other saturated substrates.

Discrepancies are found in the literature on the composition of the carbonylation products of many alcohols (4,21,26). They are probably attributable to difficulties existing up to a few years ago in the separation, identification, and quantitative determination of these compounds in complex mixtures.

The results obtained by Reppe and co-workers on the carbonylation of propyl and *sec*-butyl alcohols (4) and by Simons on the carbonylation of butyl and *sec*-butyl alcohols (26) all clearly show that this reaction does not lead, as first supposed (21), to the formation of only one product, but to a mixture of isomeric compounds. The carboxyl group generally adds to the carbon atom carrying the hydroxyl group in the starting alcohol; complications arise mainly from the formation of olefins due to dehydration of the alcohol.

Secondary alcohols react faster than the corresponding primary compounds (21). Tertiary alcohols are mainly dehydrated to the corresponding olefins (21,26).

Dicarboxylic acids may be obtained from diols; for example, pimelic acid is obtained in 94% yield from 1,5-pentanediol (4). Other diols tested have given much lower yields of dicarboxylic acids, with monocarboxylic acids, esters, and ethers formed as by-products.

The carbonylation of alcohols yields, besides carboxylic acids, a number of other products, whose importance and relevance depend on reaction conditions and on the catalysts used. The problem is general, but it has been considered in greater detail in the case of the synthesis of acetic acid from methanol and CO in the presence of $Co_2(CO)_8$ and I_2, where a selectivity of 90% has been claimed.

The main side reactions are listed below:

$$CH_3COOH + CH_3OH \rightleftharpoons CH_3COOCH_3 + H_2O \tag{3}$$

$$2CH_3OH \rightleftharpoons CH_3OCH_3 + H_2O \tag{4}$$

$$CH_3OH + CO \rightleftharpoons HCOOCH_3 \tag{5}$$

$$CO + H_2O \rightleftharpoons CO_2 + H_2 \tag{6}$$

Esterification of the carboxylic acid with unreacted alcohol (Eq. 3) is important in the absence of added water; it has considerable influence on carbonylation of the alcohol, if much of the alcohol is used up in forming the ester.

Another secondary reaction is the formation of ethers (Eq. 4) (14,16). These compounds are formed mainly when working in the presence of strong acids and at high temperatures. Ethers may be formed both from primary monofunctional alcohols and from diols (4,26).

Formic acid is another secondary product. It may be formed according to Eq. 5 followed by hydrolysis of the ester (14,16). The formation of methyl formate from methanol and CO is thermodynamically possible under conditions used for the synthesis of acetic acid (see Section II-A).

The reactions shown in Eqs. 3–5 are reversible and, therefore, are not a problem from a practical point of view. Indeed, the formation of esters and ethers may be reduced by addition of water to the reaction medium; the small amount of secondary products still obtained may be separated and returned to the reactor for transformation into the desired compounds (14,16).

However, the water gas shift reaction (Eq. 6) produces H_2 which, in turn, is consumed by reactions leading to formation of aldehydes and hydrocarbons (14,16); these cannot be recycled and are undesirable by-products.

Finally, in the carbonylation of glycols, the decomposition of dicarboxylic acids, which are primary reaction products, leads to formation of monocarboxylic acids and to at least part of the CO_2 detected at the end of the reaction (26).

The catalytic system used by Reppe (4) was formed from halide ions and a carbonyl derivative of a group VIII metal or from a compound of this metal which added CO during the reaction. The elements showing the best catalytic activity are Co (10,11,14), Ni (4,21), and Rh (17,19, 25,30,32–34). They may be used either as metals, carbonyls, halides, or as salts of organic acids. Cobalt hydrocarbonyl ($HCo(CO)_4$) (14,16) and the

TABLE 3
Carbonylation of Alcohols

Alcohol	Catalyst	*P* (atm)	Temp. (°C)	Products	Yield (%)	Ref.
Methanol	$Co(OAc)_2$, CoI_2	680	250	Acetic acid	90	14
Methanol	Cu_2I_2 on	200–260	300	Acetic acid	87	4
	Ni-kieselguhr			Ethyl acetate	6	
Methanol	Rh_2O_3, HI	1–15	175	Acetic acid	99	17
Ethanol	Cu_2I_2, Ni	200	320	Propionic acid	63	4
				Ethyl propionate	6	
1-Propanol	$Ni(CO)_4$, NiI_2	200	—	Butyric acid	63	4
				Isobutyric acid	17	
2-Propanol	Ni, $Ni(NH_3)_6Cl_2$, NaI	300	300	Isobutyric acid	68	7
Allyl alcohol	$PdCl_2$	100	120	Allyl 3-butenoate	44	28
Allyl alcohol	$PdCl_2(PPh_3)_2$	300	100	Methyl 3-butenoate	51	27
1-Butanol	$Ni(CO)_4$, $NiCl_2$, HCl	60	300	2-Methylbutanoic acid	47	21
1-Butanol	$Ni(CO)_4$, I_2	200	275	Pentanoic acid	59	26
				2-Methylbutanoic acid	20	
2-Butanol	$Ni(CO)_4$, $NiCl_2$, HCl	43	300	2-Methylbutanoic acid	70	21
2-Butanol	$Ni(CO)_4$, I_2	200	265	2-Methylbutanoic acid	64	26
				Pentanoic acid	22	
2-Methyl-1-propanol	$Ni(CO)_4$, NiI_2	200	250	Trimethylacetic acid	6	4
				3-Methylbutanoic acid	52	
tert-Butyl alcohol	$Ni(CO)_4$, $NiCl_2$, HCl	260	280	Trimethylacetic acid	13	26
				2-Methylbutanoic acid	6	
1-Pentanol	$NiCl_2$, $Ni(CO)_4$, HCl	45	300	2-Methylpentanoic acid	16	21
Neopentyl alcohol	$NiCl_2$, $Ni(CO)_4$, HCl	45	300	C_6-Acids	21	21
2-Methyl-2-butanol	$NiCl_2$, $Ni(CO)_4$, HCl	65	275	2,2-Dimethylbutanoic acid	35	21

1-Hexanol	$NiCl_2$, $Ni(CO)_4$, HCl	65	300	2-Methylhexanoic acid	55	21
2-Ethyl-1-butanol	$NiCl_2$, $Ni(CO)_4$, HCl	45	300	2-Ethyl-2-methylbutanoic acid	40	21
1-Heptanol	$NiCl_2$, $Ni(CO)_4$, HCl	55	300	2-Methylheptanoic acid	33	21
2-Heptanol	$NiCl_2$, $Ni(CO)_4$, HCl	50	300	2-Methylheptanoic acid	70	21
1-Octanol	$NiCl_2$, $Ni(CO)_4$, HCl	50	300	2-Methyloctanoic acid	30	21
2-Octanol	$NiCl_2$, $Ni(CO)_4$, HCl	60	300	2-Methyloctanoic acid	76	21
3-Cyclohexyl-1-propanol	$NiCl_2$, $Ni(CO)_4$, HCl	45	300	3-Cyclohexyl-2-methylpropanoic acid	49	21
Cyclopentanol	$NiCl_2$, $Ni(CO)_4$, HCl	50	275	Cyclopentanecarboxylic acid	84	21
4-Methyl-1-cyclohexanol	$NiCl_2$, $Ni(CO)_4$, HCl	65	275	4-Methylcyclohexanecarboxylic acid	53	21
2-Hydroxydecahydronaphthalene	$NiCl_2$, $Ni(CO)_4$, HCl	50	275	1- and 2-Decahydronaphthalene-carboxylic acids	77	21
2-Phenylethanol	$NiCl_2$, $Ni(CO)_4$, HCl	50	300	Ethylbenzene	12	21
3-Phenyl-1-propanol	$NiCl_2$, $Ni(CO)_4$, HCl	50	300	*n*-Propylbenzene	40	21
4-Phenyl-1-butanol	$NiCl_2$, $Ni(CO)_4$, HCl	50	300	*n*-Butylbenzene	40	21
1,2-Ethanediol	$Ni(CO)_4$, I_2	200	250–260	Succinic acid	15	4
1,3-Butanediol	$Ni(CO)_4$, BiOI	200	270	Adipic acid	—	4
				2-Methylglutaric acid	—	
1,4-Butanediol	$Ni(CO)_4$, I_2	200	260	Adipic acid	69	4
1,4-Butanediol	$Ni(CO)_4$, I_2	220	260	Adipic acid	44	26
				2-Methylglutaric acid	28	
				Monocarboxylic acids	10	
2,3-Butanediol	$Ni(CO)_4$, I_2	220	260	2-Butanone	55	26
				Dicarboxylic acids	3	
				Monocarboxylic acids	11	
1,5-Pentanediol	$Ni(CO)_4$, I_2	200	250	Pimelic acid	94	4
1,5-Pentanediol	$Ni(CO)_4$, I_2	200	260	Pimelic acid	70	26
				Monocarboxylic acids	16	

TABLE 3 (*contd.*)

Alcohol	Catalyst	*P* (atm)	Temp. (°C)	Products	Yield (%)	Ref.
2,2-Dimethyl-1,3-propanediol	$Ni(CO)_4$, I_2	200	260	3,3-Dimethylglutaric acid	15	26
				Monocarboxylic acids	52	
1,6-Hexanediol	$Ni(CO)_4$, I_2	200	260	Suberic acid	90	4
1,6-Hexanediol	$NiCl_2$, $Ni(CO)_4$, HCl	50	300	2-Methylhexanoic acid	30	21
1,10-Decanediol	$Ni(CO)_4$, I_2	200	260	Dodecanedioic acid	57	4
				Monocarboxylic acids	23	
1,12-Dodecanediol	$Ni(CO)_4$, I_2	200	260	Tetradecanedioic acid	60	4
1,14-Tetradecanediol	$Ni(CO)_4$, I_2	200	260	Hexadecanedioic acid	64	4

Ag and Hg salts of the hydrocarbonyls of Co and Fe (35) have also been suggested as catalysts.

Also Pd, Pt, Ir, Os, and Ru were used as catalysts in the carbonylation of alcohols to acids (20,25,27,28) at temperatures not exceeding 200°C and at CO pressures of 10–15 atm.

The halide ions used to activate the system may be introduced either as halides of Cu, alkali metals and transition metal salts, or as the free halogens (4).

Bhattacharyya and Sourirajan (8,10), when studying the vapor phase carbonylation of methanol, used catalysts supported on silica (Fe, Co, Ni, Cl, Br, I).

The results obtained by Reppe (4), by Simons (26), and by Bhattacharyya (10) all agree that catalytic activity increases along the series Fe, Co, Ni and Cl, Br, I.

A process developed by Badische Anilin & Soda-Fabrik (BASF) uses Co catalysts with iodine promoters for the synthesis of acetic acid from methanol and CO; a plant with a capacity of 30,000 tons went onstream in 1966 in the United States (14,16). The BASF process uses high pressures (~700 atm) and a temperature of about 210°C.

A plant employing iodide-promoted Rh compounds as catalysts went onstream in the United States in 1970; 300 million pounds of acetic acid per year are now being produced by the Monsanto Company in this way. Another plant for 440 million pounds of acetic acid per year using technology licensed from Monsanto should be completed by mid 1977. This new process may be carried out at temperatures as low as 175°C and CO pressures as low as one atmosphere with a selectivity as high as 99%. Similar results, namely high rates of carbonylation at low pressures, have also been obtained with homogeneous catalyst systems based upon Ir complexes.

The catalytic system involved in the synthesis of acetic acid from methanol is complex, and in the case of Co and Ni catalysts information on the influence of reaction variables is meager. However, the results obtained by both Mizoroki (11) and Bhattacharyya (8) indicate that this reaction is first order with respect to methanol concentration.

The pressure range claimed as suitable to carry out this reaction is very wide (50–1500 atm) (4,5,21).

In the presence of $Co_2(CO)_8$ and I_2, the conversion of methanol to acetic acid increases as the pressure is raised to reach a limiting value at very high pressures (14,16); in fact, the BASF synthesis of acetic acid is carried out at about 700 atm.

In the presence of Rh catalysts, conversion of methanol is practically complete at as low a CO pressure as 1 atm.

The reaction temperature, like the pressure, depends on the catalytic system adopted: In the presence of nickel–iodine catalytic systems, it is in the range of 250–280°C, while with Co (14) and Rh (17,30) derivatives it may be 100°C lower.

The conversion of methanol to acetic acid in the presence of Co catalysts increases to a maximum as the temperature is increased. At higher temperatures the amount of hydrocarbons formed becomes important; this is probably due to decomposition of the alcohol (14).

The amount of water present is another important factor in the carbonylation of alcohols. As seen earlier, the formation of esters and that of ethers are equilibrium reactions (Eqs. 3,4) and the presence of water may decrease their importance. In fact, addition of water from a few percent up to a considerable concentration has been indicated as beneficial in the synthesis of acetic acid (14) with iodine-promoted Co catalysts.

The original interpretation of the reaction mechanism given by Reppe (4) has been reformulated by von Kutepow and co-workers (14,16) in view of work carried out by Heck and Breslow (36,37) on the chemistry of cobalt carbonyl derivatives. As may be seen in the reaction sequence below, $Co_2(CO)_8$ and HI, if not introduced as such, are probably formed according to Eqs. 7 and 8. Subsequent steps in the mechanism are the formation of the alkyl halide (Eq. 9) and that of the alkylcobalt carbonyl (Eq. 10).

$$2CoI_2 + 2H_2O + 10CO \longrightarrow Co_2(CO)_8 + 4HI + 2CO_2 \tag{7}$$

$$Co_2(CO)_8 + H_2O + CO \longrightarrow 2HCo(CO)_4 + CO_2 \tag{8}$$

$$CH_3OH + HI \longrightarrow CH_3I + H_2O \tag{9}$$

$$CH_3I + HCo(CO)_4 \longrightarrow CH_3Co(CO)_4 + HI \tag{10}$$

$$CH_3Co(CO)_4 \longrightarrow CH_3COCo(CO)_3 \underset{-CO}{\overset{+CO}{\rightleftharpoons}} CH_3COCo(CO)_4 \tag{11}$$

$$CH_3COCo(CO)_4 + H_2O \longrightarrow CH_3COOH + HCo(CO)_4 \tag{12}$$

The carboxylic acid is formed from the alkylcobalt carbonyl through a sequence of reactions similar to those that occur in the hydroformylation of olefins; here water instead of H_2 reacts with the acylcobalt carbonyl. There is no direct evidence to date to support this hypothesis, but the observed influence of variables on the reaction rate is not contrary to the suggested scheme.

The following considerations are against the hypothesis, offered earlier by Adkins and Rosenthal (21), that the carbonylation of alcohols to acids proceeds through the formation of an intermediate olefin, which would then be hydrocarboxylated to the corresponding acid.

(*i*). Acetic acid is obtained in excellent yields from methanol which cannot form an olefin by dehydration; the methylcobalt carbonyl intermediate of this reaction has a relatively high stability attributed to the impossibility of alkene elimination from this compound.
(*ii*). Although tertiary alcohols are easily dehydrated to olefins, these alcohols give poor yields of acids (26).
(*iii*). Iso- and *n*-propyl alcohol both yield propylene on dehydration but give different amounts of isomeric butyric acids (4,7).

Further work is necessary to elucidate the mechanism of this reaction.

A considerable amount of information has recently been published on the rhodium-catalyzed synthesis of acetic acid from methanol and CO (17,31–34,100).

Kinetic investigations carried out in the liquid phase with a homogeneous catalyst and in the vapor phase with a heterogeneous catalyst have given identical reaction order dependencies suggesting that a similar reaction mechanism is probably operative in both cases. The reaction is first order with respect to iodide concentration and zero order with respect to methanol concentration and CO partial pressure. The rate was also found proportional, in the liquid phase, to Rh concentration.

A new interesting rhodium-catalyzed conversion of methyl formate to acetic acid has been reported; it takes place under conditions analogous to those for the carbonylation of methanol (39). By using ^{14}C-labeled reagents, evidence has been obtained suggesting that the reaction proceeds through the simultaneous decarbonylation of methyl formate and carbonylation of methyl iodide.

The data collected are consistent with the mechanism suggested for the carbonylation of methyl iodide by Roth and co-workers (32) which will be discussed at the end of the chapter.

Another synthesis of acetic acid by carbonylation of formaldehyde with CO and H_2O in the presence of halogen-promoted Fe, Co, or Ni catalysts has also been reported (40).

Finally a novel cobalt carbonyl-catalyzed synthesis of *N*-acylamino acids by carbonylation of a mixture of an aldehyde and amide has been reported (41). This reaction might be interpreted as proceeding via the carbonylation of the corresponding *N*-acylhydroxy amide:

$$RCHO + R'CONH_2 + CO \longrightarrow \underset{RCHCO_2H}{\overset{NHCOR'}{|}} \qquad (13)$$

C. Synthesis of Formates

Formates, HCOOR, are best prepared by reacting CO with alcohols in the presence of sodium alcoholate (42); this synthesis is used industrially.

These esters are also formed, although in minor amounts, during the reaction of alcohols with synthesis gas (Eq. 14). Wender and co-workers (43) were the first to detect the formation of methyl formate (2%) during the homologation of methanol with synthesis gas (1 H_2:1 CO) in the presence of $Co_2(CO)_8$ at 185°C and 180 atm. Formates have also been found among the hydroformylation products of propylene when the

TABLE 4

Reaction of Aldehydes and Alcohols with CO *and* H_2 *in the Presence of* $Co_2(CO)_8$ (45)[a]

Substrate	P_{CO} (atm)	Product composition (%)		
		Aldehyde	Alcohol	Formate
Propionaldehyde	100	21.7	43.3	35.0
Propyl alcohol	100	3.3	93.5	3.2
Hexahydrobenzaldehyde	200	12.1	57.7	11.5
Hexahydrobenzyl alcohol	200	0.7	74.4	0.5

[a] Temp 160°C; P_{H_2} 100 atm; reaction time 5 hr.

reaction was carried out at particularly high pressures and temperatures (225–300°C; 350–1500 atm) (44).

$$ROH + CO \rightarrow HCOOR \tag{14}$$

An investigation of the synthesis of formates by the reaction of alcohols and aldehydes with CO and H_2 (200–300 atm, 160°C) in the presence of $Co_2(CO)_8$ has shown that these esters are formed in much better yields from aldehydes than from alcohols (45) (Table 4).

The reaction temperature has a large influence on the amount of formate produced; the best yields of hexahydrobenzyl formate from hexahydrobenzaldehyde were obtained at 160°C. Good conversions of methanol to methyl formate (55%) were obtained at 120°C in the presence of a catalytic system formed by $Co_2(CO)_8$, PBu_3, and acetylene (46).

Butyl formate has been obtained in 26% yield by carbonylation of butanol at 250°C with CO and H_2 (2000–3000 atm) in the presence of cobalt acetate (47).

D. Homologation of Alcohols

In 1949, Wender and co-workers (48) found that an alcohol could be converted, by reaction with CO and H_2, to the primary alcohol containing one more carbon atom:

$$ROH + CO + 2H_2 \rightarrow RCH_2OH + H_2O \qquad (15)$$

This reaction has been tried with primary, secondary, and tertiary aliphatic alcohols (43,48–50). In some cases aldehydes (or their acetals) with one more carbon atom than the starting alcohol have been obtained (5,51,52,102).

The catalyst normally used is $Co_2(CO)_8$, a derivative or a cobalt salt which under reaction conditions may be transformed into a carbonyl (43, 48,49). The addition of I_2 or an alkali metal iodide as promoter increases the reaction rate but it decreases specificity (13,53). Iron (54) derivatives have also been used as catalysts for this reaction.

Mixtures of CO and H_2 (synthesis gas) are normally used in ratios ranging from 2 to 0.5 under a pressure of 200–400 atm, although in some patents, pressures from 20 to 1000 atm have been claimed (47,54). The best temperature range for this reaction is 180–250°C. The results of experiments with a number of alcohols are reported in Table 5.

The reaction of methanol with CO and H_2 has received particular attention (13,50,53). The conversion of methanol to ethanol, however, does not usually exceed 40%. The ethanol is accompanied by substantial quantities of methane, methyl and ethyl acetates, higher homologs like propanol and butanol, methyl formate, acetaldehyde, and its dimethyl acetal and water.

Higher straight-chain alcohols may also undergo homologation, although at a lower rate (49). A mixture of butyl and isobutyl alcohol is formed from propyl alcohol. Isopropyl alcohol gives the same products but at a higher rate; secondary alcohols, in general, react more rapidly than do primary ones (43). *Tert*-butyl alcohol reacts even more rapidly (48).

Benzyl alcohols show a different reactivity on homologation: The reaction leading to formation of the corresponding hydrocarbon prevails (50).

$$PhCH_2OH \xrightarrow{CO+H_2} \begin{cases} PhCH_3 \\ PhCH_2CH_2OH \end{cases} \qquad (16)$$

From benzyl alcohol, 32% of 2-phenylethanol may be obtained together with 63% of toluene (50); however, from alcohols of the type

TABLE 5
Homologation of Alcohols

Alcohol	*P* (atm)	Temp (°C)	Catalyst	Product	Yield (%)	Ref.
Methanol	200–340	180	$Co_2(CO)_8$	Ethanol	38.8	43
				Propanol	4.7	
				Butanol	0.9	
				Methyl formate	2.0	
				Methyl acetate	9.0	
				Ethyl acetate	6.3	
				Propyl acetate	0.1	
				Methane	8.5	
Methanol	400	190–210	$Co(OAc)_2$, I_2, and $(NH_4)_2HPO_4$	Ethanol	30.6	55
Ethanol	900	225	Co	1-Propanol	—	49
				1-Butanol	—	
				3-Methylbutanol	—	
1-Propanol	220	180	$Co(OAc)_2$	1-Butanol	—	48
				2-Methylpropanol	—	
				1-Pentanol	—	
				3-Methylbutanol		
2-Propanol	220	180	$Co(OAc)_2$	1-Butanol } 2-Methyl-1-propanol }	11	48
tert-Butyl alcohol	267	200	$Co_2(CO)_8$	3-Methyl-1-butanol	60	52
				2,2-Dimethyl-1-propanol	4.1	
				Isobutane	3.2	
				Isobutene	3.0	
				Higher boiling materials	26.3	

tert-Butyl alcohol	200	130	$Co(OAc)_2$	3-Methyl-1-butanal	51	56
				2,2-Dimethyl-1-propanal	10	
Cyclohexanol	1000	225	Co–Fe	Cyclohexylcarbinol	44	49
2,3-Dimethylbutan-2,3-diol	220	185	$Co_2(CO)_8$	3,4-Dimethyl-1-pentanol	26	57
				3,3-Dimethyl-2-butanone	17	
				3,3-Dimethyl-2-butanol	4	
Benzyl alcohol	250	185	$Co_2(CO)_8$	2-Phenylethanol	32	50
				Toluene	63	
p-Methylbenzyl alcohol	238	185	$Co_2(CO)_8$	2(*p*-Tolyl)ethanol	24	58
				p-Xylene	58	
m-Methylbenzyl alcohol	238	185	$Co_2(CO)_8$	2(*m*-Tolyl)ethanol	36	58
				m-Xylene	52	
p-(*tert*-Butyl)-benzyl alcohol	238	185	$Co_2(CO)_8$	2(*p*-*tert*-Butylphenyl)ethanol	28	58
				4-*tert*-Butyltoluene	54	
2,4,6-Trimethylbenzyl alcohol	238	185	$Co_2(CO)_8$	2(2,4,6-Trimethylphenyl)ethanol	18	58
				1,2,3,5-Tetramethylbenzene	58	
p-Hydroxymethylbenzyl alcohol	238	185	$Co_2(CO)_8$	2-(*p*-Tolyl)-ethanol	39	58
				1,4-bis(Hydroxyethyl)-benzene *p*-	12	
				p-Xylene	27	
p-Methoxybenzyl alcohol	238	185	$Co_2(CO)_8$	2(*p*-Anisyl)ethanol	44	58
				p-Methoxytoluene	16	
m-Methoxybenzyl alcohol	238	185	$Co_2(CO)_8$	*m*-Methoxybenzaldehyde	4	58
				m-Methoxytoluene	23	
p-Chlorobenzyl alcohol	238	185	$Co_2(CO)_8$	2(*p*-Chlorophenyl)ethanol	16	58
				p-Chlorotoluene	41	
m-Trifluoromethylbenzyl alcohol	238	185	$Co_2(CO)_8$	*m*-Trifluoromethyltoluene	5	58
p-Carbethoxybenzyl alcohol	238	185	$Co_2(CO)_8$	Ethyl *p*-toluate	27	58
p-Nitrobenzyl alcohol	238	185	$Co_2(CO)_8$	Polymeric *p*-aminobenzyl alcohol	—	58

TABLE 6
Relative Reaction Rates of Substituted Benzyl Alcohols (58)

Substituent[a]	Relative rate
p-Methoxy	10,000
p-Methyl	200
m-Methyl	50
p-tert-Butyl	50
Hydrogen	1
p-Chloro	0.8
p-Carbethoxy	0.4
m-Methoxy	0.3
m-Trifluoromethyl	0.01

[a] Temperature was 188–190°C in all cases except *p*-methoxy (92°C) and *p*-methyl (166°C).

$PhCR_1R_2OH$, where R_1 and R_2 are H, alkyl and aryl groups, the corresponding hydrocarbons are formed in excellent yields (50).

The relative rates of reaction of substituted benzyl alcohols with CO and H_2 have been determined semiquantitatively by Wender and coworkers (58) (Table 6). These data show that electron-releasing substituents in the *meta* and *para* positions markedly increase the rate of both the reduction and homologation reactions. The proportion of homologation product, as compared to reduction product, also increases in the order in which the substituent is capable of releasing electrons.

The importance of secondary reactions leading to formation of esters, acids, anhydrides, and higher alcohols is influenced mainly by gas composition, reaction temperature and catalyst used (13,49).

While no clear data exist on the influence of gas composition on the yield of the homologous alcohol (49), temperatures exceeding 220°C appear to favor reduction of the alcohol to the corresponding hydrocarbon (13). Halogens and halides, while speeding up the reaction, increase the formation of esters and acids (13,51).

Benzhydrol is only hydrogenated under hydroformylation conditions (60). This reaction is first order with respect to benzhydrol concentration. Its rate, however, is independent of CO partial pressure but increases with H_2 partial pressure and amount of catalyst used. Electron-releasing substituents in the *para* position enhance the reaction rate as they do in the homologation of substituted benzyl alcohols.

Two mechanisms have been proposed for the homologation of alcohols, one by Wender in order to rationalize the results of his work on benzyl alcohols and the other by Ziesecke to be applied to alcohols that can dehydrate. Wender, considering that $Co_2(CO)_8$ under reaction conditions is rapidly transformed into $HCo(CO)_4$ which in polar solvents behaves as a strong acid (59), originally suggested a mechanism in which $HCo(CO)_4$ reacts with the alcohol to give a benzyl carbonium ion (48). This ion by subsequent reaction with CO and H_2 then would form the aldehyde and finally the alcohol.

Ziesecke (49), considering alcohols that can dehydrate, suggested that the first step of this reaction involves dehydration of the alcohol to the corresponding olefin which then, by hydroformylation followed by hydrogenation, gives the alcohols.

In agreement with findings on the chemistry of alkyl- and acylcobalt carbonyls and with what was suggested by Sternberg and Wender (61), the homologation of alcohols may be rationalized as proceeding through formation of alkylcobalt carbonyls as intermediates. From these intermediates, with the same steps suggested for the hydroformylation of olefins, aldehydes with one more carbon atom are formed which, by hydrogenation, finally give the alcohols. The formation of hydrocarbons during the homologation may be explained by hydrogenation of the intermediate alkylcobalt tetracarbonyl.

The formation of the alkylcobalt carbonyl in the homologation of alcohols may take place by reaction of the alcohol with $HCo(CO)_4$ (49,61,85):

$$ROH + HCo(CO)_4 \rightarrow H_2O + RCo(CO)_4 \qquad (17)$$

In the case of benzhydrols and triphenylcarbinols it has been suggested that this reaction may proceed via a carbonium ion (60).

As for alcohols that can form olefins by dehydration, an olefin intermediate cannot be excluded. Addition of an olefin to $HCo(CO)_4$ is another possible route to alkylcobalt tetracarbonyls (reaction 18), and this may explain the formation of isomeric products, as in the hydroformylation of olefins (second chapter).

$$RCH{=}CH_2 + HCo(CO)_4 \rightarrow \underset{\displaystyle Co(CO)_4}{RCHCH_3} + RCH_2CH_2Co(CO)_4 \qquad (18)$$

Of interest are the results obtained by Burns (62) on the homologation of methanol labeled with ^{14}C. Methanol gives, among other products, 40 mole % of ethanol and 5 of 1-propanol. In the ethanol, all the activity was in the methyl group ($^{14}CH_3CH_2OH$). In the case of 1-propanol, the $-CH_2OH$ group had no activity while the C-2 and C-3 carbons each had half the activity of the original methanol.

These results clearly show that propanol is formed through a symmetrical intermediate. This intermediate is undoubtedly ethylene, formed either by dehydration of ethanol or by decomposition of ethylcobalt carbonyl that is formed from ethanol and $HCo(CO)_4$. The second route to ethylene is more probable, since alkylcobalt carbonyls are much more reactive than primary alcohols.

III. CARBONYLATION OF ETHERS

A. Noncyclic Ethers

Only a few symmetrical lower members of this class of compounds have been treated with CO (4,7,63–67) (Table 7); none of these compounds has yet been reported to react with CO and H_2. The main product, under appropriate conditions, is the ester derived from addition of CO to the ether:

$$ROR + CO \rightarrow RCOOR \qquad (19)$$

The carboxylic acid corresponding to the ester is always found among the reaction products; its formation seems related to the amount of water present (63,64,67).

Nickel, either as metal or carbonyl or in the form of other derivatives, has been used, in the presence of halogens, as a catalyst for this reaction in solution (4,7,63,67). Cobalt and iron derivatives supported on silica, pumice, or kieselguhr, have also been employed (64,66).

Bhattacharyya and Palit (64) have studied the carbonylation of ethyl ether and the effect of various conditions on the yield of ester. The catalytic activity of the various metals and halogens tested decreases in the order Ni>Co>Fe and I>Br>Cl. Both reaction temperature and CO pressure have a considerable influence on formation of ethyl propionate: Maximum yields are obtained at 230°C and at 240 atm. The formation of free propionic acid, however, increases continuously; its presence has been attributed (64) to the cumulative effect of the reactions reported below:

$$C_2H_5COOC_2H_5 + CO \rightarrow (C_2H_5CO)_2O \xrightarrow{H_2O} 2C_2H_5COOH \qquad (20)$$

$$C_2H_5COOC_2H_5 + H_2O \rightarrow C_2H_5COOH + C_2H_5OH \qquad (21)$$

$$C_2H_5OH + CO \rightarrow C_2H_5COOH \qquad (22)$$

$$C_2H_5COOC_2H_5 \rightarrow C_2H_5COOH + C_2H_4 \qquad (23)$$

TABLE 7

Carbonylation of Noncyclic Ethers

Substrate	Catalyst	Temp (°C)	*P* (atm)	Products	Yield (%)	Ref.
Dimethyl ether	Ni, I_2	190	200	Methyl acetate	37	4
				Acetic acid	1.6	
Diethyl ether	NiI_2, H_2O	260	110–120	Ethyl propionate	55	63
				Propionic acid	27	
Diethyl ether	FeI_2 on SiO_2	230	240	Ethyl propionate	28.8	64
				Propionic acid	10.1	
Dipropyl ether	NiI_2 on SiO_2	—	—	Propyl butyrate	21.2	65
				Butyric acid	9.1	
Diisopropyl ether	FeI_2 on SiO_2	230	270	Isopropyl isobutyrate	12.3	66
Dipentyl ether	NiI_2 on SiO_2	—	—	Pentyl hexanoate	14.2	65
				Hexanoic acid	2.4	

B. Cyclic Ethers

1. Three-Membered Rings (Epoxides)

The carbonylation of epoxides has already been dealt with in Vol. 1, p. 384; we will therefore discuss only the catalytic and stoichiometric reaction with synthesis gas ($CO+H_2$). However, to complete the picture, we will also list in a separate table some examples of the carbonylation of epoxides.

Several epoxides have been treated with synthesis gas; the reaction leads mainly to the formation of 3-hydroxyaldehydes.

$$\underset{\text{(epoxide: O bridging the two CH)}}{RCHCHR} + CO + H_2 \rightarrow R\overset{\displaystyle CHO}{\overset{|}{C}}HCHOHR \tag{24}$$

Under severe conditions, by dehydration of the primary reaction product, the unsaturated aldehydes formed may then be hydrogenated to saturated aldehydes or alcohols.

Dicobalt octacarbonyl (68–72) and $Co_2(CO)_6(PBu_3)_2$ (73) have mainly been used as catalysts. Many experiments, however, have also been carried out with stoichiometric amounts of $HCo(CO)_4$ (71,74–77). Moreover, the potassium salts of several iron carbonyl hydrides have been reacted, in stoichiometric amounts, with epoxides (78).

The reaction of ethylene oxide or propylene oxide with a stoichiometric amount of $HCo(CO)_4$, followed by treatment with excess PPh_3, leads to isolation of the corresponding 3-hydroxyacylcobalt tricarbonyl triphenylphosphine derivatives (74).

Room temperature or lower is used for stoichiometric reactions, when working under N_2 or CO at atmospheric pressure. Catalytic reactions are best performed at about 100°C and synthesis gas pressures of 150–200 atm. Epoxides, when heated with $Co_2(CO)_8$, isomerize to the corresponding carbonyl derivatives: ethylene oxide to acetaldehyde, propylene oxide to acetone (79). Side reactions may become preponderant during the reaction of epoxides with synthesis gas, if the temperature is not kept in the range of 80–100° C (68).

The rate of carbonylation of epoxides with $HCo(CO)_4$ may be greatly increased by adding small amounts of inorganic compounds (77), such as Cu_2O, CuO, Ag_2O, and organic compounds, such as ethanol, butanol, acetone, or ether (76).

The order of reactivity of some epoxides with $HCo(CO)_4$ has been found to be the following (numbers in parentheses indicate relative reactivities): cyclohexene oxide ($\sim$5) and styrene oxide $>$ propylene oxide (1) $>$ ethylene oxide $>$ epichlorohydrin (0.05–0.025) (75).

An interesting feature of the carbonylation of the oxides of terminal olefins catalyzed by $Co_2(CO)_8$ is the preponderant attachment of the formyl group to the terminal carbon atom of the epoxide. In the carbonylation of isobutylene oxide with stoichiometric amounts of $HCo(CO)_4$, the products obtained consist, after work-up, of methyl 3-hydroxy-3-methyl butyrate (93%) and methyl 3-hydroxy-2,2-dimethyl propionate (7%) (74).

The major product of the reaction of cyclohexene oxide with $HCo(CO)_4$ is the dimer of the expected hydroxyaldehyde (71).

From the catalytic carbonylation of 1,2-propylene oxide, the main product is 3-hydroxybutyraldehyde (68).

Table 8 gives results obtained from the carbonylation of various epoxides; Table 9 gives results of carboxylation reactions (68,80–83).

2. *Higher Cyclic Ethers*

Apparently, oxacyclobutane (trimethylene oxide) and its 3,3-dimethyl derivative are the only four-membered ring cyclic ethers that have been subjected to carbonylation to date (74,84). By this reaction, catalyzed by cobalt derivatives, lactones are formed in good yields (Table 10) by insertion of a mole of CO per mole of substrate (Eq. 25). The reaction pressure and especially the temperature are higher than those needed for carbonylation of epoxides.

$$\text{oxacyclobutane} + CO \longrightarrow \text{(five-membered lactone, C=O)} \qquad (25)$$

Tetrahydrofuran and its 2-methyl- and 2,5-dimethyl- derivatives are the only five-membered ring ethers studied to date (85). The carbonylation of THF has been the subject of particular attention as a possible route to adipic acid. As catalysts, Co and Ni have been used in the form of metals, carbonyls, salts, alone or together with halogens or halogen-containing compounds (85).

The reaction of THF with synthesis gas carried out in the presence of a cobalt salt (Table 10) gives mainly butanol, δ-valerolactone, and C_5–C_6 diols.

In the presence of Ni catalysts, I_2 or an iodine-releasing compound and water, the reaction gives adipic acid in good yield together with smaller amounts of valeric acid and δ-valerolactone. By working in the presence of nickel–iodine catalysts and varying the amount of water and CO

TABLE 8
Carbonylation of Epoxides

Substrate	Catalyst	Reactants	Temp (°C)	*P* (atm)	Products	Yield (%)	Ref.
Ethylene oxide	$Co_2(CO)_8$	$CO+H_2$	80	150	3-Hydroxypropionaldehyde	13	69
Ethylene oxide	$Co_2(CO)_8$	$CO+H_2$	90–100	150	Acetaldehyde	16	68
					Acrolein	3	
Ethylene oxide	$HCo(CO)_4$[a]	CO	0	1	Ethyl 3-hydroxypropionate[b]	50	75
Ethylene oxide	$Co_2(CO)_6(PBu_3)_2$	$CO+H_2$	—	100	Trimethylene glycol	92	73
1,2-Propylene oxide	$Co_2(CO)_8$	$CO+H_2$	90–95	150	3-Hydroxybutyraldehyde	56	68
					Acetone	6	
					Crotonaldehyde	5	
1,2-Propylene oxide	$Co_2(CO)_8$	$CO+H_2$	115–120	150	Acetone	51	68
					Isobutyraldehyde	5	
					Butyraldehyde	16	
					2-Methyl-1-propanol	2	
					1-Butanol	2	
					3-Hydroxybutyraldehyde	3	
1,2-Propylene oxide	$HCo(CO)_4$[a]	CO	0	1	Ethyl 3-hydroxybutyrate[b]	60	75
1,2-Propylene oxide	$Co_2(CO)_6(PBu_3)_2$	$CO+H_2$	—	100	1,3-Butanediol	—	73
1,2-Propylene oxide	$Co_2(CO)_8$	$CO+H_2$	135	135	Crotonaldehyde	25	70
					2-Methylacrylaldehyde	16	
Epichlorohydrin	$HCo(CO)_4$[a]	CO	0	1	Ethyl 4-chloro-2-hydroxybutyrate[b]	50	75
Isobutylene oxide	$HCo(CO)_4$[a]	CO	0	1	Methyl 3-hydroxy-3-methylbutyrate[b]	93	74
					Methyl 3-hydroxy-2,2-dimethylpropionate[b]	7	
Styrene oxide	$HCo(CO)_4$[a]	CO	0	1	1-Phenylethanol	19	75
					2-Phenylethanol	9	
Cyclohexene oxide	$Co_2(CO)_8$	$CO+H_2$	110–115	150	Dimer of 2-hydroxycyclohexanecarboxaldehyde	43	71

[a] $HCo(CO)_4$ used in stoichiometric amounts.
[b] After treatment of the crude reaction product with I_2 and either CH_3OH or C_2H_5OH.

TABLE 9
Carboxylation of Epoxides

Substrate	Catalyst	Reactants	Temp (°C)	*P* (atm)	Products	Yield (%)	Ref.
Ethylene oxide	Co on kieselguhr	$CO+H_2O$	70	110	Monoethyleneglycol propionate	76	80
Ethylene oxide	$NaCo(CO)_4$	$CO+CH_3OH$	65	140	Methyl 3-hydroxypropionate	55	74
1,2-Propylene oxide	$Co_2(CO)_8$	CO	160	400	Crotonic acid	81	81
1,2-Propylene oxide	$Co_2(CO)_8$	$CO+CH_3OH$	130	240	Methyl 3-hydroxybutyrate	40.3	82
					1-Methoxy-2-propanol	0.9	
					2-Methoxy-1-propanol	0.7	
Epichlorohydrin	$Co_2(CO)_6(PBu_3)_2$	$CO+CH_3OH$	70	400	Methyl 3-hydroxy-4-chlorobutyrate	46	83
Isobutylene oxide	$NaCo(CO)_4$	$CO+CH_3OH$	50	3	Methyl 3-methyl-3-hydroxybutyrate	30	74

TABLE 10
Carbonylation of Some Cyclic Ethers

Substrate	Catalyst	Reactants	Temp (°C)	*P* (atm)	Products	Yield (%)	Ref.
Oxetane (trimethylene oxide)	$Co(OAc)_2$	$CO+H_2O$	200	250	γ-Butyrolactone	55	84
3,3-Dimethyloxetane	$Co(OAc)_2$	$CO+H_2O$	200	250	3,3-Dimethyl-γ-butyrolactone	60	84
Tetrahydrofuran	$Co(OAc)_2$	$CO+H_2$	200	200	1-Butanol	5–10	85
					γ-Valerolactone	35–45	
					C_5-C_6 Diols	35–45	
					2-Hydroxymethyltetrahydropyran	5–10	
Tetrahydrofuran	$Ni(CO)_4$, $I(CH_2)_4I$	$CO+H_2O$	250	60	Adipic acid	74	85
					Valeric acid	17	
					γ-Valerolactone	9	
2-Methyltetrahydrofuran	$NiBr_2$, I_2	$CO+H_2O$	270	100	2-Methyladipic acid	—	85
2,5-Dimethyltetrahydrofuran	$Ni(CO)_4$, I_2	$CO+H_2O$	270	100	2,5-Dimethyladipic acid	—	85

present, evidence was obtained that suggested that the reaction occurred in the following steps (67):

$$\text{(tetrahydrofuran)} + CO \longrightarrow \text{(δ-valerolactone)} \quad (26)$$

$$\text{(δ-valerolactone)} + H_2O \longrightarrow HOCH_2CH_2CH_2CH_2COOH \quad (27)$$

$$\text{(δ-valerolactone)} + CO + H_2O \longrightarrow HOOCCH_2CH_2CH_2CH_2COOH \quad (28)$$

With Co catalysts, as the amount of water is decreased, the yield of lactone formed increases while that of adipic acid decreases. Above 260°C the formation of valeric acid is enhanced at the expense of adipic acid. Minor amounts of α-methylbutyrolactone, α-methylglutaric acid, and acetylsuccinic acid are always obtained, however.

Good conditions for the synthesis of adipic acid from THF in the presence of NiI_2 and water appear to be 240–270°C and a CO pressure of 200 atm (85).

The carbonylation of both linear and cyclic ethers in the presence of $Co_2(CO)_8$ and synthesis gas may be interpreted by considering that the catalyst and gases used are essentially the same as those employed in the hydroformylation of olefins. The chief differences are in the initial stages of the reactions, since the substrates differ (ethers versus olefins).

As already seen for epoxides, reaction of these oxygenated substrates with $HCo(CO)_4$, generally accepted to be the active catalytic species, leads to acylcobalt carbonyls (74):

$$\text{(oxetane)} + HCo(CO)_4 \xrightarrow{CO} HOCH_2CH_2CH_2COCo(CO)_4 \quad (29)$$

An analogous pathway was suggested by Reppe (85) years before, on a less firm experimental basis, to rationalize the reaction of THF with synthesis gas in the presence of Co catalysts.

The corresponding aldehyde is formed by reduction of the acyl derivative, as suggested by Heck (74). If water is present instead of H_2, a carboxylic acid is formed.

The only data on the stereochemistry of the carbonylation of ethers are those gained by reaction of (74) cyclohexene oxide with $HCo(CO)_4$; after work-up, methyl 2-hydroxycyclohexanecarboxylate was obtained. This result was confirmed by Roos and co-workers (71). The mechanism most consistent with these data is an ionic acid-catalyzed opening of the epoxide ring (74). The *trans* addition of $HCo(CO)_4$ to epoxides has been suggested as a two-step reaction in which the epoxide is first protonated and then attacked by $[Co(CO)_4]^-$ from the back side (86).

$$\text{cyclohexene oxide} \xrightarrow{H^+} \text{protonated epoxide } (O^+\text{–H}) \longrightarrow \text{2-hydroxycyclohexanecarbaldehyde (OH, CHO trans)} \qquad (30)$$

IV. CARBONYLATION OF DIALKYL ACETALS

Dialkyl acetals may be carbonylated under appropriate conditions (85,87,88). In the presence of Co catalysts these compounds react with CO and H_2, thus:

$$H_2C(OR)_2 + CO + H_2 \xrightarrow{-ROH} H_2C(OR)(CHO) \qquad (31)$$

This reaction is generally carried out by using as solvent the alcohol derived from the alkoxy group. The product obtained is normally the dialkyl acetal of the 2-alkoxyaldehyde (Eq. 32).

$$RCH(OR')CHO + 2R'OH \longrightarrow RCH(OR')CH(OR')_2 + H_2O \qquad (32)$$

Examples of this carbonylation are summarized in Table 11. Reaction conditions are fairly drastic: Temperatures are in the range of 160–200°C and pressures are about 500–900 atm.

The mechanism originally suggested by Reppe (85), (Eq. 33,34) is in line with those subsequently proposed for the carbonylation of the saturated substrates previously discussed.

$$RHC(OCH_3)OCH_3 + HCo(CO)_4 \longrightarrow RHC(OCH_3)Co(CO)_4 + CH_3OH \qquad (33)$$

$$RHC(OCH_3)Co(CO)_4 + CO + H_2 \longrightarrow RHC(OCH_3)CHO + HCo(CO)_4 \qquad (34)$$

TABLE 11
Carbonylation of Dialkyl Acetals

Substrate	Catalyst	Reactants	Temp (°C)	*P* (atm)	Products	Yield (%)	Ref.
Dimethoxymethane	Co	$CO + H_2$	—	—	2-Methoxyethanal 1,1,2-Trimethoxyethane	— —	85
Dimethoxymethane	CoO	$CO + H_2$ CH_3OH	160–170	600	1,1,2-Trimethoxyethane[a]		87
Diethoxymethane	CoO	$CO + H_2$ C_2H_5OH	200	700	2-Ethoxyethanol 1,1,2-Triethoxyethane	9 6	87
Dimethoxyethane	CoO	$CO + H_2$	200	950	2-Methoxy-1-propanal 2-Methoxy-1-propanol	— —	88
1,1-Dimethoxypropane	CoO	$CO + H_2$	200	800–900	2-Methoxy-1-butanal 2-Methoxy-1-butanol	— —	88

[a] All the reacted substrate (50%) was converted to 1,1,2-trimethoxyethane.

V. CARBONYLATION OF ORTHO ESTERS [R'C(OR)$_3$]

The $Co_2(CO)_8$-catalyzed synthesis of aldehydes from ortho esters, CO and H_2 was discovered in 1960 by Piacenti, Cioni, and Pino (89). Ortho esters react in the presence of $Co_2(CO)_8$ with CO and H_2 ($CO:H_2 = 1$; 80–200 atm) at 80–130°C according to Eq. 35. An aldehyde is formed having one carbon atom more than the alcohol of the ortho ester.

$$R'C(OR)_3 + CO + H_2 \rightarrow R'COOR + ROH + RCHO \tag{35}$$

By working with stoichiometric amounts of $Co_2(CO)_8$ and under a high H_2 pressure, the reaction may be carried out at room temperature when the substrate is an orthoformate (90). An analogous reactivity is shown by triethyl orthoacetate, orthopropionate, orthobenzoate, and similar compounds; tetraethyl orthocarbonate, $C(OEt)_4$, also reacts (91). More recently, isopropyl orthotitanate has been reported to give the same reaction (92) (Table 12).

Since orthoformates undergo this reaction with particular ease, this technique has proved useful in preparing a number of aldehydes (93–95) otherwise difficult to obtain (Table 13).

Ortho esters of secondary alcohols form small quantities of isomeric aldehydes; small amounts of olefins have also been detected among the reaction products. The formation of isomeric aldehydes depends not only on the nature of the substrate but also on the temperature; isomeric aldehyde formation decreases as the temperature is decreased (less olefin is formed).

TABLE 12
Carbonylation of Ethyl Esters of Ortho Acids[a]

	Mole ratios	
Ortho ester	Gas absorbed/ ortho ester	Aldehyde/ ortho ester
Orthoformate	1.0	0.5
Orthoacetate	1.6	0.7
Orthopropionate	1.6	0.7
Orthocarbonate	1.6	0.8
Orthobenzoate[b]	1.0	0.4
Orthotitanate[c]	0.6	0.2

[a] Temp 100°C; $Co_2(CO)_8$, 10 g/l; P_{CO}, 80 atm; P_{H_2} 80 atm.
[b] $Co_2(CO)_8$, 42.6 g/l.
[c] Temp 140°C.

TABLE 13

Reaction of Alkyl Orthoformates, $HC(OR)_3$, *with* CO *and* H_2 [a]

Alkyl group	Temp (°C)	P_{max} (atm)	Reaction products	Yield (mole %)	Remarks	Ref.
Methyl	100	178	1,1-Dimethoxyethane	49	No solvent	93
Ethyl	100	166	1,1-Diethoxypropane	50	No solvent	93
Propyl	100	144	1,1-Dipropoxybutane	47	No solvent	93
Butyl	100	185	1,1-Dibutoxypentane	46	No solvent	93
2-Ethyl-1-butyl	110	196	1,1-Bis(2-ethylbutoxy)-3-ethylpentane	50	No solvent	93
(*S*)-2-Methyl-1-butyl	100	184	(+)-{1,1-Bis[(2*S*)-2-methylbutoxy]-(3*S*)-3-methylpentane}	50	Optical yield, 86%	93
Isopropyl	100	173	2-Methylpropanal	57	After hydrolysis	93
			Butanal	8		
sec-Butyl	100	176	2-Methylbutanal	64	After hydrolysis	93
			Pentanal	16		
(*S*)-*sec*-Butyl	90	150	(*R*)-(−)-2-Methylbutanal	69	Optical yield, 74.3% [b]	96
			Pentanal	15	Xylene as solvent	
(*S*)-1,2-dimethyl-1-butyl	110	240	(3*S*)-2,3-Dimethylpentanal	26	Benzene as solvent	94
			(4*S*)-4-Methylhexanal	35		
			3-Ethylpentanal	6		

[a] Catalyst $Co_2(CO)_8$; $1H_2:1CO$ used.

[b] Taking into account racemization of 2-methylbutanal by heating during the reaction and its oxidation as well as the presence of racemic 2-methylbutanal due to hydroformylation of butenes, the optical yield becomes 93±4%.

When orthoformic esters of primary alcohols are reacted with synthesis gas in the absence of solvents or in the presence of the alcohol from which the ester is derived, the aldehyde formed immediately reacts with the ortho ester still present to give the corresponding acetal:

$$RCHO + HC(OR)_3 \rightarrow RCH(OR)_2 + HCOOR \tag{36}$$

The free aldehyde is thus no longer found among the reaction products; only its dialkyl acetal is recovered in a quantity equal to 0.5 mole per mole of ortho ester reacted (Table 13). The overall reaction is the following (90,93):

$$2HC(OR)_3 + CO + H_2 \rightarrow RCH(OR)_2 + 2HCOOR + ROH \tag{37}$$

Under similar conditions, orthoformates of secondary alcohols give mixtures of the free aldehyde and its dialkyl acetal: Yields of the carbonyl compounds may reach 80% (89).

With hydrocarbons as solvents, acetal formation is greatly reduced and the yield of carbonyl compounds may reach 90% even from orthoformates of primary alcohols (90,93).

The mild conditions generally needed to carbonylate ortho esters, the good yields normally obtained, and the high purity of the products formed are distinctive for this reaction; they establish the main difference from the carbonylation of ethers and of acetals. These two last classes of compounds are carbonylated at much higher temperatures and pressures and yield substantial amounts of secondary products.

The reaction of ethyl orthoformate with synthesis gas in benzene is first order with respect to both the ortho ester and to the H_2 partial pressure (90). The reaction rate is approximately proportional to the concentration of dissolved cobalt and inversely proportional to the CO partial pressure at CO pressures above 10 atm. The conversion of ethyl orthoformate to propionaldehyde is plotted against CO pressure in Fig. 1. This gives a curve very similar to the corresponding one obtained from the hydroformylation of olefins.

On keeping the transformation of the ortho ester constant, a temperature increase of 10°C in the range 100–120°C results in a 2.4-fold decrease in reaction time. An apparent activation energy of 25 kcal/mole was estimated from these experiments (90).

There are a number of analogies between this reaction and the hydroformylation of olefins, but undoubtedly the ortho ester does not decompose to an olefin which then is hydroformylated (89).

The difference in the two reaction mechanisms is found in the initial step leading to formation of an alkylcobalt carbonyl intermediate. The mechanism suggested for the carbonylation of ortho esters, which takes

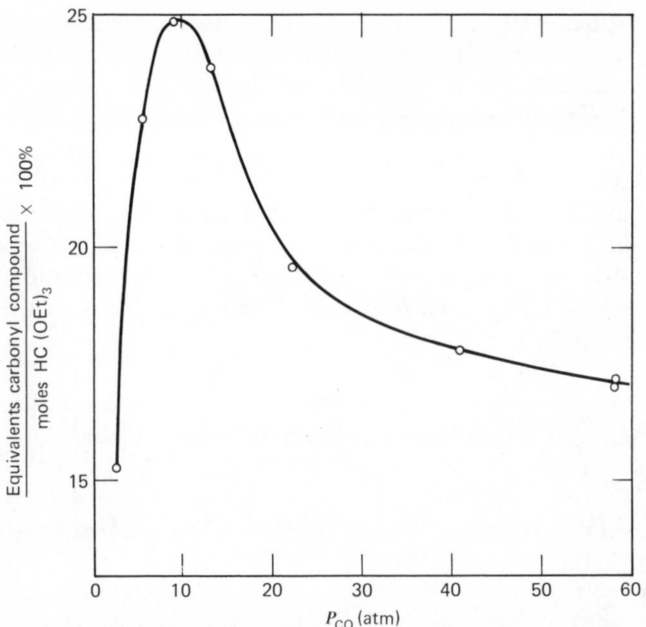

FIG. 1. Ethyl orthoformate conversion to propionaldehyde as a function of P_{CO}. Isochronic exp.: reaction time 60 min at 100°C; $Co_2(CO)_8$ 0.5 g.

into consideration the formation of isomeric aldehydes, is given below using *n*-propyl orthoformate as substrate:

$$HC(OCH_2CH_2CH_3)_3 + HCo(CO)_4 \rightarrow [Complex] \quad (38)$$

$$[Complex] \rightarrow HCOOC_3H_7 + C_3H_7OH + CH_3CH_2CH_2Co(CO)_4 \quad (39)$$

$$CH_3CH_2CH_2Co(CO)_4 \rightarrow \begin{cases} CH_3CH{=}CH_2 + HCo(CO)_4 \\ CH_3CH_2CH_2COCo(CO)_3 \end{cases} \quad (40)$$

$$CH_3CH_2CH_2COCo(CO)_3 + CO \rightleftharpoons CH_3CH_2CH_2COCo(CO)_4 \quad (41)$$

$$CH_3CH_2CH_2COCo(CO)_3 + H_2 \rightarrow CH_3CH_2CH_2CHO + HCo(CO)_3 \quad (42)$$

$$CH_3CH{=}CH_2 \xrightarrow[Co_2(CO)_8]{CO+H_2} \begin{cases} CH_3CH_2CH_2CHO \\ CH_3CH(CH_3)CHO \end{cases} \quad (43)$$

Investigation of the reaction of (+)-tris[(*S*)-1-methylpropoxy]methane with CO and H_2 has provided an insight into the steric course of this reaction and on the interaction between the orthoformate and the catalytically active cobalt carbonyl species (96). From direct reaction of the ortho ester with CO and H_2, (−)-(*R*)-2-methylbutanal has been obtained with an optical yield of about 95%. This clearly shows that the reaction takes place with inversion of configuration and with almost no racemization at the carbon atom bound to the oxygen. An S_N2 reaction on the saturated carbon is suggested.

To identify the nucleophile responsible for this reaction, $NaCo(CO)_4$ was added to a solution of tris[1-methylpropoxy]methane in the presence of CO and H_2; no reaction was observed either at 0°C or at 100°C. However, a rapid reaction took place at 0°C in the presence of $HCo(CO)_4$; thus, on addition of PPh_3 to the reaction mixture, *sec*-$C_4H_9COCo(CO)_3PPh_3$ was recovered. These results suggest the following mechanism (scheme **I**).

$$HC(OR)_2\text{—O—}\overset{CH_3}{CH}(S)\text{—}C_2H_5 + HCo(CO)_4 \longrightarrow \left[HC(OR)_2\text{—}\underset{H}{O}\text{—}\overset{CH_3}{CH}(S)\text{—}C_2H_5\right]^+ [Co(CO)_4]^- \rightarrow$$

$$\left[HC(OR)_2\text{—}\underset{H}{O}\cdots\cdots\overset{CH_3}{\underset{C_2H_5}{CH}}\cdots\cdots Co(CO)_4\right] \xrightarrow{CO} HCOOR + ROH + \underset{C_2H_5}{\overset{CH_3}{}}\!\!>\!CH\text{—}COCo \ (R)$$

$$\xrightarrow{H_2} (R)\ \underset{C_2H_5}{\overset{CH_3}{}}\!\!>\!CHCHO$$

$$R = \text{—}\overset{CH_3}{\underset{C_2H_5}{CH}}(S)$$

Scheme (I)

The preparation in good optical yield of (+)-(*S*)-3-methylpentanal and of (3*S*)-2,3-dimethylpentanal by reacting the orthoformic esters of respectively (−)-(*S*)-2-methyl-1-butanol (96) and (−)-(3*S*)-3-methyl-2-pentanol (94,95) with CO and H_2 has shown that the reaction does not involve the carbon atom in the α position with respect to the reaction center.

VI. CARBONYLATION OF LACTONES

Lactones have been carbonylated mainly to find out the role they may play during carbonylation of cyclic ethers. The carbonylations of δ-valerolactone, γ-butyrolactone and of 2-methyl-γ-butyrolactone have been studied (85) (Table 14). The main reaction, shown for δ-valerolactone (Eq. 44), leads to formation of a dicarboxylic acid with one more carbon atom.

The catalytic system generally used is either the $Ni(CO)_4$–I_2–H_2O system or a Co catalyst.

$$\text{(δ-valerolactone)} + CO + H_2O \longrightarrow HOOC(CH_2)_4COOH \qquad (44)$$

Formation of the dicarboxylic acid seems directly related to the amount of water present; good yields are obtained when water is present in stoichiometric amounts. Temperatures of 260–270°C are generally used, while the most appropriate CO pressure is about 150 atm.

The main product of the carbonylation of THF in the absence of water is δ-valerolactone. In the presence of water, both THF and the lactone carbonylate to give the same products. This has led to the suggestion that δ-valerolactone is an intermediate in the carbonylation of THF (85). A mechanism for the formation of adipic acid from THF that takes this into account has been suggested by Reppe. In the presence of NiI_2, CO, and H_2O, nickel carbonyl and H_2 are initially formed according to the following equations:

$$NiI_2 + 5CO + H_2O \rightarrow Ni(CO)_4 + 2HI + CO_2 \qquad (45)$$

$$2HI + Ni(CO)_4 \rightarrow NiI_2 + 4CO + H_2 \qquad (46)$$

The reaction of the organic substrate then proceeds as shown below:

$$\text{(THF)} \xrightarrow{HI} I(CH_2)_4OH \xrightarrow[CO+H_2O]{-HI} HOOC(CH_2)_4OH \xrightarrow{-H_2O}$$

$$\text{(δ-valerolactone)} \xrightarrow{HI} I(CH_2)_4COOH \xrightarrow[CO+H_2O]{-HI} HOOC(CH_2)_4COOH$$

The formation of monocarboxylic acids may be attributed to hydrogenation of the iodocarboxylic acid:

$$ICH_2(CH_2)_3COOH + H_2 \rightarrow CH_3(CH_2)_3COOH + HI \qquad (47)$$

An alternative mechanism based on cobalt carbonyl catalysts will be considered in the general discussion of the mechanism of carbonylation of saturated oxygenated substrates.

TABLE 14
Carbonylation of Lactones

Lactone	Catalyst	Reactants	Temp (°C)	P_{CO} (atm)	Products	Yield (%)	Ref.
β-Propiolactone	$Co_2(CO)_8$	$CO+H_2$ (2:1)	150	115	γ-Butyrolactone	45	97
					Succinic acid	38	
β-Propiolactone	$Co_2(CO)_8$	CO	150	100	Acrylic acid	19	97
					Succinic anhydride	29	
γ-Butyrolactone	Co	$CO+H_2O$			Glutaric acid	—	85
δ-Valerolactone	$Ni(CO)_4$, I_2, BiI_3	$CO+H_2O$	265	100	Adipic acid	67–75	85
					Pentanoic acid	—	—
2-Methyl-γ-butyrolactone	$Ni(CO)_4$, I_2, BiI_3	$CO+H_2O$	265	200	2-Methylglutaric acid	—	85

VII. CARBONYLATION OF ESTERS

Only the carbonylation of methyl acetate, of ethyl propionate, and of heptyl caprylate have been reported (64,98,99,103). Acetic anhydride (43% yield) is obtained from carbonylation of methyl acetate, while ethyl propionate yields propionic anhydride (40% yield) (98,99).

$$CH_3COOCH_3 + CO \rightarrow CH_3COOCOCH_3 \qquad (48)$$

The synthesis of acetic anhydride was achieved by using $(Ph_3PBu)_2NiBr_4$ as catalyst, while the synthesis of propionic anhydride was carried out in the presence of $CoBr_2$ and Et_4NBr. In both cases the reaction temperature was 180–190°C and the pressure was 650–700 atm.

The synthesis of anhydrides from esters during carbonylation of ethers has been indicated by Bhattacharyya and Palit (64) as a possible explanation for the formation of carboxylic acids. Since the carbonylation of ethers is generally carried out in the presence of water, acids could be formed by hydrolysis of the anhydrides.

VIII. DISCUSSION OF REACTION MECHANISMS

The reaction of saturated oxygenated substrates with CO catalyzed by metal carbonyls and related complexes requires, as we have seen, quite different reaction conditions depending on the nature of the substrate and the presence of promoters and of other reactants, such as H_2 and water. But quite likely the mechanisms involved are similar and have at least some steps in common.

Attempts to formulate a general interpretation of the mechanism for these reactions may be done by considering cobalt carbonyls as catalysts, although $Ni(CO)_4$ in some instances is as good a catalyst, if not better. Cobalt carbonyl, however, has the advantage over $Ni(CO)_4$ in being an effective catalyst for all these reactions; in addition, intermediates that can be isolated and studied are sometimes obtained with Co.

The carbonylation reactions catalyzed by the Co derivatives considered in this chapter may be divided into two groups: those carried out in the presence of H_2 and those carried out in the absence of H_2, but in the presence of water. The first set of reactions includes the homologation of alcohols, the carbonylation of epoxides, of THF, of acetals, and of ortho esters, and the synthesis of formates. The second group includes the synthesis of acids from alcohols, ethers, lactones, and esters.

The reactions performed in the presence of H_2 proceed at a good rate below 180°C; they do not need promoters and they form aldehydes as primary products. The aldehydes may then be transformed into acetals, alcohols, or formates.

The formation of aldehydes suggests a reaction scheme analogous to that proposed for the hydroformylation of olefins catalyzed by cobalt carbonyl. With ethers, for instance, we may write the following scheme:

$$ROR + HCo(CO)_4 \rightarrow RCo(CO)_4 + ROH \quad (49)$$

$$RCo(CO)_4 \rightleftharpoons RCOCo(CO)_3 \quad (50)$$

$$RCOCo(CO)_3 + CO \rightleftharpoons RCOCo(CO)_4 \quad (51)$$

$$RCOCo(CO)_3 + H_2 \rightarrow RCHO + HCo(CO)_3 \quad (52)$$

$$HCo(CO)_3 + CO \rightleftharpoons HCo(CO)_4 \quad (53)$$

That the carbonylation of epoxides and of orthoformates proceeds according to this scheme is supported by experimental evidence. By addition of PPh_3 to the crude product obtained by reacting an epoxide or an orthoformate with $HCo(CO)_4$, the corresponding PPh_3-substituted acylcobalt carbonyl derivative can be recovered.

While the steps given in Eqs. 49–53 may be common to this set of reactions, the interaction between the catalytically active cobalt carbonyl species and the particular substrate must be the distinctive step. According to this hypothesis, the conditions required for carbonylation of the various substrates are determined by those necessary to promote the initial stage.

Information about the first steps has been obtained from the carbonylation of epoxides and of orthoformates. As for the epoxides, the mechanism most consistent with the data is an acid-catalyzed opening of the oxide which is more sensitive to steric than to electronic factors. Orthoformates, as previously reported, react with $HCo(CO)_4$ according to an S_N2 mechanism. This path, that is, protonation of the substrate followed by nucleophilic attack of the cobalt carbonyl anion on the protonated species, may reasonably be extended to the carbonylation of alcohols, ethers and acetals.

The different minimum temperatures necessary to carry out these carbonylations, increasing in the order: epoxides $\simeq$ orthoformates $\ll$ alcohols $<$ ethers, may be related to the increasing stability of the protonated intermediates with respect to the cobalt carbonyl necleophile. It is worth mentioning, in this connection, that the decreasing stability of the alkycobalt carbonyls, which decrease with rise in temperature, is probably responsible for some side reactions during carbonylation of substrates such as alcohols and ethers, which take place at relatively high temperatures. One of these side reactions is the formation of olefins by decomposition of the alkycobalt carbonyls.

Two pathways have been formulated (61) for the synthesis of formates. The first involves formation of hypothetical formylcobalt carbonyl intermediate (Eqs. 54 and 55):

$$HCo(CO)_4 + CO \rightarrow HCOCo(CO)_4 \tag{54}$$

$$HCOCo(CO)_4 + CH_3OH \rightarrow HCOOCH_3 + HCo(CO)_4 \tag{55}$$

The second scheme (Eqs. 56–58) involves alkoxycobalt carbonyls as intermediates.

$$CH_3OH + Co_2(CO)_8 \rightarrow CH_3OCo(CO)_4 + HCo(CO)_4 \tag{56}$$

$$CH_3OCo(CO)_4 + CO \rightarrow CH_3OCOCo(CO)_4 \tag{57}$$

$$CH_3OCOCo(CO)_4 + H_2 \rightarrow HCOOCH_3 + HCo(CO)_4 \tag{58}$$

It has been suggested that the synthesis of formates by carbonylation of aldehydes proceeds through formation of the alkoxy intermediate according to Eq. 59; this would then be carbonylated as shown above (61).

$$RCHO + HCo(CO)_4 \rightarrow RCH_2OCo(CO)_4 \tag{59}$$

The mechanism of the carboxylation of alcohols, ethers, and esters in the presence of Co or Ni catalysts (in the absence of H_2) is less well understood.

The mechanism suggested to rationalize the synthesis of acids from alcohols in the presence of Co catalysts (II-B, Eqs. 7 to 12) has many steps in common with that of the homologation of alcohols.

The role of promoters is particularly obscure. v. Kutepow et al. (14,16) postulate that the halogen provides the reagent capable of attacking the substrate. According to Mizoroki (38), however, the halogen ion coordinates to Co to provide a "modified" halogenocobalt carbonyl catalyst. The increasing activity of promoters in the order $Cl < Br < I$ could, however, reasonably be explained in terms of a nucleophilic attack of a metal complex on an alkyl halide.

A clearer insight into the mechanism of carbonylation of methanol has been achieved for the iodine promoted rhodium-catalyzed reaction on the basis of the kinetic evidence discussed before and the knowledge of the chemical behavior of Rh complexes (32) (reactions 60–64).

$$CH_3OH + HI \rightleftharpoons CH_3I + H_2O \tag{60}$$

$$CH_3I + [\text{Rh complex}] \xrightarrow{\text{slow}} \left[\begin{array}{c} CH_3 \\ | \\ \text{Rh complex} \\ | \\ I \end{array} \right] \tag{61}$$

$$\left[\begin{array}{l} CH_3 \\ | \\ \text{Rh complex} \\ | \\ I \end{array}\right] + CO \longrightarrow \left[\begin{array}{l} CH_3 \\ | \\ \text{Rh complex} \cdot CO \\ | \\ I \end{array}\right] \tag{62}$$

$$\left[\begin{array}{l} CH_3 \\ | \\ \text{Rh complex} \cdot CO \\ | \\ I \end{array}\right] \longrightarrow \left[\begin{array}{l} CH_3 \\ | \\ CO \\ | \\ \text{Rh complex} \\ | \\ I \end{array}\right] \tag{63}$$

$$\left[\begin{array}{l} CH_3 \\ | \\ CO \\ | \\ \text{Rh complex} \\ | \\ I \end{array}\right] \xrightarrow{H_2O} CH_3COOH + [\text{Rh complex}] + HI \tag{64}$$

After the initial conversion of the alcohol to CH_3I, the oxidative addition of CH_3I to a Rh(I) complex (d^8 square planar) is assumed to take place to form the six-coordinated (d^6) alkylrhodium(III) species. Carbon monoxide is then inserted to give the acyl derivative as suggested in the cases of alkylmanganese and alkylcobalt carbonyls. The last step involves the reaction of the acylrhodium derivative species with water to give acetic acid with restoration of the d^6 rhodium complex. The favored path for this last step is the reductive elimination of acetyl iodide followed by its hydrolysis.

This mechanism is consistent with the results of the kinetic investigation if the oxidative addition of CH_3I is the rate-determining step. Recently Forster (101), following the various stages of the reaction between Rh(I) halogen carbonyl complexes and methyl iodide by an I.R. technique under pressure, was able to confirm and supply details on the mechanism previously suggested by Roth, showing that the intermediate present, when carrying out the carbonylation of CH_3I in the presence of $RhCl_3 \cdot 3H_2O$, is $[RhI_2(CO)_2]^-$. This species reacts with CH_3I at room temperature giving rise to the formation of the acetyl species, $[RhI_3CO(CH_3CO)]^-$, probably through the isomeric precursor $[CH_3RhI_3(CO)_2]^-$. Under CO pressure, $[RhI_3(CO)_2(CH_3CO)]^-$ is subsequently formed which spontaneously, by reductive elimination of acetyl iodide, restores the starting Rh complex $[Rh(CO)_2I_2]^-$.

The difference in rate dependence between the cobalt-catalyzed and the rhodium-catalyzed reactions suggest that the two catalytic systems either operate by different mechanisms or involve different rate-determining steps.

REFERENCES

1. Ya. T. Eidus and K. V. Puzitskii, *Russ. Chem. Rev.*, **33,** 438 (1964).
2. H. Koch and W. Gilfert, *Brennstoff-Chem.*, **36,** 321 (1955).
3. O. Hecht and H. Kröper, in K. Zeigler, Ed., *Naturforschung und Medizin in Deutschland* 1939–1946, Vol. 36, *Präparative organische Chemie,* Part I, Dieterich'sche Verlagsbuchhlandlung, Wiesbaden, 1948, p. 115.
4. W. Reppe, H. Kröper, N. v. Kutepow, and H. J. Pistor, *Ann. Chem.*, **582,** 72 (1953).
5. R. E. Brooks, U.S. Pat. 2,457,204 (1948); through *Chem. Abstr.*, **43,** 3443 (1949).
6. S. K. Bhattacharyya and S. Sourirajan, *J. Sci. Ind. Res.* (*India*), **11B,** 123 (1952); through *Chem. Abstr.*, **47,** 10475 (1953).
7. H. J. Hagemeyer, U.S. Pat. 2,739,169 (1956); through *Chem. Abstr.*, **50,** 16835 (1956).
8. S. K. Bhattacharyya and S. Sourirajan, *J. Appl. Chem.*, **6,** 442 (1956).
9. G. Natta and P. Pino, Swiss Pat. 306,638 (1955); through *Chem. Abstr.*, **51,** 2022 (1957).
10. S. K. Bhattacharyya and S. Sourirajan, *J. Appl. Chem.*, **9,** 126 (1959).
11. T. Mizoroki, M. Nakayama, and M. Furumi, *Tokyo Kogyo Shikensko Hokoku,* **58,** 302 (1963); through *Chem. Abstr.*, **62,** 6388 (1965).
12. M. Furumi, T. Mizoroki, M. Nakayama, and Y. Ando, *Tokyo Kogyo Shikensko Hokoku,* **58,** 293 (1963); through *Chem. Abstr.*, **62,** 6388 (1965).
13. T. Mizoroki and M. Nakayama, *Bull. Chem. Soc. Japan,* **37,** 236 (1964).
14. N. v. Kutepow, W. Himmele, and H. Hohenschutz, *Chem.-Ing. Tech.*, **37,** 383 (1965).
15. T. Mizoroki and M. Nakayama, *Bull. Chem. Soc. Japan,* **39,** 1477 (1966).
16. H. Hohenschutz, N. v. Kutepow, and W. Himmele, *Hydrocarbon Process.*, **45,** 141 (1966).
17. F. E. Paulik and J. F. Roth, *Chem. Commun.*, **1968,** 1578.
18. P. Ellwood, *Chem. Eng.* (*New York*), **76,** 148 (1969).
19. R. G. Schultz and P. D. Montgomery, *J. Catal.*, **13,** 105 (1969).
20. F. E. Paulik, A. Hershman, J. F. Roth, J. H. Craddock, and D. Forster, Fr. Pat. 1,573,130 (1969); through *Chem. Abstr.*, **72,** 100054 (1970).
21. H. Adkins and R. W. Rosenthal, *J. Am. Chem. Soc.*, **72,** 4550 (1950).
22. Ya. Yu. Aliev and Ya. I. Isakov, *Issled. Mineral'n i Rast Syr'ya Uzebekistana, Akad. Nauk Uz. S.S.R., Inst. Khim.*, **1962,** 95; through *Chem. Abstr.*, **59,** 3767 (1963).
23. S. Sourirajan, *Advances in Catalysis,* Vol. IX, Academic Press, New York, 1957, p. 618.
24. W. Reppe, N. v. Kutepow, and E. Titzenthaler, Ger. Pat. 922,231 (1955); through *Chem. Abstr.*, **51,** 16517 (1957).
25. F. E. Paulik, A. Hershman, W. R. Knox, and J. F. Roth, Ger. Offen. 1,941,449 (1970); through *Chem. Abstr.*, **72,** 110807 (1970).
26. R. H. Simon, Ph.D. Dissertation, Yale University (1957).
27. N. v. Kutepow, K. Bittler, and D. Neubauer, Ger. Pat. 1,221,224 (1966).
28. J. Tsuji, J. Kiji, and M. Morikawa, *Tetrahedron Letters,* **1963,** 1811.
29. G. K. N. Reddy and M. R. A. Rao, *Indian Inst. Sci. Golden Jubilee Res. Vol., 1909–1959,* 30 (1959); through *Chem. Abstr.*, **55,** 24561 (1961).
30. F. E. Paulik, A. Hershman, J. F. Roth, J. H. Craddock, W. R. Knox, and R. G. Schultz, Fr. Pat. 1,573,131 (1968).
31. H. D. Grove, *Hydrocarbon Process.*, **51** (11), 76 (1972).
32. J. F. Roth, J. H. Craddock, A. Hershman, and F. E. Paulik, *Chem. Technol.*, **23,** 600 (1971).

33. K. K. Robinson, A. Hershman, J. H. Craddock, and J. F. Roth, *J. Catal.*, **27,** 389 (1972).
34. A. Hershman, K. K. Robinson, J. H. Craddock, and J. F. Roth, Symposium on Advances in Petrochemical Technology, A.C.S. Meeting, New York, August 1972.
35. Badische Anilin- & Soda-Fabrik, AG., Brit. Pat. 713,515 (1954); through *Chem. Abstr.*, **50,** 6501 (1956).
36. R. F. Heck and D. S. Breslow, Paper presented at 2nd Int. Congr. on Catalysis, Paris (1960).
37. R. F. Heck and D. S. Breslow, *J. Am. Chem. Soc.*, **83,** 4023 (1961).
38. T. Mizoroki and M. Nakayama, *Bull. Chem. Soc. Japan*, **41,** 1628 (1968).
39. F. J. Bryant, W. R. Johnson, and T. C. Singleton, A. C. S. Meeting, Dallas, April 1973.
40. J. Kato, R. Iwanaga, and H. Wakamatsu, Ger. Pat. 1,135,884 (1962); through *Chem. Abstr.*, **58,** 4430 (1963).
41. H. Wakamatsu, J. Uda, and N. Yamakami, *Chem. Commun.*, **1971,** 1540.
42. A. Stähler, *Chem. Ber.*, **47,** 580 (1914).
43. I. Wender, R. A. Friedel, and M. Orchin, *Science*, **113,** 206 (1951).
44. R. C. Schreyer, U.S. Pat. 2,564,130 (1951); through *Chem. Abstr.*, **47,** 142 (1953).
45. L. Markó and P. Szabó, *Chem. Technol.*, **13,** 482 (1961).
46. F. Piacenti, P. Pino, and M. Bianchi, unpublished results.
47. E. I. du Pont de Nemours and Co., Brit. Pat. 665,698 (1952); through *Chem. Abstr.*, **46,** 11232 (1952).
48. I. Wender, R. Levine, and M. Orchin, *J. Am. Chem. Soc.*, **71,** 4160 (1949).
49. K. H. Ziesecke, *Brennstoff-Chem.*, **33,** 385 (1952).
50. I. Wender, H. Greenfield, and M. Orchin, *J. Am. Chem. Soc.*, **73,** 2656 (1951).
51. W. Reppe and H. Friederich, Ger. Pat. 897,403 (1953); through *Chem. Abstr.*, **50,** 16830 (1956).
52. I. Wender, J. Feldman, S. Metlin, B. H. Gwynn, and M. Orchin, *J. Am. Chem. Soc.*, **77,** 5760 (1955).
53. J. Berty, L. Markó, and D. Kalló, *Chem. Technol.*, **8,** 260 (1956).
54. G. Marullo and A. Baroni, Ital. Pat. 484,182 (1953); through *Chem. Abstr.*, **50,** 4194 (1956).
55. A. D. Riley and W. O. Bel, Jr., Belg. Pat. 625,939 (1963); through *Chem. Abstr.*, **59,** 9793 (1963).
56. H. Kröper, H. Hauber, and W. Hagen, Ger. Pat. 921,936 (1955); through *Chem. Abstr.*, **53,** 222 (1959).
57. I. Wender, S. Metlin, and M. Orchin, *J. Am. Chem. Soc.*, **73,** 5704 (1951).
58. I. Wender, H. Greenfield, S. Metlin, and M. Orchin, *J. Am. Chem. Soc.*, **74,** 4079 (1952).
59. H. W. Sternberg, I. Wender, R. A. Friedel, and M. Orchin, *J. Am. Chem. Soc.*, **75,** 2717 (1953).
60. Y. C. Fu, H. Greenfield, S. Metlin, and I. Wender, *J. Org. Chem.*, **32,** 2837 (1967).
61. H. W. Sternberg and I. Wender, *Proc. Int. Conf. Coord. Chem.*, Chem. Soc. Spec. Publ. No. 13, 35 (1959).
62. G. R. Burns, *J. Am. Chem. Soc.*, **77,** 6615 (1955).
63. Ya. Yu. Aliev, I. M. Kovina, and M. I. Monakov, *Dokl. Akad. Nauk SSR*, **9,** 37 (1960); through *Chem. Abstr.*, **56,** 5824 (1962).
64. S. K. Bhattacharyya and S. K. Palit, *J. Appl. Chem.*, **12,** 174 (1962).
65. S. K. Palit, M. K. Saxena, and N. C. Tripathy, *Indian J. Technol.*, **7,** 329 (1969).
66. S. K. Bhattacharyya and S. K. Palit, *Brennstoff-Chem.*, **43,** 169 (1962).

67. Ya. Yu. Aliev and I. B. Romanova, *Neftekhim., Akad. Nauk Turk. SSR*, **1963,** 204; through *Chem. Abstr.*, **61,** 6913 (1964).
68. C. Yokokawa, Y. Watanabe, and Y. Takegami, *Bull. Chem. Soc. Japan*, **37,** 677 (1964).
69. Y. Takegami, C. Yokokawa, and Y. Watanabe, Japan. Pat. 69 07,690 (1969); through *Chem. Abstr.*, **71,** 30066 (1969).
70. W. D. Niederhauser, U.S. Pat. 3,054,813 (1962); through *Chem. Abstr.*, **58,** 3319 (1963).
71. L. Roos, R. W. Goetz, and M. Orchin, *J. Org. Chem.*, **30,** 3023 (1965).
72. A. Rosenthal and G. Kan, *Tetrahedron Lett.*, **1967,** 477.
73. C. W. Smith, G. N. Schrauzer, R. J. Windgassen, and K. F. Koetitz, U.S. Pat. 3,463,819 (1969); through *Chem. Abstr.*, **71,** 112407 (1969).
74. R. F. Heck, *J. Am. Chem. Soc.*, **85,** 1460 (1963).
75. Y. Takegami, C. Yokokawa, Y. Watanabe, and H. Masada, *Bull. Chem. Soc. Japan*, **37,** 672 (1964).
76. Y. Takegami, C. Yokokawa, and Y. Watanabe, *Bull. Chem. Soc. Japan*, **37,** 935 (1964).
77. Y. Takegami, C. Yokokawa, Y. Watanabe, and H. Masada, *Bull. Chem. Soc. Japan*, **38,** 1649 (1965).
78. Y. Takegami, Y. Watanabe, H. Masada, and I. Kanaya, *Bull. Chem. Soc. Japan*, **40,** 1456 (1967).
79. J. L. Eisenmann, *J. Org. Chem.*, **27,** 2706 (1962).
80. M. Seon and J. Leleu, U.S. Pat. 2,782,226 (1957); through *Chem. Abstr.*, **51,** 10564 (1957).
81. W. A. McRae and J. L. Eisenmann, U.S. Pat. 3,024,275 (1962); through *Chem. Abstr.*, **57,** 2077 (1962).
82. J. L. Eisenmann, R. L. Jamartino, and J. F. Howard, Jr., *J. Org. Chem.*, **26,** 2102 (1961).
83. J. D. McClure, U.S. Pat. 3,259, 649 (1966); through *Chem. Abstr.*, **65,** 8767 (1966).
84. H. J. Nienburg and G. Elchnigg, Ger. Pat. 1,066,572 (1959); through *Chem. Abstr.*, **55,** 10323 (1961).
85. W. Reppe, H. Kröper, H. J. Pistor, and O. Weissbarth, *Ann. Chem.*, **582,** 87 (1953).
86. M. Orchin and W. Rupilius, *Catal. Rev.*, **6,** 85 (1972).
87. W. F. Gresham and R. E. Brooks, U.S. Pat. 2,449,470 (1948); through *Chem. Abstr.*, **43,** 675 (1949).
88. J. D. C. Wilson, II, U.S. Pat. 2,555,950 (1951); through *Chem. Abstr.*, **45,** 10257 (1951).
89. F. Piacenti, C. Cioni, and P. Pino, *Chem. Ind.* (*London*), **1960,** 1240.
90. P. Pino, F. Piacenti, and P. P. Neggiani, *Chim. Ind.* (*Milan*), **44,** 1367 (1962).
91. F. Piacenti and P. P. Neggiani, *Chim. Ind.* (*Milan*), **44,** 1396 (1962).
92. F. Piacenti, unpublished results.
93. F. Piacenti, *Gazz. Chim. Ital.*, **92,** 225 (1962).
94. R. Rossi, P. Pino, F. Piacenti, L. Lardicci, and G. Del Bino, *J. Org. Chem.*, **32,** 842 (1967).
95. R. Rossi, P. Pino, F. Piacenti, L. Lardicci, and G. Del Bino, *Gazz. Chim. Ital.*, **97,** 1194 (1967).
96. F. Piacenti, M. Bianchi, and P. Pino, *J. Org. Chem.*, **33,** 3653 (1968).
97. Y. Mori and J. Tsuji, *Bull. Chem. Soc. Japan*, **42,** 777 (1969).
98. W. Reppe, H. Friederich, N. v. Kutepow, and W. Morsch, U.S. Pat. 2,729,651 (1956); through *Chem. Abstr.*, **50,** 13081 (1956).

99. W. Reppe and H. Friederich, U.S. Pat. 2,730,546 (1956); through *Chem. Abstr.*, **50,** 13081 (1956).
100. J. Hjortkjaer and V. W. Jensen, *Ind. Eng. Chem., Prod. Res. Develop.*, **15,** 46 (1976).
101. D. Forster, *J. Am. Chem. Soc.*, **98,** 846 (1976).
102. G. Albanesi, *Chim. Ind. (Milan)*, **55,** 319 (1973).
103. C. Hewlett, Ger. Offen. 2,441,502 (1975); through *Chem. Abstr.*, **83,** 96448 (1975).

Reactions of Carbon Monoxide and Hydrogen with Olefinic Substrates: The Hydroformylation (OXO) Reaction

P. PINO, *Swiss Federal Institute of Technology, Department of Industrial and Engineering Chemistry, Zurich, Switzerland*

F. PIACENTI and M. BIANCHI, *Institute of Organic Chemistry, University of Florence, Italy*

I. INTRODUCTION

The hydroformylation or oxo reaction, one of the oldest carbonylation reactions, was discovered by Roelen in 1938 (1). It is still the most important carbonylation reaction industrially, and it has been the most extensively investigated.

As indicated by its name, the oxo reaction leads to the synthesis of carbonyl compounds (aldehydes and ketones) from olefins, CO, and H_2 according to reactions 1 and 2.

$$\text{>C=C<} + CO + H_2 \rightarrow -\overset{|}{\underset{H}{\overset{|}{C}}}-\overset{|}{\underset{|}{C}}-CHO \qquad (1)$$

$$2\ \text{>C=C<} + CO + H_2 \rightarrow -\overset{|}{\underset{H}{\overset{|}{C}}}-\overset{|}{\underset{|}{C}}-\underset{O}{\overset{\|}{C}}-\overset{|}{\underset{|}{C}}-\overset{|}{\underset{H}{\overset{|}{C}}}- \qquad (2)$$

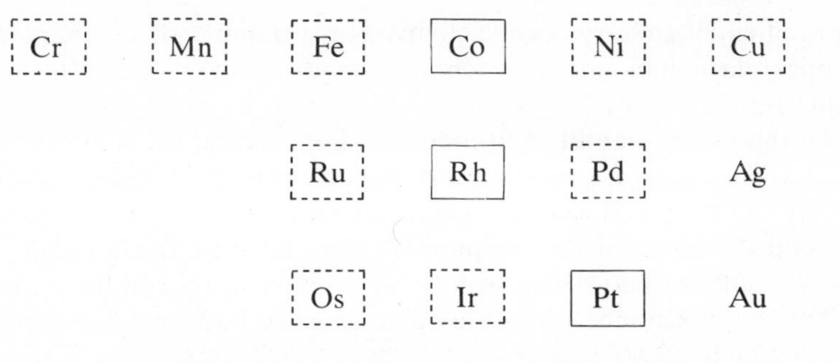

FIG. 1. Catalytic activity of various transition metals in hydroformylation.

However, as the aldehyde synthesis is much more important than the ketone synthesis, which occurs in high yield only in special cases, the oxo reaction is now more usually termed hydroformylation, as first used by Adkins and Krsek (2).

Reaction 1 is exothermic and thermodynamically favored (3) up to about 200°C. Although hydrogenation of the olefinic carbon–carbon double bond and of CO to methane and water are thermodynamically much more favored than hydroformylation, they are either suppressed or greatly diminished owing to the high selectivity of the catalytic system.

Cobalt and its compounds were originally used as catalysts.* They showed a remarkable activity accompanied by a high selectivity and seemed suitable for industrial applications. Although research on the oxo reaction has for more than 20 years been based on the use of cobalt as catalyst, a thorough investigation of the catalytic activity of different metals and of their compounds has more recently been in progress (5). The newer findings can be summarized as shown in Fig. 1.

Rhodium and its derivatives were found more active and at least as selective as cobalt, whereas ruthenium and iridium appeared fairly active but less selective. Recently hydridic platinum complexes were found to have catalytic activity comparable to that of cobalt (289). Other metals like iron, osmium, nickel, palladium, and their derivatives tested to date

* The oxo reaction was discovered by Roelen (1) while investigating the origin of the oxygenated compounds present in the products of the Fischer–Tropsch synthesis. Repeating an experiment originally carried out by Smith, Hawk, and Golden (4), he observed that propionaldehyde was formed by adding ethylene to the gaseous feed under typical conditions of the Fischer–Tropsch synthesis; this formation did not seem to be a part of the Fischer–Tropsch synthesis itself. Roelen obtained propionaldehyde and diethyl ketone in high yields by increasing the pressure and the ethylene concentration.

showed slight to very slight activity. For a number of other metals like copper, chromium, and tungsten the catalytic activity was not investigated sufficiently.

Owing to its scientific and industrial importance, the oxo reaction has been the subject of a very large number of patents, scientific papers, and reviews (1,6–10). Since a complete review of this extensive literature is beyond the scope of this chapter, we have tried to give a picture of the present state of knowledge of the oxo reaction mainly on the basis of the scientific publications without emphasizing the historical development of the research.

We shall discuss separately the synthesis of aldehydes (hydroformylation) and the synthesis of ketones. Furthermore, in order to avoid undue generalization on the catalytic behavior of different metals, the first two sections of the chapter deal with the most widely used cobalt-catalyzed hydroformylations and with the rhodium-catalyzed reaction, the importance of which is growing rapidly. A third section concerns hydroformylation reactions catalyzed by other metals.

II. HYDROFORMYLATION CATALYZED BY COBALT COMPOUNDS

Cobalt derivatives have been the first and for many years practically the only hydroformylation catalysts; even today they are still the most widely used in industry. Although at the outset heterogeneous catalysts were employed (1,11), it was soon realized that the reaction could be catalyzed in homogeneous phase by cobalt compounds soluble in the reaction medium and that cobalt carbonyls—mainly $Co_2(CO)_8$, and cobalt hydrocarbonyl, $HCo(CO)_4$—should play important roles in the catalysis (2,10–12,300). Hydroformylation of olefins by using stoichiometric amounts of $HCo(CO)_4$ (13) or $Co_2(CO)_8$ plus H_2 (14) was then successfully achieved but it never reached, in contrast to stoichiometric hydrocarboxylation of olefins with $Ni(CO)_4$ and acids (see the third chapter), practical importance even as a laboratory method. The use of $HCo(CO)_4$ is difficult because of its instability at room temperature, whereas the stoichiometric use of $Co_2(CO)_8$ requires high H_2 pressures and, therefore, does not offer enough practical advantage with respect to the catalytic procedure. However, studies of the stoichiometric hydroformylation of olefins have contributed much to the present knowledge of the reaction mechanism and will be discussed first.

In discussing the catalytic reaction we will first consider some general aspects and the types of soluble cobalt catalysts used, then the classes of olefins that have been hydroformylated and the limitations of the reaction. We will later discuss the secondary reactions and the factors which mainly control composition of the reaction products. Finally after a brief survey of the kinetic aspects of the reaction, we will discuss the most recent views on its mechanism.

A. Stoichiometric Hydroformylation

The first stoichiometric hydroformylations were carried out at room pressure and temperature using $HCo(CO)_4$ as reactant. Cyclohexene and 1-hexene were hydroformylated according to a stoichiometry that was complicated and not completely clarified (13). These very important experiments showed that a cobalt carbonyl or a cobalt carbonyl hydride was able to activate the olefin and that CO and hydrogen bound to the cobalt could react with the substrate to yield aldehydes. When 1-hexene was used in large excess, the unreacted olefin was isomerized to 2-hexene and 3-hexene.

Another stoichiometric hydroformylation at room temperature was later performed involving the reaction of the olefin with stoichiometric amounts of $Co_2(CO)_8$ under a high hydrogen pressure (14). The reaction in hydrocarbon solvents proceeds smoothly according to the following stoichiometry (reaction 3). $Co_4(CO)_{12}$ is quantitatively recovered after the reaction.

$$Co_2(CO)_8 + 2\,{>}C{=}C{<} + 2H_2 \rightarrow 1/2\,Co_4(CO)_{12} + 2\,{-}\underset{\mathrm{H}}{\overset{|}{\underset{|}{C}}}{-}\underset{\mathrm{CHO}}{\overset{|}{\underset{|}{C}}}{-} \qquad (3)$$

Cyclohexene, propylene, and isobutylene were hydroformylated and the expected aldehydes were formed in good yields; the isomeric composition of the reaction products from propylene was butyraldehyde/isobutyraldehyde = 7/3 (16).

In the absence of CO the reaction proceeded promptly at room temperature after the addition of H_2; but in the presence of CO a long induction period was noticed even at a low partial pressure. Since a retarding effect of CO under the same conditions was also noticed in the synthesis of $HCo(CO)_4$ from $Co_2(CO)_8$ and H_2 in the absence of olefins (14*b*, 77) an equilibrium as reported in reactions 4–7 was assumed to exist in solution. According to this hypothesis, $[Co_2(CO)_7]$ and not $Co_2(CO)_8$ would be responsible for the activation of hydrogen with subsequent formation of $HCo(CO)_4$. In reactions 4–7 the compounds indicated in brackets have been only hypothesized.

$$Co_2(CO)_8 \rightleftarrows [Co_2(CO)_7] + CO \quad (4)$$

$$[Co_2(CO)_7] + H_2 \rightleftarrows [H_2Co_2(CO)_7] \quad (5)$$

$$[H_2Co_2(CO)_7] \rightleftarrows HCo(CO)_4 + [HCo(CO)_3] \quad (6)$$

$$[HCo(CO)_3] + CO \rightleftarrows HCo(CO)_4 \quad (7)$$

This last species, as said earlier, would then hydroformylate the olefin.

Kinetic evidence for equilibrium in Eq. 4 was obtained by Ungvary (15). He confirmed that, as supposed in preliminary experiments (14*b*), the retarding effect of CO is due to the fact that, at low CO pressure, the rate of $HCo(CO)_4$ formation is inversely proportional to the CO concentration in solution. The strong inhibiting effect of CO reported (14*b*) in the stoichiometric hydroformylation of olefins, however, should not be attributed only to the retarding effect of CO on $HCo(CO)_4$ formation, since it might also arise from an inhibiting effect of CO on a subsequent hydroformylation step.

Finally, the stoichiometric hydroformylation of olefins at room temperature may be accomplished with $Co_4(CO)_{12}$ as the CO source under a higher pressure of H_2 (16). The reaction does not take place if hydrocarbons are used as solvents, but readily occurs in ethanol according to the stoichiometry shown in reaction 8.

$$Co_4(CO)_{12} + 4\,\rangle C{=}C\langle + 4H_2 \xrightarrow{\text{EtOH}} 4\ \begin{matrix} | & | \\ -C- & -C- \\ | & | \\ H & CHO \end{matrix} + \frac{1}{n}[Co_4(CO)_8(EtOH)_x]_n \quad (8)$$

The resulting cobalt complex could not be isolated and characterized. Since the solvent used in this experiment was ethanol, a disproportionation of $Co_4(CO)_{12}$ to $Co_2(CO)_8$ and metallic cobalt or other complexes with CO/Co ratios lower than 3 cannot in principle be excluded; in such a case no substantial difference would exist between a stoichiometric hydroformylation with $Co_2(CO)_8$ or with $Co_4(CO)_{12}$. However, the difference between the isomeric composition of the hydroformylation products obtained in the two cases (butyraldehyde/isobutyraldehyde from $Co_2(CO)_8 = 70/30$, from $Co_4(CO)_{12} = 48/52$) seems to indicate that the organometallic intermediates involved are different.

Various workers reinvestigated the reaction between olefins and $HCo(CO)_4$ under different conditions in the presence and in the absence of CO. On adding an excess of 1-hexene to a solution of $HCo(CO)_4$ under 1 atm of CO, it was found that 1 mole of gas was absorbed for every 2 moles of $HCo(CO)_4$ (17). The existence of an intermediate complex $\{[HCo(CO)_4]_2(CH_3(CH_2)_3CH{=}CH_2)(CO)\}$ was postulated by Kirch and Orchin (17), and some infrared evidence was produced in support of its existence.

Breslow and Heck (18), by mixing 1-octene with $HCo(CO)_4$, observed the formation of an acylcobalt carbonyl (see the fourth chapter of Vol. 1). They formulated the stoichiometry given in reaction (**a**) of Scheme **I** for the reaction in the presence of carbon monoxide at 25°C:

$$>C{=}C< + 2HCo(CO)_4 + CO \longrightarrow -\underset{H}{\overset{|}{C}}-\underset{CHO}{\overset{|}{C}}- + Co_2(CO)_8 \qquad \textbf{(a)}$$

$$\left.\begin{array}{l} >C{=}C< + HCo(CO)_4 \longrightarrow -\underset{H}{\overset{|}{C}}-\underset{|}{\overset{|}{C}}Co(CO)_4 \rightleftarrows -\underset{H}{\overset{|}{C}}-\underset{|}{\overset{|}{C}}COCo(CO)_3 \xrightarrow{CO} \\ -\underset{H}{\overset{|}{C}}-\underset{|}{\overset{|}{C}}COCo(CO)_4 \xrightarrow{HCo(CO)_4} -\underset{H}{\overset{|}{C}}-\underset{|}{\overset{|}{C}}-CHO + Co_2(CO)_8 \end{array}\right\} \textbf{(b)}$$

Scheme **I**

The reaction should occur according to the steps indicated in (**b**) of Scheme **I**. No reaction seems to occur at 0°C, owing to stabilization of $HCo(CO)_4$ by CO. Since acylcobalt tetracarbonyls are rather stable at room temperature in the presence of CO, addition of $HCo(CO)_4$ to a large excess of olefin results in consumption of most of $HCo(CO)_4$ due to formation of the acyl intermediate. As a consequence, only a small quantity of $HCo(CO)_4$ is available for reducing the acyl derivative to aldehyde. The yield of aldehyde under these conditions decreases on increasing the olefin concentration (19). When the CO atmosphere used for the reaction between $HCo(CO)_4$ and excess of olefin is replaced by N_2, the aldehyde yield increases considerably. A disproportionation of the acylcobalt carbonyls to aldehydes and olefins was suggested to explain this behavior (20).

Also the addition of a polar solvent such as ether to the pentane solution seems to affect the aldehyde yield. As shown in Table 1, the yield increases considerably with increase of the ether concentration (20).

TABLE 1
Stoichiometric Hydroformylation of 1- and cis-2-Pentene in Pentane. Effect of Ethyl Ether (20)[a]

Ether (ml)	Pentene	Aldehydes Yield (%)	St[b] (%)	Br[c] (%)
0	1-	42	43	57
0	2-	30	41	59
4	1-	51	53	47
4	2-	38	47	53
7	1-	62	58	42
7	2-	41	50	50

[a] 1.8 mmoles $HCo(CO)_4$; 10 mmoles pentene; total solvent 10 ml; 4 hr under N_2 and then quenched with PPh_3.
[b] Straight-chain.
[c] Branched-chain.

The isomeric composition of the aldehyde mixture formed depends strongly on the concentration of the olefin and on the CO pressure. The branched-chain isomer prevails at high olefin concentration and in the absence of CO. When using an olefin to $HCo(CO)_4$ ratio of two at 25°C, practically only the branched-chain aldehyde results in the absence of carbon monoxide; in the presence of CO the normal isomer prevails. In the first case the residual olefin was largely isomerized, while in the second case practically no isomerization had taken place (19). The presence of ether also seems to increase the ratio of straight- to branched-chain aldehydes (Table 1).

Takegami and co-workers (21) have determined the ratio of straight- to branched-chain acylcobalt carbonyls obtained by reacting 1-pentene with $HCo(CO)_4$ at 10°C when using toluene or toluene–ether (1:1) as solvent (Table 2). The ratio was determined by reacting the acyl derivatives with iodine and ethyl alcohol; the amount of branched isomer increased from 28% to 50% when using the toluene–ether mixture instead of toluene alone.

The rate of the stoichiometric hydroformylation is higher for terminal than for internal olefins (17); it is greatly enhanced by addition of nucleophiles such as C_6H_5CN, the maximum rate being achieved for a ratio of C_6H_5CN to $HCo(CO)_4$ of about two (25). The addition of C_6H_5CN favors formation of the straight-chain isomer and decreases the rate of double-bond isomerization (25). Stronger nucleophiles such as

TABLE 2
Interaction of $HCo(CO)_4$ *with Various Olefins*

Substrate	Temp (°C)	P_{CO} (atm)	Isolated products[a]	Distribution (%)	Ref.
Propylene[b]		0	Butyraldehyde	36	22
			Isobutyraldehyde	64	
Propylene[b]		5	Butyraldehyde	60	22
			Isobutyraldehyde	40	
Isobutylene	0	0.6	Methyl trimethylacetate	100	23
1-Pentene	0	[c]	Methyl hexanoate	70–50	23
			Methyl 2-methylpentanoate	30–50	
1-Pentene	15	1	Ethyl hexanoate	75	24
			Ethyl 2-methylpentanoate	25	
1-Pentene	0	1	Ethyl hexanoate	75	24
			Ethyl 2-methylpentanoate	25	
1-Pentene	−10	1	Ethyl hexanoate	70	24
			Ethyl 2-methylpentanoate	30	
1-Pentene	10	1	Ethyl hexanoate	72	21
			Ethyl 2-methylpentanoate	28	
1-Pentene[d]	10	1	Ethyl hexanoate	50	21
			Ethyl 2-methylpentanoate	50	
1-Pentene[e]	10	1	Ethyl hexanoate	34	21
			Ethyl 2-methylpentanoate	66	
Styrene	25	1	Ethyl 2-phenylpropionate	30	24
			Ethyl 3-phenylpropionate	70	
Styrene	−20	1	Ethyl 2-phenylpropionate	97	24
			Ethyl 3-phenylpropionate	3	
Butyl vinyl ether	25	1	Ethyl 2-butoxypropionate	3	24
			Ethyl 3-butoxypropionate	97	
Butyl vinyl ether	−20	1	Ethyl 2-butoxypropionate	97	24
			Ethyl 3-butoxypropionate	3	
Ethyl acrylate	25	1	Diethyl methylmalonate	3	24
			Diethyl succinate	97	
Ethyl acrylate	−10	1	Diethyl methylmalonate	97	24
			Diethyl succinate	3	
Ethyl acrylate	0	[c]	Diethyl methylmalonate	30	24
			Diethyl succinate	70	
Ethyl acrylate	−10	[c]	Diethyl methylmalonate	40	24
			Diethyl succinate	60	

[a] By treating the products of the reaction between $HCo(CO)_4$ and olefins with I_2 and alcohols, except for propylene.
[b] Vapor phase.
[c] $P_{N_2} = 1$ atm.
[d] 1:1 toluene–diethyl ether as solvent.
[e] 1:1 toluene–acetonitrile as solvent.

pyridine and triphenylphosphine inhibit the reaction (25). This was later confirmed by the observation that $HCo(CO)_3PBu_3$ does not react with ethylene at room temperature (26). In these experiments it was shown that $CH_3Co(CO)_3PBu_3$ reacts easily with CO at room temperature and that the acyl derivative thus obtained easily reacts at the same temperature with H_2 to yield acetaldehyde (26). The strong nucleophiles, therefore, seem to hinder the reaction between the substituted hydrocarbonyls and the olefin. Ethyl vinyl ether, however, was reported to react at 25°C with trialkyl- and triphenylphosphine-substituted cobalt tricarbonyl hydrides to yield the corresponding acylcobalt carbonyl derivatives (305).

Stoichiometric hydroformylation with $HCo(CO)_4$ was also attempted by using α,β-unsaturated carbonyl compounds (27) and nonconjugated diolefins such as 1,4-pentadiene (28,29). In the first case, only hydrogenation of the double bond was observed. In the second case, 5-hexenal and 2-methylcyclopentanone were obtained at room temperature, but on heating the mixture at 120°C, 1-formylcyclohexene was obtained; this compound probably arose from an intramolecular aldol condensation of heptanedial, which should be the primary reaction product.

In connection with stoichiometric hydroformylation, a series of investigations was carried out involving only the interaction between $HCo(CO)_4$ and olefins. These reactions were conducted by mixing solutions of $HCo(CO)_4$ and olefin and decomposing the acylcobalt carbonyl derivatives thus obtained with iodine and methanol. Methyl esters were formed in this manner (Scheme **II**).

$$>C{=}C< + HCo(CO)_4 \longrightarrow -\overset{|}{\underset{H}{\overset{|}{C}}}-\overset{|}{\underset{|}{C}}-CO-Co(CO)_3 \xrightarrow{I_2}$$

$$-\overset{|}{\underset{H}{\overset{|}{C}}}-\overset{|}{\underset{|}{C}}-CO-I \xrightarrow{CH_3OH} -\overset{|}{\underset{H}{\overset{|}{C}}}-\overset{|}{\underset{|}{C}}-CO-OCH_3$$

Scheme **II**

The kinetics of the reaction of $HCo(CO)_4$ with 1-hexene was investigated by Gankin and Dvinin (295).

From results reported in Table 2 it appears that even slight changes in reaction conditions have a large influence on the isomeric composition of the reaction products. According to Takegami and co-workers (30) this is due not only to a change in the mode of addition of $HCo(CO)_4$ to the double bond but, as clearly demonstrated by the reaction between ethyl acrylate and $HCo(CO)_4$ under different conditions, to a possible isomerization of the acylcobalt carbonyls. No isomerization of acylcobalt carbonyls was found under oxo conditions (80°C, $P_{CO} = 80$ atm, $P_{H_2} =$ 80 atm) (288).

B. Catalytic Hydroformylation

Although stoichiometric hydroformylation can be accomplished even at room temperature and atmospheric pressure, the catalytic reaction with cobalt as catalyst, on the other hand, generally requires temperatures between 80 and 160°C, CO pressures above 10 atm, and H_2 pressures over 50 atm.

The mildest conditions may be used when $Co_2(CO)_8$ is the catalyst precursor. Metallic cobalt or cobalt salts are commonly employed since they are more easily available than $Co_2(CO)_8$; but when using them, higher temperatures and pressures are necessary to avoid long induction periods connected with the transformation of the cobalt compounds into the catalytic species. The synthesis of cobalt carbonyls is accelerated by preformed $Co_2(CO)_8$ and Lewis bases (290).

Common reaction conditions are 120–140°C and 200 atm total pressure of a CO and H_2 mixture (1:1). With these conditions, a suitable reaction rate is achieved and formation of secondary products is minimized.

During hydroformylation, the cobalt catalyst unfortunately also catalyzes secondary reactions that lead to the loss of the primary reaction products; aldehydes must therefore be separated from the reaction mixture as quickly as possible. In laboratory practice this can be achieved by rapidly cooling the reaction vessel and distilling the products at the lowest possible temperature to separate aldehydes from high-boiling products and catalysts. The aldehyde then can be rectified under N_2 without significant loss. In industrial practice, separation of the aldehyde and recovery and recirculation of the catalyst are among the most delicate steps. There is a great deal of patent literature (31–37) and some reviews (3,10) specifying operations intended to avoid losses of catalyst and of useful reaction products.

In laboratory practice the yield of aldehydes can be improved by tying up the carbonyl groups as soon as formed by suitable reagents such as orthoformic esters (38), acetic anhydride (39), or ethylene glycol (40). When orthoformates are used, the hydroformylation is accompanied by the formation of aldehydes according to reaction 9 (see the first chapter).

$$HC(OC_2H_5)_3 \xrightarrow[CO+H_2]{Co_2(CO)_8} CH_3CH_2CHO + HCOOC_2H_5 + C_2H_5OH \quad (9)$$

1. Catalysts

Since the oxo reaction was discovered during a study of the Fischer-Tropsch process, the first hydroformylation catalysts were typical Fischer-Tropsch catalysts (1): $Co/ThO_2/MgO$/kieselguhr = 100/5/8/200. It was soon recognized that ThO_2 and MgO had no favorable effect on catalyst activity in hydroformylation: therefore, cobalt on kieselguhr or Raney cobalt was commonly used between 1945 and 1950. Since under usual reaction conditions part of the cobalt used was always dissolved in the organic reaction medium, Roelen (12) and then Adkins and Krsek (11) suggested that the true catalyst was a soluble cobalt carbonyl complex. Since then an increasing use of $Co_2(CO)_8$ as catalyst precursor dissolved in the reaction medium had appeared in the literature. But after 1954, practically all laboratory investigations on hydroformylation were carried out by using soluble cobalt compounds as catalyst precursors.

Such use of soluble catalysts, however, does not exclude the possibility of carrying out the reaction almost exclusively on heterogeneous catalysts. For instance, the hydroformylation of octadecene can be carried out at a total pressure of 20 atm in the presence of cobalt on kieselguhr, which gives rise to practically colorless reaction products that apparently contain only traces of soluble cobalt compounds (41–43). If the hydroformylation is conducted without careful control of temperature and concentration of CO in solution, catalytically active insoluble products can form. In this case, heterogeneous and homogeneous catalysis of hydroformylation can coexist and cause fluctuations in the reaction rate and in composition of the reaction products.

Research on heterogeneous catalysts has not been abandoned, as revealed by the patent literature (e.g., ref. 44). The main problem with heterogeneous catalysis is to find a support that does not release cobalt under hydroformylation conditions, in order to avoid losses of cobalt in the gas phase and the separation of the dissolved catalyst from the reaction products. Different inorganic materials such as SiO_2, Al_2O_3, and metal oxides (45) have been suggested as supports. Cobalt and rhodium complexes covalently anchored to silica were described (46). We shall not review the research in this field, since it is beyond the scope of this chapter. More recently attempts have been made to support cobalt and rhodium on crosslinked polymers containing phosphorous or nitrogen in the hope that the metal remains attached to the polymer (47,48). Polyvinylpyridine is not suitable as a support because it releases cobalt under reaction conditions (49). Interesting results might arise both from this approach and from the use of supported liquid phase catalysts as suggested for rhodium (50).

A few attempts to carry out the hydroformylation of propylene in the gaseous phase were reported (51,52). The formation of a smaller amount of secondary products was claimed for gas-phase operation than for reaction under conventional conditions. Formation of aldehydes with 99% selectivity was obtained for propylene hydroformylation in the presence of cobalt zeolites (53). Vapor phase hydroformylation of pentenes and hexenes was also attempted (54).

Despite doubts raised on the homogeneous nature of the hydroformylation reaction (55), much of the experimental findings shows that under appropriate conditions hydroformylation occurs substantially in the liquid phase.

Different cobalt compounds soluble in the reaction medium have been used as catalyst precursors for liquid-phase hydroformylation; it is generally assumed that under reaction conditions they are completely transformed into $Co_2(CO)_8$ or $HCo(CO)_4$. In fact, as already reported (Vol. 1, first chapter), the synthesis of the above catalytic precursors or intermediates occurs at temperatures and pressures of CO and H_2 similar to those generally chosen for the hydroformylation.

Different methods have been proposed to accelerate formation of $Co_2(CO)_8$ and $HCo(CO)_4$ from cobalt salts, such as the use of noble metals (56,57) or addition of $Fe(CO)_5$ (58,59). A great deal of literature exists also on promoters and inhibitors of the oxo reaction. The inhibiting effect of sulfur compounds was studied mainly by Markó and co-workers, who synthesized a group of sulfur-containing cobalt carbonyls and investigated their catalytic activity (60–68).

The effect of bases on the hydroformylation reaction was also examined (6,13,69–72). Strong bases like trialkylamines inhibit the reaction, whereas an acceleration was noticed in the presence of pyridine or lutidine when used in equimolar amount with $Co_2(CO)_8$.

The inhibiting effect of acids was discussed by Macho and Komora (73); water has no inhibiting effect on hydroformylation (70) and it was even used as a solvent (74,75). The inhibiting effect of water in the hydroformylation of unsaturated esters was attributed to formation of free acids by hydrolysis (76).

The possibility of modifying the activity of the cobalt catalyst by using ligands capable of displacing one or more molecules of CO under hydroformylation conditions evidently depends on the equilibria shown in reactions10–16 (26,78–83,296). These regulate the relative concentrations of the different cobalt complexes.

$$[Co(CO)_3L_2]^+[Co(CO)_4]^- \rightleftarrows Co_2(CO)_6L_2 + CO \qquad (10)$$

$$Co_2(CO)_6L_2 + CO \rightleftarrows Co_2(CO)_7L + L \qquad (11)$$

$$Co_2(CO)_6L_2 + H_2 \rightleftarrows 2HCo(CO)_3L \quad (12)$$

$$Co_2(CO)_7L + H_2 \rightleftarrows HCo(CO)_4 + HCo(CO)_3L \quad (13)$$

$$HCo(CO)_3L + CO \rightleftarrows HCo(CO)_4 + L \quad (14)$$

$$HCo(CO)_3L + L \rightleftarrows HCo(CO)_2L_2 + CO \quad (15)$$

$$HCo(CO)_2L_2 + L \rightleftarrows HCo(CO)L_3 + CO \quad (16)$$

In order to stabilize carbonylation catalysts, several ligands were used by Reppe and co-workers as early as 1941 in their investigation of the synthesis of acrylic esters (84). The use of ligands capable of displacing CO in cobalt complexes used as hydroformylation catalysts, however, was systematically investigated only after 1960, when it was discovered that improved selectivity for linear products could be achieved by operating in the presence of trialkyl and triarylphosphines and other ligands (5,78,85–95,301–303).

Since the pioneering work of Slaugh and Mullineaux (85,87), many investigations in this field were carried out by different groups. The following characteristics of the catalytic systems containing trisubstituted phosphines became apparent: Hydroformylation and olefin isomerization rates and formation of high-boiling products decrease, and selectivity toward the synthesis of straight-chain compounds increases. But due to the high hydrogenating power of the new catalytic systems, selectivity toward aldehyde formation with respect to olefin hydrogenation and alcohol formation decreases; catalyst stability is largely improved and low CO and H_2 pressures are sufficient to stabilize the catalytic complexes up to 200°C.

Owing to inadequate knowledge of the equilibria shown in reactions 10–16 at high temperatures and high CO and H_2 pressures, and because only qualitative spectroscopic evidence of the existence of different catalytic species is available (79,82), this research is mainly empirical and it is still not possible to tailor the catalyst best suited to obtain the product specifically desired from an olefin. On the other hand, asymmetric hydroformylation (96) can be used as a tool to establish whether at least one of the catalytic species responsible for hydroformylation contains a ligand other than an olefin, hydrogen, or CO. The effect of ligands different from CO and hydrogen on reaction kinetics, isomeric composition of the reaction products, and secondary reactions will be discussed in Sections II-B-6, 4, and 3, respectively.

2. *Substrates*

In general when an olefinic substrate is hydroformylated, the formyl group can add to either carbon of the double bond. Substitution of a formyl group on an initially saturated carbon atom accompanied by a hydrogen shift has also been observed experimentally. Three cases are therefore possible:

(i). Only one aldehyde is formed. This occurs when the carbon atoms of the double bond are equivalent, as for ethylene and unsubstituted cycloolefins, in which double-bond shifts cannot give more than one isomer.

(ii). Two aldehydes are obtained. This occurs when no double-bond shift can take place (as for styrene, *tert*-butylethylene, methyl methacrylate) or when only two different positions exist where a formyl group can be added (as for propylene, butene, 1,4-diphenyl-2-butene, etc.).

(iii). More than two aldehydes can be formed. This occurs with pentenes, hexenes, crotonic acid esters, and so on. Here, through a hydrogen shift (Section II-B-3), more than two different positions exist for attack by the formyl group.

The formation of isomeric reaction products, often difficult to separate, is one of the main drawbacks of the hydroformylation reaction. Control of the composition of the isomeric products, which has been accomplished at least up to a certain extent, will be discussed in Section II-B-4.

A survey of the olefins that undergo hydroformylation was the subject of several reviews (3,6,8,10).

We shall mention here only some of the most typical classes of unsaturated substrates that undergo hydroformylation in the presence of cobalt catalysts, in order to give an idea of the usefulness of the reaction as a synthetic tool.

a. Alkenes (Aliphatic and Cyclic Olefins). From investigations on the hydroformylation of aliphatic and cyclic olefins (97,98), it was found that the formyl group preferentially adds to a terminal carbon atom and that addition to a tertiary carbon atom occurs very slowly. The interpretation of results obtained in hydroformylation reactions is complicated by the simultaneous occurrence of hydroformylation and double-bond isomerization.

Preferential attack of the formyl group on terminal carbon atoms occurs with linear olefins, as shown in Table 3. For internal linear olefins of higher molecular weight, virtually all the possible isomers were detected when operating at 150°C (101,102). In the hydroformylation of some internal-branched olefins, essentially only attack of the formyl group at the initially saturated terminal carbon atoms was detected (97). Interpretations of these facts will be discussed later.

Olefins containing aryl groups have been relatively little investigated; some data are reported in Table 4. Aryl-substituted olefins in which the

TABLE 3
Hydroformylation of Olefins

Olefin	Temp (°C)	P_{CO}[a] (atm)	P_{H_2}[a] (atm)	Products	Distribution (%)	Ref.
1-Butene	110	2	80	Pentanal	53	99
				2-Methylbutanal	47	
1-Butene	110	140	80	Pentanal	79	99
				2-Methylbutanal	21	
cis-2-Butene	110	2	80	Pentanal	53	99
				2-Methylbutanal	47	
cis-2-Butene	110	140	80	Pentanal	71	99
				2-Methylbutanal	29	
1-Pentene	100	1.7	80	Hexanal	56	99
				2-Methylpentanal	36	
				2-Ethylbutanal	9	
1-Pentene	100	90	80	Hexanal	82	99
				2-Methylpentanal	16	
				2-Ethylbutanal	3	
2-Pentene	100	1.7	80	Hexanal	56	99
				2-Methylpentanal	36	
				2-Ethylbutanal	9	
2-Pentene	100	93	80	Hexanal	76	99
				2-Methylpentanal	19	
				2-Ethylbutanal	5	
1-Hexene	100	75	100	Heptanal	82	100
				2-Methylhexanal	14	
				2-Ethylpentanal	4	
2-Hexene	100	75	100	Heptanal	69	100
				2-Methylhexanal	21	
				2-Ethylpentanal	10	
1-Octene	150	100[b]	100[b]	Nonanal	65	101
				2-Methyloctanal	22	
				2-Ethylheptanal	7	
				2-Propylhexanal	6	
trans-4-Octene	150	100[b]	100[b]	Nonanal	55	101
				2-Methyloctanal	22	
				2-Ethylheptanal	11	
				2-Propylhexanal	12	

[a] Experiments at constant pressure.
[b] Initial pressure at room temperature.

double bond is conjugated with the aryl group are mainly hydrogenated to the corresponding phenylalkane. The possibility that hydrogenation occurs by a radical mechanism has been discussed (291). When the double bond is not conjugated with the aryl group, the expected aldehydes are formed. However, in the case of safrole (109) and eugenol (110) extensive hydrogenation or cyclization to tetrahydronaphthalene derivatives also takes place. The cyclization mechanism was investigated by Markó et al. (109). Silicon-containing olefins were also hydroformylated (111).

Optically active aldehydes have been obtained by starting with optically active olefins in the presence of conventional catalysts (112,113) and from prochiral olefins or finally from racemic olefins by using optically active cobalt complexes as catalysts or by adding chiral ligands to conventional cobalt catalysts (96).

The hydroformylation of optically active olefins has been very useful in the investigation of the reaction mechanism. Aldehydes having the same optical purity as the starting olefin or partially or completely racemized aldehydes were obtained from (*S*)-3-methyl-1-pentene, depending on the reaction conditions used (112). When the formyl group adds to an originally saturated carbon atom, partial racemization was observed (Table 5).

Up to now the most successful asymmetric hydroformylations in the presence of cobalt catalysts (Table 6) have been obtained when using styrene as substrate and $Co_2(CO)_8$ plus bis(N-α-methylbenzyl)salicylaldimine. When ethyl orthoformate was present, 2-phenylpropanal diethyl acetal having an optical purity (Table 6) of about 20% was obtained (96). The low optical yield in the absence of ethyl orthoformate was attributed to racemization of the free aldehyde. Much lower optical purity is shown by aldehydes obtained with the same catalytic systems starting with α-methylstyrene and α-ethylstyrene. Very low optical activity was exhibited by the chiral aldehydes produced from 1-butene or in the partial hydroformylation of racemic 3-methyl-1-pentene when using the same catalyst (156).

b. Diolefins and Polyolefins. The problem of preparing dialdehydes or polyaldehydes by hydroformylation of diolefins or polyolefins has not yet been successfully solved with cobalt catalysts. Conjugated diolefins yield practically only saturated monoaldehydes (118); nonconjugated diolefins or polyolefins generally yield a mixture of monoaldehydes and polyaldehydes (119–122,298). In the few cases when dialdehydes formed, they were immediately hydrogenated and finally isolated as glycols (Table 7).

TABLE 4
Hydroformylation of Aryl-substituted Olefins

Olefin	Temp (°C)	P_{CO} [a] (atm)	P_{H_2} [a] (atm)	Products	Yield (%)	Ref.
Styrene	120	100–150	100–150	2-Phenylpropanal	30	2
Styrene	120	40	40	2-Phenylpropanal	27	96
				3-Phenylpropanal	19	
				Ethylbenzene	52	
α-Methylstyrene	130	70	70	3-Phenyl-3-methylpropanal	40	103
				Isopropylbenzene	60	
α-Ethylstyrene	120	50	50	3-Phenylpentanal	11	104
				4-Phenylpentanal	7–9	
				2-Methyl-3-phenylbutanal	0.7	
				2-Phenylbutane	80	
1-Phenylpropene	140	100	100	2-Phenylbutanal	2.7	105
				2-Benzylpropanal	1.5	
				4-Phenylbutanal	1.0	
				Propylbenzene	87	
Allylbenzene	140	80	80	4-Phenylbutanal	37.5	106
				2-Benzylpropanal	9.5	
				2-Phenylbutanal	7	
				Propylbenzene	46	

1-Phenyl-1-butene	140	110	110	Butylbenzene	90	107
3-Phenyl-1-butene	130	110	110	4-Phenylpentanal	30	107
2-Phenyl-2-butene	145	108	108	*sec*-Butylbenzene	95	107
1-Phenyl-2-methylpropene	120	80	80	3-Methyl-4-phenylbutanal	11.5	108
				2-Phenyl-3-methylbutanal	4	
				Isobutylbenzene	84	
1,1-Diphenylpropene	120	80	80	1,1-Diphenylpropane	100	108
1-Vinylnaphthalene	120–125	100–150	100–150	2-(1-Naphthyl)-propanal	29	2
Safrole	150	115	115	Dihydrosafrole	48	109
				5,6,7,8-Tetrahydro-2,3-methylenedioxynaphthalene	24	
Eugenol	180	150	150	3-Methoxy-4-hydroxy-1-propylbenzene	41	110
				5,6,7,8-Tetrahydro-3-methoxy-2-naphthol	37	
				4-(3-Methoxy-4-hydroxyphenyl)-1-butanol	5	

[a] Initial pressure at room temperature.

TABLE 5
Hydroformylation of Optically Active Olefins[a]

Olefin	Temp (°C)	P_{CO} (atm)	P_{H_2} (atm)	Products[b]	Yield (%)	Optical yield (%)	Ref.
(+)-(*S*)-3-Methyl-1-pentene	100	102	80	(+)-(*S*)-4-Methylhexanal	81.0	98.0	113
				Threo-(2*R*,3*S*)-2,3-dimethylpentanal	1.9	98.9[c]	
				Erythro-(2*S*,3*S*)-2,3-dimethylpentanal	1.0	98.9[c]	
				3-Ethylpentanal	3.9	—	
(+)-(*S*)-3-Methyl-1-pentene	100	2.2	80	(+)-(*S*)-4-Methylhexanal	61.4	25.6	113
				2,3-Dimethylpentanal diastereomers	4.7	n.d.	
				3-Ethylpentanal	8.8	—	
(+)-(*S*)-3-Methyl-1-hexene	110	100	100	(*S*)-4-Methylheptanal	74.4	n.d.	114
				2,3-Dimethylhexanal diastereomers	2.7	n.d.	
				(*R*)-3-Ethylhexanal	2.9	70.0	
(−)-(*S*)-4-Methyl-1-hexene	100	80	80	(*S*)-5-Methylheptanal	69.7	95.0	115
				2,4-Dimethylhexanal diastereomers	8.0	n.d.	
				(*S*)-3-Ethylhexanal	1.3	72.0	

(+)-(*S*)-4-Methyl-2-hexene	110	80	80	(*S*)-5-Methylheptanal	60.0	94.0	116, 117
				2,4-Dimethylhexanal diastereomers	13.3	n.d.	
				(*S*)-3-Ethylhexanal	3.9	72.0	
				(*R*)-4-Methylheptanal	2.5	n.d.	
(+)-(*S*)-5-Methyl-1-heptene	100	80	80	(*S*)-6-Methyloctanal	61.6	94.0	115
				2,5-Dimethylheptanal diastereomers	15.1	n.d.	
				2-Ethyl-4-methylhexanal	1.9	n.d.	
				(*S*)-3-Ethylheptanal	1.0	74.0	
(+)-(*S*)-2,2,5-Trimethyl-3-heptene	110	95	95	(*S*)-3-Ethyl-6,6-dimethylheptanal	40.3	74.0	116, 117
				(*R*)-4,7,7-Trimethyloctanal	29.4	62.0	
				2,3,6,6-Tetramethylheptanal diastereomers	1.4	n.d.	
(+)-(*S*)-1-Phenyl-3-methyl-1-pentene	150	115	146	(+)-(*S*)-1-Phenyl-3-methylpentane	95.0	100	117
				Aldehydic products	1.0	n.d.	

[a] Benzene solvent; catalyst $Co_2(CO)_8$; experiments at constant pressure.
[b] Product oxidized to acids and determined as methyl esters.
[c] Relative to (3*S*) carbon atom.

TABLE 6

Hydroformylation of Various Olefins[a] *Using the Catalytic System* $Co_2(CO)_8$/(SAL-R)[b] (96, 104)

Olefin	Solvent	Temp (°C)	Products	Yield (%)	Asymmetric aldehyde: Prevailing chirality	Asymmetric aldehyde: Optical purity (%)
Styrene	Benzene	120	Ethylbenzene	52		
			2-Phenylpropanal	27	(*S*)	1.9
			3-Phenylpropanal	19		
Styrene	Ethanol	90	Ethylbenzene	63		
			2-Phenylpropanal	22	(*R*)	0.2
			3-Phenylpropanal	3		
Styrene	Ethanol + ethyl orthoformate	90	Ethylbenzene	65		
			2-Phenylpropanal diethyl acetal	20	(*R*)	17.1
			3-Phenylpropanal diethyl acetal	10		
α-Methylstyrene	Benzene	120	Cumene	89		
			3-Phenylbutanal	11	(*S*)	2.5
α-Methylstyrene	Ethanol + ethyl orthoformate	90	Cumene	91		
			3-Phenylbutanal diethyl acetal	5	(*R*)	0.5
α-Ethylstyrene	Benzene	120	2-Phenylbutane	94	—	0
			3-Phenylpentanal	2.5	(*S*)	1.4
			4-Phenylpentanal	3	(*R*)	0.07
			2-Methyl-3-phenylbutanal	1	n.d.	n.d.

[a] $P_{CO} = P_{H_2} = 40$ atm at 20°C.

[b] SAL-R = (+)-(*S*)-*N*-α-methylbenzylsalicylaldimine.

TABLE 7

Hydroformylation of Diolefins

Diolefin	Temp (°C)	P_{CO}[a] (atm)	P_{H_2}[a] (atm)	Products	Yield (%)	Ref.
Butadiene	150–175	95	95	Pentanal	12	118
				2-Methylbutanal	12	
Isoprene	150	90	90	Isomeric C_6 aldehydes	26	118
2,3-Dimethylbutadiene	175	145	145	3,4-Dimethylpentanal	45	118
3-Methyl-1,3-pentadiene	145	153	153	Isomeric C_7 aldehydes	33	118
Cyclopentadiene	145–155	126	126	Formylcyclopentane	37	118
Vinylcyclohexene	120–134	480–720[b]		Mono and dialdehydes	65	123
1,5-Cyclooctadiene	130	50	125	Hydroxymethylcyclooctane	70	124
				Bis(hydroxymethyl)cyclooctanes	19	
1,5-Cyclooctadiene	[c]			Hydroxymethylcyclooctane	26	125
				Bis(hydroxymethyl)cyclooctanes	63	
Cyclododecatriene	135–145	150–300[b]		Hydroxymethylcyclododecane	37	126

[a] Initial pressures at room temp.

[b] Operating pressures.

[c] Hydroformylation at 110°C and 200 atm (at 20°C) $CO + H_2 (1:1)$, followed by hydrogenation at 200°C and 300 atm.

Only monoaldehydes were obtained from cyclododecatriene; polyaldehydes were obtained by hydroformylating polybutadiene and styrene–butadiene copolymers (120).

The hydroformylation of aromatic nuclei does not occur under usual hydroformylation conditions, but under drastic conditions some positive results were reported (127,128). *N*-Formylpiperidine was obtained from pyridine (129).

c. Halogen-Containing Olefins. Hydroformylation of halogen-containing olefins presents some difficulties because hydrogen halide forms and then inactivates the catalyst (130). The best results were reported for the hydroformylation of hexafluoropropene (131) and of 2,4-dichloro- and 4-chlorophenyl allyl ethers (130).

Vinyl chloride gives α-chloropropionaldehydes (132), but 1,1-dichloroethylene could be hydroformylated only in the presence of stoichiometric amounts of $HCo(CO)_4$ (132).

d. Oxygenated Substrates. Unsaturated oxygenated compounds may be hydroformylated, with the exception of α,β-unsaturated aldehydes and ketones, which are mainly hydrogenated to the corresponding saturated carbonyl compounds or alcohols (2,27).

Generally the yields of hydroformylation products are lower with oxygenated substrates than with alkenes. No clear indications have been found of directing effects of the oxygen-containing group on the composition of the hydroformylation products.

Rather few examples are known of hydroformylation of unsaturated alcohols in the presence of cobalt catalysts (Table 8). The yields are usually low, although a mixture of bis-(hydroxymethyl)cyclohexanes in 70% yield was obtained by Falbe and others from the hydroformylation of tetrahydrobenzyl alcohol followed by hydrogenation of the resulting aldehyde (134). The hydroformylation of coniferyl alcohol is unusual in that a double bond conjugated with a phenyl ring is hydroformylated rather than hydrogenated (135).

Vinyl and allyl ethers and esters (121, 168) and acetals of α,β-unsaturated aldehydes were hydroformylated in rather low yields (Table 9).

In α,β-unsaturated cyclic ethers, the oxygen seems to direct addition of the formyl group to the α-position (Table 10).

For furan and 5,6-dihydro-γ-pyrans the formyl group adds to position 2 or to position 6, respectively; if these positions bear substituents, then the formyl group adds to positions 3 or 5. When the 3-substituent is a carbomethoxy group, the reaction takes place only at higher temperatures and the olefin is mainly hydrogenated. The same type of reaction was

TABLE 8

Hydroformylation of Unsaturated Alcohols

Substrate	Temp (°C)	P^a (atm)	Products	Yield (%)	Ref.
$CH_2{=}CHCH_2OH$	100	150	Propanal	7	133
			2-Formylpropanol, 3-Formylpropanol	17	
$CH_3CH{=}CHCH_2OH$	100–150	100–200	3-Formylbutanol	n.d.	38
$C_6H_9CH_2OH$	[b]	[b]	1,3- and 1,4-Bis(hydroxymethyl)-cyclohexane (52:48)	70	134
$HO-C_6H_3(OCH_3)-CH{=}CHCH_2OH$	170	100	3-(3-Methoxy-4-hydroxyphenyl)-tetrahydrofuran	25	135

[a] $CO:H_2 = 1:1$; initial pressure at room temp.

[b] Hydroformylation at 130°C and 200 atm $CO + H_2$ (1:2); initial pressure at reaction temp; followed by *in situ* hydrogenation at 240°C and 300 atm.

TABLE 9
Hydroformylation of Unsaturated Ethers and Acetals

Ether	Products	Yield (%)	Ref.
$CH_3OCH{=}CH_2$	3-Methoxypropionaldehyde	n.d.	136
n-$C_4H_9OCH{=}CH_2$	2-Butoxypropionaldehyde	31	2
$C_2H_5OCH_2CH{=}CH_2$	3-Ethoxyisobutyraldehyde	30	2
	2-Methylacrolein	6	
	4-Ethoxybutyraldehyde	4	
2,4-$Cl_2C_6H_3$-$OCH_2CH{=}CH_2$ (Cl–C₆H₃(Cl)–OCH₂CH=CH₂, structure)	γ-(2,4-Dichlorophenoxy)-butyraldehyde[a]	70	137
$C_6H_5OCH_2CH{=}CH_2$	4-Phenoxybutyraldehyde	84	138
$(C_2H_5O)_2CHCH{=}CH_2$	1,1-Diethoxy-3-formylpropane	16	139
	1,1-Diethoxy-2-formylpropane	4	
	1,1-Diethoxypropane	7	
$(C_2H_5O)_2CHCH{=}CHCH_3$	Methylsuccindialdehyde bis-diethyl acetal	50	38

[a] Characterized as the acid.

observed in the hydroformylation of some unsaturated sugar derivatives investigated principally by Rosenthal and his co-workers (148). Surprisingly, by using conventional hydroformylation conditions they obtained mainly alcohols. Only by very careful control of the amount of gas reacted was it possible to isolate the intermediate aldehydes, but in rather low yields. Under hydroformylation conditions no inversion of configuration of the asymmetric carbon atoms takes place, and in general two diastereomers were obtained when starting with trisubstituted dihydropyrans. From a tetrasubstituted dihydropyran only one product was obtained.

Unsaturated esters readily undergo hydroformylation in reasonably good yields. Only for cinnamic and other phenyl-substituted esters does double-bond hydrogenation largely occur under hydroformylation conditions (108,149). Some examples of the hydroformylation of esters are given in Table 11. Hydroformylation of C_{18} mono-, di-, and triunsaturated esters was studied extensively (43,154,155). The hydroformylation of methyl oleate showed that the formyl group is mainly bound to carbon atoms C_5 to C_{13}. However, a surprisingly large amount (5–15%) of linear

TABLE 10
Hydroformylation of Cyclic Unsaturated Compounds Containing Oxygen

Substrate	Products	Yield (%)	Ref.
 O	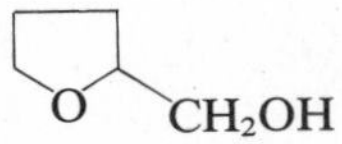 O CH_2OH	19–35	140–142
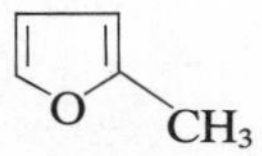 O CH_3	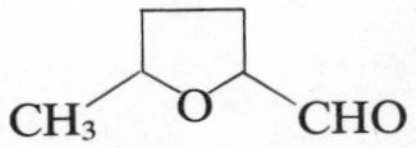 CH_3 O CHO	8	142
	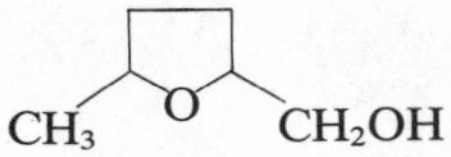 CH_3 O CH_2OH	15	
	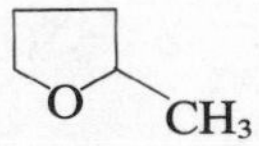 O CH_3	21	
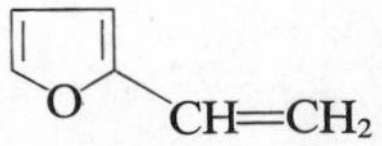 O $CH{=}CH_2$	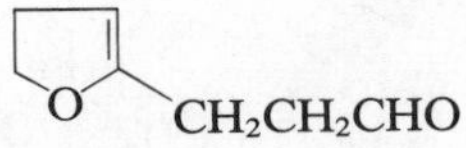 O CH_2CH_2CHO	28	142
	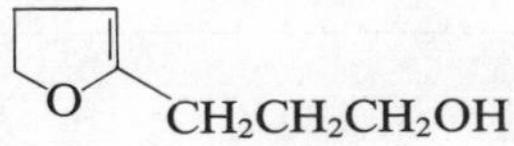 O $CH_2CH_2CH_2OH$	8	
	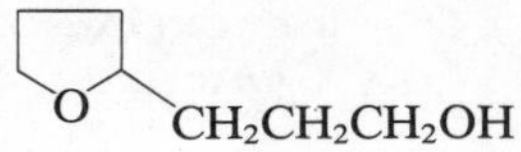 O $CH_2CH_2CH_2OH$	6	
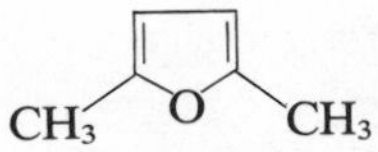 CH_3 O CH_3	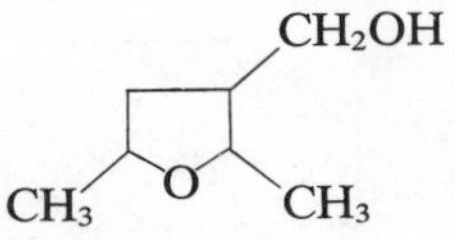 CH_2OH CH_3 O CH_3	23	141

TABLE 10 (*contd.*)

Substrate	Products	Yield (%)	Ref.
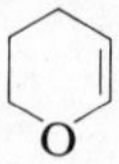 (3,4-dihydro-2H-pyran; O)	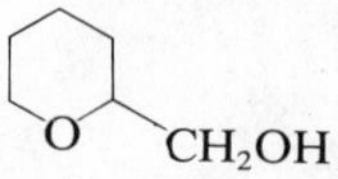 (tetrahydropyran-2-yl, O) CH_2OH	78	143
	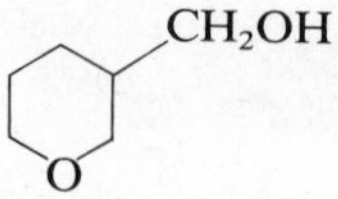 CH_2OH (tetrahydropyran-3-yl, O)	8	
	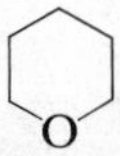 (tetrahydropyran, O)	3	
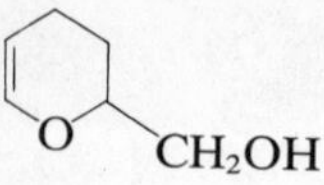 (O) CH_2OH	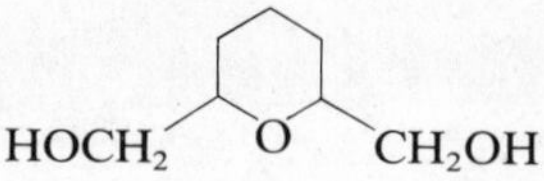 $HOCH_2$ (O) CH_2OH	77	143
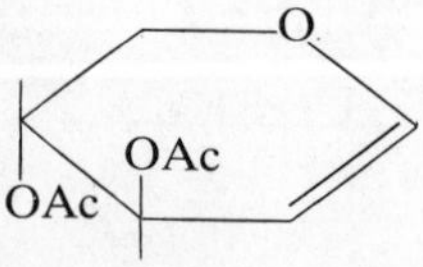 (O, OAc, OAc)	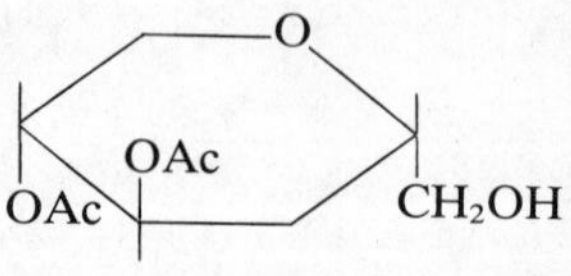 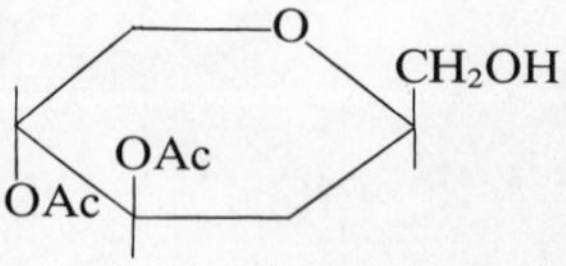 (O, OAc, OAc, CH_2OH); (O, CH_2OH, OAc, OAc)	>90	144, 145

TABLE 10 (*contd.*)

Substrate	Products	Yield (%)	Ref.
O; OAc; OAc	O; CH_2OH; OAc; OAc O; OAc; CH_2OH; OAc	>90	146
CH_2OAc; O; OAc; OAc	CH_2OAc; O; OAc; CHO; OAc CH_2OAc; O; CHO; OAc; OAc	70[a]	145
CH_2OAc; O; OAc; OAc; OAc	CH_2OAc; O; H_2COAc; OAc; OAc; OAc CH_2OAc; O; H_2COAc; OAc; OAc	~50	147

[a] The epimeric alcohols resulting from the accompanying reduction of the aldoses were obtained in about 30% yield; total yield, therefore, was essentially quantitative.

TABLE 11
Hydroformylation of Unsaturated Esters

Substrate	Products	Yield (%)	Ref.
Ethyl acrylate	Ethyl β-formylpropionate	57[a]	150
	Ethyl α-formylpropionate	18[a]	
Ethyl crotonate	Ethyl γ-formylbutyrate	59[a]	150
	Ethyl β-formylbutyrate	12[a]	
	Ethyl α-formylbutyrate	9[a]	
Methyl methacrylate	Methyl β-formylisobutyrate	60–65[b]	151
Methyl tiglate	Methyl α-methyl-γ-formylbutyrate	29	108
	Methyl α-ethyl-β-formylpropionate	19	
	Methyl α-methyl-β-formylbutyrate	1[c]	
Methyl cinnamate	Methyl β-formyldihydrocinnamate	29	108
Ethyl β-phenylcrotonate	Ethyl β-phenyl-γ-formylbutyrate	18	108
Ethyl β,β-diphenylacrylate	Ethyl β,β-diphenylpropionate	100	108
Methyl oleate	Isomeric formyl esters	78	108,152
	Isomeric alcohol esters	12	
	Methyl stearate	10	
Diethyl maleate	Diethyl α-formylsuccinate	62	153
Diethyl fumarate	Diethyl α-formylsuccinate	51	2
Diethyl itaconate	Diethyl α-(formylmethyl)succinate	56	153

[a] Reaction carried out in the presence of triethyl orthoformate; compounds isolated as diethyl acetals.

[b] As dimethyl acetal.

[c] Erythro/threo = 1 : 1.

product in which the formyl group was added to C_{18} is present when PBu_3 is added to the catalyst (154).

When the hydroformylation of unsaturated esters is carried out at higher temperatures, lactones are formed by aldehyde group reduction, followed by ring closure (149,157). This reaction was studied by Falbe and co-workers who examined a variety of substrates and showed that the hydroformylation of unsaturated esters substantially follows the same rules of hydroformylation as those for ordinary olefins. The formyl group is only rarely bound to a quaternary carbon atom; addition of the formyl group to the terminal methyl group is preferred over hydroformylation of internal double bonds. With conjugated double bonds, as in ethyl sorbate, only one formyl group is added, while the other double bond is hydrogenated (157).

Unsaturated compounds containing nitrogen do not react easily under typical hydroformylation conditions (129), probably because of formation of fairly stable complexes with the cobalt catalyst (158) (Table 12).

A very well known reaction is the hydroformylation of acrylonitrile (161–166) which yields up to 81% of β-formylpropionitrile (159); hydroformylation of other unsaturated nitriles was reported (155), but mainly in the patent literature (140,169). 2-Vinylpyridine was hydroformylated at 125°C and 300 atm to give 3-(2-pyridyl)propionaldehyde in good yield, but 4-vinylpyridine gave mainly 4-ethylpyridine (107). Finally, *N,N*-diacylallylamines are readily hydroformylated to give mainly the corresponding linear aldehydes (160).

The reactions of compounds with carbon-nitrogen and nitrogen-nitrogen double bonds with CO are discussed in Vol. 1, fifth chapter.

Hydroformylation of unsaturated organic compounds containing sulfur was studied (141,170), but only hydrogenation products were obtained.

TABLE 12
Hydroformylation of Unsaturated Compounds Containing Nitrogen

Substrate	Temp (°C)	P_{CO}[a] (atm)	P_{H_2}[a] (atm)	Products	Yield (%)	Ref.
Acrylonitrile	130	125	125	β-Formylpropionitrile[b]	81	159
				α-Formylpropionitrile	8	
				Propionitrile	3	
2-Vinylpyridine	125	150	150	3-(2-Pyridyl)-propionaldehyde	78	107
4-Vinylpyridine	125	150	150	4-Ethylpyridine	80	107
				1,4-di-(4-Pyridyl)-butane	10	
N,N-Diacetylallylamine	120	85	85	*N,N*-Diacetyl-4-amino-butyraldehyde	75	160
				N,N-Diacetyl-3-amino-isobutyraldehyde	11	
N-Vinylphthalimide	120	80	80	3-Phthalimidopropionaldehyde	69	160
				2-Phthalimidopropionaldehyde	28	
N-Allylphthalimide	120	83	83	4-Phthalimidobutyraldehyde	87	160
N-Styrylphthalimide	120	81	81	*N*-Phenethylphthalimide	66	160
				Isomeric aldehydes	7	

[a] Initial pressure at room temp., except for the vinylpyridines where the pressure is at reaction temp.

[b] Mainly as the acetal.

Despite the large number of substrates investigated, only very few are currently used in industry: The most important single olefin used is propylene followed by ethylene and the butenes. Largely used are also mixtures of C_7 olefins which are hydroformylated and then reduced to the corresponding C_8 alcohols.

One of the hydroformylation products of acrylonitrile can be used as starting material for the production of glutamic acid (171).

A better knowledge of the factors controlling the hydroformylation of more sophisticated substrates will probably favor greater use of this reaction both as a laboratory tool and as a useful synthetic method in industry. It will also help us better understand the mechanism of homogeneous catalysis by transition metal complexes and the related heterogeneous reactions.

3. *Secondary Reactions*

Because of the high reactivity of the starting materials and of their products, the hydroformylation of olefins is generally accompanied by both concurrent and successive reactions that often affect the yield of aldehydes. The most important concurrent reactions (172) are the hydrogenation and isomerization of the olefinic substrates as well as formation of ketones, carboxylic acids, and esters (reactions 17–21). No rearrangement of the carbon skeleton of olefinic substrates has been reported under typical oxo conditions, but this reaction cannot be excluded a priori since, for example, pinacol rearranges in the presence of CO, H_2, and $Co_2(CO)_8$ to give pinacolone and other products (173).

$$>C{=}C< + H_2 \longrightarrow H{-}\overset{|}{\underset{|}{C}}{-}\overset{|}{\underset{|}{C}}{-}H \tag{17}$$

$$CH_3CH_2CH{=}CH_2 \longrightarrow CH_3CH{=}CHCH_3 \tag{18}$$

$$2\ {>}C{=}C{<} + CO + H_2 \longrightarrow \left[H{-}\overset{|}{\underset{|}{C}}{-}\overset{|}{\underset{|}{C}}{-}\right]_2 CO \tag{19}$$

$$>C{=}C< + CO + HOH \longrightarrow H{-}\overset{|}{\underset{|}{C}}{-}\overset{|}{\underset{|}{C}}{-}COOH \tag{20}$$

$$>C{=}C< + CO + ROH \longrightarrow H{-}\overset{|}{\underset{|}{C}}{-}\overset{|}{\underset{|}{C}}{-}COOR \tag{21}$$

The most important of the successive reactions are hydrogenation of aldehydes to alcohols, condensation (aldolization), aldehyde polymerization, and acetal formation. Other reactions of less importance are the formation of formates, carbonylation of acetals, homologation of alcohols, and hydrogenation of the aldehydes to hydrocarbons (reactions 22–30).

$$RCHO + H_2 \longrightarrow RCH_2OH \qquad (22)$$

$$2RCH_2CHO \longrightarrow RCH_2CH(OH)CH(R)CHO \qquad (23)$$

$$RCHO + 2ROH \longrightarrow RCH(OR)_2 + H_2O \qquad (24)$$

$$3RCHO \longrightarrow \text{(cyclic trimer: 2,4,6-trialkyl-1,3,5-trioxane, ring of three O and three CHR)} \qquad (25)$$

$$RCHO + CO + H_2 \longrightarrow HCOOCH_2R \qquad (26)$$

$$RCH(OR')_2 + CO + H_2 \longrightarrow RCH(OR')CHO + R'OH \qquad (27)$$

$$RCH_2OH + CO + 2H_2 \longrightarrow RCH_2CH_2OH + H_2O \qquad (28)$$

$$RCH_2OH + H_2 \longrightarrow RCH_3 + H_2O \qquad (29)$$

$$2RCHO \longrightarrow RCOOCH_2R \qquad (30)$$

Reactions 27 and 28 are discussed in the first chapter; those aspects of the remaining reactions connected with hydroformylation will be discussed here.

a. Concurrent Reactions. (*1*) *Hydrogenation of olefinic substrates.* An exhaustive account of the hydrogenation of unsaturated substrates under hydroformylation conditions is given in an early publication of the Bureau of Mines (174). In the presence of $Co_2(CO)_8$ and a mixture of CO and H_2 (1 : 1) under a relatively high pressure, monoolefinic hydrocarbons are partially hydrogenated at fairly high temperatures. With highly branched hydrocarbons, hydrogenation may prevail over hydroformylation. Fluoroolefins are more easily hydrogenated than ordinary olefins (131,175). Conjugated diolefins undergo simultaneous hydrogenation and hydroformylation with formation of saturated aldehydes (Table 7).

When an olefinic double bond is conjugated with a phenyl ring, hydrogenation is surprisingly large (Table 4). However, with styrene below 150°C, hydroformylation is still the most important reaction, whereas under the same conditions more than 90% of hydrogenation products are obtained from α-methylstyrene and α-ethylstyrene (Table 4).

Anthracene undergoes only hydrogenation (174); other aromatic polycyclic hydrocarbons either do not react or undergo hydrogenation, but they are never hydroformylated (Table 13). Furan is both hydroformylated and hydrogenated giving rise to tetrahydrofurfuryl alcohol. α-β-Unsaturated aldehydes and ketones are mainly hydrogenated to the corresponding saturated carbonyl compounds (2,27). α-β-Unsaturated esters, however, are mainly hydroformylated (Table 11), although in some cases a small percentage of the corresponding saturated ester may be detected among the reaction products. Saturated esters were detected among the hydroformylation products of fatty acid esters (154,177).

Little is known of the hydrogenation of nitrogen-containing unsaturated compounds: Acrylonitrile is substantially hydroformylated (159,161,162), whereas hydrogenation is decreased when operating at low temperature and concentration of substrate (161); *N*-styrylphthalimide is substantially hydrogenated (160) (Table 12); vinyl and allylamines have been little investigated.

Sulfur-containing unsaturated substrates like thiophene and its derivatives are only hydrogenated under oxo conditions with formation of tetrahydrothiophene and a small amount of butane (170).

The hydrogenation of an olefinic substrate undoubtedly depends not only on the operating temperature but also on the partial pressure of CO. In general, by decreasing the CO pressure, the hydrogenation of the substrate increases. Although no systematic investigation has been carried out, a rapid hydrogenation of propylene to propane was observed when this olefin was treated at 50°C with H_2 in the presence of $Co_4(CO)_{12}$ in toluene. The solution obtained was black, and it was not established whether the reaction occurred in homogeneous or heterogeneous phase (178).

Increased hydrogenation of olefinic compounds was observed when using soluble cobalt catalysts containing trisubstituted phosphines (87,88); for example by using PBu_3 as a ligand 10–15% of the olefin was hydrogenated at 180°C (179). A cobalt cluster complex such as $Co_3(CO)_6L_3$ or hydrocarbonyls of the type $HCo(CO)_2(PBu_3)_2$ or $HCo(CO)(PBu_3)_3$, which are known to be powerful hydrogenating agents, might be responsible for this hydrogenation (83,180). Possible mechan-

TABLE 13

Hydrogenation of Some Unsaturated and Aromatic Compounds under Hydroformylation Conditions

Substrate	Temp (°C)	P_{H_2}[a] (atm)	P_{CO}[a] (atm)	Products	Yield (%)	Ref.
Isobutylene	200	100[b]	200[b]	Isobutane	53	176
				Isoamyl alcohol	35	
Hexafluoropropylene[c]				Hexafluoropropane	50	131
				Alcohols	40	
				Aldehydes	5–8	
Anthracene	150	100[b]	100[b]	9,10-Dihydroanthracene	99	174
Acrolein	120–125	100–150	100–150	Propionaldehyde	40–50	2
Crotonaldehyde	120–125	100–150	100–150	*n*-Butyraldehyde	40–50	2
Methyl vinyl ketone	120–125	100–150	100–150	Methyl ethyl ketone	70–90	2
Mesityl oxide	120–125	100–150	100–150	Methyl isobutyl ketone	70–90	2
Furan	180–185	140	70	Tetrahydrofurfuryl alcohol	35	141
Thiophene	180–185	145	72	Tetrahydrothiophene	8	141
Thiophene	200	100[b]	200	Tetrahydrothiophene	16	170
				Butane	0.2	

[a] At 20°C.
[b] Initial pressure at reaction temp.
[c] Reaction conditions not reported.

isms for this reduction were reviewed by Chalk and Harrod (9) and by Halpern (291). It seems plausible that the extent of the hydrogenation is in some way connected with the equilibria 31–32; hydrogenation is favored when the equilibria are displaced toward the alkylcobalt carbonyls which, depending on their structure, might possess different tendencies to undergo hydrogenolysis with saturation of the double bond originally present in the substrate and regeneration of the cobalt carbonyl hydride.

$$RCo(CO)_4 \rightleftharpoons [RCOCo(CO)_3] \tag{31}$$

$$RCo(CO)_4 \rightleftharpoons [RCo(CO)_3] + CO \tag{32}$$

$$C_6H_5CH_2Co(CO)_3(PPh_3) \underset{-CO}{\overset{+CO}{\rightleftharpoons}} C_6H_5CH_2COCo(CO)_3(PPh_3) \tag{33}$$

Benzylcobalt carbonyls (reaction 33) synthesized by Markó and coworkers (181) appear to be particularly stable in the presence of hydrogen. Compounds that give rise to secondary or tertiary benzylcobalt carbonyls, $C_6H_5CRR'COCo(CO)_4$, as hydroformylation intermediates, are readily hydrogenated; typical examples are styrene, α-methylstyrene, α-ethylstyrene, and (+)-(*S*)-1-phenyl-3-methyl-1-pentene. The very high optical yield (Table 5) obtained in the hydrogenation of (+)-(*S*)-1-phenyl-3-methyl-1-pentene to (+)-(*S*)-1-phenyl-3-methylpentane under hydroformylation conditions is a clear indication that the hydrogenation, as the hydroformylation of similar optically active olefins, takes place without formation of isomerization products like 1-phenyl-3-methyl-2-pentene which would necessarily cause racemization.

An interesting indication that might confirm the above hypothesis comes from the asymmetric hydroformylation of α-ethylstyrene (Scheme **III**).

In this case while at least two of the resulting aldehydes are optically active, the 2-phenylbutane is inactive. This result may be interpreted by assuming that the paraffin does not arise from the hydrogenolysis of the alkylcobalt carbonyl which gives rise (reaction 34) to the optically active 3-phenylpentanal. It might arise from hydrogenolysis of the quaternary alkylcobalt carbonyl (reaction 35) which, probably because of electronic

$$C_6H_5\overset{*}{C}H(C_2H_5)CH_2Co_x(CO)_y \longrightarrow C_6H_5\overset{*}{C}H(C_2H_5)CH_2CHO \tag{34}$$

$$C_6H_5C(CH_3)(C_2H_5){-}Co_x(CO)_y \xrightarrow{H_2} C_6H_5CH(CH_3)(C_2H_5) + HCo_x(CO)_y \tag{35}$$

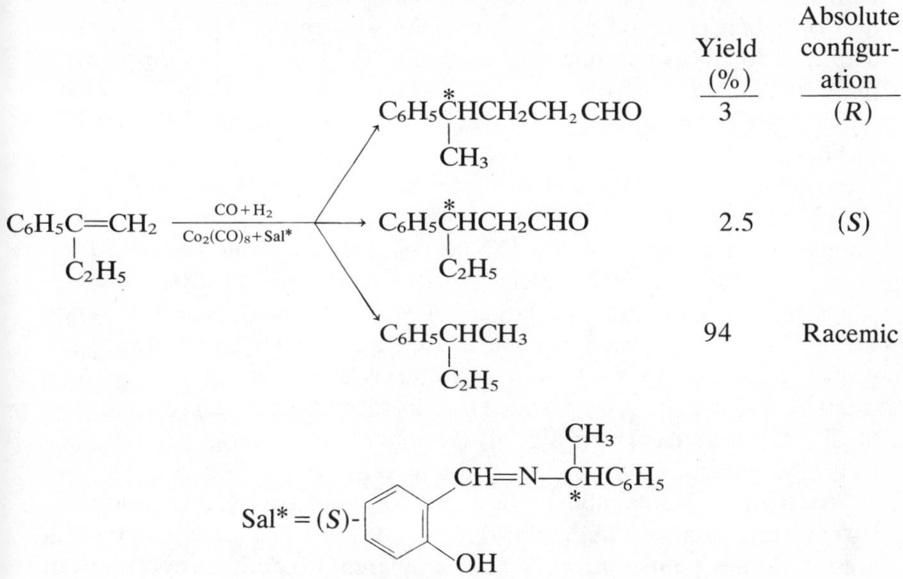

		Yield (%)	Absolute configuration
$C_6H_5C(C_2H_5)=CH_2$ $\xrightarrow[Co_2(CO)_8+Sal^*]{CO+H_2}$	$C_6H_5\overset{*}{C}H(CH_3)CH_2CH_2CHO$	3	(*R*)
	$C_6H_5\overset{*}{C}H(C_2H_5)CH_2CHO$	2.5	(*S*)
	$C_6H_5CH(C_2H_5)CH_3$	94	Racemic

Sal* = (*S*)-[2-hydroxyphenyl]–CH=N–$\overset{*}{C}H(CH_3)C_6H_5$ (OH ortho)

Scheme **III**

and steric factors, is sterically labile. This intermediate undergoes hydrogenolysis more readily than CO insertion and gives rise to completely racemized products. A free radical mechanism was suggested as an alternative explanation of the above data (291).

Stabilization of the —C—Co— bond, favoring hydrogenation with respect to —C—CO—Co— bond formation which in turn favors hydroformylation, was postulated to rationalize the hydrogenation of aldehydes and ketones having conjugated double bonds (27).

(2) *Isomerization of olefinic substrates.* Double-bond shifts in unsaturated compounds, which in the presence of $HCo(CO)_4$ occurs not only in the presence of solvents (297) but also in the vapor phase (182), was often observed under hydroformylation conditions (13,98,99,152,183–186). This shift was believed to be essentially responsible for the formation of aldehydes other than those expected on the basis of attachment of the formyl group to one of the two carbon atoms of the original double bond (97) and was therefore thoroughly investigated. Since this subject was reviewed by many authors (9,22,187), we shall limit our remarks to aspects closely connected with hydroformylation.

The relationship between olefin isomerization and isomeric composition of the hydroformylation products is more complicated than originally believed. Double-bond shift through a sequence of addition and elimination of cobalt carbonyl hydrides to olefin with formation of isomerized olefins (188,189) does not now seem the only mechanism responsible for formation of aldehydes formally derived by substitution of a formyl group for a hydrogen atom on a saturated carbon atom of the starting olefin.

Formation of isomerized olefins prior to hydroformylation at 80–120°C occurs only at low CO pressures (<50 atm), but at 150–190°C olefin isomerization occurs even at a CO partial pressure greater than 150 atm (99). To investigate this problem, reaction conditions must be chosen at which the rate of hydroformylation is low enough to allow equilibration of CO between the gaseous and liquid phases. Conflicting results are probably due either to a high hydroformylation rate (owing to high catalyst concentration, for instance), to inefficient temperature control or finally to insufficient stirring of the liquid that causes a diffusion-controlled concentration of CO in the liquid phase (186).

The findings stated above were obtained chiefly by comparing the hydroformylation of the pairs: 1-pentene/2-pentene and 1-butene/2-butene. At low temperatures (<120°C) different isomeric compositions of the reaction products were obtained at high CO pressure (>50 atm) in both cases, but the products from each pair of olefins was the same when using low CO pressures (~5 atm) (99).

Furthermore, in the hydroformylation of (+)-(*S*)-3-methyl-1-pentene at 120°C (112,113), where a double-bond shift yielding isomeric olefins must cause racemization, 4-methylhexanal having the same optical purity as the starting material was obtained when operating at high CO pressures, but 90% racemization was observed at low CO pressures.

Complete isomerization was observed by Asinger and Berg (190) in the hydroformylation of 1-dodecene at 150–200°C; by Johnson (186) using 4-methyl-1-pentene, particularly at high temperatures; and by Lai and others (108,152) using methyl oleate.

The different results obtained by changing the CO pressure seem to indicate the existence, under hydroformylation conditions, of at least two catalytic species having different CO/Co ratios—the one with a lower CO/Co ratio having a much higher catalytic activity for double-bond isomerization. The lower isomerization of the olefin observed at low temperatures (80–120°C) in the presence of catalysts containing PBu_3 can be explained on the same basis (91,94). In the presence of such phosphine ligands, higher temperatures (if other conditions are the same) are necessary to produce a sufficient concentration of coordinatively unsaturated catalytic species with lower CO/Co ratios. Only when operating at higher temperatures was the same reaction product composition observed from

1-octene and 2-octene (94) or 4-octene (101) in the presence or absence of phosphine ligands.

According to different research groups (22,187,191,192) the most probable mechanism of the double-bond shift in unsaturated substrates involves a series of additions and eliminations of a cobalt carbonyl hydride followed by dissociation of the olefin from the catalytic complex, but a 1–3 hydrogen shift catalyzed by cobalt compounds is possible in some cases. The isomerization of allyl alcohol to propionaldehyde (193) in the presence of $HCo(CO)_4$ might be considered as an example.

The formation of aldehydes having the formyl group attached to a carbon atom, which in the original substrate was a saturated carbon atom, takes place even when the hydroformylation is carried out under "nonisomerizing conditions," that is, conditions in which no double-bond shift is detectable by analysis of the unreacted olefins, and no racemization occurs when using optically active olefins. Some examples of this behavior are the production of straight-chain aldehydes by starting with straight-chain internal olefins, and the hydroformylation at the methyl group in branched olefins like 3-methyl-1-pentene, 3-methyl-1-hexene, and others (116,117).

To clarify this problem, the hydroformylation of optically active 3-methyl-1-hexene (114) and 3-methyl-1-hexene-3-d_1 (194) was carried out. From (+)-(*S*)-3-methyl-1-hexene was obtained (*R*)-3-ethylhexanal with only 30% racemization (reaction 36); also 3-(ethyl-1-d_1)hexanal was obtained from 3-methyl-1-hexene-3-d_1 (reaction 36a). Although the first result might in principle be explained by both a direct formylation of the methyl group in the C_3 position and a migration of the double bond with retention of configuration (114), the second result can only be accounted for by the second hypothesis (194).

$$\begin{array}{c} CH_2 \\ \| \\ CH \\ | \\ H{-}\overset{*}{C}{-}CH_3 \\ | \\ C_3H_7 \end{array} \xrightarrow[Co_2(CO)_8]{CO+H_2} \begin{array}{c} CH_3 \\ | \\ CH_2 \\ | \\ H{-}\overset{*}{C}{-}CH_2{-}CHO \\ | \\ C_3H_7 \end{array} \tag{36}$$

$$\begin{array}{c} CH_2 \\ \| \\ CH \\ | \\ D{-}C{-}CH_3 \\ | \\ C_3H_7 \end{array} \xrightarrow[Co_2(CO)_8]{CO+H_2} \begin{array}{c} CH_3 \\ | \\ CHD \\ | \\ H{-}C{-}CH_2{-}CHO \\ | \\ C_3H_7 \end{array} \tag{36a}$$

The shift of deuterium can be explained as a conventional double-bond isomerization by a series of cobalt hydrocarbonyl additions and eliminations (194) with formation of the isomerized olefins, 3-methyl-2-hexene and 2-ethyl-1-pentene. The hydroformylation, however, is carried out under nonisomerizing conditions and a preferential addition of one face of the prochiral olefin 2-ethyl-1-pentene to the catalyst, as required by the substantial retention of configuration, is very difficult to envisage. Therefore a hydrogen shift must occur in the organic group arising from the olefin in the catalytic complex without formation of free olefin; or in other words, a stereospecific isomerization of the catalytic complex takes place (117,194).

By taking into account the scheme of Heck and Breslow (23), the experimental facts were explained (22) by assuming that isomerization of σ-alkylcobalt to π-olefin complexes occurs much more rapidly than dissociation of the olefin from the cobalt complex, and that in the π-complexes the olefin is free to rotate around the olefin to cobalt bond (Scheme **IV**). In this case isomerization of the double bond occurs without

$$\begin{array}{c}[Co]\\|\\H\end{array} + \begin{array}{c}CH_2\\ \|\\ CH\\|\\R\end{array} \rightleftarrows \begin{array}{c}[Co]\\|\\H\end{array} \cdots \begin{array}{c}CH_2\\ |\\ CH\\|\\R\end{array} \rightleftarrows \left\{\begin{array}{c}[Co]{-}CH_2{-}CH_2{-}R\\ \updownarrow \\ \begin{array}{c}CH_3\\|\\ [Co]{-}CH\\|\\R\end{array}\end{array}\right.$$

Scheme **IV**

formation of free olefin, and therefore during the rearrangement of the π-complexes to σ-alkylcobalt compounds, racemization of an asymmetric tertiary carbon atom present in the substrate should not occur.

A possible confirmation of the above suppositions comes from the work of Piacenti's group on the hydroformylation of olefins containing *tert*-butyl groups (195). Here, no formylation at the methyl groups of the *tert*-butyl group was found (Table 14).

Further experimental evidence is given by hydroformylation of 1-methylcyclohexene (196) which yields, beside the expected *trans*-2-methylcyclohexanecarboxaldehyde, *trans*-3-methyl- and *trans*-4-methylcyclohexanecarboxaldehyde with only a small amount of the corresponding *cis* compounds (see Scheme **IX**, Section II-B-5).

TABLE 14
Composition of Hydroformylation Products from Olefins Containing a Quaternary Carbon Atom (116, 195)[a]

Olefin	Formyl group addition at carbon atoms of the olefin: C-1 (%)	C-2 (%)
3,3-Dimethyl-1-butene	99.2	0.8
2,3,3-Trimethyl-1-butene	99.2	0.8
3,3-Dimethyl-1-hexene	~100	Trace
4,4-Dimethyl-1-hexene	96.6	3.4
2,2,5,5-Tetramethyl-3-hexene	No reaction	

[a] Catalyst $Co_2(CO)_8$; benzene solvent; temp 110°C; $P_{H_2} = P_{CO} = 80$ atm. Experiments at constant pressure.

The hydroformylation of 1-methylcyclohexene under isomerizing conditions ($P_{CO} = 5$ atm) again yields a high *trans*/*cis* ratio for 2-methylcyclohexanecarboxaldehyde but a *trans*/*cis* ratio lower than 1 for the 3-methylcyclohexanecarboxaldehyde and a *trans*/*cis* ratio of 4 instead of more than 50 for the 4-methylcyclohexanecarboxaldehyde. Furthermore, hydroformylation of 3-methylcyclohexene yields the expected 3-methyl and 4-methylcyclohexanecarboxaldehydes with a *trans*/*cis* ratio of about 6 and of about 0.2, respectively (Table 15). These results clearly confirm that the isomeric products of hydroformylation under nonisomerizing conditions (high P_{CO}) originate mainly from isomerization of the substrate without formation of isomerized free olefin, whereas under isomerizing conditions (low P_{CO}) olefin isomerization is mainly involved.

The major criticism of this interpretation comes from the following facts: (*i*) rapid isomerization of the alkylcobalt compounds has never been clearly demonstrated. On the contrary, Falbe et al. (197) showed that, at 0°C, $CH_3CH_2CH_2Co(CO)_3(PPh_3)$ undergoes CO insertion without isomerization; similar results were reported for alkylcobalt carbonyls in the absence of phosphines at 90° (287). (*ii*) Heck and Breslow (23) obtained ethylene from $CH_3CH_2Co(CO)_4$ solutions only at 220°C; (*iii*) carbonylations of propyl or isopropyl orthoformate under typical hydroformylation conditions (120°C, 200 atm; $CO/H_2 = 1$), which seems to occur by a mechanism similar to that of hydroformylation, give rise to pure *n*-butyraldehyde and to practically pure isobutyraldehyde, respectively (199); (*iv*) complexes of olefins with cobalt carbonyl compounds have not

TABLE 15
Percentage Composition of the Cobalt-catalyzed Hydroformylation Products of 1- *and* 3-*Methylcyclohexene at* 120°*C* (196)[a]

	1-Methylcyclohexene		3-Methylcyclohexene
Primary products	P_{CO} 100 atm P_{H_2} 50 atm	P_{CO} 5 atm P_{H_2} 100 atm	P_{CO} 100 atm P_{H_2} 50 atm
Cyclohexanecarboxaldehydes			
trans-2-methyl	28	17	19
cis-2-methyl	<0.5	1	<1
trans-3-methyl	10	9	8
cis-3-methyl	4	22	43.5
trans-4-methyl	36	16	20
cis-4-methyl	<1	4	3
Formylmethylcyclohexane	20	31	5.5

[a] All pressures are initial pressures, in atm, measured at 20°C.

been described and if they exist, they must be present in imperceptibly small concentrations.

A possible way to reconcile these facts with the proposed mechanism is to admit (*1*) that at high CO pressure the equilibrium cobalt carbonyl hydride plus olefin yielding the π-complex (Scheme **IV**) is displaced toward the π-complex, and (*2*) that the dissociation rate of the π-complex is relatively low and in any case is lower than the rate by which isomerization of the catalyst–substrate complex and hydroformylation take place. In this case the fact that π-olefin complexes have never been detected in the reaction mixture could be explained by admitting that the concentration of the catalytic species (either complexed with the olefin or free) would be very small and not detectable. Moreover, since the presence of olefins does not influence the stability of alkylcobalt carbonyl complexes (292), a π-olefin-cobalt carbonyl hydride isomerization not involving formation of a true —C—Co— covalent bond must be postulated.

b. Successive Reactions. (*1*) *Hydrogenation of Aldehydes.* Aldehydes are reduced to alcohols by $HCo(CO)_4$ at 25°C (200). The formation of small amounts of alcohols during catalytic hydroformylation, especially when the reaction is carried out in the higher temperature range, was observed very early (201). This reaction can proceed in homogeneous phase (141), but its occurrence in the presence of heterogeneous catalysts cannot be excluded and the participation of heterogeneous catalysts in some steps of the reaction was proposed (202). The hydrogenation of aldehydes becomes important above 170°C; and on this account, processes have been proposed to produce alcohols in one step (1,141,201,203–206).

At low partial pressures of CO (<20 atm), the rate of hydrogenation increases when using PBu_3 at 160°C and 30 atm of H_2 (83). The ligand apparently acts as a stabilizer of the cobalt catalyst. However, since $HCo(CO)_2(PBu_3)_2$ rapidly hydrogenates aldehydes at 130°C and $HCo(CO)(PBu_3)_3$ is active even at 60°C, the enhanced hydrogenating catalytic activity of the cobalt catalysts in the presence of phosphines can perhaps be attributed to the presence of small amounts of hydrides with lower CO/Co ratios as well as to the greater hydrogenating activity of $HCo(CO)_3(PBu_3)$ compared to $HCo(CO)_4$. This increased catalytic activity of $HCo(CO)_3(PBu_3)$ has been ascribed to the greater hydridic character of the hydrogen in the cobalt carbonyl hydrides containing trisubstituted phosphines (88).

The extent of the hydrogenation activity of the $HCo(CO)_4/PR_3$ systems depends on the structure of the phosphines. Trialkylphosphines make for more active catalysts than do the triphenylphosphines (88). The extent of the hydrogenation is also influenced by the polarity of the solvent used (207,208). Hydroformylation of propylene at 160–185°C and 100 atm of synthesis gas (1 H_2 : 1 CO) in the presence of a $PBu_3/Co_2(CO)_8$ catalyst system gave an aldehyde to alcohol ratio of 2 when using heptane as solvent; the ratio was 24 when using dimethylformamide.

Ketones are also hydrogenated to alcohols with H_2 both in the presence and in the absence of CO and in the presence of $Co_2(CO)_8$ and trialkylphosphines at 200°C. When using optically active phosphines, no asymmetric induction was observed in the formation of secondary alcohols. Under the reaction conditions, a slight reduction of alcohols to hydrocarbons (up to 2%) was observed (209).

The kinetics and mechanism of hydrogenation under hydroformylation conditions have not been clarified. A maximum rate at a partial pressure of CO of about 20 atm was found by Markó (210) when hydrogenating propionaldehyde in the presence of a cobalt catalyst at 150°C and an H_2 partial pressure of 95 atm. The inhibiting effect of CO at high pressures had also been found by Aldridge and Jonassen (202). These investigators however, do not agree on either the overall kinetic equation or the mechanism of hydrogenation. Markó (210) and Orchin (200) favor a scheme involving the hydrogenolysis of a cobalt–oxygen bond (Scheme **Va**), whereas Aldridge and Jonassen prefer a hydrogenolysis of a cobalt–carbon bond (Scheme **Vb**). According to Markó, the alkoxycobaltcarbonyl is also involved in the synthesis of formates during hydroformylation (Scheme **Va**).

$$RCHO + [HCo(CO)_3] \longrightarrow [RCH_2OCo(CO)_3] \qquad \textbf{(a)}$$

$$[RCH_2OCo(CO)_3] \xrightarrow{H_2} RCH_2OH + [HCo(CO)_3]$$

$$[RCH_2OCo(CO)_3] \xrightarrow{CO} [RCH_2OCOCo(CO)_3] \xrightarrow{H_2} RCH_2O\overset{\displaystyle O}{\overset{\|}{C}}H + [HCo(CO)_3]$$

$$RCHO + [HCo(CO)_3] \longrightarrow [R\underset{\displaystyle OH}{\underset{|}{C}}HCo(CO)_3] \xrightarrow{HCoCO)_4} RCH_2OH + [Co_2(CO)_7] \qquad \textbf{(b)}$$

Scheme **V**

The hydrogenation of aldehydes under hydroformylation conditions is used industrially to produce butanol from propylene with phosphine-modified cobalt catalysts (10).

(2) *Aldolization of Aldehydes.* Aldol formation usually occurs under oxo conditions (2,211–215,304). Straight-chain aldehydes undergo the reaction more readily than the corresponding branched-chain aldehydes (216).

Discrepancies often found in the literature on the distribution of isomeric aldehydes may be attributed, at least in part, to a preferential aldolization of one of the reaction products (211). Apparently aldolization proceeds even at room temperature after the hydroformylation reaction is over: A larger amount of aldol products was found by working up the hydroformylation products of propylene after they had been standing overnight at room temperature than when the products were isolated immediately after reaction (178).

Aldol reactions are largely avoided by using phosphine-modified cobalt catalysts. This result can be attributed both to the low activity of the phosphine-containing catalysts in aldol formation and to the low concentration of aldehydes present in the reaction medium, since they are largely reduced to alcohols.

Aldol formation is often followed by dehydration and double-bond hydrogenation; for example, saturated branched aldehydes having twice as many carbon atoms as the original hydroformylation products were found (212). Such a scheme was proposed for the production of alcohols having $2n+2$ carbon atoms from the hydroformylation of olefins with n carbon atoms (Scheme **VI**) (217,218).

$$CH_3CH{=}CH_2 + CO + H_2 \xrightarrow{Co_2(CO)_8} CH_3CH_2CH_2CHO + CH_3\underset{\displaystyle CH_3}{\underset{|}{C}}HCHO$$

$$2CH_3CH_2CH_2CHO \xrightarrow[-H_2O]{aldol} CH_3CH_2CH_2CH{=}\underset{\displaystyle CH_2CH_3}{\underset{|}{C}}CHO$$

$$CH_3CH_2CH_2CH{=}\underset{\displaystyle CH_2CH_3}{\underset{|}{C}}CHO \xrightarrow{H_2} CH_3CH_2CH_2CH_2\underset{\displaystyle CH_2CH_3}{\underset{|}{C}}HCH_2OH$$

Scheme **VI**

Attempts were made to find hydroformylation conditions under which most of the primary aldehyde, RCH_2CHO, was transformed into the corresponding higher molecular weight aldehyde, $RCH_2CHRCHO$; this one-step reaction can have practical interest (219). However, the preferred way to synthesize high-boiling aldehydes and alcohols still is to isolate the primary aldehyde, carry out the aldolization separately, and then proceed to the dehydration and hydrogenation reactions (220).

(3) *Other Successive Reactions.* According to reaction 24, acetal formation ensues from alcohol formation. The water formed in the synthesis of acetals can readily react according to reaction 20 (221); small quantities of acids and esters were found in the reaction products. Interestingly, acetal formation decreases when using cobalt catalysts in the presence of phosphines, even though large amounts of alcohols are present. The hydrogenation of the aldehydes in this case may be more rapid than the formation of acetals; furthermore, cobalt complexes containing trisubstituted phosphine ligands may be poorer catalysts than $HCo(CO)_4$ for acetal formation. Part of the esters containing $2n+2$ carbon atoms found among the hydroformylation products of olefins with n carbon atoms are probably due to a Tishchenko reaction, which has been observed under hydroformylation conditions (222).

Small amounts of formates were often found among the hydroformylation products (172,223). According to Markó and Szabó (224), the formation of formates proceeds through the aldehyde and not through the corresponding alcohol. They suggest that this reaction is based on a hydroformylation of aldehydes (see Scheme **Va**). The carbonylation of the cobalt–oxygen bond, however, needs further investigation.

Finally, the trimerization of aldehydes, particularly of branched-chain and of fluorinated aldehydes (131) has been observed, especially when the hydroformylation reaction is carried out in the absence of a solvent (211,213).

4. Control of Isomeric Composition of the Reaction Products

One of the main drawbacks to the use of the hydroformylation reaction in synthetic organic chemistry is the simultaneous production of two or more isomeric aldehydes. The importance of controlling the isomeric composition of products, therefore, can hardly be overemphasized.

Research in this direction has been conducted since 1948 by several groups (97,99,123,184,225,226). Conflicting results were often obtained, probably because not all the numerous reaction variables were carefully controlled, and particularly because of the formation of large amounts (20–30% of the total products) of unidentified secondary reaction products (150,153,211,227).

As shown by the reaction of propyl and isopropyl orthoformate with CO and H_2 in the presence of cobalt catalysts (199,228,229), which very likely proceeds by a mechanism similar to that of the oxo reaction, it is possible to obtain only one aldehyde when using conditions very similar to those of the oxo reaction, such as 100 atm each of H_2 and CO in the presence of $Co_2(CO)_8$ at 110–140°C. But complete selectivity toward one of the possible aldehydes has never been achieved with unsymmetrical olefins.

The major developments in this field have followed two pathways. First, a better understanding was achieved of the effect of reaction variables, (such as temperature, CO and H_2 partial pressures, and catalyst concentration) on the composition of the isomeric reaction products. Second, it was shown that, by operating in the presence of ligands that can compete with CO, a change in the composition of the isomeric reaction products can be achieved. This change, however, is accompanied by a striking decrease in the reaction rate; therefore, the reaction must be carried out at temperatures at which the main reaction products are alcohols. These two lines of research will next be discussed separately. The results achieved are not only of practical importance but also have contributed to a deeper understanding of the hydroformylation mechanism.

a. Effect of Reaction Variables on the Composition of Isomeric Hydroformylation Products. The partial pressures of CO and H_2 and the temperature (99,184,299) definitely influence the composition of hydroformylation products, but, of course, to different extents. Other factors such as the use of certain solvents (ketones) (21) or the presence of an excess of one of the reaction products (230) and catalyst concentration (184) were also reported to influence the yield of the isomeric products obtained. These results, however, could not be entirely reproduced (178). Only the effect of solvents on the composition of the hydroformylation products has been corroborated at present (16).

The first indication of the effect of CO pressure on the composition of the products from the hydroformylation of propylene was disclosed in a patent by Gresham, Brooks, and Bruner (123) issued in 1948. These authors reported that, on operating at 108–120°C with a mixture of CO and H_2 (1:2) at 470–790 atm, the ratio between straight-chain isomer (*n*-butyraldehyde) and branched-chain isomer (isobutyraldehyde) was higher than 4; yet it was known that only S/B ratios between 1 and 3 were obtained under typical oxo conditions (123).

We are using the symbol S/B throughout to indicate the ratio between straight- and branched-chain aldehyde. In the case of the hydroformylation products of branched olefins (for example, 4-methyl-1-pentene), S/B stands for the ratio between the least branched and all other isomeric aldehydes formed.

Systematic investigations were carried out on the hydroformylation of propylene in the presence of ethyl orthoformate to avoid formation of appreciable amounts of secondary products (231). After some initial experiments, which gave erroneous results due to the presence of propionaldehyde diethyl acetal derived from ethyl orthoformate (199), it became clear that at high CO pressures, S/B values of 2.75 were readily achieved; whereas, at CO pressures of 10 atm or less, values of S/B of about 1.0 were obtained (Table 16).

When using a high vacuum technique to isolate the aldehydes immediately after completing hydroformylation, secondary products were found only in very small amounts (<5%) even in the absence of ethyl orthoformate (99,225). In this way, the effect of the CO partial pressure could easily be examined with a number of olefins. In the case of propylene, low values of S/B were always found at low CO pressures. By increasing the CO pressure, the value of S/B increases rapidly to reach, at a CO pressure of over 70 atm, a limiting value that increases only slightly even if the CO pressure is further increased (99). A decrease in the S/B ratio was noticed only at extremely high pressure (>1000 atm) (286). In the case of 2-pentene hydroformylation, the percentage of *n*-hexanal obtained reached a maximum of 75–76% at about 100 atm P_{CO} but it decreased at higher P_{CO} reaching the value of 63–64% at 400 atm P_{CO} (236). In the case of isomeric olefins (1-butene, 2-butene, and 1-pentene, 2-pentene), the isomeric composition is different at high CO pressures; but is the same at low CO pressures (Fig. 2). The lowest and the highest values of S/B reported for different olefins are presented in Table 17.

TABLE 16

Effect of P_{CO} *on Isomeric Composition of Propylene Hydroformylation Products in the Presence of Ethyl Orthoformate*[a] (225)

P_{CO}[b] (atm)	High-boiling products/ aldehydes (%)	*n*-Butyraldehyde[c] (%)
4	3.0	45.4
9	1.4	51.4
12.5	1.7	60.6
21	0.9	63.5
31	1.3	65.0
66	0.5	70.2
104	2.0	72.7
147	1.5	73.3
224	1.0	73.4

[a] Propylene 21 g; temp 108°C; $Co_2(CO)_8$ 1 g; P_{H_2} 80 atm; $HC(OC_2H_5)_3$ 90 g; C_2H_5OH 80.5 g.

[b] Experiments at constant pressure.

[c] Percentage of total C_4 aldehydes.

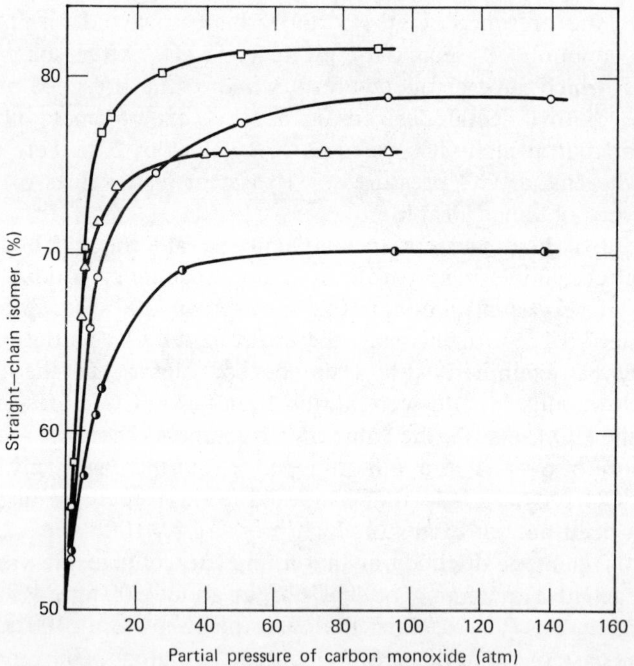

FIG. 2. Effect of partial pressure of CO on isomeric composition of hydroformylation products (99). ○ = 1-butene; ◑ = *cis*-2-butene; □ = 1-pentene; △ = 2-pentene (*cis* + *trans*).

TABLE 17
Effect of Carbon Monoxide Partial Pressure on Isomeric Distribution of Hydroformylation Products of Olefins[a] (99)

Olefin	Temp (°C)	P_{CO} (atm)	S/B[b]
Propylene	110	2.5	1.6
Propylene	110	90	4.4
1-Butene	110	2	1.1
1-Butene	110	140	3.7
cis-2-Butene	110	2	1.1
cis-2-Butene	110	138	2.4
1-Pentene	100	1.7	1.3
1-Pentene	100	90	4.5
2-Pentene	100	1.7	1.3
2-Pentene	100	93	3.1
4-Methyl-1-pentene	116	2	1.4
4-Methyl-1-pentene	116	150	8.1

[a] P_{H_2} 80 atm; solvent benzene or toluene; catalyst $Co_2(CO)_8$; experiments at constant pressure.

[b] S/B is the ratio of straight- to branched-chain aldehydes: For propylene, it is the ratio of *n*-butyraldehyde to isobutyraldehyde; for 4-methyl-1-pentene, S/B stands for the ratio of 5-methylhexanal to the other aldehydes formed.

The effect of the partial pressure of CO on S/B was observed also by Hughes and Kirschenbaum (184) who studied the hydroformylation of 1-heptene; S/B decreased from 3 to 0.8 when the CO pressure decreased from 85 to 11 atm. These authors, however, did not stress this point.

The H_2 partial pressure has a small but definite and reproducible effect on the composition of hydroformylation products as shown in Table 18, which gives data of some experiments on propylene hydroformylation in the presence and in the absence of ethyl orthoformate. Analogous results were obtained with 1-pentene. In both cases the effect of H_2 partial pressure is in the same direction as that found for CO pressure variation: The value of S/B increases when H_2 pressure is increased (99,299). Large discrepancies exist in the literature concerning the effect of temperature on the S/B ratio (Table 19).

Investigation of the effect of temperature on the isomeric composition of hydroformylation products is particularly difficult because the effect due to changes in CO concentration in the liquid phase (234), which is superimposed on true temperature effects, must be taken into account. Furthermore, since the hydroformylation rate increases greatly with rising temperature (185), the liquid phase may not become saturated with CO even if stirring is relatively efficient.

Careful examination of temperature effects, therefore, should include, for each temperature tested, a series of experiments that are conducted under various CO partial pressures to make certain that the S/B values obtained were not determined in the CO pressure interval where even small partial pressure changes greatly influence the S/B value. In addition, experiments at constant temperature and CO pressure, but at different

TABLE 18
Hydroformylation of Propylene at Various Hydrogen Partial Pressures[a]

P_{H_2} (atm)	*n*-Butyraldehyde/ total C_4 aldehydes (%)
22.5[b]	74.9
30	77.3
41	79.2
83	81.4
125	81.5
24[c]	62.2
40	65.0
55	67.7
87	70.0
170	70.4

[a] Propylene 20 g; $Co_2(CO)_8$, 1 g; conversion of propylene to aldehydes, 92.4–95.5%. Amount of high-boiling product formed (expressed as weight % of aldehydes produced), 1.2–3.9; experiments at constant pressure.

[b] For first five experiments, temp 110°C; P_{CO} 100 atm, benzene 100 g.

[c] For last five experiments, temp 90°C; P_{CO} 68 atm; C_2H_5OH, 80.5 g; $HC(OC_2H_5)_3$, 90 g.

TABLE 19
Increased Branching of Hydroformylation Products of Some Olefins with Increasing Temperature

Olefin	Temp (°C)	S/B[a]	Solvent	Ref.
Propylene	80–145	4.8–3.9	Benzene	16
Propylene	80–180	4.5–1.6	Hexane	184
1-Butene	70–180	2.7–0.8	Hexane	184
1-Pentene	80–160	5.2–3.1	Benzene	16
2-Pentene	100–160	3.1–2.0	Benzene	16
1-Heptene	70–180	4.0–0.75	Hexane	184
Isobutylene	121–200	39.0–15.0	Hexane	232
Isobutylene[b]	160–220	22.0–7.5	Methanol	233
3-Methyl-1-pentene	90–145	13.7–7.0	Dioxane	112
4-Methyl-1-pentene	100–140	8.1–6.9	Benzene	99,178
4-Methyl-1-pentene	116–175	1.3–1.1	none	186

[a] For meaning, see Table 17.
[b] Except for this experiment, all experiments were done at constant pressure.

reaction rates (that is, at different olefin or catalyst concentrations) should be performed to make certain that the CO concentration in the liquid phase is not being influenced by the rate at which CO dissolves.

An investigation along these lines was carried out for propylene and 1-pentene hydroformylation (16); the results are shown in Tables 20 and 21. In both cases an apparently large decrease in S/B with temperature was noticed at relatively high reaction rates with poor stirring (16), but only a very slight decrease in S/B with increasing temperature was obtained when the reaction rate was slow and stirring efficient.

A comparison of the results obtained by Hughes and Kirschenbaum (184) and by Macho (211) for propylene hydroformylation with the results reported in Fig. 3 reveals that variation in the isomeric aldehyde composition found by these authors corresponds to results obtained when the liquid phase was not saturated with CO. The change in S/B observed between 100 and 140°C is not very different from the change observed by Hughes and Kirschenbaum (184) between 90 and 100°C.

It may therefore be concluded that when the reaction temperature is increased from 100 to 140°C, a slight decrease in S/B occurs for propylene, 1-butene, 1-pentene, 3-methyl-1-pentene, and 4-methyl-1-pentene (Table 19). Accurate determinations of S/B for temperatures above 140°C were not possible since the hydroformylation rate becomes so high

TABLE 20
Hydroformylation of Propylene at Various Temperatures[a] (16,178)

Temp (°C)	P_{CO} (atm)	*n*-Butyraldehyde/ total C_4 aldehydes (%)
80	55	82.8
80	80	82.8
90	70	82.1
90	122	82.2
100	90	81.9
110	119	81.4
110	146	81.4
120	120	79.2
120	150	79.3
120[b]	180	81.0
135	140	75.6
135	165	75.7
135[b]	170	80.5
145	150	67.8
145	182	67.8
145[b]	170	79.8

[a] P_{H_2} 80±2 atm; 20 g propylene and 1 g $Co_2(CO)_8$ per 100 g benzene solvent; the amount of high-boiling product formed, expressed as weight percent of aldehydes, varied from 2.0 to 6.0. Experiments at constant pressure.

[b] 2.5 g propylene and 0.1 g $Co_2(CO)_8$ per 100 g solvent.

that, with the equipment available (16), it was impossible to keep the liquid phase saturated with CO during the reaction and to avoid formation of large amounts of secondary products.

In the hydroformylation of isobutylene, a large increase in pivalic aldehyde is observed when the reaction temperature is increased (232,233). Part of this effect may be due also to a variation of CO concentration in the liquid phase (Fig. 4).

In contrast to early reports (184), catalyst concentration by itself does not influence S/B (Table 22) (16,235). But when the catalyst concentration is varied, a change in product composition can occur for one or both of the following reasons: (i) increase in rate of aldolization of primary reaction products, and (ii) decrease in CO concentration in the liquid phase due to lack of saturation with CO because of increase in reaction rate. Many discrepancies among literature data may be easily explained in this way.

Although catalyst concentration does not affect S/B, the nature of the catalyst, that is the metal carbonyl used and the type of ligand added

TABLE 21

Hydroformylation of 1-Pentene at Various Temperatures[a] (16,178)

Temp (°C)	P_{CO} (atm)	g $Co_2(CO)_8$/ 100 g solvent (%)	Reaction time (min)	Hexanal/ total aldehydes (%)
80	51	1.67	200	83.9
80	100	1.48	260	83.9
100	47	1.48	120	80.7
100	99	1.48	145	81.8
110	100	1.48	75	80.5
110	152	1.48	85	80.5
120	149	1.48	40	79.7
120	200	1.48	60	79.7
140	145	1.48	15	63.8
140	202	1.48	20	76.3
140	180	0.63	50	77.9
140	220	0.63	50	77.7
160	170	1.48	8	62.8
160	220	1.48	8	64.6
160	180	0.37	35	69.0
160	220	0.37	40	69.2
160	180	0.185	80	71.1
160	220	0.185	80	71.2
160	220[b]	0.185	80	73.6
160	220[b]	0.185	80	74.4
160	180[b]	0.185	80	74.6
160	200[c]	0.092	80	75.7

[a] Experiments at constant pressure; P_{H_2} 80 ± 2 atm; benzene; 18.5 g 1-pentene per 100 g solvent. Equivalents of carbonyl compounds formed per mole of 1-pentene (wt %) varied from 87 to 96. The amount of high-boiling product formed (as wt % aldehydes) varied from 4.0 to 7.0.

[b] 3.7 g 1-pentene per 100 g benzene.

[c] 0.92 g 1-pentene per 100 g benzene.

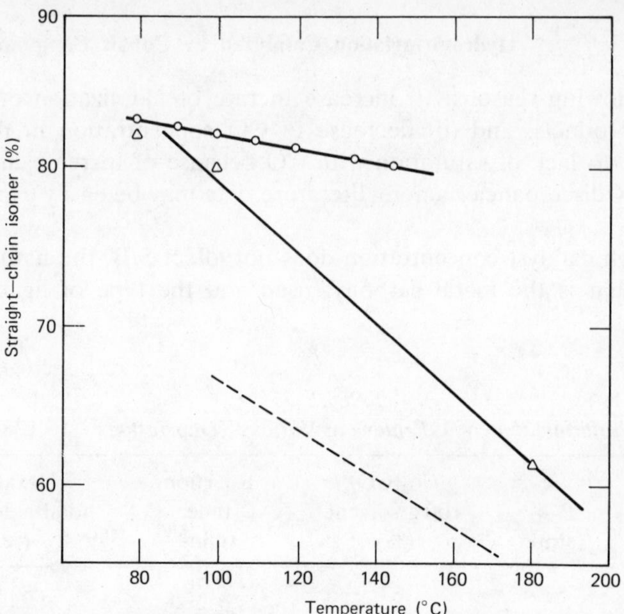

FIG. 3. Effect of temperature on distribution of products from propylene hydroformylation. ○ (16,178); △ (184); --- (211).

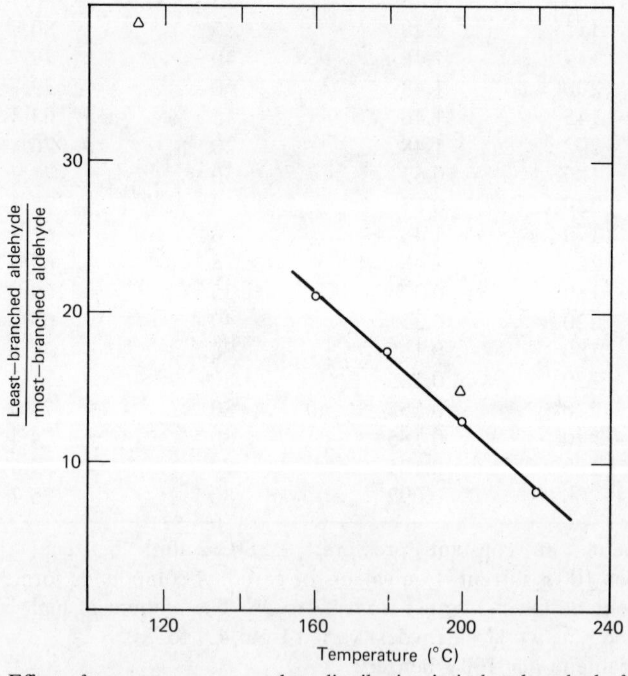

FIG. 4. Effect of temperature on product distribution in isobutylene hydroformylation. △ (232); ○ (233).

TABLE 22
Hydroformylation of Propylene and 1-Pentene Using Various Concentrations of $Co_2(CO)_8$[a] (16)

Olefin	$Co_2(CO)_8$/solvent (wt %)	S/B[b]
Propylene[c]	0.29	2.82
Propylene	0.59	2.83
Propylene	1.47	2.82
Propylene	1.47	2.82
Propylene	2.35	2.86
Propylene	2.35	2.83
1-Pentene[d]	0.56	4.4
1-Pentene	1.67	4.5
1-Pentene	2.77	4.5
1-Pentene	4.44	4.4

[a] Equivalents of carbonyl compounds per mole of olefin varied between 91 and 99%; amount of high-boiling product (as wt % aldehydes) varied from 0.6 to 4.2; experiments at constant pressure.

[b] Ratio of straight- to branched-chain aldehydes.

[c] With propylene experiments: olefin 20 g; temp 110°C; C_2H_5OH, 80 g; $HC(OC_2H_5)_3$, 90 g; P_{H_2} 80 atm; P_{CO} 230 atm.

[d] With 1-pentene: olefin 5 g; temp 100°C; benzene 27 g; P_{H_2} 80 atm; P_{CO} 100 atm.

(other than CO and olefins) can influence the composition of the hydroformylation products. Values of S/B determined for hydroformylation products of propylene when using different metal carbonyls under similar reaction conditions are given in Table 23.

The effect of solvents on the composition of hydroformylation products was studied by different authors (16,21). Experiments done with isooctane, benzene, diethyl ether, ethyl alcohol, and acetone, showed that although the solvent influences S/B (Table 24), it does not have a large effect (16,21). This effect may be connected with an interaction between solvent and catalyst as well as with the varying solubility of CO in the different media. Dissolved CO may affect equilibria among the several

TABLE 23

Hydroformylation of Propylene with Various Catalysts

Catalyst Compound	Catalyst Concentration[a]	Temp (°C)	P_{CO}[b] (atm)	P_{H_2}[b] (atm)	Time (hr)	Yield (%)	Straight-chain Aldehyde (%)	Ref.
$Co_2(CO)_8$	23.4	110	75	75	1	94	80	16
$Co_2(CO)_6(PBu_3)_2$	23.4	140	15	30	3	80	86	16
$Ru_3(CO)_{12}$	23.4	110	75	75	7	40	74	16
$(Rh(CO)_3)_n$	0.05	110	110[c]	110[c]	~1	83	53	237
$HRh(PPh_3)_3(CO)$[d]	n.d.	90–125	5	5	[e]	>95	94	238
$Os_3(CO)_{12}$	18.5	180	80	80	2	74	51	16
$IrCl(CO)(PPh_3)_2$	21.2	130	15	15	3.7	61	63	239

[a] Millimoles metal per liter of solvent; (benzene).
[b] Constant pressure at reaction temp.
[c] Measured at room temp.
[d] In molten PPh_3.
[e] 0.8–7.5 mol per mol of catalyst per min.

TABLE 24

Hydroformylation of Propylene in Various Solvents[a]
(16)

Solvent	*n*-Butyraldehyde/ isobutyraldehyde
2,2,4-Trimethylpentane	4.6
Benzene	4.5
Toluene	4.4
Ethyl ether	4.4
Ethyl alcohol, 95%	3.8
Acetone	3.6

[a] Temp 108°C; P_{CO} 140 atm; P_{H_2} 100 atm; $Co_2(CO)_8$, 1 g; propylene 20 g; solvent 100 ml; experiments at constant pressure. Equivalents of aldehydes per mole of olefin charged varied from 60 to 83%. The amount of high-boiling product (as wt % aldehydes) varied from 3.2 to 4.

catalytic complexes and hence the concentration of the catalytic species that ultimately lead to formation of the different isomeric aldehydes in various ratios (16).

Finally, the literature reports that when the oxo reaction is carried out in the presence of a large excess of one of the aldehydes produced, the amount of that aldehyde eventually decreases (226,230). Although aldehydes are known to undergo decarbonylation in the presence of group VIII metals (240,241), these results were not confirmed (299). Aldehydes do not appear to undergo decarbonylation under hydroformylation conditions. At any rate, the effect of excess of one of the aldehydes cannot be thermodynamic. In fact the thermodynamic equilibrium composition of the aldehydes is usually not reached, as shown by the fact that very different values of S/B can be obtained at a given temperature, for example, depending on the partial pressure of CO; obviously CO pressure cannot influence the equilibria between isomeric aldehydes.

b. Effect of Ligands Containing Phosphorus or Arsenic on Isomeric Composition of Hydroformylation Products. Work in the Shell laboratories (87,242) showed that the selectivity towards straight-chain hydroformylation products markedly increases for oxo reactions carried out in the presence of certain ligands. This is true when either terminal or internal olefins are used as substrates (101). This effect is clearly shown in Table 25 which compares the results of hydroformylation experiments done in the presence and the absence of ligands containing phosphorus.

The ligand structure appears to have a striking effect on the composition of the reaction products. The highest selectivity toward formation of straight-chain aldehydes or alcohols is observed when using trialkylphosphines (87,88) (Table 26). An attempt was made to relate the basicity of the phosphine to its influence on selectivity by investigating 1-hexene hydroformylation in the presence of eight trisubstituted phosphines (92). The results substantially agreed with those given in Tables 25 and 26. Since the position of the equilibria given in reactions 10–16 is not known, the above effects can arise from the structure of the phosphine present in the catalytic species and also from the relative amounts of the different catalytic species present, some with phosphine ligands and some containing only CO and hydrogen as ligands.

The second possibility was clearly shown by Piacenti et al. (95) who performed a series of experiments with propylene at 150°C with PBu_3 as a ligand ($Co/PBu_3 = 1$) at different CO pressures. It appears that the partial pressure of CO has a striking effect in decreasing the isomeric ratio (S/B) in the range between 3 and 20 atm. Above this pressure the

TABLE 25
Hydroformylation of Olefins. Decrease in Branched-Chain Product Caused by Ligands Containing Phosphorus

Olefin	Ligand	Temp (°C)	P_{CO}[a] (atm)	Straight-chain aldehyde (%)	Ref.
Propylene	—	120	50	78	178
Propylene	$P(OPh)_3$[b]	120	50	81	178
Propylene	—	120	15	61	178
Propylene	$P(OPh)_3$[b]	120	15	75	178
Propylene	PBu_3[b]	140	15	86	178
1-Octene	—	190	27	43	94
1-Octene	PBu_3	190	27	85	94
1-Octene	—	170	27	44	94
1-Octene	PBu_3	170	27	87	94
2-Octene	—	190	27	42	94
2-Octene	PBu_3	190	27	82	94
2-Octene	—	170	27	43	94
2-Octene	PBu_3	170	27	82	94

[a] Experiments at constant pressure.
[b] Moles ligand to moles $Co_2(CO)_8 = 2$; ratio in all other cases = 4.

TABLE 26
Effect of R_3P *or* R_3As *Ligands on Product Composition in the Hydroformylation of 1-Pentene with Cobalt Carbonyl Catalysts*[a] (87)

Ligand	Temp (°C)	Olefin conversion (%)	Total yield (%) Alcohols	Total yield (%) Aldehydes	Straight-chain product (%)
PEt_3	195	100	79.8	0	80.9
PBu_3	195	100	77.0	0	84.1
P(2-ethylhexyl)$_3$	195	72	65.9	3.9	67.2
$PPhBu_2$	195	100	72.0	0	80.0
PPh_3	195	40	60.7	10.3	66.0
PBu_3	150	98	83.7	1.4	91.0
$AsBu_3$	150	83	51.9	28.7	67.8

[a] Ligand : Co ratio = 2; catalysts prepared *in situ* by adding ligand to $Co_2(CO)_8$; reaction time 3 hr; $CO + H_2$ (1 : 2) 29–32 atm; experiments at constant pressure.

behavior of the system is similar to that shown by cobalt catalysts in the absence of a phosphine ligand.

Ligands including diphosphines (87), dialkylchlorophosphines, bicyclic tertiary phosphines, phosphites (243), arsines (87), and others (244) were used by various workers, but knowledge of the effect of the ligand is mainly empirical. Most authors agree that the above basic ligands favor formation of straight-chain aldehydes both by retarding double-bond isomerization in the olefin and by favoring, for steric reasons, the formation of linear alkyl- and acylcobalt derivatives; the second effect is probably most important (94,101).

By using PBu_3 as ligand, Tucci (89) investigated the effect of a number of reaction variables on the formation of *n*-butyraldehyde and butyl alcohol in the hydroformylation of propylene. Neither temperature nor catalyst concentration significantly influenced the percentage of straight-chain product formed. On the contrary, as expected from equilibria shown in reactions 10–16, the cobalt to phosphine ratio largely influenced the product composition when it was greater than 0.7 (88). CO and H_2, which influence the product composition when the cobalt/phosphine ratio is 1 (95), do not seem to have a great effect when this ratio is smaller than 0.5 (88). Only a slight effect of solvents on the percentage of straight-chain product was found for propylene hydroformylation using $Co_2(CO)_6(PBu_3)_2$ as catalyst.

5. *Stereochemistry*

Investigation of the stereochemistry of hydroformylation presents some difficulties, since the addition reaction under usual conditions is accompanied, as already discussed, by rapid isomerization of the catalyst–substrate complex and by double-bond shifts in the olefinic substrate. Furthermore, as in other catalytic reactions in which preliminary formation of a π-complex between metal and olefin is postulated, the possibility exists that attack occurs at the olefin face π-bound to the metal or at the opposite face (Scheme **VII**). This aspect is relevant in asymmetric hydroformylation as its knowledge allows relating the prevailing chirality of the aldehyde formed with the stereochemistry of the π-complex.

The most important aspects of the stereochemistry of hydroformylation have been reviewed (285). The first indication of *cis* addition of CO and H_2 to double bonds was obtained by investigating the hydroformylation of some steroids (245,246). Further evidence of *cis* addition of hydrogen or deuterium and the formyl group to a double bond in a catalytic hydroformylation was furnished by Rosenthal and co-workers in the hydroformylation of compounds **1** (147) (Scheme **VIIIa**) and **2** (144,148)

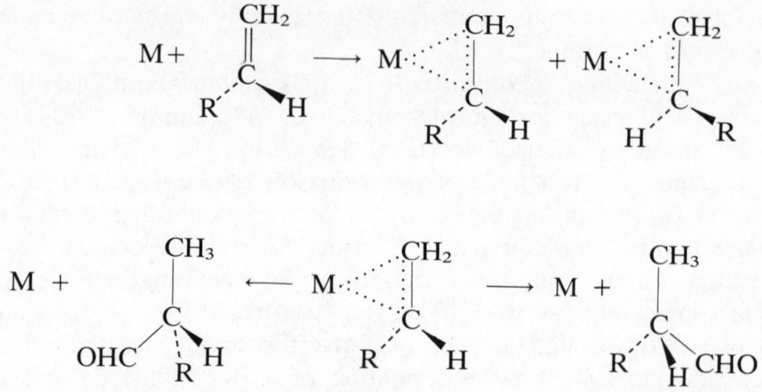

M = Metal complex; R ≠ H

Scheme **VII**

(Scheme **VIIIb**). The structure of the reaction products, demonstrated chemically for **3** and by nmr for **4**, proves that a *cis* addition to the double bond takes place. A further indication was obtained by Fichteman and Orchin (247) who, on the basis of nmr spectra, suggested that the reaction between $DCo(CO)_4$ and norbornene gives rise to compound **5**. According to these authors (248), a *cis*-addition mechanism is favored in hydroformylation, since their findings showed that in the stoichiometric reaction

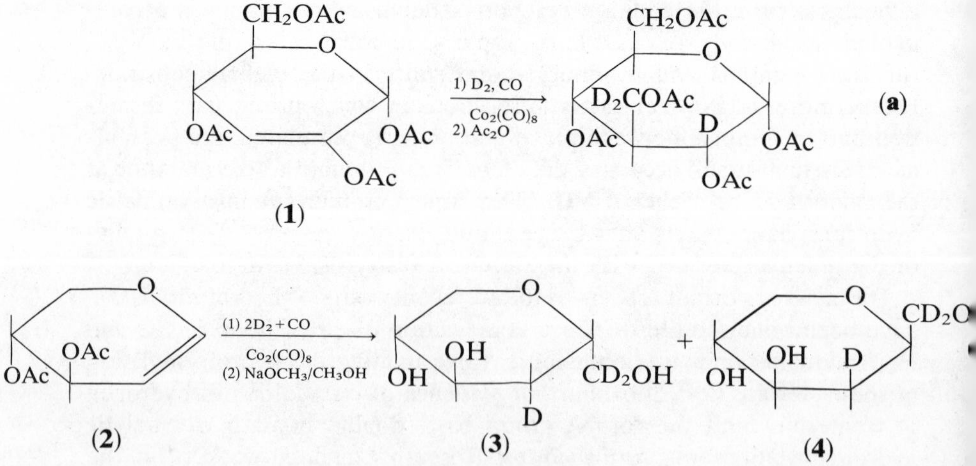

1) D_2, CO, $Co_2(CO)_8$; 2) Ac_2O (**a**)

(**1**)

(1) $2D_2 + CO$, $Co_2(CO)_8$; (2) $NaOCH_3/CH_3OH$

(**2**) (**3**) (**4**)

Scheme **VIII**

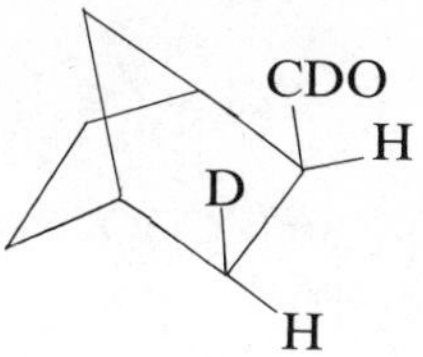

CDO H D H

(**5**)

between 1,2-diphenylcyclobutene and $HCo(CO)_4$, 96% of the 1,2-diphenylcycobutanes formed is the *cis* product, and that under catalytic oxo conditions, 90% of the same product is obtained.

An attempt to solve the problem of the stereochemistry of hydroformylation was made by investigating the stoichiometric hydroformylation products of methyl 2,3-diphenylcyclopropene-2-carboxylate (249), but no conclusive evidence was obtained (22).

Further interesting evidence in favor of substantially *cis* addition was obtained from the already cited hydroformylation of 1-methylcyclohexene and from the hydroformylation of 1,2-dimethylcyclohexene (196). These substrates allow the establishment of the stereochemistry of the process despite the formation of all possible isomers. When operating under nonisomerizing conditions with 1-methylcyclohexene, the products obtained were, besides cyclohexylacetaldehyde (20% of the aldehydes), 2-methyl-, 3-methyl-, and 4-methylcyclohexanecarboxaldehydes. The *trans* products greatly predominated over the corresponding *cis* products: *trans*/*cis* = ~13. Scheme **IX** depicts the product distribution from the hydroformylation of 1-methylcyclohexene, starting with $P_{CO} = 100$ atm and $P_{H_2} = 50$ atm at 20°C, operating at 120°C, using $Co_2(CO)_8$ as catalyst and benzene as solvent.

In the hydroformylation of 1,2-dimethylcyclohexene under non-isomerizing conditions, the products obtained, besides a mixture of dimethylcyclohexanecarboxaldehydes (44.4%), were *cis*- and *trans*-2-(methylformyl)-methylcyclohexanes; the *cis* product largely prevailed (*cis*/*trans* = 7.5). As previously discussed and as confirmed by the above hydroformylation of 1-methylcyclohexene, formation of the isomers in which the formyl group is bound to an originally saturated carbon atom occurs mostly by isomerization of the catalytic complex without dissociation to free olefin, and therefore it is stereospecific (114) (Scheme **X**, **6** → **7**). Scheme **X** gives a postulated sequence of intermediates from the hydroformylation of 1,2-dimethylcyclohexene starting with $P_{CO} =$ 100 atm and $P_{H_2} = 50$ atm at 20°C, operating at 100°C, and using $Co_2(CO)_8$ as catalyst and benzene as solvent. This representation of the hydroformylation shows that a hydrogen takes the place of the cobalt

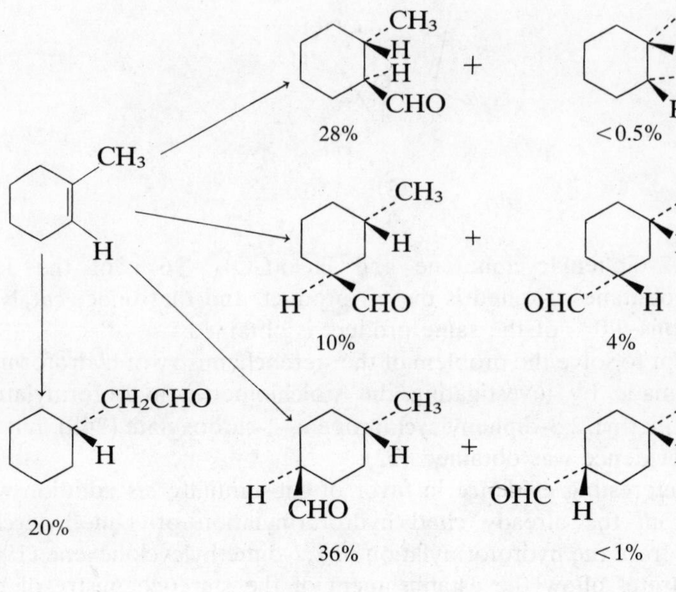

28% + <0.5%

10% + 4%

20%

36% + <1%

Scheme **IX**

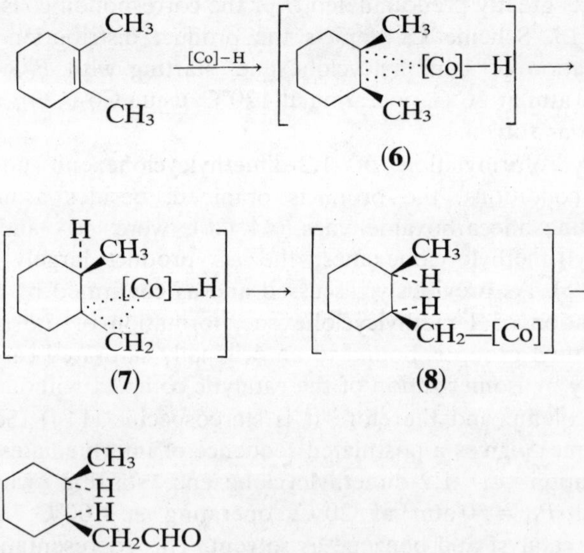

$[Co]-H$

(**6**)

(**7**) (**8**)

Scheme **X**

atom on the tertiary carbon atom in complex **7** (Scheme **X**, **7**→**8**), and thus indicates that *cis* attack of hydrogen and CO occurs at the olefin face bound to the cobalt atom in the postulated initial complex **6.**

It thus appears that *cis* addition of CO and hydrogen is not limited to a particular substrate but is probably a general phenomenon in cobalt-catalyzed hydroformylation. Furthermore, the occurrence of *cis* addition in the rhodium-catalyzed hydroformylation of 3-methyl-2-pentene (250) and of 1-methylcyclohexene (196) indicates that *cis* addition is probably a general feature of the hydroformylation reaction and is independent of the metal used as catalyst.

6. Kinetics

The kinetics of hydroformylation was investigated by many authors (6,98,185,251–253) who established some important aspects, but several points have yet to be settled. The main difficulty is to sort out the relative rates of the numerous steps (probably at least four) leading from olefins to aldehydes. The classical experiments of Natta and co-workers (252) on the hydroformylation of cyclohexene in the presence of $Co_2(CO)_8$ showed that the reaction is first order with respect to olefin over a wide range of olefin concentrations and first order with respect to H_2 partial pressure.

When the pressure of CO is increased, the reaction rate increases and reaches a maximum at a pressure which depends on the temperature, the unsaturated substrate, and probably the type of solvent.

For cyclohexene in toluene at 110–120°C, the maximum rate is observed at about 10 atm of CO. Above this pressure, the rate is inversely proportional to the CO partial pressure (Fig. 5). Substantially the same results were found by Niwa and Yamaguchi (254).

According to other investigators (6,253,255) the dependence on the H_2 partial pressure is more complicated; at constant dissolved cobalt concentration the dependence of the initial hydroformylation rate on the pressure of CO and H_2 and on the olefin concentration follows the rate law given in Eq. 37:

$$r = \frac{aP_{H_2}}{bP_{H_2} + P_{CO}}[\text{olefin}] \qquad (37)$$

Even more complicated is the effect of cobalt concentration on the reaction rate, for example, at 110°C with a high partial pressure of CO (100 atm) the rate is practically proportional to the cobalt concentration, but with a low partial pressure of CO (10–15 atm) the apparent order (with respect to cobalt concentration) is about 0.5.

Although no systematic investigation has been made, it seems, as stated before, that the partial pressure of CO at which the hydroformylation rate

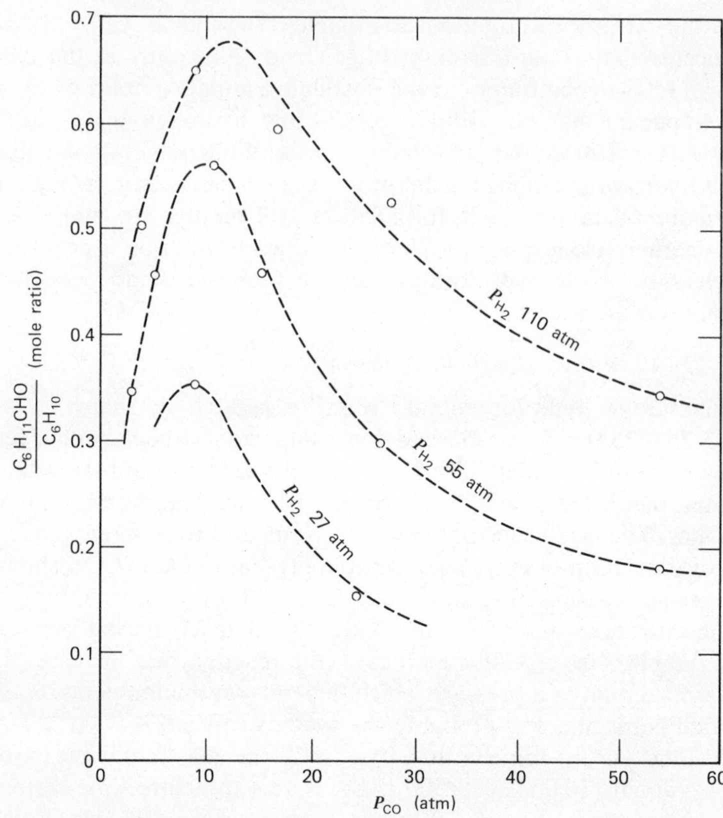

FIG. 5. Kinetics of cyclohexene hydroformylation (252). Conversion as a function of CO and H_2 pressures for isochronous experiments at constant total pressure and 110°C.

reaches a maximum depends on the type of olefin used: for propylene the partial pressure of CO was about 7 atm at 75°C (187), for 2-ethyl-1-hexene, about 12 atm at 120°C, and for methyl acrylate, about 15 atm at 100°C (256).

a. Effect of Olefin Structure on Reaction Rate. Under nonisomerizing conditions, the structure of the olefin greatly influences the hydroformylation rate as Table 27 shows. Terminal olefins are more rapidly hydroformylated than internal and cyclic olefins; the lowest rate was found for internal branched olefins.

Since the composition of the reaction products was not determined in these experiments, the ratios of the relative rates of addition of the formyl group to the terminal and internal carbon atoms are not known. These

TABLE 27
Rates of Hydroformylation of Olefins at 110°*C*[a] (98)

	Sp. reaction rate, $10^3\ k\ \text{min}^{-1}$
Straight-chain terminal olefins	
1-Pentene	68.3
1-Hexene	66.2
1-Heptene	66.8
1-Octene	65.6
1-Decene	64.4
1-Tetradecene	63.0
Straight-chain internal olefins	
2-Pentene	21.3
2-Hexene	18.1
2-Heptene	19.3
3-Heptene	20.0
2-Octene	18.8
Branched terminal olefins	
4-Methyl-1-pentene	64.3
2-Methyl-1-pentene	7.8
2,4,4-Trimethyl-1-pentene	4.8
2,3,3-Trimethyl-1-butene	4.3
Camphene	2.2
Branched internal olefins	
4-Methyl-2-pentene	16.2
2-Methyl-2-pentene	4.9
2,4,4-Trimethyl-2-pentene	2.3
2,3-Dimethyl-2-butene	1.4
2,6-Dimethyl-3-heptene	6.2
Cyclic olefins	
Cyclopentene	22.4
Cyclohexene	5.8
Cycloheptene	25.7
Cyclooctene	10.8
4-Methylcyclohexene	4.9

[a] Conditions: 0.50 mole olefin; 65 ml methylcyclohexane solvent; 2.8 g (8.2×10^{-3} mole) $Co_2(CO)_8$; 1:1 synthesis gas; 233 atm, initial pressure at room temp.

TABLE 28

Addition of Formyl Groups to Various Carbon Atoms in the Chain During Hydroformylation of Some Linear and Branched Olefins[a]

Olefin	Temp (°C)	P_{CO}[b] (atm)	P_{H_2}[b] (atm)	Ratio of addition to different carbons C_1/C_2	C_2/C_3	Ref.
Propylene	110	90	80	81.4/18.6		99
1-Butene[c]	110	93	80	78.8/21.2		99
cis-2-Butene[c]	110	94	80	71/29		99
1-Pentene	100	90	80	81.7/15.6	15.6/2.7	99
2-Pentene	100	93	80	75.8/19.2	19.2/5	99
3-Methyl-1-pentene[d]	90	95	95	92.1/3	3/~0	112
4-Methyl-1-pentene	116	150	80	89/10	10/~0	99
3-Methyl-1-hexene	110	100	100	93.8/3.1	3.1/~0	114
4-Methyl-1-hexene	100	80	80	88.1/10.2	10.2/~0	115
4-Methyl-2-hexene	110	80	80	75.3/16.6	16.6/~0	117
5-Methyl-1-heptene	100	80	80	77/18.9	18.9/2.4	115

[a] Benzene as solvent unless otherwise indicated.
[b] Experiments at constant pressure.
[c] Toluene as solvent.
[d] Dioxane as solvent.

ratios, however, were estimated from the isomeric composition of the products obtained in hydroformylation experiments done under nonisomerizing conditions for a number of olefins. These ratios seem largely dependent on the existence of branching not only at the double bond but also at the α or β position relative to the double bond (Table 28).

The data on 1-pentene and 2-pentene by Goldfarb and Orchin (183) and by Piacenti and others (99) show that the formation of hexanal is slower from 2-pentene that from 1-pentene; also *cis*-2-pentene hydroformylates more rapidly that the *trans* isomer (293). These results give interesting indications on the reaction mechanism, which will be discussed later.

b. Effect of Temperature on Reaction Rate. The effect of temperature on the hydroformylation rate is a rather complicated phenomenon; it is not only connected with the activation energy of the slowest reaction step, but also with the dependence of the solubility of CO and H_2 on temperature and with the effect of temperature on possible equilibria between different catalytically active and inactive species that contain cobalt (257,258).

Moreover, it is possible that the slowest reaction step is different at various temperatures. For instance at low temperatures (60–80°C), the activation of CO by the catalyst might be the rate-limiting step. In fact, although stoichiometric hydroformylation with either $HCo(CO)_4$ or $Co_2(CO)_8$ and H_2 occurs rapidly at room temperature, catalytic hydroformylation takes place at appreciable rates only above 60°C when using a low partial pressure of CO and $Co_2(CO)_8$ as catalyst precursor. Higher temperatures are necessary when cobalt salts are used as catalyst precursors because the formation of catalytically active cobalt carbonyl compounds from cobalt salts generally requires a temperature above 110°C (260). In many experiments conducted above 140°C, the diffusion of CO from the gas phase to the liquid phase was found slower than the actual hydroformylation reaction (16).

Higher temperatures (130–140°C) are required when strongly coordinating ligands like trialkyl- or triarylphosphines are present in quantities of either equimolar or in excess relative to the dissolved cobalt (85,86). In fact, $[Co(CO)_3PBu_3]_2$ and $HCo(CO)_3PBu_3$ react with olefins only above 100–120°C (26), in contrast to the behavior of $HCo(CO)_4$ and of $Co_2(CO)_8$ in the presence of H_2, which are active at room temperature.

From the experimental data, it appears that the hydroformylation rate increases greatly with temperature. For temperatures between 90 and 140°C, different authors reported apparent overall activation energies between 24 and 35 kcal/mole (Table 29).

TABLE 29

Apparent Activation Energies for the Hydroformylation of Some Olefins[a]

Olefin	Temp (°C)	Solvent	P_{total}[b] (atm)	Apparent overall activation energy (kcal/mole)	Ref.
Cyclohexene	100–122.5	Methylcyclohexane	200	35–27	251
1-Hexene	89–108	Toluene	250–273	27	254
1-Butene	90–140	Hexane	238	24	184

[a] $Co_2(CO)_8$ as catalyst.

[b] $H_2/CO = 1$.

c. Influence of Nucleophiles on Reaction Rate. Nucleophiles can greatly influence the hydroformylation rate; for example, strong nucleophiles in large excess relative to the cobalt catalyst cause a marked decrease in the reaction rate and may even suppress it, if the temperature is sufficiently low (13). On the other hand, a relatively strong nucleophile (as pyridine) in small concentration, or a large excess of weak nucleophiles (as ethyl alcohol or acetonitrile) can increase the reaction rate by a factor of two or more (69,76).

A systematic investigation of the effect of nucleophiles on the rate of the catalytic oxo reaction is difficult and has not been attempted. The interaction between nucleophiles and transition metals is far from being well understood, particularly when more than one nucleophile is involved. In the catalytic system during hydroformylation at least three nucleophiles are involved, namely CO, olefin, and the resulting aldehyde.

Investigation of the interaction between nucleophiles and cobalt carbonyls under typical oxo conditions is experimentally difficult: Besides the intrinsic difficulties of research on homogeneous catalysis arising from the simultaneous presence of a relatively large number of catalytically active and inactive complexes (some of which are present in very small concentration) it is necessary to operate at high temperature and pressure.

When weak nucleophiles are used as solvents, further difficulties arise. These media may influence the reaction rate for reasons that have little to do with the nucleophilicity of the solvent, for example, a high or low dielectric constant or a varying capacity for dissolving CO and H_2.

Accordingly, we shall now discuss separately (*1*) the effect on reaction rate of adding small quantities of relatively strong nucleophiles when the reaction medium is not substantially modified, and (*2*) the use of weak nucleophiles as solvents.

(*1*) *Effect of Strong Nucleophiles in Small Amounts.* The first base extensively studied as an additive to a cobalt hydroformylation catalyst was pyridine. From the hydroformylation of cyclohexene in methylcyclohexane as solvent, Wender et al. (6) found a small rate increase for pyridine/$Co_2(CO)_8$ ratios between 0.15 and 1.2, the maximum being around 0.15.

For the hydroformylation of methyl acrylate in benzene, Iwanaga (69) reported that the relative rate increased up to 3.14 times at 100°C and 4.28 times at 120°C when using pyridine/Co ratios of 10. These investigators (13,69) indicated that a large excess of pyridine inhibits or suppresses the reaction.

Small amounts of triethylamine slightly increase the rate of hydroformylation of cyclohexene, but larger amounts inhibit the hydroformylation of both cyclohexene (6) and methyl acrylate (69).

The effect of seventeen bases on the hydroformylation of methyl acrylate in benzene with a base/cobalt ratio of 10 was examined by Iwanaga (69), who found that the highest relative rate was for picoline, 3.52, and the lowest for triethylamine, 0.44. Only with pyridine, however, was the effect of the base/cobalt ratio on the rate investigated. The effect of different phosphines on the hydroformylation rate was also examined (87).

No systematic investigation has been carried out to determine the effect of the type of phosphine on the hydroformylation rate. There is no doubt, however, that tributylphosphine (with PBu_3/cobalt ratio of 2) decreases the hydroformylation rate of 1-pentene by a factor of at least 100 (87) at 110–150°C, and for 2-octene at 190°C (94) by a factor of at least 30. According to Tucci (91), the difference in catalytic activity between $HCo(CO)_4$ and $HCo(CO)_3PBu_3$ at 120°C also depends on the type of olefin used.

The hydroformylation rate in the absence of tributylphosphine under similar conditions ($HCo(CO)_4$ as catalyst) is about 230 times higher for 1-hexene, and 170 times for 2-hexene. Different phosphines have different effects on the hydroformylation rate; the literature data are conflicting in some cases. Catalysts containing PPh_3, for instance, are less active than catalysts containing PBu_3, according to Slaugh and Mullineaux (87), but are more active according to Tucci (92). The latter finds a linear relationship between the logarithm of the hydroformylation rate and the logarithm of the basicity of the ligand for phosphines containing trialkyl, triaryl, or alkyl and aryl substituents. Among the other phosphorus-containing nucleophiles, diphosphines, dialkyl chlorophosphines, cyclic, and bicyclic phosphines as well as phosphites were investigated, as discussed previously. Arsines seem to lower the catalytic activity less than do phosphines.

Unfortunately, all the above work was directed mainly toward investigating the effect of the ligands on the isomeric composition of the reaction products. No kinetically applicable information was obtained, except for the general conclusion that ligands containing nitrogen or oxygen may increase the hydroformylation rate and that ligands containing phosphorus always decrease the rate.

The effect of the relative concentrations of different phosphine-containing cobalt carbonyl species in some solvents was investigated under oxo conditions (259). An attempt was also made to determine the relative activity of these species as hydroformylation catalysts. The polarity of the solvents seems to affect both the relative concentration of the various species and their specific activity (259).

Much more work is needed to acquire an understanding of the complex phenomena involved.

TABLE 30
Effect of Solvent on Rate of Hydroformylation of Several Olefins[a]

	Specific reaction rate, v ($10^3 k$/min)				
Solvent	1-Hexene	2-Hexene	Cyclohexene[b]	Methyl acrylate[c]	Acrylonitrile
Benzene	32	9.2	6.7	41.8	12
Acetone	34	9.1	6.1	59.5	23
Methanol	54	9.2	8.9	157.0	80
Ethanol			8.7	186.0	128
Methyl ethyl ketone			5.7	39.1	

[a] Unless otherwise stated: temp 120°C; pressure 200 atm at 20°C; H_2:CO = 1:1; ref. 162.
[b] Temp 110°C; pressure 233 atm at 20°C; CO:H_2 = 1:1; refs. 98, 260.
[c] Ref. 76.

(2) *Effect of Weakly Nucleophilic Solvents on Rate.* The effect of different solvents on the reaction rate was investigated for cyclohexene (98), methyl acrylate (76), and acrylonitrile (162). As expected, rather different results were obtained, considering the different nucleophilic character of the olefinic substrates used. With cyclohexene (98,260), the effect of solvent was not large; the maximum rate found in methanol was only about 1.5 times the lowest rate determined in saturated hydrocarbons. Aliphatic ethers and ketones show a behavior similar to that of the aliphatic hydrocarbons, and aromatic hydrocarbons, such as benzene and methylnaphthalene, show a behavior intermediate between methanol and aliphatic hydrocarbons (260) (Table 30).

The hydroformylation of methyl acrylate (76) was investigated using 17 different solvents at 120°C. Under the conditions employed, no reaction was observed when using ethylene glycol or formamide as solvent. A variation in the reaction rate from one to seven was observed on changing from dioxane to ethanol. Alcohols, except *tert*-butyl alcohol, followed by acetonitrile and THF seem to give the highest reaction rates. Aromatic hydrocarbons, aliphatic ketones, and ethers are the solvents in which the lowest rates were observed.

The hydroformylation rate of acrylonitrile in ethanol and in propanol at 120°C was found to be 10 times greater than the rate in benzene or in toluene, whereas the rate in acetone was only about twice that in benzene.

The above data consistently show that the rate of hydroformylation is larger in alcohols than in hydrocarbons, and that ketones and ethers occupy an intermediate position. But no interpretation is possible on the basis of solvent nucleophilicity, since, as already mentioned, these effects might arise from many other factors.

7. *Mechanisms*

Much work on the oxo reaction was done mainly during the 1950s in academic and industrial laboratories, and the general features of the reaction mechanism are now relatively well known. A detailed knowledge of the numerous steps involved as well as of the structure of the catalytic intermediates, nevertheless, is still lacking. A great deal of work is in progress around the world stimulated by the interest devoted to homogeneous catalysis and by the growing industrial importance of the oxo and related reactions.

First we shall review the most important contributions which in 1961 led to the formulation of a mechanism still believed to be substantially sound. Then we shall examine the experimental contributions made during the ensuing ten years; finally, we shall discuss the present state of knowledge in the field.

a. Contributions to the Hydroformylation Mechanism up to 1961. Although heterogeneous catalysts were initially used to carry out hydroformylation experiments (42,43), it was soon recognized that the oxo reaction generally occurs in homogeneous phase (11,261). Only the mechanism of the homogeneously catalyzed hydroformylation will be considered here; the elementary steps, however, are probably very similar to those occurring in heterogeneously catalyzed hydroformylation.

The first attempts to rationalize the course of olefin hydroformylation involved either the formation of ketenes or cyclopropanones (84) by reaction of CO with the olefin (reaction 38).

$$\mathrm{RCH{=}CH_2 + CO \longrightarrow \underset{\substack{\diagdown\;\diagup \\ C \\ \| \\ O}}{RCH{-}CH_2} \xrightarrow{H_2} RCH_2CH_2CHO + R\underset{\substack{| \\ CH_3}}{C}HCHO} \qquad (38)$$

A competitive attack of H^+ and CO or $\overset{+}{C}HO$ on the olefin (6) or a radical mechanism (262) was successively proposed and found unsatisfactory (6). Roelen (261) and Adkins and Krsek (11) recognized $HCo(CO)_4$ as the main catalytic species having the role of activating both CO and H_2; the possibility of the addition of H^+ and $[Co(CO)_4]^-$ to the double bond was also considered (222,251).

The soundness of Adkins and Krsek's suggestion (11) was later confirmed by the success of the stoichiometric synthesis of aldehydes by interaction of olefins and $HCo(CO)_4$ at room temperature (13,263).

These experiments showed that the relatively high pressures and temperatures used for the catalytic reaction were necessary to activate CO and H_2 but not for activation of the olefin. The activation of H_2, when a sufficient pressure is used, occurs even at room temperature under a very low CO pressure, but only at higher temperature when the CO pressure is increased (Eqs. 39–42).

$$2Co + 8CO \xrightleftharpoons{150°C} Co_2(CO)_8 \tag{39}$$

$$Co_2(CO)_8 + H_2 \xrightleftharpoons{25°C} 2HCo(CO)_4 \tag{40}$$

$$4HCo(CO)_4 + 4CH_2{=}CH_2 \xrightarrow[25°C]{2H_2} 4CH_3CH_2CHO + Co_4(CO)_{12} \tag{41}$$

$$Co_4(CO)_{12} + 4CO \xrightleftharpoons{\sim 70°C} 2Co_2(CO)_8 \tag{42}$$

Definite progress in understanding the mechanism of hydroformylation was made in 1954 (252,253,255) when it was demonstrated that the lack of response of the reaction rate to the total pressure of CO and H_2 (1:1) for both heterogeneous (43) and homogeneous catalysts (251) could be considered accidental; in fact the first order dependence of the rate on the H_2 pressure was nullified by the inverse dependence of the rate on the CO pressure.

In addition, Natta, Ercoli, and Castellano (185) made a second important observation about the order with respect to the dissolved cobalt. As already mentioned, the first order dependence at high CO pressure decreased to a half-order dependence at low CO pressure. This finding, which has been ignored in most of the subsequent literature, probably indicates the existence of various catalytic species whose concentration depended on CO pressure.

$$HCo(CO)_4 + C_6H_{10} \rightleftarrows [HCo(CO)_3C_6H_{10}] + CO$$

$$[HCo(CO)_3C_6H_{10}] + HCo(CO)_4 \rightarrow C_6H_{11}CHO + [Co_2(CO)_6]$$

$$2[Co_2(CO)_6] \rightleftarrows Co_4(CO)_{12}$$

$$Co_4(CO)_{12} + 4CO \rightleftarrows 2Co_2(CO)_8$$

$$Co_2(CO)_8 + H_2 \rightleftarrows 2HCo(CO)_4$$

Scheme **XI**

On the basis of the kinetic results, two alternate mechanisms were proposed by Natta, Ercoli, and Castellano (185). The first, which might be termed an S_N2 reaction and does not correspond to an "associative mechanism" (264), involves displacement of a CO group in a $HCo(CO)_4$ molecule by the reacting olefin (Scheme **XI**).

The second, closely resembling a "dissociative mechanism" (264) involves a coordinatively unsaturated $[Co_2(CO)_6]$ species arising from dissociation of a $Co_4(CO)_{12}$ molecule (Scheme **XII**).

$$Co_4(CO)_{12} \rightleftarrows 2[Co_2(CO)_6]$$

$$[Co_2(CO)_6] + C_6H_{10} + CO \rightleftarrows [Co_2(CO)_7C_6H_{10}]$$

$$[Co_2(CO)_7C_6H_{10}] \underset{-CO}{\overset{+CO}{\rightleftarrows}} Co_2(CO)_8 + C_6H_{10}$$

$$[Co_2(CO)_7C_6H_{10}] \xrightarrow[\text{slow}]{+H_2} [Co_2(CO)_6] + C_6H_{11}CHO$$

$$2Co_2(CO)_8 \rightleftarrows 4CO + Co_4(CO)_{12}$$

Scheme **XII**

In both cases, the rate-determining step was supposed to be hydrogenation of the intermediate complexes $[HCo(CO)_3C_6H_{10}]$ or $[Co_2(CO)_7C_6H_{10}]$, respectively, either by $HCo(CO)_4$ or by H_2.

Martin (253) and Wender and co-workers (6) favored a mechanism in which a substitution reaction of the S_N2 type between $Co_2(CO)_8$ and olefin is postulated, but which does not reach equilibrium (Scheme **XIII**).

$$Co_2(CO)_8 + RCH{=}CH_2 \xrightarrow{k_1} [Co_2(CO)_7RCH{=}CH_2] + CO$$

$$[Co_2(CO)_7RCH{=}CH_2] + CO \xrightarrow{k_2} Co_2(CO)_8 + RCH{=}CH_2$$

$$[Co_2(CO)_7RCH{=}CH_2] + H_2 \xrightarrow{k_3} [Co_2(CO)_6] + RCH_2CH_2CHO$$

$$[Co_2(CO)_6] + 2CO \longrightarrow Co_2(CO)_8$$

Scheme **XIII**

These mechanisms, postulated to fit most of the features of the kinetics of hydroformylation, excluded direct attack of CO on the olefin. Furthermore, they opened several interesting and stimulating questions—the most important of which was the structure of the olefin-containing complexes $[HCo(CO)_3C_6H_{10}]$ in Scheme **XI** and $[Co_2(CO)_7C_6H_{10}]$ in Scheme **XII**, respectively. These complexes evidently are connected with the transfer of CO and H_2 to the olefin and with the nature of the hydrogenolysis process leading to the aldehydes. They were considered responsible for the inverse effect of the CO pressure on the hydroformylation rate.

Further progress was made between 1955 and 1961, particularly on the structure and reactivity of the reaction intermediates. The first indication of the mechanism of CO transfer to the substrate came from a related field: The chemistry of alkymanganese carbonyls which largely contributed to understanding the hydroformylation mechanism (265–267).

Coffield and co-workers (268) were able to show that $CH_3Mn(CO)_5$ is reversibly transformed into the corresponding acyl derivative. The CO involved in this transformation is one already coordinated to the metal (269). The transformation can be conceived either as a CO insertion through a three-membered ring intermediate or as a nucleophilic attack of an alkyl group on a coordinated CO molecule; the second path is now favored (269).

On the basis of results obtained in the stoichiometric reaction between $HCo(CO)_4$ and olefin, Kirch and Orchin (17) concluded that either $[(HCo(CO)_4)_2(RCH{=}CH_2)(CO)]$ or $[HCo(CO)_4(RCH{=}CH_2)(CO)]$ was the intermediate, but they did not formulate any hypothesis on the structures involved.

Finally, Heck and Breslow (198) studied the CO insertion reaction of $CH_3Co(CO)_4$, which was first synthesized by Hieber et al. (270).

Acylcobalt tetracarbonyls, $RCOCo(CO)_4$, the composition of which corresponded exactly to that of the complex $[HCo(CO)_4(olefin)CO]$ formulated by Kirch and Orchin (17), were then isolated.

Investigation of the chemical properties of acylcobalt carbonyls by Heck and Breslow, which is discussed in Vol. 1 (fourth chapter), showed that these carbonyls are easily reduced to aldehydes at room temperature by both $HCo(CO)_4$ and by H_2 under pressure in a reaction inhibited by CO pressure (198). On the basis of these results, the mechanism shown in Scheme **XIV** can be postulated for the hydroformylation of ethylene.

b. Remarks on the Mechanism Proposed by Heck and Breslow. Although scheme **XIV** probably offers the most satisfactory way to explain the mechanism of hydroformylation, a number of aspects need further clarification or proof.

It should be kept in mind that the retarding effect of CO pressure was noticed also in reactions **XIVa** (14b) and **XIVe** (18), as well as **XIVc** (23), and that nothing is known of its magnitude in the various steps. In this

$$Co_2(CO)_8 + H_2 \rightleftarrows 2HCo(CO)_4 \quad \textbf{(a)}$$

$$HCo(CO)_4 \rightleftarrows [HCo(CO)_3] + CO \quad \textbf{(b)}$$

$$CH_2{=}CH_2 + [HCo(CO)_3] \rightleftarrows \underset{\substack{\uparrow \\ CH_2{=}CH_2}}{[HCo(CO)_3]} \rightleftarrows [CH_3CH_2Co(CO)_3] \quad \textbf{(c)}$$

$$[CH_3CH_2Co(CO)_3] + CO \rightleftarrows CH_3CH_2Co(CO)_4 \rightleftarrows [CH_3CH_2COCo(CO)_3] \underset{-CO}{\overset{+CO}{\rightleftarrows}} CH_3CH_2COCo(CO)_4 \quad \textbf{(d)}$$

$$[CH_3CH_2COCo(CO)_3] \xrightarrow{HCo(CO)_4} CH_3CH_2CHO + [Co_2(CO)_7] \quad \textbf{(e)}$$

$$[CH_3CH_2COCo(CO)_3] \xrightarrow{H_2} CH_3CH_2CHO + [HCo(CO)_3] \quad \textbf{(f)}$$

$$[Co_2(CO)_7] + H_2 \rightleftarrows HCo(CO)_4 + [HCo(CO)_3] \quad \textbf{(g)}$$

Scheme **XIV**

connection, it would be important to determine which is the rate-determining step and whether it remains the same under different reaction conditions.

An important point not directly explainable by Scheme **XIV** is the change in reaction order from one to one-half with respect to cobalt concentration when the CO pressure is decreased.

Niwa and Yamaguchi (254) who, from kinetic experiments found the synthesis of $HCo(CO)_4$ from $Co_2(CO)_8$ and H_2 much slower than the oxo reaction, have questioned the participation of $HCo(CO)_4$ in the course of hydroformylation, although its presence at the end of the reaction was proved experimentally (271). It should be noted that the kinetics of $HCo(CO)_4$ formation was studied in a medium different from that present during hydroformylation when olefins and aldehydes are present in rather high concentration. Moreover, it is known that reactions catalyzed by $HCo(CO)_4$ are accelerated by weak Lewis bases (6,25,69).

Other unsettled questions are the possible existence of equilibria corresponding to the insertion of the olefin into the cobalt–hydrogen bond and the already-mentioned effect of bases and solvents on the reaction rate.

Finally, the explanation given for the formation of various isomeric aldehydes, for instance by hydroformylation of terminal olefins, as well as the factors influencing the composition of the isomeric products is not completely satisfactory and casts some doubt on the validity of transferring knowledge acquired from stoichiometric hydroformylation to the catalytic process. In many cases, the isomeric composition of the reaction products is completely different in the two processes.

As pointed out by Goldfarb and Orchin (183) in 1957, the hydroformylation of 1-pentene and of 2-pentene under typical oxo conditions yields isomeric ratios different enough to dismiss the possibility of a common intermediate. Double-bond isomerization before hydroformylation with formation of free isomerized olefin has to be excluded on the basis of thermodynamic and kinetic considerations (183), and therefore formation of large amounts of hexanal from 2-pentene was not understood. According to Heck and Breslow (23), the formation of hexanal from 2-pentene can be explained by assuming the existence of an equilibrium between the π-complexes postulated as first reaction intermediates. The π-complexes of terminal olefins would be much more stable and, therefore, would be present in larger concentrations than those deriving from internal olefins (Scheme **XV**). We shall discuss this point further in connection with a possible explanation of the effect of CO pressure on the composition of the oxo products.

c. Later Work on the Hydroformylation Mechanism. The general validity of the scheme proposed by Breslow and Heck was confirmed by different experiments. The second step, addition of the olefin to a Co–H bond, was confirmed by Stone and co-workers (272) who succeeded in isolating $CF_2HCF_2Co(CO)_3PPh_3$ by reacting $CF_2{=}CF_2$ with $HCo(CO)_4$ and successively with PPh_3, and by Gankin et al. (273) who isolated 2- and 3-iodohexane by treating the reaction product of $HCo(CO)_4$ and 1-hexene with iodine.

Furthermore, Nagy-Magos, Bor, and Markó (181) proved, in the case of benzylcobalt tetracarbonyl, the existence of an equilibrium between alkyl and acyl compounds (reaction 33), and showed, as with manganese derivatives, that the CO bound to the benzyl group is one of those previously coordinated to the cobalt atom. Acyl formation also occurs when one of the CO groups has been replaced by PPh_3 (181).

Investigation of the stereochemistry of the oxo reaction (Section II-B-5) yielded results in agreement with the above picture. In fact, as expected, addition of CO and hydrogen is *cis* and seems to occur at the same olefin face at which the metal is bound in the postulated initial π-complex.

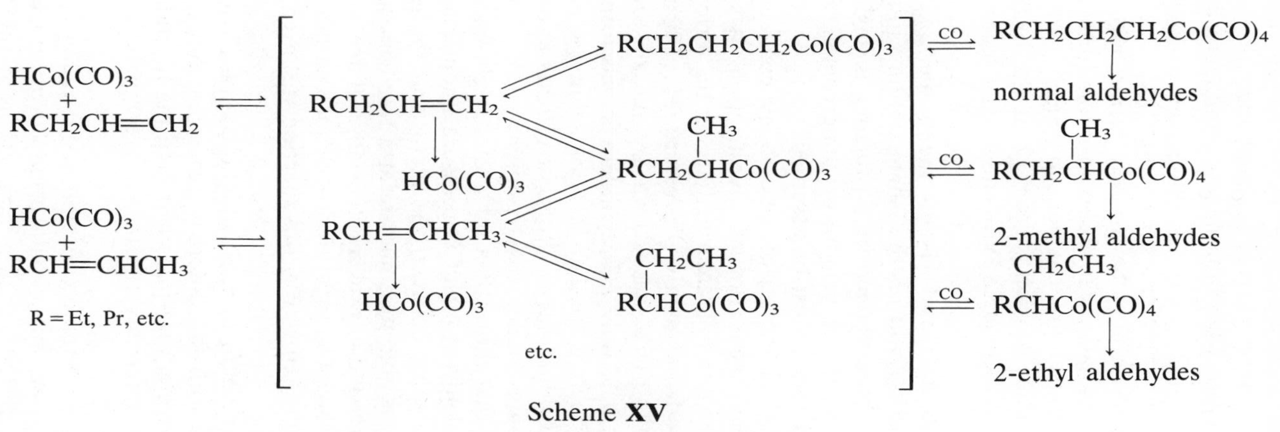
HCo(CO)3 + RCH2CH=CH2
HCo(CO)3 + RCH=CHCH3
R = Et, Pr, etc.
RCH2CH=CH2
HCo(CO)3
RCH=CHCH3
HCo(CO)3
RCH2CH2CH2Co(CO)3
CH3
RCH2CHCo(CO)3
CH2CH3
RCHCo(CO)3
etc.
CO
RCH2CH2CH2Co(CO)4
normal aldehydes
CO
CH3
RCH2CHCo(CO)4
2-methyl aldehydes
CO
CH2CH3
RCHCo(CO)4
2-ethyl aldehydes

Scheme **XV**

Other facts, although not in conflict with the mechanism of Heck and Breslow, show that some aspects of the mechanism need further clarification and that the picture given is oversimplified.

The reaction step in which the negative influence of CO partial pressure on the hydroformylation rate actually takes place is still uncertain. A possible decrease in π-complex concentration by increasing the CO partial pressure (185) cannot be the sole factor since a similar negative effect by the same variable was observed in the formylation of orthoformic esters (228,229) (see first chapter), in which no π-complexes seem to be involved. An inhibiting effect of CO in the hydrogenolysis of the acylcobalt carbonyls (Scheme **XIVd-f**) might give a consistent explanation of both facts.

The influence of CO and H_2 on the isomeric composition of the reaction products (99) and on the isomerization rate of the olefin substrate (16) cannot be explained by postulating a single catalytic species, but at least two or probably more catalytic complexes having different Co/CO and Co/H ratios must be assumed.

The stability of alkyl and acylcobalt carbonyl complexes under different reaction conditions needs further investigation. Alkyl and acylcobalt carbonyls, the most probable intermediates in the formylation of orthoesters, do not isomerize at high CO pressure (P_{CO} 100 atm) and temperatures up to 140°C. This was shown by the observations that from n-propyl orthoformate only n-butyraldehyde was formed and that from isopropyl orthoformate more than 95% of isobutyraldehyde was obtained (228). Further evidence arises from investigation of the behavior of alkylcobalt carbonyls under oxo conditions (287,292). On the other hand, the finding that more than 50% of straight-chain aldehydes were obtained from 2-olefins at high CO pressures (nonisomerizing conditions) shows that the complex between the olefin and the catalyst (perhaps a π-olefin complex) is able to isomerize at a rate comparable to or higher than the hydroformylation rate. Furthermore, different authors (20,274–277) found that at low CO pressures, acylcobalt carbonyls isomerize even at room temperature. However, this reaction is slower than conventional hydroformylation.

This fast isomerization of the catalytic complex containing the olefin can be reconciled with the different hydroformylation rates of 1-olefins and 2-olefins, if we grant as the first step of the reaction the formation of π-complexes that interconvert at a rate comparable with that of the hydroformylation.

The existence of a rapidly established equilibrium for π-complex formation from olefins and cobaltcarbonyl hydrides would introduce a further difficulty: If the rapid isomerization of the catalytic complex containing the olefin proceeds through π-complex formation, the isomerized olefin produced by rapid dissociation of the π-complex should be present in the nonhydroformylated olefins in substantial amounts, especially if the isomerized olefin is thermodynamically favored. This is not observed under nonisomerizing conditions.

A picture of the hydroformylation consistent with the existence of relatively stable alkylcobalt carbonyl complexes and with the existence of the above mentioned equilibrium can be obtained by postulating that:

a Isomerization of the π-olefin-catalyst complex does not involve formation of true σ cobalt-carbon bonds;

b The equilibrium leading from olefins to π-complex formation if it is reached, is largely displaced toward the olefin-catalyst complex.

As for point *a*, a possible olefin-catalyst complex isomerization path resulting in a 1–2 hydrogen shift (194) and not involving racemization of a tertiary asymmetric carbon atom (114) is reported in Scheme **XVa**.

Finally, further attempts to establish the rate-determining steps in the hydroformylation reaction under different conditions should be made. Since during the catalytic reaction at high CO pressures only acylcobalt carbonyls have been detected as intermediates (82,278) it seems that the rate-determining step should be hydrogenolysis of the acyl group. But in the case of substituted olefins, $HCo(CO)_4$ and not the corresponding acyl complexes were observed, which indicates that, depending on the olefin structure (279) other steps might be rate-determining.

C. Conclusions

There is little doubt that hydroformylation of olefins with cobalt catalysts involves the activation of CO, H_2, and olefins by a cobalt complex and that the transfer of CO and H_2 to the olefin takes place within this complex. This complex is finally cleaved, thus releasing the aldehyde while regenerating the catalytic metal complex.

As for H_2 activation, there are many examples of homolytic (280) and heterolytic cleavage (281) of the H_2 molecule by metal complexes. In H_2 activation by $Co_2(CO)_8$ at room temperature, a homolytic splitting seems to be favored according to reactions 4–7 (15).

During hydroformylation, hydrogen activation with formation of a cobalt carbonyl hydride might also be generated by reaction of an acylmetal carbonyl with H_2. Although no direct information is available on the course of this reaction, Scheme **XVI**, which involves oxidative addition of H_2 to a coordinatively unsaturated acylcobalt carbonyl has been suggested (22).

$-CH([Co])-CH_2-{}^{*}CD(-)-CH_2-$ $\leftarrow$ $-CH{=}CH-{}^{*}CD(-)-CH_2-$ $\rightarrow$ $[Co]-H$ $\rightarrow$ $-CH_2-CH([Co])-{}^{*}CD(-)-CH_2-$

$\rightleftharpoons$ $-CH\cdots H \,:\, CH \;[Co]\; -C\cdots D \,-\, CH_2-$ $\rightleftharpoons$

$-CH_2-CH([Co])-{}^{*}CD(-)-CH_2-$ $\leftarrow$ $-CH_2-CH{=}C(-)-CH_2-$ $\rightarrow$ $[Co]-D$ $\rightarrow$ $-CH_2-CHD-{}^{*}C(-)([Co])-CH_2-$

$\rightleftharpoons$ $-CH_2-CH\cdots D \,:\, C(-) \;[Co]\; CH\cdots H-$ $\rightleftharpoons$

$-CH_2-CHD-{}^{*}C(-)([Co])-CH_2-$ $\leftarrow$ $-CH_2-CHD-C(-){=}CH-$ $\rightarrow$ $[Co]-H$ $\rightarrow$ $-CH_2-CHD-{}^{*}CH(-)-CH([Co])-$

Scheme **XVa**

$$[RCOCo(CO)_3] \underset{}{\overset{H_2}{\rightleftharpoons}} [RCOCo(CO)_3H_2] \rightleftharpoons RCHO + [HCo(CO)_3]$$

Scheme **XVI**

In this connection, a dihydride, $[H_2Co(Ph_2CH_2CH_2Ph_2)_2]ClO_4$, was described (282), and species of the type $H_nCo_m(CO)_p$ (in which n is greater than one) might be present under CO pressure. Actually we do not know whether a monohydride or a hydride with more than one hydrogen atom bound to the same cobalt atom is present under hydroformylation conditions.

Concerning the participation of CO in the oxo reaction, one can conceive, in principle, both a bimolecular reaction between CO and a complex containing the activated olefin and H_2 or a preliminary activation of CO on the catalyst followed by intramolecular CO insertion. That the second hypothesis is correct is indicated, among other facts, by the following experiments: CO bound to cobalt in $Co_2(CO)_8$ reacts with olefins in the presence of high H_2 pressure even at room temperature (14). On the contrary, at temperatures below 70°C, the catalytic hydroformylation of olefins does not occur or occurs very slowly independent of the CO pressure used. Since H_2 and olefin are already activated at room temperature (14), the lack of catalytic hydroformylation at temperatures below 70°C might be attributed not only to a lack of H_2 activation in the presence of CO, but also to the fact that at low temperature CO is not, or very slowly, activated by the cobalt catalytic complexes. An investigation of the temperature necessary to convert $Co_4(CO)_{12}$ into $Co_2(CO)_8$, with rates comparable to the hydroformylation rate seems to confirm this view (294).

As already mentioned, the CO in $Co_4(CO)_{12}$ (16) is less reactive than CO in $Co_2(CO)_8$, but it becomes at least partly available for hydroformylation at room temperature in weak Lewis bases such as alcohols. This suggests that at least at room temperature the transfer of CO to the olefinic substrate occurs, in the absence of nucleophiles, more easily in a complex like $RCo(CO)_4$ than in a complex like $RCo(CO)_3$.

The activation of the olefin in hydroformylation is still rather puzzling. Complexes of olefins with cobalt carbonyls or cobalt carbonyl hydrides were postulated by many workers, but their existence has not been demonstrated. In principle, the activation of an olefin might occur either through a π-complex or through a four-center intermediate prior to insertion of the olefin into the Co–H bond.

Rupilius and Orchin (20), when investigating the chemical behavior of acylcobalt carbonyls, obtained evidence favoring the formation of a π-complex between olefin and the cobalt carbonyl complex at room temperature. Although results from stoichiometric experiments should not in principle be considered representative of the situation under catalytic conditions, this evidence appears very important since it may represent what happens in catalytic hydroformylation at very low CO pressures.

Admitting the formation of a π-complex, there still arises the problem of the direction of addition of CO and H_2 to the double bond. A possible way of explaining the above effects is given in Scheme **XVII**

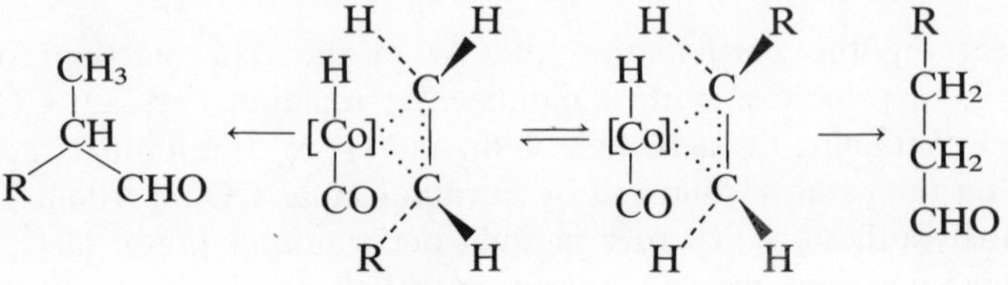

(**9**) (**10**)

Scheme **XVII**

where a conformational equilibrium in the initial π-complex is considered. Such equilibria have already been studied, for example, in square planar platinum complexes (283). According to this hypothesis, both the effect of CO and of phosphines on the isomeric composition of the reaction products can be consistently accounted for mainly on the basis of steric influences on the conformational equilibrium. However, a detailed knowledge of stereochemistry of π-complexes **9** and **10** is lacking.

Further, to put the above hypothesis on a quantitative basis, it is necessary to know the free energy difference, $\Delta G°$, between complexes **9** and **10** as well as the sign and order of magnitude of the difference $\Delta G_a^* - \Delta G_b^*$ between the free energies of activation for the transformation of **9** and **10** to the linear and branched aldehydes, respectively. If these data were evaluated, it would be possible to decide whether the prevalence of one of the isomeric hydroformylation products is mainly determined by the above-mentioned difference between complexes like **9** and **10**, by $\Delta G°$ or by the corresponding free energies of activation, $\Delta G_a^* - \Delta G_b^*$, for the transformation of **9** and **10** into aldehydes.

This approach should furnish a better understanding of the effect of different ligands and reaction conditions on the isomeric composition of the hydroformylation products and should provide qualitative and quantitative information on asymmetric induction in asymmetric hydroformylation.

As for the activation of olefinic carbon atoms not linked with the double bond in the original substrate (attack of the formyl group on originally saturated carbon atoms), it is not yet possible to establish if, besides a series of π-complex formations, other types of olefin activation play an important role. Some progress has been made in understanding the isomerization of the catalyst-substrate complexes by investigating deuterated olefins (284).

The steps following the activation of H_2, CO, and olefins, as formulated by Heck and Breslow, need no further discussion. We shall only point out that the detailed mechanism of CO transfer to the olefin is still under discussion (269). The cleavage of aldehyde from the intermediate complex by $HCo(CO)_4$ seems unlikely under typical oxo conditions, but it does occur in stoichiometric hydroformylation (20). Admitting that the aldehyde in catalytic hydroformylation is split out by H_2, it is still not clear whether H_2 activation occurs through oxidative addition of hydrogen to cobalt with formation of dihydrides or whether the Co–COR groups react directly with molecular hydrogen.

Beside these uncertainties, the main problem of oxo catalysis is the structure of the catalytically active species that react with olefins. The course of principal and secondary reactions and the order with respect to cobalt at different CO pressures (Table 31) clearly indicate that more than one catalytic species are present and that they probably operate via similar reaction paths. Their concentrations depend largely on CO and H_2 partial pressures as shown by the dependence of the composition of the reaction products on these factors. The equilibria among different catalytic species should be, in the simplest case, of the type shown in Scheme **XVIII**.

TABLE 31

Effects of High and Low Partial Pressures of Carbon Monoxide on Hydroformylation and on Secondary Reactions

Effect on	High P_{CO} 100–300 atm	Low P_{CO} 2–10 atm
Relative rate of hydroformylation	small	large
Relative rate of olefin isomerization[a]	very small	very large
Relative rate of racemization	very small	large
Order with respect to dissolved cobalt	1	0.5
Composition of isomeric products	preponderance of less branched isomers	smaller amounts of less branched isomers

[a] With formation of free isomerized olefin.

$$A \rightleftarrows nB + mCO + pH_2$$

$$n > 1 \qquad m > p$$

Scheme **XVIII**

In this scheme, A and B are complexes of the type $H_xCo_y(CO)_z$; n must be larger than one to explain the decrease of the order with respect to dissolved cobalt by decreasing CO pressure; and m must be larger than p to explain the greater variation in composition of the isomeric products with CO partial pressure than with H_2 partial pressure.

Both the change in composition of the reaction products in the presence of strong nucleophiles and the finding that *N*-substituted optically active salicylaldimines give rise to catalytic cobalt complexes with ligands still bound to cobalt (96) confirm that catalytic properties are not limited to a single cobalt carbonyl hydride or even to a cobalt carbonyl hydride containing only cobalt, CO and hydrogen, if other ligands are present.

For a better understanding of the oxo reaction, efforts should mainly be directed to acquiring a detailed knowledge of the species participating in equilibria as in Scheme **XVIII** and, if possible, determining the equilibrium constants at different temperatures. As a consequence, we may hope to achieve better control of the oxo reaction and a deeper insight into homogeneous catalysis.

REFERENCES

1. O. Roelen, Ger. Pat. 849,548 (1938); *Chem. Zentr.*, **1953,** 927.
2. H. Adkins and G. Krsek, *J. Am. Chem. Soc.*, **71,** 3051 (1949).
3. P. Pino and L. Paleari, in *Ullmanns Encyklopädie der technischen Chemie,* Vol. XIII, Urban and Schwarzenberg, München, Berlin, 1962, p. 61.
4. D. F. Smith, C. O. Hawk, and P. L. Golden, *J. Am. Chem. Soc.*, **52,** 3221 (1930).
5. F. E. Paulik, *Catal. Rev.*, **6,** 49 (1972).
6. I. Wender, H. W. Sternberg, and M. Orchin, in *Catalysis,* Vol. V, Reinhold, New York, 1957, p. 73.
7. L. F. Hatch, *Higher Oxo Alcohols,* Wiley, New York, 1957.
8. C. W. Bird, *Chem. Rev.*, **62,** 283 (1962).
9. A. J. Chalk and J. F. Harrod, in *Advances in Organometallic Chemistry,* Vol. VII, Academic Press, New York, 1968, p. 119.
10. J. Falbe, *Carbon Monoxide in Organic Synthesis,* Springer-Verlag, Berlin, 1970.
11. H. Adkins and G. Krsek, *J. Am. Chem. Soc.*, **70,** 383 (1948).
12. O. Roelen, in *The Oxo Process,* FIAT Final Report No. 1000, U.S. Dept. Commerce, Washington, D.C., 1947, p. 26.
13. I. Wender, H. W. Sternberg, and M. Orchin, *J. Am. Chem. Soc.*, **75,** 3041 (1953).

14*a*. P. L. Barrick, U.S. Pat. 2,542,747 (1951); *Chem. Abstr.*, **45,** 7584 (1951).

14*b*. P. Pino, R. Ercoli, and F. Calderazzo, *Chim. Ind.* (*Milan*), **37,** 782 (1955).

15. F. Ungvary, *J. Organometal. Chem.*, **36,** 363 (1972), and private communication.
16. P. Pino, F. Piacenti, M. Bianchi, and R. Lazzaroni, *Chim. Ind.* (*Milan*), **50,** 106 (1968).

17. L. Kirch and M. Orchin, *J. Am. Chem. Soc.*, **81,** 3597 (1959).
18. D. S. Breslow and R. F. Heck, *Chem. Ind.* (*London*), **1960,** 467.
19. G. L. Karapinka and M. Orchin, *J. Org. Chem.*, **26,** 4187 (1961).
20. W. Rupilius and M. Orchin, *Proc., Symp. Chemistry of Hydroformylation and Related Reactions*, Veszprem 1972, p. 59; *Chem. Abstr.*, **77,** 87513 (1972).
21. Y. Takegami, Y. Watanabe, H. Masada, and T. Mitsudo, *Bull. Chem. Soc. Japan*, **42,** 206 (1969).
22. M. Orchin and W. Rupilius, *Catal. Rev.*, **6,** 85 (1972).
23. R. F. Heck and D. S. Breslow, *J. Am. Chem. Soc.*, **83,** 4023 (1961).
24. Y. Takegami, C. Yokokawa, Y. Watanabe, H. Masada, and Y. Okuda, *Bull. Chem. Soc. Japan*, **37,** 1190 (1964).
25. L. Roos and M. Orchin, *J. Org. Chem.*, **31,** 3015 (1966).
26. F. Piacenti, M. Bianchi, and E. Benedetti, *Chim. Ind.* (*Milan*), **49,** 245 (1967).
27. R. W. Goetz and M. Orchin, *J. Am. Chem. Soc.*, **85,** 2782 (1963).
28. S. Husebye, H. B. Jonassen, C. L. Aldridge, and W. Senn, *Acta Chem. Scand.*, **18,** 1547 (1964).
29. J. A. Bertrand, C. L. Aldridge, S. Husebye, and H. B. Jonassen, *J. Org. Chem.*, **29,** 790 (1964).
30. S. Suga, H. Masada, H. Suda, Y. Watanabe, and Y. Takegami, *Bull. Chem. Soc. Japan*, **42,** 2920 (1969).
31. R. W. Krebs and W. E. Catterall, U.S. Pat. 2,768,974 (1956); *Chem. Abstr.*, **51,** 9976 (1957).
32. N. V. de Bataafsche Petroleum Maatschappij, Brit. Pat. 659,701 (1951); *Chem. Abstr.*, **47,** 5427 (1953).
33. S. H. Munger, U.S. Pat. 2,779,796 (1957); *Chem. Abstr.*, **51,** 9674 (1957).
34. J. K. Mertzweiller, U.S. Pat. 2,636,903 (1953); *Chem. Abstr.*, **48,** 3386 (1954).
35. T. G. Jones, M. C. Fuqua, and C. M. Downs, Jr., U.S. Pat. 2,757,200 (1956); *Chem. Abstr.*, **51,** 4414 (1957).
36. M. C. Fuqua and G. P. Hamner, U.S. Pat. 2,667,514 (1954); *Chem. Abstr.*, **49,** 1779 (1955).
37. H. Lemke, French Pat. 1,223,381 (1960); *Chem. Abstr.*, **55,** 22134 (1961).
38. P. Pino, *Gazz. Chim. Ital.*, **81,** 625 (1951).
39. Wm. W. Prichard, U.S. Pat. 2,517,416 (1950); through *Chem. Abstr.*, **45,** 648 (1951).
40. P. P. Klemchuk, Ger. Pat. 1,165,568 (1964); *Chem. Abstr.*, **61,** 2988 (1964).
41. G. Natta and E. Beati, Ital. Pats. 412,338 and 412,339 (1945).
42. G. Natta, P. Pino, and E. Beati, *Chim. Ind.* (*Milan*), **31,** 111 (1949).
43. G. Natta and E. Beati, *Chim. Ind.* (*Milan*), **27,** 84 (1945).
44. E. M. Gladrow and W. J. Mattox, U.S. Pat. 3,352,924 (1967); *Chem. Abstr.*, **68,** 77749 (1968).
45. J. W. Fitzwilliam, E. A. Naragon, and F. J. Moore, U.S. Pat. 2,830,089 (1958); *Chem. Abstr.*, **52,** 13233 (1958).
46. L. J. Boucher, A. A. Oswald, and L. L. Murrell, *Am. Chem. Soc., Div. Petrol. Chem.*, Preprints, Vol. XIX, No. 1, 1974, p. 162.
47. M. Čapka, P. Svoboda, M. Černy, and J. Hetflejš, *Tetrahedron Lett.*, **1971,** 4787.
48. K. G. Allum, R. D. Hancock, and M. J. Lawrenson, Ger. Offen. 2,022,710 (1971); *Chem. Abstr.*, **75,** 5199 (1971).
49. A. J. Moffat, *J. Catal.*, **18,** 193 (1970).
50. P. R. Rony, *J. Catal.*, **14,** 142 (1969).
51*a*. V. Yu. Gankin, N. Ya. Gordina, D. P. Krinkin, D. M. Rudkovskii, and A. G. Trifel, *Khim. i Tekhnol. Topliv i Masel*, **11,** 8 (1966); *Chem. Abstr.*, **64,** 19264 (1966).

51*b*. *Idem, Intern. Chem. Eng.*, **7,** 45 (1967).
52. V. Yu. Gankin, D. P. Krinkin, D. M. Rudkovskii, and A. G. Trifel, *Neftekhim.*, **6,** 271 (1966); *Chem. Abstr* **65,** 8704 (1966).
53. P. Centola, G. Terzaghi, R. Del Rosso, and I. Pasquon, *Chim. Ind.* (*Milan*), **54,** 775 (1972).
54. V. Yu. Gankin, D. P. Krinkin, D. M. Rudkovskii, and A. G. Trifel, *Khim. i Tekhnol. Topliv i Masel,* **11,** 10 (1966); *Chem. Abstr.*, **64,** 9474 (1966).
55. C. L. Aldridge, E. V. Fasce, and H. B. Jonassen, *J. Phys. Chem.*, **62,** 869 (1958).
56. S. Usami, T. Kondo, K. Nishimura, and Y. Koga, *Bull. Chem. Soc. Japan,* **42,** 2961 (1969).
57. S. Usami, K. Nishimura, T. Koyama, and S. Fukushi, *Bull. Chem. Soc. Japan,* **42,** 2966 (1969).
58. B. L. Booth, H. Goldwhite, and R. N. Haszeldine, *J. Chem. Soc. C,* **1966,** 1447.
59. B. L. Booth, M. J. Else, R. Fields, H. Goldwhite, and R. N. Haszeldine, *J. Organometal. Chem.*, **14,** 417 (1968).
60. L. Markó, G. Bor, and E. Klumpp, *Chem. Ind.* (*London*), **1961,** 1491.
61. V. Macho, *Chem. Zvesti,* **15,** 181 (1961); *Chem. Abstr.*, **55,** 18567 (1961).
62. L. Markó, G. Bor, and G. Almasy, *Chem. Ber.*, **94,** 847 (1961).
63. L. Markó, G. Bor, E. Klumpp, B. Markó, and G. Almasy, *Chem. Ber.*, **96,** 955 (1963).
64. L. Markó, G. Bor, and E. Klumpp, *Angew. Chem.*, **75,** 248 (1963).
65. E. Klumpp, L. Markó, and G. Bor, *Chem. Ber.*, **97,** 926 (1964).
66. L. Markó and G. Bor, *J. Organometal. Chem.*, **3,** 162 (1965).
67. L. Markó, *Acta Chim. Acad. Sci. Hung.*, **59,** 389 (1969); *Chem. Abstr.*, **70,** 118516 (1969).
68. E. Klumpp, G. Bor, and L. Markó, *J. Organometal. Chem.*, **11,** 207 (1968).
69. R. Iwanaga, *Bull. Chem. Soc. Japan,* **35,** 865 (1962).
70. V. Macho, *Chem. Zvesti,* **25,** 49 (1971); *Chem. Abstr.*, **75,** 19660 (1971).
71. V. Macho, E. J. Mistrik, and J. Stresinka, *Chem. Zvesti,* **20,** 870 (1966); *Chem. Abstr.*, **67,** 43324 (1967).
72. L. Komora, V. Macho, and O. Lustik, *Ropa Uhlie,* **14,** 281 (1972); *Chem. Abstr.*, **77,** 93382 (1972).
73. V. Macho and L. Komora, *Chem. Zvesti,* **21,** 164 (1967); *Chem. Abstr.*, **67,** 43364 (1967).
74. H. J. Nienburg, Ger. Pat. 948,150 (1956); *Chem. Abstr.*, **53,** 4136 (1959).
75. A. N. Wennerberg, R. M. Alm, and H. H. Peter, U.S. Pat. 2,894,990 (1959); *Chem. Abstr.*, **54,** 1296 (1960).
76. R. Iwanaga, *Bull. Chem. Soc. Japan,* **35,** 869 (1962).
77. R. Iwanaga, *Bull. Chem. Soc. Japan,* **35,** 774 (1962).
78. W. Rupilius, J. J. McCoy and M. Orchin, *Ind. Eng. Chem., Prod. Res. Develop.*, **10,** 142 (1971).
79. M. Bianchi, E. Benedetti, and F. Piacenti, *Chim. Ind.* (*Milan*), **51,** 613 (1969).
80. P. Szabo, L. Fekete, G. Bor, Z. Nagy-Magos, and L. Markó, *J. Organometal. Chem.*, **12,** 245 (1968).
81. G. F. Pregaglia, A. Andreetta, G. Gregorio, G. F. Ferrari and G. Montrasi, *Proc., Symp. Chemistry of Hydroformylation and Related Reactions,* Veszprem, 1972, p. 19; *Chem. Abstr.*, **77,** 113536 (1972).
82. R. Whyman, *Proc., Symp. Chemistry of Hydroformylation and Related Reactions,* Veszprem, 1972, p. 24; *Chem. Abstr.*, **77,** 113549 (1972).
83. G. F. Pregaglia, A. Andreetta, G. F. Ferrari, and R. Ugo, *J. Organometal. Chem.*, **30,** 387 (1971).
84. W. Reppe, H. Kröper, H. J. Pistor, and H. Schlenk, *Liebigs Ann.*, **582,** 38 (1953).

85*a*. L. H. Slaugh and R. D. Mullineaux, Belg. Pat. 606,408 (1962).

85*b*. *Idem*, Belg. Pat. 621,662 (1963); *Chem. Abstr.*, **59,** 11268 (1963).

86. Esso Research Engineering Co., Neth. Pat. Appl. 6,400,701 (1964); *Chem. Abstr.*, **62,** 5194 (1965).

87. L. H. Slaugh and R. D. Mullineaux, *J. Organometal. Chem.*, **13,** 469 (1968).

88. E. R. Tucci, *Ind. Eng. Chem., Prod. Res. Develop.*, **7,** 32 (1968).

89. E. R. Tucci, *Ind. Eng. Chem., Prod. Res. Develop.*, **7,** 125 (1968).

90. E. R. Tucci, *Ind. Eng. Chem., Prod. Res. Develop.*, **7,** 227 (1968).

91. E. R. Tucci, *Ind. Eng. Chem., Prod. Res. Develop.*, **8,** 215 (1969).

92. E. R. Tucci, *Ind. Eng. Chem., Prod. Res. Develop.*, **9,** 516 (1970).

93. A. Hershman and J. H. Craddock, *Ind. Eng. Chem., Prod. Res. Develop.*, **7,** 226 (1968).

94. W. Kniese, H. J. Nienburg, and R. Fischer, *J. Organometal. Chem.*, **17,** 133 (1969).

95. F. Piacenti, M. Bianchi, E. Benedetti, and P. Frediani, *J. Organometal. Chem.*, **23,** 257 (1970).

96. C. Botteghi, G. Consiglio, and P. Pino, *Chimia*, **26,** 141 (1972).

97. A. I. M. Keulemans, A. Kwantes, and T. van Bavel, *Rec. Trav. Chim.*, **67,** 298 (1948).

98. I. Wender, S. Metlin, S. Ergun, H. W. Sternberg, and H. Greenfield, *J. Am. Chem. Soc.*, **78,** 5401 (1956).

99. F. Piacenti, P. Pino, R. Lazzaroni, and M. Bianchi, *J. Chem. Soc. C*, **1966,** 488.

100. F. Piacenti and M. Bianchi, unpublished results.

101. B. Fell, W. Rupilius, and F. Asinger, *Tetrahedron Lett.*, **1968,** 3261.

102. M. A. F. El Daoushy, Thesis, Technische Hochschule, Aachen, 1964.

103. D. M. Rudkovskii and N. S. Imyanitov, *Zh. Prikl. Khim.*, **35,** 2719 (1962); *Chem. Abstr.*, **59,** 2689 (1963).

104. P. Pino, C. Salomon, C. Botteghi, and G. Consiglio, *Chimia*, **26,** 665 (1972), and unpublished results.

105. R. Lai and E. Ucciani, *C.R. Acad Sci. Paris C*, **273,** 1368 (1971); *Chem. Abstr.*, **76,** 33484 (1972).

106. R. Lai and E. Ucciani, *C.R. Acad. Sci., Paris C*, **275,** 1033 (1972); *Chem. Abstr.*, **78,** 42963 (1973).

107. A. Alberola and M. F. Brana, *An. Real Soc. Espan. Fis. Quim. Ser. B*, **63,** 683 (1967); *Chem. Abstr.*, **67,** 63939 (1967).

108. R. Lai, Thesis, University of Marseilles, 1974.

109. J. Palágyi, Z. Décsy, G. Pályi, and L. Markó, *Hung. J. Ind. Chem.*, **1,** 413 (1973).

110. F. Gaslini and L. Z. Nahum, *J. Org. Chem.*, **29,** 1177 (1964).

111*a*. C. A. Burkhard and D. T. Hurd, U.S. Pat. 2,588,083 (1952); *Chem. Abstr.*, **46,** 9120 (1952).

111*b*. *Idem, J. Org. Chem.*, **17,** 1107 (1952).

112. P. Pino, S. Pucci, and F. Piacenti, *Chem. Ind.* (*London*), **1963,** 294.

113. P. Pino, S. Pucci, F. Piacenti, and G. Dell'Amico, *J. Chem. Soc. C*, **1971,** 1640.

114. F. Piacenti, S. Pucci, M. Bianchi, R. Lazzaroni, and P. Pino, *J. Am. Chem. Soc.*, **90,** 6847 (1968).

115. F. Piacenti, M. Bianchi, and P. Frediani, *Chim. Ind.* (*Milan*), **55,** 262 (1973).

116. F. Piacenti, M. Bianchi, and P. Frediani, *Proc., Symp. Chemistry of Hydroformylation and Related Reactions*, Veszprem, 1972, p. 6; *Chem. Abstr.*, **77,** 125860 (1972).

117. F. Piacenti, M. Bianchi, and P. Frediani, *Advances in Chemistry Series*, No. 132, Am. Chem. Soc., Washington, D.C., 1974, p. 283.

118. H. Adkins and J. L. R. Williams, *J. Org. Chem.*, **17,** 980 (1952).

119. M. Morikawa, *Bull. Chem. Soc. Japan*, **37,** 379 (1964).

120. F. L. Ramp, E. J. Dewitt, and L. E. Trapasso, *J. Polym. Sci., Part A-1*, **4,** 2267 (1966).

121. J. E. Knap, in *Kirk-Othmer's Encyclopedia of Chemical Technology*, Second Supplement, Interscience, New York, 1960, p. 548.
122. F. Asinger, in *Chemie und Technologie der Monolefine*, Akademie Verlag, Berlin, 1957, p. 650.
123. W. F. Gresham, R. E. Brooks, and W. M. Bruner, U.S. Pat. 2,437,600 (1948); *Chem. Abstr.*, **42,** 4196 (1948).
124. Inventa, A-G, French Pat. 1,371,085 (1964); through *Chem. Abstr.*, **62,** 460 (1965).
125. J. Falbe and N. Huppes, *Brennstoff-Chem.*, **47,** 314 (1966).
126. G. Wilke and W. Pfohl, Ger. Pat. 1,059,904 (1959); *Chem. Abstr.*, **55,** 7321 (1961).
127. Ya. T. Eidus, B. K. Nefedov, and M. E. Vol'pin, *Izv. Akad. Nauk SSSR, Otd. Khim. Nauk*, **1963,** 548; *Chem. Abstr.*, **59,** 2701 (1963).
128. C. Bordenca, U.S. Pat. 2,619,506 (1952); *Chem. Abstr.*, **47,** 8091 (1953).
129. P. Pino and R. Ercoli, *Ricerca Scientifica*, **23,** 1231 (1953); *Chem. Abstr.*, **48,** 12106 (1954).
130. V. Macho and J. Stresinka, *Chem. Zvesti*, **22,** 656 (1968); *Chem. Abstr.*, **70,** 46730 (1969).
131. D. M. Rudkovskii, N. S. Imyanitov, and V. Yu. Gankin, *Tr., Vses. Nauchn.-Issled. Inst. Neftekhim. Protsessov*, **1960** (2), 121; through *Chem. Abstr.*, **57,** 10989 (1962).
132. G. Gut, M. H. El-Makhzangi, and A. Guyer, *Helv. Chim. Acta*, **48,** 1151 (1965).
133. V. Macho, M. Polievka, and L. Komora, *Chem. Zvesti*, **21,** 170 (1967); *Chem. Abstr.*, **67,** 21403 (1967).
134. J. Falbe, N. Huppes, and F. Korte, *Brennstoff-Chem.*, **47,** 207 (1966).
135. L. S. Nahum, *J. Org. Chem.*, **33,** 3601 (1968).
136. W. F. Gresham and R. E. Brooks, U.S. Pat. 2,497,303 (1950); through *Chem. Abstr.*, **44,** 4492 (1950).
137. J. Stresinka, V. Macho, and E. J. Mistrik, *Chem. Zvesti*, **22,** 844 (1968); *Chem. Abstr.*, **70,** 77517 (1969).
138. J. Stresinka, M. Markó, and V. Macho, *Chem. Zvesti*, **22,** 263 (1968); *Chem. Abstr.*, **69,** 76825 (1968).
139. I. Maeda and R. Yoshida, *Bull. Chem. Soc. Japan*, **41,** 2969 (1968).
140. E. I. du Pont de Nemours and Co., Brit. Pat. 614,010 (1948); *Chem. Abstr.*, **43,** 4685 (1949).
141. I. Wender, R. Levine, and M. Orchin, *J. Am. Chem. Soc.*, **72,** 4375 (1950).

142*a*. E. J. Mistrik, *Proc., Symp. Chemistry of Hydroformylation and Related Reactions*, Veszprem, 1972, p. 94; *Chem. Abstr.*, **77,** 139684 (1972).

142*b*. E. J. Mistrik and A. Mateides, *Chem. Zvesti*, **25,** 350 (1971); *Chem. Abstr.*, **76,** 71689 (1972).

143. J. Falbe and F. Korte, *Chem. Ber.*, **97,** 1104 (1964).
144. A. Rosenthal and D. Abson, *Can. J. Chem.*, **42,** 1811 (1964).
145. A. Rosenthal, D. Abson, T. D. Field, H. J. Koch, and R. E. J. Mitchell, *Can. J. Chem.*, **45,** 1525 (1967).
146. A. Rosenthal and H. J. Koch, *Can. J. Chem.*, **42,** 2025 (1964).
147. A. Rosenthal and H. J. Koch, *J. Am. Chem. Soc.*, **90,** 2181 (1968).
148. A. Rosenthal, in *Advances in Carbohydrates Chemistry*, Vol. XXIII, Academic Press, New York, 1968, p. 59.
149. J. Falbe, N. Huppes, and F. Korte, *Chem. Ber.*, **97,** 863 (1964).
150. F. Piacenti, P. Pino, and P. L. Bertolaccini, *Chim. Ind.* (*Milan*), **44,** 600 (1962).
151. C. H. McKeever and G. H. Agnew, U.S. Pat. 2,533,276 (1950); *Chem. Abstr.*, **45,** 3415 (1951).
152. R. Lai, E. Ucciani, M. Naudet, and M. O. Lai, *Bull. Soc. Chim. France*, **1969,** 793.

153. H. J. Hagemeyer, Jr. and D. C. Hull, U.S. Pat. 2,610,203 (1952); *Chem. Abstr.*, **47,** 5960 (1953).
154. E. N. Frankel, S. Metlin, W. K. Rohwedder, and I. Wender, *J. Am. Oil Chem. Soc.*, **46,** 133 (1969).
155. E. Ucciani, A. Bonfand, R. Lai, and M. Naudet, *Bull. Soc. Chim. France,* **1969,** 2826.
156. P. Pino, G. Consiglio, C. Botteghi, and C. Salomon, *Adv. in Chem. Series,* No. 132, Amer. Chem. Soc., Washington, D.C., 1974, p. 295.
157. J. Falbe, *Carbon Monoxide in Organic Synthesis*, Springer-Verlag, Berlin, 1970, p. 51.
158. W. Hieber, F. Mühlbauer, and E. A. Ehmann, *Chem. Ber.*, **65,** 1090 (1932).
159. Y. Ono, S. Sato, M. Takesada, and H. Wakamatsu, *Chem. Commun.*, **1970,** 1255.
160. S. Sato, *Nippon Kagaku Zasshi,* **90,** 404 (1969); *Chem. Abstr.*, **71,** 21828 (1969).
161. M. Honda, T. Koga, and G. Noyori, *Kogyo Kagaku Zasshi,* **70,** 1346 (1967); *Chem. Abstr.*, **68,** 40049 (1968).
162. J. Kato, H. Wakamatsu, R. Iwanaga, and T. Yoshida, *Kogyo Kagaku Zasshi,* **64,** 2139 (1961); *Chem. Abstr.*, **57,** 2064 (1962).
163. P. L. Barrick and A. A. Pavlic, U.S. Pat. 2,506,571 (1950); *Chem. Abstr.*, **44,** 7344 (1950).
164. R. Iwanaga, I. Iwamoto, and T. Yoshida, U.S. Pat. 3,010,994 (1961); *Chem. Abstr.*, **57,** 9952 (1962).
165. T. Koga and G. Noyori, *Kogyo Kagaku Zasshi,* **70,** 1172 (1967); *Chem. Abstr.*, **68,** 28966 (1968).
166. S. Sato, Y. Ono, S. Tatsumi, and H. Wakamatsu, *Nippon Kagaku Zasshi,* **92,** 178 (1971); *Chem. Abstr.*, **76,** 33755 (1972).
167. W. F. Gresham, R. E. Brooks, and W. M. Bruner, U.S. Pat. 2,549,454 (1951); through *Chem. Abstr.*, **45,** 8551 (1951).
168. W. F. Gresham and R. E. Brooks, U.S. Pat. 2,402,133 (1946); *Chem. Abstr.*, **40,** 6093 (1946).
169. R. C. Schreyer, U.S. Pat. 2,564,131 (1951); *Chem. Abstr.*, **46,** 9583 (1952).
170. S. A. Khattab and L. Markó, *Acta Chim. Acad. Sci. Hung.*, **40,** 471 (1964); *Chem. Abstr.*, **62,** 1493 (1965).
171. *Chem. Eng. News,* **43** (Mar. 22), 48 (1965).
172. M. Polievka and E. J. Mistrik, *Ropa Uhlie,* **11,** 665 (1969); *Chem. Abstr.*, **72,** 110500 (1970).
173. I. Wender, S. Metlin, and M. Orchin, *J. Am. Chem. Soc.*, **73,** 5704 (1951).
174. I. Wender, H. W. Sternberg, R. A. Friedel, S. J. Metlin, and R. E. Markby, *The Chemistry and Catalytic Properties of Cobalt and Iron Carbonyls,* Bulletin 600, U.S. Bureau of Mines, Washington, D.C., 1962.
175. H. Bestian and K. Rehn, Ger. Pat. 1,007,771 (1957); *Chem. Abstr.*, **53,** 14971 (1959).
176. L. Markó, *Chem. Ind.* (*London*), **1962,** 260.
177. R. Lai, M. Naudet, and E. Ucciani, *Rev. Franc. Corps Gras,* **15,** 15 (1968); *Chem. Abstr.*, **68,** 70410 (1968).
178. F. Piacenti, M. Bianchi, P. Frediani, and P. Pino, unpublished results.
179. J. Falbe, *Ullmanns Encyklopädie der technischen Chemie, Ergänzungsband,* 1969, p. 87.
180. G. F. Pregaglia, A. Andreetta, G. F. Ferrari, G. Montrasi, and R. Ugo, *J. Organometal. Chem.*, **33,** 73 (1971).
181. Z. Nagy-Magos, G. Bor, and L. Markó, *J. Organometal. Chem.*, **14,** 205 (1968).
182. P. Taylor and M. Orchin, *J. Am. Chem. Soc.*, **93,** 6504 (1971).

183. I. Goldfarb and M. Orchin, *Advances in Catalysis*, Vol. IX, Academic Press, New York, 1957, p. 609.
184. V. L. Hughes and I. Kirshenbaum, *Ind. Eng. Chem.*, **49,** 1999 (1957).
185. G. Natta, R. Ercoli, and S. Castellano, *Chim. Ind.* (*Milan*), **37,** 6 (1955).
186. M. Johnson, *J. Chem. Soc.*, **1963,** 4859.
187. M. Orchin, *Advances in Catalysis*, Vol. XVI, Academic Press, New York, 1966, p. 1.
188. V. Yu. Gankin, D. P. Krinkin, and D. M. Rudkovskii, *Zh. Org. Khim.*, **2,** 45 (1966); *Chem. Abstr.*, **64,** 14076 (1966).
189*a*. V. Yu. Gankin, V. A. Dvinin, and V. A. Rybakov, in N. S. Imyanitov, Ed., *Gidroformilirovanie* [Hydroformylation], Izd. Khimiya, Leningrad, 1972, p. 47; *Chem. Abstr.*, **77,** 139126 (1972).
189*b*. *Ibid.*, p. 57; *Chem. Abstr.*, **77,** 151265 (1972).
190. F. Asinger and O. Berg, *Chem. Ber.*, **88,** 445 (1955).
191. B. Fell, P. Krings, and F. Asinger, *Chem. Ber.*, **99,** 3688 (1966).
192. R. Cramer and R. V. Lindsey, *J. Am. Chem. Soc.*, **88,** 3534 (1966).
193. R. W. Goetz and M. Orchin, *J. Am. Chem. Soc.*, **85,** 1549 (1963).
194. C. P. Casey and C. R. Cyr, *J. Am. Chem. Soc.*, **93,** 1280 (1971).
195. M. Bianchi, P. Frediani, and F. Piacenti, *Chim. Ind.* (*Milan*), **55,** 798 (1973).
196. A. Stefani, G. Consiglio, C. Botteghi, and P. Pino, *J. Amer. Chem. Soc.*, in press.
197. J. Falbe, H. Feichtinger, and P. Schneller, *Chem. Ztg.*, **95,** 644 (1971).
198. R. F. Heck and D. S. Breslow, *Actes congr. intern. catalyse, 2^e, Paris*, **1960,** p. 671; *Chem. Abstr.*, **55,** 24545 (1961).
199. F. Piacenti, *Gazz. Chim. Ital.*, **92,** 225 (1962).
200. R. W. Goetz and M. Orchin, *J. Org. Chem.*, **27,** 3698 (1962).
201. O. Roelen, in K. Ziegler, Ed., *Präparative organische Chemie* (*Naturforschung und Medizin in Deutschland, Vol. XXXVI*), Part I, Dietrich, Wiesbaden, 1948, p. 157.
202. C. L. Aldridge and H. B. Jonassen, *J. Am. Chem. Soc.*, **85,** 886 (1963).
203. R. W. Rosenthal and E. A. Naragon, *U.S.* Pat. 2,670,385 (1954); *Chem. Abstr.*, **49,** 4008 (1955).
204. E. I. du Pont de Nemours & Co., Brit. Pat. 665,705 (1952); *Chem. Abstr.*, **46,** 10190 (1952).
205. Standard Oil Development Co., Brit. Pat. 668,963 (1952); *Chem. Abstr.*, **46,** 10191 (1952).
206. J. Berty, E. Oltay, and L. Markó, *Chem. Tech.* (*Berlin*), **9,** 283 (1957); *Chem. Abstr.*, **54,** 254 (1960).
207. G. F. Pregaglia, *Chim. Ind.* (*Milan*), **51,** 1043 (1969).
208. G. F. Pregaglia, *Chim. Ind.* (*Milan*), **51,** 509 (1969).
209. L. Markó, B. Heil, and S. Vastag, *Advances in Chemistry Series*, No. 132, Am. Chem. Soc., Washington, D.C., 1974, p. 27.
210. L. Markó, *Proc. Chem. Soc.*, **1962,** 67; *Chem. Abstr.*, **61,** 2926 (1964).
211. V. Macho, *Chem. Prumysl*, **12,** 240 (1962); *Chem. Abstr.*, **58,** 1338 (1963).
212. P. Guyer and E. Bosshard, *Chimia*, **18,** 131 (1964).
213. A. Matsuda and H. Uchida, *Tokyo Kogyo Shikensko Hokoku*, **57,** 50 (1962); *Chem. Abstr.*, **62,** 7625 (1965).
214. M. Polievka, *Ropa Uhlie*, **15,** 366 (1973); *Chem. Abstr.*, **79,** 125500 (1973).
215. K. A. Alekseeva, D. L. Libina, D. M. Rudkovskii, and A. G. Trifel, *Neftekhim.*, **6,** 458 (1966); *Chem. Abstr.*, **65,** 10451 (1966).
216. F. Piacenti and P. P. Neggiani, *Atti Soc. Tosc. Sci. Nat. Ser. A*, **67,** 27 (1960); *Chem. Abstr.*, **57,** 69 (1962).
217. Esso Research & Engineering Co., Brit. Pat. 761,024 (1956); *Chem. Abstr.*, **51,** 15548 (1957).

218. R. B. Mason, U.S. Pat. 2,811,567 (1957); *Chem. Abstr.*, **52,** 4677 (1958).
219. H. Uchida and A. Matsuda, *Bull. Chem. Soc. Japan*, **36,** 1351 (1963).
220. E. Guccione, *Chem. Eng.*, **72** (May 24), 90 (1965).
221. R. Ercoli, G. Signorini, and E. Santambrogio, *Chim. Ind.* (*Milan*), **42,** 587 (1960).
222. G. Natta, P. Pino, and E. Mantica, *Gazz. Chim. Ital.*, **80,** 680 (1950).
223. M. Polievka and E. J. Mistrik, *Chem Zvesti*, **26,** 149 (1972); *Chem. Abstr.*, **77,** 113388 (1972).
224. L. Markó and P. Szabo, *Chem. Tech.* (*Berlin*), **13,** 482 (1961); *Chem. Abstr.*, **56,** 7102 (1962).
225. P. Pino, F. Piacenti, and P. P. Neggiani, *Chem. Ind.* (*London*), **1961,** 1400.
226. H. J. Hagemeyer, Jr. and D. C. Hull, U.S. Pat. 2,790,832 (1957); *Chem. Abstr.*, **51,** 15552 (1957).
227. P. Pino, E. Mantica, and C. Paleari, *Ann. Chim.* (*Rome*), **40,** 237 (1950); *Chem. Abstr.*, **46,** 435 (1952).
228. F. Piacenti, C. Cioni, and P. Pino, *Chem. Ind.* (*London*), **1960,** 1240.
229. P. Pino, F. Piacenti, and P. P. Neggiani, *Chim. Ind.* (*Milan*), **44,** 1367 (1962).
230. V. Macho, *Chem. Zvesti*, **16,** 667 (1962); *Chem. Abstr.*, **59,** 429 (1963).
231. P. Pino, Paper presented at the 15th Congress of Pure and Applied Chemistry, Zurich, 1955.
232. I. Wender, J. Feldman, S. Metlin, B. H. Gwynn, and M. Orchin, *J. Am. Chem. Soc.*, **77,** 5760 (1955).
233. J. E. Knap, N. R. Cox, and W. R. Privette, *Chem. Eng. Progr.*, **62,** 74 (1966).
234*a*. K. A. Alekseeva, V. S. Dragunskaya, D. M. Rudkovskii, and A. G. Trifel, *Khim. i Tekhnol. Topliv i Masel*, **4,** 24 (1959); *Chem. Abstr.*, **53,** 21608 (1959).
234*b*. U. K. Dietler, Thesis, ETH Zürich (1974).
235. J. Falbe, H. Tummes, and J. Weber, *Brennstoff-Chem.* **50,** 46 (1969).
236. D. A. v. Bézard, G. Consiglio, and P. Pino, *Chimia*, **29,** 30 (1975).
237. H. Wakamatsu, *Nippon Kagaku Zasshi*, **85,** 227 (1964); *Chem. Abstr.*, **61,** 13173 (1964).
238. C. K. Brown, and G. Wilkinson, *J. Amer. Chem. Soc. A*, **1970,** 2753.
239. L. Benzoni, A. Andreetta, C. Zanzottera, and M. Camia, *Chim. Ind.* (*Milan*), **48,** 1076 (1966).
240. J. Tsuji and K. Ohno, *Tetrahedron Lett.*, **1965,** 3969.
241. L. Vaska, *J. Am. Chem. Soc.*, **86,** 1943 (1964).
242. W. W. Spooncer, A. C. Jones, and L. H. Slaugh, *J. Organometal. Chem.*, **18,** 327 (1969).
243. E. R. Tucci, *Ind. Eng. Chem.*, *Prod. Res. Develop.*, **8,** 286 (1969).
244. D. M. Rudkovskii and N. S. Imyanitov, in N. S. Imyanitov, Ed., *Gidroformilirovanie* [Hydroformylation], Izd. Khimiya, Leningrad, 1972, p. 5; *Chem. Abstr.*, **77,** 138991 (1972).
245. A. L. Nussbaum, T. L. Popper, E. P. Oliveto, S. Friedman, and I. Wender, *J. Am. Chem. Soc.*, **81,** 1228 (1959).
246. P. F. Beal, M. A. Rebenstorf, and J. E. Pike, *J. Am. Chem. Soc.*, **81,** 1231 (1959).
247. W. L. Fichteman and M. Orchin, *J. Org. Chem.*, **34,** 2790 (1969).
248. W. L. Fichteman and M. Orchin, *J. Org. Chem.*, **33,** 1281 (1968).
249*a*. J. G. Fish, Thesis, University of Cincinnati, 1968.
249*b*. S. Horgan, Thesis, University of Cincinnati, 1971.
250. A. Stefani, G. Consiglio, C. Botteghi, and P. Pino, *J. Am. Chem. Soc.*, **95,** 6504 (1973).
251. G. Natta and R. Ercoli, *Chim. Ind.* (*Milan*), **34,** 503 (1952).
252. G. Natta, R. Ercoli, S. Castellano, and F. H. Barbieri, *J. Am. Chem. Soc.*, **76,** 4049 (1954).

253. A. R. Martin, *Chem. Ind.* (*London*), **1954,** 1536.
254. M. Niwa and M. Yamaguchi, *Shokubai,* **3,** 264 (1961).
255. H. Greenfield, S. Metlin, and I. Wender, *Abstracts, 126th Meeting Am. Chem. Soc.*, New York, September 1954.
256. R. Iwanaga, *Bull. Chem. Soc. Japan,* **35,** 778 (1962).
257. L. Markó, *Magy. Asvanyolaj Foldgaz Kiserl. Int. Kozlemen.*, **1961,** 228; *Chem. Abstr.*, **56,** 4603 (1962).
258. J. Falbe, *Carbon Monoxide in Organic Synthesis,* Springer-Verlag, New York, 1970, p. 14.
259. G. F. Pregaglia, A. Andreetta, G. Gregorio, G. F. Ferrari, G. Montrasi, and R. Ugo, *Proc., Symp. Chemistry of Hydroformylation and Related Reactions,* Veszprem, 1972, p. 12; *Chem. Abstr.*, **77,** 113523 (1972).
260. I. Wender, *Petroleum Refiner,* **35,** 197 (1956).
261. *British Intelligence Objectives Sub-Committee, Final Report No. 447* [On work of Otto Roelen], London, 1946, 63 pp.
262. P. H. Groggins in *Unit Process in Organic Synthesis,* 4th ed., McGraw-Hill, New York, 1951, p. 583.
263. L. Kirch and M. Orchin, *J. Am. Chem. Soc.*, **80,** 4428 (1958).
264. D. Evans, J. A. Osborn, and G. Wilkinson, *J. Chem. Soc. A,* **1968,** 3133.
265. F. Calderazzo and F. A. Cotton, *Inorg. Chem.*, **1,** 30 (1962).
266. F. Calderazzo and K. Noack, *Coord. Chem. Rev.*, **1,** 118 (1966).
267. R. J. Mawby, F. Basolo, and R. G. Pearson, *J. Am. Chem. Soc.*, **86,** 5043 (1964).
268. T. H. Coffield, J. Kozikowski, and R. D. Closson, *J. Org. Chem.*, **22,** 598 (1957).
269. K. Noack and F. Calderazzo, *J. Organometal. Chem.*, **10,** 101 (1967).
270. W. Hieber, O. Vohler, and G. Braun, *Z. Naturforsh.*, **13b,** 192 (1958).
271*a*. M. Orchin, L. Kirch, and I. Goldfarb, *J. Am. Chem. Soc.*, **78,** 5450 (1956).
271*b*. M. V. Boven, N. H. Alemdaroğlu, and J. M. L. Penninger, *Ind. Eng. Chem., Prod. Res. Develop.*, **14,** 259 (1975).
272. J. B. Wilford, A. Forster, and F. G. A. Stone, *J. Chem. Soc.*, **1965,** 6519.
273. V. Yu. Gankin, V. L. Klimenko, V. A. Ribakov, and V. A. Dvinin, *Proc., Symp. Chemistry of Hydroformylation and Related Reactions,* Veszprem, 1972, p. 32; *Chem. Abstr.*, **77,** 113528 (1972).
274. Y. Takegami, C. Yokokawa, Y. Watanabe, and Y. Okuda, *Bull. Chem. Soc. Japan,* **37,** 181 (1964).
275. Y. Takegami, Y. Watanabe, H. Masada, Y. Okuda, K. Kubo, and C. Yokokawa, *Bull. Chem. Soc. Japan,* **39,** 1495 (1966).
276. Y. Takegami, Y. Watanabe, H. Masada, and C. Yokokawa, *Bull. Chem. Soc. Japan,* **39,** 1499 (1966).
277. Y. Takegami, C. Yokokawa, Y. Watanabe, H. Masada, and Y. Okuda, *Bull. Chem. Soc. Japan,* **38,** 787 (1965).
278. L. Markó, G. Bor, G. Almasy, and P. Szabo, *Brennstoff-Chem.*, **44,** 184 (1963).
279*a*. R. Whyman, *J. Organometal. Chem.*, **66,** C23 (1974).
279*b*. R. Whyman, *Ibid.* **81,** 97 (1974).
280. J. A. Osborn, F. H. Jardine, J. F. Young, and G. Wilkinson, *J. Chem. Soc. A,* **1966,** 1711.
281. J. Halpern, J. F. Harrod, and B. R. James, *J. Am. Chem. Soc.*, **88,** 5150 (1966).
282. A. Sacco and R. Ugo, *J. Chem. Soc.*, **1964,** 3274.
283. J. Ashley-Smith, Z. Douek, B. F. G. Johnson, and J. Lewis, *J. C. S. Dalton,* **1974,** 128.
284*a*. D. A. v. Bézard, G. Consiglio, and P. Pino, *Chimia,* **28,** 610 (1974).
284*b*. M. Bianchi, U. Matteoli, and F. Piacenti, *Intern. Symposium on Metals in Organic Chemistry,* Venice, Sept. 1974.

285. F. J. McQuillin, *Tetrahedron,* **30,** 1661 (1974).
286. S. Brewis, *J. Chem. Soc.*, **1964,** 5014.
287. F. Piacenti, M. Bianchi, P. Frediani, and U. Matteoli, *J. Organometal. Chem.*, **87,** C54 (1975).
288. F. Piacenti, M. Bianchi, G. Bimbi, P. Frediani, and U. Matteoli, *Coord. Chem. Reviews,* **16,** 9 (1975).
289. C.-Y. Hsu and M. Orchin, *J. Am. Chem. Soc.*, **97,** 3553 (1975).
290*a*. P. Chini, *Chim. Ind.* (*Milan*), **42,** 133 (1960).
290*b*. P. Chini, *Chim. Ind.* (*Milan*), **42,** 137 (1960).
291. H. M. Feder and J. Halpern, *J. Am. Chem. Soc.*, **97,** 7186 (1975).
292. M. Bianchi, P. Frediani, U. Matteoli, A. Girola, and F. Piacenti, *Chim. Ind.* (*Milan*), **58,** 223 (1976).
293. D. A. v. Bézard, Thesis, ETH Zürich, 1976.
294. G. Bor, U. K. Dietler, and P. Pino, *7th Intern. Conf. on Organometal. Chem.*, Venice, Sept. 1975.
295*a*. V. Yu. Gankin and V. A. Dvinin, *Kinet. Katal.*, **14,** 191 (1973); *Chem. Abstr.*, **78,** 135274 (1973).
295*b*. V. Yu. Gankin, V. A. Dvinin, V. P. Novikov, V. A. Rybakov, and S. L. Skop, *Kinet. Katal.*, **14,** 1149 (1973); *Chem. Abstr.*, **80,** 81937 (1974).
295*c*. V. Yu. Gankin, V. A. Dvinin, and V. A. Rybakov, *Kinet. Katal.*, **15,** 60 (1974); *Chem. Abstr.*, **80,** 119864 (1974).
296. M. v. Boven, N. Alemdaroğlu, and J. M. L. Penninger, *J. Organometal. Chem.*, **84,** 65 (1975).
297. M. V. McCabe, J. F. Terapane, and M. Orchin, *Ind. Eng. Chem., Prod. Res. Develop.*, **14,** 281 (1975).
298. L. Komora and V. Macho, *Chem. Prum.*, **19,** 359 (1969); *Chem. Abstr.*, **72,** 11777 (1970).
299. K. A. Alekseeva, D. L. Libina, D. M. Rudkovskii, and A. G. Trifel, *Neftekhim.*, **6,** 276 (1966); *Chem. Abstr.*, **65,** 3731 (1966).
300. R. Kummer, H. J. Nienburg, H. Hohenschutz, and M. Strohmeyer, *Advances in Chemistry Series,* No. 132, Am. Chem. Soc., Washington, D.C., 1974, p. 19.
301. J. K. Mertzweiller and H. M. Tenney, U.S. Pat. 3,351,666 (1967); *Chem. Abstr.*, **68,** 42004 (1968).
302. W. W. Spooncer, A. C. Jones, and L. H. Slaugh, *J. Organometal. Chem.*, **18,** 327 (1969).
303. BASF, Fr. Demande 2,008,554 (1970); *Chem. Abstr.*, **73,** 34780 (1970).
304. D. M. Rudkovskii, A. G. Trifel, K. A. Alekseeva, and D. L. Libina, *Khim. Prom.*, **42,** 95 (1966); *Chem. Abstr.*, **64,** 19357 (1966).
305. M. Tanaka, Y. Watanabe, T. Mitsudo, and Y. Takegami, *Bull. Jpn. Pet. Inst.*, **17,** 83 (1975); *Chem. Abstr.*, **83,** 177863 (1975).

III. HYDROFORMYLATION BY RHODIUM CATALYSTS

A. Introduction

The first indications of the possibility of using rhodium or rhodium compounds instead of cobalt as hydroformylation catalysts appeared in patents issued between 1950 and 1960 (1–4). It was claimed that rhodium had greater catalytic activity than cobalt (1), gave mostly branched-chain aldehydes by hydroformylation of linear terminal olefins (2), gave unsaturated aldehydes from compounds containing two double bonds (2), and was active in the hydroformylation of unsaturated aldehydes (3). In the early 1960s the first papers appeared that compared cobalt- and rhodium-catalyzed hydroformylation (5–7). Since 1965 three main trends became apparent in the investigation of rhodium-catalyzed hydroformylation. The first was directed at clarifying the mechanism of the hydroformylation reaction using rhodium complexes; it was pursued mainly by Wilkinson and his co-workers in England. The second was directed chiefly at resolving synthetic problems, which could not be adequately solved by using cobalt catalysts; for example, the hydroformylation of dienes and unsaturated compounds containing functional groups. The third, mainly followed by industrial laboratories, was directed at investigating the question whether its activity accompanied by good selectivity toward aldehyde formation would make rhodium a promising catalyst for use in large-scale hydroformylation plants. In two short reviews on the use of rhodium in hydroformylation by Markó (8) and by Rudkovskii (9*a*) it is shown that even if some general features of the catalytic activity of rhodium seem rather clear, the detailed mechanism of the catalysis, particularly in the presence of different ligands, is not well understood. More recent reviews (9*b*, *c*), only partially devoted to rhodium-catalyzed hydroformylation, confirm this view.

B. Stoichiometric Hydroformylation

Stoichiometric hydroformylation by reaction of unsubstituted rhodium carbonyl hydrides with olefins has not yet been reported. In fact, there is only little experimental evidence (10) on unsubstituted rhodium carbonyl hydrides.

In contrast to $HCo(CO)_3PBu_3$, which does not react with olefins below 100°C (11), $HRh(CO)(PPh_3)_3$ or, as shown by the retarding effect of excess PPh_3, $HRh(CO)(PPh_3)_2$, is able to react with olefins to give the corresponding alkylrhodium derivatives (Scheme **XIXb**) as shown by isomerization and deuterium exchange experiments (12). But when only one CO group is present in the complex, CO insertion does not take place. To obtain a stoichiometric hydroformylation, the following steps seem necessary: First, a phosphine molecule must be displaced by CO forming the dimeric species $[Rh(CO)_2(PPh_3)_2]_2$ (Scheme **XIXc**). Then by adding olefin and displacing CO by hydrogen, the aldehyde is formed probably according to Scheme **XIXd**.

$$HRh(CO)(PPh_3)_3 \leftrightharpoons HRh(CO)(PPh_3)_2 + PPh_3 \qquad \textbf{(a)}$$

$$HRh(CO)(PPh_3)_2 + RCH{=}CH_2 \leftrightharpoons RCH_2CH_2Rh(CO)(PPh_3)_2 \qquad \textbf{(b)}$$

$$HRh(CO)(PPh_3)_2 + CO \leftrightharpoons 1/2[Rh(CO)_2(PPh_3)_2]_2 + 1/2H_2 \qquad \textbf{(c)}$$

$$[Rh(CO)_2(PPh_3)_2]_2 + 2RCH{=}CH_2 + 3H_2 \rightarrow$$
$$2HRh(CO)(PPh_3)_2 + 2RCH_2CH_2CHO \qquad \textbf{(d)}$$

Scheme **XIX**

By this type of reaction, at 25°C and with 1 atm of H_2, essentially only linear aldehydes were obtained from 1-pentene and 1-hexene (linear aldehyde/branched aldehyde $= \sim 20$), whereas from internal olefins (*cis*- and *trans*-2-pentene and *trans*-2-hexene) no linear aldehydes were obtained. The reaction rate is much higher for terminal than for internal olefins. Similar results were obtained by Haszeldine and co-workers (13). When using $[Rh(CO)_2(AsPh_3)_2]_2$ they obtained a straight-chain/branched-chain aldehyde ratio of about 20 from 1-hexene.

$Rh_4(CO)_{12}$ reacts with olefins in the presence of H_2 more easily than either $Co_2(CO)_8$ (14) or $Co_4(CO)_{12}$ (15). This reaction with propylene occurs at room temperature even at H_2 pressures lower than 1 atm according to reaction 43 (16); the ratio for straight-chain to branched isomer is about one (16) or lower, depending on the propylene partial pressure (17).

$$3Rh_4(CO)_{12} + 4CH_3CH{=}CH_2 + 4H_2 \rightarrow 2Rh_6(CO)_{16} + 4C_3H_7CHO \qquad (43)$$

The reaction rate decreases greatly in the presence of CO but part of the gaseous CO is also transformed. From propylene, 3.5–4 moles of aldehyde were obtained per mole of $Rh_4(CO)_{12}$ when using a CO partial pressure of about 5 mm Hg (16*b*,17).

In contrast to cobalt carbonyl complexes, rhodium complexes are able to activate CO even at room temperature. Acylrhodium complexes seem to be formed under the above conditions and can be considered intermediates in these reactions; in fact, in the presence of acetone and water and in the absence of H_2, the anion $[C_3H_7CORh_6(CO)_{15}]^-$ was isolated as the tetraethylammonium salt. The acyl groups were characterized by infrared and nmr analysis and by decomposition of the anion with iodine and methanol to the methyl esters of the corresponding normal and isobutyric acids (16*a*).

The kinetics of the stoichiometric synthesis of aldehydes from $Rh_4(CO)_{12}$, olefins and H_2 was investigated (17). When using a low CO partial pressure (6 mm Hg), the reaction rate, at constant propylene partial pressure, is proportional to the concentration of $Rh_4(CO)_{12}$ and of H_2; the initial rate seems inversely proportional to the CO pressure, at least below a CO pressure of 10 mm Hg.

C. Catalytic Hydroformylation

The rhodium-catalyzed hydroformylation of olefins occurs under milder conditions than the corresponding reaction with cobalt (5,6,18). The reaction can be carried out at total pressures ($CO/H_2 = 1$) of 1 atm or lower at 25°C (19), but temperatures between 50 and 150°C and total pressures of 50–100 atm are generally used (94). A wide range of CO and H_2 ratios can be used for the reaction. A large excess of H_2 should be avoided, since it can cause hydrogenation of the olefin, whereas a high CO partial pressure seems to decrease the reaction rate. A systematic investigation of these factors is lacking.

Besides rhodium metal, a large number of rhodium derivatives (78,94) were used as catalysts: simple compounds like $RhCl_3$ or Rh_2O_3, Rh(I) complexes like $RhCl(CO)(PPh_3)_2$, $RhH(CO)(PPh_3)_3$ and rhodium carbonyls like $Rh_4(CO)_{12}$ (Table 32). Rhodium complexes coordinatively bound to a polymer containing phosphine groups were also proposed (28).

TABLE 32
Rhodium Compounds as Catalysts or Catalyst Precursors in Hydroformylation

Rhodium compounds	Olefins	P_{CO} (atm)	P_{H_2} (atm)	Temp (°C)	Yield[a] (%)	Ref.
Rh_2O_3	*cis*-2-Butene	100[b]	100[b]	100	90–95	20
$RhCl_3(PPh_3)_3$	1-Hexene	45[c]	45[c]	55	90	21
$Rh(NO_3)_3$	1-Octene	100	100	120	95	22
$Rh(CO)(PBu_3)_2(acac)$	1-Hexene	7–21[b]	7–21[b]	78–84	86	23
$RhCl(PPh_3)_3$	1-Pentene	50[c]	50[c]	70	~100	12
$RhH(CO)(PPh_3)_3$	1-Hexene	0.5[d]	0.5[d]	25	n.d.	24
$Rh+P(OPh)_3$	1-Octene	2.5–3.5[d]	2.5–3.5[d]	90	93	25
$Rh_4(CO)_{12}$	1-Butene	100[b]	100[b]	60	94	26
$Rh_4(CO)_{12}$	Propylene	110[b]	110[b]	110	80	6
$Rh_6(CO)_{16}+PPh_3$	1-Hexene	21[c]	21[c]	82	~100	27

[a] Moles of aldehydes per mole of olefin reacted.
[b] Initial pressure at room temp.
[c] Initial pressure at reaction temp.
[d] At constant pressure.

For similar substrates the reaction conditions are not very different whatever rhodium compounds are employed as catalysts. It may well be that under reaction conditions in the absence of phosphines, the same catalytically active species are always formed. After starting with $RhCl_3 \cdot 3H_2O$, Heil and Markó (8) detected, by infrared spectroscopy of the reaction mixture, the compounds $Rh_4(CO)_{12}$ and $Rh_6(CO)_{16}$, besides $[Rh(CO)_2Cl]_2$. Infrared spectra of mixtures from hydroformylation experiments at 80°C, in which $RhCl(CO)(PPh_3)_2$ and triethylamine or $HRh(CO)(PPh_3)_3$ were used as catalyst precursors, indicated that here too the same predominant catalyst form apparently was present (29).

Besides milder reaction conditions, a major advantage in the use of rhodium catalysts is a remarkable decrease in secondary reactions. With olefinic hydrocarbons, little or no aldol reactions or acetal formation of the aldehydes produced was observed (30) even when employing heterogeneous catalysts for vapor-phase hydroformylation (31).

1. Substrates

Here we shall mainly discuss the hydroformylation of substrates that give poor results or are unreactive with cobalt catalysts. Olefinic hydrocarbons will be considered in connection with the factors affecting the isomeric composition of the products. We may say beforehand that in the presence of rhodium compounds the hydroformylation of simple olefins is usually conducted under mild conditions (60–120°C; 50–200 atm; $CO/H_2 = 1$) and gives the expected products with a low linear to branched-chain isomer ratio.

When aryl-substituted olefins are hydroformylated in the presence of rhodium catalysts, they give the expected aldehydes in very good yields (Table 33) and, unlike the reaction using cobalt catalysts, only minor amounts of the substrate are hydrogenated.

Styrene forms 2-phenylpropanal and 3-phenylpropanal in almost equal amounts when operating at high temperature and pressure (32,37), but 2-phenylpropanal predominates (90%) when operating at room temperature and atmospheric pressure (24).

α-Methylstyrene yielded preferentially the less-branched product (33). From allyl- and propenylbenzene, a substantial amount of the aldehydes, arising from carbonylation of the initially saturated carbon atoms, are also formed (34,35) when using Rh/Al_2O_3 catalyst. In the presence of a phosphine-containing catalytic system, allylbenzene formed only the expected isomeric aldehydes with prevalence (80%) of the more branched compound (35).

Conjugated diolefins (Table 34) can be successfully hydroformylated in the presence of rhodium catalysts. From butadiene, the hydroformylation

TABLE 33

Hydroformylation of Aryl-substituted Olefins

Olefin	Catalyst	Temp (°C)	P_{CO} (atm)	P_{H_2} (atm)	Products	Yield (%)	Ref.
Styrene	$[RhCl(1,5\text{-hexadiene})]_2$ and (+)-$PhCH_2P(CH_3)Ph$	140	50	50	2-Phenylpropionaldehyde	45	32
					3-Phenylpropionaldehyde	48	
Styrene	$HRh(CO)(PPh_3)_3$	25	0.5^a	0.5^a	2-Phenylpropionaldehyde[b]	n.d.	24
					3-Phenylpropionaldehyde		
α-Methylstyrene	$RhH(CO)(PPh_3)_3$ and (−)DIOP[c]	102	43	43	3-Phenylbutanal	69	33
					2-Phenyl-2-methylpropanal	15	
α-Ethylstyrene	Rh/Al_2O_3	120	80	80	3-Phenylpentanal	63	34
					2-Methyl-3-phenylbutanal	5^d	
					4-Phenylpentanal	4	34
Allylbenzene	Rh/Al_2O_3	140	79	79	4-Phenylbutanal	40	35
					2-Methyl-3-phenylpropanal	40	
					2-Phenylbutanal	14	

Propenylbenzene	Rh/Al_2O_3	140	99	99	4-Phenylbutanal	9	36
					2-Methyl-3-phenylpropanal	34	
					2-Phenylbutanal	42	
Propenylbenzene	$Rh/Al_2O_3/PPh_3$	90	99	99	2-Phenylbutanal	80	36
					2-Methyl-3-phenylpropanal	12	
Isobutenylbenzene	Rh/Al_2O_3	120	80	80	2-Phenyl-3-methylbutanal	53	34
					3-Methyl-4-phenylbutanal	31	
2-Phenyl-2-butene	Rh/Al_2O_3	120	80	80	3-Phenylpentanal	42	34
					4-Phenylpentanal	26	
					2-Methyl-3-phenylbutanal[e]	22	
3-Phenyl-1-butene	Rh/Al_2O_3	90	80	80	4-Phenylpentanal	84	34
					2-Methyl-3-phenylbutanal[f]	17	
1,1-Diphenylpropene	Rh/Al_2O_3	160	80	80	2-Methyl-3,3-diphenylpropanal	19	34
					2-Methyl-3,3-diphenylpropanol	48	

[a] At constant pressure; all other values are for initial pressures measured at room temp.
[b] Moles 2-phenylpropionaldehyde/moles 3-phenylpropionaldehyde = 9.
[c] DIOP = (−)-2,3-0-isopropylidene-2,3-dihydroxy-1,4-bis(diphenylphosphino)butane.
[d] Threo/erythro = 3.3/1.8.
[e] Threo/erythro = 14.2/7.4.
[f] Erythro/threo = 10.4/6.6.

TABLE 34

Hydroformylation of Conjugated and Nonconjugated Di- and Polyolefins with Rhodium Catalysts

Olefin	Catalyst	Temp (°C)	P_{CO}[a] (atm)	P_{H_2}[a] (atm)	Products	Yield (%)	Ref.
Butadiene	$Rh_2O_3 + PBu_3$	125	67	134	Pentanal	50	38
					2-Methylbutanal	2	
					2-Methylpentanedial	24	
					2-Ethylbutanedial	12	
					Hexanedial	4	
Butadiene	Rh_2O_3	150	100	100	Pentanal	36	20
					2-Methylbutanal	48	
1,3-Pentadiene	$Rh_2O_3 + PBu_3$	125	67	134	2-Methylpentanal	46	38
					Hexanal	2	
					2,4-Dimethylpentanedial	16	
					2-Ethylpentanedial	9	
					2-Methylhexanedial	7	
					2-Ethyl-3-methylbutanedial	6	
					2-Propylbutanedial	1	

1,4-Pentadiene	Rh_2O_3	60–100	75	75	C_6 aliphatic aldehydes	13	7
					1-Cyclohexene-1-carboxaldehyde	9	
					1,7-Heptanedial	16	
					Branched-chain C_7 dialdehydes	8	
1,4-Pentadiene	$Rh_2O_3 + PBu_3$	135	100	100	1,7-Heptanedial	25	20
					2-Methylhexanedial	29	
					2-Ethylpentanedial	4	
					2,4-Dimethylpentanedial } 2-Ethyl-3-methylbutanedial }	17	
1,5-Hexadiene	Rh_2O_3	60–100	75	75	C_7 aliphatic aldehydes	17	7
					1,8-Octanedial	7	
					2-Methylheptanedial	16	
4-Vinylcyclohexene	$Rh + P(OAr)_3$	90–105	40	120	3-(3-Cyclohexenyl)propanal	38	39
					2-(3-Cyclohexenyl)-2-methylhethanal	9	
					Dialdehydes	19	
trans,trans,cis-1,5,9-Cyclododecatriene	$Rh_2O_3 + PBu_3$	145–220	70	140	Tris(hydroxymethyl)cyclododecane	80	20
1,5-Cyclooctadiene	Rh_2O_3	90–210	500–600[b]	500–600[b]	Hydroxymethylcyclooctane	13	40
					Bis(hydroxymethyl)cyclooctane	81	
Dicyclopentadiene	Rh_2O_3	115	66[b]	134[b]	Tricyclododecanedicarboxaldehyde	61	41

[a] Measured at room temp.
[b] Pressure at reaction temp.

product consisted of saturated monoaldehydes (~50%), mainly *n*-pentanal; a mixture of dialdehydes (~40%), mainly 2-methylpentanedial and 2-ethylbutanedial (11%); and a small amount of hexanedial (38). The formation of these dialdehydes has been explained by postulating a 1–4 and a 1–2 hydroformylation followed by a second hydroformylation of the nonconjugated double bond whose migration occurs at a lower or comparable rate than hydroformylation. From 1,3-pentadiene (38), about 50% of C_6-monoaldehydes, mainly 2-methylpentanal, and ~40% of dialdehydes, mainly 2,4-dimethylpentanedial, were obtained.

Nonconjugated di- and polyolefins (20,40,42,43) were also successfully hydroformylated to give aldehydes containing one or more formyl groups, depending on the substrate and type of catalyst used (Table 34). By using a catalytic system containing trisubstituted phosphines with 1,4-pentadiene (20), hydroformylation products were obtained in 84% yield. From 1,5-cyclooctadiene (40), bis(hydroxymethyl)cyclooctane was obtained as the main product (81%) when using a very high pressure. Tris(hydroxymethyl)cyclododecane was formed from *trans, trans, cis*-1,5,9-cyclododecatriene (20).

Unsaturated alcohols and their esters (20,45–48) can be hydroformylated with yields up to 80% (Table 35).

The two expected oxyaldehydes are obtained from allyl alcohol (20) with the γ-isomer predominating to some extent. Furthermore, high selectivity for the terminal position was observed (45) in the hydroformylation of a series of α-substituted allyl alcohols; the products were isolated in yields of 70–80% as glycols after reduction with $LiAlH_4$. Also, from vinyl acetate only one isomer is formed due to addition of a formyl group to the α-position (47) (Table 35).

3-Formylbutanol has been obtained in 75% yield from 1,4-butenediol (20); from 3,4-diacetoxy-1-butene the two expected formyl derivatives were formed (49) with prevalence of the more-branched compound.

The hydroformylation of conjugated unsaturated aldehydes and ketones with rhodium catalysts has not been reported. Nonconjugated unsaturated aldehydes, however, give the expected products in good yields (8,46) (Table 36). Diethyl acetal and the diacetate of acrolein (50) yield mixtures of formyl derivatives with the more-branched products predominating, whereas α-alkyl substituted acrolein diethyl acetals (45,51) give high yields (>80%) of products resulting from formylation of the terminal carbon atom.

Monounsaturated esters (25,34,52–54) can be hydroformylated in fairly good yields; hydrogenation of the substrate is in the range of 20–25% (Table 37).

In the presence of Rh_2O_3 or rhodium carbonyls, a predominance of β-formyl esters is formed by hydroformylation of α, β-unsaturated esters; the extent of β-addition decreases from ethyl cinnamate to ethyl crotonate, ethyl acrylate, and methyl methacrylate. The large amount of the α-formyl ester from methyl methacrylate is surprising in view of the formation of only 3-methylbutanal from isobutylene (58). From ethyl crotonate and ethyl cinnamate, the corresponding butyrolactones were the main reaction products when operating at high temperature. From methyl methacrylate α-methyl-γ-butyrolactone may be obtained in good yield by operating at higher temperature in the presence of tertiary amines (59).

Under approximately the same reaction conditions with cobalt catalysts, 70% of the product from ethyl crotonate derives from insertion of the formyl group at the γ-position, but it is only 7% in the presence of rhodium. Nonconjugated unsaturated esters, such as methyl 3- and 5-hexenoates (54) and methyl oleate (34,53,60), all give high yields (~80%) of products arising from addition of the formyl group to the original olefinic carbon atoms when using phosphine-containing rhodium catalysts.

Polyunsaturated esters were also hydroformylated. α,β-Unsaturated esters, such as methyl sorbate (20), give mixtures of saturated monoformyl derivatives due to addition of the formyl group to all carbon atoms (C_2 to C_6) in the substrate; the presence of PBu_3 suppresses the formation of some of these products. Interesting results were obtained from hydroformylation of octadecadienoates in the presence of the Rh/PPh_3 catalytic system (56). With a 9,11- or 10,12-conjugated dienic system, a complex mixture of unsaturated monoformyl and diformyl esters are obtained in which the former predominate, whereas with a nonconjugated dienic system (9,12- or 9,15-), the opposite is true. From methyl 9,12,15-octadecatrienoate, a mixture of formyl unsaturated esters is obtained as the major product (85%) together with monoformyl diunsaturated esters (12%) and a smaller amount of triformyl derivatives (56). Among other less common substrates hydroformylated in the presence of a rhodium catalyst, it is of interest to consider the nitrogen-containing unsaturated compounds like nitriles (61,62), amides (63), imides (63), and nitroolefins (26) (Table 38).

Undoubtedly rhodium catalysts have widely extended the scope of the hydroformylation reaction, especially for the class of unsaturated compounds containing conjugated double bonds. Moreover, in the presence of rhodium catalysts under appropriate conditions, hydrogenation and isomerization of the double bond may be suppressed. Because of the mild conditions used, even compounds containing rather reactive groups like

TABLE 35
Hydroformylation of Unsaturated Alcohols and Their Derivatives with Rhodium Catalysts

Substrate	Catalyst	Temp (°C)	P_{CO} [a] (atm)	P_{H_2} [a] (atm)	Products	Yield (%)	Ref.
$CH_2{=}CHOAc$	$Rh_4(CO)_{12}$	60–80	83	83	$AcOCHCHO$ | CH_3	72	47
$CH_2{=}CHCH_2OH$	$Rh_2O_3 + PBu_3$	115–120	110	110	$OHCCH_2CH_2CH_2OH$ CH_3CHCH_2OH | CHO	49 35	20
$CH_2{=}CCH_2OH$ | CH_3	$RhCl(CO)(PPh_3)_2$	80	40	40	$HOCH_2CH_2CHCH_2OH$ [b] | CH_3	70	45
$CH_2{=}CCH_2OH$ | $HC(CH_3)_2$	$RhCl(CO)(PPh_3)_2$	80	40	40	$HOCH_2CH_2CHCH_2OH$ [b] | $HC(CH_3)_2$	70	45
$CH_2{=}CCH_2OH$ | $C(CH_3)_3$	$RhCl(CO)(PPh_3)_2$	80	40	40	$(CH_3)_3CCH{-}CH_2{-}O$ | | $CH_2{-}CH{-}O$ (ring) | OH	80	45
$CH_2{=}CCH_2OH$ [c] | $CH_3CH_2CH^*$ | CH_3	$RhCl(CO)(PPh_3)_2$	80	40	40	CH_3 | $CH_3CH_2CHCHCH_2OH$ [b,d] | CH_2CH_2OH	75	45

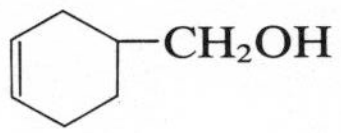 (3-cyclohexenyl)CH_2OH	Rh_2O_3	*e*	67	133	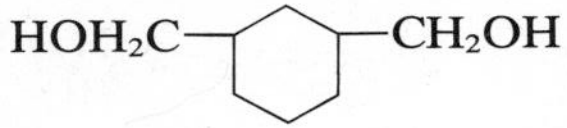 HOH_2C–(1,3-cyclohexylene)–CH_2OH	39	46
					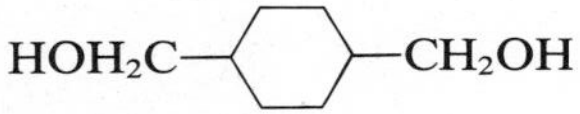 HOH_2C–(1,4-cyclohexylene)–CH_2OH	37	
$HOCH_2CH{=}CHCH_2OH$	$Rh_2O_3 + PBu_3$	150	100	100	$OHCCH(CH_2CH_2OH)CH_3$	75	20
$CH_2{=}CHCH(OAc)CH_2OAc$	$RhCl_3 \cdot 3H_2O$	100	300[f]	300[f]	$AcOCH_2CH(AcO)CH(CH_3)CHO$	71	49
					$AcOCH_2CH(OAc)CH_2CH_2CHO$	24	

[a] Initial pressure at room temp.
[b] After $LiAlH_4$ reduction of the reaction products.
[c] (*S*) antipode; optical purity, 96%.
[d] Racemization of initial asymmetric carbon atom was less than 4%.
[e] Hydroformylation at 130°C followed by *in situ* hydrogenation at 240°C.
[f] At constant pressure.

TABLE 36

Hydroformylation of Unsaturated Aldehydes and Their Derivatives with Rhodium Catalysts

Substrate	Catalyst	Temp (°C)	P_{CO} (atm)	P_{H_2} (atm)	Products	Yield (%)	Ref.
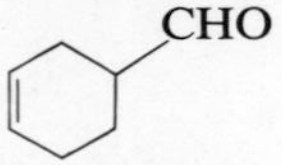 CHO	Rh_2O_3	100	120[a]	120[a]	Dialdehydes	71	3
					Hydroxymethylformyl-cyclohexanes	14	
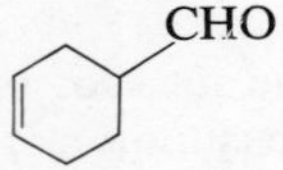 CHO	Rh_2O_3		—[b]		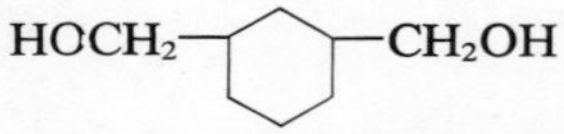 $HOCH_2$—CH_2OH	17	46
					$HOCH_2$ CH_2OH[c]	38	
					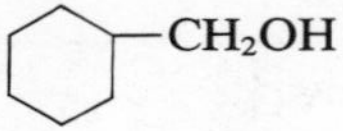 CH_2OH	6	
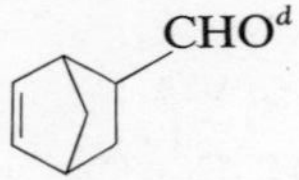 CHO[d]	Rh_2O_3		—[e]		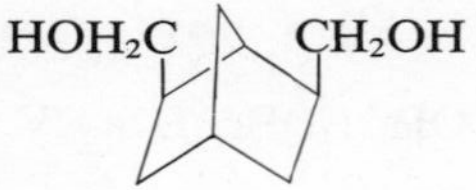 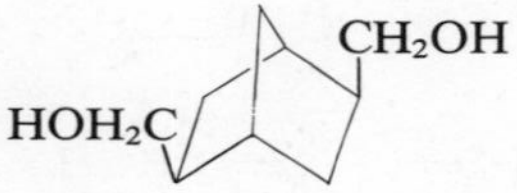 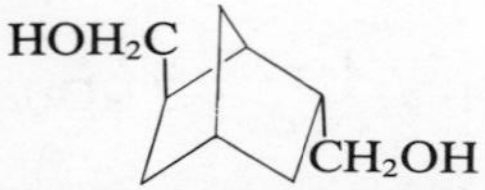 HOH_2C CH_2OH		
					HOH_2C CH_2OH	63	41
					HOH_2C CH_2OH		

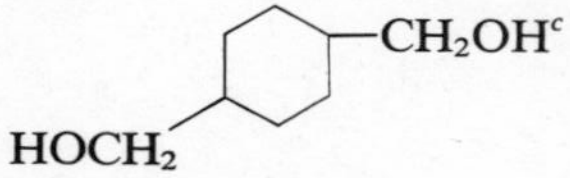

Substrate	Catalyst	Temp. (°C)	H_2	CO	Product	Yield (%)	Ref.
					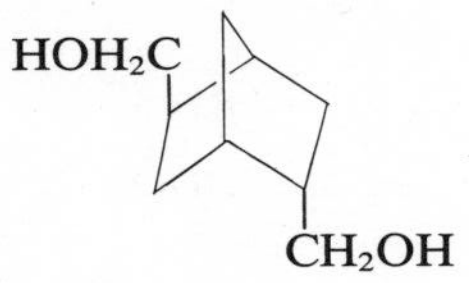 HOH_2C–(bicyclic)–CH_2OH		
$CH_2{=}CHCH(OEt)_2$	Rh_2O_3	110	100[f]	100[f]	$CH_3CH(CHO)CH(OEt)_2$	40	50
					$OHCCH_2CH_2CH(OEt)_2$	22	
$CH_2{=}CHCH(OAc)_2$	Rh_2O_3 in EtOH	100	100[f]	100[f]	$CH_3CH(CHO)CH(OAc)_2$	n.d.	50
					$OHCCH_2CH_2CH(OAc)_2$	n.d.	
$CH_2{=}C(CH_3)CH(OEt)_2$	$RhCl(CO)(PPh_3)_2 + Et_3N$	80	50[f]	50[f]	$CH_3C(CH_3)(CH_2CHO)CH(OEt)_2$	82	45*a*, 51
$CH_2{=}C(^*CH(CH_3)CH_2CH_3)CH(OEt)_2$[g]	$RhCl(CO)(PPh_3)_2 + Et_3N$	80	50[f]	50[f]	$CH_3CH_2\overset{*}{C}H(CH_3)CH(CH_2CHO)CH(OEt)_2$[h]	85–88	45*a*, 51

[a] At constant pressure.
[b] Hydroformylation at 120°C and 200 atm $CO+H_2$ (1:1) followed by *in situ* hydrogenation at 220°C and 300 atm.
[c] *Cis*/*trans* = 1/2.1.
[d] Mixture (6:4) of *endo*- and *exo*-.
[e] Hydroformylation at 130°C and 200 atm $CO+H_2$ (1:2) followed by *in situ* hydrogenation at 240°C and 300 atm.
[f] Initial pressure at room temp.
[g] (*S*) antipode; optical purity, 96%.
[h] Racemization of initial asymmetric carbon atom was less than 5%.

TABLE 37
Hydroformylation of Unsaturated Esters and Anhydrides with Rhodium Catalysts

Substrate	Catalyst	Temp (°C)	P_{CO} (atm)	P_{H_2} (atm)	Products	Yield (%)	Ref.
$CH_2{=}CHCOOC_2H_5$	$[Rh(CO)_3]_n$	110	120[a]	80[a]	$OCHCH_2CH_2COOC_2H_5$	47	55
					$CH_3CH(CHO)COOC_2H_5$	42	
$CH_2{=}C(CH_3)COOCH_3$	Rh_2O_3	130	100[b]	100[b]	$OHCCH_2CH(CH_3)COOCH_3$	63	52
					$CH_3C(CH_3)(CHO)COOCH_3$	17	
					$CH_3CH(CH_3)COOCH_3$	2	
$CH_3CH{=}CHCOOC_2H_5$	Rh_2O_3	—[c]	—[c]	—[c]	3-methyl-γ-butyrolactone (CH_3-substituted five-membered lactone, C=O)	47	52
					δ-valerolactone (six-membered lactone, C=O)	7	
					$CH_3CH_2CH(CH_2OH)COOC_2H_5$	3	
					$CH_3CH_2CH_2COOC_2H_5$	14	

$C_6H_5CH{=}CHCOOC_2H_5$	Rh_2O_3	—[d]	—[d]	—[d]	[structure: C_6H_5-substituted lactone ring, ring O, =O]	73	52
					$C_6H_5CH_2CH_2COOC_2H_5$	26	
$CH_3CH{=}CHCH{=}CHCOOCH_3$	Rh_2O_3	150	100[a]	100[a]	$OHCCH_2(CH_2)_4COOCH_3$	18	20
					$CH_3\underset{CHO}{CH}(CH_2)_3COOCH_3$	25	
					$CH_3CH_2\underset{CHO}{CH}CH_2CH_2COOCH_3$	12	
					$CH_3CH_2CH_2\underset{CHO}{CH}CH_2COOCH_3$	12	
					$CH_3CH_2(CH_2)_2\underset{CHO}{CH}COOCH_3$	0.7	
$CH_3CH{=}CHCH{=}CHCOOCH_3$	Rh_2O_3/PBu_3	150	100[a]	100[a]	$CH_3\underset{CHO}{CH}(CH_2)_3COOCH_3$	13	20
					$CH_3CH_2\underset{CHO}{CH}CH_2CH_2COOCH_3$	33	
					$CH_3CH_2CH_2\underset{CHO}{CH}CH_2COOCH_3$	42	

TABLE 37 (Continued)
Hydroformylation of Unsaturated Esters and Anhydrides with Rhodium Catalysts

Substrate	Catalyst	Temp (°C)	P_{CO} (atm)	P_{H_2} (atm)	Products	Yield (%)	Ref.
$CH_3(CH_2)_7CH{=}CH(CH_2)_7COOCH_3$	Rh/PPh_3	110	70[a]	70[a]	$CH_3(CH_2)_7CH_2CH(CHO)(CH_2)_7COOCH_3$ and $CH_3(CH_2)_7CH(CHO)CH_2(CH_2)_7COOCH_3$	84	53
cis,cis-9,12-$CH_3(CH_2)_4$-$CH{=}CHCH_2CH{=}CH(CH_2)_7COOCH_3$	Rh/PPh_3	100	70[a]	70[a]	Unsaturated monoformyl esters	23	56
					Diformyl esters	75	
cis,cis,-9,11-$CH_3(CH_2)_5$-$CH{=}CHCH{=}CH(CH_2)_7COOCH_3$ and *trans,cis*-10,12-$CH_3(CH_2)_4$-$CH{=}CHCH{=}CH(CH_2)_8COOCH_3$	Rh/PPh_3	100	70[a]	70[a]	Unsaturated monoformyl esters	59	56
					Diformyl esters[e]	19	
cis,cis-9,15-CH_3CH_2-$CH{=}CH(CH_2)_4CH{=}CH(CH_2)_7COOCH_3$	Rh/PPh_3	100	70[a]	70[a]	Unsaturated monoformyl esters	10	56
					Diformyl esters[e]	73	
$CH_3CH_2CH{=}CHCH_2CH{=}CHCH_2$-$CH{=}CH(CH_2)_7COOCH_3$	Rh/PPh_3	100	70[a]	70[a]	Monoformyl diunsaturated esters	12	56
					Diformyl unsaturated esters	85	
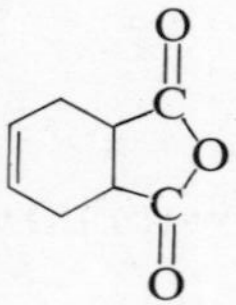 [structure]	Rh_2O_3	100	125[b]	125[b]	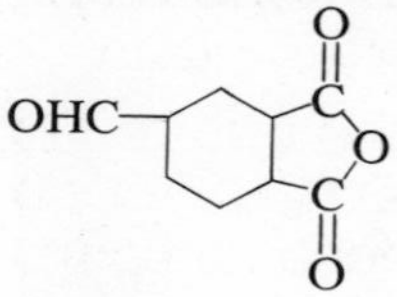 OHC— [structure]	74	57

[a] Initial pressure at room temp.
[b] Initial pressure at reaction temp.
[c] Hydroformylation at 130°C and 200 atm $CO+H_2$ (1:1) followed by *in situ* hydrogenation at 220°C and 300 atm.
[d] Hydroformylation at 120°C and 200 atm $CO+H_2$ (1:1) followed by *in situ* hydrogenation at 230°C and 300 atm.

nitro or acylamido groups can be hydroformylated to the corresponding formyl products.

2. *Asymmetric Hydroformylation*

An interesting development in the hydroformylation of olefins is the synthesis of optically active aldehydes by using rhodium catalysts in the presence of optically active ligands (32,33,64,65–67). There are three possibilities of obtaining asymmetric hydroformylation starting with non-chiral or racemic olefins: (*i*) addition of CO to an unsaturated mono- or disymmetrically disubstituted carbon atom (Scheme **XXa**), (*ii*) addition of hydrogen to a disymmetrically disubstituted unsaturated carbon atom (Scheme **XXb**), and (*iii*) preferential partial hydroformylation of a racemic olefin (Scheme **XXc**). All three possibilities of obtaining asymmetric hydroformylation were explored successfully.

Both the aromatic (Table 39) and aliphatic (Table 40) substrates were investigated. By using the same reaction conditions and the same asymmetric ligand, the highest asymmetric induction was obtained with *cis*-2-butene. Similar asymmetric inductions were obtained with styrene and aliphatic vinyl α-olefins ($RCH{=}CH_2$), but a much lower optical activity was obtained with *trans*-2-butene and 1,1-disubstituted ethylenes ($R_1R_2C{=}CH_2$). The type of ligand (Table 41), the reaction conditions (67), and the ratio of moles of ligand to moles of rhodium (32) all influence the optical yield. The best results were obtained by using diphosphines like (−)DIOP, low temperatures and pressures, and a high ratio of ligand to rhodium. The relationship between prevailing chirality of the reaction products and the chirality of the ligand was considered: For a given chiral ligand, a simple model was proposed (67) for the prediction of the prevailing chirality of products from vinyl and internal olefins. The significance of the opposite prevailing chirality found for the 2-methylbutanal obtained from 1-butene and *cis*-2-butene will be discussed later in connection with the hydroformylation mechanism (Section III-C-6).

3. *Secondary Reactions*

The selectivity of rhodium catalysts in the hydroformylation of olefins is greater than that observed for cobalt or other metals. Under hydroformylation conditions, rhodium compounds seem to have very little activity as catalysts for aldolization, olefin hydrocarboxylation and, with some exceptions, hydrogenation of carbon–carbon double bonds. The only relevant secondary reactions are hydrogenation of aldehydes and isomerization of the olefinic double bond.

TABLE 38

Hydroformylation of Nitrogen-containing Unsaturated Substrates with Rhodium Catalysts

Substrate	Catalyst	Temp (°C)	P_{CO} [a] (atm)	P_{H_2} [a] (atm)	Products	Yield (%)	Ref.
$CH_2{=}CH_2CN$	Rh_2O_3	127	100	100	$CH_3CH(CN)CH_2OH$	16	61
					CH_3CH_2CN	29	
$CH_2{=}CHCH_2CH_2CH_2CN$	$[Rh(CO)_2Cl]_2$	85	85	85	$OHCCH_2CH_2CH_2CH_2CH_2CN$	67	62
$C_6H_4(C{=}O)_2NCH{=}CH_2$	$[Rh(CO)_3]_n$	100	85	85	$C_6H_4(C{=}O)_2NCH_2CH_2CHO$	81	63
$C_6H_4(C{=}O)_2NCH_2CH{=}CH_2$	$[Rh(CO)_3]_n$	70	85	85	$C_6H_4(C{=}O)_2NCH_2CH(CH_3)CHO$	79	63
					$C_6H_4(C{=}O)_2NCH_2CH_2CH_2CHO$	16	

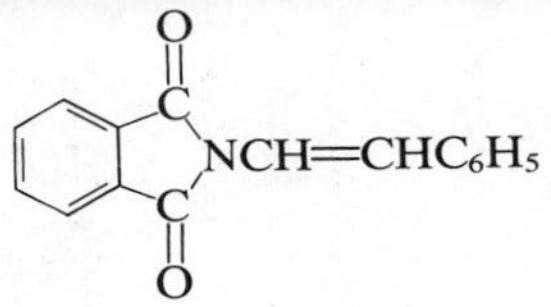 $C_6H_4(CO)_2NCH{=}CHC_6H_5$	$[Rh(CO)_3]_n$	120	85	85	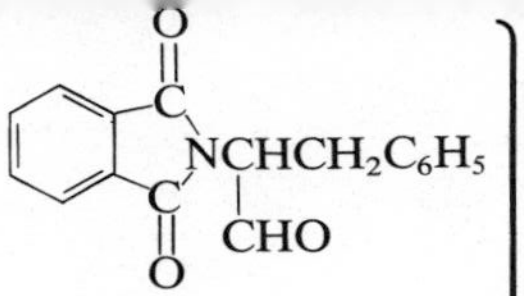 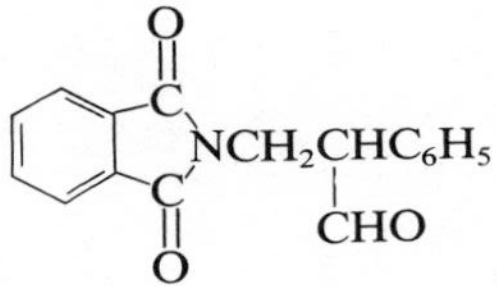 $C_6H_4(CO)_2NCH(CHO)CH_2C_6H_5$ and $C_6H_4(CO)_2NCH_2CH(CHO)C_6H_5$	75	63
$CH_3CONHCH_2CH{=}CH_2$	$[Rh(CO)_3]_4$	100	85	85	$CH_3CONHCH_2CH(CH_3)CHO$	79	63
					$CH_3CONHCH_2CH_2CH_2CHO$	8	
$O_2NCH_2CH{=}CH_2$	$[Rh(CO)_3]_4$	60	100[b]	100[b]	$O_2NCH_2CH(CH_3)CHO$	96	26
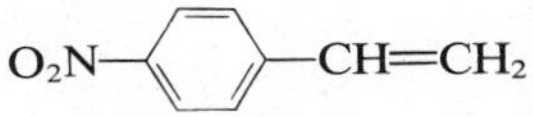 $O_2N\text{-}C_6H_4\text{-}CH{=}CH_2$	$[Rh(CO)_3]_n$	60–70	101[b]	101[b]	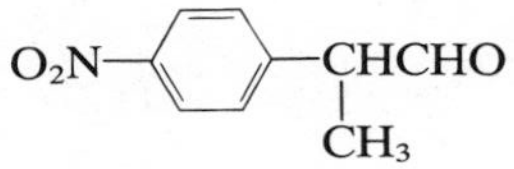 $O_2N\text{-}C_6H_4\text{-}CH(CH_3)CHO$	85	26
					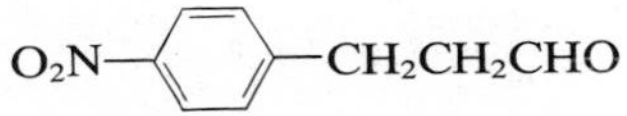 $O_2N\text{-}C_6H_4\text{-}CH_2CH_2CHO$	11	

[a] Initial pressure at room temp except where noted.
[b] Pressure at reaction temp.

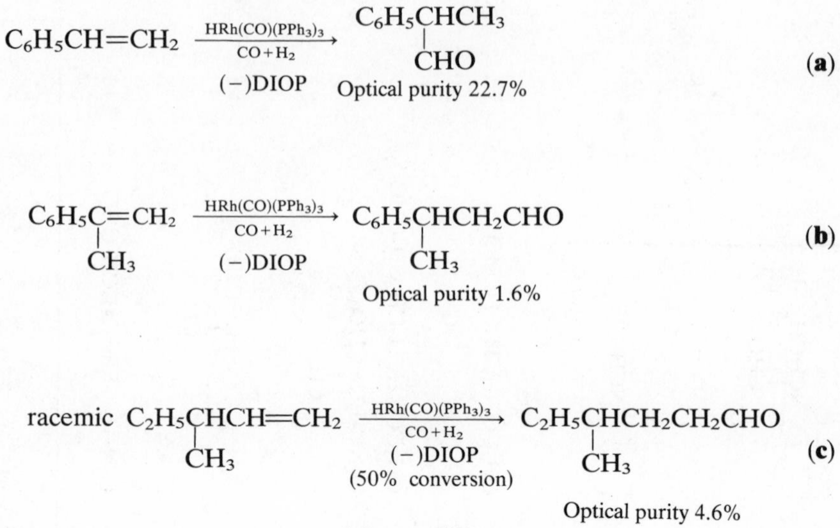

$$C_6H_5CH{=}CH_2 \xrightarrow[\text{CO}+\text{H}_2]{\text{HRh(CO)(PPh}_3)_3} \underset{\text{CHO}}{C_6H_5\overset{}{C}HCH_3} \quad \text{(a)}$$

(−)DIOP Optical purity 22.7%

$$\underset{\text{CH}_3}{C_6H_5C{=}CH_2} \xrightarrow[\text{CO}+\text{H}_2]{\text{HRh(CO)(PPh}_3)_3} \underset{\text{CH}_3}{C_6H_5CHCH_2CHO} \quad \text{(b)}$$

(−)DIOP Optical purity 1.6%

$$\text{racemic } \underset{\text{CH}_3}{C_2H_5CHCH{=}CH_2} \xrightarrow[\text{CO}+\text{H}_2]{\text{HRh(CO)(PPh}_3)_3} \underset{\text{CH}_3}{C_2H_5CHCH_2CH_2CHO} \quad \text{(c)}$$

(−)DIOP
(50% conversion)

Optical purity 4.6%

Scheme **XX**

a. Hydrogenation. The hydrogenation of aldehydes under oxo conditions was investigated by Heil and Markó (68,69) (Table 42) using $RhCl_3 \cdot 3H_2O$ as catalyst precursor. At 170–200°C the corresponding alcohols were formed in 57–78% yield. The same results were obtained with *n*-butyraldehyde in the presence of $Rh_4(CO)_{12}$. When using $RhCl_3 \cdot 3H_2O$ as a catalyst precursor, hydrogenation takes place even in the absence of CO; in this case, it was assumed that the rhodium metal acts as a heterogeneous catalyst (68).

Similar results were obtained by Falbe and Huppes (40,41), who isolated the hydroformylation products of 1,5-cyclooctadiene and dicyclopentadiene as alcohols in very good yields; they conducted the reaction in two steps, the second under rather drastic conditions: 210°C and 1200 atm of CO and H_2 (1:1).

Only with α,β-unsaturated aldehydes or ketones does hydrogenation of the olefinic double bond largely prevail during hydroformylation (8). Taking into consideration the extremely high activity of rhodium catalysts in hydroformylation, these results can be explained in terms of a more favorable ratio between the rates of hydroformylation and hydrogenation when using rhodium than when using cobalt or ruthenium catalysts. Table 43 summarizes some examples of the hydrogenation of olefinic double bonds with rhodium catalysts.

Propylene is hydrogenated to a large extent at 95°C when using diphenyl ether as solvent or when employing a very high H_2/CO ratio

(30,72), but only 5.4% of 1-hexene is hydrogenated to hexane even at 200°C (70).

A careful investigation of the hydroformylation of styrene at 1 atm and 25°C with $RhH(CO)(PPh_3)_3$ showed that practically no ethylbenzene forms when the H_2/CO ratio is one, whereas about a third of the styrene is hydrogenated when an H_2/CO ratio of two is used in the absence of excess PPh_3 (24).

Hydrogenation of the olefinic double bond in conjugated and nonconjugated diolefins is closely associated with double-bond isomerization. In all cases the first step is introduction of the formyl group, which generally gives rise to a nonconjugated aldehyde. The latter may then undergo either hydroformylation with formation of a dialdehyde or a double bond shift with formation of an α,β-unsaturated aldehyde. The conjugated double bond, as already stated, is then mainly hydrogenated under oxo conditions. This explanation is based on the fact that the yield of diformyl derivatives is much larger if nonisomerizing catalysts (for instance, PR_3-containing catalysts) are used in the hydroformylation (Table 34). This deduction, however, does not have general validity since the hydroformylation of methyl sorbate first gives a monounsaturated ester which then is hydroformylated (20).

With ethyl cinnamate, 73% of the product is due to hydroformylation, whereas 26% is due to hydrogenation of the substrate to ethyl phenylpropionate (52). From ethyl crotonate, the hydrogenation product (ethyl butyrate) increases from 18 to 30% when the temperature is increased from 70 to 180°C while using Rh_2O_3 as catalyst precursor at 200 atm ($CO/H_2 = 1$) (52).

b. Isomerization. Since the early work of Wakamatsu, rhodium carbonyls of the type $[Rh(CO)_3]_n$, probably $Rh_4(CO)_{12}$ according to the method of preparation, appeared to isomerize olefins under hydroformylation conditions (6) more rapidly than did $Co_2(CO)_8$.

Comparative hydroformylation experiments were done with butenes at 120°C and 220 atm ($CO/H_2 = 1$) in the presence of either 1 g cobalt or 5.3 mg rhodium per liter, and they were interrupted after about half of the substrate had been hydroformylated. The residual olefin from the experiments with cobalt was unchanged, whereas 29 and 15% of the isomerized olefin was recovered from the experiments done with 1-butene and *trans*-2-butene in the presence of rhodium catalysts (6).

The isomerizing activity of rhodium compounds in the presence of CO was confirmed by Asinger and co-workers (54). By heating 1-octene at 140°C for 4 hrs under pressure of CO in the presence of a hexane solution of rhodium carbonyls—prepared from Rh_2O_3, CO, and H_2—they

TABLE 39

Asymmetric Rhodium-catalyzed Hydroformylation of Aromatic Substrates

Olefin	Asymmetric ligand	Chiral aldehyde obtained: Type	Chiral aldehyde obtained: α_D^{25} ($l = 1$) (degrees)	Chiral aldehyde obtained: Asymmetric induction (%)	Ref.
α-Ethylstyrene	PPh_2 (neomenthyl*)	3-Phenylpentanal	+0.070[a]	1	64
Phenyl vinyl ether	PPh_2 (neomenthyl*)	2-Phenoxypropanal	−0.051[b]	0.3	64
Indene	(−)-$\overset{*}{P}CH_3Ph(n\text{-}Pr)$	1-Formylindane	+0.25	n.d.	65
Isoeugenol	(−)-$\overset{*}{P}CH_3Ph(n\text{-}Pr)$	2-(3-Methoxy-4-hydroxyphenyl)butanal (I); 2-Methyl-3-(3-methoxy-4-hydroxyphenyl) propanal (II)	−0.25[c]	n.d.	65

Cinnamyl alcohol	(−)-$\overset{*}{P}CH_3Ph(n\text{-}Pr)$	2-Hydroxy-3-phenyltetrahydrofuran	−0.05	n.d.	65
Cinnamaldehyde propyleneglycol acetal	(−)-$\overset{*}{P}CH_3Ph(n\text{-}Pr)$	2-Phenylsuccinaldehyde monopropyleneglycol acetal (III) / Benzylmalonaldehyde monopropyleneglycol acetal (IV)	−0.1[d]	n.d.	65
α-Methylstyrene	(−)-DIOP	3-Phenylbutanal	−0.93	1.6	33
α-Ethylstyrene	(−)-DIOP	3-Phenylpentanal	−0.80	1.8	33
Allylbenzene	(−)-DIOP	2-Benzylpropanal	+1.89	15.5	33
trans-β-Methylstyrene	(−)-DIOP	2-Phenylbutanal	−13.80	14.4	33

[a] Temp 19°C.
[b] Temp 18°C.
[c] Measured on the mixture: (I)/(II) = 85/15.
[d] Measured on the mixture: (III)/(IV) = 55/45.

TABLE 40

Hydroformylation of Aliphatic Olefins with $RhH(CO)(PPh_3)_3$ *and* (−)-DIOP *in Aromatic Solvent*[a] (67*a*)

Olefin	Total pressure,[b] (atm)	Temp (°C)	Isomeric composition of the aldehyde mixture	Isolated optically active compounds: Compound	α_D^{25} ($l=1$) (neat)	Chirality	Optical purity (%)
1-Butene	1[c]	25	12.5[d]	2-Methylbutanal	−5.27°	*R*	18.8
1-Pentene	1	25	13.5[d]	2-Methylpentanoic acid	−3.62°	*R*	19.7
2-Ethyl-1-hexene	81	100	>98% 3-Ethylheptanal	3-Ethylheptanal	−0.02°	*R*	1.1
3-Methyl-1-butene	1	25	13.5[e]	2,3-Dimethylbutanal	−4.38°	*R*	15.2
1-Octene	1	25	9[d]	2-Methyloctanal	−4.91°	*R*	15.2
cis-2-Butene	1[c]	20		2-Methylbutanal	+7.57°	*S*	27.0
cis-2-Butene	84	95	>98% 2-Methylbutanal	2-Methylbutanal	+2.28°	*S*	8.1
cis-2-Butene[f]	86	95		2-Methylbutanal	+2.22°	*S*	7.9
trans-2-Butene	84	98		2-Methylbutanal	+0.89°	*S*	3.2
cis-2-Hexene	82	95	1.5[g]	2-Methylhexanoic acid	+1.68°	*S*	7.6
				2-Ethylpentanoic acid	−0.22°	*R*	5.8
trans-2-Hexene	82	95	1.4[g]	2-Methylhexanoic acid	+0.31°	*S*	1.4
				2-Ethylpentanoic acid	−0.11°	*R*	2.9
3-Methyl-1-pentene[h]	1	40	>95% 4-Methylhexanal	4-Methylhexanal	−0.44°	*R*	4.6

[a] Moles olefin/moles of rhodium metal = 3.0×10^2 to 1.6×10^3; $H_2/CO = 1$.

[b] Initial pressure at room temp.

[c] Partial pressure of the olefin: 0.33 atm.

[d] Straight-chain aldehyde/branched aldehyde ratio.

[e] 4-Methylpentanal/2,3-dimethylbutanal ratio.

[f] Rh_2O_3 as catalyst precursor.

[g] 2-Methylhexanal/2-ethylpentanal ratio.

[h] The recovered olefin had $\alpha_D^{25} = +0.840°$ (neat, $l = 1$) (optical purity 3.3%).

TABLE 41

Rhodium-catalyzed Hydroformylation of Styrene in Aromatic Solvents Using Different Asymmetric Ligands

Catalyst	Asymmetric ligand	Rh/L[a]	Temp (°C)	P_{CO} (atm)	P_{H_2} (atm)	2-Phenylpropanal: Yield (%)	2-Phenylpropanal: Prevailing chirality	2-Phenylpropanal: Asymmetric induction, (%)	Ref.
$[Rh(CO)_2Cl]_2$	(−)-P*(Ph)(*n*-Pr)(Me)	1/2.7	80	100	100	n.d.	*S*	21.1	65
$[Rh(C_6H_{10})Cl]_2$	(+)-P*(Ph)(CH_2Ph)(Me)	1/10	120	50	50	80	*S*	17.5	32
(+)-$RhCl(CO)L_2$	L = P(Ph)(Ph)(neomenthyl*)	1/1	75	50	50	29	*S*	0.8	64
$HRh(CO)(PPh_3)_3$	(−)-DIOP	1/2	25	0.5	0.5	69	*R*	22.7	33

[a] Moles rhodium metal/moles ligand.

TABLE 42
Hydrogenation of Aldehydes under Oxo Conditions[a] (68)

Aldehyde	Temp (°C)	Products	Yield (%)
Propanal	175	Propanol	57
Butanal	200	Butanol	78
Butanal	170	Butanol	70
Benzaldehyde	200	Benzyl alcohol	64

[a] Catalyst precursor $RhCl_3 \cdot 3H_2O$; initial pressure ($CO:H_2 = 1$) 230–240 atm at room temp.

obtained the mixture of octenes expected at thermodynamic equilibrium. Yamaguchi (73), however, showed that at 90°C the isomerization rate of the olefin rapidly decreases with increasing CO pressure and that the rate is very low for a CO partial pressure higher than 150 atm.

According to Falbe and Huppes (40), the results from hydroformylation of 1,5-cyclooctadiene when using Rh_2O_3 as catalyst precursor under different pressures, show, as for the $Co_2(CO)_8$ catalyzed reaction (74), that an increase in total pressure (hence of the partial pressure of CO) decreases the extent of double-bond isomerization. Similar results were reported by Gankin et al. (75) for the hydroformylation of 2-methyl-1-pentene with CO pressures between 20 and 140 atm, a H_2 pressure of 100 atm, a temperature of 130°C, and $[Rh(CO)_2Cl]_2$ and other rhodium complexes as catalyst precursors. In these experiments, they found a change in the isomeric composition of the products with conversion because of olefin isomerization.

Trisubstituted phosphines apparently strongly inhibit the double-bond shift: When PBu_3 was added to the rhodium carbonyl solution prepared by Asinger and co-workers as described above, the isomerization of 1-octene was suppressed (54).

The effect of phosphines on the catalytic activity of rhodium compounds for isomerization was confirmed by Rupilius (20) and by Wilkinson's group (12). While 2,3-dimethyl-2-butene is readily hydroformylated to 3,4-dimethylpentanal at 140°C with Rh_2O_3 and a CO and H_2 pressure of 160 atm, no reaction is observed when PBu_3 is added even under higher CO pressures. This lack of hydroformylation was attributed to the fact that the necessary initial shift of the double bond is inhibited by the added phosphine (20).

TABLE 43

Hydrogenation of Olefins under Oxo Conditions Using Rhodium Catalysts

Olefin	Catalyst	Temp (°C)	P_{CO} (atm)	P_{H_2} (atm)	Hydrogenation products[a] (%)	Hydroformylation products[b] (%)	Ref.
1-Hexene	$Rh(CO)(Cl)(PPh_3)_2/PPh_3$ [c]	200	17[d]	17[d]	5.4	67	70
1-Octene	$HRh(CO)[(PhCH_2)_3N]_3$	25	0.5	0.5	10	61	71
Styrene	$HRh(CO)(PPh_3)_3$	25	0.33[d]	0.66[d]	22	43	24
α-Methylstyrene	Rh carbonyl	130	150[d]	150[d]	—	99	5
Propenylbenzene	Rh/Al_2O_3	140	99[e]	99[e]	13	86	36
Propenylbenzene	$Rh/Al_2O_3/PPh_3$ [f]	90	99[e]	99[e]	2	92	36
Ethyl crotonate	Rh_2O_3	70–90	100[e]	100[e]	18	37	52
Ethyl cinnamate	Rh_2O_3	120	100	100	26	73	52
Acrylonitrile	Rh_2O_3	127	100	100	29	16	61

[a] Moles of hydrogenation product/100 moles of initial olefin.
[b] Moles of hydroformylation products/100 moles of initial olefin.
[c] Moles Rh metal/moles $PPh_3 = 1/400$.
[d] At constant pressure.
[e] Initial pressure at reaction temp.
[f] Moles Rh metal/moles $PPh_3 = 1/18$.

A substantial inhibition of double-bond migration by PPh_3 was also observed when using rhodium on alumina catalysts in the hydroformylation of the methyl esters of some di- and triunsaturated linear carboxylic acids having 18 carbon atoms (56). The presence, in the methyl linoleate hydroformylation products, of substantial amounts of formyl groups at C-11, originally a CH_2 group, was explained on the basis of a peculiar reactivity of the 1,4-pentadiene system (56). Thus, in the hydroformylation of 1,4-pentadiene with the Rh_2O_3/PBu_3 catalytic system, a significant amount (>5%) of hydroformylation occurs at the methylene group (20).

Triphenylphosphine inhibits the isomerizing activity of a rhodium hydroformylation catalyst like $RhH(CO)(PPh_3)_3$ (12,76). CO under similar conditions also exerts a strong inhibiting effect on double-bond isomerization (12). The same effect of CO under pressure was found in the isomerization of 1-pentene in the presence of $RhCl(CO)(PPh_3)_2$ (12).

According to Wilkinson and co-workers (77), the isomerization process occurs via addition of olefin to the Rh–H bond and formation of an alkylrhodium complex followed by olefin elimination. Attempts by these authors to prove the existence of alkylrhodium species failed. However, they were able to demonstrate the existence of square planar rhodium complexes of the type $Rh(CF_2CHF_2)(CO)(PPh_3)_2$ (77).

Despite these experiments, the problem of the mechanism of double-bond isomerization during hydroformylation with rhodium catalysts, as with cobalt catalysts, cannot be considered entirely clarified. Both a hydrogen shift within the olefin-catalyst complex not involving dissociation of the isomerized olefin and a true rhodium hydride addition and elimination are, as with cobalt, significant. However, a double-bond shift in the substrate occurs more easily when using rhodium catalysts than when using a cobalt catalyst.

4. *Isomeric Composition of Hydroformylation Products*

Because of their theoretical and practical importance, the factors influencing the isomeric composition of the products from hydroformylation of olefins in the presence of rhodium catalysts have been extensively investigated. This research appears particularly difficult since the number of factors involved is large and the results are complicated by concurrent double-bond shifts in the substrate. Existing data on this subject can hardly be compared because of different reaction conditions used by the various research groups.

The factors influencing the isomeric composition of the products, which were mainly investigated, are: type of rhodium complex used as catalyst precursor, CO and H_2 partial pressures, temperature, catalyst concentration, and solvent. For each of these factors, propylene hydroformylation will be discussed first, since with this olefin the double-bond isomerization does not affect product composition which is therefore directly related to the nature of the hydroformylation intermediates.

a. Type of catalyst. Even since the appearance of the first papers on the hydroformylation of olefins with rhodium catalysts (6), it was evident that the linear or less branched reaction products formed in smaller amounts than when using cobalt catalysts (Table 44). The effect of catalyst type on the isomeric ratio for different unsaturated substrates will be discussed separately.

(*1*) *Propylene.* The great difference observed in product composition from the hydroformylation experiments on propylene done with rhodium or cobalt catalysts (Table 44) clearly indicates that, besides double-bond isomerization, other factors, presumably connected with the nature of the catalyst, are responsible for these results.

TABLE 44
Hydroformylation of Linear Olefins Catalyzed by Cobalt and Rhodium Carbonyls[a] (6)

Olefin	Catalyst	Yield (%)	Straight-chain aldehyde/ total aldehydes %
Propylene	Co	83	80.0
	Rh	83	53.2
1-Butene	Co	89	81.3
	Rh	87	55.2
1-Pentene	Co	84	78.7
	Rh	82	54.3
1-Hexene	Co	85	79.4
	Rh	87	54.6
2-Butene	Co	79	70.4
	Rh	88	13.6
2-Pentene	Co	71	67.1
	Rh	79	11.8
2-Hexene	Co	68	63.0
	Rh	78	9.2

[a] Temp 110°C; $CO/H_2 = 1$; 220 atm at room temp; benzene solvent.

In consideration of the large influence of various ligands on the isomeric distribution of products obtained with cobalt catalysts, much work was devoted to investigating rhodium catalysts modified either by addition of ligands or by using different catalyst precursors. Nevertheless, it is still difficult to draw a general picture of the effect of different ligands on the isomeric composition of the hydroformylation products.

As shown in Table 45, a striking effect on the isomer ratio is achieved if an excess of a ligand like PPh_3 is present in solution (24,72,78,79). Brown and Wilkinson (24) obtained 95% of the linear isomer by using a low total pressure of CO and H_2 with PPh_3 as solvent. But in the absence of excess ligand, no substantial changes were noted when using $RhH(CO)(PPh_3)_3$ instead of rhodium carbonyl (6,72).

Other authors (13) found that when operating at 70°C and 120 atm of CO and H_2 (1:1) the use of excess ligand (4:1), such as PPh_3 and $P(Bu)_3$, relative to the rhodium catalyst does not markedly increase the percentage of straight-chain aldehyde from the hydroformylation of propylene or of 1-butene.

The type of ligand has some influence on the isomeric product composition as shown by Olivier and Booth (72) who, on investigating the effect of adding different ligands to the $RhH(CO)(PPh_3)_3$ catalytic system, varied the percentage of the normal isomer from 60% when using pyrazine to 70% when using diphenylbutylphosphine.

An interesting result was reported by Robinson et al. (31) who obtained 80% of the linear isomer when using $(AsPh_3)_2Rh(CO)Cl$ supported on carbon as catalyst: The percentage of the linear product, however, decreased with time, probably because of a modification of the catalyst under reaction conditions.

(2) Higher molecular-weight linear terminal and internal olefins. With the same catalytic system, the composition of hydroformylation products from higher olefins is greatly influenced by the position of the double bond in the molecule when the reaction is carried out under appropriate conditions (Table 46). This is one of the main differences between rhodium and cobalt-catalyzed hydroformylation, since in the presence of cobalt a preponderance of the least branched isomer is formed regardless of the initial position of the double bond (see Table 3).

The data in the literature on the effect of phosphines on the hydroformylation of higher linear olefins seem conflicting at first glance (24,25,70). With terminal olefins the effect of the phosphine is similar to that observed in the hydroformylation of propylene, that is, an increase in the formation of linear product (Fig. 6); but with internal olefins the opposite is true (20). As Table 46 shows, addition of excess PBu_3 causes

TABLE 45

Isomeric Composition of Products from The Hydroformylation of Propylene with Rhodium Catalysts

Catalyst	Moles ligand/ moles Rh metal	Temp (°C)	$P_{H_2}+P_{CO}$ [a] (atm)	Solvent	*n*-Butyraldehyde/ total C_4 aldehydes (%)	Ref.
$Rh_4(CO)_{12}$	—	110	220	Benzene	53	6
$HRh(CO)(PPh_3)_3$	3	95	47	Toluene	55	72
$HRh(CO)(PPh_3)_3+PPh_3$	176	95	47	Toluene	71	72
$HRh(CO)(PPh_3)_3$	—	90–125	—[b]	PPh_3	94–95	24
$Rh(CO)Cl(PPh_3)_2$	2	80	100[c]	Benzene	64	12
$Rh(CO)Cl(PPh_3)_2$ on Al_2O_3	2	148	—[d]	Vapor phase	64	31
$Rh(CO)Cl(PPh_3)_2+PPh_3$	194	100	36[c]	Benzene	70	78
$RhCl_3+PPh_3$	5.3	90	40	2-Methyl-2-butanol	74	79
π-$C_5H_5Rh(CO)PBu_3$	1	82–88	90[c]	Heptane	50	80

[a] $H_2/CO=1$, initial pressure at room temp if not otherwise stated.
[b] Propylene/CO/$H_2=1$, total pressure at reaction temp 7.5–16 atm.
[c] Initial pressure at reaction temp.
[d] Propylene/CO/$H_2=1/1.5/1.5$, total pressure at reaction temp 49 atm.

TABLE 46

Effect of PBu_3 *on the Isomeric Product Composition from Hydroformylation of Internal Olefins*[a] (20)

Olefin	Moles PBu_3/ moles Rh metal	Linear aldehyde (%)	Branched aldehydes	%
cis-2-Butene	—[b]	49	2-Methylbutanal	51
cis-2-Butene	16	22	2-Methylbutanal	78
cis-2-Butene	31	1	2-Methylbutanal	99
trans-4-Octene	—[b]	4	2-Propylhexanal	45
			Other C_9 aldehydes	51
trans-4-Octene	12.5	1	2-Propylhexanal	81
			Other C_9 aldehydes	18
trans-4-Octene	50	—	2-Propylhexanal	99

[a] Rh_2O_3 catalyst; 200 atm of 1:1 synthesis gas measured at room temp; temp 140°C.

[b] Operating temp 100°C.

an increase in the yield of isovaleraldehyde from 2-butene of up to about 100% and up to about 99% in the yield of 2-propylhexanal from 4-octene.

The results in Fig. 6 and Table 46 may be explained by postulating that phosphine complexes not only favor formylation of the terminal position

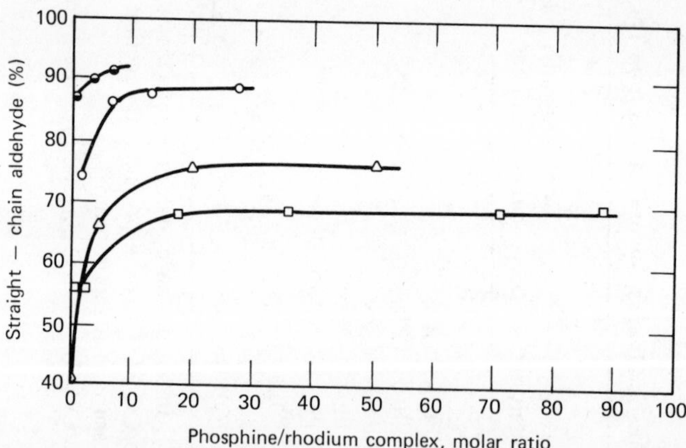

FIG. 6. Effect ot phosphine/rhodium complex molar ratio on isomeric composition of products from hydroformylation of terminal olefins ($CO/H_2 = 1$). ◒ 1-Hexene; $HRh(CO)(PPh_3)_3$: 25°C; 1 atm; PPh_3 used (24). ○ 1-Octene; Rh/C; 90°C; 5–7 atm; $P(OPh)_3$ used (25). △ 1-Hexene; $Rh(CO)(PPh_3)_2Cl$; 100°C; 34 atm; PPh_3 used (70). □ Propylene; $HRh(CO)(PPh_3)_3$; 95°C; 47 atm; PPh_3 used (72).

but also hinder double-bond isomerization. The second effect explains the data from the hydroformylation of internal olefins; this is certainly operative also for higher terminal olefins, where the addition of phosphines affects the isomer ratio more than in the hydroformylation of propylene (54,70). Similar effects were obtained by Pruett and Smith (25) in the hydroformylation of 1-octene using rhodium supported on carbon as catalyst; by adding increasing quantities of triphenylphosphite, the percentage of straight-chain aldehyde varied from 31 to 89%.

The effect of different ligands on the composition of the isomeric products was investigated mainly by Wilkinson's group (12), Pruett and Smith (25), and Craddock and co-workers (70).

Using the same type of catalyst precursor (rhodium on charcoal or soluble Rh(I) complexes), the type of ligand markedly influences the composition of the isomeric products (Table 47). But in a series of experiments with triphenylphosphines, no change in the composition of

TABLE 47

Isomeric Composition of Products from Hydroformylations Using Rhodium Catalysts with Various Ligands[a]

Olefin	Rhodium compounds	Straight-chain aldehyde (%)	Ref.
1-Pentene	$RhCl(CO)(PEt_3)_2$	72	12
	$RhCl(CO)(AsPh_3)_2$	57	12
1-Octene	$Rh/C+PBu_3$	71	25
	$Rh/C+PPh_3$	82	25
	$Rh/C+P(OPh)_3$	86	25
	$Rh/C+P[O-C_6H_3(CH_3)_2]_3$ (2,5-dimethylphenyl)	47	25
	$Rh/C+P[O-C_6H_4-Cl]_3$	93	25

[a] Reactions with 1-pentene conducted at 70°C and 100 atm CO/H_2(1/1); those with 1-octene at 90°C and at 5.5–7 atm CO/H_2 (1/1).

TABLE 48
Isomeric Composition of Products from Hydroformylation of Terminal Olefins Using Various Rhodium Complexes

Olefin	Catalyst	Straight-chain aldehyde (%)	Ref.
1-Pentene[a]	$[RhCl(CO)_2]_2$	41	12
	$RhCl(CO)(PPh_3)_2$	73	12
	$RhCl(CO)[P(p\text{-}CH_3C_6H_4)_3]_2$	74	12
	$RhCl(CO)[P(p\text{-}CH_3OC_6H_4)_3]_2$	73	12
	$RhCl(CO)[P(p\text{-}FC_6H_4)_3]_2$	73	12
1-Hexene[b]	$[Rh(CO)_2Cl]_2$	35	70
	$[(C_5H_5)_2RhCl]_2$	35	70
	$RhCl(CO)(PPh_3)_2$	52	70

[a] 1-Pentene experiments at 70°C and 100 atm measured at room temp; $CO/H_2 = 1$.
[b] 1-Hexene experiments at 100°C and constant pressure of 34 atm; $CO/H_2 = 1$.

the isomeric products was found when the phenyl group had a methyl, a methoxy, or a fluorine substituent (Table 48).

(3) Branched olefins and diolefins. Little information is available on the hydroformylation of branched olefins. Isobutylene is easily hydroformylated in 99% yield to 3-methylbutanal (58). 2,3-Dimethyl-2-butene yields exclusively 2,3-dimethylpentanal when using rhodium carbonyls in the absence of phosphines (20), but in the presence of excess PBu_3 at 140°C, that is, under nonisomerizing conditions, this olefin does not react, at least under the conditions used (20).

Phosphines also affect the composition of the hydroformylation products of olefins having conjugated double bonds, such as styrene. When using $HRh(CO)(PPH_3)_3$ (24) at 1 atm and 25°C, the fraction 3-phenylpropanal was about 12% in the absence of added ligand, while it was 21% in the presence of excess phosphine (20 : 1). A larger amount of 3-phenylpropanal was obtained by using $[RhCl(1,5\text{-hexadiene})]_2$ as a catalyst precursor and operating in the presence of 1.5 moles of $PhCH_2PCH_3(n\text{-}Pr)$ per mole of rhodium (32).

The composition of the products from the hydroformylation of diolefins is strongly influenced by the presence of phosphines (Table 34). When an appropriate concentration of these ligands is present, the formation of saturated monoaldehydes is decreased, if not entirely suppressed, while formation of dialdehydes is increased. This has been attributed to the suppression of a double-bond shift that does not allow formation of α,β-unsaturated aldehydes that are easily hydrogenated under the conditions used. With butadiene in the presence of PBu_3 (38), 1,3 and 1,2 additions mostly prevailed; but surprisingly with 1,3-pentadiene, 2,4 addition predominated over either 1,3 or 1,4 addition. In the hydroformylation of nonconjugated diolefins (such as 1,4-pentadiene and 1,5-hexadiene) in the absence of phosphines, Morikawa (7) found that 1,5 addition predominated.

(4) *Oxygenated compounds.* Interesting results were obtained by Rupilius (20) in the hydroformylation of allyl alcohol. In the presence of Rh_2O_3 and in the absence of phosphines, 40% of the alcohol isomerized to propanal; the hydroformylation products were in a 63/37 ratio in favor of the linear product. When PBu_3 was added (molar ratio of Rh_2O_3 to $PBu_3 = 0.005/1.1$), practically no propanal formed and the only secondary product was isobutyraldehyde. The ratio between the two isomeric formylpropanols is 78/22 in favor of the linear product; the ratio, however, decreases to 61/39 after taking into account that the isobutyraldehyde is due mainly to hydrogenation of the methacrolein formed by dehydration of 2-formylpropanol under hydroformylation conditions.

The effect of various ligands on hydroformylation of unsaturated esters was studied extensively (Tables 49,50). The most important results from the hydroformylation of 3- and 5-hexenoates (Table 49) is the finding that when using Rh_2O_3 a mixture of all the theoretically possible isomeric reaction products is produced. But in the presence of a phosphine, hydrogen and formyl group add only to the carbon atoms of the original double bond. The percentage of α-formyl ester from hydroformylation of methyl methacrylate found by Falbe and Huppes (52) (Table 50) varied from 5%, when operating in the presence of *N*-ethylpyrrolidine, up to 100% in the presence of triethylphosphite or 1-phenylphospholine-2. These data point to a very complex effect of ligands on the hydroformylation catalysts.

b. Effect of carbon monoxide and hydrogen partial pressure. The partial pressure of CO but not that of H_2 greatly influences the isomeric composition of the hydroformylation products when using cobalt catalysts. These effects have been little studied with rhodium catalysts in the absence of ligands other than CO and hydrogen. But numerous experiments were carried out by using rhodium catalysts containing one or more ligands different from hydrogen and CO.

In the case of propylene using rhodium tris(acac) as catalyst precursor,

TABLE 49

Isomeric Distribution of Products from Hydroformylation of Methyl Hexenoates Using Rh_2O_3 and Various Ligands[a] (54)

Ester	Catalyst	Moles ligand/ moles Rh metal	Temp (°C)	Products[b] (%) *I*	*II*	*III*	*IV*	*V*
Methyl 3-hexenoate	Rh_2O_3	—	120	1	36	28	28	8
Methyl 3-hexenoate	$Rh_2O_3+PBu_3$	15	160	—	66	34	—	—
Methyl 3-hexenoate	$Rh_2O_3+P(C_6H_{11})_3$	15	120	1	66	32	1	—
Methyl 5-hexenoate	Rh_2O_3	—	120	Trace	3	12	41	44
Methyl 5-hexenoate	$Rh_2O_3+PBu_3$	15	160	—	—	—	39	61

[a] 200 atm, $CO/H_2=1$, measured at room temp.

[b] (*I*) $CH_3(CH_2)_3\underset{\displaystyle CHO}{\underset{|}{C}H}COOCH_3$; (*II*) $CH_3(CH_2)_2\underset{\displaystyle CHO}{\underset{|}{C}H}CH_2COOCH_3$; (*III*) $CH_3CH_2\underset{\displaystyle CHO}{\underset{|}{C}H}(CH_2)_2COOCH_3$; (*IV*) $CH_3\underset{\displaystyle CHO}{\underset{|}{C}H}(CH_2)_3COOCH_3$; (*V*) $\underset{\displaystyle CHO}{\underset{|}{C}H_2}(CH_2)_4COOCH_3$.

no substantial change in the isomeric composition of the reaction products was observed when operating at 110°C and 40 atm H_2 partial pressure and when varying the CO partial pressure between 10 and 196 atm. The percentage of *n*-butyraldehyde increased slightly from 42% at the lowest CO pressure to 44% at the highest pressure (81).

TABLE 50

Isomeric Distribution of Products from Hydroformylation of Methyl Methacrylate Using Rh_2O_3 and Various Ligands[a] (52)

Catalyst	Moles ligand/ moles Rh metal	Temp (°C)	Yield[b] (%)	Methyl α-formylisobutyrate (%)
Rh_2O_3	—	130	80	21
Rh_2O_3 + N-ethylpyrrolidine	103	130–150	79	5
Rh_2O_3 + pyridine	75	130	67	12
$Rh_2O_3+PBu_3$	31	130	75	87
Rh_2O_3 + 1-phenylphospholine-2	19	80–130	86	100
$Rh_2O_3+PPh_3$	24	80	82	89
$Rh_2O_3+P(OEt)_3$	38	120–150	58	100

[a] 200 atm of $CO+H_2$ (1:1), initial pressure at reaction temp.

[b] Moles methyl α-formyl + methyl β-formylisobutyrate/initial moles methyl methacrylate.

In the absence of ligands different from CO and hydrogen, Rudkovskii's group (58) found that, at 100 atm of CO, a variation in H_2 pressure between 30 and 150 atm does not affect the product composition from the hydroformylation of 1-hexene. On the other hand, an increase in total pressure between 100 and 150 atm at 110–150°C caused an increase in the amount of branched-chain aldehydes, but an increase of total pressure between 150 and 300 atm caused a decrease. Very little compositional changes were noted when isobutylene (58*a*) was reacted under different pressures of CO and H_2 (Table 51).

In the hydroformylation of ethyl acrylate with $[Rh(CO)_3]_n$ (55), the branched product (α-carboethoxypropionaldehyde) increased from 16% to 44% when P_{CO} was increased from 40 to 160 atm; an analogous increase was found when increasing P_{H_2} (Table 52).

Similar results were obtained by Falbe and Huppes (52) in the hydroformylation of methyl methacrylate with Rh_2O_3 as catalyst precursor. On increasing the total pressure of CO and H_2 from 200 to 1000 atm and operating at 80°C, the percentage of α-formylisobutyrate increased from 50 to 84% (Table 52).

TABLE 51
Effect of Carbon Monoxide and Hydrogen Pressures on Isomeric Composition of Products from Hydroformylation of Olefins with $[Rh(CO)_3]_n$[a] (58*a*)

Olefin	P_{CO} (atm)	P_{H_2} (atm)	More-branched isomers (%)
1-Hexene[b]	100	30	51[c]
	100	150	51
	50	50	58
	75	75	62
	150	150	44
Isobutylene[d]	60	100	0.5[e]
	100	100	0.8
	150	100	1.1
	50	50	1.0
	120	100	0.6
	150	150	1.0

[a] Experiments at constant pressure.
[b] Temp 110°C.
[c] The more-branched isomers from 1-hexene consisted of 2-methylhexanal and 2-ethylpentanal.
[d] Temp 150°C, catalyst concentration 0.5 mg Rh in 100 g. of solution.
[e] 2,2-Dimethylpropanal from isobutylene in all cases.

TABLE 52

Effect of Carbon Monoxide and Hydrogen Pressures on Isomeric Composition of Products from Hydroformylation of Ethyl Acrylate and Methyl Methacrylate with Rhodium Catalysts

Substrate	Catalyst	Temp (°C)	P_{CO} (atm)	P_{H_2} (atm)	More-branched isomer (%)	Ref.
Ethyl acrylate	$[Rh(CO)_3]_n$	110	80[a]	40[a]	32	55
		110	80	160	61	
		120	40	80	16	
		120	160	80	44	
Methyl methacrylate	Rh_2O_3	80	100[b]	100[b]	50	52
		80	200	200	58	
		80	250	250	62	
		80	500	500	84	

[a] With ethyl acrylate, initial pressures at room temp.
[b] With methyl methacrylate, initial pressures at reaction temp.

The influence of P_{CO} and P_{H_2} on the isomeric composition of the products from the hydroformylation of propylene was also studied with phosphorous-containing ligands. No marked effect of P_{CO} and P_{H_2} at constant total pressure was found by Craddock's group (82). However, a decrease in the ratio of linear to branched-chain aldehyde from 6.5 to 2.5 was found by Olivier and Booth (30) by increasing P_{CO} from 1.5 to 15 atm while using $HRh(CO)(PPh_3)_3$ as catalyst. An increase in straight-chain aldehyde was found when increasing the total pressure ($CO/H_2 = 1$) from 35 to 70 atm in the hydroformylation of 1-hexene at 100°C with $RhCl(CO)(PPh_3)_2$ in the absence of excess PPh_3, but no effect of P_{CO} and P_{H_2} was noted when excess phosphine was present under the same operating conditions (70).

The most systematic approach to understanding the influence of P_{CO} and P_{H_2} on the isomeric composition of hydroformylation products was made by Brown and Wilkinson (24) who, using $HRh(CO)(PPh_3)_3$ as catalyst, conducted experiments over a wide interval of pressures. With 1-hexene at 25°C and a total pressure of 1 atm, the branched-chain isomer increased slightly when the CO/H_2 ratio exceeded 1; whereas with a CO/H_2 ratio of 0.8, over 90% of the linear aldehyde was found (Table 53). A decrease in the total pressure of the equimolar mixture of CO and H_2 from 1 to 0.7 atm also caused a slight increase in the linear aldehyde. The same effect was observed with 1-hexene by increasing the total pressure to 27 atm at 25°C, a large decrease in the percentage of the linear isomer was observed. The ratio of straight- to branched-chain

TABLE 53

Effect of Carbon Monoxide to Hydrogen Ratio at Constant Total Pressure on Isomeric Composition of Products from Hydroformylation of 1-Hexene with $HRh(CO)(PPh_3)_3$ [a] (24)

CO/H_2	Temp (°C)	Total pressure (atm)	Heptanal (%)
2	25	1	85
1	25	1	86
0.8	25	1	91
2	40	1	85
1	40	1	88
0.8	40	1	95
0.5	40	1	97
1	40	1	93[b]
0.5	40	1	98[b]
1	25	27	74
0.5	25	27	80

[a] Catalyst: 30 mmoles/l in experiments at 1 atm, 15 mmoles/l in experiments at 27 atm; benzene solvent.

[b] $PPh_3/HRh(CO)(PPh_3)_3$ molar ratio = 3.

isomer was 2.9 when using a 1 : 1 mixtures of CO and H_2, and the ratio was four when using a 1 : 2 mixture.

Similar results were obtained by Pruett and Smith (25) in the hydroformylation of 1-octene when using rhodium supported on carbon in the presence of triphenylphosphite and changing the total pressure (CO/H_2 = 1) between 8 and 170 atm (Fig. 7). The hydroformylation of methyl methacrylate with rhodium catalyst in the presence of either triphenylphosphite (25) or PBu_3 (52) at varying total pressures (CO/H_2 = 1) gave results qualitatively similar to those reported for hydroformylation in the absence of added ligands. These findings and the effect of total pressure on the distribution of the isomeric products are summarized in Fig. 7. With one exception (70), the percentage of the straight-chain aldehyde was found to decrease when the total pressure was increased (24,25,30).

These results can be rationalized by taking into account that the ratio of the less-branched to the more-branched products is larger in the presence than in the absence of phosphines, and assuming that a displacement equilibrium as indicated in reaction 44 exists under the hydroformylation conditions.

$$-\underset{|}{\overset{|}{Rh}}PR_3 + CO \rightleftarrows -\underset{|}{\overset{|}{Rh}}CO + PR_3 \tag{44}$$

R = alkyl or aryl group

c. Effect of temperature. The effect of temperature on isomeric product composition with rhodium catalysts is quite complicated because of simultaneous hydroformylation and double-bond isomerization reactions and, when phosphine-containing catalysts are used, because of the possible dissociation of the ligand from the catalytic complexes.

In the hydroformylation of propylene with rhodium carbonyl, where double-bond isomerization cannot play a role, a slight increase of the linear aldehyde was observed on increasing the temperature (6) (Table 54). These results are opposite to those found with cobalt (15), but the lack of other experimental data precludes rationalization of these findings.

An increase of the less-branched aldehyde with increasing temperature was also observed in the hydroformylation of acrylates (55) and methacrylates (52) while the opposite is apparently true with isobutylene (58).

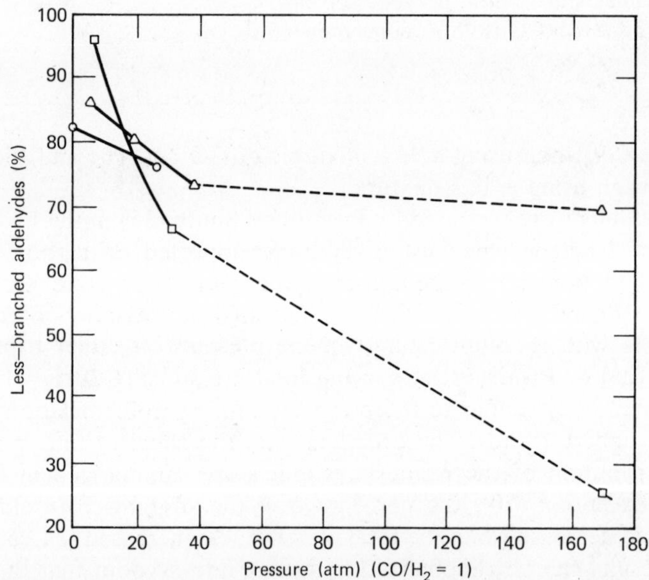

FIG. 7. Effect of total pressure on selectivity in hydroformylation of olefins with rhodium catalysts. ○ 1-Hexene; 15 mmoles/l $HRh(CO)(PPh_3)_3$; 25°C (24). □ Methyl methacrylate; 3.25 mmoles/l Rh/C; $P(OPh)_3$ 23 mmoles/l; 100–110°C (25). Δ 1-Octene; 36.4 mmoles Rh/C/l; $P(OPh_3)_3$ 241 mmoles/l; 90°C (25).

TABLE 54

Effect of Temperature on Isomeric Composition of Products from Hydroformylation with Rhodium Catalysts

Olefin	Catalyst	Temp (°C)	Pressure[a] (atm)	Less-branched isomers/ total aldehydes (%)	Ref.
Propylene	$[Rh(CO)_3]_n$	40	200	44	6
		110	200	50	
Propylene	$Rh_4(CO)_{12}$	70	120	51	13
Isobutylene	$[Rh(CO)_3]_n$	110	300[b]	99	58*a*
		170	300[b]	97	
1-Hexene	$[Rh(CO)_3]_n$	110	300[b]	57	58*a*
		140	300[b]	51	
		170	300[b]	43	
Ethyl acrylate	$[Rh(CO)_3]_n$	90	160	22	55
		120	160	69	
Methyl methacrylate	Rh_2O_3	80	1000[c]	15	52
		150	1000[c]	87	
2-Butene	$[Rh(CO)_3]_n$	50	200	~0	6
		70	200	~0	
		90	200	7	
2-Hexene	$[Rh(CO)_2Cl]_2$	80	80[c,d]	25	70
		120	80[c,d]	36	
Propenylbenzene	Rh/Al_2O_3	90	200	2	36
		140	200	9	

[a] $H_2/CO = 1$; initial pressure at room temp if not otherwise stated.
[b] Experiment at constant pressure.
[c] Initial pressure at reaction temp.
[d] $H_2/CO = 5$.

The increase in less-branched aldehyde with temperature for internal olefins probably has a different origin since double-bond isomerization seems to have a higher activation energy than hydroformylation. The most systematic research was done with ethyl crotonate. Here formylation at C_2 and C_3 occurred to about the same extent at 90°C but no linear product was formed at this temperature. At 180°C, however, formylation did not take place at C_2 and about the same amount of 3-formyl and 4-formyl derivatives were formed. At 130°C, a rather high selectivity toward formation of the 3-formyl derivative was observed. When the

TABLE 55

Effect of Temperature on Isomeric Composition of Products from Hydroformylation with Phosphine-containing Rhodium Catalysts

Olefin	Catalyst	Temp (°C)	Total pressure[a] (atm)	Less-branched isomers / Total aldehydes %	Ref.
Propylene	$HRh(CO)(PPh_3)_3 + PPh_3$	79	48[b]	70	72
		175	48[b]	58	
1-Hexene	$Rh(CO)Cl(PPh_3)_2 + PPh_3$	75	34	72	70
		100	34	76	
		125	34	70	
		150	34	67	
		200	34	41	
1-Hexene	$HRh(CO)(PPh_3)_3$	25	1	90	24
		50	1	93	
2-Hexene	$Rh(CO)Cl(PPh_3)_2$	150	34	37	70
		200	34	39	
2-Hexene	$Rh(CO)Cl(PPh_3)_2 + PPh_3$	150	34	16	70
		200	34	43	
1-Octene	$HRh(CO)[N(CH_2Ph)_3]_3$	20	1	93	71
		30	1	83	

[a] $CO/H_2 = 1$.
[b] Initial pressure at room temp; all other experiments at constant pressure.

hydroformylation of propylene was conducted in the presence of phosphines (Table 55), a decrease of linear product was observed on increasing the temperature from 79 to 175°C (72). A possible rationalization of these results, which apparently are opposite to those obtained in the absence of phosphines, can be found by supposing that the displacement of phosphines by CO is favored at higher temperatures and, therefore, some aldehydes are formed via complexes not containing phosphine ligands. Consequently, products with a lower ratio of straight- to branched-chain isomers are formed.

With higher olefins as 1-hexene, it seems that below 80°C a slight increase in the linear isomer occurs when the temperature is increased, whereas at temperatures above 100°C the opposite is true, probably because of increasing double-bond isomerization.

Finally, when using tribenzylamine as ligand between 20 and 30°C, a decrease in straight-chain aldehyde was observed on increasing the temperature (71).

d. Solvent effects. The effect of solvents on isomeric product composition was examined in a few cases. In the hydroformylation of propylene with $HRh(CO)(PPh_3)_3$ and excess PPh_3, a number of solvents were tested: toluene, dioxane, *n*-butanol, diphenyl ether, methylcyclohexane, DMF, and isobutyronitrile. Only diphenyl ether had some effect on the isomeric composition; with this solvent 80% of the linear isomer was found, instead of 67% as observed with the other solvents (72). In this case a large amount of propylene hydrogenation takes place.

When $Rh(CO)Cl(PPh_3)_2$ (70) was used as the catalyst with no excess of phosphine present, a striking effect of the solvent was observed in the hydroformylation of 1-hexene as the data in Table 56 show. The results can be rationalized by admitting that different ratios of phosphine-containing and phosphine-free catalytic species exist when operating in different solvents.

TABLE 56

Hydroformylation of 1-Hexene with $RhCl(CO)(PPh_3)_2$ *in Some Solvents*[a] (70)

Solvent	Straight-chain aldehyde (%)
Benzene	42
Dimethylformamide	71
Dioctyl phthalate	52

[a] Temp 100°C; $CO/H_2 = 1$; experiments at constant pressure (35 atm).

e. Catalyst concentration. The effect of catalyst concentration on the isomeric composition of hydroformylation products seems to have been little investigated. A small increase in the percentage of the branched isomer from ethyl acrylate was observed when increasing the concentration of $[Rh(CO)_3]_n$ (55). A dramatic effect was observed by Brown and Wilkinson (24) in the hydroformylation of 1-hexene at 25°C and 1 atm ($CO/H_2 = 1$). By increasing the concentration of $HRh(CO)(PPh_3)_3$ from 5 to 50 mmoles, the linear to branched isomer ratio increased from about 3.5 to about 9. The same behavior was also found at 40 and 50°C, and this can be explained on the basis that the presence of an increasing amount of free phosphine, which arises during formation of the catalyst, causes a change in the reaction mechanism (24). Fell and Müller (71) observed an analogous increase in the percentage of straight-chain aldehyde when increasing the catalyst concentration in the hydroformylation of 1-octene with $HRh(CO)[N(CH_2Ph)_3]_3$ at ambient pressure and temperature with 1:1 synthesis gas.

5. *Kinetics*

First, we shall discuss the kinetics of the hydroformylation of olefins catalyzed by rhodium compounds in the absence of added ligands; then, the kinetics of the reaction in the presence of various ligands; and finally, we'll compare the results obtained in the two cases. In doing so, we'll consider the effect of CO and H_2 pressures, of concentration and structure of the olefinic substrate, of temperature, and of solvent.

a. Hydroformylation with unsubstituted rhodium carbonyls. The most complete kinetic investigations of hydroformylations in the absence of added ligands were conducted by Heil and Markó (18) and by Yamaguchi (73). Their results unfortunately were obtained at different temperatures (75 and 110°C), and different pressures (at constant and at varying total pressures, respectively). In the second case (73), only the initial reaction rates were considered. In the first case (18), $Rh_4(CO)_{12}$ was used as catalyst, whereas rhodium nitrate was used in the second. Therefore, it is not surprising that the resulting kinetic expressions from both sets of experiments differ. These authors agree that the hydroformylation rate (with 1-heptene in the first case and 1-octene in the second) is first order with respect to H_2 pressure and varies inversely with the CO partial pressure between 50 and 170 atm total pressure (Fig. 8).

According to Rudkovskii's group (58), who investigated the hydroformylation of 1-hexene in the presence of rhodium carbonyls at 150°C, these orders of reaction are valid at CO pressures above 100 atm. At 150°C the reaction rate increases greatly on increasing the total pressure ($CO/H_2 = 1$) from 50 to 200 atm. These facts agree well when one takes into account that, according to both Markó and Rudkovskii, the hydroformylation rate increases to a maximum with increasing CO partial pressure and then decreases, as with cobalt catalysts. The CO pressure corresponding to the maximum rate increases with increasing temperature and is about 40 atm at 75°C and 100 atm at 110°C.

Heil and Markó (18) and Yamaguchi (73) found that the reaction order with respect to the concentration of dissolved rhodium for concentrations up to 100 mg/l is about one between 75 and 150°C and total pressures of 90–140 atm. A different dependence of rate on Rh concentration was found when $Rh_4(CO)_{12}$ or $Rh_6(CO)_{16}$ was used as catalyst precursor (93). Rudkovskii's group (58), however, when reacting 1-hexene at 110° and

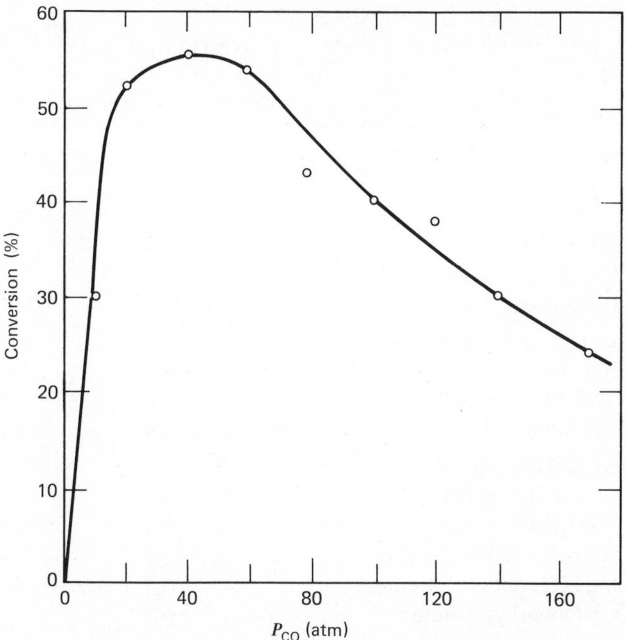

FIG. 8. Effect of P_{CO} on hydroformylation rate (18). 1-Heptene 0.361 mole/l; catalyst 0.030 mmole $Rh_4(CO)_{12}$/l; temp 75°C; P_{H_2} 40 atm; reaction time 50 min; hexane solvent.

150 atm total pressure ($CO/H_2 = 1$), found a reaction order of 0.4 with respect to rhodium concentration in the range of 5–50 mg/kg.

The greatest discrepancies relate to the qualitative and quantitative aspects of the effect of the olefin structure and concentration on the hydroformylation rate. A systematic investigation of the influence of olefin structure on the hydroformylation rate was performed by Heil and Markó (83) using $Rh_4(CO)_{12}$ at 75°C in toluene (Table 57). Unfortunately, the isomeric composition of the reaction products was not reported and therefore the data cannot be easily interpreted. The data of Heil and Markó (83) are not in accord with the observation of Wakamatsu (6) that, with $Rh_4(CO)_{12}$ as catalyst precursor, linear internal olefins are hydroformylated more rapidly than linear terminal olefins, nor do the data of Heil and Markó agree with the data of Rudkovskii's group (58) who found very similar hydroformylation rates in toluene at 150°C and 150 atm ($CO/H_2 = 1$) for 1-hexene, isobutylene, and cyclohexene. Further work by Markó's group (93) confirmed that the order with respect to the olefin concentration depends on olefin structure. Therefore,

TABLE 57
Hydroformylation of Olefins with $Rh_4(CO)_{12}$[a] (83)

Olefin	Relative reaction rate[b]
Styrene	20.3
Linear terminal olefins	
1-Hexene	9.1
1-Heptene	8.9
1-Octene	8.2
1-Decene	6.6
Linear internal olefins	
2-Hexene	5.6
cis-2-Heptene	6.6
3-Heptene	6.9
Branched olefins	
2-Methyl-1-pentene	4.2
2-Methyl-2-pentene	2.4
trans-4-Methyl-2-pentene	4.9
2,4,4-Trimethyl-2-pentene	0.5
2,3-Dimethyl-2-butene	0.1
Cyclic olefins	
Cyclohexene	1
Cyclopentene	2.7
1-Ethylcyclohexene	0.5

[a] Catalyst $53 \cdot 10^{-3}$ mmoles/l in toluene; temp 75°C; $CO/H_2 = 1$; pressure, 130 atm.

[b] Calculated from the reported data for k, rate constant.

isomerization of the substrate can greatly influence kinetic results unless initial reaction rates are measured. Further studies, however, are needed for a better understanding of the influence of olefin structure on hydroformylation rate.

The effect of olefin concentration on the hydroformylation rate was also investigated. Heil and Markó (18) found a first order dependence at 75°C; Rudkovskii's group (58) found that the olefin concentration influences the time necessary to obtain 50% conversion; whereas Yamaguchi (73), using a higher olefin concentration at 110°C, noted that there was practically no dependence of the initial reaction rate on olefin concentration in the case of 1-octene. Markó and co-workers (93) reexamined the kinetics of the hydroformylation of 1-heptene and cyclohexene. In the case of 1-heptene, the initial rate was found independent of olefin concentration in agreement with the data of Yamaguchi. With cyclohexene (93) and cycloheptene (97), however, a first order dependence on the olefin concentration was confirmed. Systematic kinetic studies at different temperatures are needed to clarify this interesting point, which might indicate a change in the rate-determining step with temperature.

Considering the probable change in the concentration of catalytic species with reaction conditions as well as the different concentrations of dissolved CO and H_2 at various temperatures, one cannot expect significant relationships between apparent activation energy and mechanism. For the apparent activation energy, a value of 28 kcal/mol between 66° and 90° was found for 1-heptene (18) and 21 kcal/mol between 90 and 110°C for 1-octene (73).

Very little is known about the effect on the hydroformylation rate of solvents that do not form stable complexes with rhodium carbonyls. According to Rudkovskii's group (58), the following relative times are necessary for 50% conversion of 1-hexene at 100°C and 150 atm total pressure ($CO/H_2=1$) in the presence of 5 mg of rhodium: ethanol 1, acetone 0.8, toluene, isooctane, and dioxane each 0.4.

b. Hydroformylation in the presence of added ligands. In general, the presence of nitrogen- or phosphorus-containing compounds that can form stable complexes with rhodium derivatives largely decreases the activity of these catalysts. A semiquantitative evaluation of the retarding effect of pyridine on the rate of hydroformylation of 1-hexene was made by Rudkovskii's group (58); their results are given in Table 58.

An analogous effect was observed in the hydroformylation of octenes in the presence of tributyl or tricyclohexylphosphine (20). The influence of

TABLE 58
Effect of Pyridine on Rate of Hydroformylation of 1-Hexene with $[Rh(CO_3)]_n$[a] (58)

Pyridine/ catalyst (molar ratio)	Olefin conversion (%)	Time for 50% conversion (min)
0	96	54
75	90	58
150	93	109
300	47	183

[a] Temp 110°C; 150 atm of $CO+H_2$ (1:1); experiments at constant pressure.

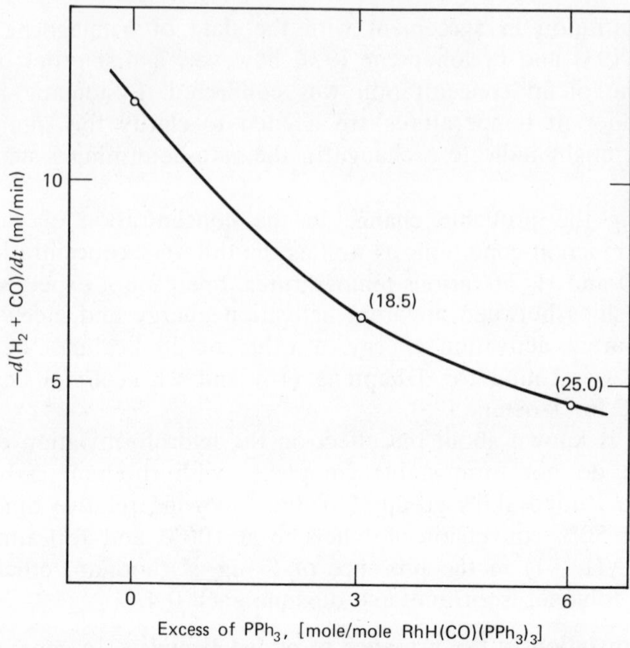

FIG. 9. Rate of hydroformylation of 1-hexene (1*M*) in benzene at 25°C as function of added excess PPh_3 at 50 cm gas pressure ($CO/H_2 = 1$). Data in parentheses give the heptaldehyde to 2-methylhexaldehyde ratios formed (24).

excess PPh_3 on the rate of hydroformylation of 1-hexene in the presence of $HRh(CO)(PPh_3)_3$ at low pressure, which was determined by Brown and Wilkinson (24), is in agreement with the above data (Fig. 9).

Surprisingly, for the hydroformylation of propylene at medium pressure in the presence of $HRh(CO)(PPh_3)_3$, Olivier and Booth (30,72) found a maximum rate for a molar ratio of 20–35 between the phosphine and the rhodium complex (Fig. 10).

According to Brown and Wilkinson (24), the kinetics of low-pressure hydroformylation in the presence of $HRh(CO)(PPh_3)_3$ is quite complicated. The rate increases linearly with concentration of rhodium above 6 mmoles/1, whereas a greater dependence on rhodium concentration was found at lower concentrations. This is probably due to changes in the equilibrium among active species. Also the dependence of the reaction rate on substrate concentration and on CO pressure is complex; an inhibiting effect of CO pressure was clearly shown. A first order dependence with respect to catalyst concentration from 2.5 to 20 mmoles/1 was also found with $HRh(CO)[N(CH_2Ph)_3]_3$ (71).

Preliminary data on propylene hydroformylation were obtained in a small continuous unit (72). A first order dependence was found with respect to catalyst concentration, H_2 pressure, and propylene concentration, but no dependence on CO partial pressure was found in the relatively small interval of pressures investigated between 79 and 107°C. On the other hand, an inhibiting effect of CO was observed in the hydroformylation of 1-pentene with *trans*-$RhCl(CO)(PPh_3)_2$ (12) as catalyst precursor.

The influence of the structure of the catalyst precursor on the rate of hydroformylation of 1-pentene was investigated by Wilkinson's group for complexes of the type $Rh(X)(CO)(PR_3)_2$ (12). By changing X, the rate decreased in the order Cl > Br > I; by changing R it decreased in the order *p*-$MeOC_6H_4 > C_6H_5 >$ *p*-FC_6H_4.

The effect of different ligands on the rate of hydroformylation of propylene was investigated (72) by adding a large excess (20–70 moles per mole of rhodium complex) of various ligands to a solution of

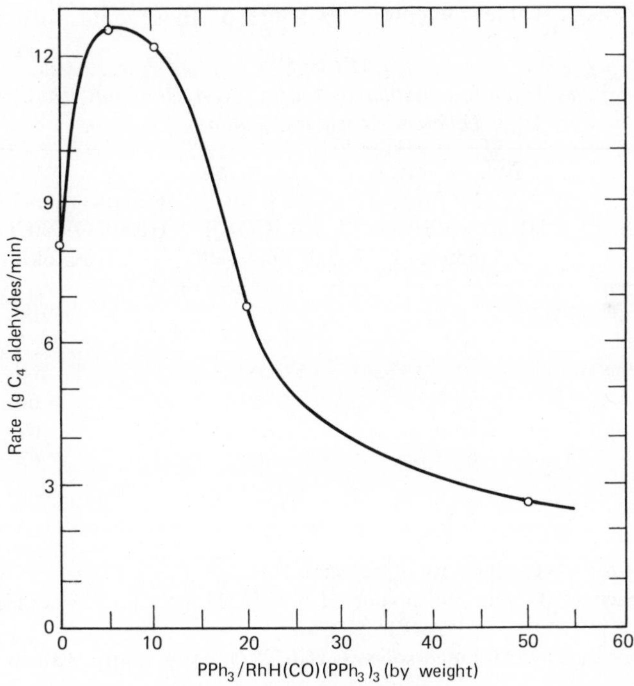

FIG. 10. Rate of hydroformylation of propylene. Temp 90°C; $CO/H_2/C_3H_6 = 1$; 34 atm; toluene solvent (30).

$RhH(CO)(PPh_3)_3$. The rate varied in the order PPh_3 > diphenylbutylphosphine > triphenylphosphite ≃ pyrazine > PBu_3 ≃ tri(*N*,*N*-dimethylamino methyl)phosphine.

The effect of solvents on the rate of hydroformylation in the presence of rhodium complexes has received little attention. In the hydroformylation of propylene with $HRh(CO)(PPh_3)_3$ and excess PPh_3, the rate decreased in the order: toluene > methylcyclohexane > DMF ≃ *n*-butanol > dioxane ≃ diphenyl ether (72). A lower reaction rate in polar solvents was also found in a low-pressure hydroformylation using $HRh(CO)[N(CH_2Ph)_3]_3$ as catalyst (71).

The effect of solvents on the rate of hydroformylation of 1-hexene and of 2-hexene using $RhCl(CO)(PPh_3)_2$ as catalyst was investigated by Craddock et al. (70) who observed a larger reaction rate in *n*-butyraldehyde than in benzene. In the former solvent, the rate was first order with respect to olefin concentration.

Finally, the effect of olefin structure on the rate of hydroformylation was examined by Brown and Wilkinson (19,24) for a large number of substrates in benzene at 50 mm Hg and 25°C with CO and H_2 (1 : 1) and $RhH(CO)(PPh_3)_3$. Table 59 compares some of these data with similar

TABLE 59

Relative Rates[a] *of Hydroformylation of Olefins with Rhodium Catalysts under Different Reaction Conditions*

	Relative rate (24) for $HRh(CO)(PPh_3)_3$, 2.5 mmoles/l[b]	Relative rate (83) for $Rh_4(CO)_{12}$, 0.053 mmole/l[c]	Relative rate (71) for $HRh(CO)[N(CH_2Ph)_3]_3$, 2.0 mmoles/l[d]
2-Methyl-1-pentene	1	4.2	n.d.
cis-2-Pentene	n.d.	n.d.	7
cis-2-Heptene	2	6.6	n.d.
1-Heptene	58.3	8.9	69
1-Hexene	58.6	9.1	67
1-Pentene	62.3	n.d.	65
Styrene	72	20.3	n.d.
Cyclohexene	n.d.	1	1

[a] Based on reported values for *k*, reaction rate.

[b] Olefin concn., 1*M*; benzene solvent; H_2/CO (1 : 1); pressure, 50 cm Hg; temp 25°C.

[c] Olefin concn., 0.4*M*; toluene solvent; H_2/CO (1 : 1); pressure, 130 atm; temp 75°C.

[d] Olefin concn., 3.2*M*; benzene solvent; H_2/CO (1.05 : 1); pressure, 1 atm; temp 25°C.

data obtained by Heil and Markó (83) at 130 atm with $Rh_4(CO)_{12}$ as catalyst. Neither of these groups of workers reported the composition of their reaction products; the relevance of secondary reactions under their reaction conditions therefore cannot be inferred. Table 59 clearly shows the dramatic effect exerted by phosphines on the relative reactivities of the various olefins. The reactivity of internal and 2-methyl olefins with respect to linear terminal olefins is much lower with the phosphine-containing catalyst. Suprisingly, styrene is by far the most reactive substrate.

Fell and Müller (71) determined relative reaction rates of several olefins in the presence of $HRh(CO)[N(CH_2Ph)_3]_3$ at 25°C with CO and H_2 at 1 atm (Table 59). As previously found with cobalt and other rhodium catalysts, the relative reactivity decreased in this order: linear terminal olefins > linear internal olefins > branched internal olefins ≃ cycloolefins.

6. *Mechanisms*

The mechanism of rhodium-catalyzed hydroformylation was investigated in the absence and in the presence of added ligands, particularly of trisubstituted phosphines. In both cases the mechanism seems more complicated than that postulated in the case of cobalt because of the larger number of complexes that could act as catalytic species with rhodium. First, we shall discuss the relatively few data available on the hydroformylation reaction carried out in the absence of strong ligands; then, we'll discuss some aspects of the hydroformylation reaction catalyzed by phosphine-containing rhodium complexes.

a. Rhodium carbonyls as catalyst precursors. The main investigations in this field were conducted by Wakamatsu (6), Heil and Markó (18,84,93), and Chini and co-workers (16). On the basis of previous knowledge about hydroformylation catalyzed by cobalt carbonyls, attempts were made to isolate or to detect by physical methods the existence, under reaction conditions, of rhodium hydrides, and of alkylrhodium and acylrhodium complexes.

The main carbonyl species detectable under hydroformylation conditions are $Rh_4(CO)_{12}$ and $Rh_6(CO)_{16}$ (18) and not acyl derivatives as observed for cobalt. Direct evidence, however, was found by Chini et al. (16) for the existence of acylrhodium carbonyls isolated as $NH_4Rh_6(CO)_{15}(COR)$ in the stoichiometric hydroformylation of propylene with $Rh_4(CO)_{12}$ and H_2. Indirect evidence was obtained by Markó's group (84), who detected complexes of the type $[Rh(CO)_2(OCOEt)_2]_2$ in ethylene hydroformylation using $Rh_4(CO)_{12}$ and synthesis gas containing small amounts of oxygen.

Since acyl formation implies formally an insertion of an olefin into a rhodium–hydrogen bond and CO insertion into a rhodium–carbon bond, and although the actual mechanism of this acyl formation has not been fully elucidated, it is generally assumed that rhodium-catalyzed hydroformylations occur via a mechanism similar to that proposed by Heck and Breslow for cobalt-catalyzed hydroformylation (Section II-B-6 of this chapter).

Investigation of the hydroformylation rate as a function of CO pressure (18,73), which showed that the maximum rate was temperature dependent, indicates some analogies but also important differences between cobalt and rhodium-catalyzed hydroformylation. These are: (*a*) for the rhodium catalyst, the maximum rate occurs at a much higher CO pressure than with cobalt; (*b*) in propylene hydroformylation, where double-bond isomerization cannot influence the isomeric composition of the products, a large increase of branched-chain aldehydes was found by decreasing the CO pressure in the case of cobalt; but with rhodium, preliminary results (81) indicate that the product composition remains practically unchanged (*n*-butyraldehyde $= 45 \pm 3\%$) for reactions at 110°C with various CO pressures between 7 and 152 atm. Tentatively, these differences can be explained by proposing that for rhodium there exists a catalytic species with a relatively low CO/Rh ratio, the concentration of which diminishes slowly both by increasing or decreasing the CO pressure with respect to the CO pressure corresponding to the maximum rate. This species induces, as with cobalt catalysts, a relatively high reaction rate and a low ratio of linear to branched-chain aldehydes. The species formed at high CO pressure might have very small catalytic activity and hence, even if they favor the formation of the less-branched isomer, they cannot greatly influence the isomeric composition of the reaction products.

As already mentioned, kinetic investigations of rhodium-catalyzed hydroformylation (18,73,93) yielded data that in part are conflicting and so no definite conclusions can be drawn now. All the authors, however, agree that the catalytic activity of rhodium is 10^3 (6,85) to 10^4 (18) times greater than that of cobalt. The difference may be connected not only with the intrinsic chemical properties of the two metals, but also with a different distribution of the metal atoms among complexes having different catalytic activity or being catalytically inactive. In cobalt-catalyzed hydroformylation, the presence of acylcobalt derivatives in the steady state was taken as an indication that the reaction of the acyl derivatives with H_2 is the rate-determining step (96). Taking into account that no detectable amount of acylrhodium carbonyls was found during hydroformylation, the large increase in reaction rate might be connected with a lower activation energy for H_2 activation (probably via oxidative addition) by rhodium than by cobalt. The effect on the reaction rate of the facile activation of CO by rhodium carbonyls, as shown by "semicatalytic" propylene hydroformylation in the presence of $Rh_4(CO)_{12}$ at room temperature (17), has never been thoroughly discussed but might be important, especially at low temperatures.

In conclusion, a mechanism for rhodium-catalyzed hydroformylation similar to that proposed for the cobalt-carbonyl catalyzed reaction seems likely. However, the details of the mechanism as well as the structure of the true catalytic species are not yet well understood.

b. Rhodium–phosphine complexes as catalyst precursors. The mechanism of the stoichiometric and of the catalytic hydroformylation in the presence of rhodium catalysts modified by phosphorous- or nitrogen-containing ligands has been thoroughly investigated since 1966. Thanks

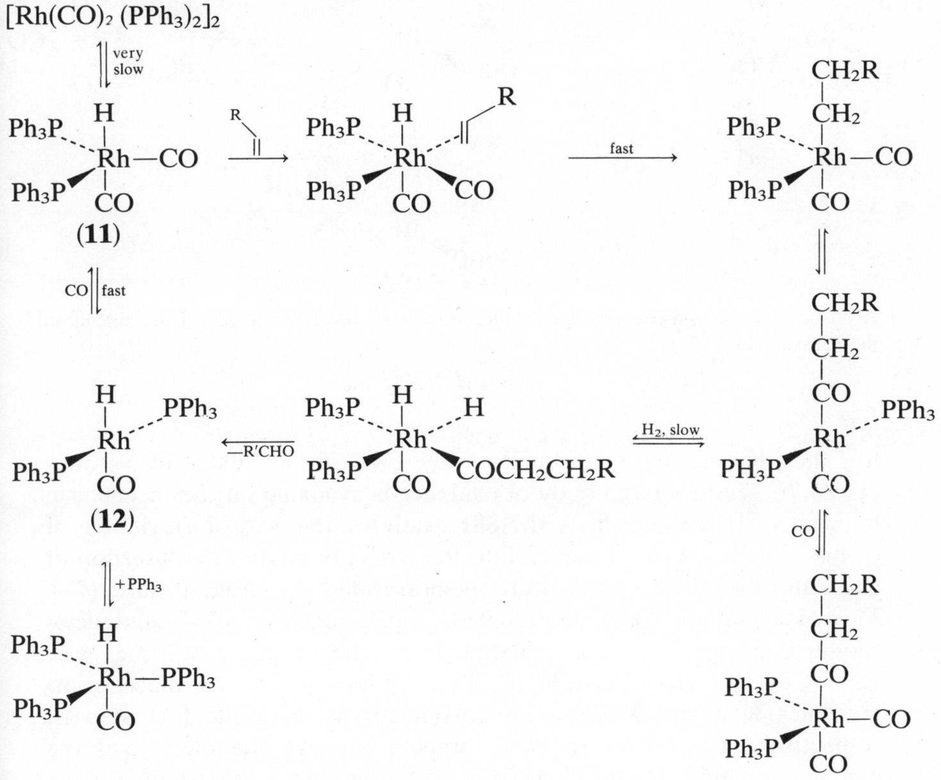

$[Rh(CO)_2(PPh_3)_2]_2$

very slow

$RhH(CO)_2(PPh_3)_2$ (**11**) —(R-CH=CH₂)→ $RhH(CO)_2(PPh_3)_2(CH_2{=}CHR)$ —fast→ $Rh(CH_2CH_2R)(CO)_2(PPh_3)_2$

CO fast

$RhH(CO)(PPh_3)_2$ (**12**) ←(−R′CHO)— $RhH_2(COCH_2CH_2R)(CO)(PPh_3)_2$ ⇌(H₂, slow)— $Rh(COCH_2CH_2R)(CO)(PPh_3)_2$

+PPh₃ (from **12**)

$RhH(CO)(PPh_3)_3$

CO (from the acyl complex)

$Rh(COCH_2CH_2R)(CO)_2(PPh_3)_2$

Suggested mechanism for hydroformylation of alkenes using an associative pathway for attack of the alkene on $RhH(CO)_2(PPh_3)_2$.

Scheme **XXI**

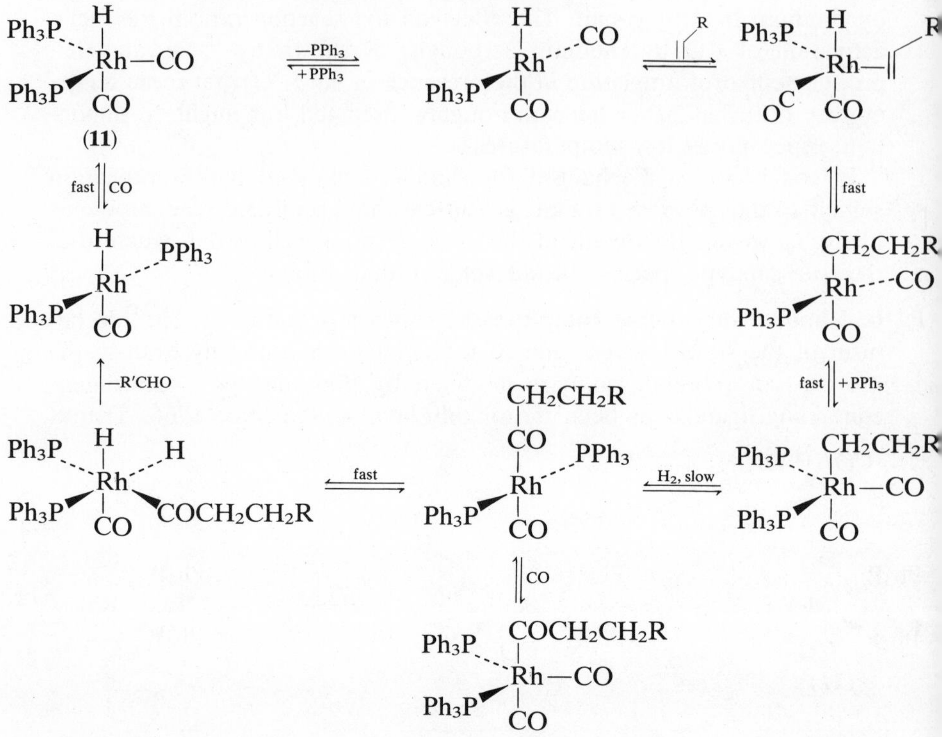

A possible mechanism for hydroformylation of alkenes preceding attack of the alkene on the Rh complex.

Scheme **XXII**

to the important contributions by Wilkinson and his groups (12,24,76,77,86) a large body of evidence is available on the mechanism by which $HRh(CO)(PPh_3)_3$ (87,88) catalyzes the hydroformylation of olefins. The insertion of olefins into the Rh–H bond and the insertion of CO into the Rh–C bond were demonstrated spectroscopically (77). Moreover, square, pentacoordinated, and octahedral alkyl- and acyl-rhodium complexes were isolated from the reaction starting with $CF_2{=}CF_2$ and $HRh(CO)(PPh_3)_3$ (77). On this basis, two mechanisms (Schemes **XXI** and **XXII**) were postulated; in these the first step is, respectively, the formation of a complex between the olefin and the pentacoordinated $HRh(PPh_3)_2(CO)_2$ (**11**) (86) or a dissociation of a molecule of phosphine from **11** followed by the formation of a complex from the olefin and the square planar $HRh(CO)_2PPh_3$ (**12**). The formation of **11** from $HRh(CO)(PPh_3)_3$ should occur (24) according to Scheme

XXIII, the first two steps of which were demonstrated spectroscopically (24,29).

$$HRh(CO)(PPh_3)_3 \rightleftharpoons HRh(CO)(PPh_3)_2 + PPh_3$$

$$2HRh(CO)(PPh_3)_2 \overset{CO}{\rightleftharpoons} [Rh(CO)_2(PPh_3)_2]_2 + H_2$$

$$[Rh(CO)_2(PPh_3)_2]_2 + H_2 \rightleftharpoons \underset{\textbf{(11)}}{2HRh(CO)_2(PPh_3)_2}$$

Scheme **XXIII**

The presence of **11** was also demonstrated (86) in solutions of $HRh(CO)(PPh_3)_3$ treated with CO.

The two steps in Schemes **XXI** and **XXII** that could not be proved experimentally, are complex formation of the olefin with **11** or **12**, and oxidative addition of H_2 to $RCORh(PPh_3)_2(CO)$. However, olefin π-complexes with rhodium are known (89) and an oxidative addition of H_2 to similar rhodium complexes was shown experimentally (90).

The *cis* addition of the formyl group and hydrogen to the double bond implied by these mechanisms was proved recently by hydroformylating the two isomeric 3-methyl-2-pentenes (91).

The rate-determining step in the hydroformylation is believed to be the oxidative addition (92) of H_2 to the complex $RCORh(PPh_3)_2(CO)$, in keeping with the first order with respect to hydrogen pressure found in the rhodium-catalyzed hydroformylation.

The inhibiting effect of CO and excess phosphine observed in hydroformylations in the presence of these complexes can be explained according to the equilibria in reactions 45 and 46 which lead to pentacoordinated species that cannot undergo oxidative additions.

$$RCORh(CO)(PPh_3)_2 + CO \rightleftarrows RCORh(CO)_2(PPh_3)_2 \tag{45}$$

$$RCORh(CO)(PPh_3)_2 + PPh_3 \rightleftarrows RCORh(CO)(PPh_3)_3 \tag{46}$$

Increase of the less-branched isomer in propylene hydroformylation due to increasing the PPh_3 concentration was attributed to both steric and electronic factors (24); however, no detailed model was given to justify this interpretation.

The model proposed to explain the asymmetric hydroformylation of vinyl and internal olefins using the catalytic system, $HRh(CO)(PPh_3)_3$/DIOP (mole Rh metal/mole DIOP = 0.33) suggests an explanation for the origin of the isomeric composition of the hydroformylation products. As for the asymmetric hydroformylation, the critical steps would be formation of a π-complex that is octahedral in the associative mechanism and could be a trigonal bipyramid in the dissociative mechanism. An octahedral π-complex is shown in Scheme **XXIV**.

Considering the four allowed conformations of the olefin around the axis of the olefin–rhodium bond, the one in which the R group is located between hydrogen and CO or between hydrogen and the phosphine group is preferred to the one in which R is located between two CO groups or between a CO group and a phosphine group. Correspondingly, more linear than branched alkyl groups are formed by insertion of the olefin into the Rh–H bond: The predominance of the less-branched isomer in the hydroformylation of vinylidene olefins ($R_1R_2C{=}CH_2$) can be explained on the same basis.

A much deeper knowledge of conformational analysis in penta- and hexacoordinated transition metal π-complexes of olefins would be necessary to judge the validity of these models. Nevertheless, they correctly predict the prevailing chirality and the prevailing isomer in the asymmetric products from the hydroformylation of vinyl and internal olefins. A more complicated situation arises when polar groups capable of interacting with the metal atom are present in the olefinic substrate.

Asymmetric hydroformylation using the same type of rhodium-DIOP catalytic system has yielded other interesting information on the hydroformylation mechanism. The asymmetric hydroformylation of *cis*-2-butene having a C_{2v} symmetry is in keeping with the structure of the catalytic complexes formulated by Wilkinson's group (12) and indicates that diastereomeric complexes can exist in which the metal atom is one of the asymmetric centers. Furthermore, the different prevailing chirality of 2-methylbutanal obtained from 1-butene and from *cis*-2-butene, arising in both cases from the *sec*-butylrhodium complex, indicates that asymmetric induction occurs at least in part before alkyl formation. Alkyl formation, therefore, even if reversible, does not reach equilibrium during hydroformylation.

The most likely step in which asymmetric induction occurs is in the formation or equilibration between the two diasteromeric π-complexes differing in the chirality of the asymmetric carbon atom bound to the rhodium atom (Scheme **XXV**). These complexes can be postulated on the basis of proposed hydroformylation mechanisms (24) and their behavior, as it concerns asymmetric induction, can be foreseen on the basis of the investigation of suitable, square planar Pt(II) olefin complexes (44). They, however, have never been detected under hydroformylation conditions.

The large amount of data accumulated on hydroformylation with $HRh(CO)(PPh_3)_3$ has contributed greatly to understanding the general lines of the reaction mechanism. But it has also shown the numerous complications arising from the existence of many complexes influencing

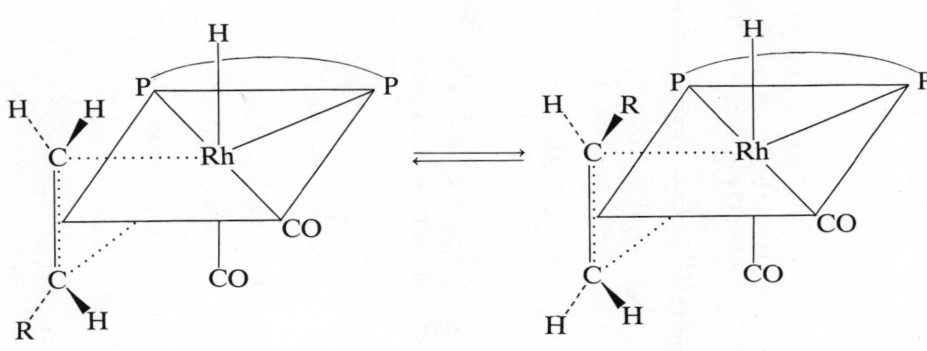

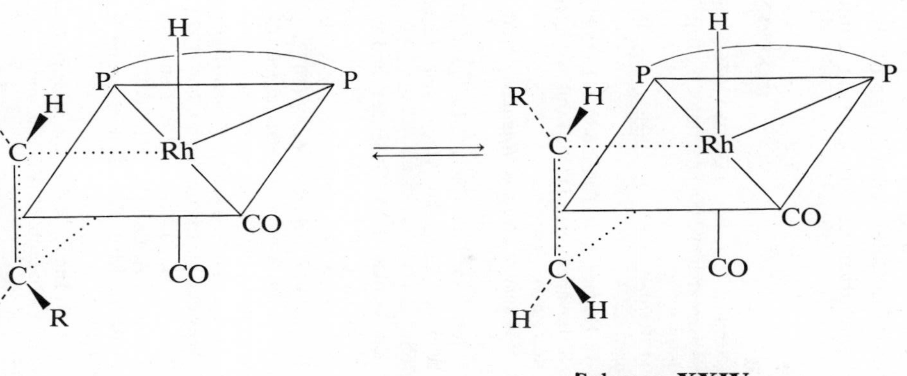

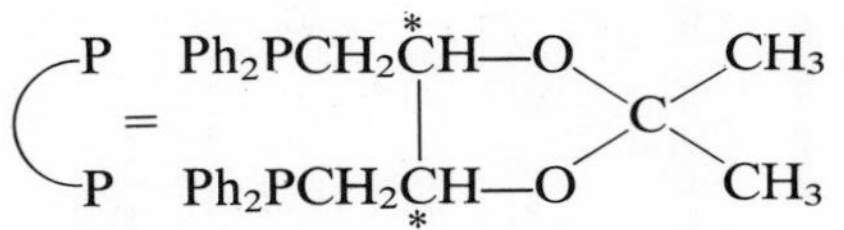

Scheme **XXIV**

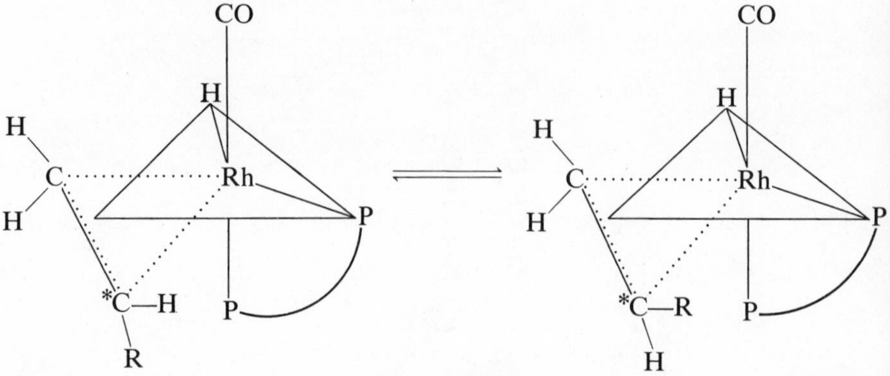

Scheme **XXV**

the reaction in different ways. These complications hinder the acquisition of a more detailed knowledge of the reaction path and make a quantitative explanation of the reaction kinetics difficult.

REFERENCES

1. G. Schiller, Ger. Pat. 953,605 (1956); *Chem. Abstr.*, **53,** 11226 (1959).
2. Esso Research and Engineering Co., Brit. Pat. 801,734 (1958); *Chem. Abstr.*, **53,** 7014 (1959).
3. J. H. Bartlett and V. L. Hughes, U.S. Pat. 2,894,038 (1959); *Chem. Abstr.*, **54,** 2216 (1960).
4. V. L. Hughes, U.S. Pat. 2,880,241 (1959); *Chem. Abstr.*, **53,** 14938 (1959).
5. N. S. Imyanitov and D. M. Rudkovskii, *Neftekhim.*, **3,** 198 (1963); *Chem. Abstr.*, **59,** 7396 (1963).
6. H. Wakamatsu, *Nippon Kagaku Zasshi*, **85,** 227 (1964); *Chem. Abstr.*, **61,** 13173 (1964).
7. M. Morikawa, *Bull. Chem. Soc. Japan*, **37,** 379 (1964).
8. B. Heil and L. Markó, *Magy. Kem. Lapja*, **23,** 669 (1968); *Chem. Abstr.*, **70,** 56861 (1968).

9*a*. V. Yu. Gankin, L. S. Genender, and D. M. Rudkovskii, in D. M. Rudkovskii, Ed., *Carbonylation of Unsaturated Hydrocarbons* [in Russian], Izd. Khimiya, Leningrad, 1968, p. 57; *Chem. Abstr.*, **71,** 2653 (1969).

9*b*. F. E. Paulik, *Catal. Rev.*, **6,** 49 (1972).

9*c*. L. Markó in *Aspects of Homogeneous Catalysis*, R. Ugo ed., D. Reidel Publ. Comp., Vol. 2, p. 3 (1974).

10*a*. W. Hieber and H. Lagally, *Z. anorg. allgem. Chem.*, **251,** 96 (1943).

10*b*. R. Whyman, *J. Chem. Soc. Dalton*, **1972**, 1375.

11. F. Piacenti, M. Bianchi, and E. Benedetti, *Chim. Ind.* (*Milan*), **49,** 245 (1967).
12. D. Evans, J. A. Osborn, and G. Wilkinson, *J. Chem. Soc. A*, **1968,** 3133.
13. B. L. Booth, M. J. Else, R. Fields, and R. N. Haszeldine, *J. Organometal. Chem.*, **27,** 119 (1971).
14. P. Pino, R. Ercoli, and F. Calderazzo, *Chim. Ind.* (*Milan*), **37,** 782 (1955).
15. P. Pino, F. Piacenti, M. Bianchi, and R. Lazzaroni, *Chim. Ind.* (*Milan*), **50,** 106 (1968).

16*a*. P. Chini, S. Martinengo, and G. Garlaschelli, *Proc., Symp. Chemistry of Hydroformylation and Related Reactions*, Veszprem, 1972, p. 68; *Chem. Abstr.*, **77,** 125858 (1972).
16*b*. G. Csontos, B. Heil, L. Markó, and P. Chini, *Hung. J. Ind. Chem.*, **1,** 53 (1973); *Chem. Abstr.*, **79,** 41608 (1973).
17. G. Csontos, B. Heil, L. Markó, and P. Chini, *Proc., Symp. Chemistry of Hydroformylation and Related Reactions*, Veszprem, 1972, p. 75; *Chem. Abstr.*, **77,** 113344 (1972).
18. B. Heil and L. Markó, *Chem. Ber.*, **101,** 2209 (1968).
19. C. K. Brown and G. Wilkinson, *Tetrahedron Lett.*, **1969,** 1725.
20. W. Rupilius, Dissertation, Aachen, Germany, 1969.
21. J. A. Osborn, G. Wilkinson, and J. F. Young, *Chem. Commun.*, **1965,** 17.
22. M. Yamaguchi, T. Onada, C. Nakajima, M. Kitatama, and Y. Yo, Ger. Offen. 1,920,960 (1969); through *Chem. Abstr.*, **72,** 31230 (1970).
23. M. J. Lawrenson and G. Foster, Ger. Offen. 1,812,504 (1969); through *Chem. Abstr.*, **71,** 101313 (1969).
24. C. K. Brown and G. Wilkinson, *J. Chem. Soc. A*, **1970,** 2753.
25. R. L. Pruett and J. A. Smith, *J. Org. Chem.*, **34,** 327 (1969).
26. M. Takesada and H. Wakamatsu, *Bull. Chem. Soc. Japan*, **43,** 2192 (1970).
27. G. Foster and M. J. Lawrenson, Ger. Offen. 1,901,145 (1969); *Chem. Abstr.*, **71,** 123572 (1969).
28*a*. M. Čapka, P. Svoboda, M. Černy, and J. Hetflejš, *Tetrahedron Lett.*, **1971,** 4787.
28*b*. C. U. Pittman, Jr. and G. O. Evans, *Chem. Technol.*, **1973,** 560.
28*c*. G. O. Evans, C. U. Pittmar , Jr., R. McMillan. R. T. Beach, and R. Jones, *J. Organometal. Chem.*, **67,** 295 (1974)
28*d*. J. C. Bailar, Jr., *Catal. Rev.*, **10,** 17 (1974).
28*e*. Z. M. Michalska and D. E. Webster, *Chem. Technol.*, **1975,** 117.
28*f*. H. Arai, T. Kaneko, and T. Kunugi, *Chem. Lett.*, **1975,** 265.
28*g*. E. Bayer and V. Schurig, *Angew. Chem.*, **87,** 484 (1975).
29. D. E. Morris and H. B. Tinker, *Chem. Technol.*, **1972,** 554.
30. K. L. Olivier and F. B. Booth, *Hydrocarbon Processing*, **49** (4), 112 (1970).
31. K. K. Robinson, F. E. Paulik, A. Hershman, and J. F. Roth, *J. Catal.*, **15,** 245 (1969).
32. I. Ogata and Y. Ikeda, *Chem. Lett.*, **1972,** 487.
33. C. Salomon, G. Consiglio, C. Botteghi, and P. Pino, *Chimia*, **27,** 215 (1973).
34. R. Lai, Thesis, University of Marseilles, 1974.
35*a*. R. Lai and E. Ucciani, *C.R. Acad. Sci., Paris, Ser. C*, **276,** 425 (1973); *Chem. Abstr.*, **78,** 147494 (1973).
35*b*. R. Lai and E. Ucciani, *Proc., Symp. Chemistry of Hydroformylation and Related Reactions*, Veszprem, 1972, p. 41; *Chem. Abstr.* **77,** 113525 (1972).
36. R. Lai and E. Ucciani, *C.R. Acad. Sci. Paris, Ser. C*, **273,** 1368 (1971); *Chem. Abstr.*, **76,** 33484 (1972).
37. I. Ogata, Y. Ikeda, and T. Asakawa, *Kogyo Kagaku Zasshi*, **74,** 1839 (1971); *Chem. Abstr.*, **75,** 140422 (1971).
38. B. Fell and W. Rupilius, *Tetrahedron Lett.*, **1969,** 2721.
39. R. L. Pruett and K. O. Groves, U.S. Pat. 3,499,933 (1970); *Chem. Abstr.*, **73,** 14291 (1970).
40. J. Falbe and N. Huppes, *Brennstoff-Chem.*, **47,** 314 (1966).
41. J. Falbe and N. Huppes, *Brennstoff-Chem.*, **48,** 182 (1967).
42. T. Alderson and R. V. Lindsey, Jr., U.S. Pat. 3,081,357 (1963); *Chem. Abstr.*, **59,** 8594 (1963).

43. J. Falbe, Ger. Pat. 1,618,384 (1971); *Chem. Abstr.*, **75,** 129416 (1971).
44. R. Lazzaroni, P. Salvadori, and P. Pino, *J. Organometal. Chem.*, **43,** 233 (1972).
45*a*. C. Botteghi, G. Consiglio, G. Ceccarelli, and A. Stefani, *J. Org. Chem.*, **37,** 1835 (1972).
45*b*. C. Botteghi, G. Ceccarelli, and G. Consiglio, *J. prakt. Chem.*, **314,** 840 (1972).
46. J. Falbe, N. Huppes, and F. Korte, *Brennstoff-Chem.*, **47,** 207 (1966).
47. Ajinomoto Co., Inc., French Pat. 1,361,797 (1964); *Chem. Abstr.*, **61,** 11894 (1964).
48. W. Himmele and W. Aquila, Ger. Offen. 1,945,479 (1971); *Chem. Abstr.*, **74,** 111589 (1971).
49. W. Himmele, F. J. Müller, and W. Aquila, Ger. Offen. 2,039,078 (1972); *Chem. Abstr.*, **76,** 112700 (1972).
50. I. Maeda and R. Yoshida, *Bull. Chem. Soc. Japan*, **41,** 2969 (1968).
51. C. Botteghi and L. Lardicci, *Chim. Ind.* (*Milan*), **52,** 265 (1970).
52. J. Falbe and N. Huppes, *Brennstoff-Chem.*, **48,** 46 (1967).
53. E. N. Frankel, *J. Am. Oil Chem. Soc.*, **48,** 248 (1971).
54. B. Fell, W. Rupilius, and F. Asinger, *Tetrahedron Lett.*, **1968,** 3261.
55. Y. Takegami, Y. Watanabe, and H. Masada, *Bull. Chem. Soc. Japan*, **40,** 1459 (1967).
56*a*. E. N. Frankel and F. L. Thomas, *J. Am. Oil Chem. Soc.*, **49,** 10 (1972).
56*b*. E. H. Pryde, E. N. Frankel, and J. C. Cowan, *ibid.*, **49,** 451 (1972).
57. Deutsche Gold-und Silber-Scheideanstalt vorm. Roessler, Neth. Appl. 6,516,193 (1966); *Chem. Abstr.*, **66,** 2215 (1967).
58*a*. V. Yu. Gankin, L. S. Genender, and D. M. Rudkovskii, in D. M. Rudkovskii, Ed., *Carbonylation of Unsaturated Hydrocarbons* [in Russian], Izd. Khimiya, Leningrad, 1968, p. 61; *Chem. Abstr.*, **71,** 2913 (1969).
58*b*. Idem, *Zh. Prikl. Khim.*, **40,** 2029 (1967); *Chem. Abstr.*, **68,** 77592 (1968).
59. F. Falbe and N. Huppes, U.S. Pat. 3,318,913 (1967); *Chem. Abstr.*, **67,** 53715 (1967).
60. J. P. Friedrich, G. R. List, and V. E. Sohns, *J. Am. Oil Chem. Soc.*, **50,** 455 (1973).
61. D. G. Kuper, U.S. Pat. 3,520,914 (1970); through *Chem. Abstr.*, **73,** 66071 (1970).
62. Imperial Chemical Industries Ltd., French Pat. 1,509,547 (1968); *Chem. Abstr.*, **70,** 77337 (1969).
63. S. Sato, M. Takesada, and H. Wakamatsu, *Nippon Kagaku Zasshi*, **90,** 579 (1969); *Chem. Abstr.*, **71,** 49178 (1969).
64. M. Tanaka, Y. Watanabe, T. Mitsudo, K. Yamamoto, and Y. Takegami, *Chem. Lett.*, **1972,** 483.
65. W. Himmele, H. Siegel, W. Aquila, and F. J. Müller, Ger. Offen. 2,132,414 (1973); *Chem. Abstr.*, **78,** 97328 (1973).
66. G. Consiglio, C. Botteghi, C. Salomon, and P. Pino, *Angew. Chem.*, **85,** 663 (1973).
67*a*. P. Pino, G. Consiglio, C. Botteghi, and C. Salomon, *Advances in Chemistry Series*, N. 132, Am. Chem. Soc., Washington, D.C., 1974, p. 295.
67*b*. M. Tanaka, Y. Watanabe, T. Mitsudo, and Y. Takegami, *Bull. Chem. Soc. Japan*, **47,** 1698 (1974).
67*c*. R. Stern, A. Hirschauer, and L. Sajus, *Tetrahedron Lett.*, **1973,** 3247.
67*d*. M. Tanaka and I. Ogata, *Chem. Commun.*, **1975,** 735.
67*e*. M. Tanaka, Y. Ikeda, and I. Ogata, *Chem. Lett.*, **1975,** 1115.
68. B. Heil and L. Markó, *Chem. Ber.*, **99,** 1086 (1966).
69. B. Heil and L. Markó, *Acta Chim. Acad. Sci. Hung.*, **55,** 107 (1968); *Chem. Abstr.*, **68,** 77430 (1968).
70. J. H. Craddock, A. Hershman, F. E. Paulik, and J. F. Roth, *Ind. Eng. Chem., Prod. Res. Develop.*, **8,** 291 (1969).

71*a*. B. Fell and E. Müller, *Proc., Symp. Chemistry of Hydroformylation and Related Reactions*, Veszprem, 1972, p. 105; *Chem. Abstr.*, **77,** 113524 (1972).
71*b*. Idem, *Monatsh. Chem.*, **103,** 1222 (1972); *Chem. Abstr.*, **78,** 97792 (1973).
72. K. L. Olivier and F. B. Booth, *Am. Chem. Soc., Div. Petrol. Chem., Preprints*, Vol. XIV, 1969, p. A-7; *Chem. Abstr.*, **75,** 5160 (1971).
73. M. Yamaguchi, *Kogyo Kagaku Zasshi*, **72,** 671 (1969); *Chem. Abstr.*, **71,** 38263 (1969).
74. F. Piacenti, P. Pino, R. Lazzaroni, and M. Bianchi, *J. Chem. Soc. C*, **1966,** 488.
75. V. Yu. Gankin, L. S. Genender, L. N. Dzeniskevich, D. M. Rudkovskii, V. A. Rybakov, and M. A. Khalyutina, in N. S. Imyanitov, Ed., *Gidroformilirovanie* [Hydroformylation], Izd. Khimiya, Leningrad, 1972, p. 66; *Chem. Abstr.*, **77,** 151202 (1972).
76. M. Yagupsky and G. Wilkinson, *J. Chem. Soc. A*, **1970**, 941.
77. G. Yagupsky, C. K. Brown, and G. Wilkinson, *J. Chem. Soc. A*, **1970,** 1392.
78. F. E. Paulik, J. F. Roth, and K. K. Robinson, French Pat. 1,560,961 (1969); through *Chem. Abstr.*, **72,** 54762 (1970).
79. F. B. Booth, U.S. Pat. 3,511,880 (1970); through *Chem. Abstr.*, **73,** 14191 (1970).
80. M. J. Lawrenson, Brit. Pat. 1,254,222 (1971); *Chem. Abstr.*, **76,** 33787 (1972).
81. C. Botteghi and P. Pino, unpublished results.
82. A. Hershman, K. K. Robinson, J. H. Craddock, and J. F. Roth, *Ind. Eng. Chem., Prod. Res. Develop.*, **8,** 372 (1969).
83. B. Heil and L. Markó, *Chem. Ber.*, **102,** 2238 (1969).
84. B. Heil, L. Markó, and G. Bor, *Chem. Ber.*, **104,** 3418 (1971).
85. N. S. Imyanitov and D. M. Rudkovskii, *Kinet. Katal.* **8,** 1240 (1967); *Chem. Abstr.*, **68,** 86867 (1968).
86. D. Evans, G. Yagupsky, and G. Wilkinson, J. Chem. Soc. A, **1968,** 2660.
87. S. J. LaPlaca and J. A. Ibers, *J. Am. Chem. Soc.*, **85,** 3501 (1963).
88. S. S. Bath and L. Vaska, *J. Am. Chem. Soc.*, **85,** 3500 (1963).
89*a*. V. Schurig and E. Gilav, *Chem. Commun.*, **1971,** 650.
89*b*. R. Cramer, *J. Am. Chem. Soc.*, **88,** 2272 (1966).
90. C. O'Connor and G. Wilkinson, *J. Chem. Soc. A*, **1968,** 2665.
91. A. Stefani, G. Consiglio, C. Botteghi, and P. Pino, *J. Am. Chem. Soc.*, **95,** 6504 (1973).
92. G. Wilkinson, *Bull. Soc. Chim. France*, **1968,** 5055.
93. G. Csontos, B. Heil, and L. Markó, *Ann. N.Y. Acad. Sci.*, **239,** 47 (1974).
94. B. Cornils, R. Payer, and K. C. Traencker, *Hydrocarbon Processing*, **54** (6), 83 (1975).
95. D. A. v. Bézard, Thesis, E. T. H. Zürich, 1976.
96. R. Whyman, *J. Organometal. Chem.*, **81,** 97 (1974).
97. S. Toros, *Magy. Kem. Lapja*, **29,** 543 (1974); *Chem. Abstr.*, **83,** 177870 (1975).

IV. HYDROFORMYLATION CATALYZED BY METAL COMPLEXES NOT CONTAINING COBALT OR RHODIUM

A. Introduction

A large number of patents exist on catalytic activity in the hydroformylation reaction of complexes of metals other than cobalt and rhodium, but the work has been sporadic rather than systematic (1). Very little is known about the catalytic activity of metals of groups I–V; also copper has received some attention (82, 83). An interesting comparison of the

TABLE 60
Hydroformylation of Cyclohexene with Different Metal Carbonyls[a] (2)

Metal	Metal carbonyl (moles metal/l)	Temp. (°C)	Olefin conversion to aldehyde (%)	Olefin conversion total (%)
Cr	0.08	250	~2	~2
Mo	0.08	250	≤1	~0
W	0.08	250	<1	~0
Mn	0.08	225	40[b]	74
Co	0.003	150	80	82
Rh	—[c]	150	97	~97
Ir	0.003	150	2	10[d]
Fe	0.08	250	7	~7
Ni	0.08	250	<2	5

[a] P_{CO} 150 atm and P_{H_2} 100 atm (at 20°C); dioxane solvent; 1 mole olefin/1; reaction time 3 hr.

[b] Aldehyde plus alcohol; over 20% hydrogenation to cyclohexane.

[c] Traces.

[d] Main reaction is hydrogenation of cyclohexene to cyclohexane.

reactivities of different metal compounds was published by Imyanitov and Rudkovskii (2), who examined the hydroformylation of cyclohexene in the presence of carbonyls of metals of groups VI, VII, and VIII (Table 60); see also the work of Hieber and co-workers (3).

The carbonyls of group VI metals show little or no activity. As for group VII metal compounds, manganese shows some activity at high temperature (4–6,8*b*), but there is apparently none with some rhenium compounds (7). The question of the catalytic activity of rhenium, however, is not settled, since other authors (8*a*) reported that 1-pentene, in the presence of a complex obtained by heating NH_4ReO_4 with PBu_3, is transformed into C_6 alcohols at 250°C and 100 atm of CO and H_2; similar results were found for 1-octene using $Re_2(CO)_{10}$—PR_3 catalytic systems (8*b*).

All group VIII metal carbonyls show at least some catalytic activity. Research in this field is still in progress. Since the reactivity of the different metals can be greatly influenced by the presence of different ligands, a fair comparison of the catalytic activity of the different metals is not possible, but a tentative comparison is shown in Table 61.

In the first and in the second vertical triad it seems that highest catalytic activity is shown by metal complexes of the second large period (Ru and Rh), followed by elements of the first large period (Co and Fe), and

TABLE 61
Hydroformylation of Propylene with Different Catalysts[a]

Catalyst	Concn.[b]	Temp (°C)	Pressure[c]	Time (hr)	Yield (%)	*n*-Butyraldehyde (%)	Ref.
$Co_2(CO)_8$	17	110	220	2	83	80	9
$[Rh(CO)_3]_n$	0.05	110	220	~1	83	53	9
$Rh_6(CO)_{16}$	0.18	70	120	3	51	51	10
$IrCl(CO)(PPh_3)_2$	21.2	130	30[d]	3.7	61	63	11
$[C_5H_5Fe(CO)_2]_2$	233	150	100	40	70	32	14
$Ru_3(CO)_{12}$	23.4	110	150[d]	7	40	74	12
$Os_3(CO)_{12}$	18.5	180	160[d]	2	74	51	12
Pd/C	—	100	100	3	38	75	13*b*
$PtCl_2(PPh_3)_2/SnCl_2$[e]	4.4	66	89	6	90	87	76

[a] $CO/H_2 = 1$, benzene solvent.
[b] Mmoles metal/l solvent.
[c] Initial pressure at room temp.
[d] At constant pressure.
[e] Pt/Sn = 1:5; methyl isobutyl ketone as solvent.

finally by elements of the last large period (Ir and Os). Furthermore, the elements of the second vertical triad show greater catalytic activity than that of the other triads.

Little is known about the catalytic activity of Ni, Pd, and Pt in hydroformylation. Poor results were reported up to now when palladium catalysts were used (16,68,69,79). The best results up to now were obtained by using platinum complexes modified with phosphines (8*a*,15) or phosphites and $SnCl_2$ (66,80). Also palladium compounds seem more active than originally estimated (13,16), particularly in the presence of phosphines and diphosphines (18).

Since hydroformylations in the presence of nickel, osmium (8*a*,12), and palladium complexes have been little studied up to now, they are not discussed in this chapter. But hydrocarbonylations with iron, ruthenium, iridium, and platinum compounds are considered separately.

B. Hydrocarbonylations with Iron Catalysts

1. Hydroformylation

Apart from numerous patents (19–22), relatively little has been published on the catalytic activity of iron compounds in the hydroformylation reaction. Haszeldine's group (23) using $Fe(CO)_5$ as catalyst precursor at 120°C in toluene under a pressure of 130 atm of CO and H_2, obtained only a 29% hydroformylation of ethylene; part of the aldehyde was hydrogenated to *n*-propanol. According to these authors, the catalytic activity of $Fe(CO)_5$ is greatly enhanced by traces of cobalt. Imyanitov and Rudkovskii (2) hydroformylated cyclohexene in the presence of an unspecified iron carbonyl at 250°C, 115 atm of CO, and 100 atm of H_2. After 3 hr, 10% of the cyclohexene disappeared; of this, 20% had been hydroformylated and 80% hydrogenated to cyclohexane. It was shown earlier, however, that iron pentacarbonyl is 10^{-6} times less active than cobalt carbonyl (24,25). The low activity of iron catalysts was attributed to the low acidity of the $H_2Fe(CO)_4$ that supposedly takes part in the catalytic process (2,26–28). It is interesting to note that acetic acid is necessary for the decomposition of the acyliron derivatives in the stoichiometric synthesis of aldehydes from alkyl halides (29); for this transformation the steps shown in Scheme **XXVI** were suggested (30)

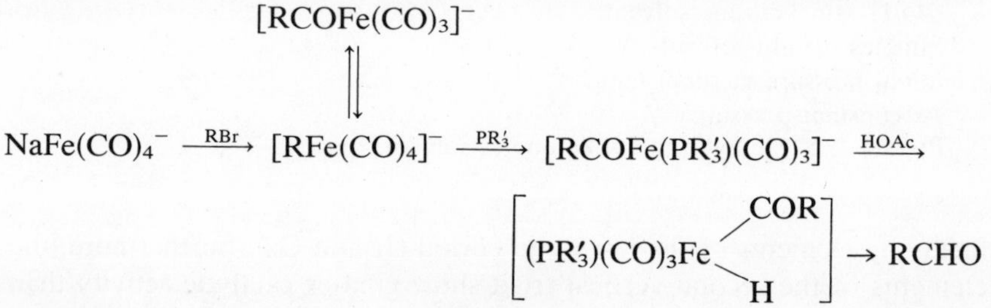

$$
\begin{array}{c}
[RCOFe(CO)_3]^- \\
\updownarrow \\
NaFe(CO)_4^- \xrightarrow{RBr} [RFe(CO)_4]^- \xrightarrow{PR'_3} [RCOFe(PR'_3)(CO)_3]^- \xrightarrow{HOAc} \\
\left[(PR'_3)(CO)_3Fe \begin{array}{l} \diagup COR \\ \diagdown H \end{array} \right] \rightarrow RCHO
\end{array}
$$

Scheme **XXVI**

According to Wilkinson's group (31) the addition of PPh_3 to $Fe(CO)_5$ when operating at 100°C and 100 atm total pressure ($CO/H_2 = 1$) apparently increases the hydroformylation rate of 1-pentene without marked effect on the ratio of straight- to branched-chain isomers (~2.7) (Table 62).

TABLE 62
Hydroformylation of 1-Pentene with Iron Complexes[a] (31)

Catalyst	Catalyst/alkene (molar ratio)	Conversion (%)
$Fe(CO)_5$	0.11	4
$Fe(CO)_5 + PPh_3$	0.11	31
$Fe(CO)_4(PPh_3)$	0.0044	30
$Fe(CO)_3(PPh_3)_2$	0.0028	37

[a] Temp 110°C; 100 atm $CO + H_2$ (1:1) at room temp; benzene solvent; reaction time 16 hr.

In the hope that binuclear carbonyl complexes would be better activators than mononuclear carbonyls for olefins, CO and H_2, Tsuji and Mori (14) tried $[C_5H_5Fe(CO)_2]_2$ as catalyst precursor. The hydroformylation of propylene and of butenes in the presence of this catalyst was carried out (in a mixture of benzene and toluene) under a total pressure of 100–150 atm of CO and H_2 ($CO/H_2 = 1$); the best yield was obtained at 150°C (Table 63). The 2-methyl substituted aldehyde prevailed over the straight-chain aldehyde.

TABLE 63
Hydroformylation of Olefins in the Presence of $[C_5H_5Fe(CO)_2]_2$[a] (14)

Olefin	Conversion in 40 hr (%)	More-branched aldehyde (%)
Propylene	70	68
1-Butene	54	67
cis-2-Butene	25	72
trans-2-Butene	17	75

[a] Temp 150°C; 100 atm of $CO + H_2$ (1:1) at room temp; benzene solvent.

2. *Reppe Synthesis of Alcohols*

If in a hydroformylation with an iron catalyst, hydrogen gas is replaced by an alkaline aqueous medium, a rapid hydroformylation under mild conditions occurs followed by aldehyde hydrogenation to the corresponding alcohol along with the formation of CO_2 (Scheme **XXVII**).

$$CH_2{=}CH_2 + 2CO + H_2O \xrightarrow[NR_3]{Fe(CO)_5} CH_3CH_2CHO + CO_2$$

$$CH_3CH_2CHO + CO + H_2O \xrightarrow[NR_3]{Fe(CO)_5} CH_3CH_2CH_2OH + CO_2$$

Scheme **XXVII**

This reaction was discovered by Reppe and co-workers (32) in 1943. They first reacted $Fe(CO)_5$ and ethylene stoichiometrically at 95–100°C in the presence of NaOH and obtained a mixture of the salts of propionic and formic acids and of propanol together with other organic compounds insoluble in water. By using methanol as solvent and triethylamine as base, they reacted ethylene, propylene, butenes, and octenes, respectively, at 100–175°C under 5–200 atm of CO and obtained alcohols with one more carbon atom in rather good yields. From propylene and butenes (33), a mixture of the two isomeric alcohols was obtained analogous to the isomeric aldehydes formed by hydroformylation of the same olefins. Piperylene and 1,5-hexadiene under similar conditions yielded, respectively, saturated hexanols and heptanols (33).

These syntheses can be carried out in the absence of CO by reacting $Fe(CO)_5$ with olefins at 110°C in the presence of *N*-alkylpyrrolidine (amine/$Fe(CO)_5 = 1.5$) and of water and methanol. From propylene was obtained an 84% yield (based on $Fe(CO)_5$) of *n*-butanol and isobutanol (*n*-butanol/isobutanol = 3.3) (34).

The main difficulty in making the above process catalytic was to find a reagent that would fulfill both the following requirements: Give labile salts with CO_2 and be basic enough to transform $Fe(CO)_5$ into iron carbonyl anions like $Fe(CO)_4^{2-}$ or $[HFe_3(CO)_{11}]^-$. These anions are believed to be the catalytically active species.

Many nitrogen-containing compounds were tried, such as triethylamine and tributylamine (34), potassium salts of *N,N*-dimethylaminoacetic acid or *N,N*-dimethylaminopropionic acid and other tertiary amines (33,35,36). Good results, however, were obtained with *N*-alkylpiperidines; after optimization of reaction conditions, ethylene and propylene could be catalytically transformed at 90–110°C and 10–15 atm total pressure into propyl, butyl, and isobutyl alcohols respectively in about 90% yield based on the olefin (37). When the reaction was carried out in the presence of H_2, an increase in rate was observed and the induction period was shortened (38). The most favorable results were obtained for H_2 pressures below 10 atm (34). As in many other carbonylation reactions, an increase in rate was observed when decreasing the CO pressure from 27 to 3 atm (34).

The *n*-butanol/isobutanol ratio from the reaction of propylene (37) in the presence of $Fe(CO)_5$ and base is higher than that for the corresponding aldehydes from the conventional oxo process with cobalt catalysts. No systematic investigation of the influence of reaction variables on the isomeric composition of the products has yet been reported. On conducting the reaction of propylene in the presence of H_2, a decrease in the percentage of *n*-butanol was observed, whereas CO seemed to have little effect; however, a maximum in the reaction rate was found at a CO pressure of 20 atm when operating at 120°C (70) and a favorable effect of the presence of hydrogen was claimed (72). The isomeric composition of the products seems to depend also on the type of amine used (34). An increase in straight-chain over branched alcohols is claimed to be achieved by addition of ethers and thioethers to the reaction mixture

(39); best results were obtained when using DMF as solvent (71). Among the secondary reactions, hydrogenation of the substrate was observed, but hydrogenation can be less than 5% with propylene (37).

The first attempt to interpret the mechanism of the Reppe alcohol synthesis was made by Kröper (40) who formulated the reaction as shown in Scheme **XXVIII**.

$$H_2Fe(CO)_4 + CH_2{=}CH_2 \rightarrow \begin{cases} [CH_3CH_2]^+[HFe(CO)_4]^- \\ [CH_2CH_2CHO]^+[HFe(CO)_3]^- \end{cases}$$

$$\xrightarrow[2H_2O]{3CO} CH_3CH_2CH_2OH + H_2Fe(CO)_4 + 2CO_2$$

Scheme **XXVIII**

According to Kröper, the $H_2Fe(CO)_4$ formed by reaction of $Fe(CO)_5$ with water in the presence of bases is the actual catalyst and no free aldehyde should be formed.

Wender's group (41) investigated the properties of the alkaline solution of $H_2Fe(CO)_4$ and concluded that $HFe(CO)_4^-$, which is present in large concentration if there is no large excess of alkali, slowly dimerizes at room temperature to form a hydrogen-containing anion that loses H_2 and gives rise to $[Fe_2(CO)_8]^{2-}$, which was independently prepared by Hieber and Brendel (42) (Scheme **XXIX**). It was indicated that the hydrogen-containing dimeric anion could be responsible for carbonylation as well as

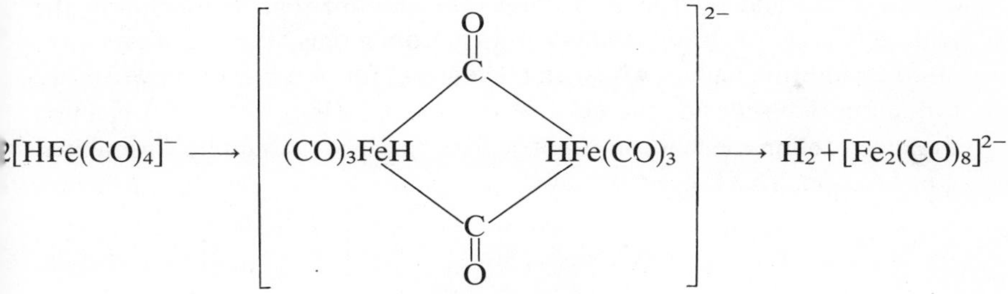

$$2[HFe(CO)_4]^- \longrightarrow \left[(CO)_3FeH(\mu\text{-}C{=}O)_2HFe(CO)_3 \right]^{2-} \longrightarrow H_2 + [Fe_2(CO)_8]^{2-}$$

Scheme **XXIX**

related reactions such as double-bond isomerization and aldehyde hydrogenation (41).

According to these same authors (41) the synthesis of alcohols occurs in two steps. In the first step, the olefin reacts with $[H_2Fe_2(CO)_8]^{2-}$ giving rise to a complex which decomposes to form the aldehyde and $[Fe_2(CO)_6]^{2-}$. In the second step, the aldehyde is hydrogenated to the corresponding alcohol (Scheme **XXX**). To confirm this scheme, cyclopentanecarboxaldehyde was synthesized by reacting $NaHFe(CO)_4$ with excess cyclopentene (41). Also, benzaldehyde was easily hydrogenated by $NaHFe(CO)_4$ (41).

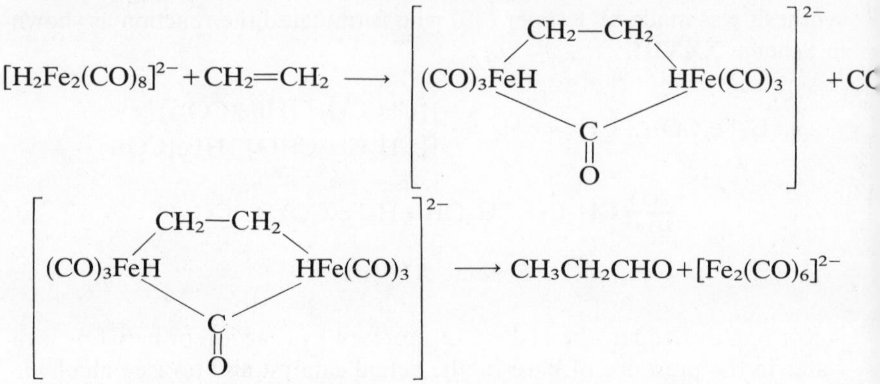

$$[H_2Fe_2(CO)_8]^{2-} + CH_2{=}CH_2 \longrightarrow \left[\begin{array}{c} CH_2{-}CH_2 \\ (CO)_3FeH \qquad HFe(CO)_3 \\ C{=}O \end{array}\right]^{2-} + CO$$

$$\left[\begin{array}{c} CH_2{-}CH_2 \\ (CO)_3FeH \qquad HFe(CO)_3 \\ C{=}O \end{array}\right]^{2-} \longrightarrow CH_3CH_2CHO + [Fe_2(CO)_6]^{2-}$$

$$[Fe_2(CO)_6]^{2-} + H_2O + 3CO \rightarrow CO_2 + [H_2Fe_2(CO)_8]^{2-}$$

$$[H_2Fe_2(CO)_8]^{2-} + CH_3CH_2CHO \rightarrow CH_3CH_2CH_2OH + [Fe_2(CO)_8]^{2-}$$

Scheme **XXX**

On the basis of the observation that complexes of the type $(NR_3H)^+(HFe_3(CO)_{11})^-$ (43) react stoichiometrically with olefins to give alcohols, Kutepow and Kindler (37) concluded that the $[HFe_3(CO)_{11}]^-$ should be the catalytically active complex since it might supply both activated CO and hydrogen. Taking into account that in this anion the hydrogen forms a bridge between two iron atoms, these authors proposed an intermediate in which an alkyl group forms a bridge between two iron atoms. Considering the most recent views on the $[HFe_3(CO)_{11}]$ anion (44) the reaction mechanism proposed by Kutepow should be formulated as in Scheme **XXXI.**

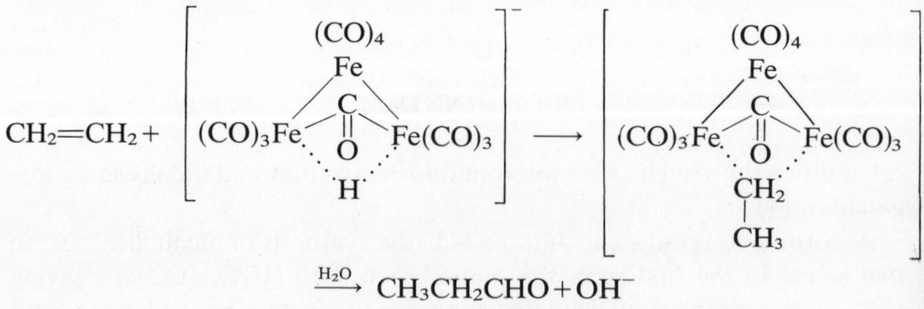

$$CH_2{=}CH_2 + \left[\begin{array}{c} (CO)_4 \\ Fe \\ (CO)_3Fe \overset{C=O}{\cdots} Fe(CO)_3 \\ H \end{array}\right]^- \longrightarrow \left[\begin{array}{c} (CO)_4 \\ Fe \\ (CO)_3Fe \overset{C=O}{\cdots} Fe(CO)_3 \\ CH_2 \\ CH_3 \end{array}\right]^-$$

$$\xrightarrow{H_2O} CH_3CH_2CHO + OH^-$$

Scheme **XXXI**

The aldehyde thus formed is then hydrogenated by an iron carbonyl hydride species. Rudkovskii's group (34) found that $H_2Fe_3(CO)_{11}$ reacts with ethylene at room temperature to give *n*-propanol, whereas in the presence of a tertiary amine the reaction proceeds much more slowly. According to these workers, the reaction, therefore, should follow the path shown in Scheme **XXXII**.

$$3Fe(CO)_5 + NR_3 + 2H_2O \rightarrow (NR_3H)^+(HFe_3(CO)_{11})^- + 2CO + 2CO_2 + H_2$$

$$(NR_3H)^+(HFe_3(CO)_{11})^- \rightleftarrows H_2Fe_3(CO)_{11} + NR_3$$

$$CH_2{=}CH_2 + 2H_2Fe_3(CO)_{11} + 9CO \rightarrow CH_3CH_2CH_2OH + 6Fe(CO)_5$$

Scheme **XXXII**

Considering the evidence on the existence of acylcarbonylferrates (45–47), a mechanism similar to that formulated for cobalt-catalyzed hydroformylation can tentatively be adopted as a working hypothesis. The alkyl and acyl groups are probably bound to an iron atom that is part of a complex carbonyl molecule or anion. The work of Collman and co-workers (29) on the chemistry of alkyl- and acyltetracarbonylferrates as intermediates in the synthesis of aldehydes and ketones is of particular interest in this connection. But more work is needed to clarify the structure of the reaction intermediates in the Reppe alcohol synthesis and in related iron-catalyzed reactions.

C. Hydroformylation with Ruthenium Catalysts

Ruthenium was mentioned as a hydroformylation catalyst in some patents of the 1960s (17,48–52). The synthesis of propionaldehyde from ethylene, CO, and H_2 in the presence of ruthenium was mentioned in a paper by Pichler and co-workers (53) who found that the hydroformylation of ethylene proceeds much faster than the reduction of CO to polymethylene. They attributed the rate difference to the fact that the hypothetical ruthenium acyls, which should be the intermediates in both cases, are much more rapidly cleaved when ethylene is complexed with the catalyst.

A deeper investigation of the hydroformylation catalyzed by ruthenium derivatives has since been carried out with $Ru_3(CO)_{12}$ as catalyst precursor (54). Some linear and branched olefins besides cyclohexene were hydroformylated (Table 64). The reaction is rather slow even at 150°C. Branched-chain olefins and cyclohexene react more slowly than linear olefins. The reaction rate reaches a maximum at a CO pressure of 20 atm, whereas at a CO pressure of 150 atm the reaction rate is about a fifth of that corresponding to the maximum. The isomeric composition of the

TABLE 64
Hydroformylation of Olefins with $Ru_3(CO)_{12}$[a] (54)

Olefin						Yield (%)		
	Mmoles	Catalyst (g)	P_{CO} (atm)	P_{H_2} (atm)	Time (min)	Total aldehydes	Paraffins	Less-branched[b] aldehyde (%)
Propylene	131	0.12	24	85	150	47	42	70
Propylene	119	0.12	98	63	150	41	20	69
1-Butene	96	0.20	29	45	210	51	26	76
1-Butene	128	0.20	90	75	210	12	8	77
1-Pentene	106	0.20	22	53	300	39	30	75
1-Pentene	94.5	0.20	93	61	300	24	8	80
4-Methyl-1-pentene	71.4	0.20	24	54	300	38	23	88
4-Methyl-1-pentene	71.4	0.20	90	60	300	26	10	91
Cyclohexene	97.6	0.10	20	55	120	23	5	—
Cyclohexene	95.5	0.10	93	60	120	5	0.5	—

[a] Temp 150°C; toluene (28 g) as solvent; reactions at constant pressure.
[b] Aldehydes formed by addition of —CHO to the terminal carbon atom.

products changes very little with CO pressure in contrast to results obtained with cobalt. The ratio of linear to branched-chain isomers is intermediate between that found when using $Co_2(CO)_8$ at high CO partial pressures and that found when using $Rh_4(CO)_{12}$. Similar results were found for propylene hydroformylation using different catalyst precursors (55). A careful investigation of the secondary reaction products was carried out, and it was shown that irradiation with γ-rays accelerates the reaction by a factor of 3 (55).

One of the main characteristics of hydroformylations catalyzed by ruthenium carbonyls is the competitive hydrogenation of the olefinic substrate, which depends largely on its structure, and, for the same olefin, on the CO pressure. The hydrogenation rate decreases more than the hydroformylation rate with increasing CO pressure.

The concomitant occurrence of hydroformylation and hydrogenation hinders investigation of the reaction kinetics. Experiments carried out by varying the olefin concentration, H_2 pressure, or the catalyst concentration showed that with cyclohexene both hydroformylation and hydrogenation are first order reactions with respect to olefin concentration. The ratio of the amount of olefin hydroformylated to that hydrogenated varies with changing H_2 pressure.

The rate of hydroformylation when operating under high CO pressure increases proportionally to the ruthenium concentration. At low CO pressure, the ratio of the amount of olefin hydroformylated to the amount hydrogenated, other reaction conditions being the same, varies with changing concentration of dissolved ruthenium. To rationalize the above rather complicated picture, it was assumed that hydrogenation and hydroformylation are catalyzed by different complexes present in concentrations depending on the partial pressures of CO and H_2 and, at constant CO and H_2 pressures, on catalyst concentration. The fact that with varying CO partial pressure a change in hydroformylation rate results in no change in isomer composition, can tentatively be explained by the oversimplified assumption that, in contrast with the postulation for cobalt catalysis, only one complex is active as the hydroformylation catalyst and that its concentration depends on CO pressure.

Phosphine-substituted ruthenium catalysts were studied by Wilkinson and his groups (31,56), who obtained an 80% conversion of 1-pentene (straight- to branched-chain product ratio of ~2.7) at 110°C and 100 atm ($CO/H_2 = 1$) in 16 hr with $Ru(CO)_3(PPh_3)_2$ as catalyst. Other workers (57) obtained from propylene a linear to branched-chain aldehyde ratio of 1.8 by operating at 120°C and 102 atm with $Ru(CO)_3[P(CH_3)_2Ph]_2$. Much lower rates were obtained by using $RuCl_3(PPh_3)_3{\cdot}CH_3OH$ or $RuCl_2(PPh_3)_3$ as catalysts (31). In the last case, the low catalytic activity is attributed to formation of the insoluble $RuCl_2(CO)_2(PPh_3)_2$.

Some catalytic systems containing phosphines and other ligands were investigated in detail by Wilkinson and coworkers (73). Addition of 3 moles of PPh_3 and triphenylphosphite to $Ru_3(CO)_{12}$ causes a large increase in the reaction rate, whereas addition of PR_3 or 1,2-bis(diphenylphosphino)ethane has the opposite effect. The type of ligand largely affects the isomeric composition of the reaction products: The highest percentage of linear aldehyde is found when 1-hexene is used as substrate and operating in the absence of added ligands.

When using $Ru(CO)_3(PPh_3)_2$ as catalyst precursor, the effects of catalyst concentration, H_2 pressure and CO pressure were found (73) to be similar to those observed (54) when using $Ru_3(CO)_{12}$. The former catalytic system seems to be more sensitive than the latter to substrate structure; cyclohexene and butadiene were not hydroformylated at 120°C and 100 atm ($P_{CO}/P_{H_2} = 1$).

The addition of PPh_3 to $Ru(CO)_3(PPh_3)_2$ causes in the case of 1-hexene an increase in the percentage of linear aldehyde produced accompanied by a strong decrease of the reaction rate.

In conclusion, from a practical point of view, the use of ruthenium catalysts in hydroformylation allows facile recovery of catalyst and avoids formation of high molecular-weight condensation products. Ruthenium, however, has a lower catalytic activity than cobalt or rhodium and, at least in the absence of phosphines, causes a rather rapid hydrogenation of the olefinic substrate.

From the viewpoint of reaction mechanism, the isolation of an acyl derivative of ruthenium (58) can be taken as indicative that the general path of ruthenium-catalyzed hydroformylations might be similar to that proposed for the cobalt-catalyzed reaction. The strong inhibiting effect of CO pressure has been taken as an indication that, at least when $Ru(CO)_3(PPh_3)_2$ is used as catalyst precursor, the rate-determining step is the formation of $RuH_2(CO)_2(PPh_3)_2$ by reaction of the precursor with hydrogen which is known (73) to involve CO elimination (74).

D. Hydroformylation with Iridium Catalysts

Iridium was proposed as a hydroformylation catalyst since 1956 (59); the stoichiometric hydroformylation of olefins by $Ir_2(CO)_6(PPh_3)_2$ was mentioned (60). The catalytic activity of iridium, however, is lower than that of cobalt or rhodium and, therefore, has been much less investigated (2,27,61,62). "$IrCl_4$" in the presence of PBu_3 (8*a*) is apparently a much more active catalyst than "$IrCl_4$" alone. The hydroformylation of propylene in the presence of iridium catalysts was investigated by Andreetta et al. (11) using $IrCl(CO)L_2$ ($L = PBu_3$ or PPh_3). The hydroformylation occurs rather slowly at 130°C and constant pressure ($P_{CO} = P_{H_2} = 75$ atm). A mixture of isomeric butyraldehydes containing 58% of the linear isomer was obtained and 25% of the propylene was hydrogenated to propane. The reaction rate decreased on adding excess phosphine, but increased on decreasing the CO partial pressure from 75 to 15 atm, although the H_2 pressure was simultaneously decreased from 75 to 15 atm.

The hydrogenation of propylene can be decreased by lowering both the H_2 partial pressure and the reaction temperature as well as by addition of excess phosphine. As with rhodium catalysts, the isomeric composition of the products does not change substantially on decreasing the CO pressure; the percentage of linear isomer increases slightly on adding excess phosphine. Similar results are reported for 1-octene hydroformylation in the presence of $Ir_4(CO)_{12}$ and PBu_3 (75). The hydroformylation of 1-pentene (31) proceeds slowly at 70°C and 100 atm total pressure ($CO/H_2 = 1$) in the presence of $HIr(CO)(PPh_3)_3$ (10% conversion in 18 hr). Cyclohexene gives good yields of cyclohexanecarboxaldehyde at 150–160°C under 250 atm pressure ($CO/H_2 = 1$) and using "$IrCl_4$" supported on kieselguhr as catalyst (59).

The rather poor catalytic activity of iridium complexes has allowed detection and, in most cases, isolation of reaction intermediates (63). By using $HIr(CO)(PPh_3)_2$ in the reaction of ethylene with CO, the pentacoordinated propionyl derivative, $CH_3CH_2COIr(CO)_2(PPh_3)_2$, was isolated (Scheme **XXXIII**). The reaction can be reversed by heating a benzene solution of the complex under nitrogen. The propionyl derivative reacts with H_2 yielding propionaldehyde and the pentacoordinated hydride $HIr(CO)_2(PPh_3)_2$. The probable intermediate [$H_2Ir(CO)_2(PPh_3)$-($COCH_2CH_3$) or $H_2Ir(CO)(PPh_3)_2(COCH_2CH_3)$] has not been isolated. But by reacting the pentacoordinated propionyl complex with HCl, the unstable complex $CH_3CH_2COIr(H)(Cl)(CO)(PPh_3)_2$ was obtained; this complex readily decomposes to propionaldehyde and $IrCl(CO)(PPh_3)_2$. Finally, after reacting the pentacoordinated $HIr(CO)(PPh_3)_3$ and $HIr(CO)_2(PPh_3)_2$ with ethylene, spectroscopic evidence for the existence of the corresponding ethyliridium complexes was obtained; in the second case, the presence of a pentacoordinated olefin complex of the type $CH_3CH_2COIr(CO)_2(C_2H_4)PPh_3$ was observed spectroscopically. The chemistry of alkyl, aryl and acyl complexes of iridium was also investigated by Kubota and Blake (65). An infrared spectral study of the reaction of phosphine-substituted derivatives of $Ir_4(CO)_{12}$ with CO and H_2 under pressure was accomplished (64*a*). $IrH(CO)_2P(i\text{-}C_3H_7)_3$ was

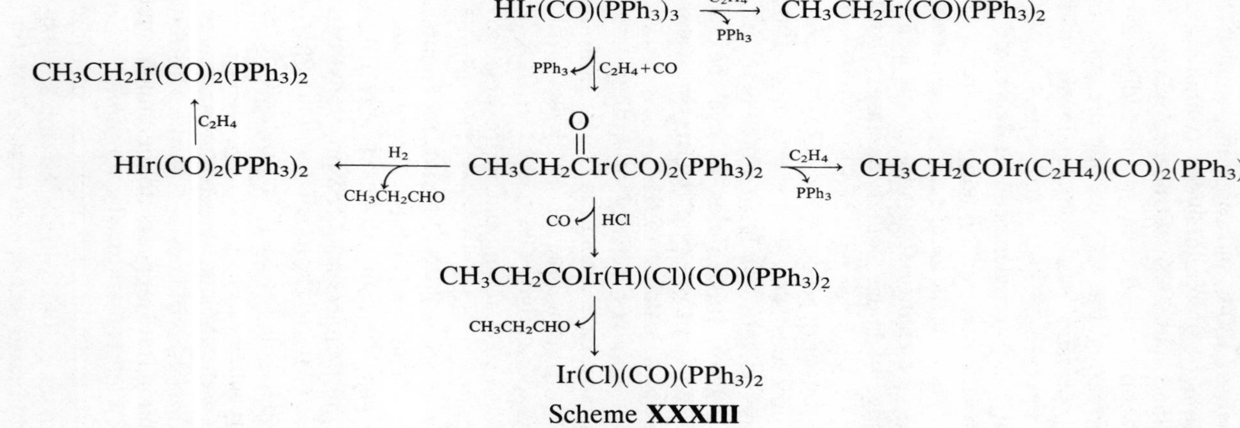

Scheme **XXXIII**

then reacted with C_2H_4, CO and H_2 (64*b*); the existence of alkyl and acyl complexes was confirmed and the steps subsequent to the alkyl complex formation were shown to be reversible. Some evidence was obtained in favor of the existence of the trihydride $IrH_3(CO)_3P(i\text{-}C_3H_7)_3$ (64*b*).

These intermediates can be used as models for investigating the reactivity of the corresponding π-olefin, alkyl-, and acylrhodium complexes which are much more unstable and difficult to characterize. The investigation of iridium complexes confirms that the general features of the mechanism in hydroformylations catalyzed by rhodium and iridium are very similar, but further work on differences between these catalysts would be most informative.

E. Hydroformylation with Platinum Catalysts

The possibility of using Pt complexes as catalysts in the hydroformylation of olefins was ignored for a long time, although platinum chloride in the presence of $SnCl_2$ was successfully used in synthesizing esters from olefins and CO (77). Indeed square planar complexes of Pt(II) were shown to have rather slight catalytic activity in hydroformylation. However, high temperature is necessary to obtain reasonable conversions and under these conditions selectivity toward formation of aldehydes is small (67,78,8*a*). A real breakthrough in this field occurred when $SnCl_2$ was added to the Pt(II) complexes (76). In this case not only a very large increase in the reaction rate was observed, but a selectivity toward aldehydes exceeding 90% was obtained, accompanied by a very large prevalence of linear aldehydes.

Besides the above cited patents the literature on this subject is very scarce. The best known catalyst precursor is $PtH(SnCl_3)(CO)(PPh_3)_2$; when using this complex the hydroformylation of 1-pentene, carried out at 100°C under 200 atm total pressure ($CO/H_2 = 1$), yields predominantly *n*-hexanal, with the ratio between straight-chain and branched-chain aldehydes being very high (about 19). Similar results were obtained using *trans*-$PtH(SnCl_3)(PPh_3)_2$ or *trans*-$PtH(Cl)(PPh_3)_2$ in the presence of excess $SnCl_2 \cdot 2H_2O$ (66,80).

Branched, internal, and cyclic olefins can also be hydroformylated in the presence of these systems; according to preliminary experiments (76,80) the reaction rate appears to be much smaller in the case of internal or cyclic olefins than for the terminal olefins.

Some experiments carried out while using Pt complexes containing optically active phosphines such as (-)DIOP or neomenthyldiphenylphosphine showed that from 2-methyl-1-butene optically active 3-methyl-1-pentanal can be obtained with optical yields up to 9.4% (80).

Further experiments (81) carried out using $PtCl_2$ (-)DIOP in the presence of $SnCl_2{\cdot}2H_2O$ (Sn/Pt = 5) showed that the prevailing chirality of the hydroformylation products of terminal olefins is opposite to that observed when using Rh complexes and the same ligand. This fact shows that asymmetric induction in hydroformylation is not merely connected with a substrate-chiral ligand interaction. Furthermore, the products obtained by hydroformylation of 1-butene, *cis*-2-butene, and *trans*-2-butene have the same chirality and the same optical purity.

If the general lines of the hydroformylation mechanism are the same when using Rh or Pt catalysts, the above facts can be interpreted by admitting that asymmetric induction, which in the case of Rh catalysts was shown to occur before or during formation of the alkyl rhodium complexes, takes place with platinum catalysts after the formation of metal alkyls.

The data available are not sufficient to allow evaluation of the importance of platinum catalytic systems in hydroformylation. It appears that they have an activity intermediate between Rh- and Co-containing catalysts and that they show an unexpected selectivity toward formation of linear aldehydes. However, the complexity of the Pt catalytic systems used up to now (which are prepared from platinum halides, trisubstituted phosphines and $SnCl_2$) can hardly compete in industrial applications with the traditional cobalt catalysts.

REFERENCES

1. The most significant patents are reported by J. Falbe, *Carbon Monoxide in Organic Synthesis*, Springer-Verlag, Heidelberg, New York, 1970, p. 15, by F. E. Paulik, *Catal. Rev.*, **6,** 49 (1972), and L. Markó in *Aspects of Homogeneous Catalysis*, R. Ugo Ed., D. Reidel Publ. Comp., 1974, Vol. 2, p. 3.
2. N. S. Imyanitov and D. M. Rudkovskii, *J. prakt. Chem.*, **311,** 712 (1969).
3*a*. W. Hieber, R. Nast, and J. Sedlmeier, *Angew. Chem.*, **64,** 465 (1952),
3*b*. W. Hieber, W. Beck, and G. Braun, *ibid.*, **72,** 795 (1960).
4. T. A. Weil, S. Metlin, and I. Wender, *J. Organometal. Chem.*, **49,** 227 (1973).
5. Ethyl Corp., Brit. Pat. 863,277 (1958); through *Chem. Abstr.*, **56,** 9969 (1962).
6. O. E. H. Klopfer, U. S. Pat. 3,050,562 (1962); through *Chem. Abstr.*, **57,** 13217 (1962).
7. T. G. Selin, *U. S. Dept. Com., Office Tech. Serv., PB Rept.* **133796** (1960); *Chem. Abstr.*, **56,** 4142 (1962).
8*a*. L. H. Slaugh and R. D. Mullineaux, U.S. Pat. 3,239,571 (1966); *Chem. Abstr.*, **65,** 618 (1966).
8*b*. B. Fell and J. Shanshool, *Chem.-Ztg.*, **99,** 231 (1975).
9. H. Wakamatsu, *Nippon Kagaku Zasshi*, **85,** 227 (1964); *Chem. Abstr.*, **61,** 13173 (1964).
10. B. L. Booth, M. J. Else, R. Fields, and R. N. Haszeldine, *J. Organometal. Chem.*, **27,** 119 (1971).

11. L. Benzoni, A. Andreetta, C. Zanzottera, and M. Camia, *Chim. Ind.* (*Milan*), **48,** 1076 (1966).
12. P. Pino, F. Piacenti, M. Bianchi, and R. Lazzaroni, *Chim. Ind.* (*Milan*), **50,** 106 (1968).
13*a*. J. Tsuji, O. Iwamoto, and M. Morikawa, Japan. Pat. 67 23,005 (1967); through *Chem. Abstr.*, **69,** 35431 (1968),
13*b*. *Idem,* Japan. Pat. 67 23,006 (1967); *Chem. Abstr.*, **69,** 51600 (1968),
13*c*. J. Tsuji and O. Iwamoto, Japan. Pat. 67 23,007 (1967); *Chem. Abstr.*, **69,** 51601 (1968).
14. J. Tsuji and Y. Mori, *Bull. Chem. Soc. Japan,* **42,** 527 (1969).
15. L. H. Slaugh and R. D. Mullineaux, U. S. Pat. 3,239,570 (1966); *Chem. Abstr.*, **64,** 19420 (1966).
16. J. Tsuji, N. Iwamoto, and M. Morikawa, *Bull. Chem. Soc. Japan,* **38,** 2213 (1965).
17. G. Wilkinson, French Pat. 1,459,643 (1966); *Chem. Abstr.*, **67,** 53652 (1967).
18. C. Botteghi, unpublished results.
19*a*. O. Roelen, Ger. Pat. 103,362 (1938),
19*b*. *Idem,* U. S. Pat. 2,327,066 (1943); *Chem. Abstr.*, **38,** 550 (1944).
20. W. F. Gresham and R. E. Brooks, U. S. Pat. 2,497,303 (1950); *Chem. Abstr.*, **44,** 4492 (1950).
21. W. F. Gresham and R. E. Brooks, Brit. Pat. 637,999 (1950); *Chem. Abstr.*, **44,** 9473 (1950).
22. B. H. Gwynn and J. H. Hirsch, U. S. Pat. 2,734,922 (1956); *Chem. Abstr.*, **50,** 16830 (1956).
23. B. L. Booth, H. Goldwhite, and R. N. Haszeldine, *J. Chem. Soc. C,* **1966,** 1447.
24. N. S. Imyanitov and D. M. Rudkovskii, in D. M. Rudkovskii, Ed., *Carbonylation of unsaturated Hydrocarbons* [in Russian], Izd. Khimiya, Leningrad, 1968, p. 28; *Chem. Abstr.*, **71,** 105708 (1969).
25. N. S. Imyanitov and D. M. Rudkovskii, *Kinet. Katal.*, **8,** 1051 (1967).
26. N. S. Imyanitov and D. M. Rudkovskii, *Kinet. Katal.*, **8,** 1240 (1967); *Chem. Abstr.*, **68,** 86867 (1968).
27. N. S. Imyanitov and D. M. Rudkovskii, *Zh. Prikl. Khim.*, **40,** 2020 (1967); *Chem. Abstr.*, **68,** 95367 (1968).
28*a*. N. M. Bogoradovskaya and N. S. Imyanitov, in N. S. Imyanitov, Ed., *Gidroformilirovanie* [Hydroformylation], Izd. Khimiya, Leningrad, 1972, p. 138; *Chem. Abstr.*, **77,** 152579 (1972).
28*b*. N. S. Imyanitov, *Zh. Obshch. Khim.*, **45,** 1344 (1975); *Chem. Abstr.*, **83,** 96336 (1975).
29*a*. J. P. Collman, S. R. Winter, and D. R. Clark, *J. Am. Chem. Soc.*, **94,** 1788 (1972),
29*b*. W. O. Siegl and J. P. Collman, *J. Am. Chem. Soc.*, **94,** 2516 (1972).
30. M. P. Cooke, Jr., *J. Am. Chem. Soc.*, **92,** 6080 (1970).
31. D. Evans, J. A. Osborn, and G. Wilkinson, *J. Chem. Soc. A,* **1968,** 3133.
32. W. Reppe, Ger. Pat. 890,942 (1953); *Chem. Zentr.*, **1954,** 5395.
33. W. Reppe and H. Vetter, *Liebigs Ann. Chem.*, **582,** 133 (1953).
34. N. M. Bogoradovskaya, N. S. Imyanitov, and D. M. Rudkovskii, in N. S. Imyanitov, Ed., *Gidroformilirovanie* [Hydroformylation], Izd. Khimiya, Leningrad, 1972, p. 146; *Chem. Abstr.*, **77,** 163947 (1972).
35. Yu. I. Berezina, P. A. Moshkin, and L. L. Klinova, *Khim. Prom.* (*Moscow*) **44,** 890 (1968); *Chem. Abstr.*, **70,** 57058 (1969).
36. N. v. Kutepow, H. Kindler, K. Eisfeld, K. Dettke, H. Jenne, and H. Detzer, Ger. Pat. 1,114,796 (1961); *Chem. Abstr.*, **57,** 2076 (1962).
37. N. v. Kutepow and H. Kindler, *Angew. Chem.*, **72,** 802 (1960).

38. H. W. B. Reed and P. O. Lenel, U. S. Pat. 2,911,443 (1959); cf. Brit. Pat. 794,067 (1958); *Chem. Abstr.*, **53,** 218 (1959).
39. N. v. Kutepow, H. Kindler, K. Eisfeld, K. Dettke, H. Jenne, and H. Detzer, U. S. Pat. 3,013,538 (1963); cf. Ger. Pat., ref. 36.
40. H. Kröper, in *Houben-Weyl Methoden der Organischen Chemie*, Georg Thieme Verlag, Stuttgart 1955, Vol. IV, Part 2, p. 392.
41*a*. H. W. Sternberg, R. Markby, and I. Wender, *J. Am. Chem. Soc.*, **78,** 5704 (1956).
41*b*. H. W. Sternberg, R. Markby, and I. Wender, *ibid.*, **79,** 6116 (1957).
41*c*. I. Wender, H. W. Sternberg, R. A. Friedel, S. J. Metlin, and R. E. Markby, *The Chemistry and Catalytic Properties of Cobalt and Iron Carbonyls*, Bulletin 600, U. S. Bureau of Mines, Washington, 83 pp., 1962.
42. H. Hieber and G. Brendel, *Z. anorg. allg. Chem.*, **289,** 324 (1957).
43. M. Heintzeler and N. v. Kutepow, Ger. Pat. 948,058 (1956); *Chem. Abstr.*, **53,** 6054 (1959).
44. L. F. Dahl and J. F. Blount, *Inorg. Chem.*, **4,** 1373 (1965).
45. V. Kiener and E. O. Fischer, *J. Organometal. Chem.*, **42,** 447 (1972).
46. Y. Watanabe, T. Mitsudo, M. Tanaka, K. Yamamoto, T. Okajama, and Y. Takegami, *Bull. Chem. Soc. Japan*, **44,** 2569 (1971).
47. G. M. Whitesides and D. J. Boschetto, *J. Am. Chem. Soc.*, **91,** 4313 (1969).
48. P. Smith and H. H. Jager, Brit. Pat. 966,482 (1961); *Chem. Abstr.*, **61,** 10593 (1964).
49. P. Smith and H. H. Jager, Ger. Pat. 1,159,926 (1963); *Chem. Abstr.*, **60,** 14389 (1964).
50. L. H. Slaugh and R. D. Mullineaux, U. S. Pat. 3,239,566 (1966); *Chem. Abstr.*, **64,** 15745 (1966).
51. T. Alderson and R. V. Lindsey, Jr., U. S. Pat. 3,081,357 (1963); *Chem. Abstr.*, **59,** 8594 (1963).
52. A. B. Stiles, U. S. Pat. 3,244,644 (1966); *Chem. Abstr.*, **65,** 82 (1966).
53. H. Pichler, B. Firnhaber, and D. Kioussis, *Brennstoff-Chem.*, **44,** 337 (1963).
54. G. Braca, G. Sbrana, F. Piacenti, and P. Pino, *Chim. Ind.* (*Milan*), **52,** 1091 (1970).
55. H. F. Schultz and F. Bellstedt, *Ind. Eng. Chem., Prod. Res. Develop.*, **12,** 176 (1973).
56. D. Evans, J. A. Osborn, F. H. Jardine, and G. Wilkinson, *Nature*, **208,** 1203 (1965).
57. M. J. Lawrenson and M. Green, Ger. Offen. 2,026,926 (1971); *Chem. Abstr.*, **74,** 124822 (1971).
58. G. Braca, G. Sbrana, and E. Benedetti, *Proc., Symp. Chemistry of Hydroformylation and Related Reactions*, Veszprem, **1972,** p. 127; *Chem. Abstr.*, **77,** 113238 (1972).
59. G. Schiller, Ger. Pat. 953,605 (1956); *Chem. Abstr.*, **53,** 11226 (1959).
60. L. Malatesta, M. Angoletta, and F. Conti, *J. Organometal. Chem.*, **33,** C43 (1971); *Chem. Abstr.*, **76,** 41448 (1972).
61. N. S. Imyanitov and D. M. Rudkovskii, *Neftekhim.*, **3,** 198 (1963); *Chem. Abstr.*, **59,** 7396 (1963).
62. M. Yamaguchi, *Shokubai*, **11,** 179 (1969); through *Chem. Abstr.*, **73,** 13787 (1970).
63. G. Yagupsky, C. K. Brown, and G. Wilkinson, *J. Chem. Soc. A*, **1970,** 1392.
64*a*. A. J. Drakesmith and R. Whyman, *J. Chem. Soc., Dalton Trans.*, **1973,** 362.
64*b*. R. Whyman, *J. Organometal. Chem.*, **90,** 303 (1975).
65. M. Kubota and D. M. Blake, *J. Am. Chem. Soc.*, **93,** 1368 (1971).
66. C. Y. Hsu and M. Orchin, *J. Am. Chem. Soc.*, **97,** 3553 (1975).
67. G. A. Rowe, Brit. Pat. 1,368,434 (1974); *Chem. Abstr.*, **82,** 142590 (1975).
68. S. Usami, K. Nishimura, and S. Fukushi, Japan. Pat. 72 11,410 (1972); *Chem. Abstr.*, **77,** 19182 (1972).
69. G. A. Rowe, Brit. Pat. 1,368,802 (1974); *Chem. Abstr.*, **82,** 72542 (1975).
70. A. L. Lapidus, E. Z. Gil'denberg, and Ya. T. Eidus, *Kinet. Katal.*, **16,** 252 (1975); *Chem. Abstr.*, **82,** 155180 (1975).

71. Ya. T. Eidus, A. L. Lapidus, and E. Z. Gil'denberg, *Kinet. Katal.*, **14,** 598 (1973); *Chem. Abstr.*, **79,** 78000 (1973).
72. N. M. Bogoradovskaya, N. S. Imyanitov, and D. M. Rudkovskii, *Zh. Prikl. Khim.*, **46,** 616 (1973); *Chem. Abstr.*, **79,** 4914 (1973).
73. R. A. Sanchez-Delgado, J. S. Bradley, and G. Wilkinson, *J. Chem. Soc. Dalton*, **1976,** 399.
74. F. L'Eplattenier and F. Calderazzo, *Inorg. Chem.*, **7,** 1290 (1968).
75. J. A. Thomas, Brit. Pat. 1,367,623 (1974); *Chem. Abstr.*, **82,** 142591 (1975).
76. I. Schwager and J. F. Knifton, Ger. Offen. 2,322,751 (1973); *Chem. Abstr.*, **80,** 70327 (1974).
77. E. L. Jenner and R. V. Lindsey, Jr., U.S. Pat. 2,876,254 (1959); *Chem. Abstr.*, **53,** 17906 (1959).
78. J. J. Mrowca, U.S. Pat. 3,876,672 (1975); *Chem. Abstr.*, **84,** 30432 (1976).
79. W. H. Bradler, Jr., S. B. Cavitt, and R. M. Gipson, French Pat. 1,530,136 (1968); *Chem. Abstr.*, **71,** 60714 (1969).
80. C. Y. Hsu, Thesis, Cincinnati, 1974; *Chem. Abstr.*, **82,** 154899 (1975).
81. G. Consiglio and P. Pino, *Helv. Chim. Acta*, **59,** 642 (1976).
82. W. F. Gresham and A. McAlevy, U. S. Pat. 2,564,104 (1951); *Chem. Abstr.*, **46,** 4561 (1952).
83. R. H. Bennett and W. R. Deever, U. S. Pat. 3,839,459 (1974); *Chem. Abstr.*, **81,** 169126 (1974).

V. SYNTHESIS OF KETONES FROM UNSATURATED SUBSTRATES

A. Introduction

By reacting unsaturated substrates with CO and H_2 in the presence of metal carbonyls or their precursors, ketones were first obtained by Roelen (1) according to reaction 47.

$$2CH_2{=}CH_2 + CO + H_2 \rightarrow CH_3CH_2COCH_2CH_3 \quad (47)$$

Instead of molecular hydrogen, other hydrogen donors such as alcohols (2,3*b*,35), water (3,4), or aniline (5) can be used for this unusual hydrocarbonylation reaction; sometimes, better selectivity was achieved in the absence of gaseous hydrogen. In certain cases, when very little H_2 is available, polyketones can be obtained in one step as shown by Reppe and Magin (4) and Iwashita and Sakuraba (3*b*,35) using ethylene (reaction 48), by Dokiya and Bando (6) using butadiene, and by Belluco's group (7) using norbornadiene.

$$n(CH_2{=}CH_2) + (n-1)CO + H_2 \rightarrow CH_3CH_2(COCH_2CH_2)_{n-2}COCH_2CH_3 \quad (48)$$

Finally, ketones have been obtained by reacting olefins with CO in the absence of H_2 or hydrogen donors. Under these conditions, the carbonylation is followed by cyclization as shown by the reaction between cyclohexene and CO giving rise to perhydrofluorenone (8) (reaction 49).

$$2\,\text{(cyclohexene)} + CO \longrightarrow \text{(dodecahydrofluorenone)} \quad (49)$$

Despite its potential synthetic value, the ketone synthesis has scarcely been investigated. This is probably due to the low selectivity of the reaction—generally it leads to a mixture of ketones, aldehydes, and other oxygenated products. The problem of controlling the isomeric composition of the product from the ketone synthesis is even more difficult than in the hydroformylation reaction; thus, even a simple substrate like 1-butene gives, with excess of $HCo(CO)_4$, a mixture of three isomeric ketones (reaction 50) (9).

$$CH_3CH_2CH{=}CH_2 \xrightarrow{HCo(CO)_4} \begin{cases} CH_3(CH_2)_3CO(CH_2)_3CH_3 \\ CH_3(CH_2)_3COCH(CH_3)CH_2CH_3 \\ CH_3CH_2CH(CH_3)COCH(CH_3)CH_2CH_3 \end{cases} \quad (50)$$

The ketone synthesis has been reviewed (10–12). On account of its small practical importance, only those aspects of the stoichiometric and catalytic synthesis of ketones that may contribute to an overall understanding of the carbonylation reaction will be considered here.

B. Stoichiometric Synthesis of Ketones

1. Hydrocarbonylation

Both monoolefins and diolefins undergo stoichiometric carbonylation to yield ketones: Ethylene on reaction with $HCo(CO)_4$ gave diethyl ketone as well as *n*-propanol (13). 1-Butene on treatment in pentane with gaseous $HCo(CO)_4$ at −6 to −8°C and then heating at 130°C for 3 hr gave a ketone/aldehyde ratio of 9/1; all the expected isomeric products were present (9). Similarly, butadiene gave 5-nonanone and 3-nonen-5-one; and 1,4-pentadiene yielded 2-methylcyclopentanone as the main product (9,14). Conjugated dienes can react with acylcobalt carbonyls to form unsaturated ketones (Vol. 1, p. 388).

Nonconjugated 1,5-dienes react smoothly with $Ni(CO)_4$ and dilute HCl in acetone at 60°C to give ketonic products in relatively high yields (~70%) instead of the expected esters (15): 1,5-Hexadiene gave 2,5-dimethylcyclopentanone and 2-methylcyclohexanone (reaction 51); 1,5-cyclooctadiene gave bicyclo[3.3.1]nonan-9-one in 60% yield; but 1,5-octadiene gave only carboxyl derivatives.

$Ni(CO)_4$, HCl; CH_3, CH_3, +, O, CH_3 (51)

1 mole 0.45 mole 0.24 mole

$Ni(CO)_4$ can also be used in stoichiometric amounts to synthesize ketones from cyclic olefins and from diolefins. The first synthesis of this kind was performed by Bird and co-workers (16,17), who obtained the ethyl ester of bicyclo[2.2.1]heptanecarboxylic acid (**13**) and dibicyclo[2.2.1]hept-2-yl ketone (**14**) from bicyclo[2.2.1]heptene-2, norbornene (reaction 52). Bicyclo[2.2.1]heptadiene gave the corresponding doubly unsaturated ketone.

$Ni(CO)_4$, H_2O; C_2H_5OH, HOAc; $CO_2C_2H_5$ (**13**) (52)

O, C (**14**)

2. *Carbonylation*

By reacting bicyclo[2.2.1]heptadiene with $Fe(CO)_5$, the ketones **15**, **16**, and **17** were isolated (16,18). These compounds are formed by addition of one molecule of CO to two or three molecules of diolefin, or by addition of two CO molecules to three molecules of diolefins (16,19)

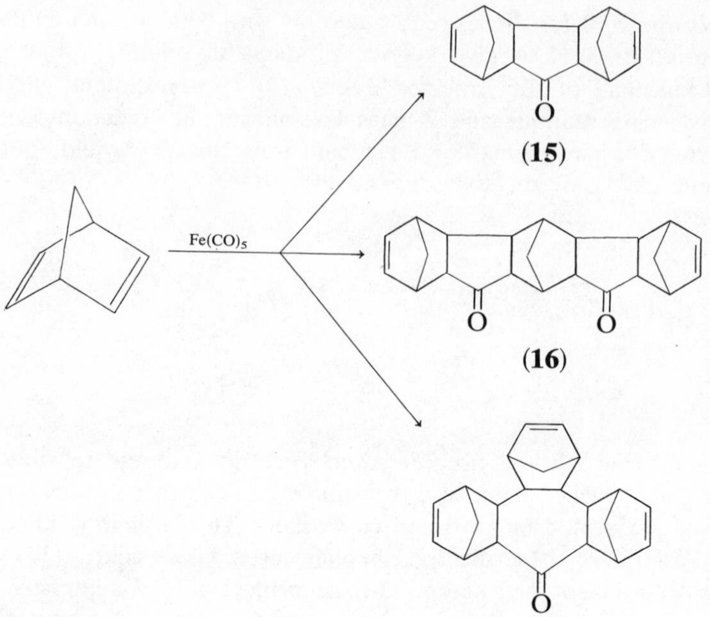

$\xrightarrow{Fe(CO)_5}$ (15) + (16) + (17) (53)

(reaction 53). More complicated reactions occur when bicyclo[2.2.1]heptene reacts with CO in the presence of $Co_2(CO)_8$ (20).

Irradiation of 2-substituted cyclopropylethylenes at room temperature in the presence of $Fe(CO)_5$ or $Fe_2(CO)_9$ resulted in the formation of substituted cyclohexenones as the major products (21) (reaction 54).

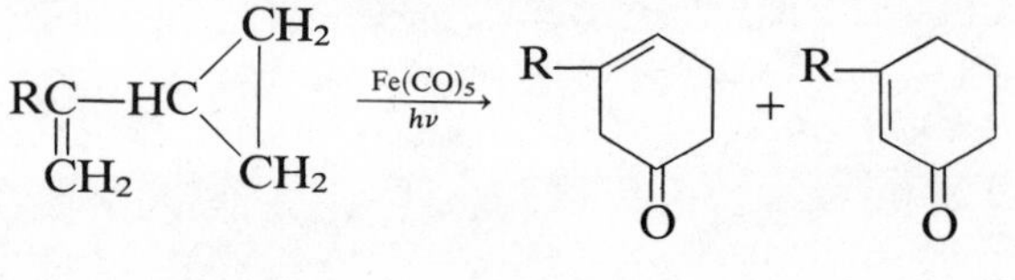

$$RC(=CH_2)-HC(CH_2)_2 \xrightarrow[h\nu]{Fe(CO)_5} \text{R-cyclohexenone} + \text{R-cyclohexenone} \quad (54)$$

R = $-CH_3$, —Ph, *p*-$CH_3OC_6H_4$-, *p*-ClC_6H_4-, 2-thienyl

Finally, 11-vinyl-3,7-cycloundecadien-1-one was obtained by reacting the bis-π-allyl nickel complex $C_{12}H_{18}Ni$ with CO (22) (reaction 55).

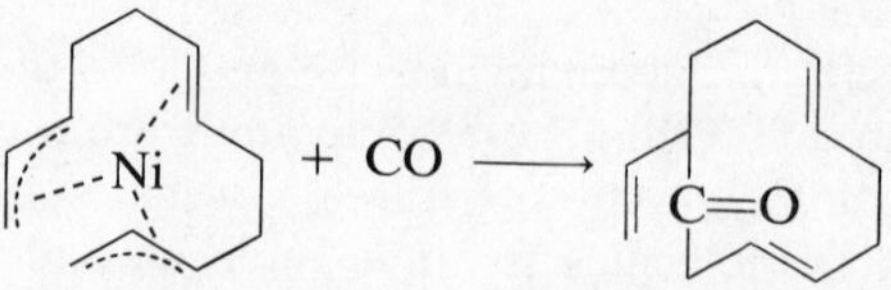

$$\text{Ni complex} + CO \longrightarrow \text{C=O} \quad (55)$$

C. Catalytic Synthesis of Ketones

1. *Hydrocarbonylation*

The catalytic synthesis of ketones will be discussed with emphasis on the hydrogen donor employed. Besides gaseous hydrogen, alcohols, water, and other compounds containing active hydrogen can be used. On reacting olefins with CO and H_2 under oxo conditions, only ethylene gives rise to a ketone (diethyl ketone) in substantial yield. For instance, from propylene only 3.5% conversion to a mixture of ketones with seven carbon atoms was obtained when using $RuCl_3$ as catalyst precursor (23).

Attempts to improve this synthesis of ketones were made by different authors (24–26). Good yields were obtained by using a heterogeneous cobalt catalyst at 65°C and 21 atm total pressure; for example, use of these conditions with ethylene/CO/H_2 ratios of 2.4/1/1.1 or 3.8/1/1 gave yields of up to 57% diethyl ketone and ratios of diethyl ketone to propionaldehyde, in weight, approaching 40/1 (24*a*).

A systematic investigation of the effect of some variables on the synthesis of diethyl ketone from ethylene, CO, and H_2 in the presence of $Co_2(CO)_8$ was carried out by Dokiya and Bando (27). They studied the effect of temperature, of ethylene, and of H_2 and CO partial pressures, of $Co_2(CO)_8$ concentration, and of addition of bases. Molar ratios of diethyl ketone to propionaldehyde up to about 3 were obtained at 120°C by using $Co_2(CO)_8$ as catalyst (pyridine/$Co_2(CO)_8$ molar ratio = 6) and a H_2 partial pressure of 11 atm. The CO pressure in the interval between 25 and 100 atm does not seem critical, but high concentrations of $Co_2(CO)_8$ and ethylene seem to favor the formation of diethyl ketone. Addition of base influenced both the reaction rates and the selectivity of the reaction—pyridine appeared the best compromise.

The highest yield (99%) and selectivity (99%) from the presence of an organic base were reported by Swakon and Field (28). They interacted ethylene, CO (17.6 atm) and H_2 (3.5 atm) in the presence of $Co_2(CO)_8$ and benzylamine or benzonitrile. Using nitriles and amines as ligands, Kenzie (26) obtained conversions of 90% and selectivity toward diethyl ketone of 95%. Phosphine-containing cobalt catalysts were also used (29,30).

The synthesis of dibutyl ketone from butadiene, CO and H_2 in the presence of $Co_2(CO)_8$ at 150–200°C was described by Adkins and Williams (31), who isolated the ketone in low yield as the semicarbazone from the reaction mixture containing aldehydes. Dokiya and Bando (6) also isolated 3-methyloctanone and polyketones of the type $C_4H_9CO(C_4H_8CO)_nC_4H_9$ from the products of the hydroformylation of butadiene.

The possibility of obtaining ketones from olefins in the absence of molecular hydrogen but in the presence of various hydrogen donors was demonstrated in the early 1950s (2,32). When secondary alcohols were used as hydrogen donors, the corresponding ketone was formed by dehydrogenation (Scheme **XXXIV**). But when methanol was used, no formaldehyde was found among the reaction products, hence the hydrogen necessary must have derived either directly from decomposition of the methanol or from the water arising from methanol by dehydration to form dimethyl ether.

The yield of ketones depends strongly on both the olefin structure and the ratio of olefin to hydrogen donor as well as on the type of alcohol used as hydrogen donor.

The best selectivity for the formation of ketones was found in the reaction of ethylene in the presence of isopropanol and $Co_2(CO)_8$ at high temperatures (220–250°C) and pressures (>350 atm) (12,33). Under these conditions, yields of up to 95% of diethyl ketone (based on starting ethylene) were obtained (Table 65); the main secondary product was isopropyl propionate. Lower yields were generally obtained when using primary alcohols as hydrogen donors (Table 65). High yields of diethyl

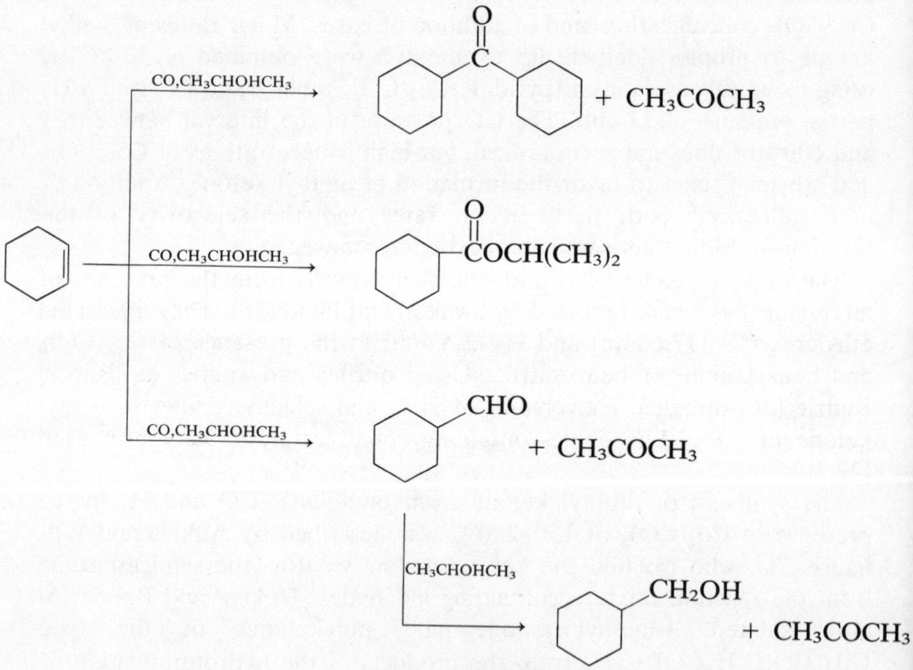

$CO, CH_3CHOHCH_3$ → + CH_3COCH_3

$CO, CH_3CHOHCH_3$ → $COCH(CH_3)_2$

$CO, CH_3CHOHCH_3$ → CHO + CH_3COCH_3

$CH_3CHOHCH_3$ → CH_2OH + CH_3COCH_3

Scheme **XXXIV**

TABLE 65
Synthesis of Diethyl Ketone from Ethylene, Carbon Monoxide, and Alcohols[a] (33)

Alcohol	Temp (°C)	Yield[b] (%)
Isopropanol	210	95
Isopropanol	220	88
Isopropanol	240	79
Isopropanol	250	83
Methanol	220	61
Ethanol	220	81
Propanol	220	84

[a] Catalyst $Co_2(CO)_8$; initial P_{CO} 160 atm at 20°C

[b] Moles diethyl ketone per 2 moles of starting ethylene.

ketone were reported using rhodium catalysts in the presence of isopropanol at 175°C and high CO pressures (>200 atm) (34). The same reaction was carried out at lower pressure in the presence of CH_3OH/H_2O mixtures as hydrogen donors, using $Rh(H)(CO)(PPh_3)_3$ (57).

As mentioned earlier, selectivity in the ketone synthesis is greatly influenced by olefin structure (2). The yields of ketones are low in the case of propylene and of cyclohexene (Table 66), and only traces of ketones were found when using isobutylene as substrate (2). In these reactions the main products are esters (from hydrocarboxylation of the olefin), aldehydes (from olefin hydroformylation), and alcohols (from reduction of aldehydes) (Scheme **XXXIV**). As expected from this scheme, the yield of diethyl ketone from ethylene would decrease and the yield of the alcohol derived from the olefin would increase as the ratio of isopropanol to olefin increased (2).

Ketoester formation was observed when using methanol as the hydrogen donor. Thus, methl 4-ketohexanoate was obtained in good yields from the reaction of ethylene in the presence of cobalt acetate and methanol or isopropanol at 200–220°C and 300 atm of CO (reaction 56) (37).

$$2CH_2{=}CH_2 + 2CO + CH_3OH \xrightarrow{CO} CH_3CH_2COCH_2CH_2COOCH_3 \qquad (56)$$

The same kind of reaction was obtained with lower selectivity under milder conditions (200–300 atm, 80–200°C) when using nickel catalysts

TABLE 66

Formation of Ketones from Olefins by Hydrocarbonylation with Alcohols as Hydrogen Donors

Olefin	Catalyst	Hydrogen donor	Temp (°C)	P_{CO}[a] (atm)	Yield[b] (mole %) Carboxylic acid derivatives	Ketones	Alcohols	Ref.
Ethylene	Rh_2O_3	Acetic acid, Methanol	130	~500	25[c]	32[d]	—	35
Propylene	Raney Co	Methanol	200	250	33	26	3[e]	2
Cyclohexene	Raney Co	Isopropanol 0.5[f]	210	250	15	12	4	2
		Isopropanol 1.3[f]	210	250	30	6	14	2
		Isopropanol 10[f]	210	250	31	Trace	40	2
Methyl acrylate	$Co_2(CO)_8$ + pyridine	Methanol	110	150[g]	78	13[h]	—	36

[a] Initial pressure at room temp.
[b] Based on olefin charged.
[c] Methyl homolevulinate.
[d] Octa-3,6-dione 26%, undeca-3,6,9-trione 6%.
[e] Aldehydes.
[f] Alcohol/olefin, molar ratio.
[g] Pressure at reaction temp.
[h] Dimethyl γ-ketopimelate.

containing nitrile groups. In this case, esters and polyketoesters were also obtained (38).

Under milder conditions (100°C and 140 atm) a similar synthesis was observed as a secondary reaction during the hydrocarboxylation of ethylene in the presence of palladium and HCl (39); with ethanol as solvent, ethyl 4-ketohexanoate was found in the reaction products. Ketoesters were also obtained from 1,5-hexadiene and from 1,4-pentadiene in the presence of $PdI_2(PBu_3)_2$ at 150°C and 1000 atm of CO (reaction 57) (40). The reaction rate increases when the acidity of the medium is increased; with *p*-toluenesulfonic acid present, the reaction proceeds even at 250 atm.

$$CH_2{=}CHCH_2CH_2CH{=}CH_2 + 2CO + CH_3OH \xrightarrow{PdI_2(PBu_3)_2}$$

$$CH_3\text{—}(\text{cyclopentanone ring, C=O})\text{—}CH_2COOCH_3 \quad (57)$$

Ketoesters in relatively low yields were obtained by carbonylation of methyl acrylate in the presence of a cobalt catalyst (36) (Table 66) and also with ruthenium and rhodium catalysts (3*a*); in some cases, methyl γ-ketopimelate was isolated. Methyl γ-ketocaproate was found in the products of the reaction of methyl acrylate with CO, methanol, and ethylene in the presence of rhodium or ruthenium catalysts (3*a*).

Water has been used as the hydrogen donor in the synthesis of ketonic compounds in the presence of various catalysts (Table 67). By using an aqueous solution of $CoSO_4$ and $FeSO_4$ as catalyst in the carbonylation, ethylene gave equal amounts of propionic acid and diethyl ketone (41). In the presence of $Ni(CN)_2$, a mixture of ketonic compounds, including polyketones containing the $(CH_2CH_2CO)_n$ entity, were obtained from ethylene (4). Metallic cobalt can also be used as catalyst (43). At much higher CO pressures (1000 atm) and 190°C in the presence of ruthenium or rhodium catalysts, ethylene gave mainly diethyl ketone and octan-3,6-dione (3). Under similar conditions, propylene gave the three expected ketones containing seven carbon atoms (3).

Nonconjugated diolefins react with CO and water in the presence of cobalt catalysts to yield cyclic ketones. A rather good yield was reported from the reaction of 1,5-hexadiene, which gave mainly 2,5-dimethylcyclopentanone together with small amounts of 2,5-dimethylcyclopent-2-enone (42) (reaction 58). Finally tetraphenylbutatriene reacted with CO and water in the presence of $Co_2(CO)_8$ to form 2-(β,β-diphenylethyl)-3-phenylindanone (44) (reaction 59).

TABLE 67

Synthesis of Ketones from Olefins by Hydrocarbonylation with Water as Hydrogen Donor

Olefin	Catalyst	Temp (°C)	P_{CO} (atm)	Main reaction products	Ref.
Ethylene	$CoSO_4 \cdot 7H_2O$, $FeSO_4 \cdot 7H_2O$, $MgSO_4 \cdot 7H_2O$	197	90[a]	Et_2CO CH_3CH_2COOH CH_3CH_2CHO	41
Ethylene	$K_2[Ni(CN)_4]$	150	75[b]	Et_2CO $EtCOCH_2CH_2COEt$ $EtCO(CH_2CH_2CO)_2Et$ CH_3CH_2COOH $EtCOCH_2CH_2COOH$	4
Ethylene	$Ru(acac)_3$	190	500[b]	Et_2CO $EtCOCH_2CH_2COEt$	3*a*
Ethylene	$Rh(acac)_3$	190	500[b]	Et_2CO $EtCOCH_2CH_2COEt$	3*a*
Propylene	$Rh(acac)_3$	190	1000[b]	$(CH_3CH_2CH_2)_2CO$ $(CH_3CH)_2CO$ (CH with CH_3 branch) $CH_3CH_2CH_2COCHCH_3$ (CH with CH_3 branch)	3*a*
1,4-Pentadiene	$Co_2(CO)_8$	165	110[a]	2-Methylcyclopentanone	42
1,5-Hexadiene	$Co_2(CO)_8$	165	110[a]	2,5-Dimethylcyclopentanone 2,5-Dimethylcyclopent-2-enone	42
1,6-Heptadiene	$Co_2(CO)_8$	165	110[a]	2-Methyl-5-ethylcyclopentanone	42

[a] Initial pressure at room temp.

$$\text{1,5-hexadiene} \xrightarrow[CO]{Co_2(CO)_8} \text{CH}_3\text{-cyclopentenone-CH}_3 \xrightarrow[Co_2(CO)_8]{CO, H_2O} \text{CH}_3\text{-cyclopentanone-CH}_3 \quad (58)$$

$$(C_6H_5)_2C{=}C{=}C{=}C(C_6H_5)_2 \xrightarrow[Co_2(CO)_8]{CO, H_2O} \text{indanone with } C_6H_5, H, H, CH_2CH(C_6H_5)_2 \quad (59)$$

Attempts have been made to use in the catalytic synthesis of ketones some hydrogen source other than molecular hydrogen, water, or alcohols. By reacting propylene with CO in the presence of $Co_2(CO)_8$ and small amounts of aniline (aniline/propylene = 0.011), the products were dipropyl ketone and other ketonic compounds (5) (reaction 60).

$$2CH_3CH{=}CH_2 + 2CO + 2PhNH_2 \rightarrow (CH_3CH_2CH_2)_2CO + (PhNH)_2CO \quad (60)$$

Tetralin was also used as a hydrogen donor in the synthesis of ketones from propylene in the presence of cobalt catalysts (45).

2. *Carbonylation*

The synthesis of ketones also occurs in the absence of hydrogen donors, but generally higher temperatures are required and lower yields are obtained. Cobalt and palladium have mainly been used as catalysts for this synthesis, although platinum may have some catalytic activity. For example, cyclohexene reacts with CO at 700 atm and 300–350°C to yield a mixture of stereoisomeric perhydrofluorenones (8); this reaction, carried out in the absence of catalysts in a platinum-lined autoclave, has been described as a thermal carbonylation. Dimethylcyclopropene, in the presence of $Pd(PPh_3)_4$ at 30°C and 70 atm CO pressure, yields compound **17*a*** (56).

CH₃ CH₃ / CH₃ CH₃ C=O / CH₃ CH₃

17a

With $PdI_2(PBu_3)_2$ at 150°C and a pressure of 1000 atm of CO, 1,5-cyclo-octadiene was transformed into bicyclo[3.3.1]non-2-en-9-one; the residual olefin was 1,3-cyclooctadiene (46). Under similar conditions

(1000 atm, 200°C) nonconjugated diolefins like 1,4-pentadiene, 1,5-hexadiene, and 1,7-heptadiene gave a mixture of lactones and cyclopentenones according to reaction 61 (47,48). In this case, the reaction might

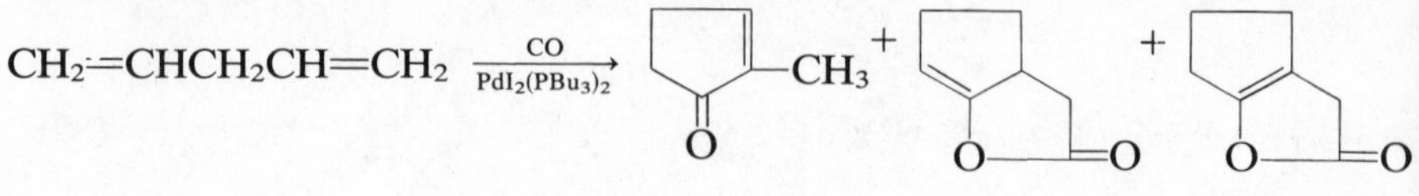

$$CH_2{=}CHCH_2CH{=}CH_2 \xrightarrow[PdI_2(PBu_3)_2]{CO}$$ CH₃ + O O=O + O O=O (61)

formally be only a carbonylation not involving hydrogen. In fact, the investigators postulated that the catalyst is a palladium hydride complex of the type $HPdL_2I$ ($L = PBu_3$) which adds to the olefin and is eliminated after the reaction. These authors did not explain how the hydride forms when the reaction is carried out in the absence of water or methanol.

Cumulenes react with CO in the presence of $Co_2(CO)_8$, even in the absence of water (49). At 230–250°C and 150 atm of CO, tetraphenylbutatriene gives 70% of 2-(β,β-diphenylvinyl)-3-phenylindone; and tetraphenylallene gives the expected 2-diphenylmethyl-3-phenylindone along with the isomeric 2,2,4-triphenylnaphthalenone (reaction 62).

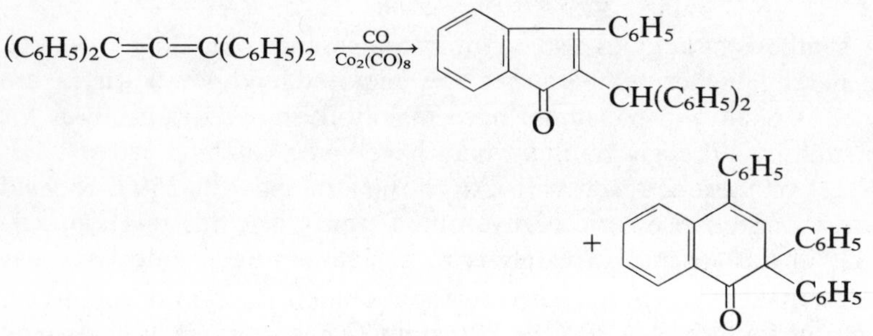

$$(C_6H_5)_2C{=}C{=}C(C_6H_5)_2 \xrightarrow[Co_2(CO)_8]{CO}$$ C_6H_5, $CH(C_6H_5)_2$, O + C_6H_5, C_6H_5, C_6H_5, O (62)

Norbornadiene in benzene solution reacts smoothly with CO in the presence of catalytic amounts of $PdCl_2$ (50). After 18 hr at 50°C with CO at 100 atm, there was obtained a 1:1 copolymer of CO and norbornadiene for which the structure given in reaction 63 was proposed from

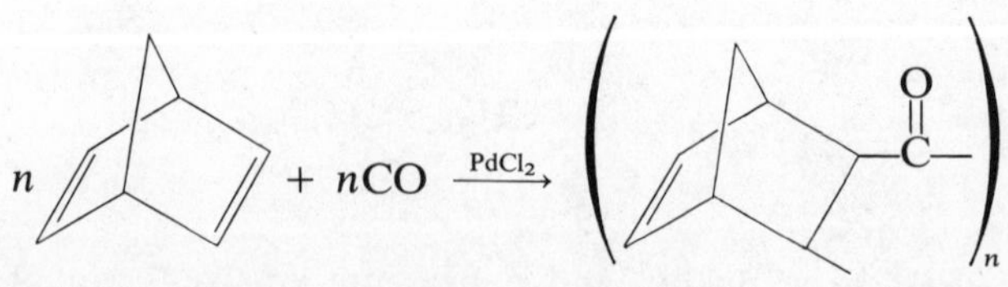

$$n \; + nCO \xrightarrow{PdCl_2}$$ O, C, n (63)

nmr evidence; apparently no attempts were made to determine the nature of the end groups in this product.

The use of copper (58) or ruthenium (59) catalysts was proposed for synthesizing unsaturated ketones from ethylene and CO.

3. *Mechanism*

Little is known about the mechanism of the synthesis of ketones by carbonylation of olefinic substrates, and no detailed kinetic investigations on the reaction have been made. The stoichiometric carbonylation of olefins to ketones by $HCo(CO)_4$ (9,14) and the decomposition of acylcobalt carbonyls, however, give some hints on the reaction path that might lead from olefins to ketones in the presence of cobalt catalysts.

For the hydroformylation of olefins, there is little doubt that acylcobalt carbonyls play an important role as reaction intermediates. Assuming that acylcobalt carbonyls are also formed in the ketone synthesis, Jonassen and his group (9) and Heck (51) proposed two different paths leading from the cobalt-containing intermediate to the ketonic compound (Scheme **XXXV**).

$$RCOCo(CO)_x \xrightarrow{R'Co(CO)_x} RCOR' + Co_2(CO)_{2x} \qquad \textbf{(a)}$$

$$RCOCo(CO)_x \xrightarrow{>C=C<} RCO\overset{|}{\underset{|}{C}}\overset{|}{\underset{|}{C}}Co(CO)_x \qquad \textbf{(b)}$$

$$\downarrow H_2$$

$$RCO\overset{|}{\underset{|}{C}}\overset{|}{\underset{|}{C}}H + HCo(CO)_x$$

Scheme **XXXV**

According to the first path, **a**—inspired by the remarks of Hieber et al. (52) that acetone forms in the decomposition of methylcobalt carbonyl—the acylcobalt carbonyl reacts with an alkylcobalt carbonyl to yield the ketone and a dimeric cobalt carbonyl. This reaction was interpreted as an oxidative addition of the C–Co bond of the alkyl intermediate to the coordinatively unsaturated cobalt atom of the acyl intermediate.

As for the second path, **b**—essentially based on the isolation from diolefins of complexes of type **18**, which by thermal decomposition yield saturated and unsaturated cyclic ketones (51)—insertion of a carbon–carbon double bond (already coordinated to cobalt) between the cobalt and the acyl group was postulated (51) (Scheme **XXXVI**).

The high steric requirements of ketone synthesis are in accord with both schemes, since it is known that insertion of 2,2-disubstituted or

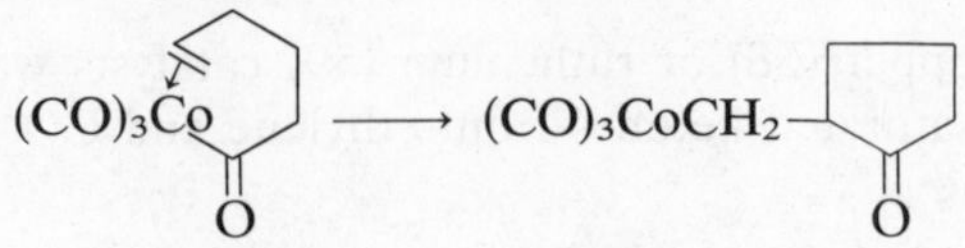

$$(CO)_3Co \text{(cyclic acyl, C=O)} \longrightarrow (CO)_3CoCH_2\text{–(cyclopentanone)}$$

(18)

Scheme **XXXVI**

trisubstituted olefinic double bonds into a metal to carbon bond is very difficult and that branched alkylcobalt carbonyls, on the other hand, are unstable (53). Possibly path **b** (Scheme XXXV) prevails when a nonconjugated diolefin is reacted, since this forms a ring of reasonable size, such as **18** (Scheme **XXXVI**) whereas with monoolefins or conjugated diolefins, a route resembling path **a** is preferred. The formation of a polyketone like $CH_3CH_2CO(CH_2CH_2CO)_nCH_2CH_3$ (4), which is obtained from ethylene when operating in the presence of water, and the formation of methyl 4-ketohexanoate from ethylene, CO and methanol (37), seem to favor path **b** (Scheme **XXXV**) for the ketone synthesis from ethylene.

With cobalt catalysts, CO addition is always accompanied by addition of hydrogen or hydrogen-donor compounds; hence, formation of a metal alkyl by addition of a probably already π-complexed olefin to a metal–hydrogen bond was postulated as the first step of the reaction. However, both in the case of iron (16) (reaction between $Fe(CO)_5$ and bicyclo-[2.2.1]heptadiene yielding **15**, **16**, and **17**) and of palladium (46) (catalytic carbonylation of 1,5-cyclooctadiene to bicyclo[3.3.1]non-2-en-9-one) the carbonylation products do not contain added hydrogen; therefore activation of the double bond without insertion into a metal–hydrogen bond cannot be excluded. This possibility was considered both for the hydrocarboxylation and the hydrocarbonylation of olefins with nickel catalysts, where a cyclic intermediate such as **19** has not been completely

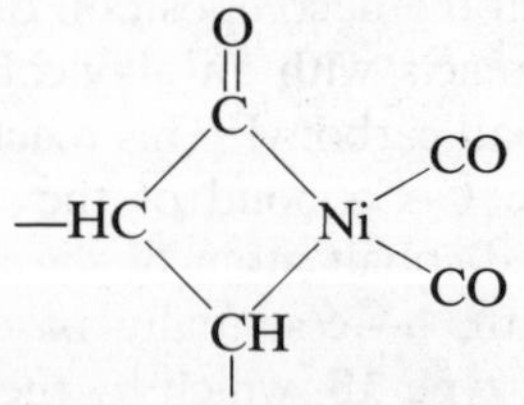

(19)

discarded (17), and in the case of palladium for which insertion of carbon–carbon olefinic bonds into Pd–Cl or Pd–COCl bonds was hypothesized (54).

Although catalytic species containing Pd-H bonds are postulated in the carbonylation of cyclic diolefins in the absence of H_2 (46,55), there is no experimental evidence of the need to activate the olefinic double bond by insertion into a metal–hydrogen bond in order to obtain acylmetal species such as **19.**

Obscure at present are the factors affecting the competition between olefin and hydrogen or hydrogen donors in the reaction with the acyl-metal derivative in the mechanism suggested for carbonylation of ethylene (Scheme **XXXVb**) and of butadiene. Of course, the ketone synthesis is favored when using a catalyst with poor ability to activate molecular hydrogen. Hydrogen from water (or from other hydrogen sources) seems readily utilized in this reaction. Equally unknown are the factors affecting the competition between CO addition and hydrogenolysis of a carbon–metal bond causing formation of mono- and polyketones.

The addition of an olefin (possibly already complexed to the metal) to the acyl group must be assumed to explain formation of polyketones from ethylene and from butadiene. Preferential formation of polyketones versus olefin or diolefin dimerization or polymerization may be due to a lower reactivity of the M–CH_2 bond with the olefin than that of the M–COR bond, or due to a much lower concentration of the alkyl derivative under reaction conditions.

REFERENCES

1*a*. O. Roelen, U. S. Pat. 2,327,066 (1943); *Chem. Abstr.*, **38,** 550 (1944),

1*b*. O. Roelen, *Naturforschung und Medezin in Deutschland*, Vol. XXXVI, Dietrich, Wiesbaden, 1948, p. 166.

2*a*. G. Natta, P. Pino, and R. Ercoli, *J. Am. Chem. Soc.*, **74,** 4496 (1952);

2*b*. *Idem, Proc. Third World Petroleum Congress, The Hague, 1951, Section V*, E. J. Brill, Leiden, Netherlands, 1951, p. 1.

3*a*. T. Alderson and J. C. Thomas, U. S. Pat. 3,040, 090 (1962); *Chem. Abstr.*, **57,** 16407 (1962).

3*b*. Y. Iwashita, M. Sakuraba, and A. Nakamura, Japan. Pat. 74 48,406 (1974); *Chem. Abstr.*, **83,** 9207 (1975).

4*a*. W. Reppe and A. Magin, U. S. Pat. 2,577,208 (1951); *Chem. Abstr.*, **46,** 6143 (1952),

4*b*. *Idem*, Ger. Pat. 880,297 (1953); *Chem. Abstr.*, **52,** 11907 (1958).

5. P. Pino and R. Magri, *Chim. Ind.* (*Milan*), **34,** 511 (1952).

6*a*. M. Dokiya and K. Bando, *Bull. Chem. Soc. Japan*, **41,** 1741 (1968),

6*b*. *Idem, Nippon Kagaku Kaishi*, **1973,** 1523; *Chem. Abstr.*, **79,** 125365 (1973).

7. M. Graziani, G. Carturan, and U. Belluco, *Chim. Ind.* (*Milan*), **53,** 939 (1971).

8. P. T. Lansbury and R. W. Meschke, *J. Org. Chem.*, **24,** 104 (1959).

9. J. A. Bertrand. C. L. Aldridge, S. Husebye, and H. B. Jonassen, *J. Org. Chem.*, **29,** 790. (1964).

10. J. Falbe, *Carbon Monoxide in Organic Synthesis*, Springer-Verlag, Heidelberg, New York, 1970.

11. C. W. Bird, *Transition Metal Intermediates in Organic Chemistry*, Logos Press, London, 1967, p. 205.
12. N. S. Imyanitov, D. M. Rudkovskii, and M. V. Khokhlova, in N. S. Imyanitov, Ed., *Hydroformylation* [in Russian], Khimiya (Leningrad), 1972, p. 161; *Chem. Abstr.*, **77,** 163914 (1972).
13. W. Reppe and H. Vetter, *Liebigs Ann. Chem.*, **582,** 133 (1953).
14. S. Husebye, H. B. Jonassen, C. Aldridge, and W. Senn, *Acta Chem. Scand.*, **18,** 1547 (1964).
15. B. Fell, W. Seide, and F. Asinger, *Tetrahedron Lett.*, **1968,** 1003.
16. C. W. Bird, R. C. Cookson, and J. Hudec, *Chim. Ind.* (*London*), **1960,** 20.
17. C. W. Bird, R. C. Cookson, J. Hudec, and R. O. Williams, *J. Chem. Soc.*, **1963,** 410.
18. R. Pettit, *J. Am. Chem. Soc.*, **81,** 1266 (1959).
19. R. C. Cookson, J. Henstock, and J. Hudec, *J. Am. Chem. Soc.*, **88,** 1059 (1966).
20. Y. Colleuille and P. Perras, French Pat. 1,352,841 (1964); through *Chem. Abstr.*, **61,** 593 (1964).
21. R. Victor, R. Ben-Shoshan, and S. Sarel, *Tetrahedron Lett.*, **1970,** 4253.
22*a*. G. Wilke et al., *Angew. Chem. Intern.* (*Eng.*.) *Ed.*, **5,** 151 (1966),
22*b*. B. Bogdanovic, M. Kroener, and G. Wilke. *Liebigs Ann. Chem.*, **699,** 1 (1966);
23. H. F. Schulz and F. Bellstedt, *Ind. Eng. Chem., Prod. Res. Develop.*, **12,** 176 (1973).
24*a*. E. A. Naragon, A. J. Millendorf, and J. H. Vergilio, U. S. Pat. 2,699,453 (1955); *Chem. Abstr.*, **50,** 1893 (1956),
24*b*. J. W. Fitzwilliam, E. A. Naragon, and F. J. Moore, U. S. Pat. 2,830,089 (1958); *Chem. Abstr.*, **52,** 13233 (1958).
25. W. F. Gresham, R. E. Brooks, and W. M. Bruner, U. S. Pat. 2,549,454 (1951); *Chem. Abstr.*, **45,** 8551 (1951).
26. N. Kenzie, Ger. Offen. 2,061,798 (1971); *Chem. Abstr.*, **75,** 63135 (1971).
27. M. Dokiya and K. Bando, *Kogyo Kagaku Zasshi*, **71,** 1866 (1968); *Chem. Abstr.*, **70,** 77247 (1969).
28. E. A. Swakon and E. Field, U. S. Pat. 3,257,459 (1966); *Chem. Abstr.*, **65,** 8764 (1966).
29. L. H. Slaugh, Belg. Pat. 621,662 (1963); *Chem. Abstr.*, **59,** 11268 (1963).
30*a*. L. H. Slaugh, French Pat. 1,300,404 (1963),
30*b*. *Idem*, Fr. Addn., 82,181 (1964); *Chem. Abstr.*, **61,** 1761 (1964).
31. H. Adkins and J. L. R. Williams, *J. Org. Chem.*, **17,** 980 (1952).
32. W. F. Gresham and R. E. Brooks, U. S. Pat. 2,526,742 (1950); *Chem. Abstr.*, **45,** 2017 (1951).
33. N. S. Imyanitov, D. M. Rudkovskii, and M. V. Khokhlova, *Neftekhim.*, **11,** 63 (1971).
34. V. L. Hughes and R. S. Brodkey, U. S. Pat. 2,839,580 (1958): *Chem. Abstr.*, **52,** 17109 (1958).
35. Y. Iwashita and M. Sakuraba, *Tetrahedron Lett.*, **1971,** 2409.
36. A. Matsuda, *Bull. Chem. Soc. Japan*, **42,** 571 (1969).
37. W. F. Gresham and R. E. Brooks, U. S. Pat. 2,542,767 (1951); *Chem. Abstr.*, **46,** 5615 (1952).
38. W. Reppe, H. Friederich, and K. Wiebusch, Ger. Pat. 892,893 (1953); *Chem. Abstr.*, **52,** 16211 (1958).
39. J. Tsuji, M. Morikawa, and J. Kiji, *Tetrahedron Lett.*, **1963,** 1437.
40. S. Brewis and P. R. Hughes, *Chem. Commun.*, **1965,** 489.
41. Ruhrchemie A.-G., Brit. Pat. 787,258 (1957); *Chem. Abstr.*, **52,** 10150 (1958).
42. P. P. Klemchuk, U. S. Pat. 2,995,607 (1959); *Chem. Abstr.*, **56,** 1363 (1962).
43. W. Reppe and H. Kroper, Ger. Pat. 860,350 (1952); *Chem. Zentr.*, **1954,** 1587.

44. P. J. Kim and N. Hagihara, *Mem. Inst. Sci. Ind. Res., Osaka Univ.*, **24,** 133 (1967); *Chem. Abstr.*, **68,** 2982 (1968).
45. J. H. Staib, W. R. F. Guyer, and O. C. Slotterbeck, U. S. Pat. 2,864,864 (1958); *Chem. Abstr.*, **53,** 9063 (1959).
46. S. Brewis and P. R. Hughes, *Chem. Commun.*, **1966,** 6.
47. S. Brewis and P. R. Hughes, *Chem. Commun.*, **1967,** 71.
48. S. Brewis and P. R. Hughes, *Abstracts, 157th Meeting Am. Chem. Soc.*, April 1969, PETR 53.
49. P. J. Kim and N. Hagihara, *Bull. Chem. Soc. Japan*, **38,** 2022 (1965).
50. J. Tsuji and S. Hosaka, *J. Polym. Sci., Polym. Lett. Ed.*, **3,** 703 (1965).
51. R. F. Heck, *J. Am. Chem. Soc.*, **85,** 3116 (1963).
52. W. Hieber, O. Vohler, and G. Braun, *Z. Naturforsch.*, **13b,** 192 (1958).
53. R. F. Heck and D. S. Breslow, *Chem. Ind. (London)*, **1960,** 467.
54. J. Tsuji, *Accounts Chem. Res.*, **2,** 144 (1969).
55. J. Falbe, *Angew. Chem. Intern. (Eng.) Ed.* **5,** 435 (1966).
56. P. Binger and U. Schuchardt, *Angew. Chem.*, **87,** 715 (1975).
57. H. Hara, Ger. Offen. 2,321,191 (1973); *Chem. Abstr.*, **80,** 26757 (1974).
58. J. D. McClure, Ger. Offen. 2,054,307 (1971); *Chem. Abstr.*, **75,** 19704 (1971).
59. J. D. McClure, Ger. Offen. 2,046,060 (1971); *Chem. Abstr.*, **75,** 5260 (1971).

Acknowledgements

We wish to thank Drs. G. Consiglio and A. Stefani (ETH, Zürich), P. Frediani and U. Matteoli (Florence University), and Prof. C. Botteghi (Sassari University) for their valuable help in careful reading of the manuscript and the proofs; we wish also to thank Dr. G. Bor (ETH, Zürich), Prof. G. Chiusoli (Parma University), and G. F. Pregaglia (Montedison S.p.A., Milan) for useful suggestions.

Hydrocarboxylation of Olefins and Related Reactions

P. PINO, *Swiss Federal Institute of Technology, Department of Industrial and Engineering Chemistry, Zurich, Switzerland*

and

F. PIACENTI AND M. BIANCHI, *Institute of Organic Chemistry, University of Florence, Florence, Italy*

I. INTRODUCTION

The synthesis of carboxylic acids or their esters by reaction of an olefinic substrate with CO and water or alcohols, catalyzed by phosphoric acid or other Lewis acids, has been known since 1931 (1,2); branched aliphatic acids and esters are obtained. In the presence of H_2SO_4, the reaction may be carried out under milder conditions (3). This synthesis, however, has serious limitations due both to experimental conditions and isomeric composition of the products which are mostly branched.

Much more interesting is the carboxylation of olefinic substrates conducted in the presence of metal carbonyls or metal compounds which, under reaction conditions, may be transformed into metal complexes containing CO groups. This reaction, discovered by Reppe and Kröper around 1940 (4), consists in the addition of hydrogen and a carboxyalkyl, thiocarboxyalkyl, amide, or a similar group to an olefinic substrate, and may therefore be defined as a hydrocarboxylation (Eq. 1).

$$>C{=}C< + CO + ZH \longrightarrow -\underset{\underset{\underset{Z}{|}}{CO}}{\overset{|}{\underset{|}{C}}}-\overset{|}{\underset{\underset{H}{|}}{C}}- \tag{1}$$

Here Z = $-$OR; $-$NHR; $-NR_2$; $-$OOCR′; $-$SR′. R = $-$H, alkyl, aryl; and R′ = alkyl.

Acids, esters, amides, anhydrides, or thioesters may be obtained by using water, alcohols, amines, carboxylic acids, or mercaptans as ZH reagents.

Numerous olefins were carboxylated in this way by using metal carbonyls as CO donors (stoichiometric reaction) or by employing CO under pressure in the presence of appropriate derivatives of transition metals as catalysts (catalytic reaction).

Other carboxylation reactions were found which do not involve addition of hydrogen (5,11,16) or involve addition of other groups [$-$Cl (6,59), $-CCl_3$ (7–10), $-$OR (11–13), CH_3COO- (5*a*, 14, 15, 59)] to the double bond instead of hydrogen as shown in Scheme **I**. These reactions, with the exception of trichlorocarboxylation, if carried out stoichiometrically involve reduction of the metal component; they can be carried out catalytically only in the presence of oxygen or other oxidizing agents. The chemical reactions involved are somewhat different from the hydroformylation and hydrocarboxylation reactions, and will not be discussed in this chapter.

The syntheses of acids and esters, very similar and widely studied, will be treated first, then the less studied syntheses of other functional derivatives of carboxylic acids such as amides, anhydrides, and thioesters will be treated.

The hydrocarboxylation of olefinic substrates, as already mentioned, is an addition of a hydrogen and a carboxyl group to an olefinic double bond. It is therefore analogous, formally, to the hydroformylation of

$$
\mathrm{H_2C{=}CH_2 + CO} \longrightarrow
\begin{cases}
\xrightarrow{\mathrm{PdCl_2}} \mathrm{CH_2ClCH_2COCl} \\
\xrightarrow[\mathrm{PdCl_2}]{\mathrm{CCl_4}} \mathrm{Cl_3CCH_2CH_2COCl} \\
\xrightarrow[\mathrm{PdCl_2,\,O_2}]{\mathrm{ROH}} \mathrm{ROOCCH_2CH_2COOR} \\
\xrightarrow[\mathrm{PdCl_2}]{\mathrm{C_2H_5OH}} \mathrm{C_2H_5OCH_2CH_2COOC_2H_5} \\
\xrightarrow[\mathrm{Pd,\,O_2}]{\mathrm{CH_3COOH}} \mathrm{CH_3COOCH_2CH_2COOH} \\
\xrightarrow[\mathrm{Pd,\,O_2,\,CuCl_2}]{\mathrm{ROH}} \mathrm{CH_2{=}CHCOOR}
\end{cases}
$$

Scheme **I**

olefins, where a hydrogen and a formyl group are added to the same type of substrate.

Thermodynamically, hydrocarboxylation is more favored than hydroformylation (18). The hydrocarboxylation of olefins, however, is slower than hydroformylation and must, as a rule, be carried out at higher temperatures. When water or methanol is used as hydrogen donor, the yields are generally satisfactory due to the relatively low importance of competitive reactions and to the stability of the reaction products under operating conditions.

A serious limitation to the use of this reaction as a synthetic method is, as in hydroformylation of olefins, formation of at least two isomeric products when an unsymmetrical olefin is used as substrate (reaction 2). Up to now, no general method of controlling the isomeric composition of the products has been found.

$$
\mathrm{RCH{=}CH_2 + CO + ZH} \longrightarrow
\begin{cases}
a\,\mathrm{RCH_2CH_2COZ} \\
b\,\mathrm{RCH(CH_3)COZ}
\end{cases}
\qquad (2)
$$

$$a + b = 1;\ \mathrm{R} = \text{alkyl};\ \mathrm{Z} = -\mathrm{OH}, -\mathrm{OR}$$

Molecules containing both olefinic unsaturation and, at a suitable distance, a group having an active hydrogen, may, under certain conditions, react with CO to give cyclic compounds (19–24). Lactones may therefore be obtained (reaction 3) from unsaturated alcohols (20), lactams from unsaturated amides (21*a*), and cyclic imides from unsaturated amides (21*b*).

We shall designate this type of reaction as "intramolecular" hydrocarboxylation even if the transfer of the hydrogen atom from the functional group to the double bond has not been demonstrated.

$$CH_2{=}CHCH_2OH \xrightarrow[Co_2(CO)_8]{CO} \text{γ-butyrolactone} \qquad (3)$$

The hydrocarboxylation of olefins may be an interesting synthetic method not only in the laboratory but also in industry. Propionic acid is, in fact, manufactured by hydrocarboxylation of ethylene (25), and the hydrocarboxylation of 1,5,9-cyclododecatriene reached, at least, the pilot plant stage (26).

Several reviews were published on hydrocarboxylation (27–30). We shall limit ourselves to discussing only those aspects useful to characterize the hydrocarboxylation of olefins.

II. SYNTHESIS OF ACIDS AND ESTERS

The first synthesis of acids and esters in the presence of transition metal complexes was performed by Reppe (4) who, as an extension of research on the chemistry of acetylene, used stoichiometric amounts of $Ni(CO)_4$ (31–33). The synthesis of esters in the presence of metallic cobalt was then investigated by Dupont et al. (34); Natta and Pino (35) independently observed the formation of esters during the hydroformylation of α-pinene in methanol when using the same catalyst. This observation generated a series of investigations on this synthesis in the presence of cobalt catalysts.

Emphasis was put on the use of derivatives of noble metals, such as ruthenium (36–38), rhodium (39–42,178), palladium (26,43,44), platinum (45,46), iridium (41,42), and osmium (45) as hydrocarboxylation catalysts; but most of this information comes from the patent literature. A systematic investigation has been carried out only with palladium (26,47), which in the presence of phosphorus-containing ligands seems more active than cobalt or nickel.

A. Stoichiometric Hydrocarboxylation

In Table 1 are summarized some results obtained in the hydrocarboxylation of olefins with stoichiometric amounts of $Ni(CO)_4$ and excess water or alcohols. The reaction seems to be promoted by ultraviolet light (50), by the presence of halogen ions (48,49), and, especially by acids (50–52). In the presence of acids, high conversions may be obtained when operating at 50–60°C, whereas in their absence temperatures of 250–300°C are necessary.

The hydrocarboxylation of ethylene has also been achieved with stoichiometric amounts of $Co_2(CO)_8$ under CO pressure (48). Methyl esters were obtained by reacting acylcobalt tetracarbonyls with methanol at 50°C, and by reacting triphenylphosphine acylcobalt tricarbonyls with methanol and iodine (55). Acylcobalt carbonyls may be prepared by reacting an excess of olefin with a solution of $HCo(CO)_4$ at 0°C (54). (See Vol. 1, p. 373).

By reacting unsaturated esters under CO with stoichiometric amounts of $KHFe(CO)_4$, the corresponding diesters were obtained after treatment with iodine in ethanol. This synthesis probably proceeds via acylmetal carbonyl intermediates (56). As shown by the data of Table 1, the stoichiometric hydrocarboxylation of olefins with $Ni(CO)_4$ has not been so deeply and systematically investigated as the corresponding reaction of acetylenic substrates.

Olefin and diolefin complexes of palladium are easily carbonylated with CO; in general, however, chlorocarboxy compounds or diesters are the main reaction products (5*d*,6*b*,57–59,85), but their formation is not discussed in this chapter.

From a preparative point of view the stoichiometric hydrocarboxylation of olefins does not offer practical advantages even on a laboratory scale, in comparison with the corresponding catalytic synthesis.

B. Catalytic Hydrocarboxylation

The first studies on catalytic hydrocarboxylation were carried out by Reppe (32) using $Ni(CO)_4$ and nickel halides as catalysts. High temperature (270°C) and pressure (200 atm) were employed to achieve reasonable conversions of the olefins.

Cobalt catalysts were later introduced and studied by Natta et al. (18,35,60), Gresham and Brooks (61), and Dupont et al. (34). Besides cobalt and nickel compounds, palladium, which was more recently investigated mostly by Tsuji et al. (62), von Kutepow's group (26), and by Fenton (47), deserves particular attention (180). Derivatives of other metals were used, such as iron, ruthenium, rhodium, iridium, and platinum, but their catalytic activity has not been systematically investigated.

As substrates, mono- and polyunsaturated hydrocarbons as well as unsaturated compounds containing functional groups were used. As hydrogen donors, water and methanol offer the advantage of minimizing

TABLE 1

Stoichiometric Hydrocarboxylation of Olefins with $Ni(CO)_4$

Olefin	Hydrogen source	Available CO[a]	Temp (°C)	Products	Yield (%) Carboxyl equiv./moles olefin	Yield (%) Carboxyl equiv./moles CO	Remarks	Ref.
1-Octene	C_2H_5OH	0.8	200	Ethyl 2-methyloctanoate	75	94	NiI_2 or I_2 added	48
1-Octene	C_2H_5OH	1.2	190	Ethyl nonanoate Ethyl 2-methyloctanoate	37	24	CH_3COOH added	49
1-Octene	H_2O	2.4	55–60	Nonanoic acid 2-Methyloctanoic acid 2-Ethylheptanoic acid 2-Propylhexanoic acid	5	2	Exposure to daylight, HCl added	50
1-Octene	H_2O	2.4	55–60	Nonanoic acid 2-Methyloctanoic acid 2-Ethylheptanoic acid 2-Propylhexanoic acid	69	29	Exposure to uv light, HCl added	50
cis-4-Octene	H_2O	2.4	55–60	2-Methyloctanoic acid 2-Ethylheptanoic acid 2-Propylhexanoic acid	15	6	Exposure to uv light, HCl added	50
1-Heptene	CH_3OH	—	45–48	Methyl octanoate Branched-chain isomeric esters	75	87	HCl added	51

Styrene	Cyclohexanol	2.3	240	Cyclohexyl 3-phenylpropanoate	27	11	CoI_2 or $CoBr_2$ added	48
Cyclopentene	C_2H_5OH	3.8	56	Ethyl cyclopentanecarboxylate	21	6	CH_3COOH and H_2O added	52
Cyclohexene	H_2O	0.3	270	Cyclohexanecarboxylic acid	8	28	—	49
Cyclohexene	H_2O	2.4	55–60	Cyclohexanecarboxylic acid	80	33	Exposure to uv light, HCl added	50
Cyclooctene	H_2O	0.7	270	Cyclooctanecarboxylic acid	31	44	200 atm CO	53
Bicyclo[2.2.1]-2-heptene	C_2H_5OD	3.1	56	Ethyl 3-*exo*-deuterobicyclo[2.2.1]-heptane-2-*exo*-carboxylate	72	23	D_2O and CH_3COOD added	52
				3-*exo*-Deuterobicyclo[2.2.1]heptane-2-*exo*-carboxylic acid	10	3		
Butadiene	H_2O	0.35	250	α-(Carboxycyclohexyl)propionic acid	29	82	Hydroquinone added,	48
				α-(Cyclohexenyl)-3-propionic acid	19	53	200 atm CO	
Cyclopentadiene	C_2H_5OH	0.61	56	Ethyl tricyclo[5.2.1.$0^{2,6}$]dec-3- or 4-ene-*exo*-8-carboxylate	69	14	H_2O and CH_3COOH added	52
1,4-Cyclohexadiene	C_2H_5OH	2.6	56	Ethyl cyclohexen-3-carboxylate	6	2	H_2O and CH_3COOH added	52
Bicyclo[2.2.2]-octadiene	C_2H_5OH	3.4	56	Diethyl bicyclo[2.2.2]octane-2,5-dicarboxylate } Diethyl bicyclo[2.2.2]octane-2,6-dicarboxylate	60	35	H_2O and CH_3COOH added	52

[a] Moles CO initially bound to metal/moles starting olefin.

secondary reactions. Experiments with other aliphatic alcohols were also performed, but with less interesting results.

The role of the solvent is very important in the synthesis of acids: The reaction occurs quite easily when the reaction medium can dissolve water, olefinic substrate, and catalyst (63). When using cobalt catalysts the only important secondary products are the corresponding carbonyl compounds that are formed in relatively small amounts (63); the reaction is less selective when operating in the presence of palladium catalysts (62*a*).

The asymmetric synthesis of acids and esters from prochiral-substituted ethylenes was obtained using $PdCl_2$ as catalyst in the presence of an optically active diphosphine (17,64); solvents exert a great effect on the optical yield. Under suitable conditions, optical yields up to 60% (17,65*b*) were obtained.

1. Catalysts

Nickel compounds must be used in the presence of activators such as hydrogen halides or halide ions (32,49) or HI (67). Cobalt derivatives are more effective and they do not require activators. No significant difference was found in the catalytic activity of metallic cobalt, cobalt salts, or $Co_2(CO)_8$ (66). Milder conditions are possible when using $Co_2(CO)_8$ (165°C, 230 atm) than when using cobalt metal or cobalt salts (200–220°C; 350 atm). Probably all cobalt compounds are transformed under reaction conditions into cobalt carbonyl complexes which constitute the true catalytic species. Pyridine increases the activity of $Co_2(CO)_8$ in the synthesis of both carboxylic acids (68) and esters (69,70,172), while strong bases suppress such activity. The concentration of pyridine in the reaction medium is not, as in hydroformylation (71), a critical factor; good yields may be obtained even when pyridine is used as solvent (68, 69,84,90,91). Complexes containing pyridine (74) or pyridine derivatives (115) were successfully used as catalysts in hydrocarboxylation.

Nickel compounds are generally more selective catalysts than cobalt derivatives due to the much larger ability of the latter to catalyze the synthesis of aldehydes and ketones from olefins, CO, and hydrogen donors. Palladium in various forms is also an active hydrocarboxylation catalyst: Satisfactory conditions are 80°C and a CO pressure of 700 atm (44), but even much lower pressures can be used (177). Palladium chloride was first investigated (62*a*), but the results were not very interesting for both catalytic activity and selectivity. Operating in the presence of phosphines, catalytic activity is improved (26) and depends largely on the type of phosphine used. Very active catalysts can be obtained by using diphosphines (17,64,65) like (–)DIOP (72). A large research effort is in progress to investigate the influence of ligand structure and the catalytic activity of palladium complexes (148): $SnCl_2$ was proposed as promoter (149,174) and catalysts containing, besides palladium, another group VIII metal like cobalt (170) or iron (171) was used. Chloroplatinic acid, when activated with $SnCl_2$, shows marked activity in oxygenated solvents at 200 atm at 80–90°C, which are particularly mild conditions for this type of synthesis (46). Also $K_2PtCl_6/SnCl_2$ systems were used in the presence of ethers (175) and ketones (176). Active catalysts were obtained from phosphines (150), phosphites (151), or arsines (173) containing Pt(II) complexes. Hydrocarboxylation with some group VIII noble metal halides was also carried out in molten salts (152).

Table 2 gives results obtained by Imyanitov and Rudkovskii (42) for relative activities of various metal carbonyls in the synthesis of acids and esters. These results, however, were obtained under conditions tending to give rather low selectivity for these reactions, and, therefore, must be considered with care. Nevertheless, it may be said that the activity of rhodium, iridium, iron, and nickel carbonyls is rather low. The carbonyls of osmium (45) and of ruthenium (45,38), although not extensively investigated, were reported to be poor catalysts for these reactions.

Data on hydrocarboxylation of propylene and cyclohexene in the presence of various catalysts are assembled in Tables 3 and 4. These results give some idea of the yields of carboxylic acids and their esters obtainable under appropriate conditions with different catalysts. But since the activity of the particular catalyst usually is greatly affected by reaction variables, a systematic investigation is necessary to determine optimum operating conditions. Such investigations were conducted for $Co_2(CO)_8$ (63,79), $Ru_3(CO)_{12}$ (38), and $PdCl_2$ in the presence of phosphines (42,47). Acids generally increase the hydrocarboxylation rate when $Pd(PPh_3)_2Cl_2$ is used as catalyst (80); in the presence of *p*-toluenesulfonic acid the CO pressure may be reduced from 1000 to 250 atm (80).

TABLE 2

Relative Catalytic Activities of Metal Carbonyls in Hydrocarboxylation of Cyclohexene (42)

Hydrogen source	$Co_2(CO)_8$	$Fe(CO)_5$	$Ni(CO)_4$	$Rh_6(CO)_{16}$	$[Ir(CO)_3]_4$
H_2O	1	$\leq 10^{-7}$	$<10^{-7}$	10^{-3}	10^{-5}
CH_3OH	1	$<10^{-5}$	$<10^{-5}$	10^{-2}	$<10^{-4}$

TABLE 3
Catalytic Hydrocarboxylation of Propylene in the Presence of Metal Complexes

Catalyst	Hydrogen source	Temp (°C)	P_{CO} (atm)	Time (hr)	Yield (%)	Ref.
$Co_2(CO)_8$	H_2O	130	123[a]	5	86	75
$Ni(CO)_4$	H_2O	250	50[a]	5	76	75
$Co_2(CO)_8$	CH_3OH	145	150[a]	10	93	76
$H_4Ru_4(CO)_{12}$	CH_3OH	190	22[b]	5	35	38
$Ru_3(CO)_{12}$	CH_3OH	190	24[b]	5	12	38
$Pd(PPh_3)_2Cl_2$	CH_3OH	90	700[b]	4	~100	77
$H_2PtCl_6/SnCl_2$	CH_3OH	80	800[c]	16	74	45

[a] Measured at room temp.
[b] Constant pressure.
[c] Initial pressure at reaction temp.

2. *Hydrogen Donors*

Water, primary and secondary open-chain alcohols (66), cyclohexanol (48), and phenol (49) as well as polyols (49,154) were used as hydrogen donors. Little work was reported on the effect of water or structure of the alcohol on the hydrocarboxylation reaction.

From Tables 3 and 4 it appears that for the hydrocarboxylation of both cyclohexene and propylene, in the range of CO pressure and temperature investigated, the reaction is more rapid when using water than when using

TABLE 4
Catalytic Hydrocarboxylation of Cyclohexene in the Presence of Some Metal Complexes

Catalyst	Hydrogen source	Temp (°C)	P_{CO} (atm)	Time (hr)	Yield (%)	Ref.
$Co_2(CO)_8$	H_2O[a]	165	197[b]	3	89	63
$Ni(CO)_4$	H_2O	285	200[b]	16	n.d.	78
$Pd(PPh_3)_2Cl_2$	H_2O[c]	120	700[d]	14	78	44
$Co_2(CO)_8$	CH_3OH	165	242[b]	6	75	66
$[Ru(CO)_3Cl_2]_2$	CH_3OH	190	20[d]	5	21	38

[a] Dioxane as solvent.
[b] Initial pressure at reaction temp.
[c] In the presence of HCl.
[d] Constant pressure.

methanol. Similar results were found in the hydrocarboxylation of 1-hexene and 2-ethylhexene (63) in the presence of cobalt catalysts.

With metallic cobalt as catalyst, hydrocarboxylation was investigated by using different alcohols as hydrogen donors. The best yields were obtained with primary alcohols, whereas very poor yields were obtained with tertiary alcohols (79*b*). There was, however, no great difference in the yields obtained with different primary alcohols; the yields with secondary alcohols were lower (79*b*).

Table 5 shows the results of representative experiments carried out with $Co_2(CO)_8$ or $H_2PtCl_6/SnCl_2$ as catalysts. The synthesis of acids in the presence of $Co_2(CO)_8$ and acetone as solvent proceeds at a higher rate than the synthesis of esters and shows good selectivity, since only a few percent of the corresponding aldehydes are formed. Surprisingly in the presence of $H_2PtCl_6/SnCl_2$ catalysts, the rate of the synthesis of esters seems higher than that of acids (46).

Selectivity in the synthesis of esters when using $Co_2(CO)_8$ at 165°C and 230–250 atm seems to decrease when passing from primary to secondary alcohols and with increasing molecular weight of the alcohol (66). On the other hand, according to Gankin et al. (69), when the reaction temperature is raised to 190°C, the selectivity is unexpectedly high (90%), at least with primary alcohols, and the reaction is not influenced by the molecular weight of the alcohol. The reaction rate, according to the same authors, increases when passing from methanol to ethanol.

In the case of the palladium/(−)DIOP catalytic system, an investigation was carried out on the effect of different hydrogen donors

TABLE 5

Hydrocarboxylation of Olefins with Different Hydrogen Donors

Olefin	Catalyst	Hydrogen donor	Temp (°C)	P_{CO}[a] (atm)	Time (hr)	Olefin conversion (%)	Yield[b] (%)	Ref.
Cyclohexene	$Co_2(CO)_8$	H_2O	165	197	3	—	89	63
Cyclohexene	$Co_2(CO)_8$	CH_3OH	165	246	6	72	81	66
Cyclohexene	$Co_2(CO)_8$	C_4H_9OH	165	230	7	83	45	66
Cyclohexene	$Co_2(CO)_8$	$(CH_3)_2CHOH$	165	226	6	69	27	66
1-Hexene	$Co_2(CO)_8$	CH_3OH	190	300	3.5	58	90	69
1-Hexene	$Co_2(CO)_8$	C_2H_5OH	190	300	2	90	82	69
1-Hexene	$Co_2(CO)_8$	$(CH_3)_2CHCH_2OH$	190	300	~2	65	58	69
1-Dodecene	$H_2PtCl_6/SnCl_2$[c]	H_2O	90	200	1–2	100	80	46
1-Dodecene	$H_2PtCl_6/SnCl_2$[c]	CH_3OH	90	200	1–2	100	80	46

[a] Initial pressure at reaction temp.

[b] Equivalents of acid/moles olefin charged × 100.

[c] Molar ratio 1:5; acetone as solvent.

TABLE 6
Hydrocarboxylation of α-Methylstyrene in the Presence of $PdCl_2$ *and* (-)DIOP *Using Different Hydrogen Donors* (17,153)

Hydrogen donor	Temp (°C)	P_{CO}[a] (atm)	Time (hr)	Yield[b] (%)	Optical yield[c] (%)
H_2O[d]	146	431	24	50	13.4
CH_3OH	100	385	118	40	3.0
C_2H_5OH	100	382	142	67	6.3
sec-C_3H_7OH	100	397	118	75	14.2
sec-C_4H_9OH	100	386	41	n.d.	8.2
tert-C_4H_9OH	100	392	93	63	19.6
(-)(*S*)-$C_2H_5\overset{*}{C}H(CH_3)CH_2OH$[e]	100	390	46	70	0.2

[a] Initial pressure at the reaction temp.
[b] α-Phenylpropionate/β-phenylpropionate ≃ 99/1; when using $(PPh_3)_2PdCl_2$ as catalyst the ratio is 20/80.
[c] Prevailing absolute configuration: (*S*).
[d] Using THF as solvent; molar ratio H_2O/substrate = 1.1.
[e] Using $PdCl_2(PPh_3)_2$ as catalyst.

on the optical yield obtained in the hydrocarboxylation of α-methylstyrene (Table 6). The selectivity toward the expected carboxylic derivatives is relatively good in all cases; the most important secondary products are the corresponding saturated ethers (64). On using an alcohol as solvent, the highest optical yield was obtained with *tert*-butanol. With optically active 2-methylbutanol an asymmetric synthesis was achieved even if the ligand was not chiral (PPh_3 instead of (−)DIOP) (153).

3. *Substrates*

Many unsaturated substrates were subjected to hydrocarboxylation: monoolefins, conjugated and nonconjugated diolefins, allenes, and unsaturated molecules containing other functional groups. The main interest was in determining the nature of the reaction products. For the monoolefins, however, a reactivity scale was also determined (63,69).

a. Monoolefins. As shown in Table 7, monoolefins on hydrocarboxylation generally give rise to an isomeric mixture of carboxylic acids or their esters. In the presence of palladium catalysts, however, only methyl

TABLE 7

Catalytic Hydrocarboxylation of Olefins

Olefin	Hydrogen donor	Catalyst	Temp (°C)	P_{CO} (atm)	Products	Yield (%)	Ref.
Ethylene	H_2O	$Co_2(CO)_8$	285	200[a]	Propionic acid	85	78
Ethylene	CH_3OH	$H_2PtCl_6/SnCl_2$	80	800[a]	Methyl propionate	93	45
Propylene	H_2O	$Co_2(CO)_8$	130	123[b]	Butyric acid	64	75
					Isobutyric acid	20	
Propylene	CH_3OH	$Co_2(CO)_8$	145	150[b]	Methyl butyrate	74	76
					Methyl isobutyrate	19	
1-Butene	H_2O	$Co_2(CO)_8$	180	150[b]	Pentanoic acid	52	81
					2-Methylbutanoic acid	23	
1-Butene	CH_3OH	$Co_2(CO)_8$	180	150[b]	Methyl pentanoate	56	81
					Methyl 2-methylbutanoate	19	
2-Butene[d]	H_2O	$Co_2(CO)_8$	210	250[b]	Pentanoic acid	24	68
					2-Methylbutanoic acid	13	
2-Butene[d]	CH_3OH	$(PPh_3)_2PdCl_2/HCl$	70	700[c]	Methyl 2-methylbutanoate	95	26
Isobutylene	H_2O	$Ni(CO)_4/NiI_2$	250	200[a]	3-Methylbutanoic acid	37	49
					Trimethylacetic acid	6	

[a] Initial pressure at reaction temp.
[b] Initial pressure at 20°C.
[c] Constant pressure.
[d] *Cis,trans* mixture.

TABLE 7 (contd.)
Catalytic Hydrocarboxylation of Olefins

Olefin	Hydrogen donor	Catalyst	Temp (°C)	P_{CO} (atm)	Products	Yield (%)	Ref.
Isobutylene	C_2H_5OH	$(PPh_3)_2PdCl_2$/HCl	100	700[c]	Ethyl trimethylacetate	60	26
					Ethyl 3-methylbutanoate	30	
1-Pentene	H_2O	$Co_2(CO)_8$	145	180[c]	Hexanoic acid	52	82
					2-Methylpentanoic acid	17	
					2-Ethylbutanoic acid	5	
1-Pentene	H_2O	$Ni(OAc)_2$/HI	300	400[c]	Hexanoic acid	41	67
					2-Methylpentanoic acid	46	
1-Pentene	CH_3OH	$Co_2(CO)_8$	150	127[c]	Methyl hexanoate	58	82
					Methyl 2-methylpentanoate	18	
					Methyl 2-ethylbutanoate	5	
2-Pentene[d]	H_2O	$Co_2(CO)_8$	145	180[c]	Hexanoic acid	49	82
					2-Methylpentanoic acid	17	
					2-Ethylbutanoic acid	6	
2-Pentene[d]	H_2O	$Ni(OAc)_2$/HI	300	400[c]	Hexanoic acid	39	67
					2-Methylpentanoic acid	49	
2-Pentene[d]	CH_3OH	$Co_2(CO)_8$	150	140[c]	Methyl hexanoate	55	82
					Methyl 2-methylpentanoate	20	
					Methyl 2-ethylbutanoate	6	
2-Methyl-2-butene	H_2O	$Ni(OAc)_2$/HI	300	400[c]	3-Methylpentanoic acid	41	67
					4-Methylpentanoic acid	20	
					2,3-Dimethylbutanoic acid	20	

1-Hexene	H_2O	$Ni(OAc)_2$	300	400[c]	Heptanoic acid	32	67
					Branched C_7 acids	48	
1-Hexene	CH_3OH	$Co_2(CO)_8$	175	300[a]	Methyl heptanoate	75	69
					Methyl 2-methylhexanoate	19	
					Methyl 2-ethylpentanoate	6	
1-Octene	H_2O	Ni on silica	285	200[a]	Nonanoic acid 2-Methyloctanoic acid	84	49
1-Octene	C_2H_5OH	NiI_2/AcOH	220	70[b]	Ethyl 2-methyloctanoate	95	4*b*
2-Ethyl-1-hexene	H_2O	$Co_2(CO)_8$	165	165[a]	C_9 acids	57	63
2-Ethyl-1-hexene	CH_3OH	$Co_2(CO)_8$	165	160[a]	Methyl esters of C_9 acids	20	63
1-Dodecene	H_2O	Ni on silica	285	200[a]	2-Methyldodecanoic acid Tridecanoic acid	28	49
1-Dodecene	CH_3OH	$H_2PtCl_6/SnCl_2$	90	200[a]	Methyl tridecanoate	63	46
					Methyl esters of C_{13} branched acids	13	
1-Hexadecene	C_2H_5OH	NiO/NiO_2	220	100[a]	Ethyl esters of C_{17} acids	61	49
1-Octadecene	H_2O	Ni	285	200[a]	2-Methyloctadecanoic acid	67	49
1-Octadecene	C_2H_5OH	NiI_2	210–250	100[a]	Ethyl 2-methyloctadecanoate	65	49
Cyclohexene	H_2O	$Co_2(CO)_8$	165	197[a]	Cyclohexanecarboxylic acid	89	63
Cyclooctene	H_2O	$(PPh_3)_2PdCl_2$	120	700[c]	Cyclooctanecarboxylic acid	61	44
Cyclododecene	C_2H_5OH	$PdCl_2$/HCl	120	100[c]	Ethyl cyclododecanecarboxylate	93	83

[a] Initial pressure at reaction temp.
[b] Initial pressure at 20°C.
[c] Constant pressure.
[d] *Cis,trans* mixture

2-methylbutanoate was obtained from 2-butene (26). The reaction rate is greatly influenced by the nature of the olefinic substrate both in the synthesis of acids (63) and of esters (63,69).

According to Ercoli (63) the reaction rate in the presence of $Co_2(CO)_8$ decreases in the order: linear terminal olefins > cyclic olefins > terminal β-disubstituted olefins (vinylidene olefins). On the basis of the data reported by Gankin et al. (69) on the relative reactivity of various olefins in the synthesis of methyl esters with $Co_2(CO)_8$ as catalyst (Table 8), the reactivity sequence indicated by Ercoli may be enlarged as follows: linear α-olefins > internal olefins > cyclic olefins > vinylidene olefins. It may be added that, in the presence of $Co_2(CO)_8$, reactivity decreases as the molecular weight of the α-olefin increases. A similar sequence of reactivities was noticed by Levering and Glasebrook (67) in the synthesis of acids in the presence of $Ni(CO)_4$ promoted by HI. With the $H_2PtCl_6/SnCl_2$ catalyst, propylene seems to react more slowly than higher olefins (46).

Significant asymmetric induction in the hydrocarboxylation of prochiral olefins has been obtained up to now only by using palladium derivatives as catalysts. Table 9 gives some results obtained in asymmetric hydrocarboxylation of several terminal and internal olefins. The highest asymmetric

TABLE 8
Effect of Olefin Structure on Reaction Rate in the Synthesis of Methyl Esters[a] (69)

Olefin	Relative reaction rates
Propylene	1.00
1-Hexene	0.80
1-Heptene	0.77
2-Heptene	0.31
1-Octene	0.68
1-Decene	0.62
Isobutylene	0.21
2,4,4-Trimethyl-1-pentene	0.14
2-Methyl-1-pentene	0.17
2-Methyl-2-pentene	0.07
Cyclohexene	0.19

[a] $Co_2(CO)_8/CH_3OH$ (g/g) = 0.028; 0.25 mole of olefin; moles CH_3OH/moles olefin = 5; pyridine addition, 25% by weight; temp. 190°C; P_{CO} 300 atm, constant.

TABLE 9

Asymmetric Hydrocarboxylation of Olefins[a] (64, 17)

Olefin	Hydrogen donor	P_{CO}[b] (atm)	Products	Yield (%)	Prevailing configuration	Optical yield (%)
1-Butene	C_2H_5OH	300	Ethyl pentanoate	51		
			Ethyl 2-methylbutanoate	17	(*S*)	7.6
cis-Butene	C_2H_5OH	345	Ethyl pentanoate	16		
			Ethyl 2-methylbutanoate	64	(*R*)	7.2
trans-Butene	CH_3OH	340	Methyl pentanoate	39		
			Methyl 2-methylbutanoate	41	(*S*)	1.6
Styrene	CH_3OH	350	Ethyl 3-phenylpropanoate	18		
			Ethyl 2-phenylpropanoate	72	(*S*)	3.2
3-Methyl-1-butene	CH_3OH	285	Methyl 4-methylpentanoate	63		
			Methyl 2,3-dimethylbutanoate	7	(*S*)	10.3
3,3-Dimethyl-1-butene	CH_3OH	280	Methyl 4,4-dimethylpentanoate	86		
			Methyl 2,3,3-trimethylbutanoate	4	(*S*)	2.0
2-Methyl-1-butene	*t*-C_4H_9OH[c]	300	*t*-Butyl 3-methylpentanoate	60	(*R*)	4.3
2,3-Dimethyl-1-butene	*t*-C_4H_9OH[c]	300	*t*-Butyl 3,4-dimethylpentanoate	75	(*S*)	4.6
2,3,3-Trimethyl-1-butene	*t*-C_4H_9OH[c]	300	*t*-Butyl 3,4,4-trimethylpentanoate	65	(*S*)	19.5
α-Ethylstyrene	*t*-C_4H_9OH[c]	300	*t*-Butyl 3-phenylpentanoate	80	(*S*)	44.7

[a] Using $PdCl_2$ and (–)-DIOP (72) in molar ratio 1:2, temp 100°C.

[b] Initial pressure at room temp.

[c] Using benzene as solvent; moles alcohol/moles olefin = 1.5.

induction was obtained when using benzene as solvent and α-ethylstyrene as substrate (17,64).

b. Diolefins. Nonconjugated diolefins give rise to dicarboxylic acids and unsaturated monocarboxylic acids (Table 10). Saturated monocarboxylic acids were obtained from 1-5 hexadiene when using $Co_2(CO)_8$ as catalyst (84). In the presence of $Pd(PBu_3)_2I_2$, a formal hydrocarboxylation followed by an intramolecular cyclization takes place (80) (reaction 4).

$$\begin{array}{l} CH_2CH{=}CH_2 \\ | \\ CH_2CH{=}CH_2 \end{array} \xrightarrow[Pd(PBu_3)_2I_2]{CO+CH_3OH} \text{2-methyl-5-(}CH_2COOCH_3\text{)cyclopentanone } (CH_3,\ {=}O,\ CH_2COOCH_3) \qquad (4)$$

Data for the carboxylation of 1,5,9-cyclododecatriene are summarized in Table 11. In this case, as for nonconjugated linear diolefins, mixtures of tricarboxylic acids, monounsaturated dicarboxylic acids, and diunsaturated monocarboxylic acids are formed. With palladium catalysts, even small changes in reaction temperature (26) or type of ligand bound to palladium (85) markedly influence the composition of the reaction products.

The most used catalysts for hydrocarboxylation of conjugated diolefins are $Co_2(CO)_8$ (41,88) and palladium derivatives (43). The carbonyls of iridium, rhodium (39,156), iron, and nickel are very poor catalysts (41), but with $Ni(CO)_4$, good results were obtained in stoichiometric reactions when using cyclopentadiene as substrate (52). From butadiene (Table 12) with $Co_2(CO)_8$ present, 3-pentenoic acid and valeric acid (or their esters) are formed along with a mixture of dicarboxylic acids, mainly adipic acid (Scheme **II**).

The solvent used has a great effect on the proportion of dicarboxylic derivatives formed: Pyridine seems to determine the preferential formation of these compounds (90–92,157), whereas with other solvents like dioxane, acetone, or isoquinoline (88), the main product is 3-pentenoic acid. Valeric acid, formed in good yield in the synthesis of acids from butadiene, is practically absent, both as free acid and as ester, among the reaction products in the corresponding synthesis of esters. The hydrogenation of 3-pentenoic acid during the acid synthesis might be due to the H_2 formed by reaction of water with CO (the water–gas shift). Under hydrocarboxylation conditions, butadiene may also dimerize to vinylcyclohexene (40,43,88), which in turn may be hydrocarboxylated (48). The synthesis of ethyl nona-3,8-dienoate from butadiene through dimerization and hydrocarboxylation in the presence of $Pd(acac)_2$ and PPh_3 was reported (155).

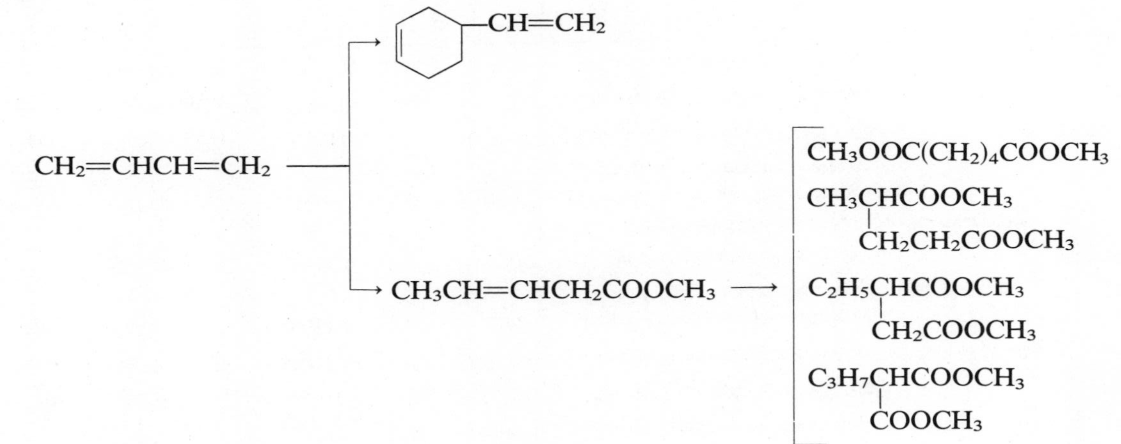

CH₂=CHCH=CH₂
CH=CH₂
CH₃CH=CHCH₂COOCH₃
CH₃OOC(CH₂)₄COOCH₃
CH₃CHCOOCH₃
CH₂CH₂COOCH₃
C₂H₅CHCOOCH₃
CH₂COOCH₃
C₃H₇CHCOOCH₃
COOCH₃

Scheme **II**

TABLE 10
Hydrocarboxylation of Nonconjugated Dienes

Diene	Catalyst	Hydrogen donor	Temp (°C)	P_{CO} (atm)	Products	Yield (%)	Ref.
1,5-Hexadiene	$Co_2(CO)_8$[a]	H_2O	210	250[b]	Suberic acid	34	84
					Saturated C_7-monocarboxylic acids	26	
1,5-Hexadiene	$Ni(CO)_4$	H_2O	180	n.d.	2-Methyl-4-vinylbutyric acid	20	49
1,5-Hexadiene	$Pd(PBu_3)_2I_2$[c]	CH_3OH	150	1000[d]	Methyl 2-keto-3-methylcyclopentylacetate	50	80
4-Vinylcyclohexene	$Pd(PPh_3)_2Cl_2$	CH_3OH	60	300–700[e]	Methyl 2-(3-cyclohexenyl)propionate	85	26
4-Vinylcyclohexene	$Pd(PPh_3)_2Cl_2$	CH_3OH	120	300–700[e]	Methyl 2-(methoxycarbonylcyclohexyl)propionate	80	26
1,5-Cyclooctadiene	PdI_2	C_2H_5OH	100	100[b]	Ethyl 4-cyclooctene-1-carboxylate	5	85
					Diethyl cyclooctane-1,4(and 1,5)-dicarboxylate	95	
1,5-Cyclooctadiene	$Pd(acac)_2$/HCl	C_2H_5OH	100	100[b]	Ethyl 4-cyclooctene-1-carboxylate	75	85
					Diethyl cyclooctane-1,4(and 1,5)-dicarboxylate	4	
1,5-Cyclooctadiene	$Pd(PBu_3)_2I_2$[c]	CH_3OH	150	1000[d]	Methyl 4-cyclooctene-1-carboxylate	45	80
					Dimethyl cyclooctanedicarboxylate	30	
1,5-Cyclooctadiene	$Pd(PPh_3)_2Cl_2$	C_2H_5OH	60	300–700[e]	Ethyl 4-cyclooctene-1-carboxylate	95	26
1,5-Cyclooctadiene	$Pd(PPh_3)_2Cl_2$	C_2H_5OH	100	300–700[e]	Diethyl cyclooctanedicarboxylate	80	26
1,5-Cyclooctadiene	$Pd(PPh_3)_2Cl_2$	H_2O	130	400[e]	4-Cyclooctene-1-carboxylic acid	51	44

[a] In the presence of pyridine.
[b] Initial pressure at 20°C.
[c] In the presence of *p*-toluenesulfonic acid.
[d] Pressure at reaction temp.
[e] Constant pressure.

TABLE 11
Hydrocarboxylation of 1,5,9-*Cyclododecatriene*

Catalyst	Hydrogen source	Temp (°C)	P_{CO} (atm)	Products	Yield (%)	Ref.
$PdCl_2/HCl$	C_2H_5OH	100	100[a]	Ethyl cyclododecadienecarboxylate	Trace	83
				Diethyl cyclododecenedicarboxylate	85	
$Pd(PPh_3)_2Br_2/HCl$	CH_3OH	80	700	Methyl 5,9-cyclododecadiene-1-carboxylate	40	86
				Dimethyl 9-cyclododecene-1,5-dicarboxylate	8	
$Pd(PPh_3)_2Cl_2/HCl$	C_2H_5OH	35–50	300–700	Ethyl 4,8-cyclododecadiene-1-carboxylate	>90	26
$Pd(PPh_3)_2Cl_2/HCl$	C_2H_5OH	50–70	300–700	Diethyl 5-cyclododecenedicarboxylate	70–80	26
$Pd(PPh_3)_2Cl_2/HCl$	C_2H_5OH	> 70	300–700	Triethyl cyclododecanetricarboxylate	70–80	26
$Pd(PPh_3)_2Cl_2$	H_2O	130	700	5,9-Cyclododecadiene-1-carboxylic acid	16	44
Co/CoS	CH_3OH	175	275	Methyl cyclododecanecarboxylate	57	87

[a] Initial pressure at 20°C; other experiments at constant pressure.

TABLE 12
Hydrocarboxylation of Conjugated Diolefins

Diene	Hydrogen donor	Catalyst	Temp (°C)	P_{CO} (atm)	Products	Yield (%)	Ref.
Butadiene	H_2O	$Co_2(CO)_8$	140	3000^a	*trans*-3-Pentenoic acid	75.0	89
Butadiene	H_2O	$Co_2(CO)_8$ [b]	220	250^a	3-Pentenoic acid	<1	90
					Adipic acid	16	
					α-Methylglutaric acid	7.5	
					Ethylsuccinic acid	25.5	
					α,α'-Dimethylsuccinic acid	≤3	
					Propylmalonic acid	<0.5	
					Valeric acid	20	
					α-Methylbutyric acid	<2	
Butadiene	CH_3OH	$Co_2(CO)_8$ [b]	140	600^d	Methyl 3-pentenoate	58	88
					4-Vinylcyclohexene	Small	
Butadiene	CH_3OH	$Co_2(CO)_8$ [b]	210	400^c	Methyl 3-pentenoate	23	91
					Methyl adipate	43	
					Methyl α-methylglutarate	7	
					Methyl ethylsuccinate	3	
					Methyl propylmalonate	<0.1	
Butadiene	CH_3OH	Rh_2O_3	150	60–70[c]	Methyl 4-pentenoate	27.6	40
					Methyl 3-pentenoate	18.8	
					Methyl 2-pentenoate	1.9	
					4-Vinylcyclohexene	13	
					Butenes	2.8	

Butadiene	H_2O	$Pd(PPh_3)_2Cl_2/HCl$	120–140	700[c]	3-Pentenoic acid	70	44
Butadiene	CH_3OH	$Pd(PBu_3)_2Br_2$	150	1000[d]	Methyl 3-pentenoate	73	43
					Methyl 4-pentenoate	1	
					4-Vinylcyclohexene	1–2	
1,3-Pentadiene	CH_3OH	Na_2PdI_4	70	1000[d]	Methyl 2-methylpentenoate	34	43
Isoprene	CH_3OH	Na_2PdI_4	70	1000[d]	Methyl 3-methyl-3-pentenoate	15	43
					Methyl 4-methyl-3-pentenoate	38	
					Methyl 4-methyl-4-pentenoate	10	
Cyclopentadiene	CH_3OH	Na_2PdI_4	70	1000[d]	Methyl 2-cyclopentenecarboxylate	73	43
2,3-Dimethyl-1,3-butadiene	CH_3OH	Na_2PdI_4	70	1000[d]	Methyl 3,4-dimethyl-3-pentenoate	50	43
1,3-Cyclohexadiene	C_2H_5OH	$PdCl_2$	100	100[a]	Ethyl 2-cyclohexenecarboxylate	80	85
1,3-Cyclooctadiene	CH_3OH	$Pd(PBu_3)_2I_2$[e]	150	1000[d]	Methyl 2-cyclooctenecarboxylate	14	80
					Saturated diesters	3	
1,3-Cyclooctadiene	C_2H_5OH	$PdCl_2$	100	100[a]	Ethyl 2-cyclooctenecarboxylate	9.5	85
					3-Ethoxycyclooctene	1.3	
Cyclooctatetraene	H_2O	$Ni(CO)_4$	270	200[c]	Dicarboxylic acids of dimeric cyclooctatetraene	51	53

[a] Initial pressure measured at 20°C.
[b] In the presence of pyridine.
[c] Constant pressure.
[d] Initial pressure at reaction temp
[e] In the presence of *p*-toluenesulfonic acid.

For the hydrocarboxylation of other conjugated diolefins, palladium was mainly used as catalyst in the presence of methanol or ethanol while using relatively low temperatures (56–150°C) and very high CO pressures (1000 atm). Unsaturated monocarboxylic esters were obtained; the yields were higher (70–80%) with cyclic diolefins like cyclopentadiene (43) or 1,3-cyclohexadiene (85).

The rate of reaction decreases in this order: butadiene, isoprene, 1,5-hexadiene > piperylene, 2,4-hexadiene > 2,3-dimethylbutadiene-1,3 (93). With butadiene the synthesis of acids in the presence of $Co_2(CO)_8$ is faster than that of esters (94).

Propadiene is practically the only allene studied in the hydrocarboxylation reaction (36,45,95–97). In the presence of platinum and nickel, and also ruthenium catalysts, while operating at 110–150°C with 24–900 atm of CO, the expected methacrylic esters are formed when methanol or ethanol is present (36,45,95) (Table 13). At higher temperatures in the presence of iron and ruthenium catalysts, tarry materials are formed in greater amounts, and fairly low yields of the methacrylic esters or acids are obtained along with the products formed by carboxylation of cyclic dimers or trimers of propadiene (36). However a high yield of methyl methacrylate was claimed from the reaction of allene with $Fe(CO)_5$ and CH_3OH in the presence of α-picoline (158). Low yields are reported in the hydrocarboxylation of 1,2-hexadiene with $Ni(CO)_4$ and methanol.

c. Unsaturated Substrates Containing Functional Groups. A fairly large number of unsaturated substrates containing various functional groups were hydrocarboxylated in the presence of nickel, cobalt, and palladium catalysts. The functional groups in the unsaturated molecules (Table 14) are mainly halogens, carboxy, carboalkoxy, and cyano groups. The products formed in these reactions are mixtures of isomeric esters or acids as expected in this type of synthesis.

Particularly good yields are obtained by reacting vinyl chloride in the presence of $Pd(PPh_3)_2Cl_2$ and HCl (26). The same catalyst precursor was used in the hydrocarboxylation of polyunsaturated fatty acids and esters (159). Pyridine and hydrogen gas help obtaining good results in the hydrocarboxylation of substrates like acrylonitrile (98) and methyl acrylate (99) in the presence of $Co_2(CO)_8$.

With $Co_2(CO)_8$ as catalyst, crotonic acid yields glutaric acid in addition to the expected methylsuccinic acid. Undecylenic and oleic acids, in the presence of $Ni(CO)_4$, yield mixtures of dicarboxylic derivatives with 12 or 19 carbon atoms, respectively (Table 14).

When the functional group in the unsaturated molecule is a hydroxyl

TABLE 13
Hydrocarboxylation of Allenes

Substrate	Hydrogen donor	Catalyst	Temp (°C)	P_{CO}[a] (atm)	Products	Yield (%)	Ref.
Allene	CH_3OH	$PtCl_2/SnCl_2$	90	975[b]	$CH_2{=}C(CH_3){-}COOCH_3$	39	45
Allene	CH_3OH	$Ni(CO)_4$[c]	140	24	$CH_2{=}C(CH_3){-}COOCH_3$	62	95
Allene	CH_3OH	$Fe(CO)_5$	180–190	700–900	$CH_2{=}C(CH_3){-}COOCH_3$	25	36
					3-methylene-1-methylcyclobutane-1-COOCH$_3$ ($CH_2{=}$cyclobutane with CH_3 and $COOCH_3$)	Small amounts	
Allene	CH_3OH	"$Ru_2(CO)_9$"	135–150	700–900	$CH_2{=}C(CH_3){-}COOCH_3$	50	36

[a] Constant pressure.
[b] Initial pressure measured at room temp.
[c] Methacrylic acid as cocatalyst.
[d] $RuCl_3{\cdot}3H_2O/RuCl_4{\cdot}H_2O$; pyridine added.
[e] CH_3COOH and H_2O as cocatalysts.

TABLE 13 (cont'd.)
Hydrocarboxylation of Allenes

Substrate	Hydrogen donor	Catalyst	Temp (°C)	P_{CO}[a] (atm)	Products	Yield (%)	Ref.
Allene	CH_3OH	"$Ru_2(CO)_9$"	140–200	700–1000	$CH_2{=}C(CH_3){-}COOCH_3$	18	36
					$CH_3OOC{-}C(CH_3)_2{-}CH_2{-}C({=}CH_2){-}COOCH_3$	11	
Allene	H_2O	$RuCl_x$[d]	120	700–1000	$CH_2{=}C(CH_3){-}COOH$	20	36
Allene	H_2O	$RuCl_x$[d]	175	800–1000	$CH_2{=}C(CH_3){-}COOH$	5	36
					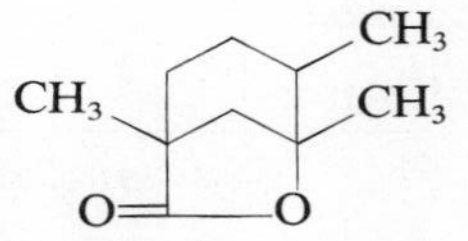 bicyclic lactone (structure: CH_3, CH_3, CH_3, $O{=}C{-}O$)	40	

					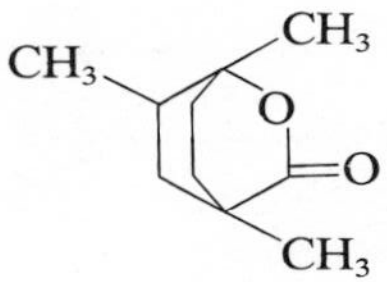 bicyclic lactone (CH_3, CH_3, CH_3, O, =O)	8	
Allene	CH_3OH	$RuCl_x$[d]	175	1000	$CH_2{=}C(CH_3){-}COOCH_3$	10	36
					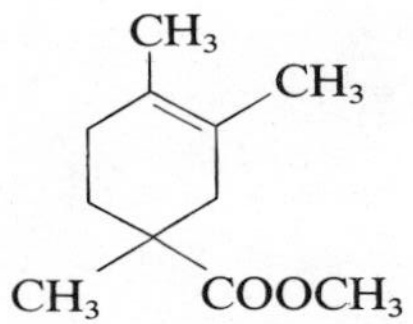 cyclohexene ring (CH_3, CH_3, CH_3, $COOCH_3$)	12	
1,2-Hexadiene	CH_3OH	$Ni(CO)_4$[e]	155	28	$CH_3CH_2CH_2CH{=}C(CH_3){-}COOCH_3$	12	95
					$CH_3CH_2CH_2CH_2{-}C({=}CH_2){-}COOCH_3$	3	

[a] Constant pressure.
[b] Initial pressure measured at room temp.
[c] Methacrylic acid as cocatalyst.
[d] $RuCl_3 \cdot 3H_2O/RuCl_4 \cdot H_2O$; pyridine added.
[e] CH_3COOH and H_2O as cocatalysts.

TABLE 14

Hydrocarboxylation of Olefinic Substrates Containing Functional Groups

Substrate	Hydrogen donor	Catalyst	Temp (°C)	P_{CO}[a] (atm)	Products	Yield (%)	Ref.
Vinyl chloride	C_2H_5OH	$Pd(PPh_3)_2Cl_2$, HCl	90–120	700	Ethyl α-chloropropionate	80	26
					Ethyl β-chloropropionate	5	
Acrylonitrile	CH_3OH	$Co_2(CO)_8$	94	140	Methyl α-cyanopropionate	79	98
					Methyl β-cyanopropionate	10	
Methyl acrylate	CH_3OH	$Co_2(CO)_8$	110	160	Dimethyl methylmalonate	3	99
					Dimethyl succinate	75	
					Dimethyl γ-ketopimelate	13	
3,3-Dimethoxypropene	CH_3OH	$Co_2(CO)_8$	180	250[b]	Methyl 4,4-dimethoxybutanoate	41	100
Crotonic acid	H_2O	$Co_2(CO)_8$	150	200[c]	Glutaric acid, Methylsuccinic acid	52	63
Methyl 3-butenoate	CH_3OH	$Pd(PPh_3)_2Cl_2/HCl$	120	700	Dimethyl methylsuccinate	67	86
Methyl 3-butenoate	CH_3OH	$Pd(PPh_3)_2Cl_2/HCl$	90–120	700	Dimethyl methylsuccinate	80	26
					Dimethyl glutarate	10	
Diethyl maleate	C_2H_5OH	$Pd(PPh_3)_2Cl_2$	140	700	Triethyl 1,1,2-ethanetricarboxylate	70	26
3-(β-cyanoethoxy)propene	CH_3OH	$Co_2(CO)_8$	120	200	Methyl 4-(β-cyanoethoxy)butyrate	52	101
					Methyl 2-methyl-3-(β-cyanoethoxy)propionate	2	
					Propionaldehyde dimethylacetal	28	
Phenyl vinylsulfonate	C_2H_5OH	$Pd(PPh_3)_2Cl_2$, HCl, $Pd(Pyr)(PPh_3)Cl_2$	90	700	Phenyl 3-ethoxycarbonylethane sulfonate	36	102
Ethyl 10-undecenoate	C_2H_5OH	$Ni(CO)_4$, CuI	270	200	Diethyl dodecanedicarboxylate	38	49
10-Undecenoic acid	H_2O	$Co_2(CO)_8$	145	210[c]	Dodecanedicarboxylic acid	76	63
Methyl oleate	CH_3OH	$Ni(CO)_4$	280	200	Dimethyl nonadecanedicarboxylate	36	49
Diethyl tetrahydrophthalate	C_2H_5OH	$Pd(PPh_3)_2Cl_2/HCl$	90	700	Triethyl cyclohexane-1,2,4-tricarboxylate	86	86
Diethyl bicyclo[2.2.1]4-heptene-1,2-dicarboxylate	C_2H_5OH	$Pd(PPh_3)_2Cl_2/HCl$	90	700	Triethyl bicyclo[2.2.1]heptane-1,2,4-tricarboxylate	52	86

[a] Constant pressure except when noted otherwise.
[b] Initial pressure at 20°C.
[c] Initial pressure at reaction temp.

group, hydrocarboxylation does not give the products normally expected but gives lactones due to an "intramolecular" reaction (Scheme **III**). Five-membered and six-membered rings are formed, the former generally being preponderant. The presence of an alkyl substituent in position 3 favors formation of the six-membered ring lactone. Reppe and Kröper (49) first described this reaction, which was later thoroughly investigated by Falbe and others (20) and by Matsuda (23).

$$CH_3CH{=}CHCH_2OH \xrightarrow[Co_2(CO)_8]{CO} \text{(five-membered lactone)} + \text{(six-membered lactone)}$$

$$\downarrow$$

$$CH_3CH_2CH_2CHO$$

$$\swarrow \qquad \searrow$$

$$CH_3CH_2\underset{\underset{\displaystyle C_3H_7}{|}}{\underset{CH}{\underset{\|}{C}}}CHO \qquad\qquad C_3H_7COOC_4H_9$$

Scheme **III**

Table 15 summarizes results of typical experiments. In the hydrocarboxylation of allyl alcohol, a 60% yield of α-butyrolactone may be obtained with $Co_2(CO)_8$ in acetonitrile as solvent (23), but the yield was only 2% when benzene was used (20). Rhodium (20) and nickel derivatives (49) were also used as catalysts, but $Co_2(CO)_8$ seems the best so far.

The main side reaction, which may affect the yield of lactones considerably, is the isomerization of primary and secondary unsaturated alcohols to saturated aldehydes and ketones, respectively (Scheme **III**). Unsaturated tertiary alcohols tend to dehydrate to give diolefins which then are hydrogenated to monoolefins (20).

4. *Secondary Reactions*

Several secondary reactions may take place during the hydrocarboxylation of olefinic substrates (reactions 5–13) especially when the reaction is carried out at high temperatures while using alcohols as hydrogen donors.

$$2CH_3OH \rightarrow CH_3OCH_3 + H_2O \tag{5}$$

$$CH_3{-}\overset{\displaystyle CH_3}{\underset{\displaystyle CH_3}{\overset{|}{\underset{|}{C}}}}{-}OH \rightarrow CH_2{=}\overset{\displaystyle CH_3}{\underset{\displaystyle CH_3}{\overset{|}{\underset{|}{C}}}} + H_2O \tag{6}$$

TABLE 15
Hydrocarboxylation of Unsaturated Alcohols[a] (20)

Substrate	Products	Yield, %
Allyl alcohol[b]	γ-Butyrolactone	60
	Propionaldehyde	19
	Isobutyraldehyde	4
Methallyl alcohol	β-Methyl-γ-butyrolactone	2
	Isobutyraldehyde	17
Crotyl alcohol	α-Methyl-γ-butyrolactone	2
	δ-Valerolactone	0.5
	Butyraldehyde	23
	2-Ethyl-2-hexenal	3
	Butyl butyrate	14
2,2-Dimethyl-3-buten-1-ol	α,β,β-Trimethyl-γ-butyrolactone	51
	γ,γ-Dimethyl-δ-valerolactone	14
2,2,3-Trimethyl-3-buten-1-ol	α,α,β,β-Tetramethyl-γ-butyrolactone	3
	β,γ,γ-Trimethyl-δ-valerolactone	25
2-Methyl-2-ethyl-3-buten-1-ol	α,β-Dimethyl-β-ethyl-γ-butyrolactone	40
	γ-Methyl-γ-ethyl-δ-valerolactone	13
4-Hydroxymethylcyclohexene[c]	2-Hydroxymethylcyclohexanecarboxylic acid γ-lactone	6
	Hydroxymethylcyclohexane	5
	Cyclohexylmethyl cyclohexanecarboxylate	10
	Cyclohexanecarboxaldehyde	13
1-Hexen-4-ol	α-Methyl-γ-ethyl-γ-butyrolactone	2
	Ethyl propyl ketone	73
1-Hepten-4-ol	α-Methyl-γ-propyl-γ-butyrolactone	2
	Dipropyl ketone	70
2-Methyl-4-penten-2-ol	α,γ,γ-Trimethyl-γ-butyrolactone	10
	δ,δ-Dimethyl-δ-valerolactone	2
3-Methyl-5-hexen-3-ol	α,α-Dimethyl-γ-ethyl-γ-butyrolactone	29
	δ-Methyl-δ-ethyl-δ-valerolactone	6
4,6-Dimethyl-1-hepten-4-ol	α,γ-Dimethyl-γ-isobutyl-γ-butyrolactone	10
	δ-Methyl-δ-isobutyl-δ-valerolactone	2

[a] P_{CO} 300 atm, constant; temp. 240°C; catalyst $Co_2(CO)_8$; benzene solvent.

[b] P_{CO} 400 atm (including 4% H_2, constant; temp. 130°C; acetonitrile solvent; pyridine added (23).

[c] Temp. 210°C.

$$CH_3CHOHCH_3 \rightarrow CH_3COCH_3 + H_2 \qquad (7)$$

$$H_2O + CO \rightarrow CO_2 + H_2 \qquad (8)$$

$$>C=C< + H_2 \rightarrow >CH{-}CH< \qquad (9)$$

$$>C=C< + H_2 + CO \rightarrow -\overset{|}{\underset{|}{C}}H{-}\overset{|}{\underset{|}{C}}{-}CHO \qquad (10)$$

$$2>C=C< + H_2 + CO \rightarrow >CH\overset{|}{\underset{|}{C}}CO\overset{|}{\underset{|}{C}}CH< \qquad (11)$$

$$>C=C< + ROH \rightarrow >CHCOR< \qquad (12)$$

$$ROH + CO \rightarrow RCOOH \qquad (13)$$

In the synthesis of esters the alcohol may undergo the reactions shown in equations 5–7, 12, and 13. In some cases a good part of the alcohol gives, through reactions 5–7, a relatively large amount of water or H_2. Hydrogen may be formed also by the water–gas shift reaction (reaction 8). The olefinic substrate, in the presence of H_2 and CO, then may give either the corresponding paraffin (reaction 9), ketone (reaction 11), or, by hydroformylation, aldehyde (reaction 10). The importance of these side reactions depends on the catalyst used, the reaction temperature and pressure, the nature of the solvent, and the alcohol and olefin employed. Reactions 5–11 are mainly observed when using nickel or cobalt catalysts at high temperature; reaction 12 was observed in the presence of palladium catalysts (64).

In the hydrocarboxylation of 1-octene in the presence of nickel and ethanol, Reppe and Kröper (49) obtained a conversion of 25% of the alcohol to propionic acid and 12.5% to ethyl propionate.

An unusually large hydrogenation of the substrate was observed during hydrocarboxylation of propylene in the presence of methanol when using $[Ru(CO)_3Br_2]_2$ as catalyst; on operating at 190°C with 22 atm of CO, 7% of the olefin was transformed into propane (38). In the presence of the $PdCl_2/PPh_3$ catalytic system, hydrogenation of the olefinic substrate was observed at low CO pressures (10–20 atm) (47). The effect of added hydrogen under these conditions was investigated by Fenton (103). The possibility that conjugated diolefins undergo hydrogenation before hydrocarboxylation was discussed by Rudkovskii et al. (93,160).

Aldehydes were always detected among the reaction products in the synthesis of acids and esters when using either cobalt (66,75,76,82) or ruthenium (38) catalysts. In hydrocarboxylation of cyclohexene at 165°C, 5–10% of the olefin is transformed to aldehyde in the presence of water, 11–13% in the presence of methanol, and about 20% in the presence of isopropanol (66). In the synthesis of esters, the yield of aldehydes increases with decreasing CO pressure.

Diethyl ketone is the most important secondary product of the hydrocarboxylation of ethylene and may even become the main reaction product (64%) (37,61,104–108). Ketones (104) were found (12%) among the hydrocarboxylation products of cyclohexene, of propylene, and of methyl acrylate; cyclic ketoesters were obtained from dienes (69) (reaction 4). Detailed discussions of the side reactions encountered in this synthesis in the presence of nickel or cobalt catalysts were published by Reppe and Kröper (49), and by Pino and Ercoli (66).

Besides the previously discussed secondary reactions that mainly involve different kinds of intermolecular hydrogen transfer processes, other secondary reactions were observed that are concerned with intramolecular hydrogen shifts in the unsaturated substrate.

Double-bond shifts during hydrocarboxylation have been little investigated. In the synthesis of esters and acids from 1-pentene and 2-pentene, the variation of the isomeric composition of the reaction products with temperature in the presence of cobalt catalyst was attributed to an isomerization of the substrate preceding hydrocarboxylation.

At a CO pressure of 200 atm, isomerization is slower than hydrocarboxylation at 140–150°C, but it is faster at 200°C (109). During the hydrocarboxylation of 1-pentene at 140°C, the unreacted substrate is rapidly isomerized at a CO pressure of 45 atm, but less than 10% isomerization is observed at a CO pressure of 186 atm (82) (Table 16).

TABLE 16
Olefin Isomerization During Hydrocarboxylation of 1-*Pentene Using* $Co_2(CO)_8$ (82)

Hydrogen source	Temp (°C)	P_{CO}[a] atm	Residual olefin composition[b] (%) 1-Pentene	2-Pentene[c]
H_2O	140	45	52	48
	140	186	92	8
CH_3OH	150	30	44	56
	150	210	93	7

[a] Constant pressure.
[b] Olefin conversion ~20%.
[c] Mixture of *cis* and *trans* isomers.

This behavior is similar to that shown by the same olefins under hydroformylation conditions and was discussed in the second chapter.

Very little is known about substrate isomerization under hydrocarboxylation conditions with other catalysts. With palladium in the presence of an excess of PPh_3, very little or no double-bond shift occurs (26,47,73). In the presence of (–)-DIOP, however, products arising from attack of CO on carbon atoms not part of the original double bond were isolated (65*a*). The same observation was made on using nickel and platinum catalysts and different olefinic substrates (Table 7). In most of these cases it is not known whether an isomerization of substrate takes place with formation of isomerized olefins or whether an intramolecular isomerization of the substrate–catalyst system takes place without formation of free isomerized olefins.

5. Isomeric Composition of the Reaction Products

As with hydroformylation, the main drawback of hydrocarboxylation of olefins, from the viewpoint of laboratory as well as of industrial applications, is formation of two or more isomeric products. Many attempts therefore were made to determine and to control the composition of the isomeric products from hydrocarboxylation. Although these efforts were not very successful, we shall briefly review some aspects of the problem closely connected with the reaction mechanism.

a. Effect of Type of Metal Used as Catalyst and of Added Ligands. In Table 17 the isomeric product composition obtained in the synthesis of acids and esters from some terminal olefins with cobalt catalysts at high CO pressures is compared with that from hydroformylation. In all cases the linear isomer prevails, but the percentage of straight-chain isomer

TABLE 17

Straight-Chain Isomer Formation in Hydroformylation and Hydrocarboxylation of Olefins Using $Co_2(CO)_8$[a] (109)

Olefin	Yield (%)		
	By hydroformylation	By hydrocarboxylation	
		Acids	Methyl esters
Propylene	76[b]	75	79
1-Butene	74[b]	69	75
1-Pentene	74[b]	65	64

[a] P_{CO} 150 atm (constant); temp. 180°C.

[b] Extrapolated values on the basis of experiments done under nonisomerizing conditions (second chapter).

TABLE 18
Hydrocarboxylation of Olefins Using Various Catalysts

Olefin	Hydrogen source	Catalyst	Temp (°C)	P_{CO} (atm)	Linear isomer (%)	Ref.
Propylene	H_2O	$Co_2(CO)_8$	130	123[a]	75	75
Propylene	H_2O	$Ni(CO)_4$	250	50[a]	43	75
Propylene	H_2O	$Ru_3(CO)_{12}$	190	20	52	109
Propylene	CH_3OH	$Co_2(CO)_8$	145	150[a]	79	76
Propylene	CH_3OH	$Ru_3(CO)_{12}$	190	24	46	38
Propylene	C_2H_5OH	$PdCl_2$[b]	80	50[a]	33	62[a]
1-Pentene	H_2O	$Co_2(CO)_8$	240	270	63	82
1-Pentene	H_2O	$Ni(CO)_4$	250	160	61	82
1-Pentene	H_2O	$Ru_3(CO)_{12}$	200	20	50	82
1-Pentene	CH_3OH	$Co_2(CO)_8$	200	240	64	82
1-Pentene	CH_3OH	$Ru_3(CO)_{12}$	200	20	42	82

[a] Initial pressure measured at 20°C; otherwise at constant pressure.
[b] In the presence of HCl.

decreases much more in hydrocarboxylation than in hydroformylation when passing from propylene to 1-pentene. The type of metal used, in the absence of added ligands (like phosphites, phosphines, arsines, etc.), seems to have a marked effect on the isomeric composition (Table 18); however, reaction conditions (CO pressure and temperature) are in some cases quite different and the comparison has only qualitative value. In all cases examined, nickel and ruthenium carbonyls as well as palladium catalysts seem to shift, more than cobalt, the isomeric composition toward branched isomers. Branched isomers prevail in some cases; see also Tables 7–11.

The influence of added ligands on the composition of the isomeric products was investigated. The type of ligand bound to the metal seems to have a large effect.

In the stoichiometric synthesis of acids from $Ni(CO)_4$ and 1-heptene, Chiusoli and Merzoni (51) obtained in the ester mixture, 50% of branched esters when using HI, 57% when using HCl, but 94% when using either acetic, monochloroacetic or *o*-hydroxybenzoic acid.

In the synthesis of acids and esters using cobalt catalysts, the presence of pyridine increases the relative amount of linear products (69,84,110), The palladium-catalyzed hydrocarboxylation was mostly investigated in the presence of phosphines (Table 19). In the case of propylene an ethyl butyrate/ethyl isobutyrate ratio of 0.44 was found in the presence of PPh_3 (26,86). In this case acids cause a marked enhancement of the

TABLE 19

Effect of Different Ligands on Composition of Isomeric Products in Palladium-catalyzed Hydrocarboxylations

Olefin	Temp (°C)	P_{CO}[a] (atm)	Alcohol	Ligand	Linear or less-branched isomer (%)	Ref.
Propylene	90	700[b]	EtOH	PPh_3	33	26
Propylene	100	720	MeOH	DIOP	64	162
Styrene	90	700[b]	EtOH	PPh_3	95	26
Styrene	100	380	MeOH	DPE[c]	67	153
Styrene	100	460	MeOH	DIOP	80	162
1-Butene	100	460	MeOH	PPh_3	44	162
1-Butene	100	440	MeOH	DPE[c]	67	153
1-Butene	100	450	EtOH	DIOP	75	162
1-Butene	100	370	MeOH	DMOP[d]	73	153
1-Dodecene	100	410	EtOH	PPh_3	49	162
1-Dodecene	100	380	EtOH	DIOP	67	162
α-Methylstyrene	100	390	$EtCH(CH_3)CH_2OH$	PPh_3	19	153
α-Methylstyrene	100	390	EtOH	DIOP	99	162
Isobutylene[e]	90	700[b]	EtOH	PPh_3	33	26
2-Ethyl-1-hexene	100	480	MeOH	DIOP	75	153

[a] Initial pressure measured at room temp.
[b] Reactions at constant pressure.
[c] DPE = 1,2-bis(diphenylphosphino) ethane.
[d] DMOP = 2,3-O-Methylene-2,3-dihydroxy-1,4-bis(diphenylphosphino)butane.
[e] In the presence of HCl.

reaction rate but a relatively small change in the composition of the reaction products (26). In the absence of added acids, the type of phosphine greatly influences the composition of the isomeric reaction products; diphosphines favor formation of linear isomers (161,162), but only with α-methylstyrene was practically just one isomer obtained.

The effect of the type of phosphine and of the phosphine to palladium ratio on the isomeric composition of the product arising from the hydrocarboxylation of 1-octene was investigated by Fenton (47) using water as hydrogen source. At 125°C the percentage of straight-chain isomer increases from 66 to 78 by increasing the $PPh_3/PdCl_2$ molar ratio from 4 to 13. Nine different triarylphosphines were used as added ligands; the other reaction conditions being the same, the percentage of straight-chain acid varies from 50 in the presence of tris(*p*-anisyl)phosphine to 70 in the presence of tris(*o*-tolyl)phosphine. The same author also investigated the effect of other reagents on the product composition (47,103,111,112).

The addition of $SnCl_2$ to $PdCl_2(PPh_3)_2$ in the hydrocarboxylation of propylene increases the straight-chain/branched-chain ratio in the esters formed (179).

b. Effect of Reaction Variables. Systematic studies on the effect of different factors on composition of the isomeric products from hydrocarboxylation were carried out with propylene, 1-butene, 1-pentene, 2-pentene (82,109), and 1-hexene (69) by using $Co_2(CO)_8$ as catalyst and with 1-octene in the presence of $PdCl_2/PR_3$ catalytic systems (47). The variables investigated were (besides the already-discussed effect of the type of catalytic system) temperature, CO partial pressure, catalyst concentration, and concentration and type of hydrogen donor.

In Table 20 is reported the maximum percentage of linear isomer obtained in the acid and ester syntheses from 1-pentene when varying temperature, CO pressure, and hydrogen donor concentration while using $Co_2(CO)_8$ as catalyst. The studies indicated that all reaction variables except catalyst concentration have a definite effect on the isomeric composition of the hydrocarboxylation products. The CO partial pressure and reaction temperature have an effect analogous to but smaller than that observed in the hydroformylation reaction. The concentration of hydrogen donor has a small effect on product composition and in a direction opposite to that observed for hydroformylation. The concentration of $Co_2(CO)_8$ does not influence the composition of the products.

Gankin et al. (69) report that the isomeric composition of esters obtained from 1-hexene, methanol, and CO in the presence of $Co_2(CO)_8$

TABLE 20

Effect of Some Variables on Composition of Isomers from Hydrocarboxylation of 1-Pentene Using $Co_2(CO)_8$[a] (82)

Hydrogen source	Variable	Range	Hexanoic acid (%)
H_2O	Temperature	135–240°C	73–63
CH_3OH		150–230°C	72–64
H_2O	Partial pressure of CO	38–240 atm	64–71
CH_3OH		35–240 atm	63–72
H_2O	Hydrogen source (moles)/ olefin (moles)	1.1–10	67–66
CH_3OH		1.8–140	73–68

[a] Experiments at constant pressure.

is not affected by the methanol concentration, catalyst concentration, CO partial pressure, and nature of the solvent. However, they found that addition of up to 10 wt% of pyridine gave an increase in the less-branched isomer.

The results reported by Gankin et al. (69) and Piacenti et al. (82) are not really contradictory, since they were obtained at different temperatures (175°C versus 145–150°C) while the range of CO partial pressure differed. The greatest effect of CO pressure on product composition determined by Piacenti et al. was noticed between 35 and 100 atm of CO for 1-pentene and for propylene, whereas Gankin et al. reported only experiments done at 200 and 300 atm. The effect of structure of the olefin on product composition was also investigated. The position of the double bond seems to affect, although slightly, the isomeric product ratio when operating under appropriate conditions as shown by results obtained with 1-pentene and 2-pentene (82). In the acid synthesis and in the ester synthesis, conducted at 145 and 150°C, respectively, with $Co_2(CO)_8$ present, practically the same isomeric product composition was obtained from both olefins when using low CO partial pressures (35–38 atm). But a definite difference in composition was observed when a CO pressure higher than 100 atm was used. This behavior, analogous to that observed in hydroformylation of olefins, is attributable at least in part to isomerization of the olefin before its hydroformylation. This interpretation is suggested by the isomeric composition of the residual olefins in experiments interrupted at various degrees of conversion (Table 16). Isomerization of the substrate at high temperatures was also observed by Gankin and co-workers (69).

When using cobalt catalysts under nonisomerizing conditions, a predominance of hexanoic acid and of glutaric acid is obtained from 2-pentene and from crotonic acid, respectively. For a possible explanation of these results, which are similar to those observed in the hydroformylation of the same substrates, see the discussion in the second chapter. An anomalous behavior was detected in the synthesis of esters from acrylonitrile (98) and from methyl acrylate (99). The mixture of isomers contains, respectively, 11 and 82% of the less-branched isomer (Table 14). A different kind of interaction between functional groups and catalysts might explain these results.

With palladium the structure of the substrate strongly influences the isomeric composition of the products (Table 21). Although some authors (113) emphasize a large influence of electronic effects, in our opinion the available data do not allow evaluation of the relative importance of steric and electronic effects—the former certainly play, as in hydroformylation, a very important role.

TABLE 21
Synthesis of Esters in the Presence of $Pd(PPh_3)_2Cl_2$ (26)

Substrate[a]	Less-branched isomer (%)	
Vinyl chloride	Ethyl 3-chloropropionate	6
Methyl 3-butenoate	Dimethyl propane-1,3-dicarboxylate	11
Propylene	Ethyl butanoate	33
Isobutylene	Ethyl 3-methylbutanoate	33

[a] Temp. 90–120°C; P_{CO} 700 atm (constant); 3–4% HCl.

The effect of reaction variables in palladium-catalyzed hydrocarboxylation of 1-octene was investigated by Fenton (111). When using the $PdCl_2/PPh_3$ catalytic system, investigation of the temperature effect is complicated by the fact that decarbonylation takes place at an observable rate at temperatures higher than 150°C. A maximum content (66%) for straight-chain isomer was found at 125°C. An effect of water concentration was found: At 150°C the percentage of straight-chain isomer increases when decreasing water concentration. Contrary to what was observed with cobalt catalyst, the percentage of straight-chain isomer markedly decreases on increasing the CO partial pressure: 85 and 66% straight-chain acid at 100 and 800 atm, respectively. No effect on the composition of isomeric products was observed when varying the catalyst concentration.

In the presence of the $Pd(PPh_3)_2Cl_2/HCl$ catalytic system, the less-branched isomer was found in a smaller amount starting from both linear and branched terminal unsaturated substrates regardless of the electron–withdrawing or releasing effect of the substituent in the substrate (26) (Table 21).

6. *Kinetics*

Studies on the kinetics of the syntheses of acids and esters from cyclohexene, CO, and water or methanol, respectively, were made by Ercoli and co-workers (79*a*,114). These results, together with those of corresponding hydroformylation reactions, are summarized in Table 22.

The overall reaction rate, as shown also by the temperature range chosen for the investigation, decreases in this order: hydroformylation > synthesis of acids > synthesis of esters. As for hydroformylation, by varying the CO pressure, a maximum reaction rate was found that occurs in different pressure ranges depending on the hydrogen donor and type of metal used as catalyst.

TABLE 22
Kinetic Characteristics of Some Cyclohexene Carbonylations Using $Co_2(CO)_8$ (114)

Hydrogen donor	H_2 at 100–120°C	H_2O at 145–165°C	CH_3OH at 165–175°C
Apparent order with respect to hydrogen donor	1	1	1
Apparent order with respect to olefin	1	0.5	~0
Apparent order with respect to cobalt	0.5[a]–1	0.6–0.8	0.5
P_{CO} corresponding to the maximum reaction rate	~10–15 atm	~170–220 atm	~60–100 atm

[a] At CO partial pressure corresponding to the maximum reaction rate.

When using cobalt as catalyst and cyclohexene as substrate, the maximum rate, which for hydroformylation takes place at fairly low CO pressures (10–15 atm depending on temperature), occurs at 170–220 atm in the synthesis of acids and at 60–100 atm in the synthesis of methyl esters.

The effect of CO pressure on the reaction rate was also investigated by using $[Ru(CO)_3Cl_2]_2$ as catalyst and propylene as substrate at 190°C. In this case, a maximum for the reaction rate was found at 40 atm (38). As with hydroformylation and the synthesis of acids and esters, the order with respect to the hydrogen donor concentration is one and the order with respect to the dissolved cobalt concentration, in the range of CO pressure investigated, is 0.5–1.

The most important difference between the kinetics of hydroformylation and hydrocarboxylation of olefins is in the order with respect to the substrate concentration; thus, while in hydroformylation the order with respect to olefin is one, it is one-half in the synthesis of hexahydrobenzoic acid, and zero in the synthesis of methyl hexahydrobenzoate: The order with respect to the olefin decreases as the overall reaction rate decreases. The results of Ercoli and co-workers (114) were confirmed in the $Co_2(CO)_8$-catalyzed synthesis of methyl esters by Matsuda and others (70) who made the interesting observation that, when operating with pyridine present, the reaction rate is largely enhanced and an order of about one with respect to the olefin can then be observed. Changes in the reaction kinetics were also observed in the $Co_2(CO)_8$-catalyzed hydrocarboxylation of propylene in the presence of *n*-butyl alcohol by addition of pyridine (181). Other workers (69) reported that on addition of increasing amounts of pyridine, the reaction rate reaches a maximum and then decreases. When using acrylonitrile or methyl acrylate as substrates, an enhancement of the reaction rate in the synthesis of esters was also observed in the presence of small amounts of hydrogen and/or pyridine (98).

The kinetics of the hydrocarboxylation of diolefins was investigated (94,116). In the synthesis of 3-pentenoic acid from butadiene, while the dependence of the reaction rate on the CO pressure and on temperature is analogous to that observed in hydrocarboxylation of monoolefins, the reaction order appears to be between 0 and −1 with respect to the concentration of water. No explanation has yet been given of these results. The kinetics of the successive transformation of 3-pentenoic acid into butanedicarboxylic acids was also investigated (164).

7. *Mechanisms*

Since Reppe obtained the synthesis of esters from olefins, CO, and alcohols in the presence of $Ni(CO)_4$, many attempts were made to rationalize the addition of hydrogen and a carboxyl or a carboalkoxyl group to an olefin. It was first supposed that the reaction proceeds through the formation of a ketene from CO and the olefin (49) (reaction 14).

$$RCH{=}CH_2 + CO \rightarrow RCH_2CH{=}CO \xrightarrow{R'OH} RCH_2CH_2COOR' \quad (14)$$

A second suggestion involved the formation of a cyclopropanone intermediate (reaction 15) which, at least formally, could rationalize the observed results (49). This last hypothesis was criticized by Dupont et al.

$$RCH{=}CH_2 + CO \longrightarrow \underset{\text{(cyclopropanone: } RCH{-}CH_2 \text{ bridged by } C{=}O)}{RCH{-}CH_2} \xrightarrow{R'OH} \begin{cases} RCH_2CH_2COOR' \\ RCH(CH_3)COOR' \end{cases} \quad (15)$$

(34) on the grounds that in the absence of water and methanol either unsaturated aldehydes or polymeric materials should be formed from cyclopropanones. None of the substances, however, were detected in the reaction products at that time. The above criticism may not be entirely

valid, since Reppe and co-workers (107,117,118) subsequently synthesized polyketones and ketocarboxylic acids from ethylene and CO; the paper by Dupont and co-workers (34) represents, however, one of the earliest attempts to explain the hydrocarboxylation of olefins by postulating formation of an organometallic complex containing both olefin and CO (Scheme **IV**). As already mentioned, cobalt is a better catalyst than nickel

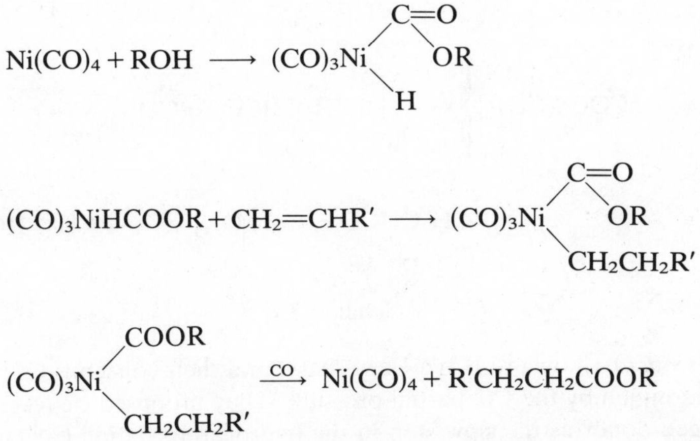

$$Ni(CO)_4 + ROH \longrightarrow (CO)_3Ni\begin{smallmatrix} C{=}O \\ \diagup \quad \diagdown \\ \quad\quad OR \\ \diagdown \\ H \end{smallmatrix}$$

$$(CO)_3NiHCOOR + CH_2{=}CHR' \longrightarrow (CO)_3Ni\begin{smallmatrix} C{=}O \\ \diagup \quad \diagdown \\ \quad\quad OR \\ \diagdown \\ CH_2CH_2R' \end{smallmatrix}$$

$$(CO)_3Ni\begin{smallmatrix} COOR \\ \diagup \\ \\ \diagdown \\ CH_2CH_2R' \end{smallmatrix} \xrightarrow{CO} Ni(CO)_4 + R'CH_2CH_2COOR$$

R = alkyl or H

Scheme **IV**

in hydrocarboxylation of olefins although small amounts of aldehydes are formed during hydrocarboxylation of olefins in the presence of $Co_2(CO)_8$. On the basis of these facts, Natta and co-workers (60) suggested that $HCo(CO)_4$, which was thought to be catalytically active in hydroformylation, was also the active species in hydrocarboxylation. In consideration of the acid character of $HCo(CO)_4$ in polar media, they suggested an ionic addition of this compound to the double bond with formation of a species closely resembling an acylcobalt carbonyl which then was cleaved to the corresponding carboxylic derivative (Scheme **V**) (60). The same concepts were discussed more recently by Imyanitov (113). A similar mechanism was then suggested by Kröper (119) to rationalize the reaction carried out in the presence of $Ni(CO)_4$ in the absence or in the presence of inorganic acids (Scheme **VI**).

Ercoli and others (79*a*) in discussing the results of their kinetic investigation of the synthesis of esters, pointed out that many cobalt carbonyl species may be present in a solution of $Co_2(CO)_8$ in Lewis bases. These species should have CO/Co ratios varying from 4 in $Co_2(CO)_8$ to 8/3 in

$$HCo(CO)_4 \rightleftarrows [Co(CO)_4]^- + H^+$$

$$RCH{=}CHR + H^+ \rightarrow RCH_2R\overset{(+)}{C}H$$

$$RCH_2R\overset{(+)}{C}H + [Co(CO)_4]^- \rightleftarrows \left\{ \begin{array}{l} CH_2R \\ | \\ CH \leftarrow [COCo(CO)_3] \\ | \\ R \end{array} \right\}$$

$$\left\{ \begin{array}{l} CH_2R \\ | \\ CH \longleftarrow [COCo(CO)_3] \\ | \\ R \end{array} \right\} \longleftrightarrow [RCH_2\underset{\displaystyle R}{\underset{|}{C}}HCOCo(CO)_3] \xrightarrow{CO}$$

$$RCH_2\underset{\displaystyle R}{\underset{|}{C}}H\overset{(+)}{C}O + [Co(CO)_4]^- \xrightarrow{R'OH} RCH_2\underset{\displaystyle R}{\underset{|}{C}}HCOOR'$$

Scheme **V**

$CoB_6[Co(CO)_4]_2$, where B is a Lewis base, and their concentration should be determined by the CO partial pressure. They proposed cleavage of the hydrogen donor as the slow step in the hydrocarboxylation (Scheme **VII**) and postulated a close analogy between hydrocarboxylation and hydroformylation of olefins: $HCo(CO)_4$ formation and cleavage of the acylcobalt carbonyls should take place according to the sequence shown in Scheme **VII**.

$$RCH{=}CH_2 \xrightarrow{Ni(CO)_4} \left[R\overset{(-)}{C}HCH_2 \longleftarrow \underset{\displaystyle O}{\underset{\|}{\overset{(+)}{C}}}Ni(CO)_3 \right] \xrightarrow[CO]{H_2O} RCH_2CH_2COOH + Ni(CO$$

$$(CH_3)_2C{=}CH_2 + H^+ \longrightarrow (CH_3)_2\overset{(+)}{C}CH_3 \rightleftharpoons (CH_3)_2CH\overset{(+)}{C}H_2$$

$$\downarrow Ni(CO)_4$$

$$[\overset{(+)}{N}i(CO)_3\underset{\displaystyle O}{\underset{\|}{C}}\rightarrow C(CH_3)_3] \qquad [\overset{(+)}{N}i(CO)_3\underset{\displaystyle O}{\underset{\|}{C}}\rightarrow CH_2CH(CH_3)_2]$$

$$\downarrow H_2O/CO \qquad \downarrow H_2O/CO$$

$$(CH_3)_3CCOOH + Ni(CO)_4 + H^+ \qquad (CH_3)_2CHCH_2COOH + Ni(CO)_4 + H^+$$

Scheme **VI**

$$[Co_2(CO)_7] + CH_3OH \rightarrow HCo(CO)_4 + [CH_3OCo(CO)_3]$$

$$HCo(CO)_4 + RCH{=}CH_2 \rightarrow [RCH_2CH_2COCo(CO)_3]$$

$$[RCH_2CH_2COCo(CO)_3] + [CH_3OCo(CO)_3] \rightarrow RCH_2CH_2COOCH_3 + 1/2\,Co_4(CO)_{12}$$

Scheme **VII**

In 1963 Heck and Breslow (55), after isolation of acylcobalt carbonyls, and investigation of their chemical and spectroscopic properties, proposed the mechanism shown in Scheme **VIII**. They pointed out that the last step, when occurring in the presence of stoichiometric amounts of acylcobalt carbonyl, is very rapid even at 50°C. The formation of $HCo(CO)_4$ in the last step was ascertained, but it was found that under the reaction conditions chosen (50°C, 1 atm CO) it decomposes very rapidly.

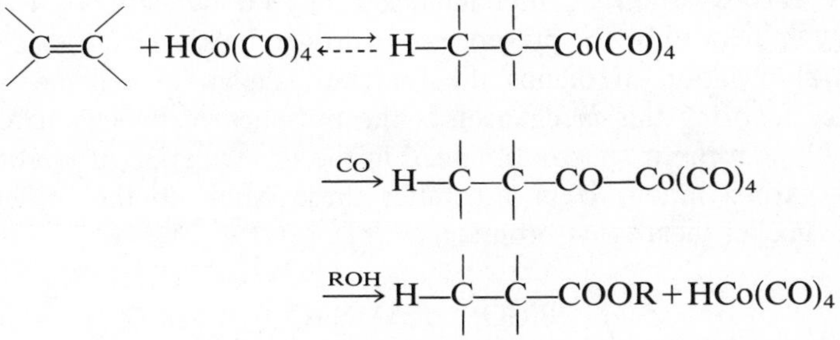

$$\rangle C{=}C\langle + HCo(CO)_4 \rightleftharpoons H{-}\overset{|}{\underset{|}{C}}{-}\overset{|}{\underset{|}{C}}{-}Co(CO)_4$$

$$\xrightarrow{CO} H{-}\overset{|}{\underset{|}{C}}{-}\overset{|}{\underset{|}{C}}{-}CO{-}Co(CO)_4$$

$$\xrightarrow{ROH} H{-}\overset{|}{\underset{|}{C}}{-}\overset{|}{\underset{|}{C}}{-}COOR + HCo(CO)_4$$

Scheme **VIII**

The results of the kinetic investigation by Ercoli's group may be reconciled with Heck and Breslow's mechanism by postulating that during the hydrocarboxylation reaction, although the equilibrium corresponding to the formation of the alkylcobalt carbonyl is largely displaced toward the reaction products, the concentration of the acylcobalt carbonyl species is extremely small and correspondingly the slowest step becomes the cleavage of the acylcobalt carbonyl by the alcohol or by another hydrogen donor. This mechanism would be in agreement with the explanation given by others (69,92,120) on the accelerating effect of pyridine on hydrocarboxylation, which is believed to arise from the facile cleavage of the acylcobalt carbonyls by pyridine (Scheme **IX**). An alternative explanation of the "pyridine effect" was given by Matsuda (98), who proposed the formation of a pyridine-containing cobalt carbonyl complex which should possess a higher catalytic activity.

$$RCOCo(CO)_4 + C_5H_5N \longrightarrow [RCONC_5H_5]^+[Co(CO)_4]^- \xrightarrow{R'OH}$$

$$RCOOR' + [HNC_5H_5]^+[Co(CO)_4]^- \rightleftharpoons HCo(CO)_4 + C_5H_5N$$

Scheme **IX**

Owing to the possible existence of many different cobalt carbonyl species having different catalytic activity, as demonstrated by the variation in reaction rate with CO pressure, the above mentioned acyl carbonyl species can be considered only as possible models of the reaction intermediates. The state of knowledge in this field is different from that on hydroformylation. While Markó and co-workers (121) were able to provide spectroscopic evidence for the presence of acylcobalt carbonyls in the reaction medium during hydroformylation, similar data are still lacking for the hydrocarboxylation of olefins.

As for the nickel-catalyzed hydrocarboxylation, Heck (122) after studying the carboxylation of allyl halides catalyzed by $Ni(CO)_4$ (which is strongly influenced by the presence of hydrogen halides) suggested for the hydrocarboxylation of olefins the reactions shown in Scheme **X**. The evidence favoring this mechanism is the presence of two bands at 4.78 and 5.72 μ in the ir spectra obtained in the investigation of π-allynickel halides carboxylation; Heck attributes these bands to the presence of butenoylnickel dicarbonyl bromide.

$$HX + Ni(CO)_4 \rightleftharpoons HNi(CO)_2X + 2CO$$

$$HNi(CO)_2X + CH_2{=}CH_2 \longrightarrow CH_3CH_2Ni(CO)_2X$$

$$CH_3CH_2Ni(CO)_2X + CO \longrightarrow CH_3CH_2CONi(CO)_2X$$

$$\xrightarrow{2CO} CH_3CH_2COX + Ni(CO)_4$$

$$CH_3CH_2COX + R'OH \longrightarrow CH_3CH_2COOR' + HX$$

$$CH_3CH_2CONi(CO)_2X + R'OH \longrightarrow CH_3CH_2COOR' + HNi(CO)_2X$$

Scheme **X**

Three mechanisms were formulated for the palladium-catalyzed hydrocarboxylation: The first scheme is similar to that already discussed for cobalt and nickel-catalyzed hydrocarboxylation (123) (Scheme **XI**). The second (26) is substantially different since palladium would activate the olefin but not the CO and no H–Pd bond would be involved. However, palladium complexes containing carbonyl groups were isolated (124) and

$$HX + Pd + nL \rightleftarrows L_nHPdX \underset{-CH_2{=}CH_2}{\overset{+CH_2{=}CH_2}{\rightleftarrows}} L_nHPdX(CH_2{=}CH_2) \rightleftarrows$$

$$CH_3CH_2PdL_nX \underset{-CO}{\overset{+CO}{\rightleftarrows}} CH_3CH_2COPdL_nX \rightleftarrows$$

$$Pd + nL + CH_3CH_2COX \xrightarrow{ROH} CH_3CH_2COOR + HX$$

Scheme **XI**

an activation of CO by the catalytic complexes as in similar reactions seems probable. The third general scheme for the hydrocarboxylation of unsaturated compounds (125) involves insertion of the olefin into a Pd–COOR bond (Scheme **XII**) as previously proposed for oxidative carboxylations like chlorocarboxylations (123) and for carboalkoxylations (126). A common scheme for the oxidative carboxylation and for hydrocarboxylation did not at first seem acceptable. In fact, in some oxidative carboxylations like alkoxycarboxylation a *trans* addition of the carboxyl group and of the alkoxy group occurs (57,127,128). On the contrary, the

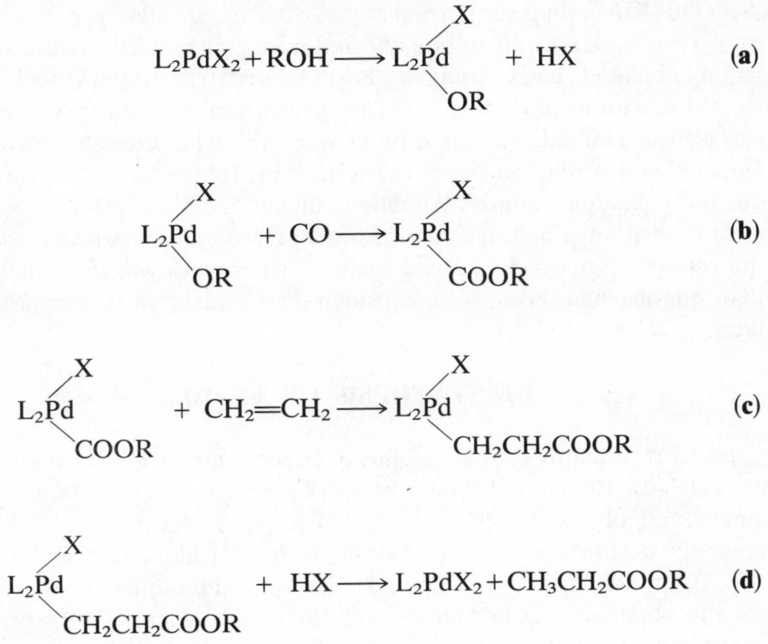

$$L_2PdX_2 + ROH \longrightarrow L_2Pd\begin{matrix}X\\OR\end{matrix} + HX \qquad \textbf{(a)}$$

$$L_2Pd\begin{matrix}X\\OR\end{matrix} + CO \longrightarrow L_2Pd\begin{matrix}X\\COOR\end{matrix} \qquad \textbf{(b)}$$

$$L_2Pd\begin{matrix}X\\COOR\end{matrix} + CH_2{=}CH_2 \longrightarrow L_2Pd\begin{matrix}X\\CH_2CH_2COOR\end{matrix} \qquad \textbf{(c)}$$

$$L_2Pd\begin{matrix}X\\CH_2CH_2COOR\end{matrix} + HX \longrightarrow L_2PdX_2 + CH_3CH_2COOR \qquad \textbf{(d)}$$

Scheme **XII**

stoichiometric hydrocarboxylation using $Ni(CO)_4$ as the CO donor was shown to proceed by *cis* addition of hydrogen and the carboxyl group to the double bond at least in the strained olefins investigated (52). Furthermore, the catalytic hydrocarboxylation of *trans*- and *cis*-3-methyl-2-pentene with a $PdCl_2/PPh_3$ catalytic system (Pd/P molar ratio = 0.5) at 100°C and 350 atm of CO showed a very large prevalence of the threo epimer in the first case (threo/erythro = 94/6) and of the erythro epimer in the second case (threo/erythro = 9/91) which clearly indicates a substantially *cis* addition of the hydrogen and of the carboxyl group to the olefin (73).

However, it was shown (129) that oxidative carboxylations can occur both with *cis* and *trans* addition to the double bond depending on whether the first group added is already bound to the metal in the catalytic complex or comes from the solution. Therefore the mechanism reported in Scheme **XII** is in principle acceptable both for oxidative carboxylation and hydrocarboxylation.

The greater hydrocarboxylation rate observed in palladium-catalyzed hydrocarboxylation in the presence of acids (26) can be explained both by an enhanced rate in the formation of a Pd–H bond (Scheme **XI**) and by an enhanced rate in the cleavage of a Pd–C bond (Scheme **XIId**). Furthermore the presence of an alkoxy group in the catalytic complex seems to be indicated by the asymmetric hydrocarboxylation obtained when using a Pd/PPh_3 catalytic system and (*S*)-2-methylbutanol as alcohol (153). The mechanism outlined in Scheme **XII** was also substantially accepted by Fenton (47), who discussed the various facts in favor of the individual reaction steps. He postulated a pentacoordinated palladium complex containing olefin, one molecule of phosphine, a chlorine atom, a hydrogen atom, and a carboxy group as intermediate in the catalytic process. It is possible that further work will show that more than one mechanism must be considered to explain all the experimental facts.

III. SYNTHESIS OF AMIDES

The first attempts to react olefins, CO, and ammonia were done in the early 1940s. Reppe and co-workers (48) reported that a mixture of the amide and of the nitrile of 2-methylstearic acid can be obtained by reacting octadecene and ammonia with stoichiometric amounts of $Ni(CO)_4$ at 290°C and 134 atm. At the same time, other workers (138) tried to obtain amines and amides via olefin carboxylation in the presence of amines by using heterogeneous catalysts containing group VIII metals. Further investigations were conducted by different authors

(19,24,91,104,130–143) using aliphatic or aromatic amines besides ammonia; in all cases the main reaction proceeds according to reaction 16.

$$CH_3CH{=}CH_2 + CO + RR'NH \begin{cases} \nearrow CH_3CH_2CH_2CONRR' \\ \searrow CH_3CH(CH_3)CONRR' \end{cases} \qquad (16)$$

Cobalt and nickel derivatives were mainly used as catalysts and the reactivity of different olefins with various aliphatic amines was investigated. However, Rh (165), Fe (142), and Ru (36) derivatives are also catalytically active.

The yields of amides were often rather good, and the reaction occurred at a rate comparable to that of the other hydrocarboxylations. This synthesis of amides, however, appears much less studied than the synthesis of other oxygenated products from olefins and CO.

A. Effect of Reaction Variables

As indicated in Table 23, nickel and cobalt are by far the most popular catalysts for the amide synthesis. Nickel has been used as metal, as $Ni(CN)_2$ or $K_2Ni(CN)_4$; cobalt as metal (Raney cobalt or cobalt supported on kieselguhr), as a salt or as $Co_2(CO)_8$. Cobalt catalysts have been more fully investigated. Some scattered experiments were done with iron ($Fe(CO)_5$) (142), Rh_2O_3 (165), and ruthenium ($RuCl_3$) (36) (Table 23).

Because of the different conditions used, it is difficult to compare the rate of the synthesis of amides with those of the other hydrocarboxylation reactions. Nevertheless, in the reaction of cyclohexene (132) and butadiene (94) with CO and aniline in the presence of metallic cobalt and of $Co_2(CO)_8$, respectively, the rate of anilide synthesis seems much lower than that of hydroformylation and intermediate between the rates of the synthesis of acids and of esters.

With cobalt catalysts, aromatic amines (particularly aniline) were more active as hydrogen sources than aliphatic amines, which, however, are more active than ammonia. Secondary amines have also been successfully used in the amide synthesis (36,130) and amides have been used as hydrogen donors (138*b*,140) particularly in "intramolecular" hydrocarboxylations (21*b*) (Table 24).

Different olefinic substrates were reacted with CO and amines to give amides; aniline and ammonia were mainly used (Table 25). Aliphatic monoolefins give good yields of the expected amides. From ethylene and

TABLE 23

Synthesis of Amides in the Presence of Various Catalysts

Catalyst	Olefin	Hydrogen source	Temp (°C)	P_{CO} (atm)	Products	Yield (%)	Ref.
Ni	Ethylene	$C_6H_5NH_2$	325	170[a]	Propionanilide	46	131
$Ni(CN)_2$	Ethylene	NH_3/H_2O	110	65[a]	Propionamide	22	136*a*
Co (Raney)	Ethylene	$CH_3CH_2CH_2NH_2$	200	100[b]	N-Propylpropionamide	58	142
$Fe(CO)_5$	Ethylene	$CH_3CH_2CH_2NH_2$	200	75[a]	*N*-Propylpropionamide	30	142
$RuCl_x$[c]	Allene	Cyclohexyl—NH_2	135	1000[a]	*N*-Cyclohexylmethacrylamide	16	36

[a] Initial pressure at reaction temp.
[b] Initial pressure at room temp.
[c] $RuCl_3 \cdot 3H_2O/RuCl_4 \cdot H_2O$

TABLE 24

Synthesis of Amides in the Presence of Various Amines or Amides

Hydrogen source	Olefin	Catalyst	Temp (°C)	P_{CO} (atm)	Products	Yield (%)	Ref.
NH_3	Cyclohexene	Co	250	200[a]	Hexahydrobenzamide	35	138
NH_3	Cyclohexene	Rh_2O_3	150	250[a]	Hexahydrobenzamide	14	165
NH_3	Ethylene[b]	Rh_2O_3	150	250[a]	Propionamide	15	165
n-$C_3H_7NH_2$	Ethylene	Co	200	100[a]	*N*-Propylpropionamide	58	142
$C_{18}H_{37}NH_2$	Cyclohexene	Co	220	400[c]	*N*-Stearylhexahydrobenzamide	n.d.	132*b*
(cyclohexyl)—NH_2	Allene	$RuCl_4$[d]	135	1000[c]	*N*-Cyclohexylmethacrylamide	16	36
$C_6H_5NH_2$	Cyclohexene	Co	205	450[c]	Hexahydrobenzanilide	n.d.	132*a*
α-$C_{10}H_7NH_2$	Cyclohexene	Co	205	450[c]	*N*-α-Naphthylhexahydrobenzamide	n.d.	132*b*
$(CH_3)_2NH$	Ethylene	Co	170–190	500–865[c]	*N,N*-Dimethylpropionamide	58	130
$HCONH_2$	Undecylenic acid	Ni	270	200[c]	Dodecanedioic acid diamide	19	49
$HCONHC_6H_5$	1-Octene	$Co_2(CO)_8$	225	370[c]	Anilides of nonanoic acids	77	140

[a] Initial pressure at room temp.
[b] 2,4,5-Triethylimidazole (yield 52%) obtained as main product. Reaction carried out in the presence of CH_3OH and H_2O.
[c] Initial pressure at reaction temp.
[d] $RuCl_3 \cdot 3H_2O/RuCl_4 \cdot H_2O$.

TABLE 25

Synthesis of Amides from Various Unsaturated Substrates

Substrate	Hydrogen source	Catalyst	Temp (°C)	P_{CO} (atm)	Products	Yield (%)	Ref.
Ethylene	NH_3	Co	195–265	460–840[a]	Propionamide	72	130
Ethylene	$C_6H_5NH_2$	Co	220	400[a]	Propionanilide	80	132*b*
Propylene	NH_3	Co	185–205	580–730[a]	Butyramide	40	130
Propylene	$C_6H_5NH_2$	$Co_2(CO)_8$	179	336[a]	Butyranilide, Isobutyranilide }	90	134
1-Butene	NH_3	Co	190–200	800–940[a]	Valeramide	10	130
Isobutylene	$C_6H_5NH_2$	Co	200	415[a]	Isovaleranilide	34	132*a*
1-Octene	$C_6H_5NH_2$	$Co_2(CO)_8$	200	350[a]	Anilides of nonanoic acids	66	140
Cyclopentene	NH_3/H_2	Co	200	75[b]	*N*-Cyclopentylmethylcyclo-pentanecarboxamide	28	142
					Cyclopentanecarboxamide	3	
Cyclohexene	$C_6H_5NH_2$	Co	220	400[a]	Hexahydrobenzanilide	n.d.	132*b*
Styrene	$C_6H_5NH_2$	Co	160	360[a]	*β*-Phenylpropionanilide	26	132*a*
Allene	C_6H_{11}—NH_2 (cyclohexylamine)	$RuCl_x$[c]	135	1000[a]	*N*-Cyclohexylmethacrylamide	16	36
Butadiene	NH_3[d]	$Ni(CN)_2$	140	80[e]	2-Methylbutyramide	29	136*a*
Butadiene	$C_6H_5NH_2$	$Co_2(CO)_8$[f]	210	400[e]	Anilides of C_6 dicarboxylic acids	61	91
2,3-Dihydrofuran	NH_3[d]	$Ni(CN)_2$	135	70[e]	Tetrahydrofurancarboxamide	14	136*a*
Methyl crotonate	$C_6H_5NH_2$	Co	200	180[g]	Methylsuccinic acid dianilide	18	104
Undecylenic acid	NH_3	Ni	270	200[e]	Dodecanedioic acid diamide	60	49

[a] Total pressure at reaction temp. [b] 75 atm of H_2 present. [c] $RuCl_3 \cdot 3H_2O/RuCl_4 \cdot H_2O$. [d] In the presence of H_2O and CH_3OH.
[e] Total pressure (constant). [f] In the presence of pyridine. [g] Initial pressure at room temperature.

cyclohexene, the anilides of propionic acid and cyclohexanecarboxylic acid were easily obtained; from terminal olefins only one product was generally isolated, although not always in quantitative yield (130). This result is probably due to a great predominance of one isomeric product which makes isolation of any others difficult.

Systematic investigations of the reaction products were apparently conducted only with propylene (134). Both *n*-butyranilide and isobutyranilide were isolated; the normal/(normal + iso) ratio was 0.83 in the range of 150–290°C with a propylene/aniline ratio of 0.1–9.1 and 3–12 hr reaction time. Also the type of solvent used (benzene, isooctane, dioxane, ethyl ether, or methanol) did not influence the product ratio. Only the anilide of dihydrocinnamic acid was isolated in low yield when styrene is used as substrate (132).

As for diolefins, only butadiene was investigated; operating at 400 atm of CO and 210°C, using $Co_2(CO)_8$ as catalyst and with pyridine present, the anilides of adipic and ethylsuccinic acids were the main reaction product. By using $Ni(CN)_2$, Reppe and Magin (136*a*) obtained from butadiene, CO, and ammonia, a low yield of 2-methylbutanamide. Olefins containing functional groups were also reacted with CO and aniline; for example, from methyl crotonate, the dianilide of methyl succinic acid was isolated. The most interesting results in this field were obtained in the intramolecular hydrocarboxylation of unsaturated amides (21*b*,141) and amines (21*a*,130). The unsaturated amines give rise to lactams (reaction 17) and the unsaturated amides form imides (reaction 18). Both

$$CH_2{=}CHCH_2NH_2 \xrightarrow[Co_2(CO)_8]{CO} \text{(2-pyrrolidinone, N–H)} \tag{17}$$

$$CH_2{=}CHCONHR \xrightarrow[Co_2(CO)_8]{CO} \text{(N-R succinimide)} \tag{18}$$

reactions were studied mainly by Falbe and Korte (21*a*) who used mainly $Co_2(CO)_8$ as catalyst, temperatures of 170–280°C, and CO pressures of 100–300 atm. These reactions have been reviewed (30).

Gresham (130) was the first to show that γ-butyrolactam can be obtained in low yield by reacting allylamine with CO in the presence of a cobalt catalyst at 300°C and 700 atm. A much higher yield was obtained by Falbe and Korte (21*a*) by using much milder conditions and $Co_2(CO)_8$ as catalyst with allylamine, *N*-monosubstituted allylamines, as well as γ-alkylallylamines. In the latter case a mixture of 3-alkyl-2-pyrrolidones and 2-piperidones were obtained. As in other cobalt-catalyzed carbonylation reactions, the attack of CO can also occur at the carbon atom adjacent to the double bond (reaction 19). As shown by experiments carried out using 1-aminomethyl-3-cyclohexene (reaction 20). CO attacks the position gamma to the nitrogen atom in the intramolecular hydrocarboxylation.

$$CH_3CH{=}CHCH_2NH_2 \xrightarrow[Co_2(CO)_8]{CO} \text{3-methyl-2-pyrrolidone} + \text{2-piperidone} \quad (19)$$

$$\text{1-aminomethyl-3-cyclohexene } (CH_2NH_2) \xrightarrow[Co_2(CO)_8]{CO} \text{bicyclic lactams (NH, O)} \quad (20)$$

Only unsubstituted amides are formed from olefins, CO, and ammonia, which shows that the NH_2 groups of amides (with the possible exception of formamides) are not very active as hydrogen donors. Unsaturated amides on the other hand undergo a very smooth intramolecular hydrocarboxylation reaction in the presence of $Co_2(CO)_8$ to form dicarboxylic acid imides as shown by Falbe and Korte (21*b*). The reaction proceeds with unsubstituted amides but, as expected, not with disubstituted amides. Only five- and six-membered ring imides of dicarboxylic acids are formed, which confirms the results of the intramolecular hydrocarboxylation of unsaturated amines. When there is a possibility for the formation of five- and six-membered ring imides of dicarboxylic acids, as from crotonic acid, the five-membered ring compound largely prevails. The opposite is true when the carbon atom in position 3 of the amide has two alkyl substituents (reaction 21).

$$CH_3CH{=}CHCONH_2 \xrightarrow[Co_2(CO)_8]{CO} \underset{68\%}{\text{2-methylsuccinimide } (CH_3)} + \underset{19\%}{\text{glutarimide}} \quad (21)$$

B. Secondary Reactions

Secondary reactions in the synthesis of amides were investigated mostly with cobalt catalysts; they generally depend on the ability of the catalyst to favor hydrogen transfer reactions (104,132,134). As discussed in Vol. 1, cobalt carbonyl catalysts react with amines (144) to form cobalt carbonyl hydrides. These compounds in turn can supply hydrogen to the unsaturated substrates and thus promote hydrogenation as well as the syntheses of aldehydes and ketones. When hydrogen donors other than amines are present, competition takes place among the different carbonylation reactions to give a mixture of products. Fortunately, the monofunctional amides, with the possible exception of formamides, are not good hydrogen donors as already mentioned. The amides themselves, therefore, do not react to give secondary products with the exception that they form nitriles above 250°C (48,130).

1. Hydrogen Transfer

The clearest example of hydrogen transfer was found in the reaction between styrene, CO, and aniline in the presence of Raney cobalt (104). In this reaction, carried out at 200°C under an initial CO pressure of 250 atm, the yield of amides (mostly β-phenylpropionanilide) based on CO was about 26%, while a large amount of diphenylurea was also formed, and ethylbenzene was found among the reaction products (reaction 22).

$$PhCH{=}CH_2 \xrightarrow[\text{Co (Raney)}]{\text{CO, PhNH}_2} \begin{cases} PhCH_2CH_2CONHPh \\ (PhNH)_2CO + PhCH_2CH_3 \end{cases} \tag{22}$$

The formation of diphenylurea from aniline in the presence of cobalt catalysts was first reported by Krzikalla and Woldan (145).

It is interesting to note that a simultaneous hydrogenation and hydrocarboxylation was observed when using $Ni(CN)_2$ as catalyst. From the reaction between butadiene, CO, and ammonia, the product 2-methylbutyramide (Table 25) was obtained by Reppe and Magin (136*a*).

Since Raney cobalt gives rise to $Co_2(CO)_8$ under reaction conditions and $Co_2(CO)_8$ can react with amines to give $HCo(CO)_4$ (144), these results can be rationalized (reactions 23–25) by assuming that a cobalt carbonyl hydride is the catalytically active species that promotes anilide formation as well as the secondary reactions.

$$PhNH_2 + Co_2(CO)_8 + CO \rightarrow (PhNH)_2CO + 2HCo(CO)_4 \tag{23}$$

$$2HCo(CO)_4 + PhCH{=}CH_2 \rightarrow PhCH_2CH_3 + Co_2(CO)_8 \tag{24}$$

$$2HCo(CO)_4 + 2C_2H_4 + CO \rightarrow Co_2(CO)_8 + (CH_3CH_2)_2CO \tag{25}$$

Besides the mentioned styrene hydrogenation observed in the cobalt-catalyzed hydroformylation of styrene, a ketone synthesis was observed when ethylene or propylene (130) was used as a substrate—particularly when a large excess of olefin relative to the amine was used (134).

2. *Synthesis of Amides in the Presence of Hydrogen or Hydrogen Donors*

Experiments in which olefins were reacted with CO and ammonia in the presence of hydrogen gas showed that, besides the expected primary amide, a secondary amide forms. The latter arises, at least formally, from the reaction of the olefin with CO and with an amine originating from the unsaturated substrate but having one carbon atom more than the olefin (142). Thus from ethylene, CO, and ammonia, *N*-propylpropionamide, 3,5-dimethyl-2-ethylpyridine, and 3,5-dimethyl-4-ethyl-pyridine were obtained, besides propionamide. The corresponding cyclopentylmethyl- and cyclohexylmethylamides were obtained from the reaction of cyclopentene and cyclohexene, respectively, in the presence of CO and H_2. In all cases, di- and trialkylamines were also formed which, starting with an olefin having n carbon atoms, contained $2n+2$ or $3n+3$ carbon atoms. The formation of pyridine derivatives from ethylene, CO, and ammonia clearly shows that propionaldehyde must be formed as a reaction intermediate; therefore, the formation of these amides and amines from ethylene in the presence of cobalt catalysts was rationalized as shown in Scheme **XIII**. The same workers (142) have also discussed the influence of $Fe(CO)_5$, which seems to favor formation of amines instead of amides.

Secondary reactions are simpler when the amide synthesis is carried out in the presence of water or of methanol. By using water, an excess of ethylene relative to ammonia, and $Ni(CN)_2$ as catalyst, the products obtained were propionic acid and propionamide (136). These results seem to indicate that under the conditions chosen the amide synthesis is faster than the simultaneously occurring acid synthesis. Similarly, in the synthesis of butyric acid anilides with Raney cobalt as catalyst and methanol as solvent, a mixture of the expected anilides and methyl esters was obtained (134). Since no methanolysis of the butyric acid anilides was observed, anilides and esters should arise from concurrent hydrocarboxylation reactions. The rate of anilide synthesis from propylene is about three times the rate of methyl ester synthesis from the same olefin. Of course, as in the case of acids, the activity of the cobalt catalyst in the ester synthesis can be modified by the presence of the amine.

A series of interesting secondary reactions were observed in the reaction between olefins, CO, and NH_3 in the presence of CH_3OH and H_2O;

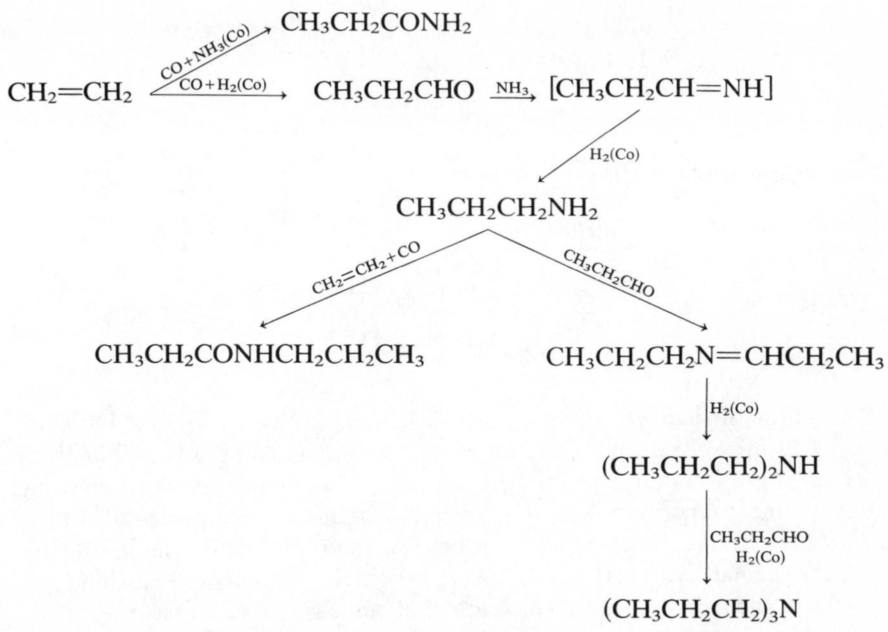

Scheme **XIII**

with ethylene, propylene, and 1-butene, 2,4,5-triethyl-, 2,4,5-tripropyl-, and 2,4,5-tributylimidazole, respectively, were formed as main reaction products; with cyclohexene, *N*-formyl-bis(hexahydrobenzyl)amine was obtained in 50% yield (165).

C. Mechanisms

The mechanism of the amide synthesis has been little investigated except for the intramolecular synthesis which leads to cyclic imides (21*b*). On the basis of analogies with the mechanism of the synthesis of acids and esters and of the possible presence of cobalt carbonyl hydride species, a plausible mechanism for the synthesis of amides involves the formation of an acylcobalt carbonyl followed by a nucleophilic attack of the amine on the acyl group (Scheme **XIV**).

The same scheme was proposed for the intramolecular hydrocarboxylation of unsaturated amines and amides. However, nothing is known about the stability of amino complexes of cobalt carbonyl species under reaction condition; these must play an important role as shown by the

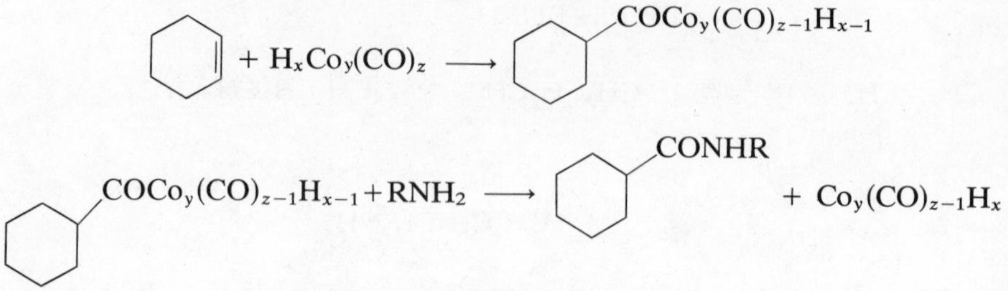

$$\text{C}_6\text{H}_{10} + H_xCo_y(CO)_z \longrightarrow C_6H_{11}COCo_y(CO)_{z-1}H_{x-1}$$

$$C_6H_{11}COCo_y(CO)_{z-1}H_{x-1} + RNH_2 \longrightarrow C_6H_{11}CONHR + Co_y(CO)_{z-1}H_x$$

$$H_xCo_y(CO)_{z-1} + CO \rightarrow H_xCo_y(CO)_z$$

Scheme **XIV**

variation of yield with temperature (21). There are no data on the influence of CO pressure on the reaction rate and product composition. For these reasons, an intramolecular elimination of an amide molecule arising from the acyl group bound to cobalt and an amine molecule already present in the complex cannot be excluded. In our opinion, the intramolecular hydrocarboxylations investigated by Falbe and Korte (21*b*), which gave cyclic imides, indicate that at least in this case the amide group is already complexed to the cobalt atom before formation of the acylcobalt carbonyl species. This can also explain why, in intramolecular hydrocarboxylations, the amide group acts as a hydrogen donor while amides do not generally act as hydrogen donors in hydrocarboxylations.

In connection with the postulation of Scheme **XIV** it must be noted that, when an α,β-unsaturated amide is reacted with CO, hydrocarboxylation occurs even at 200°C; furthermore, the reaction takes place even in the presence of hydrogen (21*b*) which, with olefins, gives rise mainly to hydroformylation when ammonia or amines are also present (142). The possibility of an intramolecular imide synthesis is also confirmed by the isolation of a complex $[Co(CO)_2CH_2{=}CHCONH_2]_x$, formed by reaction between $Co_2(CO)_8$ and acrylamide; it decomposes at 200–240°C to give succinimide (21*b*).

IV. SYNTHESIS OF THIOESTERS, ACID CHLORIDES, AND ANHYDRIDES

The hydrocarboxylation of olefins was extended by Reppe and coworkers (48,49) to the synthesis of thioesters and anhydrides in the presence of $Ni(CO)_4$ and, sometimes, nickel halides (reactions 26 and 27). When anhydrous HCl was used instead of carboxylic acids, acid chlorides were obtained in the presence of ruthenium, rhodium (39), and palladium (166,182) catalysts (reaction 28).

$$CH_2{=}CH_2 + CO + RSH \longrightarrow CH_3CH_2COSR \qquad (26)$$

$$CH_2{=}CH_2 + CO + CH_3CH_2COOH \longrightarrow (CH_3CH_2CO)_2O \qquad (27)$$

$$CH_2{=}CH_2 + HCl + CO \longrightarrow CH_3CH_2COCl \qquad (28)$$

When using a palladium catalyst, HCl, and alcohols, the final reaction products are esters, but acyl chlorides are believed to be intermediates in the reaction.

Generally, the reactions 26, 27, and 28 require relatively high pressures; little work was done after their discovery, but the most interesting aspects were reviewed (28*b*,147). The synthesis of anhydrides, mostly of propionic anhydride, was reinvestigated because of its potential industrial application (146,147). The most systematic investigations were carried out by Brooks and co-workers (146), who tried to optimize the yield of propionic anhydride. They also observed that the CO pressure greatly influences the reaction rate. The best conditions correspond, at each temperature, to a low stability of $Ni(CO)_4$ which might be transformed, at least in part, into a species having a CO to nickel ratio lower than 4. $RhCl_3$ was found a rather poor catalyst for this synthesis when operating at 150–180°C and 660–1000 atm (39). For the same synthesis Rh_2O_3 in the presence of ethyl iodide (167) and a three-component catalytic system containing $NiCl_2{\cdot}6H_2O$, copper, and sodium iodide (163) was proposed. The cyclic anhydrides of methylsuccinic and glutaric acid can be obtained by starting with allyl alcohol in the presence of $RhCl_3/CoI_2$, $RhCl_3/NiI_2$, and $RhI_3/CoCl_2$ catalytic systems (168). In the presence of rhodium catalysts and an iodine-containing promoter, ethylene and methanol simultaneously react with CO in the presence of water (169) to yield a mixture of acetic and propionic acids. Therefore the formation of cyclic anhydrides, instead of lactones (20), from allyl alcohol can be explained by admitting that, in the presence of the above catalysts, allyl alcohol carbonylation is more rapid than its hydrocarboxylation. The vinylacetic acid, which should be the intermediate product, is then hydrocarboxylated in the absence of water to cyclic anhydrides.

Nothing can be said on the mechanism of the synthesis of anhydrides because of the lack of systematic experimental research.

REFERENCES

1*a*. G. B. Carpenter, U.S. Pat. 1,981,801 (1931); *Chem. Abstr.*, **29,** 479 (1935).
1*b*. J. C. Woodhouse, U.S. Pat. 2,003,477 (1931); *Chem. Abstr.*, **29,** 4854 (1935).
1*c*. W. E. Vail, U.S. Pat. 1,979,717 (1931); *Chem. Abstr.*, **29,** 181 (1935).
2. D.V.N. Hardy, *J. Chem. Soc.*, (*a*) **1936,** 358; (*b*) **1936,** 362; (*c*) **1936,** 364.
3. H. Koch and W. Haag, *Angew. Chem.*, **70,** 311 (1958).
4*a*. W. Reppe and H. Kröper, Ger. Pat. 765,969 (1953); *Chem. Abstr.*, **51,** 13904 (1957).
4*b*. W. Reppe and H. Kröper, Ger. Pat. 879,987 (1953); *Chem. Abstr.*, **52,** 11899 (1958).
4*c*. FIAT. Report No. 273.
5*a*. D. M. Fenton and K. L. Olivier, *Chem. Technol.* **1972,** 220.
5*b*. J. B. Zachry and C. L. Aldridge, U.S. Pat. 3,176,038 (1965); *Chem. Abstr.*, **62,** 16064 (1965).
5*c*. T. Inoue and S. Tsutsumi, *J. Am. Chem. Soc.*, **87,** 3525 (1965).
5*d*. T. Yukawa and S. Tsutsumi, *J. Org. Chem.*, **34,** 738 (1969).
5*e*. D. M. Fenton and K. L. Olivier, *Amer. Chem. Soc., Div. Petr. Chem.*, Los Angeles, 1971, Prepr. B5; *Chem. Abstr.*, **78,** 29128 (1973).
5*f*. G. Biale, U.S. Pat. 3,523,971 (1970); *Chem. Abstr.*, **73,** 88433 (1970).
5*g*. M. Dokiya and K. Bando, *Kogyo Kagaku Zasshi*, **71,** 508 (1968); *Chem. Abstr.*, **69,** 51485 (1968).
5*h*. K. L. Olivier and W. D. Schaeffer, U.S. Pat., 3,461,157 (1969); *Chem. Abstr.*, **71,** 90837 (1969).
6*a*. J. Tsuji, M. Morikawa, and J. Kiji, *Tetrahedron Lett.*, **1963,** 1061.
6*b*. J. Tsuji, M. Morikawa, and J. Kiji, *J. Am. Chem. Soc.*, **86,** 4851 (1964).
6*c*. A. U. Blackham, U.S. Pat. 3,119,861 (1964); *Chem. Abstr.*, **60,** 9155 (1964).
7. T. Suzuki and J. Tsuji, *Tetrahedron Lett.* **1968,** 913.
8. K. G. Ihrman, H. A. Filbey, and E. F. Zaweski, U.S. Pat. 3,361,811 (1968); *Chem. Abstr.* **69,** 18626 (1968).
9. T. Suzuki and J. Tsuji, *J. Org. Chem.*, **35,** 2982 (1970).
10. J. Tsuji and T. Suzuki, Japan. Pat. 68 09,726 (1968); *Chem. Abstr.*, **69,** 105931 (1968).
11. T. Inoue and S. Tsutsumi, *Bull. Chem. Soc. Japan*, **38,** 2122 (1965).
12. J. Tsuji, M. Morikawa, and J. Kiji, Japan. Pat. 65 19,940 (1965); *Chem. Abstr.*, **63,** 16219 (1965).
13. J. Tsuji and H. Takahashi, Japan. Pat. 68 15,165 (1968); *Chem. Abstr.*, **70,** 67653 (1969).
14. K. L. Olivier, U.S. Pat. 3,415,871 (1968); *Chem. Abstr.*, **70,** 37200 (1969).
15. K. L. Olivier, U.S. Pat. 3,505,394 (1970); *Chem. Abstr.*, **73,** 3480 (1970).
16. D. M. Fenton, U.S. Pat. 3,397,226 (1968); *Chem. Abstr.*, **69,** 66919 (1968).
17. G. Consiglio and P. Pino, *Chimia*, **30,** 193 (1976).
18*a*. G. Natta, P. Pino, and E. Mantica, *Chim. Ind.* (*Milan*), **32,** 201 (1950).
18*b*. B. E. Kuvaev, N. S. Imyanitov, and D. M. Rudkovskii, in D. M. Rudkovskii, Ed.,

Carbonylation of Unsaturated Hydrocarbons [in Russian], Izd. Khimiya, Leningrad, 1968, p. 222; *Chem. Abstr.*, **71,** 2708 (1969).
19. J. Falbe, *Angew. Chem., Int. Ed. Engl.*, **5,** 435 (1966).
20. J. Falbe, H. J. Schulze-Steinen, and F. Korte, *Chem. Ber.*, **98,** 886 (1965).
21*a*. J. Falbe and F. Korte, *Chem. Ber.*, **98,** 1928 (1965).
21*b*. *Idem, Chem. Ber.*, **95,** 2680 (1962).
22. J. Falbe and F. Korte, *Chem. Eng. Technol.*, **36,** 158 (1964).
23. A. Matsuda, *Bull. Chem. Soc. Japan,* **41,** 1876 (1968).
24. J. Falbe and F. Korte, *Angew. Chem.*, **74,** 291 (1962).
25. J. Falbe, *Ullmanns Encyklopädie der Technischen Chemie,* 3 Auflage, Urban and Schwarzenberg, München 1970, p. 120.
26. K. Bittler, N. v. Kutepow, D. Neubauer, and H. Reis, *Angew. Chem.*, **80,** 352 (1968).
27. C. W. Bird, *Chem. Rev.*, **62,** 283 (1962).
28*a*. Ya. T. Eidus and K. V. Puzitskii, *Russian Chem. Rev.*, **33,** 438 (1964).
28*b*. Ya. T. Eidus, K. V. Puzitskii, A. L. Lapidus, and B. K. Nefedov, *Russian Chem. Rev.*, **40,** 429 (1971).
28*c*. Ya. T. Eidus, A. L. Lapidus, K. V. Puzitskii, and B. K. Nefedov, *Russian Chem. Rev.*, **42,** 199 (1973).
29. C. W. Bird, *Transition Metal Intermediates in Organic Chemistry,* Logos Press, London, 1967, Ch. 7, p. 149.
30*a*. J. Falbe, *Carbon Monoxide in Organic Synthesis,* Springer-Verlag, New York, 1970.
30*b*. J. Falbe, *Angew. Chem.*, **78,** 532 (1966).
31. W. Reppe, Ger. Pat. a.I. 65361 (1939); as cited in Ref. 32, p. 109.
32. W. Reppe, *Experientia,* **5,** 93 (1949).
33. W. Reppe, *Liebigs Ann.*, **582,** 1 (1953).
34. G. Dupont, P. Piganiol, and J. Vialle, *Bull. Soc. Chim. France,* **1948,** 529.
35. G. Natta and P. Pino, *Chim. Ind.* (*Milan*), **31,** 109 (1949).
36. J. J. Kealy and R. E. Benson, *J. Org. Chem.*, **26,** 3126 (1961).
37. T. Alderson and J. C. Thomas, U.S. Pat. 3,040,090 (1962); *Chem. Abstr.*, **57,** 16407 (1962).
38. G. Sbrana, G. Braca, F. Piacenti, G. Marzano, and M. Bianchi, *Chim. Ind.* (*Milan*), **54,** 117 (1972).
39. T. Alderson and V. A. Engelhardt, U.S. Pat. 3,065,242 (1962); *Chem. Abstr.*, **58,** 8912 (1963).
40. J. B. Zachry and C. L. Aldridge, U.S. Pat. 3,161,672 (1964); through *Chem. Abstr.*, **62,** 9018 (1965).
41. N. S. Imyanitov and D. M. Rudkovskii, *Zh. Prikl. Khim.*, **39,** 2811 (1966); *Chem. Abstr.*, **66,** 75465 (1967).
42. N. S. Imyanitov, and D. M. Rudkovskii, *Kinet. Katal.* **8,** 1240 (1967); *Chem. Abstr.*, **68,** 86867 (1968).
43. S. Brewis and P. R. Hughes, *Chem. Commun.*, **1965,** 157.
44. N. v. Kutepow, D. Neubauer, and K. Bittler, Ger. Pat. 1,229,089 (1966); *Chem. Abstr.*, **66,** 37510 (1967).
45. E. L. Jenner and R. V. Lindsey, Jr., U.S. Pat. 2,876,254 (1959); *Chem. Abstr.*, **53,** 17906 (1959).
46. L. J. Kehoe and R. A. Shell, *J. Org. Chem.*, **35,** 2846 (1970).
47. D. M. Fenton, *J. Org. Chem.*, **38,** 3192 (1973).
48. BIOS. Final Report, No. 1811 (1947).
49. W. Reppe and H. Kröper, *Liebigs Ann.*, **582,** 38 (1953).
50. B. Fell and J. M. J. Tetteroo, *Angew. Chem.*, **77,** 813 (1965).

51. G. P. Chiusoli and S. Merzoni, *Chim. Ind.* (*Milan*), **51,** 612 (1969).
52. C. W. Bird, R. C. Cookson, J. Hudec, and R. O. Williams, *J. Chem. Soc.*, **1963,** 410.
53. W. Reppe, O. Schlicting, K. Klager, and T. Toepel, *Liebigs Ann.*, **560,** 1 (1948).
54. R. F. Heck and D. S. Breslow, *J. Am. Chem. Soc.*, **83,** 4023 (1961).
55. R. F. Heck and D. S. Breslow, *J. Am. Chem. Soc.*, **85,** 2779 (1963).
56*a*. H. Masada, M. Mizuno, S. SuJa, Y. Watanabe, and Y. Takegami, *Bull. Chem. Soc. Japan*, **43,** 3824 (1970).
56*b*. H. Masada, *Asahi Garasu Kogyo Gijutsu Shoreikai Kenkyu Hokoku*, **16,** 165 (1970); *Chem. Abstr.*, **75,** 4872 (1971).
57. L. F. Hines and J. K. Stille, *J. Am. Chem. Soc.*, **94,** 485 (1972).
58. J. Tsuji and S. Hosaka, *J. Am. Chem. Soc.*, **87,** 4075 (1965).
59. D. Medema, R. van Helden, and C. F. Kohll, *Inorg. Chim. Acta*, **3,** 255 (1969).
60. G. Natta, P. Pino, and E. Mantica, *Gazz. Chim. Ital.*, **80,** 680 (1950).
61. W. F. Gresham and R. E. Brooks, U.S. Pat. 2,448,368 (1948); *Chem. Abstr.*, **43,** 669 (1949).
62*a*. J. Tsuji, M. Morikawa, and J. Kiji, *Tetrahedron Lett.*, **1963,** 1437.
62*b*. J. Tsuji, *Advances in Organic Chemistry*, Vol. VI, Interscience, 1969, p. 109.
63. R. Ercoli, *Chim. Ind.* (*Milan*), **37,** 1029 (1955).
64. C. Botteghi, G. Consiglio, and P. Pino, *Chimia*, **27,** 477 (1973).
65*a*. G. Consiglio, *Helv. Chim. Acta*, **59,** 124 (1976).
65*b*. G. Consiglio and P. Pino, unpublished results.
66. P. Pino and R. Ercoli, *Chim. Ind.* (*Milan*), **36,** 536 (1954).
67. D. R. Levering and A. L. Glasebrook, *J. Org. Chem.*, **23,** 1836 (1958).
68. N. S. Imyanitov, B. E. Kuvaev, and D. M. Rudkovskii, in D. M. Rudkovskii, Ed., *Carbonylation of Unsaturated Hydrocarbons* [in Russian], Izd. Khimiya, Leningrad, 1968, p. 176; *Chem. Abstr.*, **71,** 21650 (1969).
69. V. Yu. Gankin, M. G. Katsnel'son, and D. M. Rudkovskii, in D. M. Rudkovskii, Ed., *Carbonylation of Unsaturated Hydrocarbons* [in Russian], Izd. Khimiya, Leningrad, 1968, p. 178; *Chem. Abstr.*, **71,** 21653 (1969).
70. A. Matsuda and H. Uchida, *Bull. Chem. Soc. Japan*, **38,** 710 (1965).
71. I. Wender, H. Greenfield, S. Metlin, and M. Orchin, *Catalysis*, Vol. V, Reinhold, New York, 1957, p. 110.
72. (−)DIOP:(−)-2,3-O-Isopropylidene-2,3-dihydroxy-1,4-*bis*(diphenylphosphino)butane; H. B. Kagan and T. P. Dang, *J. Am. Chem. Soc.*, **94,** 6429 (1972).
73. G. Consiglio and P. Pino, *Gazz. Chim. Ital.*, **105,** 1133 (1975).
74. A. Matsuda, K. Bando, S. Shin, and Y. Horiguchi, *Proc. Int. Conf. High Pressure*, 4*th*, *1974* (Publ. 1975), 725; *Chem. Abstr.*, **83,** 96359 (1975).
75. F. Piacenti, P. P. Neggiani, and F. Calderazzo, *Atti Soc. Tosc. Sci. Nat.*, ***B*, 1962,** 42.
76. F. Piacenti and C. Cioni, *Atti Soc. Tosc. Sci. Nat.*, **B, 1962,** 1.
77. BASF, Neth. Appl. 6,409,121 (1965); *Chem. Abstr.*, **63,** 14726 (1965).
78. W. Reppe and H. Kröper, Ger. Pat. 863,194 (1953); through *Chem. Abstr.*, **48,** 1425 (1954).
79*a*. R. Ercoli, M. Avanzi, and G. Moretti, *Chim. Ind.* (*Milan*), **37,** 865 (1955).
79*b*. P. Pino, R. Ercoli, and E. Mantica, *Gazz. Chim. Ital.*, **81,** 635 (1951).
80. S. Brewis and P. R. Hughes, *Chem. Commun.*, **1965,** 489.
81. F. Piacenti and M. Bianchi, unpublished results.
82. F. Piacenti, M. Bianchi, and R. Lazzaroni, *Chim. Ind.* (*Milan*), **50,** 318 (1968).
83. J. Tsuji and T. Nogi, *Bull. Chem. Soc. Japan*, **39,** 146 (1966).
84. N. S. Imyanitov and D. M. Rudkovskii, *Zh. Prikl. Khim.*, **40,** 2825 (1967); *Chem. Abstr.*, **68,** 86805 (1968).

85. J. Tsuji, S. Hosaka, J. Kiji, and T. Susuki, *Bull. Chem. Soc. Japan*, **39,** 141 (1966).
86. N. v. Kutepow, K. Bittler, and D. Neubauer, Ger. Pat. 1,221,224 (1966); cf. Neth. Appl. 6,409,121 (1965); *Chem. Abstr.*, **63,** 14726 (1965).
87. T. Rull, *Bull. Soc. Chim. France*, **1964,** 2680.
88. A. Matsuda, *Bull. Chem. Soc. Japan*, **46,** 524 (1973).
89. A. M. Hyson, U.S. Pat. 2,586,341 (1952); through *Chem. Abstr.*, **46,** 8670 (1952).
90. B. E. Kuvaev, N. S. Imyanitov, and D. M. Rudkovskii, *Zh. Prikl. Khim.*, **40,** 1359 (1967); *Chem. Abstr.*, **67,** 99587 (1967).
91. N. S. Imyanitov and D. M. Rudkovskii, *Zh. Prikl. Khim.*, **39,** 2335 (1966); *Chem. Abstr.*, **66,** 10530 (1967).
92. N. S. Imyanitov and D. M. Rudkovskii, *Zh. Organ. Khim.*, **2,** 231 (1966); *Chem. Abstr.*, **65,** 2119 (1966).
93. N. M. Bogoradovskaia, N. S. Imyanitov, and D. M. Rudkovskii, *Zh. Prikl. Khim.*, **39,** 2807 (1966); *Chem. Abstr.*, **66,** 85213 (1967).
94. N. S. Imyanitov and D. M. Rudkovskii, in D. M. Rudkovskii, Ed., *Carbonylation of Unsaturated Hydrocarbons* [in Russian], Izd. Khimiya, Leningrad, 1968, p. 211; *Chem. Abstr.*, **70,** 114486 (1969).
95. S. Kunichika, Y. Sakakibara, and T. Okamoto, *Bull. Chem. Soc. Japan*, **40,** 885 (1967).
96. J. Tsuji and T. Susuki, *Tetrahedron Lett.*, **1965,** 3027.
97. T. Susuki and J. Tsuji, *Bull. Chem. Soc. Japan*, **41,** 1954 (1968).
98. A. Matsuda, *Bull. Chem. Soc. Japan*, **40,** 135 (1967).
99. A. Matsuda, *Bull. Chem. Soc. Japan*, **42,** 571 (1969).
100. Y. Takayama, M. Ikeda, and R. Sunaoka, Japan. Pat. 68 13,448 (1968); *Chem. Abstr.*, **70,** 11138 (1969).
101. A. Matsuda, *Bull. Chem. Soc. Japan*, **42,** 2596 (1969).
102. N. v. Kutepow, K. Bittler, and D. Neubauer, Ger. Pat. 1,227,023 (1966); cf. Netk. Appl., 6,409,121 (1965); *Chem. Abstr.*, **63,** 14726 (1965).
103. D. M. Fenton, U.S. Pat. 3,641,074 (1972); *Chem. Abstr.*, **76,** 99120 (1972).
104. G. Natta, P. Pino, and R. Ercoli, *J. Am. Chem. Soc.*, **74,** 4496 (1952).
105. W. F. Gresham and R. E. Brooks, U.S. Pat. 2,526,742 (1950); *Chem. Abstr.*, **45,** 2017 (1951).
106. D. G. Hedberg, Jr., U.S. Pat. 2,510,105 (1950); *Chem. Abstr.*, **44,** 7344 (1950).
107. W. Reppe and A. Magin, Ger. Pat. 880,297 (1953); *Chem. Abstr.*, **52,** 11907 (1958).
108. G. P. Chiusoli, *Chemie Industrie*, **70,** 53 (1953).
109. P. Pino, F. Piacenti, M. Bianchi, and R. Lazzaroni, *Chim. Ind.* (*Milan*), **50,** 106 (1968).
110. V. Yu. Gankin, E. A. Merkulov, and V. A. Rybakov, in N. S. Imyanitov, Ed., Gidroformilirovanie, [Hydroformylation], Izd. Khimiya, Leningrad, 1972, p. 173; *Chem. Abstr.* **77,** 163985 (1972).
111. F. M. Fenton, U.S. Pat. 3,661,949 (1972); *Chem. Abstr.*, **77,** 74837 (1972).
112. F. M. Fenton, U.S. Pat. 3,530,155 (1970); *Chem. Abstr.*, **74,** 12624 (1971).
113. N. S. Imyanitov, in N. S. Imyanitov, Ed., *Gidroformilirovanie* [Hydroformylation], Izd. Khimiya, Leningrad, 1972, p. 13; *Chem. Abstr.* **77,** 138992 (1972).
114. R. Ercoli, G. Signorini, and E. Santambrogio, *Chim. Ind.* (*Milan*), **42,** 587 (1960).
115*a*. A. Matsuda, Japan. Kokai 74 79,991 (1974); *Chem. Abstr.*, **82,** 139383 (1975).
115*b*. H. Isa, T. Inagaki, N. Kojima, and I. Kadoya, Ger. Offen. 2,503,996 (1975); *Chem. Abstr.*, **83,** 178336 (1975).
115*c*. H. Isa, T. Inagaki, Y. Kiyonaga, and M. Nagayama, Ger. Offen. 2,447,069 (1975); *Chem. Abstr.*, **83,** 58146 (1975).
115*d*. A. Matsuda, Japan. Kokai 75 46,589 (1975); *Chem. Abstr.*, **83,** 42852 (1975).
115*e*. A. Matsuda, Japan. Kokai 75 45,798 (1975); *Chem. Abstr.*, **83,** 42853 (1975).

116. B. E. Kuvaev, N. S. Imyanitov, and D. M. Rudkovskii, in D. M. Rudkovskii, Ed., *Carbonylation of Unsaturated Hydrocarbons* [in Russian], Izd. Khimiya, Leningrad 1968, p. 225; *Chem. Abstr.*, **71,** 21648 (1969).
117. W. Reppe, Ger. Pat. 894,991 (1951); *Chem. Abstr.*, **52,** 12899 (1958).
118. W. Reppe, H. Friederich, and K. Wiebusch, Ger. Pat. 892,893 (1953); *Chem. Abstr.*, **52,** 16211 (1958).
119. H. Kröper, *Anlagerung von Kohlenmonoxide und Verbindungen mit Aciden Wasserstoff* (*Carbonylierung*), Houben-Weyl, Vol. IV/2, George Thieme, Stuttgart, 1955, p. 385.
120. N. S. Imyanitov, B. E. Kuvaev, and D. M. Rudkovskii, in D. M. Rudkovskii, Ed., *Carbonylation of Unsaturated Hydrocarbons* [in Russian], Izd. Khimiya, Leningrad, 1968, p. 215; *Chem. Abstr.*, **71,** 29966 (1969).
121. L. Marko, G. Bor, G. Almasy, and P. Szabo, *Brennstoff-Chem.*, **44,** 184 (1963).
122. R. F. Heck, *J. Am. Chem. Soc.*, **85,** 2013 (1963).
123. J. Tsuji, *Accounts Chem. Res.*, **2,** 144 (1969).
124*a*. A. Misono, Y. Uchida, M. Hidai, and K. Kudo, *J. Organometal. Chem.*, **20,** 7 (1969).
124*b*. K. Kudo, M. Hidai, and Y. Uchida, *ibid.*, **33,** 393 (1971).
125. O. L. Kaliya, O. N. Temkin, N. G. Mekhryakova, and R. M. Flid, *Dokl. Akad. Nauk SSSR*, **199,** 1321 (1971); *Chem. Abstr.*, **76,** 13388 (1972).
126. R. F. Heck, *J. Am. Chem. Soc.*, **93,** 6896 (1971).
127. J. K. Stille, L. F. Hines, *J. Am. Chem. Soc.*, **92,** 1798 (1970).
128. J. K. Stille, L. F. Hines, R. W. Fries, P. K. Wong, D. E. James, and K. Lau, *Advances in Chemistry Series*, No. 132, *Am. Chem. Soc.*, Washington D.C. 1974, p. 90.
129. J. K. Stille, D. E. James, and L. F. Hines, *J. Am. Chem. Soc.*, **95,** 5062 (1973).
130. E. I. Du Pont de Nemours Co. and W. F. Gresham, Brit. Pat. 628,659 (1949); *Chem. Abstr.*, **44,** 8364 (1950).
131. D. M. Newitt and S. A. Momen, *J. Chem. Soc.*, **1949,** 2945.
132*a*. P. Pino and C. Paleari, *Gazz. Chim. Ital.*, **81,** 646 (1951).
132*b*. G. Natta, P. Pino, and R. Ercoli, *Proc. IIIrd World Petroleum Congress*, 1951, *p.* 1.
133. P. Pino, Ital. Pat. 471,913 (1952); *Chem. Abstr.*, **48,** 7627 (1954).
134. P. Pino and R. Magri, *Chim. Ind.* (*Milan*), **34,** 511 (1952).
135. A. T. Larson, U.S. Pat. 2,497,310 (1950); *Chem. Abstr.*, **44,** 4489 (1950).
136*a*. W. Reppe and A. Magin, Brit. Pat. 672,379 (1952); *Chem. Abstr.*, **47,** 5428 (1953).
136*b*. W. Reppe and H. Kröper, Ger. Pat. 868,149 (1951).
137. P. L. Barrick, U.S. Pat. 2,542,747 (1951); *Chem. Abstr.*, **45,** 7584 (1951).
138*a*. H. J. Nienburg and E. Keunecke, Ger. Pat. 863,799 (1953); *Chem. Abstr.*, **48,** 1427 (1954).
138*b*. J. F. Olin and T. E. Deger, U.S. Pat. 2,422,631 (1947); *Chem. Abstr.*, **41,** 5892 (1947).
139. BASF, Brit. Pat. 705,791 (1954); *Chem. Abstr.*, **49,** 4708 (1955).
140. B. F. Crowe and O. C. Elmer, U.S. Pat. 2,742,502 (1956); *Chem. Abstr.*, **50,** 16849 (1956).
141. Shell Internationale, Belg. Pat. 623,333 (1963); *Chem. Abstr.*, **60,** 9157 (1964).
142. A. Striegler and J. Weber, *J. Prakt. Chem.*, **29,** 281 (1965).
143. J. Falbe, H. Weitkamp, and F. Korte, *Tetrahedron Lett.*, **1965,** 2677.
144*a*. W. Hieber, G. Schulten, and R. Marin, *Z. Anorg. Allgem. Chem.*, **240,** 263 (1939).
144*b*. J. Palagi and L. Marko, *J. Organometal. Chem.*, **17,** 453 (1969).
145. H. Krzikalla and E. Woldan, Ger. Pat. 863,800 (1940); *Chem. Zentr.*, **1953,** 2674.
146. R. E. Brooks, W. F. Gresham, J. V. E. Hardy, and J. M. Lupton, *Ind. Eng. Chem.*, **49,** 2004 (1957).
147. V. Yu. Gankin, M. G. Katsnel'son, and D. M. Rudkovskii, in D. M. Rudkovskii, Ed.,

Carbonylation of Unsaturated Hydrocarbons [in Russian], Izd. Khimiya, Leningrad, 1968, p. 168; *Chem Abstr.*, **71,** 101259 (1969).
148*a*. T. Kajimoto, S. Wakamatsu, R. Nakanishi, M. Hara, K. Ohno, and J. Tsuji. U.S. Pat. 3,723,486 (1973); *Chem. Abstr.*, **79,** 41976 (1973).
148*b*. E. N. Frankel, U.S. Pat. Appl. 416,122 (1973);*Chem. Abstr.*, **83,** 62371 (1975).
148*c*. M. Yamaguchi and K. Tano, Japan. Pat. 73 13,088 (1973); *Chem. Abstr.*, **79,** 31028 (1973).
148*d*. M. Yamaguchi and K. Tano, Japan. Pat. 74 43,930 (1974); *Chem. Abstr.*, **83,** 27589 (1975).
148*e*. Y. Kajimoto, S. Wakamatsu, R. Nakanishi, M. Hara, K. Ohno, and J. Tsuji, Japan. Pat. 74 04,449 (1974); *Chem. Abstr.*, **82,** 72541 (1975).
148*f*. H. Fukutani, M. Tokizawa, and H. Okada, Japan. Pat. 75 00,645 (1975); *Chem. Abstr.*, **83,** 27606 (1975).
149. J. J. Mrowca, U.S. Pat. 3,859,319 (1975); *Chem. Abstr.*, **82,** 139393 (1975).
150*a*. F. Knifton, Ger. Offen. 2,303,118 (1973); *Chem. Abstr.*, **79,** 146008 (1973).
150*b*. K. Nozaki, Ger. Offen. 2,410,246 (1974); *Chem. Abstr.*, **82,** 3806 (1975).
151. J. J. Mrowca, U.S. Pat. 3,906,015 (1975); *Chem. Abstr.*, **84,** 4507 (1976).
152*a*. G. W. Parshall, U.S. Pat. 3,832,391 (1974); *Chem. Abstr.*, **82,** 3737 (1975).
152*b*. G. W. Parshall, *J. Am. Chem. Soc.*, **94,** 8716 (1972).
153. G. Consiglio and P. Pino, unpublished results.
154. H. Isa, T. Inagaki, Y. Kiyonaga, and M. Nagayama, Ger. Offen., 2,404,955 (1974); *Chem. Abstr.*, **83,** 27605 (1975).
155*a*. W. E. Billups, W. E. Walker, and T. C. Shields, *Chem. Commun.*, **1971,** 1067.
155*b*. H. Fukutani, M. Tokizawa, and H. Okada, Japan. Pat., 73 25,169 (1973); *Chem. Abstr.*, **79,** 104765 (1973).
156. D. G. Kuper and W. B. Hughes, U.S. Pat. 3,746,747 (1973); *Chem. Abstr.*, **79,** 104777 (1973).
157. A. Matsuda and K. Bando, Japan. Pat. 74 20,177 (1974); *Chem. Abstr.*, **82,** 58475 (1975).
158. Y. Mori and J. Tsuji, Japan. Pat. 72 37,931 (1972); *Chem. Abstr.*, **78,** 85052 (1973).
159. E. N. Frankel, F. L. Thomas, and W. K. Rohwedder, *Advances in Chemistry Series*, No. 132, Am. Chem. Soc., Washington, D.C., 1974, p. 145.
160. N. S. Imyanitov, B. E. Kuvaev, and D. M. Rudkovskii, *Zh. Prikl. Khim.*, **40,** 2821 (1967); *Chem. Abstr.*, **68,** 86804 (1968).
161. Y. Sugi, K. Bando, and S. Shin, *Chem. Ind.* (*London*), **1975,** 397.
162. G. Consiglio and M. Marchetti, *Chimia*, **30,** 26 (1976).
163. H. J. Hagemeyer, Jr., U.S. Pat. 2,739,169 (1956); *Chem. Abstr.*, **50,** 16835 (1956).
164. B. E. Kuvaev, N. S. Imyanitov, and D. M. Rudkovskii, in D. M. Rudkovskii, Ed., *Carbonylation of Unsaturated Hydrocarbons* [in Russian], Izd. Khimiya, Leningrad, 1968, p. 232; *Chem. Abstr.*, **71,** 21649 (1969).
165. Y. Iwashita and M. Sakuraba, *J. Org. Chem.*, **36,** 3927 (1971).
166. J. F. Knifton, U.S. Pat. 3,880,898 (1975);*Chem Abstr.*, **83,** 113698 (1975).
167. D. Forster, A. Hershman, and F. E. Paulik, Ger. Offen. 2,364,446 (1974); *Chem. Abstr.*, **81,** 119316 (1974).
168. D. Freudenberger, Ger. Offen. 2,339,633 (1975); *Chem. Abstr.*, **83,** 27607 (1975).
169. *Res. Discl.*, **128,** 18 (1974); *Chem. Abstr.*, **83,** 9122 (1975).
170. A. Matsuda, Japan. Pat. 72 25,327 (1972); *Chem Abstr.*, **77,** 87908 (1972).
171. D. T. Thompson and R. Jackson, Ger. Offen. 2,114,544 (1971); *Chem. Abstr.*, **76,** 13931 (1972).
172. M. G. Katsnel'son, *Zh. Prikl. Khim.*, **46,** 2745 (1973); *Chem. Abstr.*, **80,** 59394 (1974).

173. J. F. Knifton, U.S. Pat. 3,892,788 (1975); *Chem. Abstr.*, **83,** 113701 (1975).
174. M. Yamaguchi and K. Tano, Japan. Kokai 72 17,713 (1972); *Chem. Abstr.*, **77,** 139448 (1972).
175. R. A. Schell, U.S. Pat. 3,661,948 (1972); *Chem. Abstr.*, **77,** 126006 (1972).
176. R. A. Schell, U.S. Pat. 3,660,439 (1972); *Chem. Abstr.*, **77,** 126007 (1972).
177. M. Yamaguchi and K. Tano, Ger. Offen. 2,124,718 (1971); *Chem. Abstr.*, **76,** 33805 (1972).
178. G. Foster and J. R. Bethell, Ger. Pat. 2,101,909 (1972); *Chem. Abstr.*, **76,** 126399 (1972).
179. S. A. Butter, Ger. Offen. 2,106,541 (1971); *Chem. Abstr.*, **75,** 129357 (1971).
180. P. M. Maitlis, *The Organic Chemistry of Palladium*, Academic Press, New York, Vol. II, 1971.
181. M. Polievka and V. Macho, *Chem. Zvesti*, **26,** 154 (1972); *Chem. Abstr.*, **77,** 113349 (1972).
182. N. v. Kutepow, K. Bittler, D. Neubauer, and H. Reis, Ger. Pat. 1,237,116 (1967); *Chem. Abstr.*, **68,** 21550 (1968).

Organic Syntheses via Allylic Complexes of Metal Carbonyls

G. P. CHIUSOLI* AND L. CASSAR, *Istituto Ricerche Donegani, Montedison, Novara, Italy*

* Present address: Istituto di Chimica Organica, Università via D'Azeglio 85, Parma, Italy.

I. INTRODUCTION

For the most part, only reactions involving transition metal complexes that contain carbon monoxide as a ligand will be considered. Analogous reactions with complexes not containing carbon monoxide will be treated in sections concerning mechanisms, when necessary. Reactions that do not involve insertion of carbon monoxide, even though the latter is present as a ligand, are treated in a separate section of this chapter. Reactions leading to π-allylic complexes are not considered in this chapter, unless they are useful in explaining the mechanism of the organic reaction described. On the other hand, references to reviews and papers dealing with the preparation and properties of π-allylic complexes of transition metals will be given in the text.

A. Historical and General Considerations

Interest in reactions occurring via allylic complexes has arisen in the last decade as a consequence of concurrent synthetic and structural studies. Allylic compounds have long been known to be specially able to undergo substitution reactions much more easily than the corresponding saturated compounds. Further, they are frequently subject to rearrangements consisting of the migration of groups from one end of the allylic system to the other (1). Among the reactions of allyl compounds investigated were many coupling reactions, for example,

$$2RCH{=}CHCH_2X + Mg \rightarrow (RCH{=}CHCH_2)_2 + MgX_2 \qquad (1)$$

for which a metal carbonyl, that is, $Ni(CO)_4$, was first used as the coupling agent in 1943 (2). This reaction can be considered the starting point for a new topic, which, being on the borderline between organic and inorganic chemistry, has led to remarkable progress in both fields. At first the importance of complex formation in the reactions of allylic compounds with metal carbonyls was not recognized. Webb and Borcherdt (2*a*) studied the coupling reaction and observed that products resulting from coupling of terminal carbon atoms were mainly formed, even when starting from secondary allylic isomers, for example:

$$2RCH{=}CHCH_2Cl \text{ (or } RCH(Cl)CH{=}CH_2) + Ni(CO)_4 \rightarrow$$

$$RCH{=}CHCH_2CH_2CH{=}CHR + NiCl_2 + 4CO \qquad (2)$$

They found that if the reaction was not completed, the unreacted compounds were not isomerized to the same extent as the products. They were thus led to favor a radical reaction, discarding the hypothesis involving complex intermediates able to isomerize the reactants before coupling.

In 1958 Chiusoli (3,4) described the reaction of allylic halides with $Ni(CO)_4$ and proposed the formation of nickel complexes as catalysts. He also pointed out the analogy between allylic complexes and those of cyclopentadiene, butadiene, and other conjugated molecules.

An example of this kind of reaction is shown here for allyl chloride:

$$CH_2{=}CHCH_2Cl + CO + CH_3OH \xrightarrow{Ni(CO)_4} CH_2{=}CHCH_2COOCH_3 + HCl \qquad (3)$$

Here, the double bond is not affected as in the oxo or similar reactions; instead a halogen or a suitable leaving group on carbon is replaced by a carbonyl group. Formally, hydroformylation reactions may be regarded as addition carbonylation, whereas the type of reaction in Eq. 3 may be considered a replacement or substitutive carbonylation.

Reaction 3 occurs at room temperature. A slight pressure of carbon monoxide (1–2 atm) is needed; otherwise coupling products predominate. If, however, the pressure is increased to about 100 atm, the reaction slows down and a higher temperature is necessary. This behavior, analogous to that observed in hydroformylations (5–8), is consistent with formation of a complex (an intermediate in the reaction between the allylic compound and nickel carbonyl) that requires evolution of CO from the carbonyl.

The catalytic activity of allylnickel complexes did not seem limited to this kind of carbonylation. A highly specific reaction was also found, which consists of 1,2-insertion of a molecule of acetylene between the allylic group and the CO group, as shown in Eq. 4 for allyl chloride; the double bond derived from acetylene is *cis*.

$$CH_2{=}CHCH_2Cl + HC{\equiv}CH + CO + CH_3OH \xrightarrow{Ni(CO)_4}$$

$$\begin{array}{ccc} CH_2{=}CHCH_2 & & COOCH_3 \\ & \diagdown\ \ \diagup & \\ & C{=}C & \\ & \diagup\ \ \diagdown & \\ H & & H \end{array} + HCl \qquad (4)$$

It was reasonable to expect that a specific reaction of this kind could be achieved only by catalytic action of allylic complexes, in which the other reactants are also coordinated. Isolation of allylic complexes began in 1959. Tricarbonyl (π-but-1,2,3-enyl) cobalt (**1**) had been prepared as far

$$\text{(}\pi\text{-butenyl)}{-}Co(CO)_3$$

(**1**)

back as 1952 by Prichard (9), but the correct structure of this complex was not fully established until 1960 by Heck and Breslow (11) and by Jonassen and co-workers (12–14). (The nomenclature proposed by Chaudhari, Knox, and Pauson is being adopted here (10)). In 1959 Smidt and Hafner (15) reported the synthesis of bis-π-allylpalladium chloride

(2)

(2) and proposed that it has a sandwich structure analogous to ferrocene and similar compounds. Also in 1959, Moiseev, Fedorovskaya, and Syrkin (16) described allylpalladium complexes and suggested the coexistence of monomeric and dimeric forms. The situation has since been shown by Cotton (17) to involve equilibria of the following type (Eqs. 5 and 6) (where L is a basic ligand):

$$\underset{\text{dimeric } \pi\text{-complex}}{\mathbf{2}} + 2L \rightleftharpoons 2\, \underset{\text{monomeric } \pi\text{-complex}}{\underset{(\mathbf{3})}{(\pi\text{-}C_3H_5)Pd(Cl)(L)}} \tag{5}$$

$$\mathbf{3} + L \rightleftharpoons \underset{\text{monomeric } \sigma\text{-complex}}{\underset{(\mathbf{4})}{CH_2{=}CHCH_2{-}Pd(L)_2{-}Cl}} \tag{6}$$

π-Allyl (cyclopentadienyl) palladium was also reported by Shaw (18) and σ- and π-allylmanganese carbonyls by Kaesz, King, and Stone (19) in 1960. In 1961 bis(π-allyl) nickel bromide was described by Fischer (30) and bis(π-allyl) nickel by Wilke (31).

Ever since their meeting in the area of allylmetal complexes, organic and inorganic research have strongly interacted and the literature on this subject has grown rapidly. The synthesis, structure and properties of these intriguing allylic complexes have been described (20–53, 102*a–e*).

B. Structure and Bonding of Allylic Complexes

In σ-allylic complexes, the allyl radical contributes one electron to the metal, which gives rise to a conventional bond by pairing with an electron from the metal. In π-allylic complexes, three electrons are formally contributed by the allyl radical. Electroneutrality is restored by back donation of electronic charge from the metal to the allylic group (24).

The x-ray structures of several π-allylic complexes are known (e.g. 25–28). In $(\pi\text{-}C_3H_5PdCl)_2$ (**5**) the allyl groups are *anti* to each other. At

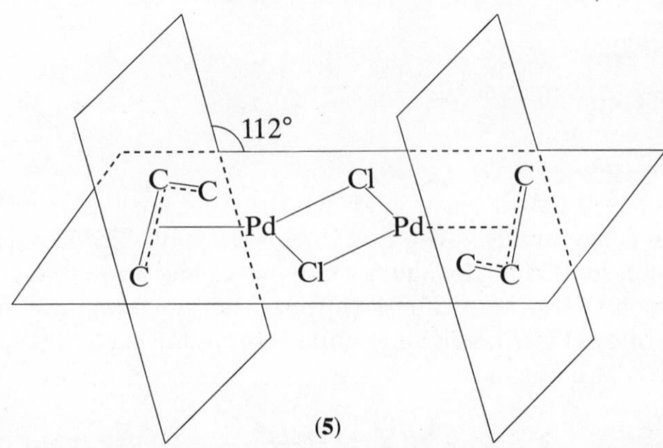

(5)

$-140°C$ their three carbon atoms form an angle of 119.8° (standard deviation 0.91°) and the plane thus defined forms an angle of $111.5 \pm 0.92°$ with the plane of Cl–Pd–Cl (28). Within the experimental error the C–C bonds are equal, thus indicating delocalization of the allyl system; the three carbon atoms are at the same distance (~ 2.11 Å) from palladium (25,25*a*). In monomeric π-$C_4H_7Pd(PPh_3)Cl$, (**5a**), the allyl plane forms an angle of 116° with the Pd–P–Cl plane (26). The Pd–C

H(3) H(4)
PPh_3
H_3C— —Pd
Cl
H(1) H(2)

(5a)

distances are 2.28, 2.22 and 2.14 Å for Pd–C (H(1),H(2)),Pd–C(CH_3) and Pd–C(H(3),H(4)) respectively. The Pd–C bond opposite Cl is thus shorter by 0.14 Å (standard deviation = 0.03 Å) than that opposite PPh_3.

Results by Mason and Wheeler (29) show that in both π-1,1,3,3-tetramethylallylpalladium chloride and in π-2-methallylpalladium chloride dimers the methyl groups are distorted from the plane of the allyl carbon atoms. The nickel complex **6** and the analogous iodo derivative have been described by Fischer and Bürger (30), who proposed a structure with the

Br
—Ni Ni—
Br

(6)

two allyl groups in planes perpendicular to the Ni$\langle^{Br}_{Br}$ plane. Such planes have the two bromine atoms in common. The dipole moment of **6** (1.31 D in benzene) as well as that of the iodo derivative (1.62 D) may be interpreted in terms of the angle formed by the two planes joined by the two bridging bromine atoms.

Wilke (31) has furnished the x-ray structure of **7** which is a sandwich structure with two methallyl groups *anti* to each other.

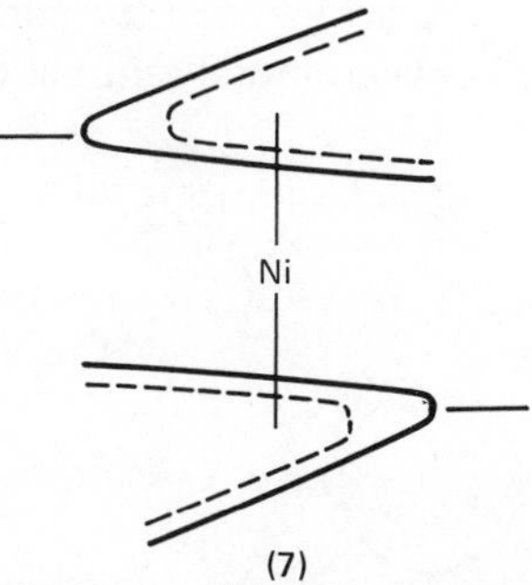

More recently some structures of pentacoordinated π-allylic complexes have been elucidated; the structure of complex **8** was found to be a distorted square pyramid (54) (Fig. 1). In this complex, the π-allyl ligand

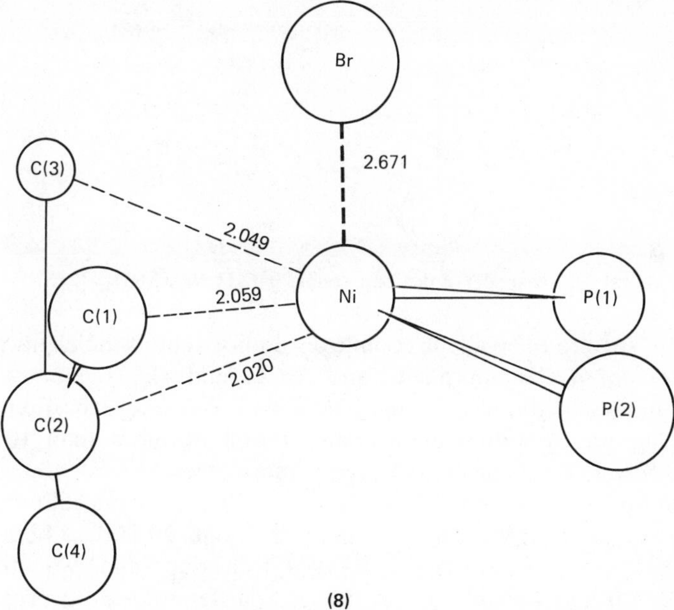

FIG. 1. Structure of π-methallyl[(bis-1,2-diphenylphosphine)ethano]bromonickel (54).

forms an angle of 106.5° with the basal coordination plane of nickel. This plane is assumed to be two-thirds the way along from the central (C(2)) to the terminal allylic carbon atoms (C(1)) and C(3)). The methyl group is distorted by 9.5° from the plane of the allyl group towards nickel. This probably is a consequence of a distortion causing a better overlapping of the metal orbitals with the orbitals of the allyl ligand.

The structure of **9** is also interesting because the complex has catalytic properties in carbonylation reactions (55). The ligands of this complex (which is ionic in solution) are arranged at the corners of a square pyramid (Fig. 2), but the distance between metal and halide (3.45 Å) exceeds that of a covalent bond.

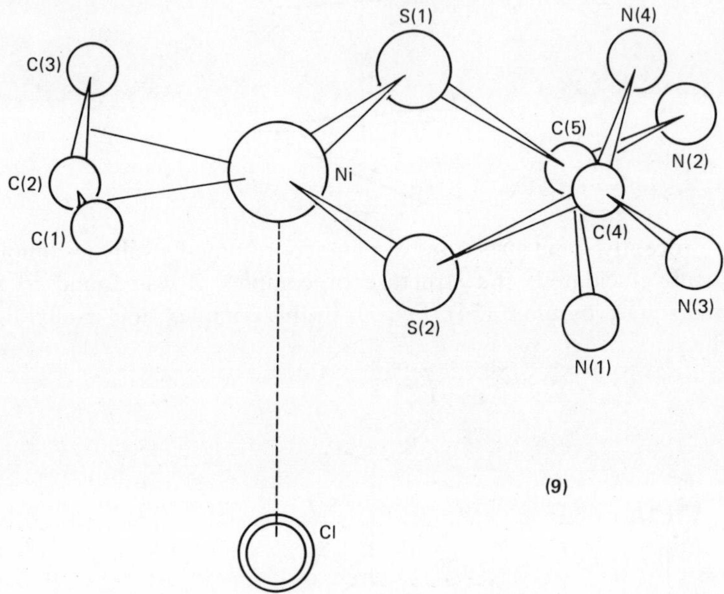

FIG. 2. Molecular model of the ionic couple π-$C_3H_5Ni[SC(NH_2)_2]_2^+,Cl^-$ (55).

Infrared spectra of π-allylic complexes do not exhibit the characteristic absorption of $=CH_2$ in plane, and $=CH$ and $=CH_2$ out of plane bending modes of the allyl group. The C=C stretching vibration of the free allyl group is shifted upon complexation by more than 100 cm^{-1} towards lower frequencies (21). For more recent work see (21*a*).

The ultraviolet and visible spectra of π-allylpalladium chloride have been investigated (50). Three bands at about 29,500, 34,000, and 40,500 cm^{-1} have been assigned to *d*–*d* transitions and two others at about 44,000 and 46,500 cm^{-1} to charge transfer transitions. The latter are due to excitations of electrons from the $p\pi$ orbital of the chloride to the d_{z^2} orbital of palladium and from the d_{xz} palladium orbital to the antibonding allyl orbital, respectively.

NMR studies have been carried out by several authors (17,20,23,34–39,41,44–49*d*). Since the behavior of the allylic protons in solution can be followed by this technique, important results relevant to both the interpretation of bonding and reactivity have been obtained. These studies of allylic complexes indicate that the allyl group can take a variety of conformations under the action of subtle steric and electronic effects (π–σ interconversions are treated in Section C of this introduction.) In most cases a π-allylic complex shows a nuclear magnetic resonance spectrum of the A_2M_2X type, owing to different shieldings of the 1,4 (*syn*) and 2,3 (*anti*) protons by the metal atom: one doublet for H_{1-4}, one doublet for H_{2-3} and a multiplet for H_5. The stereochemistry (*syn, anti*) is related to the middle

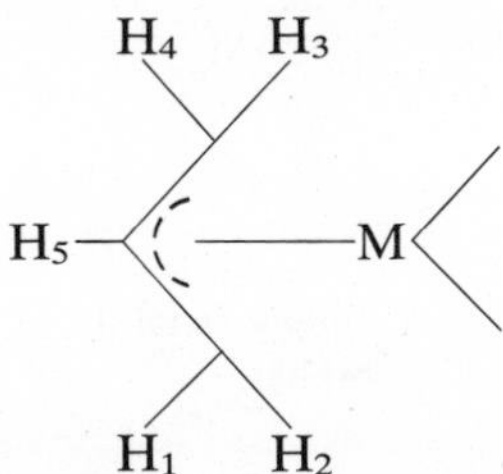

hydrogen atom, H_5, on the middle carbon atom. In the presence of basic ligands, the nuclear magnetic spectrum of a π-allylic complex indicates that the terminal protons are magnetically equivalent, resulting in an A_4X type spectrum (33). This has been referred to as a dynamic allyl (20). It should be noted, however, that the A_4X spectrum can be originated by phenomena quite different from each other (48).

Treatment of bonding in allylmetal complexes has been reported in several papers (24,42,43). The three molecular orbitals of the allyl group (Fig. 3) (ψ_1 bonding, ψ_2 nonbonding, and ψ_3 antibonding) interact with the metal orbitals by donating charge from the bonding and nonbonding orbitals and receiving back donation of charge from the metal into the antibonding orbital.

A more recent treatment of the π-allylmetal bond in a square planar system (42) shows that bonding between metal and allyl groups is essentially that between an allyl anion acting as a bidentate ligand and a metal cation. All three allyl molecular orbitals are involved in bonding, the main interaction being bonding between the nonbonding allyl ψ_2 and the vacant metal $d_{x^2-y^2}$ orbitals and back-bonding between the vacant antibonding allyl ψ_3 and the filled metal d_{xz} and d_{yz} orbitals. In order to obtain a better overlap of the orbitals involved, the plane of the allyl group is tilted to

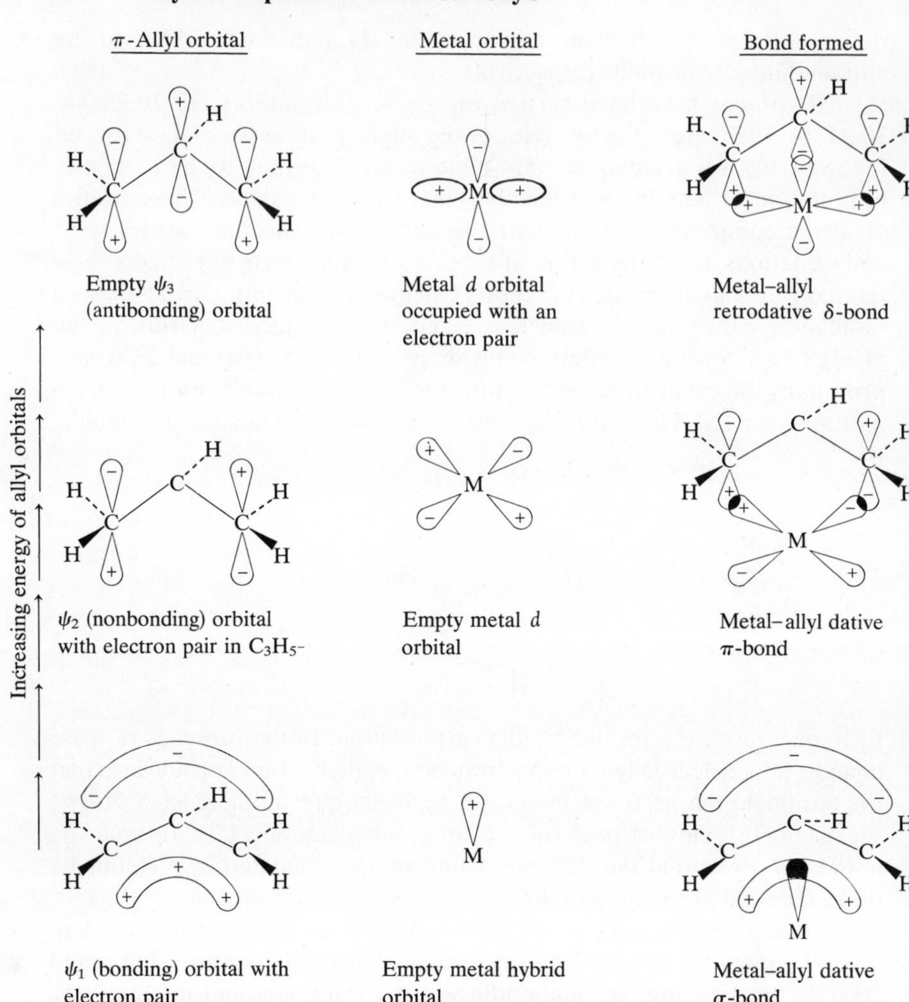

FIG. 3. Bonding of π-allyl group to a metal atom (101).

form an angle of 120° with the xy plane of the metal, as shown in Fig. 3a.

Another detailed treatment of bonding in π-allylmetal compounds appeared more recently (102a).

C. π–σ Rearrangements

As for organic syntheses, the most important property of the π-allylic complexes appears to be their interconversion with the σ-form in the presence of basic ligands.

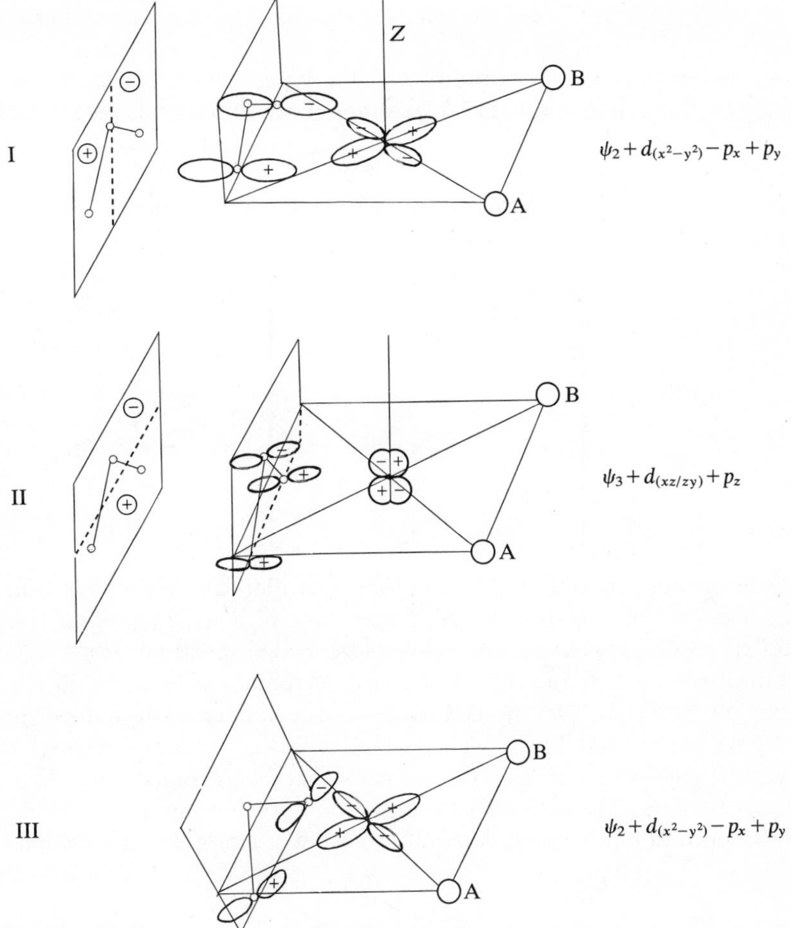

FIG. 3*a*. The groups in a square planar d_8 system (42).

An A_4X spectrum is known for the allylmagnesium halides; it has been attributed to a rapid equilibrium between the two allylic compounds:

$$CH_2{=}CH\overset{*}{C}H_2MgX \rightleftarrows \overset{*}{C}H_2{=}CHCH_2MgX \tag{7}$$

The types of transformation of π-allylic complexes that can be seen by the nmr method have been the subject of controversy (17,20,23,34–49,41,44–49). But the possibility of following the behavior of suitable complexes containing an optically active ligand has allowed elucidation of the course of the epimerization and isomerization of certain π-allylpalladium complexes (41,44–49*d*). The π–σ rearrangement appears to be primarily responsible for the changes that have been observed in these cases.

In relation to reactivity, equilibria of the types shown in Eqs. 5 and 6 appear to play a major role (17,23), although the situation is complicated by a number of fast rate processes that occur in solution (53). A possible way of equilibration in the presence of bases is shown in Eq. 8 (17,53), where L is a basic ligand such as dimethylphenylphosphine.

(8)

An interesting example, $C_3H_5NiBr(PEt_3)_2$, is offered by Walter and Wilke (56); This complex shows the A_4X spectrum of a dynamic allyl, while $C_3H_5NiBr(PPh_3)_2$ reveals a normal A_2M_2X π-allyl spectrum. It is evident that the more basic character of the triethylphosphine can cause equilibrium among the terminal allyl protons. Only a small percentage of the σ-form may be needed to effect equilibration, and it is possible that the complex with PEt_3 may be as much as 99% π (17) if the rate constants of Eqs. 5 and 6 are high.

Later findings (57) show that the situation is even more complicated by the presence of all the possible equilibria as in reaction 9. The position of

(9)

bond breakage in passing from π- to σ-allylic complexes in square planar compounds has been studied using nmr and x-ray investigations. It appears that good σ-donors such as phosphines cause the allyl carbon in the *trans* position to be more weakly bonded, while poor σ-donors, such as chlorine, favor a strong bond with the *trans* carbon atom (42,53). *Trans* effect refers to the transition state and does not necessarily imply a lengthening of the metal–carbon bond in the ground state (*trans* influence) (48,53*a*,*b*).

D. Reactivity of Allylic Complexes

From a synthetic point of view, an important consequence of the existence of equilibria as in Eq. 6 is this: The bidentate allyl ligand can pass from the π to the σ form, and the site formerly occupied by one end of the allyl groups may be taken by a reactive molecule that is in the right position to insert between the allyl group and the metal atom. A thorough study of the *cis* insertion of CO has been made by Noack and Calderazzo (58). Insertion reactions have been discussed in several papers (59,60,61,61*a*). Typical reactions of allyl halides with carbonyl complexes of nickel and, in certain cases, with other metal carbonyls are collected in Table 1. Besides structural nmr studies, further information, based on radioisotopic tracer experiments and on reactions of preformed π-complexes is available to help elucidate the nature of the intermediates in these reactions.

If allyl bromide, labeled in C-1, is made to react with tetracarbonylnickel, the carbonylation product (Eq. 4) is found to be completely scrambled, that is, C-1 possesses the same activity as C-3. (These experiments were carried out in the presence of acetylene, because the carbonylation product with insertion of acetylene was easier to obtain than that without acetylene.) The unconverted allyl bromide is also scrambled, but this isomerization reaction is independent of the carbonylation. In solution, the isomerization of allyl bromide is readily catalyzed by allylnickel bromide and the process is favored by a slight carbon monoxide pressure. If, however, allyl acetate is used, the isomerization is minimized, while the carbonylation product appears to be completely equilibrated. These facts indicate that equilibration of the C-1 and C-3 positions occurs only at the moment of formation of a π-allyl complex (62,63). If the coupling product of allyl acetate, (1,5-hexadiene) is examined, only half is found completely rearranged; the implication is that this is the π-allyl half.

As to the intermediacy of π-allylic complexes, the following reactions are relevant. Di-μ-chloro-di-π-allyldipalladium (**2**) and analogous compounds have been carbonylated (114,119):

$$\mathbf{2} + 2CO \rightarrow 2CH_2{=}CHCH_2COCl + Pd \tag{10}$$

TABLE 1
Some Typical Reactions of Allyl Halides (AX) *with Metal Carbonyls* (M)

	Ref.
(X = Cl, Br, I; R, R′, R″ = alkyl groups)	
$AX + ROH \xrightarrow{M} A{-}OR + HX$	57,71,114
$AX + PPh_3 \xrightarrow{M} A{-}PPh_3X$	57,249
$AX + CO + H_2O \xrightarrow{M} A{-}COOH + HX$	3,111,114
$AX + HC{\equiv}CH + CO + H_2O \xrightarrow{M} A{-}CH{=}CHCOOH + HX$	4
$AX + CH_2{=}CHCH{=}CH_2 + CO + H_2O \xrightarrow{M} A{-}CH_2CH{=}CHCH_2COOH + HX$	72
$AX + HX + M \rightarrow A{-}H + MX_2$	104
$AX + RCOR' + M \rightarrow A{-}C(R)(R')OMX$	10,51,105
$AX + CH_2{=}CHCN + M \rightarrow A{-}CH_2CH(CN)MX$	62,74
$AX + AX + M \rightarrow A{-}A + MX_2$	2
$AX + A'X' + M \rightarrow A{-}X' + A'{-}X + M$	52

Di-μ-bromo-di-π-allyldinickel (**6**) gives methyl but-3-enoate when treated with CO in methanol (64,65). In ether, it gives butenoyl bromide according to Eq. 11: (66).

$$\mathbf{6} + 10CO \rightarrow 2CH_2{=}CHCH_2COBr + 2Ni(CO)_4 + 2HBr \quad (11)$$

This kind of reaction implies opening of the bridge and formation of a monomeric species; CO is then inserted, possibly via a pentacoordinated intermediate (67,68) (Eq. 12):

$$\mathbf{6} + 2CO \longrightarrow (\pi\text{-}C_3H_5){-}Ni(CO)_2{-}Br \xrightarrow{CO} (CH_2{=}CHCH_2){-}Ni(CO)_3{-}Br \longrightarrow$$

$$CH_2{=}CHCH_2CO{-}Ni(CO)_2{-}Br \xrightarrow{CO} CH_2{=}CHCH_2COBr + Ni(CO)_4 \quad (12)$$

Wilke (20) reacted **10** to obtain the chloride of cyclooct-2-enecarboxylic acid:

$$[C_8H_{13}NiCl]_2 \xrightarrow[-Ni(CO)_4]{+5CO} C_8H_{13}COCl$$

(**10**)

Halogenobis(thiourea)-π-allylnickel complexes such as **9** and halogeno(thiourea)-π-allylnickel complexes have been isolated (68) and shown to be effective in catalytic carbonylations. Ionic intermediates are probably at work here in polar solvents. The reaction of **6** in the presence of acetylene gives a complex mixture of products derived from further reaction of the hexa-2,5-dienoylnickel bromide first formed (66,69).

Complex **11**, prepared by Heck, Chien, and Breslow (70), polymerizes acetylene if treated with a mixture of acetylene and CO (67). If, however, CO is fed into a benzene–methanol solution, absorption takes place with formation of the pentacoordinated complex **12** (57,67):

$(\pi\text{-}C_3H_5)Ni(I)(PPh_3)$ (**11**) $(\pi\text{-}C_3H_5)Ni(CO)(I)(PPh_3)$ (**12**)

Complex **12** reacts with CO and acetylene:

$$\mathbf{12} + HC{\equiv}CH + 3CO + CH_3OH \rightarrow$$
$$CH_2{=}CHCH_2CH{=}CHCOOCH_3 + Ni(CO)_3PPh_3 + HI \quad (13)$$

If another molecule of triphenylphosphine is added, no CO is evolved, and only polymerization of acetylene occurs. If PPh_3 is added in the absence of acetylene, an allyltriphenylphosphonium salt is obtained.

This result indicates that three coordination sites on the complex are needed for carbonylation with insertion of acetylene. The latter can occupy one of the two positions held by the π-allyl group and will then be cis to the σ-allyl group. Insertion of acetylene then occurs followed by insertion of carbon monoxide:

$(\sigma\text{-}C_3H_5)Ni(HC{\equiv}CH)(CO)(I)(PPh_3)$

When **12** is treated with CO at atmospheric pressure in methanol, it gives mainly methyl allyl ether along with small amounts of methyl but-3-enoate, according to Eqs. 14 and 15 (57):

$$\mathbf{12} + 3CO \rightarrow CH_2{=}CHCH_2OCH_3 + HI + OC{-}\underset{\displaystyle CO}{\overset{\displaystyle CO}{\underset{|}{\overset{|}{Ni}}}}{-}PPh_3 \qquad (14)$$

$$\mathbf{12} + 4CO \rightarrow CH_2{=}CHCH_2COOCH_3 + HI + OC{-}\underset{\displaystyle CO}{\overset{\displaystyle CO}{\underset{|}{\overset{|}{Ni}}}}{-}PPh_3 \qquad (15)$$

Formation of allyl ethers also occurs with cobalt complexes (71):

$$\underset{(\mathbf{13})}{(\pi\text{-}C_3H_5)Co(CO)_3} + CH_3ONa + CO \longrightarrow CH_2{=}CHCH_2OCH_3 + NaCo(CO)_4 \qquad (16)$$

The coupling reaction was observed with **12** (57):

$$\mathbf{12} + CH_2{=}CHCH_2Br \rightarrow CH_2{=}CHCH_2CH_2CH{=}CH_2 + NiBrI + PPh_3 + CO \qquad (17)$$

along with carbonylation (Eq. 15); coupling occurs with **13** (59) (Eq. 16) and other complexes such as **6**, which do not contain carbon monoxide (52).

The reaction of allyl halides with butadiene and CO on a palladium complex (**14**) was also reported (72):

$$\underset{(\mathbf{14})}{[(\pi\text{-}C_3H_5)PdCl]_2} + CH_2{=}CHCH{=}CH_2 + CO + H_2O \longrightarrow CH_2{=}CHCH_2CH_2CH{=}CHCH_2COOH + HCl + Pd \qquad (18)$$

This reaction appears analogous to the insertion reaction of acetylene on allylnickel complexes; however, unlike allylnickel complexes, substituted allylic groups react at the more substituted end.

The simple bond of norbornene was also shown to be reactive in an insertion reaction on an allylnickel complex. Norbornene inserts into π-methallylnickel acetate to give a new nickel complex, which, upon treatment with CO and water, gives an acid (73) (Eq. 19).

$$H_3C-\langle\!\!\!\!\!\!\!\!\,\,\,-Ni(Br)(C_7H_{10}) \xrightarrow[CO]{CH_3COOK} CH_2{=}C(CH_3){-}CH_2{-}C_7H_{10}{-}Ni(CO)_2OAc \xrightarrow[H_2O]{CO}$$

$$CH_2{=}C(CH_3){-}CH_2{-}C_7H_{10}{-}COOH + Ni(CO)_4 + HOAc \tag{19}$$

Addition of allylic groups to carbonyl compounds was also studied with π-allylic complexes in the absence of carbon monoxide (51) (Eq. 20):

$$\left[CH_3{-}C(CH_2)_2\,Ni{-}Br\right]_2 + R'COR'' \longrightarrow CH_2{=}C(CH_3){-}CH_2{-}C(ONiBr)(R'')(R') \tag{20}$$

Analogously, reactions of the allyl group with activated olefins were obtained with a π-allylic complex (74):

$$\mathbf{6} + 2CH_2{=}CHCN \rightarrow (CH_2{=}CHCH_2CH_2\underset{\displaystyle CN}{\underset{|}{C}}H)_2Ni + NiBr_2 \tag{21}$$

Thus, all the reactions reported in Table 1 have been observed with π-allylic complexes.

Most of these reactions can be rationalized as insertion processes (60), wherein a coordination site, formerly held by the allyl ligand, is occupied by a new molecule which then inserts between the allyl group and the metal. Pentacoordinated complexes are possible intermediates.

Sometimes the product of the reaction of the allyl group is split off; in other cases it remains attached to the complex and continues to react with other ligands, giving rise to more complex structures, which will be described later. Here also, the new group formed can be split off or retained. The reaction may be catalytic if the product is split off from the complex and if the metal is restored to an oxidation state that enables it to re-form the allylic complex. This can generally be formed by oxidative addition of the metals in its lower valence states to an allyl halide. Oxidative addition requires coordinative unsaturation and ligands able to increase the electron density on the metal (75,76,76*a*). The stereochemical course of the addition has been studied by Deeming and Shaw (77).

As for the manner by which the allyl group reacts with another ligand, there are two possibilities, depending on whether the reacting carbon is bound to the metal or to the other end of the allylic system.

Presently, the only suggestion comes from work on allylmagnesium compounds by Felkin (40,78) who proposed that crotylmagnesium halides react with carbonyl compounds or epoxides, here indicated by L, as follows:

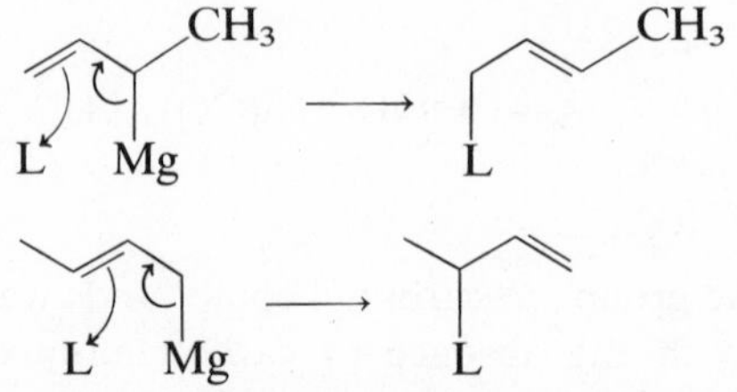

CH_3 L Mg → CH_3 L

L Mg → L

Thus, the linear product comes from the branched complex and the branched product from the linear complex. If L is considered the substituent and the metal the leaving group, this is an S_E2' type reaction. A cyclic transition state does not appear to be favored since only an S_E2' transition state and not an S_Ei' one meets the requirement that a concerted electrophilic substitution with rearrangement should be an antarafacial process (78,79). At present it is not possible to evaluate the generality of these conclusions for the transition metals.

The coordinated allyl group can be attacked by molecules or groups classifiable as nucleophiles or electrophiles. When the allyl group is attacked by electrophiles such as carbon monoxide, ketones, or activated olefins, the reaction can be regarded as an S_E type substitution, the metal being viewed as the leaving group. The above scheme of an S_E2' reaction possibly also holds for reactions on transition metals. In any case, a $\pi \rightarrow \sigma$ conversion should be operative.

Support for Felkin's mechanism comes from the hydrogenolysis rate of σ-crotyl-pentacyanocobaltate which is about one million times faster than that of the corresponding benzyl derivative (79*a*).

This behavior should be ascribed to the different protonation sites of benzyl and crotyl group, the latter being preferentially attacked by the proton and probably by other electrophiles at the noncoordinated carbon of the allylic system. See also (220).

Nucleophiles, such as Cl^-, CH_3O^-, CH_3COO^-, amines, and active methylene compounds, also attack the allylic group in certain cases, the metal being reduced to the zerovalent state (98). It is likely that the π-allylic and not the σ-allylic form is attacked, as recently shown for bis(π-allyl) nickel complexes (297).

Attack by nucleophiles on π-coordinated species generally is favored by a low electron density on the latter and by higher oxidation states of the metal, whereas the reverse is true for attack by electrophiles (76). Coordinated allyl groups generally can undergo both types of reaction. Carbonylation can be viewed as an electrophilic attack of CO on the allyl group, although one cannot exclude that certain carbonylation reactions originate by nucleophilic attack of an alkoxy group (OR) on coordinated carbon monoxide and of the resulting —COOR group on a π-coordinated ligand.

The end of the allyl group attacked by the reagents depends on the metal, the ligands, the solvent, and on the structure of the reagents and of the allylic group. Thus, the carbonylations observed on nickel so far indicate a sheer preference for the unsubstituted end in the case of an asymmetrical substituted allylic group. This is probably due to steric effects.

Reactions on palladium also show attack at the unsubstituted end during carbonylation, but addition occurs at the more substituted end during butadiene insertion. This suggests that the geometry of the transition state is more important than steric effects.

It should also be recalled that ligands able to exert a *trans* effect mobilize the opposite C–metal bond in a square planar system (42,53). As shown by nmr evidence (39) the allyl group selectively opens the bond opposite to the mobilizing ligand, thus offering the opportunity for a specific reaction at one of the two originally coordinated carbon atoms.

Nucleophiles attack the allyl group coordinated to palladium at either of the two ends depending on the structure of the allyl group. For example, an allyl group substituted at a terminal carbon by a butenyl group **15** reacts with the acetate group at the internal position, whereas it reacts at the external position if the substituent is —CH_2OCH_3 as in **16** (80). Nucleophiles attack π-allyls on the face opposite to palladium (298).

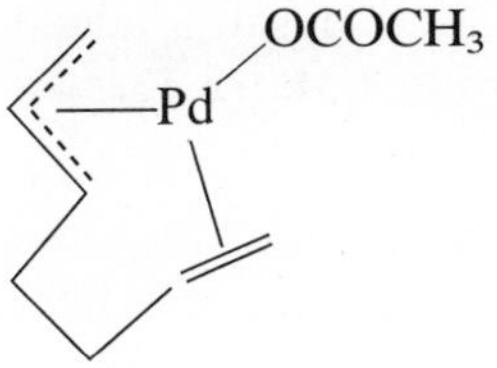

$$\text{(15)}\ \xrightarrow{CO} CH_2{=}CH\underset{\displaystyle OCOCH_3}{\underset{|}{C}H}CH_2CH_2CH{=}CH_2 \qquad (22)$$

(15)

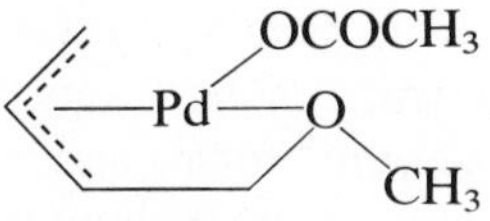

$$\text{(16)}\ \xrightarrow{CO} CH_3OCOCH_2CH{=}CHCH_2OCH_3$$

(16)

Another aspect of the reactivity of π-allylmetal complexes is the stereoselectivity that can be achieved by appropriate use of ligands and solvents (20,81–84*a*,238–240). The synthesis of 2,5-dienoic esters is a clear example of stereoselectivity concerning both the allyl group that gives rise to a *trans* double bond and acetylene that gives rise to a *cis* double bond. The stereochemistry of the allyl group also is subject to subtle steric and electronic effects on the allyl group, the ligands of the complex, and the effect of solvent (82–84). This is exemplified by the allylation reaction of π-allylic complexes (82,83,84*a*).

Allylation of complex **17** (L-ligand) in a nonpolar solvent gives mainly a *trans* product, but the methyl derivative **18** gives the *cis* product.

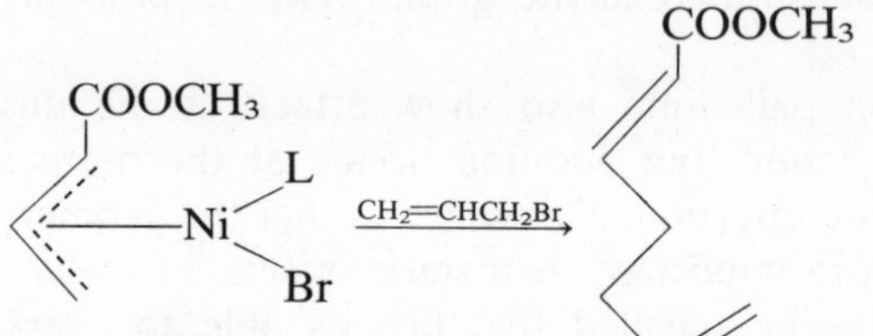

COOCH₃ … Ni(L)Br $\xrightarrow{CH_2{=}CHCH_2Br}$ COOCH₃

(17)

COOCH₃ … H₃C–…Ni(L)Br $\xrightarrow{CH_2{=}CHCH_2Br}$ H₃C– … COOCH₃

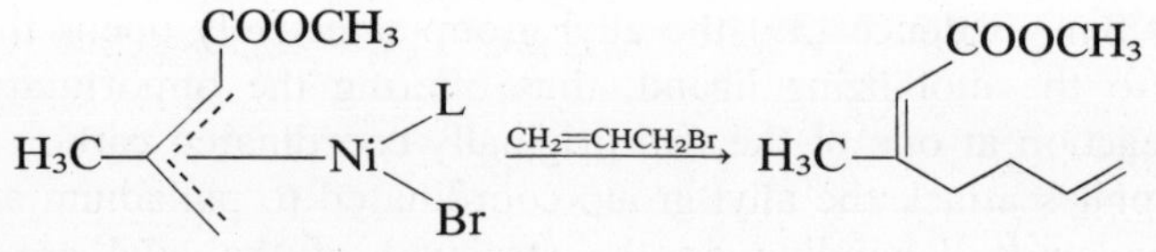

(18)

In a solution of dimethylformamide and acetonitrile containing $ZnCl_2$, **18** gives an allylation product 65–70% *trans.* If lutidine is present in **18**, the allylation product in methylformamide, although formed in low amount, is more than 90% *trans.* If a quaternary ammonium salt is added to the reaction mixtures employed, the branched product **18***a* is formed in more than 92% yield.

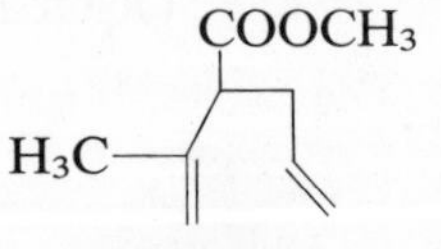

COOCH₃ H₃C–

(18a)

The steric behavior of the allylic group is not yet clear. One interpretation is that the allyl group, on passing to the σ form, can by rotation reach the energetic level corresponding to the *cis* or *trans* form; and, if the reaction is kinetically controlled, one of the two stereoisomers may be preferred.

On the other hand, if thermodynamic equilibrium is reached between the *cis* and trans forms; the reaction could also give one of the two stereoisomers in preference. The *trans* product in the above example can be considered the result of an equilibrium reaction promoted by the ligand.

Finally, a branched product can be formed in the above example, probably as a consequence of the increased carbanionic character of the allyl ligand, caused by the anion of the added salt and the highly polar medium. Carbanions of this type are known to react at the α-position.

So far we have considered the reactivity of π-allylic complexes that interconvert with the σ form.

Stable σ-allylic complexes with Mn, Fe, W, and Mo are known (22). The following reactions leading to new complexes were reported:

(*a*) protonation of σ-allylic complexes (22),
(*b*) carbonylation of allylpentacarbonylmanganese (85),
(*c*) insertion of SO_2 into allylpentacarbonylmanganese (86):

$$CH_2{=}CHCH_2Mn(CO)_5 + SO_2 \rightarrow CH_2{=}CHCH_2\overset{\overset{\Large O}{\|}}{\underset{\underset{\Large O}{\|}}{S}}{-}Mn(CO)_5 \quad (23)$$

and into 2-alkynylpentacarbonylmanganese (87):

$$HC{\equiv}CCH_2Mn(CO)_5 + SO_2 \rightarrow H_2C{=}C{=}CH{-}O{-}\underset{\underset{\Large O}{\|}}{S}{-}Mn(CO)_5 \quad (24)$$

The reactions of these complexes will not be treated further since the subject is limited to organic syntheses. For reviews see (61*a*,*b*).

Reference to allylic intermediates in organic syntheses can be found in several reviews (22,60,76,81,92–102*e*). The organic syntheses described in the following pages represent a rapidly expanding field that has already begun to branch into many areas of interest. Apart from the multitude of problems raised by the need to understand the behavior of organometallic complexes in catalysis, research is now focusing on at least three main organochemical topics. The first is related to the possibility of directing a selective series of additions of organic molecules to an allylmetal complex; the second refers to the stereoselectivity that can be achieved by coordinating allyl groups to the metal; the third, is the application of orbital symmetry rules to organic syntheses of allylic organometallics (88–91).

II. Carbonylation of the Allyl group with Carbon Monoxide and Related Reactions

A. Carbonylation Reactions of Allylic Compounds and Their Complexes

Allylic groups undergo carbonylation (as in Eq. 3) at the terminal carbon atom under mild conditions, and the double bond remains unaltered unless cyclization occurs. This behavior differentiates them from other olefinic substrates that undergo carbonylation.

1. Direct Carbonylation to Unsaturated acids and Their Functional Derivatives

a. Ni. At room or slightly higher temperature, allyl halides are coupled by $Ni(CO)_4$ (2,2*a*) (Eq. 2). Other conditions being equal, carbonylation occurs simply by increasing the carbon monoxide pressure slightly (3). In this way carbonylation is superimposed on the coupling reaction. *Trans* acids or esters are formed in the presence of water or alcohols, respectively. This reaction occurs with extraordinary ease compared to that of alkyl halides, which were shown by Reppe (103) to form esters only at high temperatures and pressures. The reaction can be expressed as follows:

$$R_2C{=}\underset{}{\overset{R}{\overset{|}{C}}}{-}\overset{R}{\overset{|}{C}}HX + CO + R'OH \xrightarrow{Ni(CO)_4} R_2C{=}\overset{R}{\overset{|}{C}}{-}\overset{R}{\overset{|}{C}}H{-}COOR' + HX \tag{25}$$

where R can be H or an electron-releasing group; R′ = H or an alkyl group; X = Cl, Br, I, OCOR″ (R″ = alkyl) or the tosyl group. Here, $Ni(CO)_4$ is viewed as behaving catalytically; secondary reactions, such as the formation of 1,5-hexadiene, are held responsible for partial transformation of $Ni(CO)_4$ into nickel halide.

The reaction is carried out by treating a solution of allyl halide in the desired alcohol or in water (the latter in mixture with a homogenizing solvent such as acetone) under CO pressure and with an excess of $Ni(CO)_4$.

Unsaturated acid chlorides are obtainable, generally by using an inert solvent. The chloride of 5-cyanopent-3-enoic acid was thus obtained from 1-chloro-4-cyanobut-2-ene by operating at 100°C under 40 atm of CO in an inert solvent such as benzene (3).

The types of compounds reacting with $Ni(CO)_4$ at low temperatures include primary and secondary allyl halides having alkyl substituents on the carbon atoms of the allylic system. Aromatic or cycloaliphatic groups can also be attached. The reaction gives better yields if alkyl substituents are on the carbon atoms of the double bond.

In contrast, some electron-withdrawing substituents prevent the reaction from occurring or strongly limit the yield. In ethereal solvents, this leads to coupling; in aqueous alcoholic solvents, substitution of the allylic halogen by hydrogen takes place (104). The reaction is described in Section III-A.

In the presence of certain ketones as solvents carbonylation products can be obtained; for instance, from methyl γ-bromocrotonate, the glutaconic half ester is obtained along with coupling products (105):

$$BrCH_2CH{=}CHCOOCH_3 + CO \xrightarrow[H_2O]{Ni(CO)_4} HOOCCH_2CH{=}CHCOOCH_3 + HBr \quad (26)$$

If allylic compounds like $CH_2{=}CHCHRX$ are employed, then complete allylic rearrangement occurs. Here the same product is obtained as from $RCH{=}CHCH_2X$ (see Table 2). In general, products of the allylic rearrangement do not form easily from compounds of formula $RCH{=}CHCR'R''X$, but an allylic rearrangement takes place when R=H. Mixtures of isomers are obtained from terminally substituted secondary allylic halides such as $RCH{=}CHCHRX$. The rearrangements observed in this carbonylation are likely influenced substantially by steric factors.

It was found (106) that carbonylations of allylic halides with $Ni(CO)_4$ can be carried out with high yields (~90% based on allyl chloride) at atmospheric pressure by using thiourea as a ligand and by neutralizing the hydrochloric acid as it is liberated in the reaction. A procedure involving the use of $NiCl_2$, thiourea, and a reducing agent such as a Mn–Fe alloy also proved effective (267). See Section II-A-3 for more details.

Benzyl bromide and iodide have been carbonylated; phenylacetic esters and dibenzyl ketone were obtained along with dibenzyl (107).

By using as catalyst a halogenotricarbonylnickel anion (formed from $Ni(CO)_4$ and halide ions of an alkaline or ammonium salt, MX), it is possible to effect a catalytic carbonylation of benzyl chloride (108) to phenylacetic acid and its esters. Yields and catalyst activity are good when using a mixture of dipolar aprotic and hydroxylic solvents and neutralizing the acids formed during the reaction with a base such as CaO (27):

$$PhCH_2Cl + CO + ROH + B \xrightarrow{M[XNi(CO)_3]} PhCH_2COOR + BHCl \quad (27)$$

Other materials reactive with $Ni(CO)_4$ are propargylic-type halides such as propargyl chloride, which gives butadienoic and itaconic acids

(4,109,110). These are regarded as oxo-type reactions (see second chapter). From a mechanistic point of view, it is not yet clear, however, whether these reactions first involve an attack of carbon monoxide on the triple bond followed by halide elimination or halide abstraction by nickel leading to a coordinated propargylic group able to react with carbon monoxide.

b. Co. The carbonylation reaction has been extended to several other metal carbonyls. Heck and Breslow (71) found that a catalytic reaction could be obtained by using cobalt carbonyl anion as a catalyst:

$$RX + CO + R'OH + B \xrightarrow{Co(CO)_4^-} RCOOR' + HBX \qquad (28)$$

where R,R′ are alkyl groups; X is Cl, Br, I, OSO_3H, OSO_2R, and so on; and B is a base. This reaction (described in the fourth chapter of Vol. 1 of this series) is general for RX compounds. Allyl compounds also react. Heck reported a 33% yield of methyl butenoate from allyl bromide in the presence of sodium methoxide and $NaCo(CO)_4$ for 20 hr at 35°C. However, he did not obtain the product from π-allylcobalt tricarbonyl (13) as would be expected from the behavior of the corresponding nonallylic complexes; instead, allyl methyl ether was formed (Eq. 16).

Benzyl chloride, a compound related to allylic halides, reacts with CO in methanol to give phenylacetic acid in over 95% yield, if cobalt carbonyl anion is formed in situ and the HCl is neutralized with calcium hydroxide (71*a*). Reduction of cobalt chloride was effected with a Mn–Fe alloy in the presence of CO (71*b*).

c. Fe. The salt $NaFe(NO)(CO)_3$ reacts with benzyl chloride to give a phenylacetyl complex that can be isolated as the triphenylphosphine derivative, $PhCH_2COFe(CO)_2(NO)(PPh_3)$, and converted into methyl phenylacetate by refluxing with sodium methoxide in methanol. The normal allylic halides do not undergo CO insertion, but give π-allylic complexes (10).

d. Pd, Rh, Pt. Palladium chloride, palladium nitrate, palladium oxide, palladium metal, rhodium chloride, or a mixture of hexachloroplatinic acid and stannous chloride were all found effective for the carbonylation of allyl chloride by Tsuji and his co-workers (111). Palladium chloride was the most effective of these catalysts. 3-Butenoic esters and their functional derivatives were obtained at temperatures between 80 and 120°C and initial pressures of 100–150 atm working in solution. Thus from allyl chloride in ethanol, ethyl but-3-enoate, and ethyl crotonate were obtained in 47 and 5% yields, respectively, whereas the chloride of but-3-enoic acid was formed in benzene. In THF the ester,

$CH_2{=}CHCH_2COO(CH_2)_4Cl$, was obtained in 88% yield. Apparently the THF ring was opened and reacted as if it were the alcohol $HO(CH_2)_4Cl$.

Allylic alcohols reacted in ethanol analogously to the corresponding chlorides. From allyl alcohol, in the absence of solvent, allyl but-3-enoate was obtained. But-2-en-1-ol and its tautomer, but-1-en-3-ol, formed ethyl pent-3-enoate in ethanol. Apparently only the formation of the same complex by an allylic rearrangement can explain this behavior, as with $Ni(CO)_4$.

Allylic esters and ethers in alcohols formed the corresponding butenoates (Eq. 29). In benzene, the anhydrides resulting from the insertion of one and two CO molecules were obtained (Eq. 30).

$$CH_2{=}CHCH_2OCOCH_3 + CO \xrightarrow{PdCl_2} \underset{58\%}{CH_2{=}CHCH_2COOCOCH_3} \qquad (29)$$

$$CH_2{=}CHCH_2OCH_2CH{=}CH_2 + CO \xrightarrow{PdCl_2} CH_2{=}CHCH_2COOCH_2CH{=}CH_2 \xrightarrow{CO} \underset{19\%}{CH_2{=}CHCH_2COOCOCH_2CH{=}CH_2} \qquad (30)$$

Allyl phenyl ethers or allyl tosylates can also be employed as starting materials for carbonylation.

In a later study, Imamura and Tsuji (112) isolated the following products from the carbonylation of 1,4-diethoxybut-2-ene in ethanol at 60–160°C:

$$EtOCH_2CH{=}CHCH_2OEt + CO + EtOH \xrightarrow{PdCl_2}$$

$$EtOOCCH_2CH{=}CHCH_2COOEt \qquad (\mathbf{19})$$

$$EtOCH_2CH{=}CHCH_2COOEt \qquad (\mathbf{20})$$

$$CH_2{=}CH\underset{\displaystyle COOEt}{\underset{|}{C}H}CH_2OEt \qquad (\mathbf{21})$$

$$CH_3CH{=}CHCH_2COOEt \qquad (\mathbf{22})$$

$$CH_2{=}CHCH{=}CHCOOEt \qquad (\mathbf{23})$$

$$EtOCH_2\underset{\displaystyle OEt}{\underset{|}{C}H}CH{=}CH_2 \qquad (\mathbf{24})$$

$$EtOCH_2CH{=}CHCH_3 \qquad (\mathbf{25})$$

$$CH_3CH{=}CHCH_3 \qquad (\mathbf{26})$$

The products, obtained in variable amounts according to the conditions, result from double carbonylation (**19**), monocarbonylation without rearrangement (**20**) and with rearrangement (**21**), monocarbonylation and hydrogenolysis (**22**), carbonylation and 1,4-elimination (**23**), rearrangement (**24**), and hydrogenolysis (**25**) and (**26**).

Tsuji showed that olefins also react with palladium chloride (113):

$$CH_2{=}CH_2 + PdCl_2 + CO \rightarrow ClCH_2CH_2COCl + Pd \qquad (31)$$

The reaction with olefins takes place at room temperature, whereas allylic carbonylation occurs at higher temperature. It is interesting to compare the behavior of allyl chloride in the two types of reaction:

$$CH_2{=}CHCH_2Cl \rightarrow CH_2{=}CHCH_2COCl$$
$$CH_2{=}CHCH_2Cl \rightarrow ClCH_2CHClCH_2COCl$$

The first is a substitutive carbonylation, that is, chlorine is replaced by the COCl group; the second is an additive one, as in Eq. 31.

Carbonylation of π-allylpalladium chloride in benzene was also shown to give vinylacetyl chloride.

Since complexes like **27** are readily obtainable from β,γ-unsaturated esters such as butenoic esters (113*a*), it is possible to start from an allylic compound and carbonylate it twice with final formation of dicarboxylic esters (Eq. 32).

$$[(\pi\text{-}C_3H_5)PdCl]_2 + CO + CH_3OH \longrightarrow CH_2{=}CHCH_2COOCH_3 + HCl$$
$$CH_2{=}CHCH_2COOCH_3 \xrightarrow{PdCl_2} [(\pi\text{-}C_3H_4COOCH_3)Pd{-}Cl]_2 \; (\mathbf{27}) \qquad (32)$$
$$[(\pi\text{-}C_3H_4COOCH_3)PdCl]_2 \xrightarrow[CH_3OH]{CO} HCl + Pd + CH_3OOC{-}CH_2{-}CH{=}CH{-}COOCH_3$$

(**27**)

A catalytic reaction of allylic chlorides involving use of palladium chloride or of palladium chloride complexes was independently described by Dent, Long, and Whitfield (114). They were able to obtain high yields of acid chlorides by treating allylic chlorides in the absence of solvents with CO at about 500 atm and 110–150°C in the presence of catalytic amounts of $PdCl_2$ or of **2**.

The site of carbonylation was the terminal allylic carbon atom; if two different positions of attack were available the least substituted was the reactive one. Thus the tautomers 1-chlorobut-2-ene and 3-chlorobut-1-ene reacted in the same way to give the chloride of pent-3-enoic acid (Eq. 33). The reaction was specific for allylic chlorides; aliphatic and aromatic chlorides did not react, and benzyl chloride formed a tar.

$$\left.\begin{array}{l} CH_3CH{=}CHCH_2Cl \xrightarrow{Pd} \\ CH_2{=}CH{-}CHClCH_3 \xrightarrow{Pd} \end{array}\right\} [(\pi\text{-}CH_3CHCHCH_2)PdCl]_2 \xrightarrow{CO} CH_3CH{=}CHCH_2COCl \quad (33)$$

The reaction of allyl halides with $PdCl_2$ in methanol was much more complicated. A comparison with the reaction with $Ni(CO)_4$ showed that several products besides methyl but-3-enoate and methyl crotonate formed—methyl isobutyrate and methyl 3-chlorocrotonate were identified among these products. Under the same conditions, $Ni(CO)_4$ gave allyl methyl ether and hexa-1,5-diene as by-products (114). Long and Whitfield (115) also studied the carbonylation of the π-allyl, π-methallyl, and π-crotylpalladium chloride complexes in benzene at 200 atm and 50°C; after hydrolysis with water they obtained the corresponding acids in 65, 57, and 88% yields.

A patent (116) claims the use of promoters such as acetylene, indene, cyclopentadiene, and fluorene in carbonylations of allyl halides catalyzed by Pd or Rh or by their complexes.

The reaction of diolefin–palladium chloride complexes is also a type of carbonylation of allylic compounds. These complexes are really π-allylic complexes. In fact, from the reaction of butadiene with $PdCl_2$ in benzene, Tsuji (117) isolated complex **28**:

$$[(\pi\text{-}ClCH_2CHCHCH_2)PdCl]_2$$

(**28**)

On treatment with carbon monoxide, **28** gave $ClCH_2CH{=}CHCH_2COCl$ and $ClCH_2CH{=}CHCH_2Cl$; in ethanol the product was

$CH_3CH{=}CHCH_2COOEt$. Also present was a small amount of β-hydromuconic ester, which, in the presence of a base, loses hydrogen with formation of muconic ester and metallic palladium.

To establish whether carbonylation could occur only at the C-1 position or at both the C-1 and C-4 positions, complex **29** was chosen:

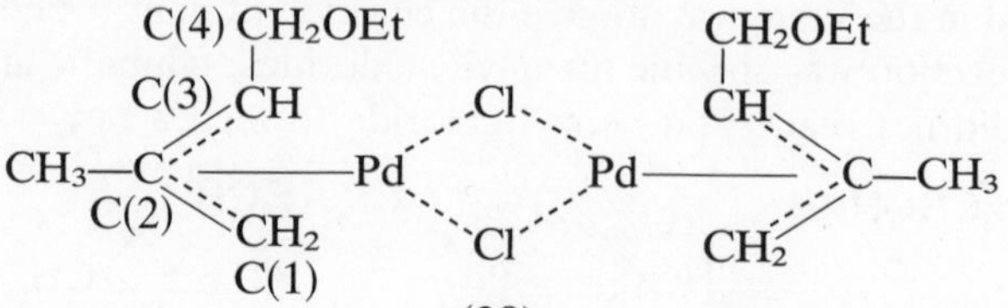

```
        C(4) CH2OEt                              CH2OEt
              |                                   |
        C(3) CH          Cl                      CH
            //  .      .    .                        \\
CH3—C  ------------Pd          Pd------------   C—CH3
   C(2) \\ .            .    .                       //
          CH2            Cl                      CH2
          C(1)
                       (29)
```

Carbonylation occurred at both positions; the products formed and the positions of attack are these:

Compound		Attack at
$EtOCH_2CH{=}C(CH_3)CH_2COOEt$	**(30)**	C-1
$CH_3C(CH_3){=}CHCH_2COOEt$	**(31)**	C-4
$EtOOCCH_2CH{=}C(CH_3)CH_2COOEt$	**(32)**	C-1 + C-4
$CH_3C(CH_3)(OEt)CH_2CH_2COOEt$	**(33)**	C-4
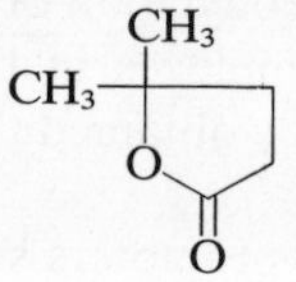 (γ,γ-dimethyl-γ-butyrolactone: ring with CH_3, CH_3, O, C=O)	**(34)**	C-4

C-1 products predominated in benzene at 20°C, whereas products resulting from attack at both C-1 and C-4 were formed at 100°C. Both C-1 and C-4 carbonylation products were accompanied by minor amounts of compounds resulting from partial substitutive hydrogenation. When carried out in ethanol, the same reaction showed mainly C-1 attack at room temperature and C-4 attack at 100°C. The latter is favored by the presence of HCl.

Bordenca and Marsico (118) reported that ethyl 2-methylpent-3-enoate is formed in 40–50% yields from reaction of *trans*-piperylene with $PdCl_2$ in ethanol at 100–110°C under a maximum CO pressure of 1700 psi. Brewis and Hughes (119) reported that the carbonylation of conjugated dienes in alcohols to β,γ-unsaturated esters is best carried out with ligand-modified catalysts. The yield and catalytic activity were shown to depend on the type of halide present. The dihalogenobis(tributylphosphine)palladium complexes, $PdX_2(PBu_3)_2$, gave a 20% yield of methyl pent-3-enoate (21 moles/gram atom of palladium) when X = Cl; 64% (54 moles) when X = Br, and 68% (61 moles) when X = I. Carbonylation of butadiene with $Co_2(CO)_8$ in isoquinoline at 120°C and 300 atm was described (119*a*).

The platinum complex containing $P(C_6H_4F)_3$ was employed by Parshall (120) for the catalytic carbonylation of allyl alcohol. The Reppe synthesis of crotonic acid from allyl alcohol, CO, and $Ni(CO)_4$ at 250 atm and 300°C (103) is likely related to the above reaction of allyl halides, as implied by Parshall's results (120). The complex $RhCl(CO)(PPh_3)_2$ (75,76) was employed by Tsuji (123) in the carbonylation of benzyl chloride at 150°C and 100 atm.

Allenes also form allyl type complexes with transition metals (124,125); complexes **35** and **38** for example have been carbonylated by Tsuji (126,127); see also (102*d*).

$$[Cl\text{—}C(CH_2)_2PdCl]_2\ (\mathbf{35}) \xrightarrow[ROH]{CO} \begin{cases} CH_2{=}C(COOR)CH_2COOR & (\mathbf{36}) \\ CH_2{=}C(Cl)CH_2COOR & (\mathbf{37}) \end{cases} \quad (34)$$

$$[CH_2{=}C(CH_2Cl)\text{—}C(CH_2)_2PdCl]_2\ (\mathbf{38}) \xrightarrow{CO,\ ROH} \quad (35)$$

(**39**) 3,3,4-trimethyl-2(5H)-furanone-type lactone (CH_3, CH_3, CH_3, O, C=O)

$$(CH_3)_2C{=}C(CH_2COOR)_2\ (\mathbf{40}) \qquad ROOCCH_2(CH_3)C{=}C(CH_2COOR)_2\ (\mathbf{41})$$

Carbonylation of **42** (written here as a monomer) in ethanol gave ethyl acetoacetate (127):

$$[CH_3COO\text{—}C(CH_2)_2PdCl]\ (\mathbf{42}) \xrightarrow[EtOH]{CO} [CH_3COO\text{—}C(CH_2COOEt){=}CH_2] \xrightarrow{H_2O} CH_3COCH_2COOEt \quad (36)$$

The carbonylation reactions described in this section are listed in Table 2.

TABLE 2

Synthesis of Unsaturated Acids and Their Functional Derivatives by Carbonylation of Allylic Compounds

Substrate	Catalyst	Products[a]	Yield[b]	Ref.
Allyl chloride	$Ni(CO)_4$	Methyl but-3-enoate	65/90	31,65,106,153,267
(π-Allyl $NiBr)_2$	—	Methyl but-3-enoate	50	64,65
1-Chlorobut-2-ene	$Ni(CO)_4$	Methyl pent-3-enoate	33	3
3-Chlorobut-1-ene	$Ni(CO)_4$	Methyl pent-3-enoate	—	3
Methallyl chloride	$Ni(CO)_4$	Methyl 3-methylbut-3-enoate Methyl 3-methylbut-2-enoate	88	3,267
Methyl γ-bromocrotonate	$Ni(CO)_4$	4-Carbomethoxybut-3-enoic acid	33	105
1,3-Dichloroprop-2-ene	$Ni(CO)_4$	Methyl 4-chlorobut-3-enoate	5	104
Methyl β-methyl-γ-bromo-crotonate	$Ni(CO)_4$	4-Carbomethoxy-3-methylbut-3-enoic acid	42	105
1-Chloro-4-cyanobut-3-ene	$Ni(CO)_4$	Methyl 5-cyanopent-3-enoate Dimethyl β-hydromuconate	30 10	3
1-Chloro-4-cyanobut-3-ene	$Ni(CO)_4$	5-Cyanopent-3-enoic acid	38	
1-Chloro-4-cyanobut-3-ene	$Ni(CO)_4$	5-Cyanopent-3-enoyl chloride	20	3
Allyl chloride	$Co(CO)_4^-$	Methyl but-3-enoate	33	71
Allyl chloride	$PdCl_2$	But-3-enoyl chloride Crotonoyl chloride	91 6	111,114
Allyl chloride	(π-Allyl $PdCl)_2$	But-3-enoyl chloride	99	114
Allyl chloride	(π-But-2-enyl $PdCl)_2$	But-3-enoyl chloride	96	114
Allyl chloride	Pd	But-3-enoyl chloride Crotonyl chloride	55 8	114
(π-Allyl $PdCl)_2$	—	But-3-enoic acid	65	115
(π-Allyl $PdCl)_2$	—	Ethyl but-3-enoate Ethyl crotonate	47 5	111

Allyl alcohol	$PdCl_2$	Ethyl but-3-enoate	42	111
Allyl ethyl ether	$PdCl_2$	Ethyl but-3-enoate Ethyl crotonate	57 8	111
Allyl phenyl ether	$PdCl_2$	Ethyl but-3-enoate Ethyl crotonate	18 10	111
Allyl acetate	$PdCl_2$	Ethyl but-3-enoate	80	111
Allyl tosylate	$PdCl_2$	Ethyl but-3-enoate Ethyl crotonate	24 26	111
Diallyl ether	$PdCl_2$	Ethyl but-3-enoate	73	111
Diallyl ether	$PdCl_2$	But-3-enoic anhydride	19	111
1-Chlorobut-2-ene	(π-Allyl $PdCl)_2$	Pent-3-enoyl chloride	81	114
(π-But-1,2,3-enyl $PdCl)_2$		Pent-3-enoic acid	88	115
(π-But-1,2,3-enyl $PdCl)_2$		Methyl pent-3-enoate	55	119
3-Chlorobut-1-ene	$PdCl_2$	Pent-3-enoyl chloride 1-Chlorobut-2-ene	80 3	114
But-2-en-1-ol	$PdCl_2$	Ethyl pent-3-enoate	42	111
But-2-en-1-ol acetate	$PdCl_2$	Ethyl pent-3-enoate	64	111
But-3-en-2-ol	$PdCl_2$	Ethyl pent-3-enoate	39	111
Pent-3-en-2-ol acetate	$PdCl_2$	Ethyl 2-methylpent-3-enoate	39	111
1-Chloro-2-methylprop-2-ene	(π-Methallyl $PdCl)_2$	3-Methylbut-3-enoyl chloride 3-Methylbut-2-enoyl chloride	70 7	114
(π-Methallyl $PdCl)_2$	—	3-Methylbut-3-enoic acid	57	115
1-Bromocyclohex-2-ene	$PdCl_2$	Ethyl cyclohex-2-ene-1-carboxylate	30	111
(1-Carbethoxy-π-allyl $PdCl)_2$	—	Diethyl glutaconate	n.d.	113*a*
Butadiene	$(PBu_3)_2PdCl_2$	Methyl pent-3-enoate	20	119
Butadiene	$(PBu_3)_2PdBr_2$	Methyl pent-3-enoate	64	119
Butadiene	$(PBu_3)_2PdI_2$	Methyl pent-3-enoate	68	119

(*contd.*)

TABLE 2 (cont'd.)

Substrate	Catalyst	Products[a]	Yield[b]	Ref.
1,3-Pentadiene	$(PBu_3)_2PdI_2$	Methyl 2-methylpent-3-enoate	34	119
Cyclopentadiene	$(PBu_3)_2PdI_2$	Methyl cyclopent-2-ene-1-carboxylate	73	119
2,3-Dimethylbuta-1,3-diene	$(PBu_3)_2PdI_2$	Methyl 3,4-dimethylpent-3-enoate	50	119
Isoprene	$(PBu_3)_2PdI_2$	Methyl 3-methylpent-3-enoate	15	119
		Methyl 4-methylpent-3-enoate	38	
		Methyl 4-methylpent-4-enoate	10	
Cyclohexa-1,3-diene	$PdCl_2$	Ethyl cyclohex-2-ene-1-carboxylate	80	260
Cycloocta-1,3-diene	$PdCl_2$	Ethyl cyclooct-2-ene-1-carboxylate	9	260
		3-Ethoxycyclooct-1-ene	1	
Butadiene-$PdCl_2$	—	5-Chloropent-3-enoyl chloride	0.6	117
		1,4-Dichlorobut-2-ene		
Butadiene-$PdCl_2$	—	Methyl 5-chloropent-3-enoate	1	117
		1,4-Dichlorobut-2-ene		
Pd complex **29**	—	**30**, **31**, **32**, **33**, **34**	—	117
Pd complex **35**	—	**36** and **37**	—	126,127
Pd complex **38**	—	**39**	24	127
		40	20	
		41	9	
Pd complex **42**	—	Ethyl acetoacetate	15	127

[a] Only methyl or ethyl esters tabulated.
[b] Only maximum yields reported if data from different authors are available.

e. Mechanisms. The direct substitutive carbonylation of allylic compounds can be carried out with various transition metals. Extensive studies, however, have only been made with nickel and palladium. As pointed out in Section I, a first step, consisting of the oxidative addition of the metal to the allyl halide, leads to the formation of an intermediate allyl complex. Carbonylation should occur through nucleophilic attack of the allyl group on the carbonyl group, a feature probably common to several types of carbonylation (128). The acyl carbonyls thus formed are subject in some cases to splitting by halide ions or other nucleophilic groups with formation of acyl halides or acids and esters if water or alcohols are present. As for $Ni(CO)_4$, the initial complex may well be a halogeno(carbonyl)-π-allylnickel complex (64–66,129), probably containing two carbonyl groups (57). The inhibiting effect of CO pressure on the reaction between allyl halides and $Ni(CO)_4$ is consistent with the necessity of liberating CO from $Ni(CO)_4$ (94,130,131) to form **43** (Eq. 37).

$$Ni(CO)_4 \rightleftarrows Ni(CO)_3 + CO$$

$$CH_2{=}CHCH_2X + Ni(CO)_3 \longrightarrow \left[CH_2{=}CHCH_2{-}Ni(CO)_3{-}X \right] \xrightarrow{-CO}$$

(**43**)

$$(\pi\text{-}C_3H_5)Ni(CO)_2X \xrightarrow{-2CO} \left[(\pi\text{-}C_3H_5)NiX \right]_2 \qquad (37)$$

(**43a**)

$$\downarrow 3CO$$

$$CH_2{=}CHCH_2COX + Ni(CO)_4$$

The first step is in agreement with the findings of Day, Basolo, and Pearson (132,133). The second step is an oxidative addition (75,76,134,135). The stereochemical course of allyl halide addition to various types of complexes was studied by Deeming and Shaw (77). They showed that allyl chloride adds *cis* (i.e., allyl and Cl are mutually *cis*) to complexes of the type *trans*-$IrCl(CO)L_2$ ($L = PMe_2Ph$ or $AsMe_2Ph$) in benzene. Isomerization then can transform the original *cis* addition to the *trans* product. Further work is needed to establish the degree of generality of this interpretation. It has been shown, however, that both types of addition are possible in principle (272).

The complex **43a** formed by oxidative addition of the species $Ni(CO)_3$, followed by loss of CO, should be pentacoordinated. This complex is probably an intermediate, in view of the isolation of complexes such as **12** (ionic in polar solutions). Pentacoordination should induce insertion (135*a*).

The carbonylation is probably a type of *cis*-ligand insertion (58,59,136,137), as observed in the carbonylation of $CH_3Mn(CO)_5$ (58). The characteristics of insertion reactions have been discussed by Cossee (61): The role of the transition metal seems also that of providing an empty *d* orbital in order to keep the migrating group bonded to the complex during the reaction, thus lowering the energy of activation.

After CO insertion, final splitting of the complex leads to formation of acyl halides (66). This was shown to occur in benzene (3) and in ether (66). Since it was not possible to detect the chloride in alcoholic solution, an investigation was carried out on the acid anhydrides, which should form analogously, by using the acetate instead of the chloride anion. Splitting of the acyl group is a feature common to both the carbonylation of allyl halides and the acrylic synthesis; because it is easy to carry out with organic acids, the latter was investigated. Acid anhydrides were detected by infrared analysis and identified chemically as anilides even in methanolic solutions (138). It was thus shown that acid anhydrides and probably acyl chlorides may be formed even in hydroxylic solvents during acyl group removal from the acyl nickel complex.

Benzyl halides are likely to follow a reaction path similar to that of allyl halides; π-benzylic type complexes of molybdenum have been isolated (268).

As for palladium, carbonylation of di-μ-chloro-di-π-allyldipalladium complexes (**2**) should occur analogously to that of the nickel complexes, through formation of a Pd–CO bond. Formation of complexes of type **2** from palladium chloride in methanol–water is favored by the presence of carbon monoxide (139). The following mechanism based on oxidation of a carbonyl group has been proposed (140) (some ligands are omitted for simplicity) (Eq. 38):

$$\begin{array}{c} \text{Cl} \\ | \\ \text{HO—Pd} \\ | \\ \text{CO} \end{array} — \begin{array}{c} \text{CH}_2 \\ \| \\ \text{CH} \\ | \\ \text{CH}_2\text{Cl} \end{array} \longrightarrow \begin{array}{c} \text{Cl} \\ | \\ \text{Pd} \\ | \\ \text{HO—C} \\ \| \\ \text{O} \end{array} — \begin{array}{c} \text{CH}_2 \\ \| \\ \text{CH} \\ | \\ \text{CH}_2\text{Cl} \end{array} \longrightarrow$$

$$\left[\begin{array}{c} \text{Cl} \\ \diagdown \\ \text{Pd} \\ \diagup \end{array} \text{———} \begin{array}{c} \text{CH}_2 \\ \vdots \\ \text{CH} \\ \vdots \\ \text{CH}_2 \end{array} \right] + \text{HCl} + \text{CO}_2 \qquad (38)$$

Primary amines may replace water in reaction 38. In this case, the amines are oxidized to ureas and oxamides via isocyanates (98,141).

The carbonylation of allyl chloride to vinylacetyl chloride in presence of di-μ-chloro-di-π-allylpalladium was studied kinetically in dimethoxyethane solvent (72). A dependence of the first order on concentration of allyl chloride and of the palladium complex and of the second order on CO pressure was found. The results were explained by double coordination of CO to the complex and attack of allyl chloride at the resulting carbonylic complex.

Ligands like PPh_3 were also found to exert a pronounced influence on reaction rate, which passes through a maximum at a phosphine/Pd molar ratio of 0.5. At higher ratios the rate decreases, the reaction being practically inhibited at ratios higher than 4.

π-Allyl complexes were also proved to be intermediates in the carbonylation of allylic ethers and esters (72,112). π-Allyl complexes are again intermediates in the carbonylation of dienes to β,γ-unsaturated esters with palladium chloride complexes in alcohols. The reaction was explained as follows (119):

$$CH_2{=}CHCH{=}CH_2 + HPdL_2X \longrightarrow \left[\begin{array}{c} CH_3 \\ | \\ CH \\ HC \cdots\cdots Pd(L)(X) \\ CH_2 \end{array}\right] \xrightarrow[CH_3OH]{CO}$$

$$CH_3CH{=}CHCH_2COOCH_3 + HPdL_2X \qquad (39)$$

where X = Cl, Br, or I; L = ligand.

2. *Carbonylation of Bis(π-allyl)nickel Complexes to Ketones*

A direct carbonylation leading to ketones was studied by Wilke (20). He started from a series of allylic compounds of nickel, prepared from allylic Grignard complexes and nickel halides. The formula for the simplest member of the series, bis(π-allyl)nickel (**44**) is given below:

$$\langle\!\!\langle \text{———Ni———} \rangle\!\!\rangle$$

(**44**)

Wilke found that, although the normal reaction of nickel diallyls with CO at low temperature is the coupling of two allylic groups (discussed later under coupling reactions), there are some allylic groups that react in part or completely with CO; for example, the π-methallyl and π-cyclooctenyl groups. The latter gave a 92% yield of the symmetrical ketone **45** with concomitant formation of $Ni(CO)_4$:

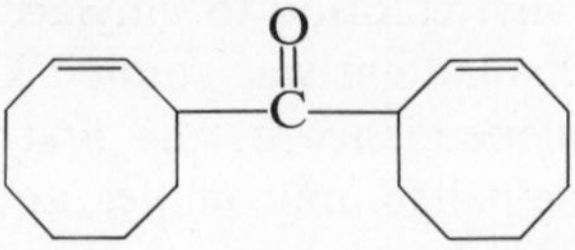

(45)

The ketone $CH_2{=}C(CH_3)CH_2COCH_2(CH_3)C{=}CH_2$ is obtained from bis(π-methallyl) nickel; vinylcycloundecadienone (**45b**) is formed from complex **45a**. The preference for insertion of carbon monoxide over the coupling reaction in certain cases has been attributed to steric effects.

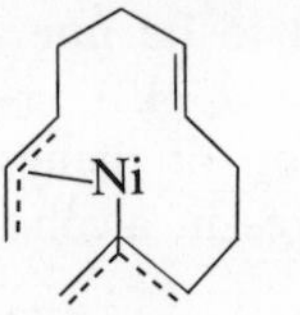

(45a) 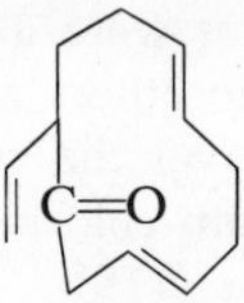 **(45b)**

Diallyl ketone has also been obtained from thermal decomposition of $C_3H_5Fe(CO)_3X$ (X = I or NO_3) in methanol (142). Dibenzyl ketone was obtained in high yield by reaction of benzyl halide with $Ni(CO)_4$ in a polar aprotic solvent at 50–60°C (107).

3. Carbonylation with Insertion of Acetylene

A reaction, taking place under surprisingly mild conditions, occurs if allylic halides are carbonylated in aqueous or alcoholic solvents in the presence of acetylene, and using $Ni(CO)_4$ as catalyst (Eq. 40):

$$R_2C{=}CRCHRX + HC{\equiv}CH + CO + R'OH \xrightarrow{Ni(CO)_4}$$

$$R_2C{=}CRCHRCH{=}CHCOOR' + HX \qquad (40)$$

where R can be H or an electron-releasing group, X = Cl, Br, or I, and R′ = H or an alkyl group (4,143). The reaction takes place readily at room temperature and atmospheric pressure and offers a simple method for obtaining dienoic acids or esters with three carbon atoms more than the original allyl group. The products contain two double bonds; one of these is conjugated with the carbonyl group and is in the *cis* form. Since it derives from the reacted acetylene, apparently *cis* addition of an allyl group and of CO to acetylene occurred. The presence of very small amounts of the *trans* form, observed mainly in the case of the acids, is likely due to subsequent isomerization occurring under the influence of the acidic conditions and of the temperature. It should be recalled that addition of hydrogen and CO to diphenylacetylene proved to be a *cis* addition (144). By-products of this synthesis were: coupling products of the allyl radicals, olefins resulting from substitution of the halogen of the allyl halide with hydrogen, plienols and cyclopentanonic compounds formed by cyclization of the hexadienoyl group.

The nature of the reaction is best ascertained by treating allyl chloride with acetylene and $Ni(CO)_4$. Evolution of CO first occurs; absorption of acetylene and CO follows. If more CO is present, less $Ni(CO)_4$ is necessary. Acetylene–carbon monoxide mixtures are passed or circulated through alcoholic solutions at a fixed ratio. A high ratio favors the formation of acrylates.

A similar procedure is effective for preparing the hexadienoic acids. Mixtures of acetone and water are useful as solvents, (e.g., 90% acetone, 10% water). Yields as high as 65–75% are obtained in dilute solution. Products obtained by analogous procedures from other allyl compounds are tabulated in Table 3.

Yields with homologous halides are generally of the order of 80%, if high dilutions (for example, 30 moles of alcohol to 1 mole of allyl halide) and low temperatures (10–30°C) are used. The consumption of $Ni(CO)_4$ is correspondingly lower, attaining values of the order of 0.1 moles/mole of reacted allyl halide. A further consumption of $Ni(CO)_4$ may occur if the hydrogen halide formed is used in the synthesis of acrylic acid or its esters, which are formed in large amounts if the temperature exceeds 40–50°C and if the acetylene is fed in excess over CO. The hydrogen halide, however, can be neutralized with MgO or with several other compounds during the reaction. Although the primary products are 2,5-dienoic compounds, they are related to members of the 3,5 and 2,4 series by the tautomerism characteristic of pentadiene systems activated by a carboxylic group (145) (Eq. 41):

$$RCH{=}CHCH_2CH{=}CHCOOR' \rightleftarrows RCH{=}CHCH{=}CHCH_2COOR' \rightleftarrows RCH_2CH{=}CHCH{=}CHCOOR' \quad (41)$$

The nature of R determines the thermodynamic stability of the tautomers. Proton shifts occur easily under alkaline conditions and with more difficulty under acidic conditions, leading in a first stage to formation of the less stable of the other two possible tautomers.

From the hexadienoate, methyl sorbate is readily obtained by adding an equal volume of a 5% methanolic sodium hydroxide solution and

TABLE 3

Synthesis of $\alpha,\beta,\delta,\varepsilon$-Unsaturated Acids and Esters

Allylic compound	Products	Yield (%)	Ref.
Allyl chloride	Hexa-2,5-dienoic acid	45, 75	4,179
	Methyl ester	69, 74	4,153,179,26
Allyl bromide	Methyl hexa-2,5-dienoate	67	205
Allyl iodide	Methyl hexa-2,5-dienoate	70	205
Allyl alcohol	Methyl hexa-2,5-dienoate	20, 67	153,160
Methyl allyl ether	Methyl hexa-2,5-dienoate	84	161
Butyl allyl ether	Methyl hexa-2,5-dienoate	84	161
Allyl acetate	Methyl hexa-2,5-dienoate	58	63,161
Allyl chloroformate	Hexa-2,5-dienoic acid	91	165
Allyl p-toluenesulfonate	Methyl hexa-2,5-dienoate	—	63
Methallyl chloride	5-Methylhexa-2,5-dienoic acid	64, 70	4,179
Methallyl chloroformate	5-Methylhexa-2,5-dienoic acid	87 }	165
	Methyl ester	68, 80 }	4,179
Methyl methallyl ether	Methyl 5-methylhexa-2,5-dienoate	68	161
Crotyl chloride	Hepta-2,5-dienoic acid	55, 71	4,179
	Methyl ester	70, 81	129,179
3-Chloro-but-1-ene	Hepta-2,5-dienoic acid	34, 77	4
Crotyl alcohol	Methyl hepta-2,5-dienoate	39	160
Methyl crotyl ether	Methyl hepta-2,5-dienoate	55	161
Crotyl acetate	Methyl hepta-2,5-dienoate	51	161
1-Chloro-3-methylbut-2-ene	Methyl 6-methylhepta-2,5-dienoate	35	4
4-Chloropent-2-ene	4-Methylhepta-2,5-dienoic acid	—	4
3-Bromohex-1-ene	Methyl nona-2,5-dienoate	62	205

3-Bromooct-1-ene	Methyl undeca-2,5-dienoate	73	205
3-Bromooctadec-1-ene	Methyl eneicosa-2,5-dienoate	48	205
3-Bromodec-1-ene	Methyl trideca-2,5-dienoate	64	205
3-Bromotetradec-1-ene	Methyl heptadeca-2,5-dienoate	45	205
1-Chloro-5,5-dimethylhex-2-ene	Methyl 8,8-dimethylnona-2,5-dienoate	55	129
1,3-Dichloropropene	Methyl 6-chlorohex-2,5-dienoate	30	104
4-Chloro-1-acetoxy-but-2-ene	Methyl 7-acetoxyhepta-2,5-dienoate	35	4
Methyl 7-acetoxyhepta-2,5-dienoate	Methyl deca-2,5,8-trien-1,10-dioate	80	167
But-2-en-1,4-diol	Methyl deca-2,5,8-trien-1,10-dioate	51	168
Butadiene monoxide	Methyl 7-hydroxyhepta-2,5-dienoate	77	168
1-Chloro-5-methoxy-pent-2-ene	7-Methoxyhepta-2,5-dienoic acid	—	165
1-Cyano-4-chlorobut-2-ene	7-Cyanohepta-2,5-dienoic acid	63	4
	Methyl ester	79	4
5-Chloropent-3-enoic acid	Octa-2,5-dien-1,8-dioic acid	32	4
1,8-Dichloroocta-2,6-diene	Methyl tetradeca-2,5,9,12-tetraen-1, 14-dioate	19	4
Cyclopent-2-enyl-1-chloride	β-(2-Cyclopenten-1-yl)acrylic acid	50	4
Cyclohex-2-enyl-1-bromide	β-(2-Cyclohexen-1-yl)acrylic acid	50	164
Cyclohex-2-en-1-ol	β-(2-Cyclohexen-1-yl)acrylic acid	16	160
Cinnamyl chloride	6-Phenylhexa-2,5-dienoic acid	32	4
Cinnamyl alcohol	Methyl 6-phenylhexa-2,5-dienoate	29	160
Methyl γ-bromocrotonate	6-Carbomethoxyhexa-2,5-dienoic acid	40	105
Methyl γ-bromo-β-methylcrotonate	6-Carbomethoxy-5-methylhexa-2,5-dienoic acid	45	105
1-Cyano-3-bromoprop-1-ene	6-Cyanohexa-2,5-dienoic acid	8	205

warming to 50°C for 2 hr. A mixture of isomers consisting mainly of *trans, trans*-methyl sorbate (ca. 90%) is formed. Methyl sorbate is separated by distillation (146,147).

Since 2,5 and 3,5-dienoic acids easily cyclize to phenols by refluxing with acetic anhydride and an acid catalyst (e.g., hexadienoic and 5-methylhexadienoic acid) or with acetic anhydride and an alkaline catalyst (e.g., ω-substituted hexadienoic acids) (148), the carbonylation of allylic compounds with insertion of acetylene affords a convenient method for synthesizing phenols (Eq. 42). Hexadienoic esters with Friedel-Crafts

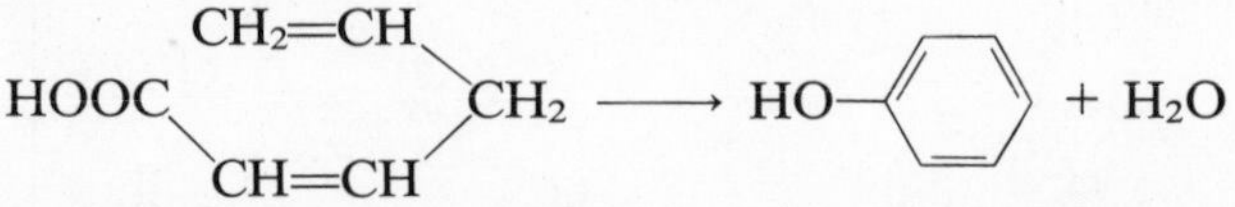

$$HOOC\begin{matrix} CH_2{=}CH \\ \\ CH{=}CH \end{matrix}CH_2 \longrightarrow HO{-}C_6H_5 + H_2O \qquad (42)$$

catalysts and acetic anhydride give the acetate of *o*-hydroxyacetophenone (**46**)(149,150). Hexadienoyl chloride and catalytic amounts of iron, pal-

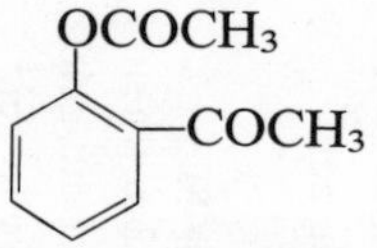

$$o\text{-}CH_3COOC_6H_4COCH_3$$

(**46**)

ladium, or other transition metals give phenol (149,150). Other reactions of hexa-2,5-dienoic acid and its methyl ester are reported in Scheme **I**.

As already noted for carbonylation with carbon monoxide alone, secondary allylic halides with terminal double bonds react as if they were the primary isomers resulting from an allylic rearrangement. Thus, $CH_3CH{=}CHCH_2Cl$ and $CH_2{=}CHCH(Cl)CH_3$ give the same product.

Mixtures of isomers are produced when starting from the other secondary allylic halides, $R^1CH{=}CHCHR^2X$, since an allylic rearrangement will give rise to two different isomers. No reaction is noted under 50°C with tertiary allylic halides, unless they undergo an allylic rearrangement and react as a primary or secondary compound. If electron-withdrawing substituents are present on the carbon atoms of the allylic system (such as CN, COOH, $CONH_2$, and COOR) the synthesis does not occur in alcoholic solvents or gives poor yields (104). Substitution of the halogen by hydrogen occurs, as already described. The use of ketonic solvents has led to better results as will be shown in Section II-A-4. With substituents such as Cl and Br a large excess of $Ni(CO)_4$ is necessary to obtain the expected product, but the yield is still low. 1,4-dichlorobut-2-ene decomposes with formation of *cis*-but-2-ene and butadiene.

Benzyl halides, which may be considered allylic halides, do not react

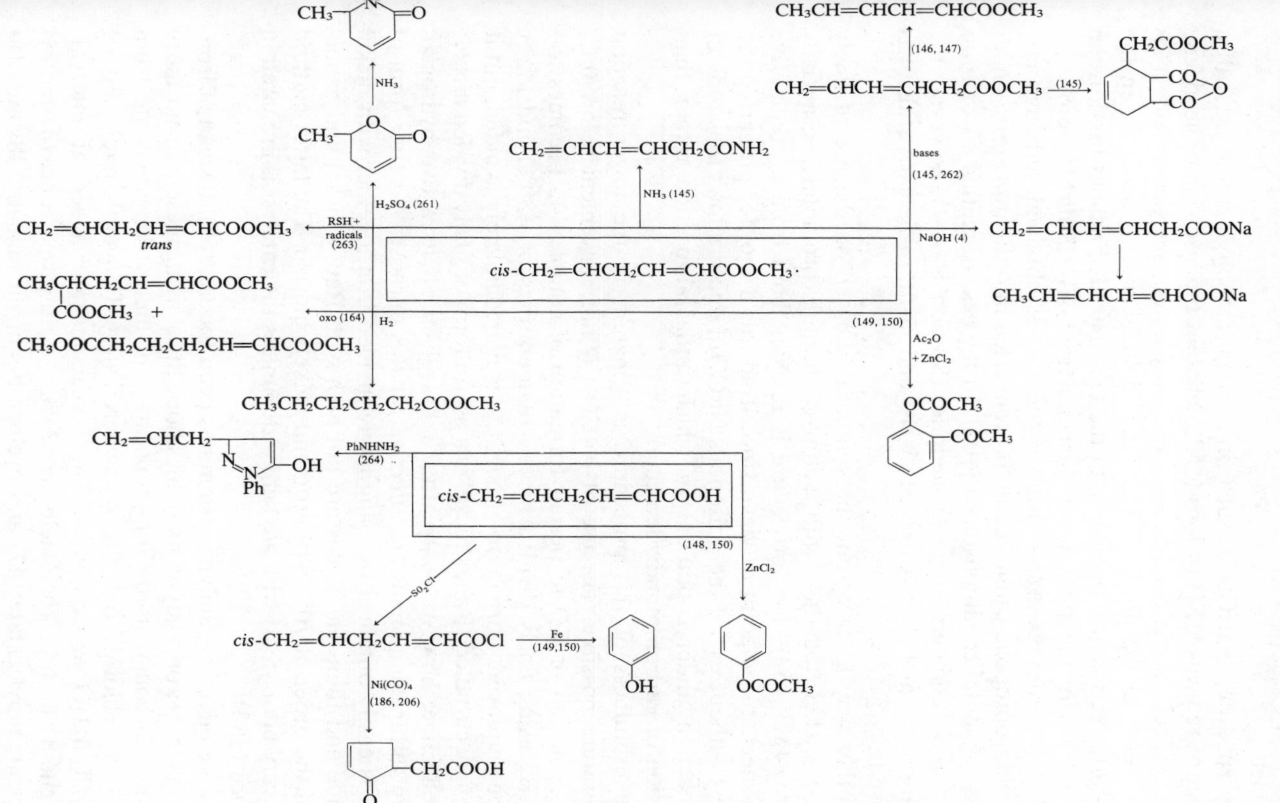

Reactions of hexa-2,5-dienoic acid and its methyl esters

Scheme **I**

with acetylene under the conditions described. Saturated halides also do not react.

It is apparent that CO as such, in a metal carbonyl, can react with a number of radicals (151) or ions (152), whereas the insertion of acetylene requires special steric conditions. As already noted, acetylene yields a *cis* double bond in the product. The other double bond is generally in the *trans* form; this is not due solely to the fact that allyl halides used in the synthesis are in the *trans* form. Starting from *cis*-1,3-dichloropropene, methyl 6-chloro-hexa-2-*cis*-5-*trans*-dienoate was obtained, although *cis*-1,3-dichloropropene cannot easily be converted into the *trans* form under the experimental conditions used (104). In the case of methyl *cis* or *trans* β-methyl-γ-bromocrotonate in acetone, (Section II-A-4), however, the formation of a 1:1 mixture of 5-*cis* and 5-*trans*-dienoic acids was observed.

Another way to carry out the synthesis is based on the use of finely divided nickel together with sulfur-containing promoters, especially thiourea (129,153). This procedure is relevant both from a preparative and a mechanistic point of view, since it affords a tool for preparing the catalyst without use of the poisonous $Ni(CO)_4$, and it allows the study of the reaction under conditions in which initial absorption rather than evolution of carbon monoxide occurs.

The usefulness of this procedure is apparent in the lower reaction temperature possible. The use of $Ni(CO)_4$ is not convenient below 0°C, since evolution of CO to permit formation of complex **43** becomes too slow. By using finely divided nickel, temperatures of −30°C and lower may be employed. Formation of $Ni(CO)_4$ is not completely avoided, but it is strongly curtailed. Raney nickel is a useful form of finely divided nickel; it is effective, however, only after it has given off its active hydrogen; certain sulfur compounds are effective for this purpose. A better form of nickel is easily obtained by using finely divided iron together with nickel chloride and thiourea. Yields are not high with Raney nickel and generally of the order of 70% with iron and nickel chloride. The latter method gives very poor yields with allyl chloride, unless the catalyst is prepared in advance (153).

A more effective catalytic system has been found (265). Instead of iron, a manganese–iron alloy containing about 80% manganese and no more than 1% carbon is used. For preparation of methyl hexadienoate from 25 g of allyl chloride, only 2 g of powdered alloy, 1.5 g of thiourea, and 4 g of $NiCl_2 \cdot 6H_2O$ are required. These reagents are added at intervals throughout 8 hr. The yield is 74%. A few percent of methyl cyclopentenonylacetate (**47**) and cyclopentenonyllevulinate (**48**) are also formed.

$$\text{(cyclopentenone)-}CH_2COOCH_3 \quad (\mathbf{47}) \qquad \text{(cyclopentenone)-}CH_2COCH_2CH_2COOCH_3 \quad (\mathbf{48})$$

In the absence of thiourea, nickel chloride is slowly reduced to metal and carbonylation does not occur appreciably. The activity of thiourea in promoting reduction of nickel salts is well known (154,155). $Ni(CO)_4$ can be prepared from iron, nickel chloride, thiourea, and CO in alcohols or water (153,156). However, other sulfur compounds such as sulfides and thiosulfates are more effective than thiourea for producing $Ni(CO)_4$ by this procedure (153). Although the latter can also be produced by reducing nickel salts with sodium hydrosulfite or formamidinsulfinic acid (157,158), the iron–nickel chloride system is a much cheaper way to produce $Ni(CO)_4$ at atmospheric pressure and room temperature.

π-Allylic complexes of nickel containing one or two molecules of thiourea as a ligand (**9**) (68) behave catalytically in carbonylations with insertion of acetylene. A synthetic method implying preparation *in situ* has been reported (266).

The synthesis of dienoic acids and esters can also be achieved by reacting "onium" salts such as allyltrimethylammonium chloride or allyl-pseudothiouronium chloride (159). Another way to effect the synthesis of dienoic acids starts from allyl alcohols (160), ethers, or esters (161,162) in the presence of HCl. For these cases the reactions can be written as shown in Eqs. 43–45.

$$CH_2{=}CHCH_2OH + HC{\equiv}CH + CO + ROH \xrightarrow[HCl]{Ni(CO)_4} CH_2{=}CHCH_2CH{=}CHCOOR + H_2O \quad (43)$$
$$(\mathbf{49})$$

$$CH_2{=}CHCH_2OR + HC{\equiv}CH + CO \xrightarrow[HCl]{Ni(CO)_4} \mathbf{49} \quad (44)$$

$$CH_2{=}CHCH_2OCOR' + HC{\equiv}CH + CO + 2ROH \xrightarrow{Ni(CO)_4} \mathbf{49} + R'COOR + H_2O \quad (45)$$

It should be stressed that HCl is used in less than stoichiometric amounts, since it is re-formed during the reaction. The synthesis with allyl alcohols seems to be less interesting from a practical point of view, since allylic esters are also formed and secondary reactions occur.

As for allylic ethers, the most useful synthesis is with allyl methyl ether, which is easily prepared from allyl chloride and alcoholic sodium hydroxide. It may seem that this is the same as starting from allyl chloride but that is not so. The production per unit of volume is increased six- to sevenfold by using the ether (yield 80–85%) (161). This is attributable to formation *in situ* of the reactive allyl group and to the lower concentration of nickel chloride; the latter strongly retards the reaction.

Allylic esters are also reactive. This method provides a useful laboratory tool since allylic esters generally are easily prepared. Nickel salts of the carboxylic acid form in large amounts, if the reaction is allowed to go further without adding HCl. Free organic acids are formed; these attack $Ni(CO)_4$ and cause formation of substantial quantities of acrylic esters. Allyl acrylate, readily obtainable from 2 moles of acrolein (163), is one of the most reactive allylic esters (162) (Eq. 46).

$$CH_2{=}CHCOOCH_2CH{=}CH_2 + HC{\equiv}CH + CO + 2CH_3OH \rightarrow CH_2{=}CHCH_2CH{=}CHCOOCH_3 + CH_2{=}CHCOOCH_3 + H_2O \qquad (46)$$

At higher temperatures, about 100–140°C, allylic esters react directly with acetylene, CO, and $Ni(CO)_4$ in alcoholic solution in the absence of HCl (63,153,164).

Allyl chloroformates have been successfully employed at 35–45°C in the absence of added acids (165).

The transfer of the allyl group directly from allylic esters to nickel without addition of acids is a type of cleavage not common among the allylic esters (1,161). It has been observed by Bauld (166) in the coupling reaction of allylic esters with $Ni(CO)_4$.

The reaction with allylic esters or ethers has allowed the synthesis of sebacic acid from 1,4-dichlorobut-2-ene (167). As already noted, dichlorobutene decomposes when treated with $Ni(CO)_4$. If, however, a chlorine atom is first replaced by the acetoxy group, the reaction with acetylene and CO occurs easily, provided the liberated HCl is not permitted to attack the acetoxy groups. By neutralizing the acid with MgO, CaO, or other agents, 7-acetoxyhepta-2,5-dienoate (**50**) can be obtained (Eq. 47). By reacting **50** with acetylene and CO, and with addition of HCl, 2,5,8-decatrien-1,10-dioate (**51**) is easily prepared (Eq. 48). Its hydrogenation affords the sebacic diester.

$$CH_3COOCH_2CH{=}CHCH_2Cl + HC{\equiv}CH + CO + CH_3OH \rightarrow \underset{\mathbf{(50)}}{CH_3COOCH_2CH{=}CHCH_2CH{=}CHCOOCH_3} + HCl \qquad (47)$$

$$\mathbf{50} + HC{\equiv}CH + CO + 2CH_3OH \rightarrow$$
$$\underset{\mathbf{(51)}}{CH_3OOCCH{=}CHCH_2CH{=}CHCH_2CH{=}CHCOOCH_3} + CH_3COOCH_3 + H_2O \quad (48)$$

The 7-hydroxy derivative that gives the acetate **50** was obtained from *cis*-but-2-ene-1,4-diol in the presence of an acid (such as formic, acetic, or phosphoric) or of an inorganic halide (such as LiCl or NaBr) which causes slow formation *in situ* of the intermediate monohalide (168). In this way formation of a dihalide, which decomposes to butadiene, is prevented. Butadiene monoxide, $CH_2{-}CH{-}CH{=}CH_2$ (with O bridging the first two carbons), behaves similarly, to give, in the presence of HCl, $HOCH_2CH(Cl)CH{=}CH_2$, that can react as if it were the primary isomer, as already shown. In the presence of acetic acid and lithium chloride, methyl 7-hydroxy-hepta-2,5-dienoate is produced in good yield. The diester **51** was not obtained from butadiene monoxide.

Vinyl lactones were also employed as starting materials. The reaction can be expressed in the simplest case by the following equation:

$$CH_2{=}CH{-}\text{(γ-butyrolactone ring, C=O)} + HC{\equiv}CH + CO + H_2O \xrightarrow{Ni(CO)_4}$$
$$HOOCCH{=}CHCH_2CH{=}CHCH_2CH_2COOH$$

The product is nona-2-*cis*-5-*trans*-dien-1,9-dioic acid. The reaction can be done in hydroxylic solvents in the presence of catalytic amounts of HCl at room or slightly higher temperature. The direct reaction with $Ni(CO)_4$, in the absence of HCl, was also achieved at higher temperature (169).

Allylic compounds also can be employed in catalytic reactions under pressure (153,170). For example, allyl methyl ether reacts in methanolic solution with acetylene and CO at 100–200°C and 30–50 atm when using a complex of nickel bromide with acetylacetone as catalyst.

All types of allylic compounds may be used. The catalysts effective in the acrylic synthesis are also effective in the synthesis of dienoic esters; however, much acrylic ester is formed. This is understandable, since the conditions are practically the same as those employed for the acrylic synthesis under pressure. This difficulty can be overcome by using lower partial pressures of acetylene and lower temperatures; formation of acrylic esters is completely eliminated and dienoic esters are formed, although with low conversions.

The complex present in the carbonylation reaction at room temperature should form from nickel bromide in the catalytic reaction under pressure. A hydride is likely to form first. Chatt and Shaw (171) have described such reductions from reactions of alcohols with group VIII metals.

a. Mechanisms. The reaction of π-allylnickel complexes like **43a** with acetylene and CO very likely involves coordination of acetylene followed by its insertion between the allyl group and Ni. Insertion of CO then takes place. Insertion of the latter into vinylnickel bonds was found to be stereospecific (see Section IV-E). The last step is hydrolysis of the acyl derivative with formation of HX (Eq. 49). A pentacoordinated intermediate complex, able to ionize in solution, is assumed, owing to the behavior of the complex of type **12** (see also Section I).

$$(\pi\text{-}C_3H_5)Ni(CO)_2X\ (\mathbf{43a}) \xrightarrow{HC\equiv CH} CH_2{=}CH{-}CH_2{-}Ni(CO)_2(HC\equiv CH){-}X \xrightarrow{CO}$$

$$CH_2{=}CH{-}CH_2{-}CH{=}CH{-}CO{-}Ni(CO)_2{-}X \xrightarrow[CH_2=CHCH_2X]{H_2O} \quad (49)$$

$$HX + CH_2{=}CH{-}CH_2{-}CH{=}CH{-}COOH + \mathbf{43a}$$

Analogous complexes should form in reactions with the catalytic system nickel halide–iron–thiourea as also shown by the behavior of allylnickel complexes containing thiourea as ligand (68). A halogeno (carbonyl)(thiourea)π-allylnickel complex displaying ionic character in methanol should be the intermediate. Insertion of acetylene between the allyl group and nickel probably precedes carbonylation as an independent step. The attempted isolation of compounds corresponding to the addition of an allylic group to a molecule of acetylene starting from the π-allylnickel complex (**6**) led to (in the presence of excess allyl bromide) formation of *cis, cis*-diallybutadiene (172). This result indicates that

$$[(\pi\text{-}C_3H_5)NiBr]_2\ (\mathbf{6}) + 4HC\equiv CH + 2CH_2{=}CHCH_2Br \longrightarrow \quad (50)$$

$$CH_2{=}CH{-}CH_2{-}CH{=}CH{-}CH{=}CH{-}CH_2{-}CH{=}CH_2 + 2NiBr_2$$

addition of the allyl group to acetylene does not require a simultaneous carbonylation. The formation of a cyclic product from the reaction of allylnickel complexes with alkynes described by Brenner, Heimbach, and

Wilke (173) suggests the same conclusion (Eq. 51). Addition of allyl

$$\text{L—Ni(C}_3\text{H}_5)_2 + CH_3C{\equiv}CCH_3 \longrightarrow \text{(dimethylcyclodecatriene)} \qquad (51)$$

halides to methyl propiolate leads to *trans*-methyl hexa-2,5-dienoate along with heavier products (Eq. 52).

$$CH_2{=}CHCH_2Br + HC{\equiv}CCOOCH_3 + Ni + H^+ \rightarrow$$
$$CH_2{=}CHCH_2CH{=}CHCOOCH_3 + NiBr^+ \qquad (52)$$

The stereochemistry of this insertion should be the same as that of acetylene and CO (*cis* insertion). A proton then takes the place that would be occupied by CO (81).

The insertion of acetylene into allylnickel complexes to give dienoic esters appears unique, although additions of alkyl and acyl groups to alkynes on Co (59,174), Cr (175,176), or other metals (59*a*) have been reported. The reaction that it resembles most is Reppe's acrylic synthesis (103). Addition of a hydrogen to acetylene (177), formally analogous to addition of an allyl group to acetylene, could be the first step. Palladium chloride gives two parallel reactions, if treated with allyl halides, acetylene, and CO in CH_3OH (178). The first between allyl halides and CO forms butenoic derivatives; the second, between acetylene and CO, gives mainly maleic and muconic diesters. Very poor yields of dienoic compounds have been obtained so far from the reaction of allyl halides, acetylene, and cobalt carbonyl (179) or iron carbonyl (164) derivatives. It is likely, however, that other metal complexes resembling the nickel complex, after coordination of the metal with the proper ligands, react similarly. Coordination of acetylene to metals is reviewed elsewhere (180,181,181*a*).

The particular reactivity of nickel is ascribable to the lower stability of vinylacetyl-type complexes. If vinylacetyl chloride is treated with $Ni(CO)_4$ in THF, CO evolves and 1,5-hexadiene forms:

$$CH_2{=}CHCH_2COCl + Ni(CO)_4 \rightarrow CH_2{=}CH(CH_2)_2CH{=}CH_2$$
$$+ NiCl_2 + 4CO \qquad (53)$$

In the presence of acetylene, addition of the allyl group to acetylene occurs (183). On the other hand, $Ni(CO)_4$ seems unsuitable for reactions with aliphatic saturated compounds.

Surprisingly enough, aromatic iodo derivatives are reactive with $Ni(CO)_4$ both in direct carbonylation (182) and in carbonylation with insertion of acetylene (183); but insertion of acetylene takes place only after carbon monoxide is added. Benzyl halides behave similarly (107,108).

The cobalt carbonyl anion appears particularly effective in promoting carbonylation of aliphatic compounds even with insertion of acetylene. Yet the acyl carbonyl formed at first is too stable to allow evolution of CO and direct attack of the resulting alkyl group on acetylene (174,183) (see Vol. 1, fourth chapter).

4. *Carbonylation and Cyclization with Insertion of Alkynes*

The use of aprotic solvents such as ethers or esters in the reactions described above leads, in general, to formation of hexadienoylnickel halides, which readily cyclize to cyclopentenone or cyclohexenone rings. The use of certain solvents, such as ketones or some aliphatic esters, not only enhances this trend, but causes further addition of acetylene and CO with formation of lactone rings (105,143). After addition of the allylic group to acetylene and CO, the resulting hexadienoyl group cyclizes and forms a cyclopentenone or cyclohexenone ring (the former is usually preferred) which can either add another CO molecule or gain or lose hydrogen. The loss or capture of hydrogen instead of another insertion of CO results in formation of phenol or cyclohexenone, as shown in Eq. 54 for allyl chloride (nonessential ligands are omitted). Equations for the

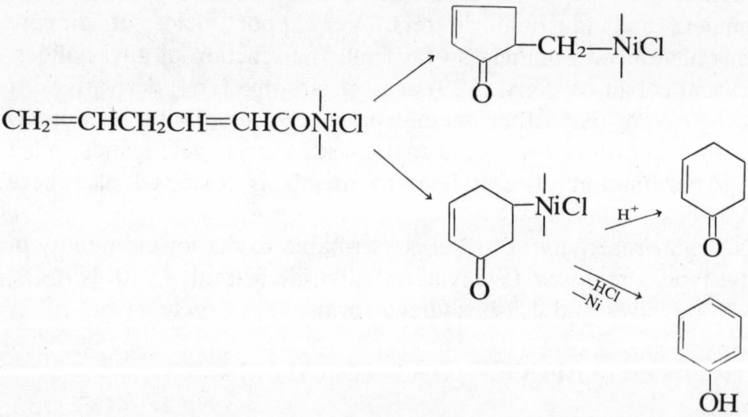

$$CH_2{=}CHCH_2CH{=}CHCONiCl \rightarrow \text{(cyclopentenone)}{-}CH_2{-}NiCl$$
$$CH_2{=}CHCH_2CH{=}CHCONiCl \rightarrow \text{(cyclohexenone)}{-}NiCl \xrightarrow{H^+} \text{cyclohexenone}; \quad \xrightarrow{-HCl,\,-Ni} \text{phenol (OH)} \qquad (54)$$

formation of phenol and cyclohexenone are:

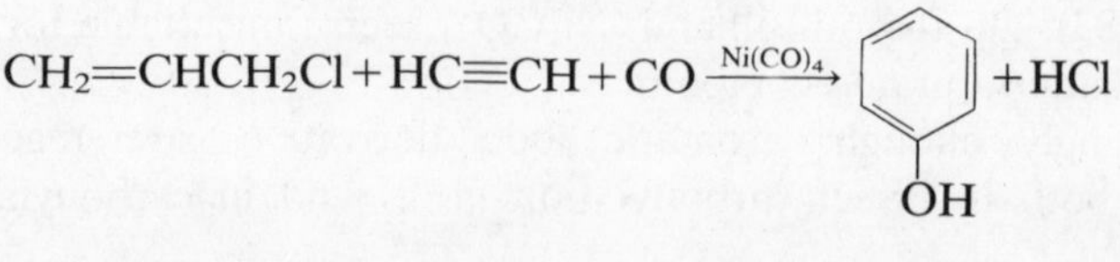

$$CH_2{=}CHCH_2Cl + HC{\equiv}CH + CO \xrightarrow{Ni(CO)_4} C_6H_5OH + HCl \qquad (55)$$

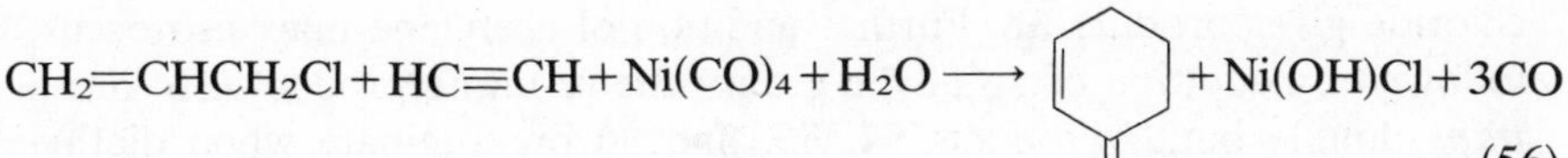

$$CH_2{=}CHCH_2Cl + HC{\equiv}CH + Ni(CO)_4 + H_2O \longrightarrow \text{(cyclohexenone)} + Ni(OH)Cl + 3CO \qquad (56)$$

The formula Ni(OH)Cl is an oversimplification, $Ni(OH)_2 + NiCl_2$ being mainly found as the result of disproportionation either of Ni(OH)Cl or of the alkylnickel halide; in the latter case to dialkylnickel and nickel chloride, followed by hydrolysis. This may be true of similar reactions in this chapter. In Eq. 56, other proton sources such as HCl or a nickel complex containing H and Cl as ligands may replace water. If carbon monoxide is inserted into the cycloalkenonylnickel complexes (Eq. 54) an acid or its functional derivative may be formed from the new acyl groups, just as with the hexadienoyl group (Eq. 57). The overall catalytic reaction

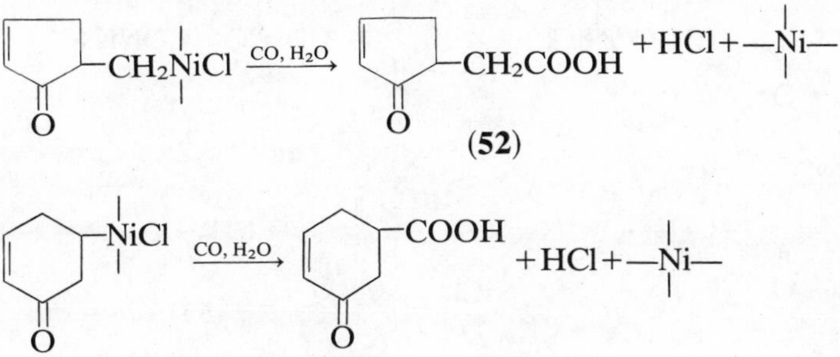

$$\text{(cyclopentenone)}{-}CH_2NiCl \xrightarrow{CO,\ H_2O} \text{(cyclopentenone)}{-}CH_2COOH\ \mathbf{(52)} + HCl + {-}Ni{-}$$

$$\text{(cyclohexenone)}{-}NiCl \xrightarrow{CO,\ H_2O} \text{(cyclohexenone)}{-}COOH\ \mathbf{(53)} + HCl + {-}Ni{-} \qquad (57)$$

leading to these types of compounds may be written for allyl chloride as follows:

$$CH_2{=}CHCH_2Cl + HC{\equiv}CH + 2CO + H_2O \xrightarrow{Ni(CO)_4} \mathbf{52} \text{ or } \mathbf{53} + HCl \qquad (58)$$

The cyclohexenone derivative is found only in small amounts. In the absence of water or alcohols, the acids are present as chlorides. The chloride of cyclopentenonylacetic acid is found mainly as the isomer **54**.

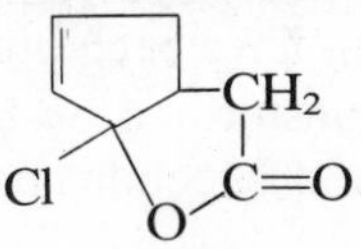

(54)

This compound is also preparable from the free acid and thionyl chloride.

Other groups, as allylic groups resulting from the reaction of allyl halides with $Ni(CO)_4$, can attack the cyclopentenonylacetyl moiety. Allyl chloride gives product **55**. Further addition of acetylene may also occur, followed by addition of an allylic group; the resulting product (**56**) has a *trans* double bond. Products **54**, **55**, and **56** predominate when diethyl

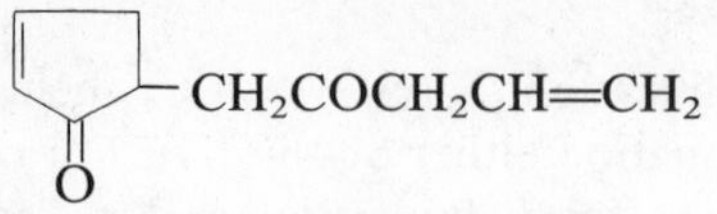

(cyclopentenone)–$CH_2COCH_2CH{=}CH_2$ (cyclopentenone)–$CH_2COCH{=}CHCH_2CH{=}CH_2$

(55) 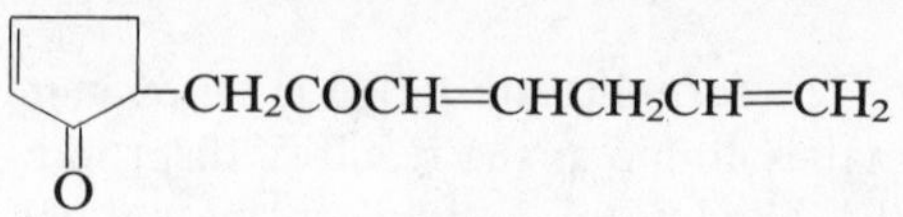 **(56)**

ether is used as the solvent at 30°C (69). The reaction proceeds further when ketones and, to a lesser extent, esters are used as solvents. After the insertion of acetylene into the cyclopentenonylacetylnickel bond, a new molecule of carbon monoxide adds to give mainly **57** (Eq. 59). The

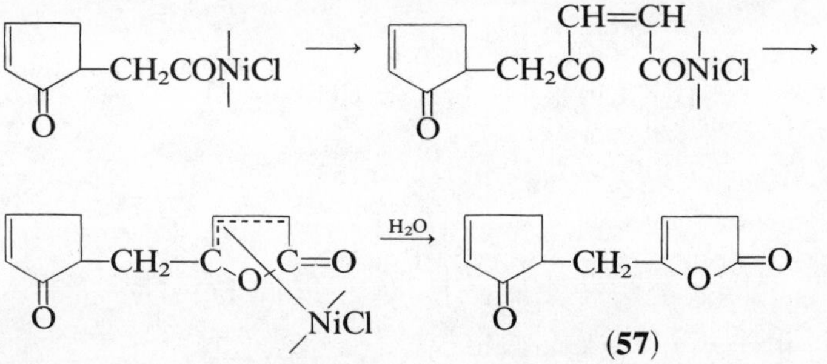

(cyclopentenone)–$CH_2CONiCl$ ⟶ (cyclopentenone)–$CH_2CO\ \ CH{=}CH\ \ CONiCl$ ⟶ (59)

(cyclopentenone)–CH_2–C(O)C=O / NiCl $\xrightarrow{H_2O}$ (cyclopentenone)–CH_2–(butenolide)=O

(57)

overall stoichiometry of this reaction can be expressed as follows:

$$CH_2{=}CHCH_2Cl + 2HC{\equiv}CH + Ni(CO)_4 + H_2O \rightarrow \mathbf{57} + NiOHCl + CO \quad (60)$$

Although all the above products are present in a normal reaction of allyl chloride, the lactone predominates when the reaction is conducted in moist acetone. This is a β,γ-unsaturated lactone, that is, it is the less thermodynamically stable of the two (α,β and β,γ) possible isomers that can form from a π-lactonyl nickel complex, the assumed intermediate. π-Lactonyl complexes of cobalt have been obtained by Heck (174).

The lactone **57** is easily transformed by hydrogenation into the saturated compound **58** or by mild saponification into the γ-keto acid **59**:

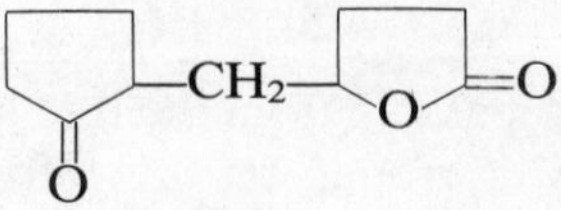

(cyclopentanone)–CH_2–(butyrolactone)=O

(58)

$$\text{(2-oxocyclopent-3-enyl)}-CH_2COCH_2CH_2COOH$$

(59)

Hydrogenolysis of **58** gives the acid **60**. Oxidative cleavage of **60** (184)

$$\text{(2-oxocyclopentyl)}-CH_2CH_2CH_2CH_2COOH$$

(60)

gives δ-ketosebacic acid (**61**). Minor amounts of products, resulting from

$$HOOC-CH_2-CH_2-CH_2-CO(CH_2)_4COOH$$

(61)

a Reformatsky-type reaction (reaction of α-halogenoesters with zinc and carbonyl compounds) of the lactone ring with ketones, are always present if the reaction is carried out in ketonic solvents. Since these products can become the major ones by further reducing the amount of water present, this reaction will be discussed in the next section. A double insertion of acetylene is observed (185) when ethers or esters are used as solvents and a low concentration of CO is maintained. An ε-lactone (**62**) containing two conjugated double bonds, is formed along with the products described above. Vinyl cyclopentenonyl ketone (**63**), a hydrogenated inter-

$$\text{(2-oxocyclopent-3-enyl)}-CH_2-C\!=\!CH-CH\!=\!CH-CH_2-CO-O\ (\text{ring})$$

(62)

mediate corresponding to the stage before formation of lactone **57**, can be prepared together with **57** by adding HCl or $ZnCl_2$ to the same reaction mixture (186). The distribution of products greatly changes with

$$\text{(2-oxocyclopent-3-enyl)}-CH_2COCH{=}CH_2$$

(63)

the nature of the allylic compound employed (105). For example, **64** is the chief product from crotyl chloride in acetone as solvent. The acid **65** is formed in small amounts. Six carbon ring compounds are mainly

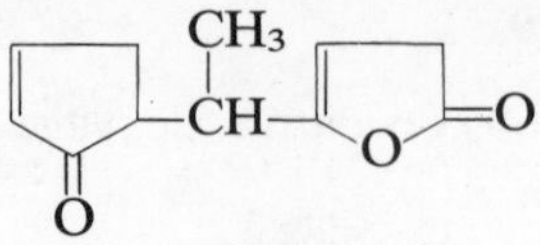

(**64**) 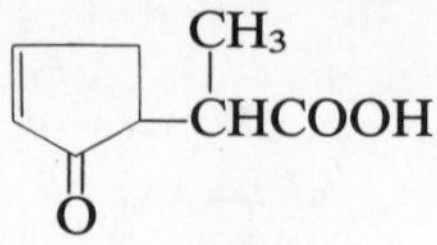 (**65**)

formed from methallyl chloride:

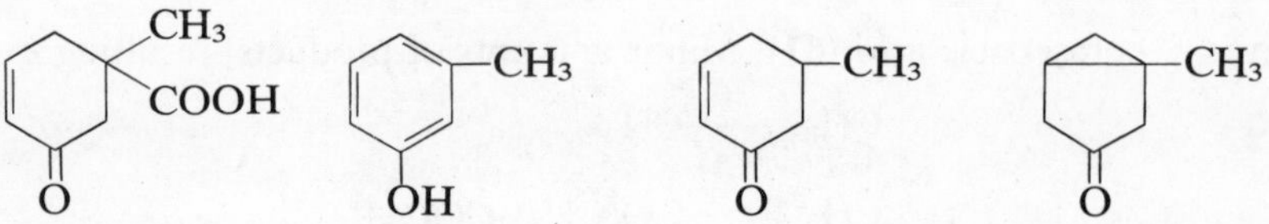

The reaction of acetylene with allylic compounds having electron-withdrawing substituents in the allyl group takes place when ketones are used as solvents; these allylic compounds undergo substitutive hydrogenation in hydroxylic solvents (Section III-A). Only some of the products are cyclic; instead, linear hexadienoic compounds are formed in significant yield. From methyl γ-bromocrotonate in acetone, for example, the linear product **66** (expected from the "normal" reaction of an allyl halide in the presence of water) is obtained in 40% yield along with methyl cyclopentenonylacetate (**47**) (105).

$$CH_3OOCCH{=}CHCH_2CH{=}CHCOOH$$

(**66**)

In this case, **47** results from abstraction of a proton from water according to the following overall reaction:

$$CH_3OOCCH{=}CHCH_2Br + HC{\equiv}CH + Ni(CO)_4 + H_2O \rightarrow \mathbf{47} + NiOHBr + 3CO \quad (61)$$

The proton can also come from the HBr originating from the normal catalytic reaction.

Coupling products are also present. Their amount depends on the type of solvent used—ranging from over 90% in benzene or ether to 9% in acetone and practically none in cyclohexanone. Products resulting from substitutive hydrogenation are also present when water is released from condensation reactions of the ketones employed. Lactonic products and Reformatsky-type products are present only in minor amounts.

It must be stressed that a slight change in the structure of the allyl compound has a large effect on the type of compound formed. For example methyl β-methyl-γ-bromocrotonate yields mainly the "normal" product **67** (2-*cis*-5-*trans* and 2-*cis*-5-*cis*, 1:1) and a lactone whose analysis and spectroscopic properties are in agreement with structure **68**.

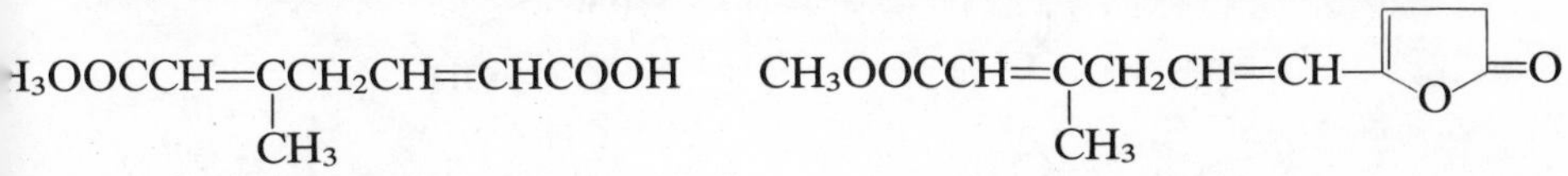

$$H_3OOCCH{=}\underset{\displaystyle CH_3}{\underset{|}{C}}CH_2CH{=}CHCOOH \qquad CH_3OOCCH{=}\underset{\displaystyle CH_3}{\underset{|}{C}}CH_2CH{=}CH\text{–}$$

(67) **(68)**

Chloro- or bromocyanopropene gives (in acetone) chiefly coupling products, and only an 8–10% yield of the expected product:

$$BrCH_2CH{=}CHCN + HC{\equiv}CH + CO \xrightarrow[H_2O]{Ni(CO)_4} NCCH{=}CHCH_2CH{=}CHCOOH \qquad (62)$$

When monosubstituted alkyl or aryl acetylenes are used, cyclic products always form (disubstituted acetylenes are poorly reactive); this is the predominant reaction in aqueous mixtures of acetone, dioxane and several other solvents (187,188).

The intramolecular reactivity of acylnickel complexes formed from higher acetylenes is apparently so high that it is not necessary to avoid the presence of large amounts of water or to use special solvents. The chief products obtained from the named reactants in acetone-water are as follows:

From methylacetylene and allyl chloride

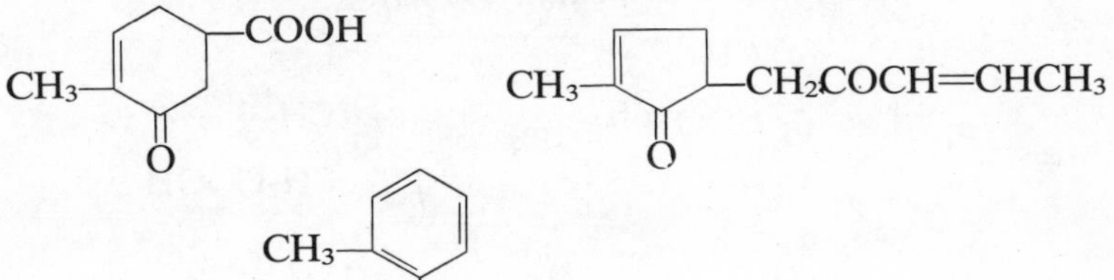

CH_3– –COOH; O

CH_3– –$CH_2COCH{=}CHCH_3$; O

CH_3– ; OH

From oct-1-yne and allyl chloride

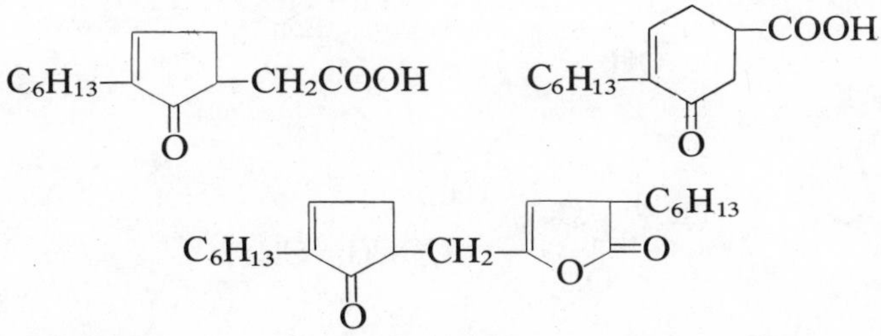

C_6H_{13}– –CH_2COOH; O

C_6H_{13}– –COOH; O

C_6H_{13}– –CH_2– –C_6H_{13}; O; =O

From hex-1-yne and methallyl chloride

CH_3 $COOH$ C_4H_9 O

CH_3 C_4H_9 OH

CH_3 C_4H_9 O

CH_3 C_4H_9 CH_2 C_4H_9 O O O

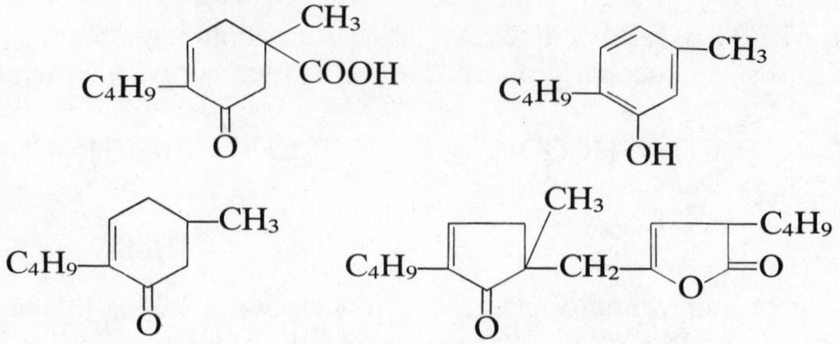

Quantitative studies were made with phenylacetylene:

From phenylacetylene and allyl chloride

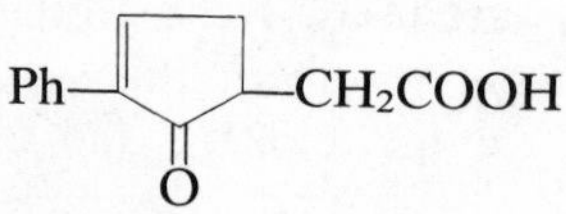

Ph CH_2COOH O 21%

COOH Ph O 4%

Ph $CH_2COCH{=}CHPh$ O 64%

Ph CH_2 Ph O O O 4%

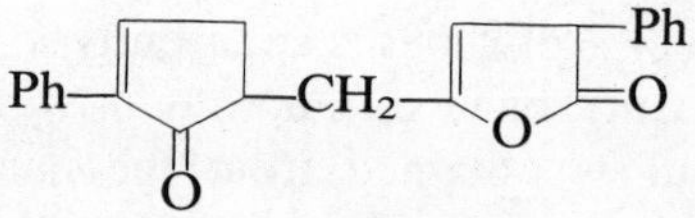

From phenylacetylene and methallyl chloride

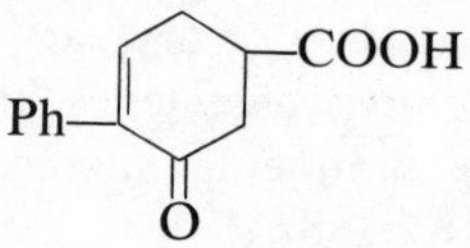

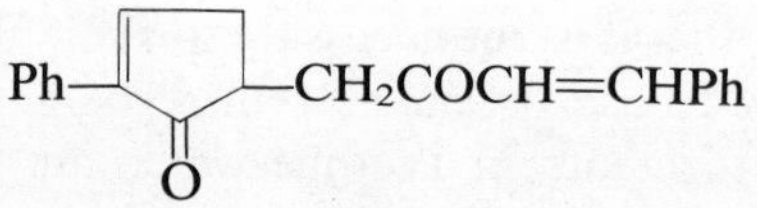

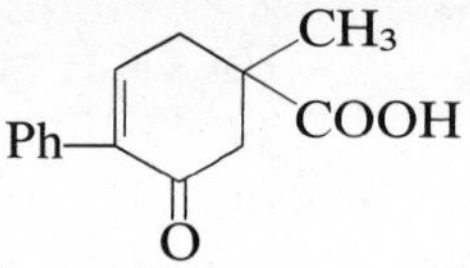

CH_3 $COOH$ Ph O ~12%

CH_3 Ph CH_2COOH O ~1%

CH_3 Ph OH 34%

CH_3 Ph O 6%

CH_3 Ph $CH_2COCH{=}CHPh$ O 37%

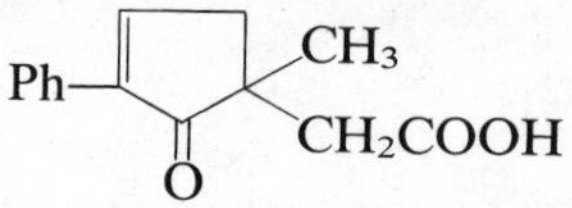

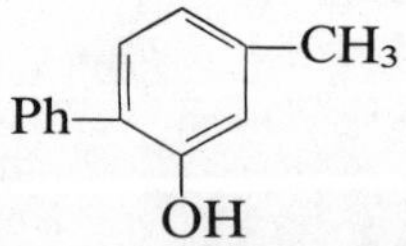

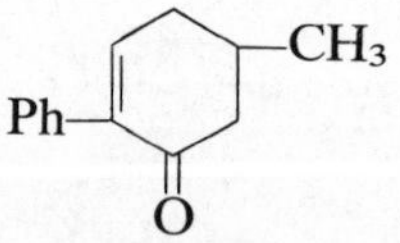

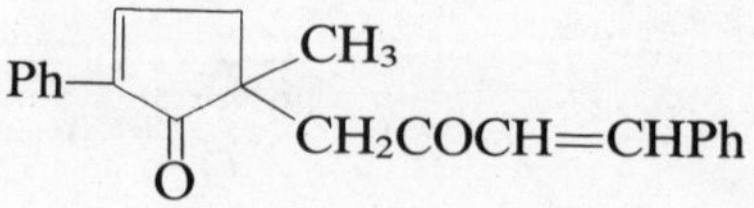

From phenylacetylene and crotyl chloride

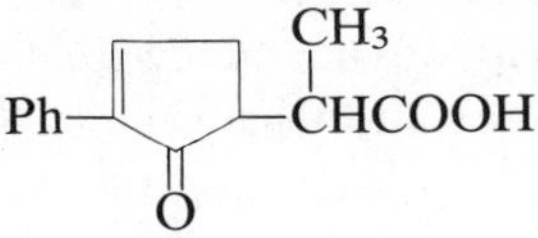

Ph–(2-cyclopentenone)–$CH(CH_3)COOH$ 20%

Ph–(2-cyclopentenone)–$CH(CH_3)COCH{=}CHPh$ 71%

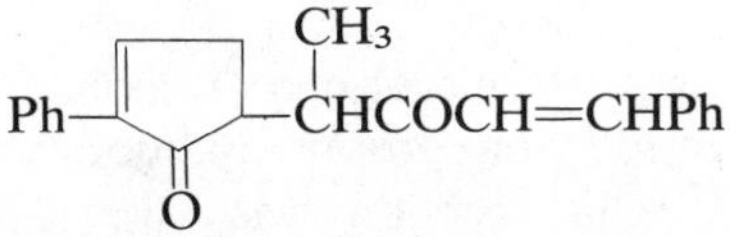

The reactions presented so far can be altered at any stage of the successive additions by exploiting steric and electronic effects in the presence of competitive ligands (189,190). For example, diphenylacetylene in an acetone–water solution reacts with allyl halides to a limited extent to give **69**. Addition of the diphenylcyclopentenonylacetyl group to diphenylacetylene to give **70** practically does not take place because the attack of water to give acid **69** (along with other carbonylation products) is favored.

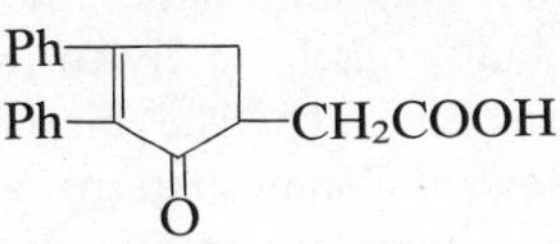

Ph,Ph–(2-cyclopentenone)–CH_2COOH **(69)**

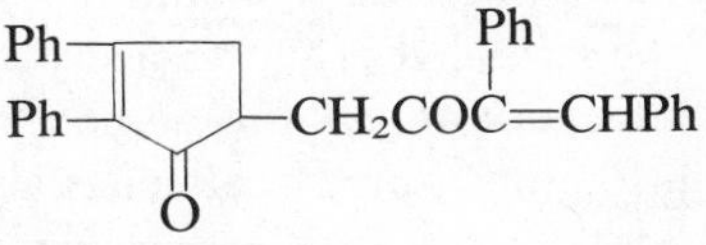

Ph,Ph–(2-cyclopentenone)–$CH_2COC(Ph){=}CHPh$ **(70)**

By using a suitably substituted allylic group, a reaction can be stopped by hydrogenation elimination, if this competes with further addition of CO. Thus, isoprene hydrochloride reacts with phenylacetylene and $Ni(CO)_4$ to give mainly a ketone derived from H elimination (189) (Eq. 63).

$$PhC{\equiv}CH + ClCH_2CH{=}C(CH_3){-}CH_3 + CO \xrightarrow[-HCl]{Ni(CO)_4} \text{Ph–(2-cyclopentenone)}{=}C(CH_3)_2 \quad (63)$$

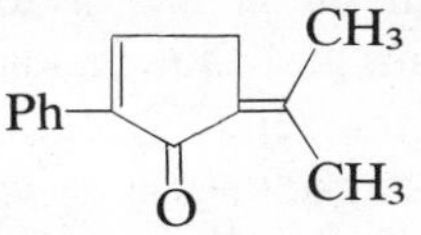

The following conclusions were drawn from the results reported here:

(*i*) Allyl groups add to the unsubstituted acetylenic carbon atom to a large extent. Allyl chloride, for instance, adds to the terminal carbon atom in 93% yield; the remaining 7% was unidentified.

(*ii*) Cyclopentenone or cyclohexenone rings always form; the former generally more easily than the latter.

(*iii*) Cyclopentenonylmethylene or a cyclohexenonyl group can add to carbon monoxide with formation of acylnickel carbonyls.

(*iv*) Alternatively, the cyclohexenone group may lose or acquire hydrogen with formation of phenols or cyclohexenones, and the cyclopentenonylalkylene group can give vinylidenic cyclopentenone.

(*v*) Acyl metal carbonyls may split to acids under the action of water.

(*vi*) Cyclopentenonic acyl metal carbonyls may further add a molecule of alkyne. At this stage, proton addition may cause splitting to unsaturated diketones.

(*vii*) Alternatively, a new molecule of carbon monoxide may be taken up, followed by lactonization and proton addition with formation of unsaturated ketolactones. This reaction was observed mainly for alkylacetylenes.

(*viii*) A methyl group on the secondary carbon atom of the allyl group favors formation of a six-membered carbon ring.

(*ix*) The 3-methyl substituted allyl group reacts at the unsubstituted side.

a. Mechanisms. Scheme **II** depicts the general pattern of the reactions for allylic halides, acetylene, and carbon monoxide as determined from the products isolated. It should be noted that addition of the allyl group and of the carbonyl group to acetylene is analogous formally to the —H and —CO addition in the acrylic synthesis (95,191). The direction of addition for substituted acetylenes also follows the same pattern, and the preference of the allyl group for the unsubstituted carbon is highly selective. It is not clear, however, to what extent steric effects are responsible for the direction of addition. The cyclization reaction can be regarded as an electrophilic addition of the acyl group to the double bond. Acyl addition to olefins catalyzed by iron powder has been reported (150). Similar behavior of the acyl groups towards alkynes is observed during reaction with $Ni(CO)_4$. Here, the unsubstituted end of the alkyne is also attacked by acyls; the other end can be attacked by a proton, and thus yields an unsaturated ketone. Instead of proton attack, the other end of the alkyne may be subject to carbon monoxide attack which produces a γ-acylacryloyl group that cyclizes to a nickel-bonded lactone ring. Proton uptake results in formation of a lactone derivative.

The relationship between five-carbon and six-carbon cycloalkenone rings deserves further comment. The preferred formation of five-ring compounds appears to agree with the behavior of acetylene and CO with cobalt and iron carbonyls, as observed by Pino and by others (192–195).

As already noted, cyclohexenonic and phenolic rings form predominantly from methallyl halides. But these types of compounds are also present in significant amounts in the products from unsubstituted or 3-substituted halides. Unsubstituted hexadienoyl halides mainly cyclize to

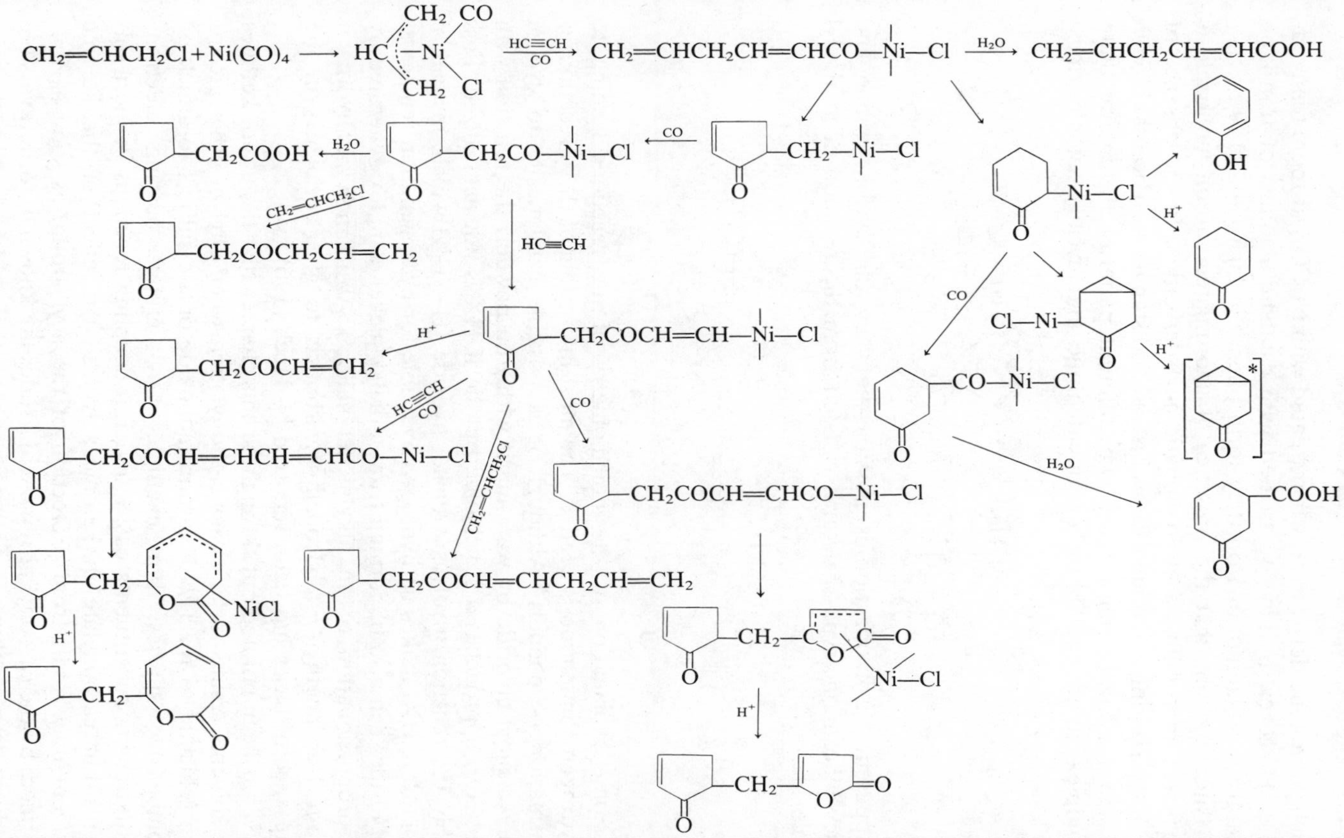

$$CH_2{=}CHCH_2Cl + Ni(CO)_4 \longrightarrow \text{(π-allyl)Ni(CO)Cl} \xrightarrow[CO]{HC\equiv CH} CH_2{=}CHCH_2CH{=}CHCO{-}Ni{-}Cl \xrightarrow{H_2O} CH_2{=}CHCH_2CH{=}CHCOOH$$

* Isolated in the case of [structure: methyl-substituted bicyclic ketone, CH_3]

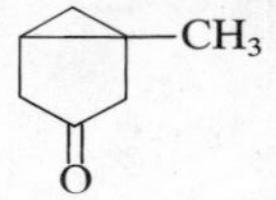

Reactions of allyl chloride with acetylene and $Ni(CO)_4$.

Scheme **II**

cyclopentenone derivatives, when treated with $Ni(CO)_4$ at room temperature. However, if they are treated with catalytic amounts of transition metals as palladium or iron they form phenol in high yield (149).

Thus, it seems that the presence of coordinated carbon monoxide is sufficient to stabilize a cycloaliphatic acyl derivative. In the absence of CO, the six-ring compound forms instead and stabilization occurs through aromatization. To allow these two different patterns, the intermediate complex could equilibrate cyclopentenonic and cyclohexenonic rings:

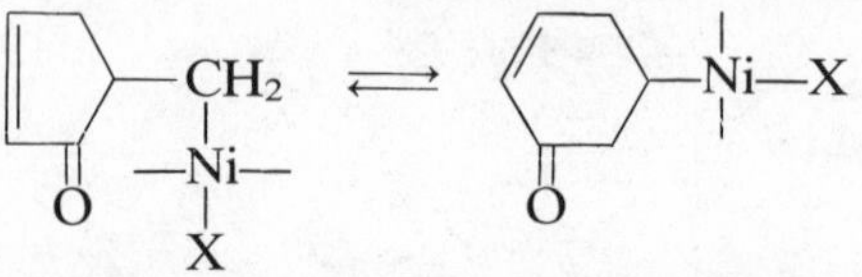

Apparently isomerization is possible between the cyclohexenone and bicyclohexanone-type structures isolated from the reaction of methallyl chloride:

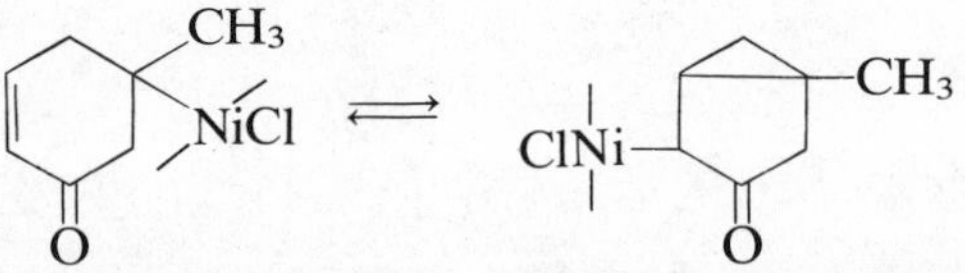

Some explanation of the stoichiometry of the reactions leading to hydrogen elimination is needed. The most interesting of these reactions is formation of phenols. Although it is possible catalytically to cyclize hexadienoyl chloride by warming it with iron powder, the reaction with $Ni(CO)_4$ approaches a stoichiometric one if HX is not neutralized. Possibly the reaction involves a nickel hydride (H–Ni–X) which, instead of liberating HX, makes its hydrogen available for hydrogenation reactions. (Hydrides can also be formed from a dialkylnickel complex generated by disproportionation of an alkylnickel halide.) Abstraction of hydrogen from 1-acylmethyl-π-allylcobalt tricarbonyls to give acyldienes in the presence of bases has been reported by Heck (196).

It has been pointed out throughout this chapter that the use of ketones as solvents results in an easier carbonylation when allyl or other groups are reluctant to react with carbon monoxide or are subject to attack by competitive species. For example, allyl groups bearing electron-withdrawing substituents easily undergo coupling reactions in inert solvents but carbonylate in ketonic solvents. The effect of the ketonic solvent may result from protection of the acyl group, as soon as it is formed, by the carbonyl group. Acid halides are known to form reversible adducts with aldehydes and ketones; moreover, aldehyde adducts have been isolated (197–199):

$$\text{PhCOCl} + \text{PhCHO} \rightleftarrows \text{PhCOOCH(Cl)Ph} \tag{64}$$

It will be shown later that, in the presence of $Ni(CO)_4$, aldehydes and certain ketones react to form products resulting from an initial attack of the acyl group on the carbonyl group (Section II-A-6).

Trapping of acyls has been noted in Friedel-Crafts reactions (200). Thus protection of the acyl group likely is obtained by reversible addition of the ketonic solvent. This addition probably occurs also in a nickel complex such as **71** (201).

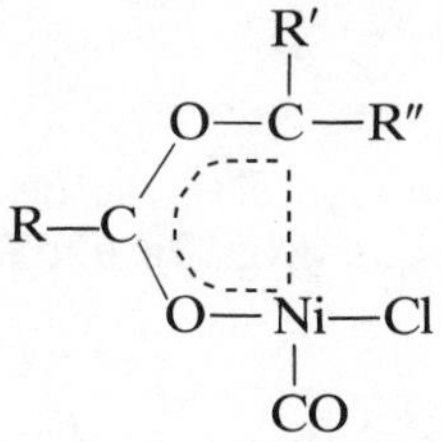

```
            R′
            |
         O—C—R″
        /  ⋮
   R—C     ⋮
        \  ⋮
         O—Ni—Cl
            |
            CO
```

(**71**)

The oxidation state of the metal in the series of reactions taking place on nickel deserves some comment. Coordinated allyl groups appear to behave generally as nucleophiles, whereas acyl groups frequently behave as electrophiles. Since both types of groups are involved in a series of additions, this different behavior may be explained by two possible patterns:

(*i*) A different electronic distribution accompanies changes in the type of ligand, the formal charge on nickel varying from 0 to 2.

(*ii*) The charge on nickel does not vary during the series of additions. The nickel-bonded carbon atom exhibits a nucleophilic or electrophilic nature depending on the adjacent groups, and its character may change on addition of new groups. This behavior parallels that of organic radicals (202,203). Thus CO addition to an alkyl group may confer an electrophilic character to the new group formed, as shown by the easy reaction with water or alcohols. It is known, however, that the acyl group can also behave as a nucleophile, for example in the addition to activated olefins (see Section II-B-3). Thus, also the type of substrate has an influence on the type of reactivity of the acyl group.

5. Carbonylation and Cyclization with Acetylene Insertion Followed by Reformatsky-type Reactions

By approaching anhydrous conditions and using certain reactive ketones as solvents, increasing amounts of products resulting from reaction between the lactone group and the ketone are formed (105). This reaction is analogous to the Reformatsky reaction (204) (See Section III-C-1). The overall reaction may be expressed as shown in Eq. 65.

$$CH_2{=}CHCH_2Cl + 2HC{\equiv}CH + Ni(CO)_4 + RCOR' + H_2O \rightarrow$$

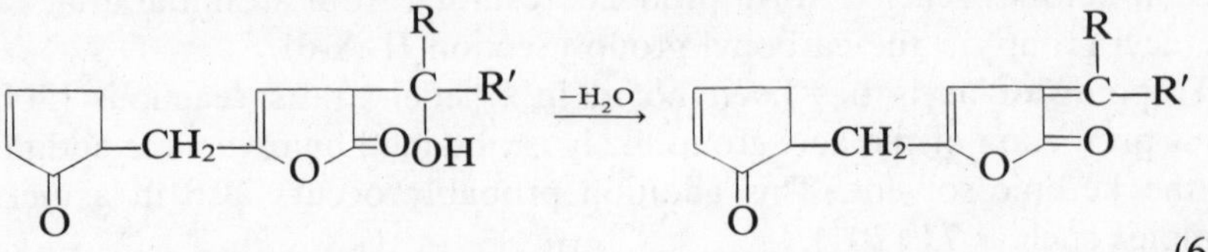

(72) $\xrightarrow{-H_2O}$ (73) (65)

Enolizable ketones are generally very reactive. Cyclopentanone gives **74** and **75**, whereas cyclohexanone gives **76** and **77**. Other ketones, such as

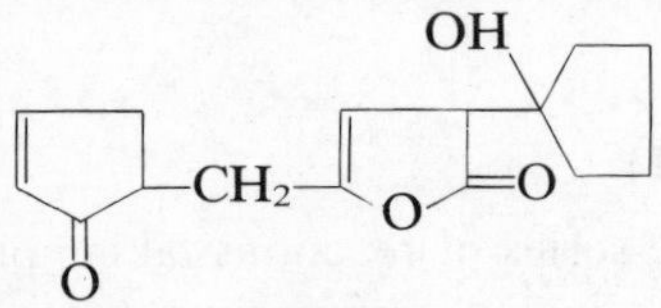

(74) 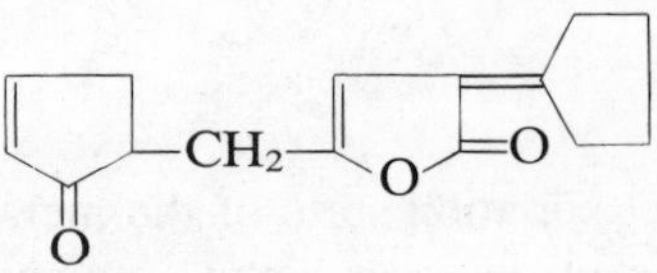 (75)

mesityl oxide, are unreactive. This provides a useful way to avoid Reformatsky-type products, if they are not desired.

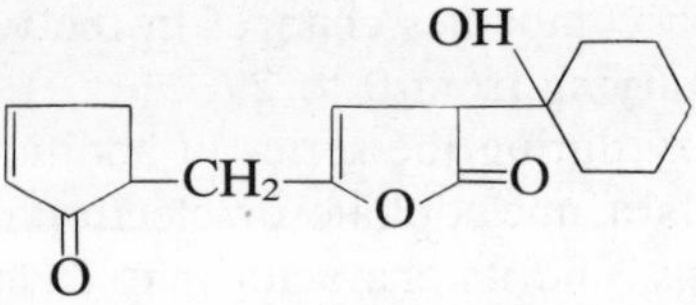

(76) 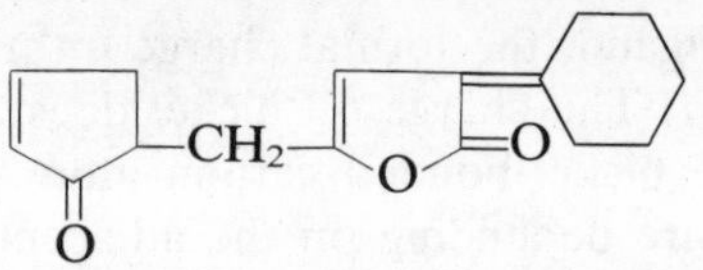 (77)

Alcohols produced by reaction with ketones are easily dehydrated. Attempts to isolate them often result in formation of an α,β-unsaturated alcohol (**78**) (isomer of **72**) which is more resistant to dehydration. Since nickel alcoholates, such as **79**, are isolatable (205) and π-lactonyl cobalt

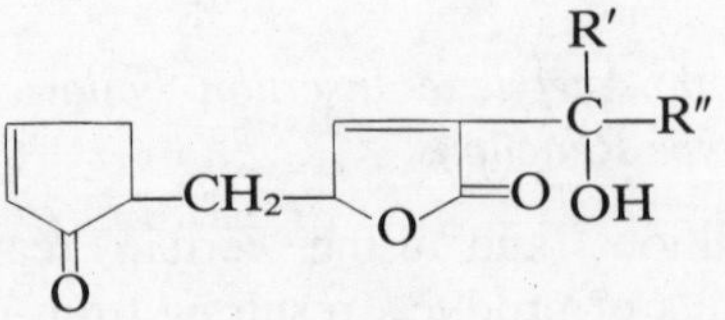

(78) 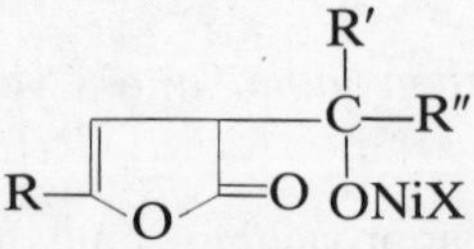 (79)

complexes have been isolated by Heck (71), the reaction should be written as shown in Eq. 66.

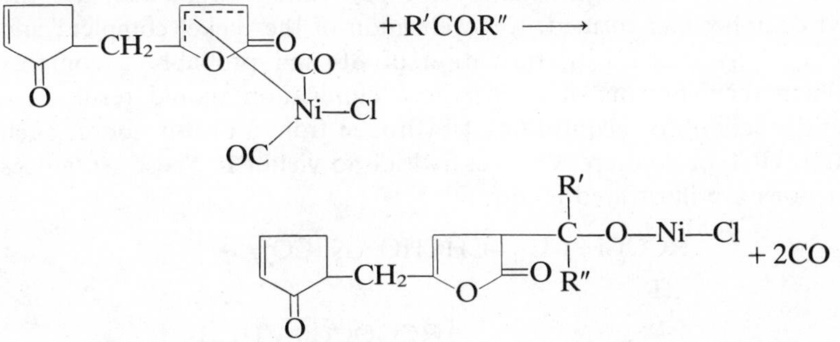

$$\text{(cyclopentenone)}\text{—CH}_2\text{—(lactone=O)}\cdot\text{Ni(CO)}_2\text{Cl} + R'COR'' \longrightarrow \text{(cyclopentenone)}\text{—CH}_2\text{—(lactone=O)—C(R')(R'')—O—Ni—Cl} + 2CO \quad (66)$$

The direct reaction of allylic compounds with carbonyl compounds is treated in Section III-1-C.

6. Carbonylation and Cyclization with Acetylene Insertion Starting from Acyl Halides and α,β-Unsaturated Aldehydes

α,β-Unsaturated aldehydes react with acyl chlorides, acetylene, and $Ni(CO)_4$, a room temperature and atmospheric pressure, giving rise to cyclopentenonic structures (186,206); with acrolein, the main products result from the following reactions:

$$RCOCl + CH_2{=}CHCHO + HC{\equiv}CH \xrightarrow{Ni(CO)_4} RCOOCH{=}\text{(cyclopentenone)} + HCl \quad (67)$$

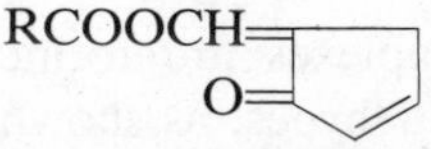

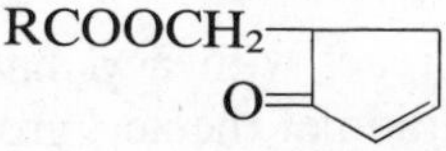

80

$$RCOCl + CH_2{=}CHCHO + HC{\equiv}CH + Ni(CO)_4 + H_2O \longrightarrow RCOOCH_2\text{—(cyclopentenone)} + Ni(OH)Cl + 3CO \quad (68)$$

81

The total amount and the ratio of these products depend on the solvent used; products **80** and **81** are best prepared in ethereal solvents. In ketonic solvents, acyl halides (and especially aroyl halides) tend to react more easily with acetylene and carbon monoxide to give lactones rather than with the unsaturated aldehydes (186). Evidently the ketonic solvent competes with the aldehyde for the acyl group, which is captured and released to acetylene rather than to the aldehyde.

a. Mechanisms. The following sequence should take place: (*1*) formation of a π-allyl from reaction of the acyl halide with $Ni(CO)_4$ and the aldehyde; (*2*) an insertion of acetylene and carbon monoxide into the π-allyl complex first formed; (*3*) cyclization of the nickel complex; and (*4*) loss of hydrogen with formation of **80** and probably a complex containing the elements of HCl (whose elimination should result in a catalytic reaction), or acquisition of hydrogen from a proton source such as water, HCl, or a complex nickel hydride to yield **81**. These sequences of reactions are illustrated in Eq. 69.

$$RCOCl + CH_2{=}CHCHO + Ni(CO)_4 \rightarrow$$

$$\text{RCOOCH}\overset{\text{H}}{\overset{|}{\text{C}}}\text{CH}_2\ (\pi\text{-allyl})\text{–Ni(Cl)(CO)} \xrightarrow{HC\equiv CH} \left[\begin{array}{c}\text{RCOOCH}{=}\text{CHCH}_2\diagdown \\ \qquad\qquad\qquad\text{CH} \\ \text{ClNiCOCH}\diagup\!\!\!=\end{array}\right] \longrightarrow$$

$$\left[\begin{array}{c}\text{RCOOCH}\text{–(cyclopentenone, O=)} \\ | \\ \text{Ni} \\ | \\ \text{Cl}\end{array}\right] \begin{array}{l} \xrightarrow{-Ni,\ -HCl} \textbf{(80)} \\ \xrightarrow[-Ni(OH)Cl]{H_2O} \textbf{(81)} \end{array} \qquad (69)$$

As to how the π-allyl complex is formed, the following observations are relevant:

π-Allyl complexes are formed from acylcobalt carbonyls and α,β-unsaturated aldehydes, as shown by Heck (207):

$$RCOCo(CO)_4 + CH_2{=}CHCHO \longrightarrow (\pi\text{-}C_3H_4\text{–OOCR})\text{–}Co(CO)_3 + CO \qquad (70)$$

Carbonyl compounds also react with acyl halides and $Ni(CO)_4$ to give products resulting from addition of the acyl group to the carbonyl oxygen. For instance, benzaldehyde reacts with benzoyl chloride and $Ni(CO)_4$ to give benzyl benzoate after hydrolysis with water, and to give the dibenzoate of 1,2-dihydroxy-1,2-diphenylethane after raising the temperature (186):

$$PhCHO + PhCOCl + Ni(CO)_4 \longrightarrow \left[\begin{array}{c}\text{PhCOOCHPh} \\ | \\ \text{OC–Ni–CO} \\ | \\ \text{Cl}\end{array}\right] \begin{array}{l}\nearrow \text{PhCOOCH}_2\text{Ph} \\ \searrow \begin{array}{c}\text{PhCOOCHPh} \\ | \\ \text{PhCOOCHPh}\end{array}\end{array} \qquad (71)$$

Bauld (182) showed that benzoyl groups from the reaction of benzoyl chloride and $Ni(CO)_4$ react with benzil to form the enolic diester:

$$2\,PhCOCl + PhCOCOPh + Ni(CO)_4 \longrightarrow \underset{\substack{|\\ OCOPh}}{PhC}{=\!=}\underset{\substack{|\\ OCOPh}}{CPh} + NiCl_2 + 4\,CO \qquad (72)$$

It was also shown later (206) that when the same reaction is carried out in moist acetone, the benzoate of benzoin forms. This result can be explained by trapping of the benzoyl group on benzil according to Eq. 73 (nonessential ligands are omitted) or also by the action of benzoyl chloride on a complex formed from benzoylnickel chloride according to

$$\begin{matrix} PhCO \\ | \\ PhCO \end{matrix} + Ni(CO)_4 \xrightarrow[-CO]{PhCOCl} \underset{\substack{|\\ OCOPh}}{\overset{\substack{PhCO\\ |}}{PhC}}{-\!-}\underset{|}{\overset{|}{Ni}}{-}Cl \xrightarrow{H_2O} \underset{\substack{|\\ OCOPh}}{PhCHCOPh} + NiClOH + CO \qquad (73)$$

Eq. 73*a*. Thus various types of reactions appear to be based on the

$$PhCO{-}\underset{|}{\overset{|}{Ni}}{-}Cl \longrightarrow PhCO{-}\underset{\substack{|\\ NiCl}}{\overset{\substack{Ph\\ |}}{C}}{-}ONiCl \xrightarrow[H_2O]{PhCOCl} PhCO\overset{\substack{Ph\\ |}}{C}HOCOPh + NiCl_2 + NiClOH \qquad (73a)$$

trapping of the acyl group by the oxygen of the carbonyl function. This leads to stable complexes in the case of α,β-unsaturated aldehydes.

7. *Carbonylation with Insertion of Dienes*

In the presence of $PdCl_2$, allylic compounds, dienes, and CO react to form octa-3,7-dienoyl chlorides according to the following equation that is analogous to the reaction involving the insertion of acetylene on nickel (72):

$$R^2CH{=}\overset{\substack{R^1\\ |}}{C}CH_2Cl + CH_2{=}CH\overset{\substack{R^3\\ |}}{C}{=}CH_2 + CO \xrightarrow{PdCl_2}$$

$$H_2C{=}\underset{\substack{|\\ R^2}}{\overset{\substack{R^1\\ |}}{C}}CHCH_2\overset{\substack{R^3\\ |}}{C}{=}CHCH_2COCl \qquad (74)$$

$$R^1, R^2, R^3 = H \text{ or } CH_3$$

The products obtained from allyl chloride and butadiene contained two double bonds; small amounts of a C_5 acid,

$$CH_3CH{=}CHCH_2COOH,$$

and a C_9 acid, $H_2C{=}CH(CH_2)_3CH{=}CHCH_2COOH$, were also obtained.

These acids are clearly derived from one or two molecules of butadiene, probably via allylic complexes. The reaction takes place in benzene solution. For example, allyl chloride reacts at 90°C with butadiene and CO under an initial pressure of 52 atm with $[(\pi\text{-}C_3H_5)PdCl]_2$ as catalyst. After 6 hr, a 71% conversion of allyl chloride is obtained. The yield of 3,7-octadienoyl chloride plus vinylacetyl chloride (molar ratio 0.8) is 90%. It is interesting to note that the conjugated diene always attacks the π-allyl group at the most substituted carbon atom.

a. Mechanism. The mechanism of this interesting reaction apparently involves insertion of dienes into an allyl–palladium bond, followed by carbonylation. The reaction rate increases with the electron-withdrawing ability of substituents present in the π-allyl segment. Kinetic studies were carried out on the reaction of isoprene with di-μ-bromo-di-π-methallylpalladium (72). The following expression was found:

$$-\frac{dc}{dt} = K_2\,[\text{palladium complex}]\,[\text{isoprene}]$$

The rate-determining step appears to be coordination of the diene to the π-allylpalladium complex. No reaction occurs in the presence of PPh_3 (1 mole per mole of palladium complex).

The direction of addition of dienes to the allylic group is opposite to that observed with acetylene. This fact is ascribed (72) to the different mode of coordination of butadiene.

The scheme proposed by Medema, van Helden, and Kohll (72) for the carbonylation of allyl chloride in the presence of butadiene is this:

$[(\pi\text{-allyl})PdCl]$ (**14**) $\xrightarrow{CO}$ $(\pi\text{-allyl})Pd(CO)Cl$ $\xrightarrow{CH_2{=}CHCH_2Cl}$ $(\pi\text{-allyl})Pd(CO)Cl(CH_2{=}CHCH_2Cl)$ $\longrightarrow$ $CH_2{=}CHCH_2COCl$ + **14**

14 $\xrightarrow{CH_2{=}CHCH{=}CH_2}$ $(\pi\text{-heptadienyl})PdCl$ $\xrightarrow{CO}$ $(\pi\text{-heptadienyl})Pd(CO)Cl$ $\xrightarrow{CH_2{=}CHCH_2Cl}$ $(\pi\text{-heptadienyl})Pd(CO)Cl(CH_2{=}CHCH_2Cl)$ $\longrightarrow$ $CH_2{=}CHCH_2CH_2CH{=}CHCH_2COCl$ + **14**

Evidence for an electrocyclic mechanism involving butadiene as monodentate ligand has been provided (220).

Based on nmr observations of the enhancement of the *trans* effect of a ligand in a π-$C_3H_5Pd(Cl)L$ complex by addition of the dimer $[(\pi\text{-}C_3H_5)PdCl]_2$ and on other kinetic observations, Volger and co-workers (271) recently proposed a mechanism involving a trinuclear complex in which the monomer is linked to the dimer through a chlorine bridge. This fact enhances the difference between the *trans* effects of L and Cl and favors the opening of the two metal–carbon bonds in the chlorine-bonded monomeric species.

B. Related Carbonylations

This section discusses some reactions related to the additions reported in the preceding sections, but not involving allylic intermediates.

1. Acyl Halides with Insertion of Acetylene

Acyl halides are used as starting materials in reactions with $Ni(CO)_4$ and acetylene, giving rise to reactions analogous to those obtained from allylic halides, acetylene, and CO through the intermediate formation of acylnickel complexes.

Acyl halides react at room temperature with $Ni(CO)_4$ and acetylene in ketones or esters as solvents (186) to give a carbonylation reaction with insertion of acetylene. The products thus obtained result from the reaction shown in Eq. 75. Water is usually present only in traces, most of it

$$RCOX + HC{\equiv}CH + Ni(CO)_4 \longrightarrow [\text{R–(lactone)=O}\cdots Ni(Cl)]\ (\mathbf{82}) \xrightarrow{H_2O} \text{R–(lactone)=O}\ (\mathbf{82a}) + NiOHCl + 2CO \quad (75)$$

being added at the end of the reaction. Halide ions have a favorable effect on the formation of lactones (**186a**). This result is to be compared with that of Heck who found that acylcobalt tetracarbonyls react with formation of π-lactonyl cobalt carbonyl complexes and that, in the presence of bases, doubly unsaturated lactones (174) could be obtained by a catalytic reaction (see Vol. 1, fourth chapter). Also with $Ni(CO)_4$, the intermediate should be a π-lactonyl complex (**82**) analogous to that shown in Eq. 59.

Water as well as other proton sources decompose the complex with formation of **82a**. It should be noted that the reaction can be stopped at the stage before the second addition of CO, thus obtaining vinyl ketones, merely by adding acid halides or zinc halides (186). The reaction can be written as shown in Eq. 76.

$$RCOX + HC{\equiv}CH + HY + Ni(CO)_4 \rightarrow RCOCH{=}CH_2 + NiXY + 4CO \quad (76)$$

Unsaturated lactones similar to **82** also form, but are isomerized mostly to the α,β-conjugated isomers. Working under the conditions of formation of **82** but with a low concentration of CO in the reaction mixture (ethers or esters as solvents, vigorous flow of acetylene), 2 moles of acetylene are inserted and seven-membered lactone rings with conjugated double bonds (**83**) are formed (185), where R can be an aryl or alkyl

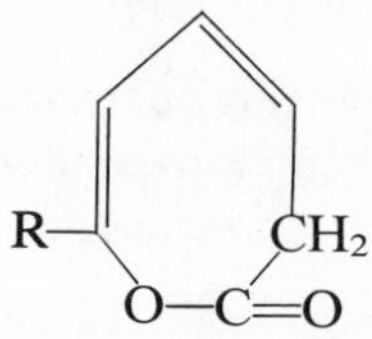

R— CH_2

O—C=O

(**83**)

group. This new class of compounds is characterized by infrared lactone absorptions at about 1770–1775 cm^{-1}. The protons of the methylene group are found in the region of 6.9–7.0τ and are nonequivalent.

Interestingly, these reactions occur with *cis*-hexadienoyl chloride. With this acyl chloride, cyclization occurs first (105) with formation of the cyclopentenonylpentenolactone (**57**). In the absence of acetylene, it is possible to obtain cyclopentenonylacetyl chloride (**54**), preferably in nonpolar solvents such as benzene and heptane. Phenol is also formed as by-product. *Trans*-hexadienoyl chloride reacts as a normal acyl chloride (69). It should be borne in mind that, in the absence of CO, hexadienoyl chloride cyclizes to phenol under the catalytic influence of transition metals (149).

Cyclization of unsaturated acylcobalt carbonyls also leads to alkylcyclopentenones. π-Hex-5-enoylcobalt carbonyl gives 2-methylcyclopentenone and cyclohexenone on decomposition. Sorbylcobalt carbonyl cyclizes to methyl-π-cyclopentenonylcobalt carbonyl(208).

The tendency to form cyclic complexes from acetylene and carbon monoxide is well known for cobalt and iron carbonyls (192,209–211); (see also Vol. 1, p. 273).

The aroyl group can also be generated *in situ* starting from aromatic iodides. Although aromatic iodo derivatives are not allylic halides, they react with $Ni(CO)_4$ under mild conditions. Bauld (182) showed that these compounds could be carbonylated by merely refluxing them with $Ni(CO)_4$ in methanol.

Acetylene was shown to react easily under pressure and even under milder conditions, giving rise mainly to γ-ketoacids or esters (183):

$$RI + HC{\equiv}CH + Ni(CO)_4 + R'OH + H_2O \rightarrow RCOCH_2CH_2COOR' + Ni(OH)I + 2CO \quad (77)$$

γ-Ketoacids are related to β,γ-unsaturated lactones from which they are obtained by hydrolysis.

An insertion reaction of acetylene between two acyl groups resulting from lithium aroyltricarbonylnickelates (212) according to Eq. 78 was observed.

$$2Li[RCONi(CO)_3] + R'C{\equiv}CH + 2H^+ \rightarrow RCOCHR'CH_2COR \quad (78)$$

This reaction was carried out at $-70°C$ and gave good yields of 1,4-diketones. At higher temperature ($-30°C$), γ-butenolactones were formed in low yields.

2. *Acyl Halides with Insertion of Acetylene Followed by Reformatsky-type Reactions*

In the presence of anhydrous enolizable ketones, reaction 75 may proceed further to form products resulting from reaction of the π-lactonyl group with ketones. Thus, from benzoyl chloride in cyclohexanone, it is possible to obtain **84** by a reaction under mild conditions which provides

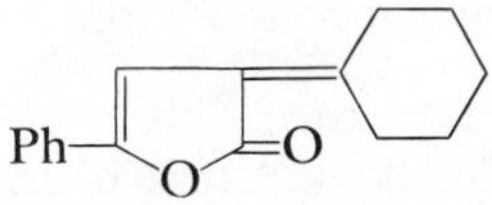

Ph— O =O

(**84**)

a very useful tool for preparing a broad class of compounds not hitherto easily accessible. Analogous Reformatsky-type reactions have been treated in Section II-A-5.

3. *Acyl Halides with Insertion of Olefins*

The acyl group generated *in situ* from iodobenzene can also add to various olefins such as styrene, acrylonitrile, and ethyl acrylate at 50–60°C (213). The products correspond to those obtained by addition of one molecule of olefin per molecule of iodobenzene and then by elimination or uptake of hydrogen, or by further addition of CO, followed by lactonization or esterification. Thus, after 100 hr of reaction with iodobenzene and $Ni(CO)_4$, styrene gave 43% of $PhCOCH_2CH_2Ph$, 1% of PhCOCH=CHPh, and 25% of **85**. The pattern of this reaction seems

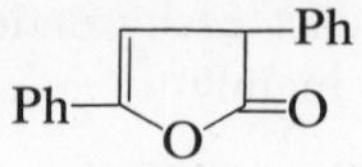

(**85**)

analogous to those already reported for the reaction of acyl groups with acetylenes.

α,β-Carbonyl compounds (such as benzalacetone, methyl cinnamate, cyclohex-2-enone, 3-methyl-3-pent-2-enone, mesityl oxide, 3-but-2-enone, and methyl crotonate) were reacted by Corey and Hegedus (214) with the acyl group generated *in situ* from butyllithium and $Ni(CO)_4$ at −50°C (Eq. 79). Excellent yields of diketones were obtained; for example,

$$[RCONi(CO)_3]^- + \;>C{=}C{-}CO{-}R' \longrightarrow \left[-\overset{|}{\underset{RCO}{C}}-\overset{|}{C}{\cdots}\overset{O}{\overset{\|}{C}}-R' \right]^- + [Ni(CO)_3] \quad (79)$$

4,4-dimethylnonane-2,5-dione was obtained in 89% yield (Eq. 80).

$$(CH_3)_2C{=}CHCOCH_3 + [n\text{-}C_4H_9CONi(CO)_3]^-Li^+ \xrightarrow{H^+} \underset{n\text{-}C_4H_9CO}{(CH_3)_2\overset{}{C}CH_2COCH_3} \quad (80)$$

4. *β-Chlorovinyl Ketones and Vinyl Halides*

Aliphatic or aromatic chlorovinyl ketones, RCOCH=CHCl, in inert solvents react with CO without insertion of acetylene to give the same products observed in the reactions of the corresponding acyl halides **82**, **83**, as well as other products as **86** formally derived from addition of the lactonyl group to the acylvinyl group. β-Chlorovinyl ketones also react

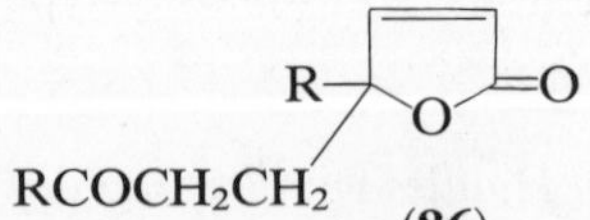

(**86**)

with $Ni(CO)_4$, in water or alcohols, to give β-acylacrylic acids or esters (215):

$$RCOCH{=}CHCl + CO + H_2O \xrightarrow{Ni(CO)_4} RCOCH{=}CHCOOH + HCl \quad (81)$$

The corresponding β-acylpropionic acids, $RCOCH_2CH_2COOH$, are present as by-products along with partially hydrogenated coupling products, $RCOCH{=}CHCH_2CH_2COR$.

Sometimes other products (**87**) are formed; these formally result from addition of the group —HC=CHCOR to the acylacrylic ester followed by hydrogen uptake.

$$\begin{array}{c} RCOCH_2CH{=}\underset{\displaystyle |}{C}CH_2COR \\ \quad COOR' \end{array}$$

(**87**)

Lactones like **82** are also formed in small amounts, unless R is an aromatic group.

In the presence of acetylene, a small percentage of esters such as **88** are found.

$$RCOCH{=}CHCH{=}CHCOOR'$$

(**88**)

Acetals, $RCOCh_2CH(OR')_2$, are also formed in the presence of alcohols.

Vinyl chlorides react slowly at 45°C and atmospheric pressure. Kröper (216) reported that drastic conditions (150°C, pressure unspecified) are necessary for vinyl halides. Vinyl bromides and iodides, however, react under mild conditions: Vinyl bromide gives acrylic acid; α-bromostyrene gives atropic acid; and β-bromostyrene gives cinnamic acid in good yields (215).

Insertion of acetylene occurs to a very small extent in some cases; β-bromostyrene, for example, gives some $PhCH{=}CHCH{=}CHCOOR'$. The occurrence of intermediate complexes, in which the aromatic ring takes part in the formation of a bidentate ligand analogous to the allylic one, cannot be excluded.

An efficient catalyst for carbonylation of vinyl halides was described by Corey and Hegedus (214). It is based on the mixture of $Ni(CO)_4$ and potassium *t*-butoxide in *t*-butyl alcohol, or on the less efficient mixture of $Ni(CO)_4$ and sodium methoxide in methanol. Nickel is known to be present as an anion in these mixtures (see Vol. 1, First chapter). High yields of esters or amides (the latter in the presence of amines) were obtained.

For related carbonylations of vinylmercury compounds *via* palladium complexes see (215*a*,*b*).

III. SUBSTITUTION AND ADDITION REACTIONS OF THE ALLYLIC GROUP ON METAL CARBONYLS WITHOUT INSERTION OF CARBON MONOXIDE

A. Substitutive Hydrogenation Reactions

This section deals with reactions of the allylic group occurring on metal carbonyls but not involving insertion of CO.

As pointed out previously, allylic halides having electron-withdrawing substituents on the allylic carbon atoms are not carbonylated in hydroxylic solvents, but undergo substitutive hydrogenation (104), that is, the replacement of the halide X with H:

$$RCH{=}CHCH_2X + Ni(CO)_4 + H_2O \rightarrow RCH_2CH{=}CH_2 + Ni(X)OH + 4CO \quad (82)$$

For example, 1-chloro-3-cyanoprop-2-ene, $ClCH_2CH{=}CHCN$, gives but-3-enonitrile, $CH_2{=}CHCH_2CN$, in almost quantitative yield. The same product is obtained from the allyl isomer, $CH_2{=}CHCH(Cl)CN$. Of the two possible isomers, but-3-enonitrile and but-2-enonitrile, the thermodynamically less stable 3-en compound is formed. From 2-chloro-4-phenylbut-3-enonitrile, there is obtained the more stable isomer, 4-phenylbut-3-enonitrile. Other typical cases have already been described, for instance, the formation of unsaturated γ-lactones by abstracting a proton from water, leading to the less stable isomer.

This substitutive hydrogenation may sometimes serve as a preparative tool. For example, it is possible, to go from an aldehyde to a nitrile having an additional carbon atom by the following sequence:

$$RCH{=}CHCHO \xrightarrow{HCN} RCH{=}CHCHOHCN \xrightarrow[HCl]{Ni(CO)_4} RCH{=}CHCH_2CN \quad (83)$$

Analogous reactions have been reported by Takegami and co-workers (218), who reacted allylic halides with potassium iron carbonylates [$K_2Fe(CO)_4$, $KHFe(CO)_4$, and $K_2Fe_2(CO)_8$] in ethyl alcohol at 30°C. For example, *trans*-1-chlorobut-2-ene gave mainly *trans*-but-2-ene along with *cis*-but-2-ene and but-1-ene.

1. Mechanism

These reactions appear to be based on protonation of allylnickel intermediates. Protonation is a general reaction for allylic complexes (22,31), σ-Allylic complexes are subject to protonation with formation of olefins (22). The olefin may remain coordinated to the metal (219) or may be displaced by the anion of the protonating acid (22).

It is possible that hydride intermediates are also involved, at least in certain instances. Bönnemann (221) has reported the following equilibrium at low temperature:

$$(\pi\text{-}C_3H_5)Ni(PF_3)H \underset{-50°C}{\overset{-40°C}{\rightleftharpoons}} CH_3CH{=}CH_2 \longrightarrow Ni{-}PF_3 \qquad (84)$$

See also (221*a*) for π-allyl-hydride exchange on Mo.

B. Coupling Reactions

Allylic halides react with $Ni(CO)_4$ to give the well-known coupling reaction (2). The solvents employed are normally alcohols, ethers, esters, or nitriles. Bauld (166) showed that allylic esters also couple in the presence of $Ni(CO)_4$ in THF.

The compounds resulting from head-to-head coupling often constitute most of the product. By starting from unsymmetrically substituted halides, the same mixture is obtained from both allylic isomers (2). If certain *cis* or *trans* allylic halides having electron-withdrawing substituents on the allyl carbons are reacted in nonhydroxylic solvents (to avoid substitutive hydrogenation) the same mixture of linear stereoisomers is obtained (84) when starting from either the *cis* or *trans* stereoisomer.

Thus, on refluxing 1-chloro-3-cyanoprop-2-ene, $ClCH_2CH{=}CHCN$, (or the bromo compound) with $Ni(CO)_4$ in ether or in benzene the products of reaction 85 were predominantly the *cis, trans* and the *cis, cis* compounds with only small amounts of the *trans, trans* isomer.

$$2ClCH_2CH{=}CHCN + Ni(CO)_4 \rightarrow CNCH{=}CHCH_2CH_2CH{=}CHCN + NiCl_2 + 4CO \qquad (85)$$

Under the same conditions, γ-bromocrotonate gave the corresponding diester, $CH_3OOCCH{=}CHCH_2CH_2CH{=}CHCOOCH_3$, as a mixture of the *cis, trans* and *trans, trans* forms, practically without the *cis, cis* isomer.

Bis(π-allyl)nickel complexes are also coupled if treated with CO (20). Bis(π-allyl)nickel treated with CO at −78°C absorbs one molecule of CO and then three other molecules between −40 and +20°C to form $Ni(CO)_4$ and hexa-1,5-diene. Dicrotylnickel behaves similarly at −20°C to yield 98% of dicrotyl (**89**), the linear coupling product. But when treated with CO at higher temperatures, it also forms the two other possible coupling products **90** and **91**.

$CH_3CH{=}CHCH_2CH_2CH{=}CHCH_3$ 58%

(89)

$CH_2{=}CHCH(CH_3)CH_2CH{=}CHCH_3$ 38%

(90)

$CH_2{=}CHCH(CH_3){-}CH(CH_3)CH{=}CH_2$ 4%

(91)

As shown by Corey (52) this type of coupling differs from that with allylic halides. He observed that cyclooctenyl bromide couples with $Ni(CO)_4$ to give only the dimer, while di-π-cyclooctenylnickel with CO gives a mixture of the dimer and dicyclooctenyl ketone.

Allylcobalt tricarbonyl was found to give hexa-1,5-diene when treated with allyl bromide (59). 2,5-Dimethylhexa-1,5-diene was obtained by reaction of CO with bis(2-methylallyl)bis(triphenylphosphine)ruthenium which was converted to $Ru(CO)_3(PPh_3)_2$ (222).

Coupling reactions have been extended to intramolecular cyclizations. Thus, Corey (223,224) coupled the dibromide **92** to obtain 1,6-dimethylcyclododeca-1,5,9-triene (**93**) (Eq. 86). Product **93** was obtained

$$BrCH_2CH{=}C(CH_3)CH_2CH_2CH{=}CHCH_2CH_2C(CH_3){=}CHCH_2Br \longrightarrow$$

(92)

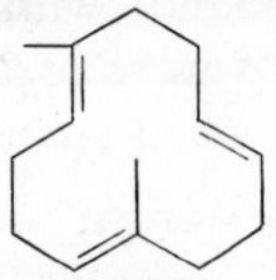

$+ NiBr_2 + 4CO$ (86)

(93)

as a 2:1 mixture of the all *trans* isomer and of the *cis*, *trans*, *trans* isomer starting from the all-*trans* **92**.

With $Ni(CO)_4$ at 50°C, 1,1-bis(chloromethyl)ethylene in THF gave 1,4,7-trimethylenecyclononane (**93a**). This synthesis apparently involves

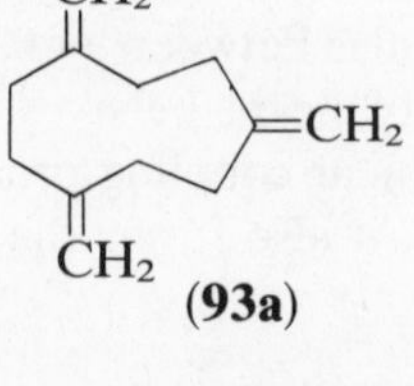

(93a)

both intermolecular and intramolecular coupling of three molecules of the starting dichloride. Cyclization product **95** corresponding to the intermediate **94** (derived from the coupling of two molecules of the reagent) was also formed in low yield.

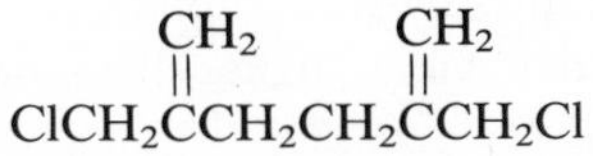

$$ClCH_2C(=CH_2)CH_2CH_2C(=CH_2)CH_2Cl$$

(94)

CH_2=(cyclohexane ring)=CH_2

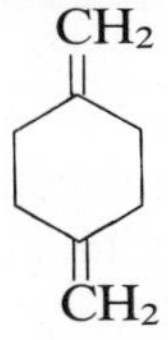

(95)

Use of $Ni(CO)_3PPh_3$ in tetraglyme in place of $Ni(CO)_4$, however, inhibited the formation of **93a** but not that of **95**. Furthermore, the reaction of **94** with $Ni(CO)_4$ gave mainly polymers and only 10% of **95**, whereas the same reaction with $Ni(CO)_3(PPh_3)$ resulted in a 60% yield of **95**. This indicates that a single PPh_3 ligand on nickel favored formation of the six-membered ring.

Cyclization of a series of dibromides of structure

$$BrCH_2CH{=}CH(CH_2)_nCH{=}CHCH_2Br$$

with $Ni(CO)_4$ in DMF at 50°C under argon gave the same type of isomers (95–98% *trans, trans* and small amounts of *cis, trans*) when starting with either *cis, cis* or *trans, trans* isomers (225). The products obtained were:

$[\ ]_n$		
$n = 6$	(cyclic diene)	59%
$n = 8$	(cyclic diene)	70–74%
$n = 12$	(cyclic diene)	76–84%

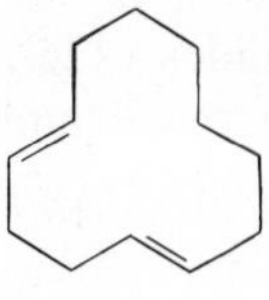

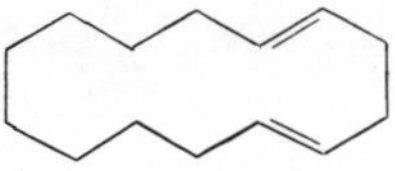

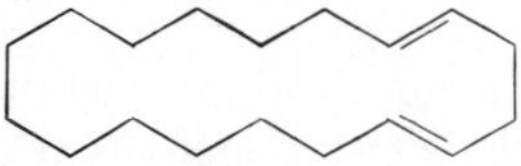

With $n = 2$ or 4, six-membered ring structures were favored, despite the general tendency to join terminal carbon atoms. 4-Vinylcyclohexene (42%) and *cis, cis*-cycloocta-1,5-diene (5%) were obtained for $n = 2$ when using $Ni(CO)_4$. The relative amount of cyclooctadiene was raised to 20% by using $Ni(CO)_3PPh_3$. With $n = 4$, only 1,2-divinylcyclohexane was obtained—mainly *cis* with $Ni(CO)_4$, and mainly *trans* with $Ni(CO)_3PPh_3$.

This method of cyclization is applicable to the synthesis of several natural products. Humulene, 2,6,6,9-tetramethylcycloundeca-1,4,8-triene, was synthesized in this manner (226). Elemol, a divinylcyclohexane sesquiterpene, was also synthesized (226*b*). Macrocyclic lactones can be obtained analogously by starting with allylic dibromides containing an O—C=O group in the chain (226*c*).

A bis(π-allyl)nickel intermediate obtained directly from butadiene also reacts with CO to give vinylcyclohexene in a yield higher than 88% (20,93) (Eq. 87).

$$\text{L—Ni(bis-}\pi\text{-allyl)} \xrightarrow{3CO} \text{vinylcyclohexene} + LNi(CO)_3 \qquad (87)$$

1. Mechanisms

The stereochemical course of the coupling reaction of allylic halides, $YCH{=}CHCH_2X$, can be explained by assuming that the two coupling molecules behave differently from each other; that is, one has a preferred configuration as a π-allyl ligand, while the other can react as the *cis* or *trans* form (82–84). The reaction should occur as shown in Eq. 88 (L is CO or other ligands) (51,52,82,84) as a type of nucleophilic substitution effected by the coordinated allyl group on the allylic carbon of a second molecule. Possible occurrence of a radical mechanism was proposed by Hegedus who observed racemization of optically active 2-iodooctane in reaction with allylnickel halides (226*a*). Probably there is more than one mechanism depending on the ligands and substituents on the allyl group.

$$(\pi\text{-YC}_3H_4)NiL_2X \longrightarrow Y\text{-CH=CH-CH}_2\text{-}NiL_2X \rightleftarrows (\text{CH}_2\text{=CH-CHY})NiL_2X \xrightarrow{YCH{=}CHCH_2X} YCH{=}CHCH_2CH_2CH{=}CHY + NiX_2 + 2L \qquad (88)$$

The occurrence of an isomerization reaction (63) can sometimes obscure the observation of the stereoselectivity. Corey and others (52) noted that the reaction leading to formation of π-allylnickel complexes from $Ni(CO)_4$ and allyl halides is reversible (Eq. 89). A rapid exchange also occurs between the π-allyl coordinated group and a new molecule of allyl

$$(\pi\text{-C}_3H_5)Ni(CO)Br \underset{-CO}{\overset{CO}{\rightleftarrows}} CH_2{=}CHCH_2Br + Ni(CO)_4 \qquad (89)$$

halide (Eq. 90). Thus, in the presence of readily isomerizable allylic groups it is not easy to follow the behavior of the two halves.

$$H_3C-\langle\!\!(-Ni(CO)Br + CH_2{=}CHCH_2Br \rightleftharpoons H_3C-\langle\!\!(-Ni(CO)(Br)(CH_2{=}CHCH_2Br) \rightleftharpoons \langle\!\!(-Ni(CO)Br + CH_2{=}C(CH_3)-CH_2Br \qquad (90)$$

Corey and Semmelhack (51) could avoid the exchange reaction by using complexes such as **96**, where R is H, CH_3, COOEt, and so on for

$$R-\langle\!\!(-Ni(\mu\text{-}Br)_2Ni-)\!\!\rangle-R$$

(**96**)

the cross-coupling reaction with nonallylic halides such as alkyl, vinyl or aryl halides, R′X. In polar, coordinating media such as DMF, *N*-methylpyrrolidone or hexamethylphosphoramide, a facile reaction occurs (Eq. 91).

$$\mathbf{96} + 2R'X \rightarrow 2RC({=}CH_2)CH_2R' + 2NiXBr \qquad (91)$$

With these solvents, however, mixtures of *cis* and *trans* isomers were obtained from allylic groups that could give rise to stereoisomers.

Stereoselectivity could be observed in cross-coupling reactions of two allylic halves by using allylic ligands containing electron-withdrawing substituents. Here the displacement reaction (90) is more difficult. Since the allylic ligand is more anionic, it more readily tends to attack the carbon of another allylic molecule, rather than be attacked by halides in a displacement reaction (82, 84*a*).

Reactions of the following complexes (written in monomeric form) with allyl bromide in aprotic solvents of low polarity gave more than 80% *trans* isomer in the case of the carbomethoxy substituent and more than 80% *cis* isomer in the case of the cyano substituent (82). Analogous

$$\langle\!\!(^{COOCH_3}-Ni\langle^{Br} \qquad \langle\!\!(^{CN}-Ni\langle^{Cl}$$

results were obtained by reacting the π-allylnickel halide complex with γ-bromocrotonate and γ-bromocrotonitrile, respectively.

Another case is offered by the synthesis of geranyl acetate. This compound was obtained with 85% stereoselectivity from the reaction of isoprene hydrochloride with the allylic complex indicated in Eq. 92 (82).

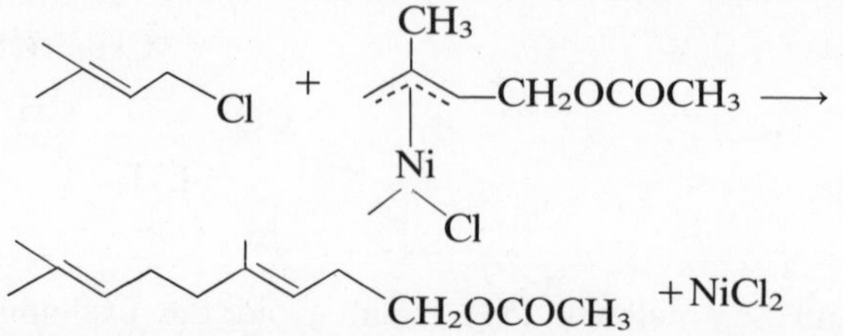

$$(CH_3)_2C{=}CHCH_2Cl + [\pi\text{-}CH_2C(CH_3)CH\text{–}CH_2OCOCH_3]NiCl \longrightarrow \text{geranyl–}CH_2OCOCH_3 + NiCl_2 \quad (92)$$

Here, displacement of the complexed allyl group can occur, but it is of no concern as to the purity of the product because the displaced group gives isoprene on reaction with the nickel complex. A better result was obtained by Sato et al. (82*a*) who started from the nickel complex of isoprene hydrochloride and 1-chloro-2-methyl-4-acetoxy-2-butene.

The use of appropriate solvents and ligands can reverse the type of stereoselectivity observed, however. These points as well as other general aspects of the stereoselectivity of the reactions of allylic groups have been treated in Section I of this chapter.

An extensive study on coupling reactions of bis(π-allyl)nickel complexes was made by Wilke and co-workers (93,173,238–240). Under the influence of a donor ligand two butadiene molecules couple on zerovalent nickel to give a bis(π-allyl)nickel complex which undergoes intramolecular coupling at the "allylic" positions. Insertion of alkenes or alkynes may occur if these species are present as ligands. Coupling after insertion leads to large cycles. Degree of conversion, temperature, type of ligands, and solvents influence the stereochemical course of these reactions. A simplified scheme is shown below (R,R′ = H, alkyl, aryl)

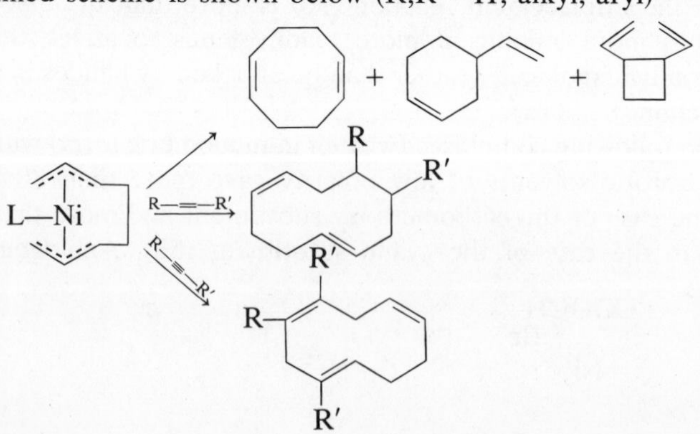

L—Ni (bis-π-allyl) → cycloocta-1,5-diene + 4-vinylcyclohexene + divinylcyclobutane

L—Ni —(R—=—R′)→ ten-membered ring with R, R′

L—Ni —(R—≡—R′)→ ten-membered ring with R, R′, R′

Analogous formation of bis(π-allyl)metal complexes is observed by using isoprene or allene in place of butadiene.

To gain further insight into this very wide subject the reader is referred to several reviews (93,102*c*,*d*). For these reactions Heimbach, Jolly, and Wilke (93,227) advanced an interpretation based on the assumption that π-ligands and the metal constitute a conjugated system to which the Woodward–Hoffmann rule on conservation of orbital symmetry is applicable.

C. Addition Reactions

1. *Carbonyl Groups*

Allylic halides with electron-withdrawing substituents on the allylic carbon add to carbonyl groups of ketones and aldehydes when treated with $Ni(CO)_4$ (105). The reaction yields a mixture of the two possible allylic isomers. From γ-bromocrotonate and acetone, for example, the following reaction occurs, possibly through the intermediates as shown in Eq. 93. The molar ratio of branched to straight-chained product is 88 : 12.

$$CH_3COOCCH{=}CHCH_2Br + CH_3COCH_3 + Ni(CO)_4 \rightarrow$$

$$\left[\text{(π-allyl with } COOCH_3\text{)}Ni(CO)_2Br\right] \rightarrow \left[CH_3OOC{-}CH{=}CH{-}CH_2{-}Ni(CO)_2Br \leftarrow O{=}C(CH_3)_2\right] \rightarrow$$

$$\left\{\begin{array}{l} CH_3OOCCH{=}CHCH_2{-}C(ONiBr)(CH_3){-}CH_3 \\ \\ CH_3OOC{-}CH(CH{=}CH_2){-}C(ONiBr)(CH_3){-}CH_3 \end{array}\right. + 4CO \qquad (93)$$

A related type of addition has been observed (10) by reaction of allyldicarbonylnitrosyliron (Eq. 94).

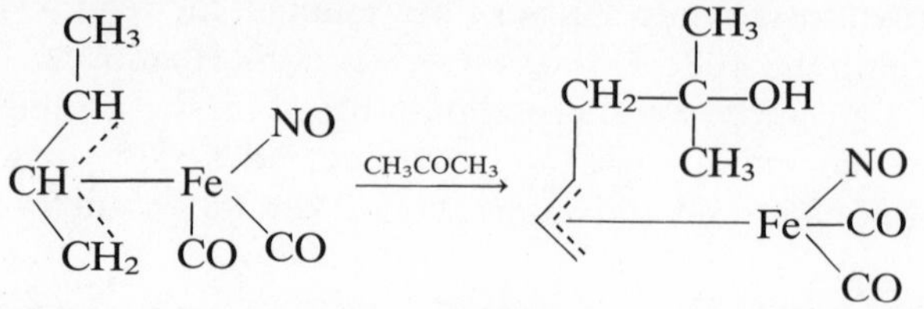

$$\begin{array}{c} CH_3 \\ | \\ CH \\ \vdots \\ CH \cdots Fe(NO)(CO)_2 \\ \vdots \\ CH_2 \end{array} \xrightarrow{CH_3COCH_3} \begin{array}{c} CH_2\text{—}C(CH_3)_2\text{—}OH \\ | \\ \text{(allyl)} \cdots Fe(NO)(CO)_2 \end{array} \tag{94}$$

a. Mechanisms. The nearest analogy is that of the Reformatsky and Grignard reaction. It is known (228–231) that for γ-halogenocrotonates and other vinylogs, the products of the reaction with zinc consist mainly of the "abnormal" branched product.

Related reactions on nickel complexes will now be reported. π-Allylnickel halides not containing CO were reacted with aldehydes, ketones and epoxides in DMF (51), as shown in Table 4, and with quinones to prepare allylquinones (231*a*). Bis–π–(2-methoxyallyl nickel bromide) was shown to be an effective reagent to introduce the acetonyl functional group (231*b*, 231*c*). α-Methylenebutyrolactones have been obtained from bis–π–(2-carbalkoxyallylnickel bromides) (231*d*) and carbonyl compounds.

π-Allynickel complexes are also likely to be intermediates in the reactions of butadiene with bis(cyclooctadiene)nickel in the presence of

TABLE 4
Reaction of π-Methallylnickel Halides with Various Substrates (51)

Substrate	Product	Yield (%)
Benzaldehyde	$CH_2{=}C(CH_3)CH_2CHOHPh$	85
Acrolein	$CH_2{=}C(CH_3)CH_2CHOHCH{=}CH_2$	80
Cyclopentanone	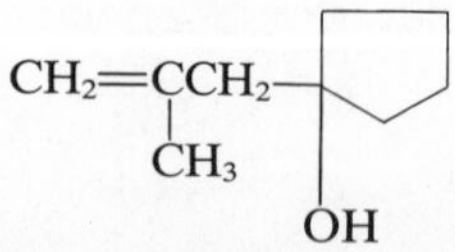 $CH_2{=}C(CH_3)CH_2\text{—}C_5H_8\text{—}OH$ (1-hydroxycyclopentyl)	50
Styrene oxide	$CH_2{=}C(CH_3)CH_2CH(Ph)CH_2OH$	60

aldehydes or ketones as described by Wilke (232). Possibly, the intermediates are of the type like **97** which, on further reaction with butadiene, give the *trans* products shown in Table 5.

[structure: allyl–Ni(L)–O–CR$_2$ chelate]

(**97**)

TABLE 5

Reactions of Bis(cyclooctadiene)nickel with Butadiene and Aldehydes or Ketones (232)

Substrate	Product (*trans*)	Yield (%)
CH_3COCH_3	$CH_3C(CH_3)(OH)CH_2CH{=}CHCH_2C(CH_3)(OH)CH_3$	60
cyclohexanone	1-hydroxycyclohexyl–$CH_2CH{=}CHCH_2$–(1-hydroxycyclohexyl)	70
$PhCOCH_3$	$PhC(CH_3)(OH)CH_2CH{=}CHCH_2C(CH_3)(OH)Ph$	68
$PhCHO$	$PhCH(OH)CH_2CH{=}CHCH_2CH(OH)Ph$	72
retinal (CHO)	[retinyl–CH(OH)–CH_2–CH=]$_2$	15

Allylnickel or allylcobalt complexes prepared *in situ* by reduction of the respective halides with a manganese-iron alloy in the presence of allyl halides react selectively with aldehydes in methanolic solution. These are Grignard-type reactions that are carried out in a protic solvent (232*a*).

2. *Activated Olefins*

At room temperature and atmospheric pressure, allylic compounds react with $Ni(CO)_4$ and some vinylic olefins activated by electron-withdrawing groups such as COOR, CN and CHO to give linear compounds with one double bond (Eq. 95) if water or alcohols are present. Products containing two double bonds (Eq. 96) predominate if the reaction is performed in an inert solvent (62,233):

$$CH_2{=}CHCH_2Cl + CH_2{=}CHCOOCH_3 + Ni(CO)_4 + H_2O \rightarrow$$
$$CH_2{=}CHCH_2CH_2CH_2COOCH_3 + NiOHCl + 4CO \qquad (95)$$

$$CH_2{=}CHCH_2Cl + CH_2{=}CHCOOCH_3 + Ni(CO)_4 \rightarrow$$
$$CH_2{=}CHCH_2CH{=}CHCOOCH_3 + [Ni(CO)_3HCl] \qquad (96)$$

The double bond conjugated with the carboxylic group is *trans*. Reaction 96 is thus stereoselective, giving rise to the same compounds obtained from allyl halides, acetylene, and CO except that the *trans* rather than the *cis* form is obtained. The same synthesis can be performed with bis(π-allyl)nickel bromide.

The reactivity of activated olefins varies greatly. If acrylonitrile is reacted in benzene, complexes such as **98** are isolated.

$$\begin{array}{r} CH_2{=}CHCH_2CH_2\underset{|}{C}HCN \\ {-}Ni{-}\;\; \\ CH_2{=}CHCH_2CH_2\overset{|}{C}HCN \end{array}$$

(98)

These complexes are polymeric; the cyano group is likely coordinated with nickel. Spectroscopic and chemical evidence show that the vinyl group is not affected by coordination. Its typical infrared and nmr absorptions remain unchanged, and the reaction of the double bond with benzonitrile oxide yields isoxazolines without destroying the complex. Thermal decomposition gives rise to $CH_2{=}CHCH_2CH{=}CHCN$ and $CH_2{=}CHCH_2CH_2CH_2CN$ along with isomeric hexadienonitriles.

The reaction may also be carried out with bis(acrylonitrile)nickel (62,74) instead of $Ni(CO)_4$.

The reaction with activated olefins in benzene is preferred to carbonylation with insertion of acetylene. If it is performed in the presence of CO and acetylene, carbonylation is prevented and addition of allylic groups to the terminal olefinic carbon occurs.

The preparation of hexenoic compounds in aqueous–alcoholic solvents was first accomplished by reacting allylic compounds with activated olefins in the presence of nickel derived from the previously described catalytic system: $NiCl_2$–iron powder–thiourea. The reagents were warmed in either an alcohol, or water, or aqueous–alcoholic mixtures in the absence of CO. The products, freed of sulfur compounds, were hexenoic compounds corresponding to those shown in Eq. 95 (233).

$Fe_3(CO)_{12}$ was employed in reactions between benzyl halides and activated olefins, $CH_2{=}CHY$, at 40–110°C in various solvents (234). As with $Ni(CO)_4$ the products were mixtures of the type $PhCH_2CH_2CH_2Y$ or $PhCH_2CH{=}CHY$. Interestingly, products resulting from the reaction of 2 moles of olefins such as $PhCH_2CH_2CH(Y)CH_2CH_2Y$ are also present. Polar solvents favor benzylation; even phenylacetylene in DMF gives 1,3-diphenylpropane and 1,3-diphenylpropene.

a. Mechanisms. The reaction on nickel probably occurs as shown in Eq. 97 (Y is an activating group). The final complex undergoes hydrogenation

$$(\text{allyl})\text{—Ni}(L)(L)(X) + CH_2{=}CHY \longrightarrow (CH_2{=}CHY)(\text{allyl})\text{Ni}(L)(L)(X) \longrightarrow$$

$$L\text{—Ni—}L\ (X)(\text{CH(Y)CH}_2\text{CH}_2\text{CH}{=}\text{CH}_2) \begin{cases} \xrightarrow{-NiHXL_2} CH_2{=}CHCH_2CH{=}CHY \\ \xrightarrow[-Ni(OH)X,\ 2L]{H_2O} CH_2{=}CHCH_2CH_2CH_2Y \end{cases} \qquad (97)$$

in hydroxylic solvents, as already described, with formation of nickel oxychloride and abstraction of hydrogen from the solvent. In benzene, it gives a mixture of hydrogenated and dehydrogenated products, RCH_2CH_2Y and $RCH{=}CHY$. Since no stoichiometric relation exists between the two products, formation of the first is not concerted with dehydrogenation of the complex intermediate from which the second is formed. A probable intermediate, H–Ni–X, derived from dehydrogenation, gives up its hydrogen partly for hydrogenation of the complex and partly for other secondary reactions. As previously pointed out, nickel hydrides can also be formed from a dialkylnickel complex generated by disproportionation of an alkylnickel halide.

Disproportionation reactions of π-allyliron carbonyl halides have been described (235,236).

Addition of allyl groups to butadiene occurring in a polymerization reaction was reported (237). The reaction is performed by treating $Ni(CO)_4$ and allylic halides with butadiene; CO also copolymerizes with butadiene.

Beside coupling of allylic ends and co-oligomerizations with alkenes and alkynes, Wilke and co-workers (93,173,238–240) also observed addition reactions of allyl groups to olefins, followed by hydrogen elimination, as in Eq. 98 for example:

Ni → Ni ⇄ Ni →

trans

Ni H → (98)

Addition of butadiene to activated olefins was reported by several authors, see, for example, (240*a*,*b*). Additions of π-allylpalladium acetate and of π-allylpalladium acetylacetonate (256) to butadiene were also observed. For reactions of complexes not containing CO, see the literature dealing with allyl additions to dienes in polymerization reactions (238,241,242).

D. Isomerization

If $Ni(CO)_4$ is reacted with an allylic halide to a limited conversion, the unreacted part is found to be isomerized to a large extent. The process is catalytic because several molecules of allyl halides are isomerized per mole of $Ni(CO)_4$ (63).

Isomerization of olefins by iron carbonyls (102,243–246) or cobalt carbonyls (102,247) was attributed to intermediate formation of π-allyl complexes, for example:

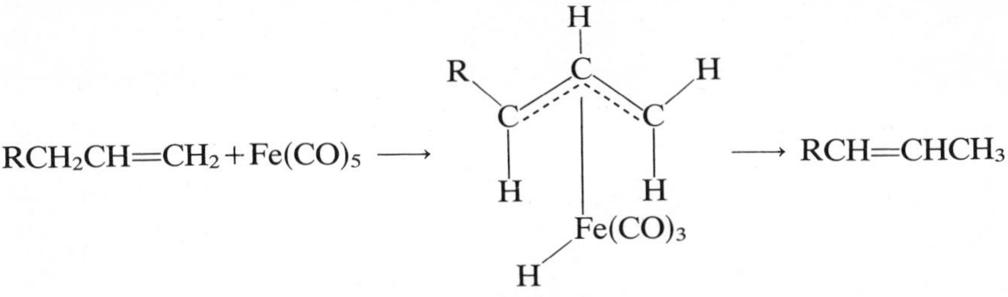

$$RCH_2CH{=}CH_2 + Fe(CO)_5 \longrightarrow \text{(99)} \longrightarrow RCH{=}CHCH_3$$

(99)

But other mechanisms also were proposed; these are given in the review by Orchin (102).

The rearrangement of allyl alcohol to propionaldehyde by iron carbonyl possibly involves π-hydroxyallyl complexes as shown in Eq. 100 (248,249). Evidence for this mechanism was provided by studies with

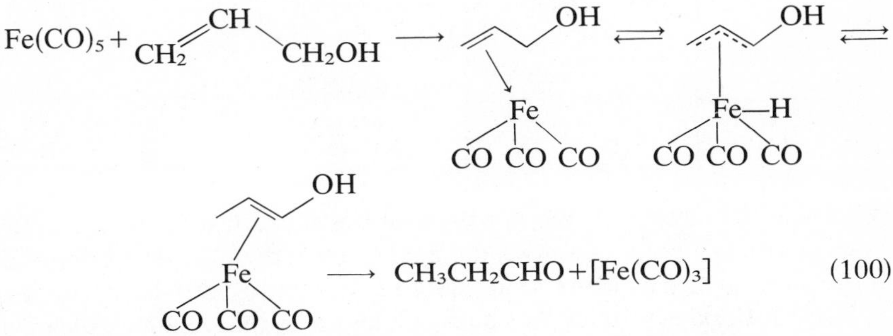

$$Fe(CO)_5 + CH_2{=}CHCH_2OH \longrightarrow \cdots \rightleftarrows \cdots \rightleftarrows \cdots \longrightarrow CH_3CH_2CHO + [Fe(CO)_3] \qquad (100)$$

deuterated allyl alcohol (250).

An interesting thermal isomerization from an acetylallyl to an oxapentadienyl complex was also observed (251,252):

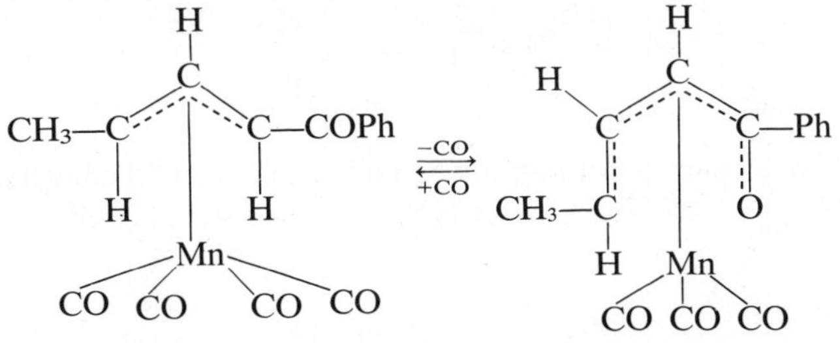

$$\xrightleftharpoons[+CO]{-CO} \qquad (101)$$

(100)

E. Hydroxylation and Reactions with Nucleophiles

π-Allyl ligands on iron carbonyls (236,253) are subject to attack by hydroxyl groups (248,249) (for instance, as Eqs. 102 and 103). In

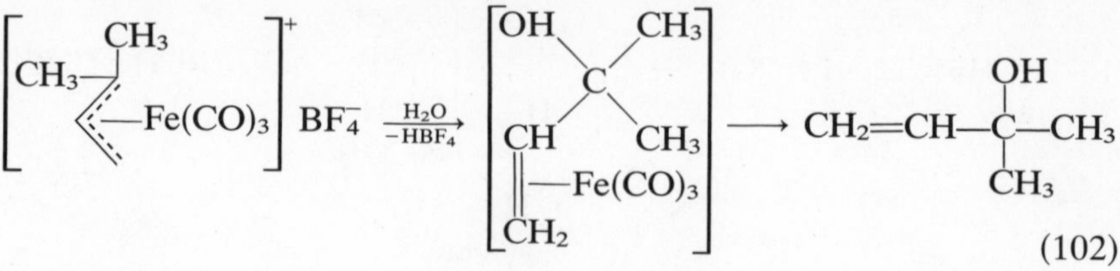

$$\left[(CH_3)_2C{=}CH{=}CH_2\text{—}Fe(CO)_3\right]^+ BF_4^- \xrightarrow[-HBF_4]{H_2O} \left[(OH)(CH_3)_2C\text{—}CH{=}CH_2\text{—}Fe(CO)_3\right] \longrightarrow CH_2{=}CH\text{—}C(OH)(CH_3)\text{—}CH_3 \quad (102)$$

$$\left[CH_3CH{=}CH{=}CH_2\text{—}Fe(CO)_3\right]^+ BF_4 \xrightarrow[-HBF_4]{H_2O} \left[(OH)(CH_3)(H)C\text{—}CH{=}CH_2\text{—}Fe(CO)_3\right] \longrightarrow$$

$$\left[(HO)(CH_3)C{=}CH(CH_3)\text{—}Fe(CO)_3\right] \longrightarrow CH_3CH_2COCH_3 \quad (103)$$

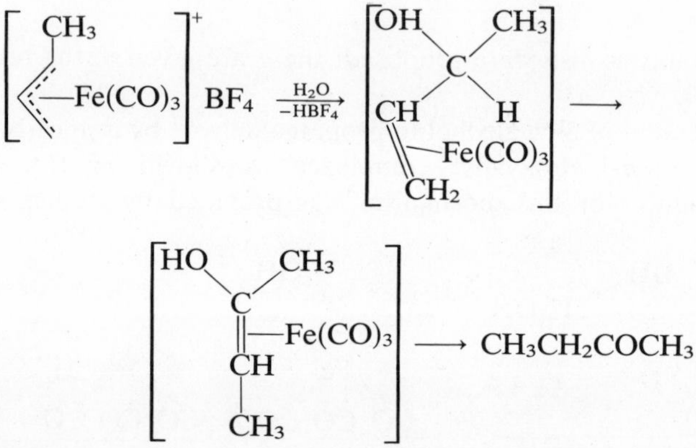

sequence 103, owing to the presence of an allylic hydrogen, a double bond shift takes place and gives the enol form of methyl ethyl ketone. It is possible that an Fe–H bond is formed during the isomerization (248,249). Besides water, other nucleophiles seem able to attack the allyl group. Thus, PPh_3 gives an allyltriphenylphosphonium salt (249). With complex **101**, instead of hydroxylation, reduction at the expense of water occurs with formation of **102**. This is another example of substituent effects determining a completely different course of a reaction.

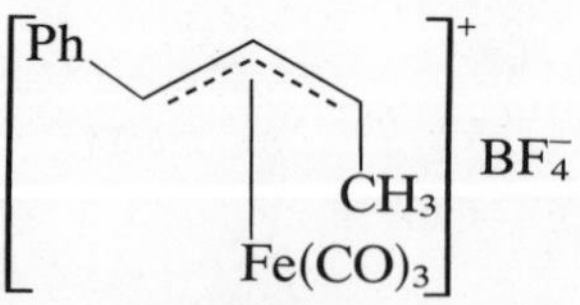

$$\left[Ph\text{—}CH{=}CH{=}CH\text{—}CH_3 \;\; Fe(CO)_3\right]^+ BF_4^-$$

(101)

$$PhCH{=}CH\text{—}CH_2\text{—}CH_3$$

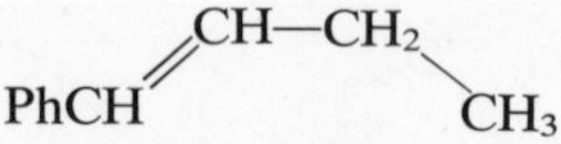

(102)

Analogous behavior of nickel complexes has been shown throughout this chapter. Allyl chloride, in CH_3OH, with $Ni(CO)_4$ gives allyl methyl ether as a secondary product, the principal product being methyl but-3-enoate; if, however, a CN or $COOCH_3$ substituent is present on the allyl group, the attacking species is not OCH_3 but H.

Nucleophilic additions to coordinated allylic group comprise a very wide subject. Only two examples involving the presence of CO on the intermediate metal complex will be reported here; for the reactions of metal complexes not containing CO, see references 98,100,254,255. The attack of acetylacetonate on the allyl group in complex **103** in the presence of CO was reported (256):

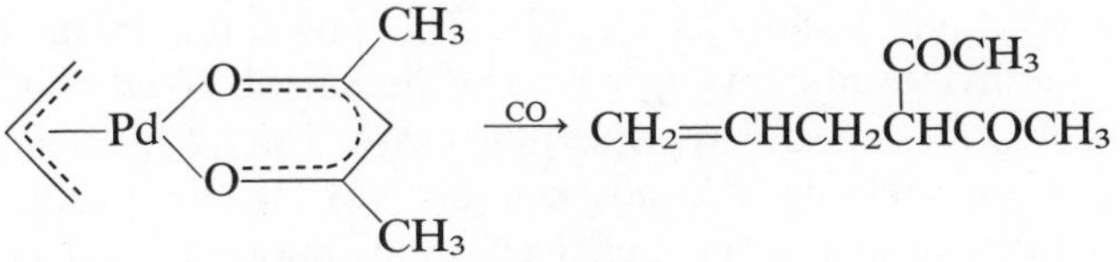

$$\text{(103)} \xrightarrow{CO} CH_2{=}CHCH_2CH(COCH_3)COCH_3 \qquad (104)$$

(103)

Bis(π-allyl)palladium acetate reacts with butadiene to give an intermediate of probable structure **104** which decomposes to **104a**, shown in Eq. 105. Besides the addition of acetate, other additions of nucleophiles such as $C_6H_5O^-$ (257) and malonate (258) with concomitant reduction of palladium complexes to the metal have been described (See also 98,100,254,255,255*a*.)

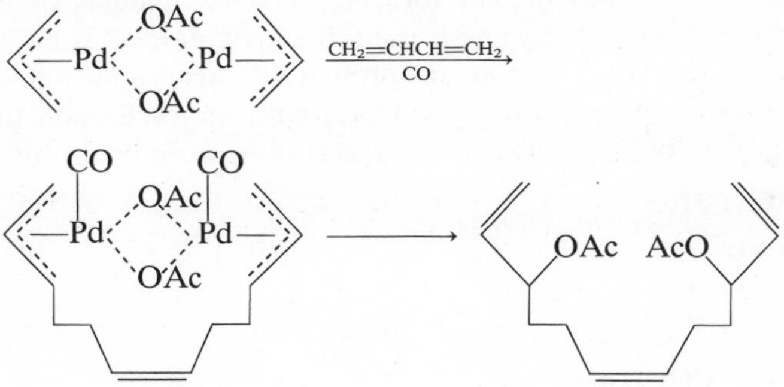

$$\xrightarrow[CO]{CH_2{=}CHCH{=}CH_2} \text{(104)} \longrightarrow \text{(104a)} \qquad (105)$$

(104) **(104a)**

IV. FURTHER DEVELOPMENTS OF INSERTION REACTIONS INVOLVING CARBON MONOXIDE

A short summary of synthetic developments not covered in preceding sections follows.

A. Synthesis of *m*-Cresol from Methallyl Chloride, Acetylene, and Carbon Monoxide

Although the reaction of methallyl chloride with acetylene and CO can lead to substantial amounts of *m*-cresol by working in acetone as solvent, the reaction has only a limited catalytic character. The reason is the hydrogen is not liberated as HCl by cyclization of the methylhexadienoyl group but serves to hydrogenate the methylcylohexenoyl–nickel bond with formation of methylcyclohexenone, as previously shown. In the presence of a catalytic complex formed by reaction of $Ni(CO)_4$ and iodide ions plus a neutralizing agent such as MgO and powdered iron or manganese–iron alloy (the latter being more efficient), cyclization easily occurs in satisfactory yield (73%)(273). The powdered metal helps to cyclize the methylhexadienoyl group. The action of powdered iron on hexadienoyl chloride was already described (150). The reaction can also be carried out completely in the absence of $Ni(CO)_4$ by producing the catalytic complex *in situ* from nickel chloride, thiourea, and powdered iron or manganese–iron alloy. The yield, however is only 35%.

B. Synthesis of Alkenylidenphthalides from Allyl Halides, Hydroxyphenylacetylene, and Carbon Monoxide

It has been shown in this chapter that higher acetylenes easily form the cyclopentenone or cyclohexenone ring on reaction with allyl halides and $Ni(CO)_4$ and that they do not form appreciable amounts of the open chain acid; but, when acetylene is used, the main product is the latter. If, however, a hydroxyl group is present in the *ortho* position of phenylacetylene, it successfully competes with the cyclization (to cyclopentenone) by forming a lactone (280) according to equation 106:

$$o\text{-}C_6H_4(C{\equiv}CH)(OH) + CH_2{=}CHCH_2X + CO \xrightarrow{Ni(CO)_4} \left[C_6H_4\text{-lactone}({=}CHCH_2CH{=}CH_2)({=}O) \right] + HCl \longrightarrow C_6H_4\text{-lactone}({=}CHCH{=}CHCH_3)({=}O) \quad (106)$$

The product is an alkenylidenphthalide, which for the simplest allyl halides, is found with two conjugated double bonds. But if the allyl halide has a terminal substituent stabilizing the double bond, the product is mainly found with nonconjugated double bonds. No product deriving from the reverse addition of allyl halide and CO to the triple bonds was formed.

This synthesis confirms that stepwise additions on nickel can be altered at any point by competition with suitable groups. Another example of competition is offered in the following section.

C. Synthesis of Lactones from Allyl Halides, Hydroxyalkylacetylenes, and Carbon Monoxide

The same principle outlined in the preceding item has been applied to the catalytic synthesis of γ-lactones from hydroxyalkylacetylenes (274). The latter are easily obtainable from aldehydes and propargyl halides. An example of reaction is the following:

$$PhCHOHCH_2C\equiv CH + ClCH_2CH{=}CH_2 + CO \xrightarrow{\text{Ni cat}}$$

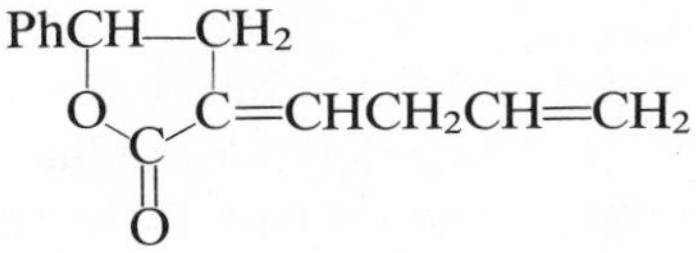

$$\begin{array}{l} PhCH\!-\!CH_2 \\ \;\;|\qquad\;\; | \\ \;\;O\diagdown_{C}\diagup C{=}CHCH_2CH{=}CH_2 \\ \qquad \| \\ \qquad O \end{array} + HCl \qquad (107)$$

D. Synthesis of Hexadienoyl Halides

Hexadienoyl halides have recently been synthesized on transition metal catalysts by reaction of allyl halides with acetylene and CO in an inert medium such as benzene (275). $PdCl_2$ has been used as catalyst under CO and acetylene pressure. Thus hexadienoyl chloride is obtained at 50 atm and 80°C in 3 days. Other halides of group VIII metals are claimed as catalysts.

E. Stereospecific Carbonylation of Styrylnickel Complexes

It was shown (275*a*) that styrylnickel complexes of the type

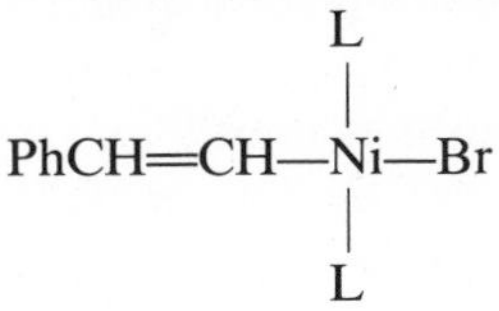

$$\begin{array}{c} \qquad\qquad\quad L \\ \qquad\qquad\quad | \\ PhCH{=}CH\!-\!Ni\!-\!Br \\ \qquad\qquad\quad | \\ \qquad\qquad\quad L \end{array}$$

($L = PPh_3$ or PEt_3) carbonylate stereospecifically by reaction with CO in methanol. Thus the *cis*-styryl complex gives *cis*-methyl cinnamate and the *trans*-styryl complex gives *trans*-methyl cinnamate.

This result is in agreement with the observed stereospecificity of acetylene insertion and supports the hypothesis that the insertion proceeds stepwise with intermediate formation of vinyl complexes.

F. Synthesis of Acids and Esters from Allyl Halides, Ethylene, and Carbon Monoxide

An ethylene insertion reaction parallel to the acetylene insertion described in Section-II-A was described (277). It is based on the use of ethylene under pressure of 30–40 atm in the presence of a low partial pressure of CO in aqueous or alcoholic media. Yields are in the range of 40–60%.

Carbonylation of allyl halides with triple-bond insertion was described in the chapter. Insertions of double bonds between the allyl group and CO had been observed only with the strained olefin norbornene. This insertion was shown to be stereospecific (276), 2-*exo*methallyl-3-*exo*carboxynorbornane being formed. Activated olefins also insert between the allyl group and nickel, but without further insertion of CO, as already shown.

$Ni(CO)_4$ as well as allylnickel complexes can be used as catalysts in the reaction of simple olefins. The reaction takes place in the range of 20–60°C. The reaction is represented by Eq. 108, which is written for the simplest term of the series of allyl halides, namely allyl chloride.

$$CH_2{=}CHCH_2Cl + CH_2{=}CH_2 + CO + ROH \xrightarrow{\text{Ni cat.}} CH_2{=}CHCH_2CH_2CH_2COOR + HCl \quad (108)$$

Products deriving from linear insertion of 2 moles of ethylene and of CO are also present such as $CH_2{=}CHCH_2CH_2CH_2COCH_2CH_2COOR$. A heavier product present in small amounts is probably

$$CH_2{=}CHCH_2CH_2CH_2COCH_2CH_2COCH_2CH_2COOR.$$

Cyclopentanonic products to be treated in the next section are also formed.

The scope and limitations of this reaction are analogous to those already mentioned for acetylene insertion and will not be discussed here. It is only to be pointed out that insertion of ethylene has proved to be a matter of competition with CO, the simple increase of ethylene partial pressure, under conditions which also allow coordination of CO, being the key to the synthesis.

G. Synthesis of Saturated Cyclopentanonic Derivatives from Allyl Halides, Olefins, and Carbon Monoxide

On passing from ethylene to higher α-olefins the yield of insertion products drops to a few percent and coupling reactions of two allyl groups or direct carbonylation to β,γ-unsaturated acids or esters predominate. Among the insertion products, those resulting from reaction 108 are only in minor amounts, whereas cyclopentanonic product deriving from reaction 109 (written for crotyl chloride) prevails (R = H or alkyl):

$$CH_3CH{=}CHCH_2Cl + 2CO + CH_2{=}CHR + R'OH \xrightarrow{\text{Ni cat.}}$$

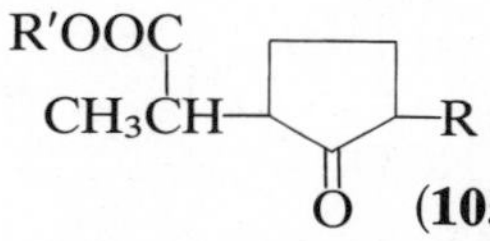

R'OOC–CH(CH$_3$)–(2-oxocyclopentyl, R-substituted) + HCl (109)

(**105**)

Neutral products containing hydrogen in place of the carboxyl are also present in minor amounts. Neither *cis* nor *trans* internal olefins seem to react appreciably.

The behavior of higher α-olefins parallels perfectly that of higher acetylenes already described in Section II-A. It is probably due to a lower rate of hydrolysis of the acylnickel derivatives containing an α-substituent (as those derived from higher acetylenes or olefins) in respect to the acylnickel derivatives not containing α-substituents (as those derived from acetylene or ethylene). The slow rate of hydrolysis favors the competitive process consisting of ring closure which also is to be regarded as a nucleophilic attack of the double bond at the acyl group. As for cyclic olefins it can be observed that the ability to insert depends on the strain of the double bond. Thus cyclohexene practically does not react with crotyl chloride under the conditions tested, but norbornene is

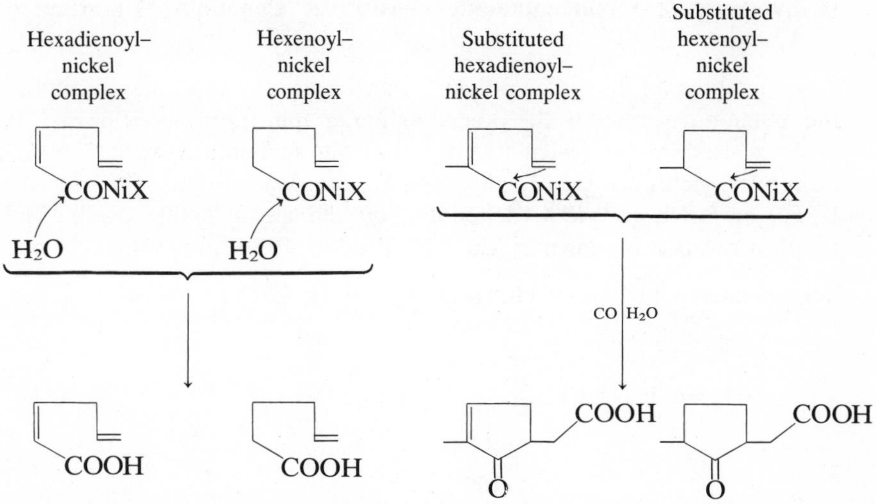

Hexadienoyl–nickel complex | Hexenoyl–nickel complex | Substituted hexadienoyl–nickel complex | Substituted hexenoyl–nickel complex

CONiX, H_2O | CONiX, H_2O | CONiX | CONiX

↓ | ↓ CO | H_2O

COOH | COOH | COOH (cyclopentanone, O) | COOH (cyclopentanone, O)

highly reactive giving rise to a series of products (**106a–f**, R = crotyl R′ = H or COOH) whose manner of formation follows the pattern already discussed throughout the chapter.

Most of the product, however, consists of short-chain polymers probably resulting from the alternate insertion of norbornene and CO. Products **106e** and **106f** probably correspond to the first step of the cooligomerization (277).

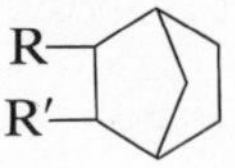

(**106**) 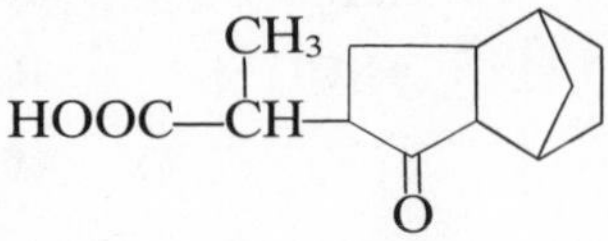 (**106a**) 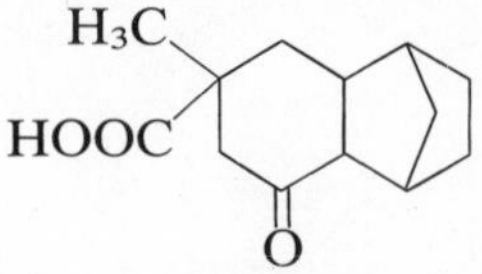 (**106b**)

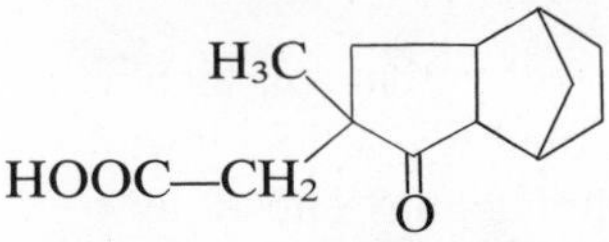

(**106c**) 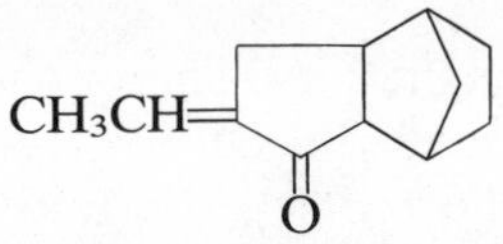 (**106d**) 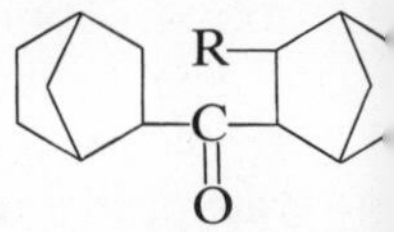 (**106e**)

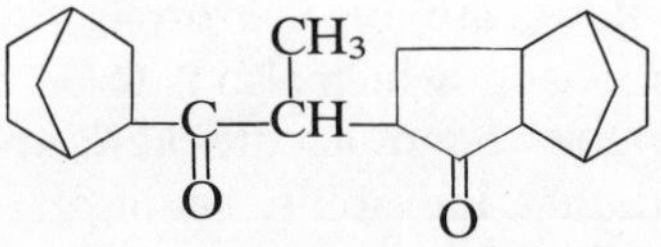

(**106f**)

H. Synthesis of Cyclopentanonic Derivatives Containing Unsaturated Chains

Nonconjugated dienes or polyenes can be used with similar results in the reaction described in the preceding paragraph. If, however, hexa-1,5-diene is used, yields (calculated for the sole acid) reach higher values. Moreover, yields are greatly improved by addition of some salts like KF, KPF_6, or $SnCl_2$ (10–40% depending on the allylic halide used). The simplest reaction is shown by Eq. 110. Product **107** is predominantly one

$$CH_2{=}CHCH_2Cl + CH_2{=}CHCH_2CH_2CH{=}CH_2 + 2CO + H_2O \xrightarrow{\text{Ni cat.}}$$

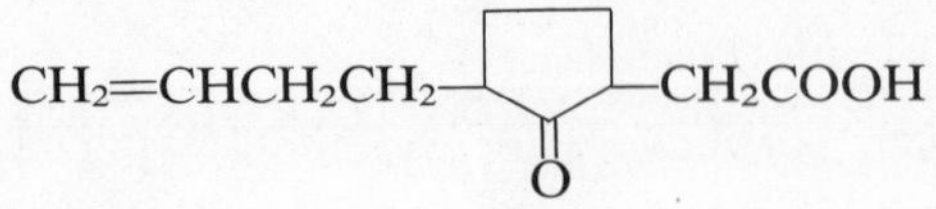

$$CH_2{=}CHCH_2CH_2\text{–}(\text{cyclopentanone})\text{–}CH_2COOH + HCl \qquad (110)$$

(**107**)

of the two possible diastereoisomers (mp 42–43°C). With a deficiency of CO increasing amounts of a ketone **107a** are formed, resulting from the stoichiometric reaction 111:

$$CH_2{=}CHCH_2Cl + CH_2{=}CHCH_2CH_2CH{=}CH_2 + CO + HCl + Ni \rightarrow$$

$$CH_2{=}CHCH_2CH_2\text{–}[\text{cyclopentanone}]\text{–}CH_3 + NiCl_2 \qquad (111)$$

(**107a**)

Also in this reaction one diastereoisomer of **107** is found to predominate (~95%); another minor product is **108**:

$$CH_2{=}CHCH_2CH_2\text{–}[\text{cyclopentanone}]\text{–}CH_2COCH_2CH{=}CH_2$$

(**108**)

With methallyl chloride the main product is the expected **109**. Among the

$$CH_2{=}C(CH_3)CH_2CH_2\text{–}[\text{cyclopentanone}]\text{–}CH_2COOH \quad \text{(m.p. 66–67°)}$$

(**109**)

neutral products, **110**, **111**, and **112** are present. The cyclohexenonic

$$CH_2{=}C(CH_3)CH_2CH_2\text{–}[\text{cyclopentanone}]\text{–}CH_3$$

(**110**)

$$H_3C\text{–}[\text{cyclohexenone}]\text{–}CH_2CH_2CH{=}CH_2$$

(**111**)

$$CH_2{=}C(CH_3)CH_2CH_2\text{–}[\text{cyclopentanone}]\text{–}CH_2COCH_2\text{—}C(CH_3){=}CH_2$$

(**112**)

derivative corresponds to the cyclization on the methallyl double bond without subsequent carbonylation (probably prevented by steric effects) but with hydrogen elimination. From crotyl chloride are obtained products **113** and **114**, corresponding to the two possible ways of cyclization of the open-chained acylnickel intermediate first formed, namely on the originally allylic double bond and on the other double bond of 1,5-hexadiene: The ratio between **113** and **114** is about 1:3. The first is a

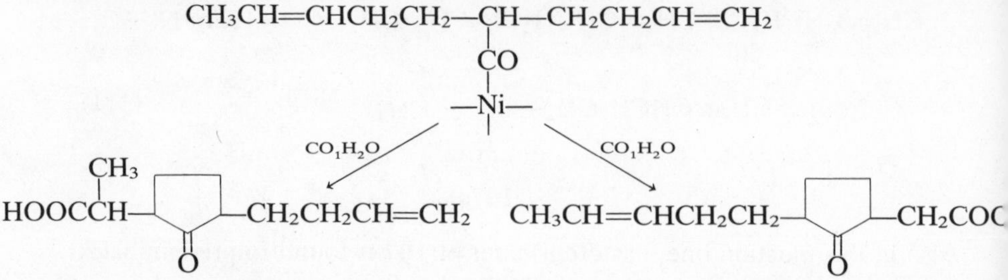

$$CH_3CH{=}CHCH_2CH_2{-}\underset{\underset{\underset{|}{-Ni-}}{|}}{\underset{CO}{\underset{|}{CH}}}{-}CH_2CH_2CH{=}CH_2$$

$\xrightarrow{CO, H_2O}$ (both branches)

$$HOOC\overset{CH_3}{\overset{|}{C}}H{-}[\text{cyclopentanone}]{-}CH_2CH_2CH{=}CH_2 \quad \textbf{(113)}$$

$$CH_3CH{=}CHCH_2CH_2{-}[\text{cyclopentanone}]{-}CH_2COOH \quad \textbf{(114)}$$

mixture of two diastereoisomers, one of which generally predominates (~90%). The second was isolated as a solid, mp 52°C.

The neutral products are also of two kinds (**115** and **116**):

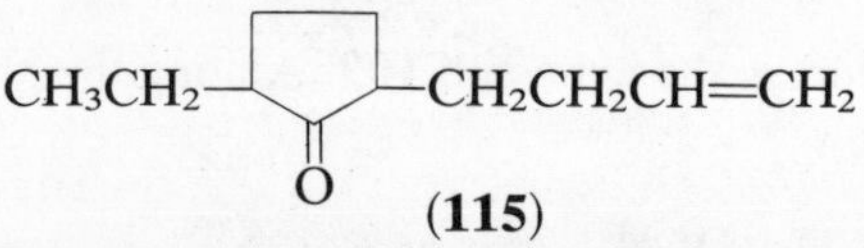

$$CH_3CH_2{-}[\text{cyclopentanone}]{-}CH_2CH_2CH{=}CH_2 \quad \textbf{(115)}$$

$$CH_3CH{=}CHCH_2CH_2{-}[\text{cyclopentanone}]{-}CH_3 \quad \textbf{(116)}$$

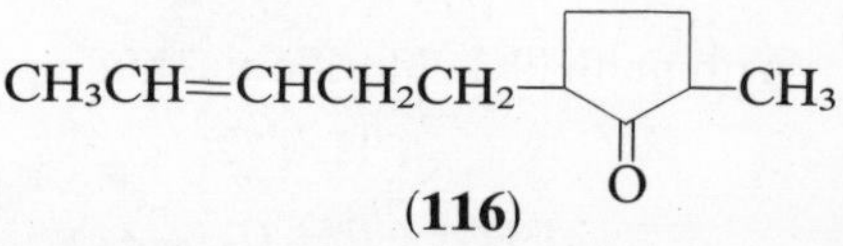

Products like **117** and **118** are also present.

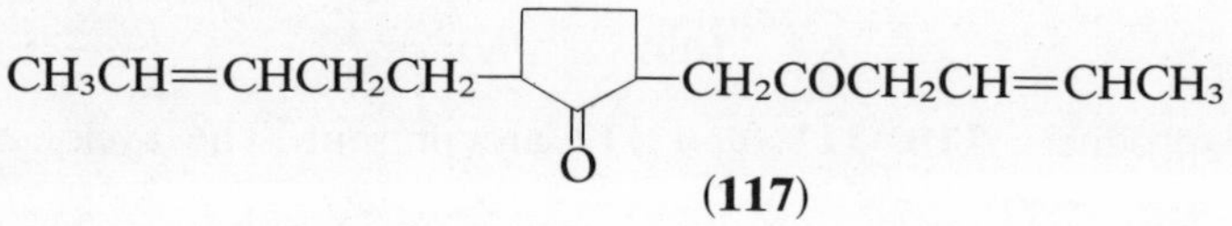

$$CH_3CH{=}CHCH_2CH_2{-}[\text{cyclopentanone}]{-}CH_2COCH_2CH{=}CHCH_3 \quad \textbf{(117)}$$

$$CH_3CH{=}CHCH_2CH_2{-}[\text{cyclohexenone}] \quad \textbf{(118)}$$

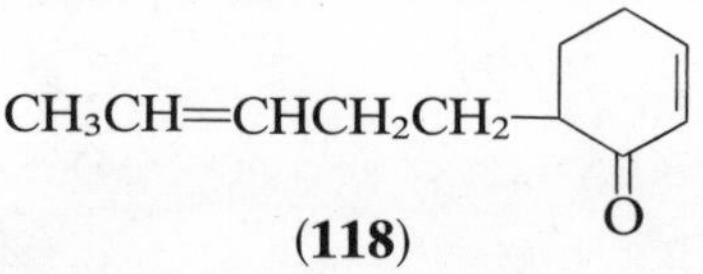

With substituted 1,5-dienes, a surprising lack of reactivity is observed. Thus, octa-2,6-diene and 2,5-dimethylhexa-1,5-diene do not react appreciably. On the contrary, 3-methylhepta-1,5-diene gives the expected products **119**, **120**, and **121**. These products easily form even if crotyl

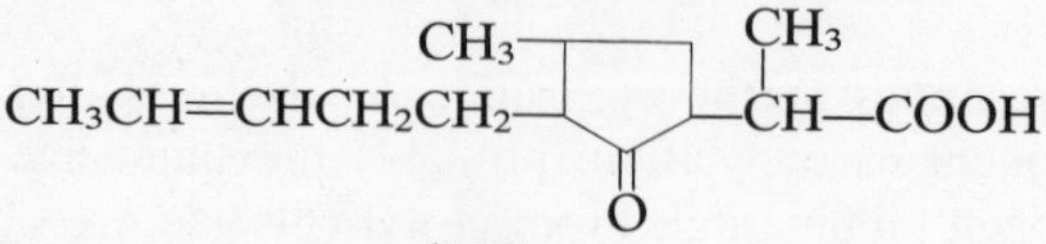

$$CH_3CH{=}CHCH_2CH_2{-}[\text{cyclopentanone}, CH_3]{-}\overset{CH_3}{\overset{|}{C}}H{-}COOH \quad \textbf{(119)}$$

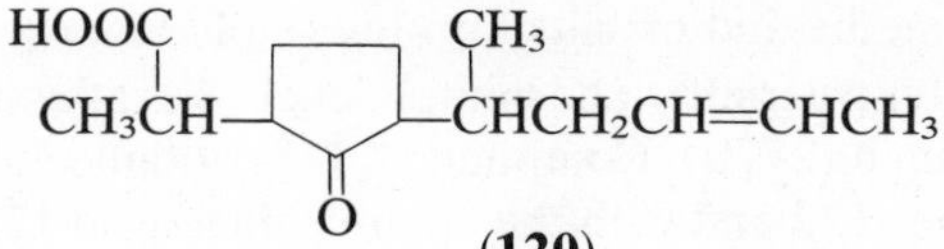

(**120**)

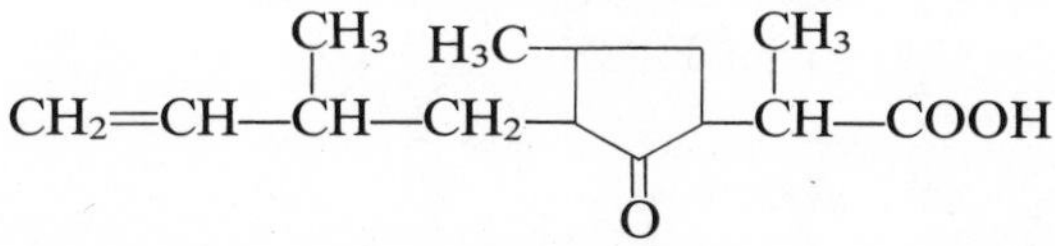

(**121**)

chloride alone is used, because the latter gives two coupling products with $Ni(CO)_4$: octa-2,6-diene (70%), which is not reactive, and 3-methylhepta-1,5-diene, which gives the above products. The same behavior is observed for allyl chloride, whose coupling product, hexa-1,5-diene, easily forms and reacts to give the cyclopentanonic derivatives.

Scheme **III** depicts the hexadiene insertion reaction. A cationic nickel complex is postulated as the intermediate on the basis of the positive

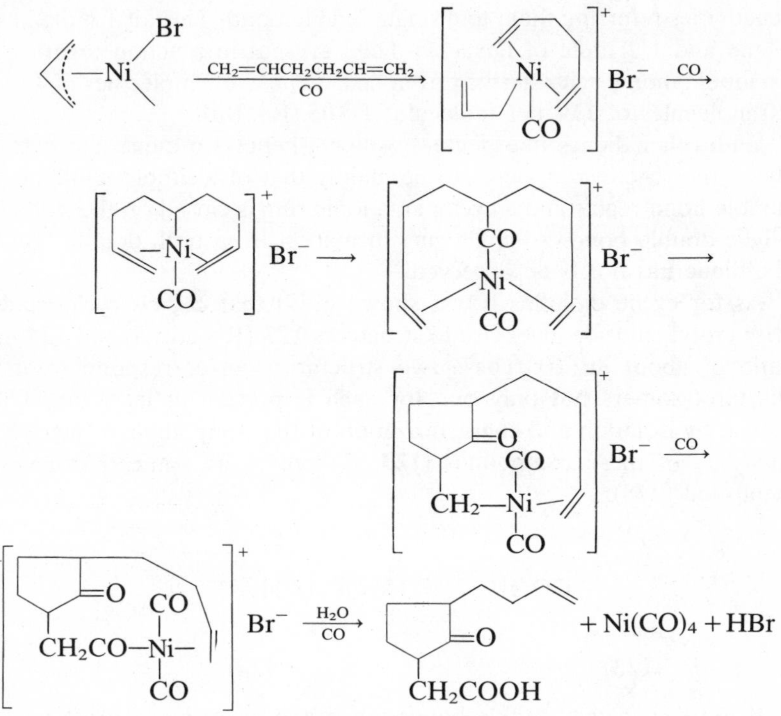

Scheme **III**

effect of polar media and of salts forming stable anions. A similar kind of complex π-allyl(tristriethylphosphite)nickel hexafluorophosphate was isolated by Tolman (277*a*). Compound **107** is obtained along with a cyclohexanonic isomer **122** and with the open-chained acid **122a** corresponding

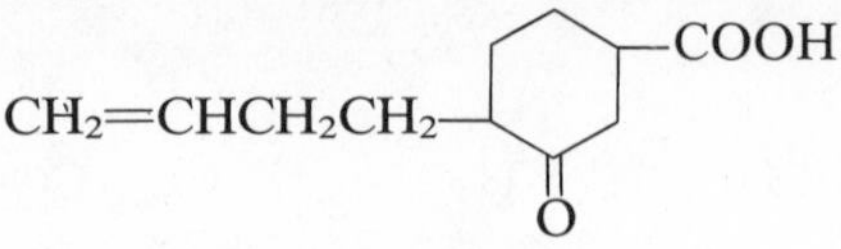

$CH_2{=}CHCH_2CH_2$– (cyclohexanone ring bearing –COOH and C=O)

(**122**)

$$CH_2{=}CHCH_2CH_2{-}CH(COOH){-}CH_2CH_2CH{=}CH_2$$

(**122a**)

to the intermediate stage before cyclization. The reactivity of the dienes formed by coupling of allyl groups always causes the formation of products in competition with those that can be derived from monoolefins. The latter are by far less reactive even if one compares the respective reactivities referring them to a single double bond. Thus, if 1 mole of hex-1-ene and 1/2 mole of hexa-1,5-diene are put in reaction together, the cyclopentanonic acids derived from hexadiene are 7 molecules of **113** and 20 molecules of **114** per molecule of **105** (R = Bu).

With other dienes like penta-1,3-diene, hepta-1,6-diene and octa-1,7-diene the reactivity appears to be mainly that of a simple α-olefin: One double bond reacts and a cyclopentanonic ring is closed on the side of the allylic double bond. A single ring formation from both double bonds of the diene has not been observed.

As for cyclic diolefins it was shown (277) that *cis, cis*-cyclooctadiene with crotyl chloride gives rise to structures **123** (R = crotyl) and **124** in the ratio of about 80:20. These two structures can correspond to several diastereoisomers but only one for each is present at least for 80% as shown by isolation and characterization of the compounds formed. X-ray analyses of these compounds (**123**, R = crotyl, as semicarbazone) were published (278).

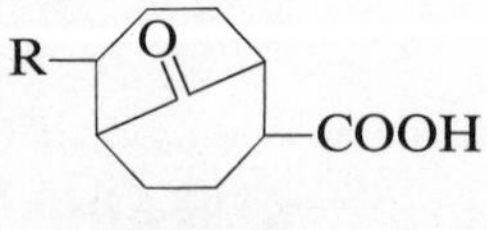

R– (bicyclic ketone) –COOH HOOC–CH(CH_3)– (bicyclic ketone with cyclooctene ring, C=O)

(**123**) 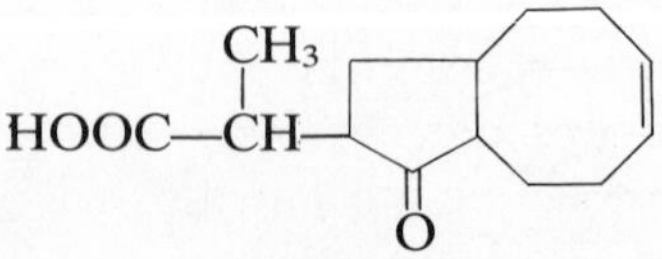 (**124**)

If more than two double bonds are in the inserting olefin, a new and interesting situation arises. The double bonds must be in the 1,5-position.

With *trans*-1,5,9-decatriene and crotyl chloride, products **125**, **125a**, **125b**, **125c**, and probably **125d** (R = crotyl) are formed. Product **125** is the major one, but the presence of substantial amounts of **125c**, are

$$RCH_2\text{-}[C_5H_7(=O)]\text{-}[C_5H_7(=O)]\text{-}CH_2\text{—}COOH$$

(**125**)

$$RCH_2\text{-}[C_5H_7(=O)]\text{-}CH(COOH)\text{—}CH_2CH_2CH\text{=}CH_2$$

(**125a**)

$$HOOC\text{—}CH(CH_3)\text{-}[C_5H_7(=O)]\text{-}CH_2CH_2CH\text{=}CHCH_2CH_2CH\text{=}CH_2$$

(**125b**)

$$(CH_2\text{=}CHCH_2CH_2\text{-})_2[C_5H_6(=O)]\text{-}CH(CH_3)\text{—}COOH$$

(**125c**)

$$CH_2\text{=}CHCH_2CH_2CH(R)\text{-}[C_5H_7(=O)]\text{-}CH_2COOH$$

(**125d**)

indicative of the special activity of the triene. The insertion of an internal double bond was not observed with 1,5-diolefins. In the case of compound **125c** it can be attributed to the higher coordination power of the triene ligand. It clearly appears from the results of the double-bond insertion reactions that the geometry of the molecules to be inserted into the allylnickel bond is of fundamental importance. This is in accord with the observed stereoselectivity.

Another important factor is also to be added to those affecting the insertion of a molecule: the assistance of another chelating group in addition to that which occupies the coordination site *cis* to the allyl group. The existence of these "assisted" insertions is clearly shown by the competition experiment cited before, in which hex-1-ene and hexa-1,5-diene were caused to react with crotyl chloride. If only the products deriving from the ring closure on the site of the allylic double bond are compared, it clearly appears that only chelation can justify the by far larger amount of product obtained from hexa-1,5-diene. Chelation effects are favored by the formation of cationic complexes of nickel. This can be achieved by the appropriate use of stable anions and of polar solvents. Also in hydrocarboxylation reactions with Ni (278*a*) and Pd (278*b*) complexes, 1,5-dienes were shown to react preferentially.

It is worth recalling the type of anchimeric assistance of a double bond or oxygen atom in the selective formation of allylic acetates from palladium complexes described in Section I-d (80). Whereas π-crotyl complexes of palladium acetate give a mixture of two isomers, the presence of a double bond or of oxygen in an appropriate position leads to the selective formation of one isomer.

I. Synthesis of Alkenylsuccinic Acids from Allyl Halides by Carbonylations with Double-Bond Insertion

It was found (279,280) that a double bond can be inserted between the allyl group and CO if present in the β,γ-position of an unsaturated acid. The reaction is carried out at 40–60°C in acetone–water in the presence of $Ni(CO)_4$ and a soluble salt of the unsaturated acid, according to reaction 112, which is written for the simplest members.

$$CH_2{=}CHCH_2Cl + CH_2{=}CHCH_2COONa + CO + H_2O \xrightarrow{Ni(CO)_4}$$

$$CH_2{=}CHCH_2CH_2\underset{\displaystyle COOH}{\underset{|}{CH}}{-}CH_2COOH + NaCl \qquad (112)$$

The reactions can be extended to higher allylic halides and to other β,γ-unsaturated acids; the α,β- and γ,δ-unsaturated acids are not reactive under the conditions tested so far.

Terminally unsubstituted β,γ-unsaturated acids are preferred to substituted ones. Products deriving from the reverse addition of the allyl group and CO to the double bond are found only in very limited amount. The allyl group always seems to prefer the γ-carbon atom.

Under the same conditions, allyl halides are able to give β,γ-unsaturated acids, and the latter can react according to reaction 112. Thus it is possible to start directly from allyl halides to form alkenylsuccinic acids (Eq. 113). The ease with which the β,γ-unsaturated acid is

$$2CH_2{=}CHCH_2Cl + 2CO + 2H_2O \xrightarrow[MgO]{Ni(CO)_4} CH_2{=}CHCH_2CH_2\underset{\displaystyle COOH}{\underset{|}{C}H}CH_2COOH + 2HCl \quad (113)$$

inserted depends greatly on the presence of substituents in the γ-position. This is clearly shown by the reaction of crotyl chloride with itself. The simple carbonylation of crotyl chloride mainly leads to pent-3-enoic acid; only limited amounts ($<0.5\%$) of the isomeric 2-methylbut-3-enoic acid are formed (Eq. 114). When, however, crotyl chloride is

$$CH_3CH{=}CHCH_2Cl + CO + H_2O \begin{cases} \nearrow CH_3CH{=}CHCH_2COOH \\ \searrow CH_2{=}CH{-}\overset{\displaystyle CH_3}{\overset{|}{C}H}{-}COOH \end{cases} \quad (114)$$

reacted in the presence of MgO to form an alkenylsuccinic acid, about 20% of the product corresponding to insertion of the branched acid is formed (reaction 115). Another aspect of reaction 115 worth mentioning

$$2CH_3CH{=}CHCH_2Cl + 2CO + 2H_2O \begin{cases} \nearrow CH_3CH{=}CHCH_2\overset{\displaystyle CH_3}{\overset{|}{C}H}{-}\underset{\displaystyle COOH}{\underset{|}{C}H}{-}CH_2COOH \quad \text{(80\% of two diastereoisomers)} \\ \searrow CH_3CH{=}CHCH_2CH_2\underset{\displaystyle COOH}{\underset{|}{C}H}{-}\overset{\displaystyle CH_3}{\overset{|}{C}H}{-}COOH \quad \text{(20\% of two diastereoisomers)} \end{cases} \quad (115)$$

is the formation of succinic anhydrides. If the reaction is carried out in a mixture of two solvents, one miscible and the other nonmiscible with water, it is possible to isolate substantial amounts of anhydrides (280,280*a*). The mechanism of the reaction can be represented for the simplest case by Scheme **IV** showing the synthesis of but-3-enylsuccinic acid.

$$CH_2{=}CHCH_2Cl + Ni(CO)_4 \longrightarrow (\pi\text{-}C_3H_5)Ni(CO)_2Cl \xrightarrow{CH_2=CHCH_2COONa}$$

$$(\pi\text{-}C_3H_5)Ni(CO)_2{-}OOCCH_2CH{=}CH_2 \longrightarrow \text{[chelate: } \sigma\text{-allyl–}Ni(CO)_2\text{–O–C(=O)CH}_2CH{=}CH_2\text{]}$$

$$\xrightarrow{CO} \text{[Ni(CO)}_2\text{ chelate of butenylsuccinyl]} \xrightarrow{CO} \text{[butenylsuccinic anhydride]} \xrightarrow{H_2O}$$

$$\xrightarrow{CO,\,H_2O} CH_2{=}CHCH_2CH_2CH(COOH)CH_2COOH + Ni(CO)_4$$

Scheme (**IV**)

According to Scheme **IV** the double bond of vinylacetic acid should occupy a coordination site previously held by one of the two ends of the allyl group. A $\pi \rightarrow \sigma$ interconversion was already postulated for insertion of acetylene (see Introduction and the recent review on nmr by Fedorov (281)). It is worth noting that the insertion of the double bond occurs even in the presence of acetylene. This suggests that the geometric conditions obtainable by a chelating group such as the vinylacetic are such that the latter can compete successfully with acetylene for insertion. The assumption that vinylacetic acid is contained in the complex as a nickel salt is supported by the reaction of allyl vinylacetate with $Ni(CO)_4$ in the same acetone–water mixture. Butenylsuccinic acid is formed in significant amount (Eq. 116.):

$$CH_2{=}CHCH_2OOCCH_2CH{=}CH_2 + Ni(CO)_4 \longrightarrow (\pi\text{-}C_3H_5)Ni(CO)_2{-}OCOCH_2CH{=}CH_2 \xrightarrow{H_2O} \tag{116}$$

$$CH_2{=}CHCH_2CH_2{-}\overset{COOH}{\overset{|}{C}H}{-}CH_2COOH$$

Another point refers to the formation of alkenylsuccinic anhydrides. Their formation in aqueous solution offers a means for elucidating a general problem connected with the hydrolysis of acyl–metal bonds. We previously had observed that anhydrides formed even in methanolic solution during hydrolysis of acryloylnickel complexes in the presence of acetic acid (138). Mixed anhydrides were formed, but it had not been possible to establish whether the anhydrides were intermediates or products of the competition of acetic acid with methanol for the acyl group. The formation of butenylsuccinic anhydride offered a means of settling this problem. This anhydride was hydrolyzed in methanol, and then transformed into methyl butyl esters (Eq. 117).

CO, Ni, O, CO — $\xrightarrow{CH_3OH}$ COOH, $COOCH_3$ **(A)**

$\xrightarrow{CH_3OH}$ (A) or (B) (117)

$COOCH_3$, COOH **(B)**

The two possible esters were in the ratio **A** : **B** = 20 : 80. The carbonylation of allyl chloride to butenylsuccinic half-esters was then carried out in methanol. The half-esters were converted into methyl butyl esters. This time their ratio was 86 : 14. If the reaction passed through the anhydride as intermediate, the same ratio should have been obtained as was found for the hydrolysis of the anhydride. On the other hand only product **A** should have been formed if the products were not derived from the anhydride. The result shows that the anhydride is formed partly as a result of competition for the acyl group between methanol and the carboxylate group. The latter is present in lower concentration, but it is favored by its proximity to the acyl group and by the tendency to form the five-membered ring of the succinic anhydride.

J. Carbonylation of Butadiene

A later article (282) describes the reaction of butadiene with CO (see Section II-A-7) in ethyl alcohol in the presence of palladium acetylacetonate and PPh_3 to give ethyl nona-3,8-dienoate (Eq. 118),

$$2CH_2{=}CHCH{=}CH_2 + CO + C_2H_5OH \xrightarrow{Pd}$$

$$CH_2{=}CHCH_2CH_2CH_2CH{=}CHCH_2COOC_2H_5 \qquad (118)$$

together with 1-ethoxyocta-2,7-diene and 3-ethoxyocta-1,7-diene.

Tsuji and co-workers (283) found that the absence of halide coordinated to palladium is essential for nonadienoate being obtained. In the presence of halide pent-3-enoate is formed. An intermediate bis(π-allyl) palladium complex formed from two molecules of butadiene is postulated:

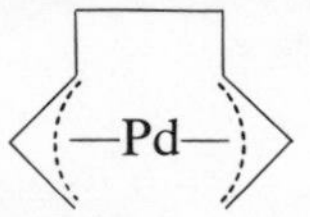

—Pd—

An alternative hypothesis is based on the attack of a coordinated carbomethoxy group on butadiene.

The reaction of allylic compounds with acetylene and CO was applied to the coordinated chain of butadiene oligomers which are known to be obtained on π-allylnickel halides. If the coordinated chain which also is a π-allylnickel complex is caused to react with CO, acetylene and water or methanol, long chain unsaturated acids or esters are formed (284).

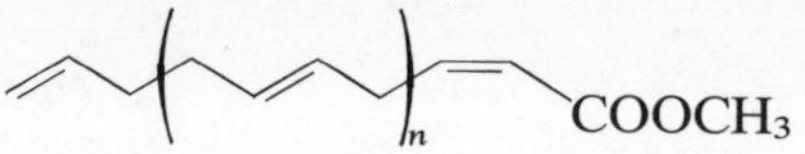

n $COOCH_3$

Preparative aspects of the carbonylation of allyl halides at atmospheric pressure were reviewed (285).

K. Synthesis of Allyl Sulfonates and of Diallyl Sulfones

Allyl halides have been caused to react with $Ni(CO)_4$ in methanol at room temperature in the presence of SO_2. Allyl sulfonates and diallyl sulfones were obtained (286) according to these equations:

$$CH_2{=}CHCH_2Cl + SO_2 + CH_3OH \xrightarrow{Ni(CO)_4} CH_2{=}CHCH_2SO_3CH_3$$

$$2CH_2{=}CHCH_2Cl + SO_2 + Ni(CO)_4 \longrightarrow$$

$$(CH_2{=}CHCH_2)_2SO_2 + NiCl_2 + 4CO$$

For a review of SO_2 insertion see (61*b*).

L. Synthesis of Benzaldoxime

An anionic complex of cobalt was caused to react with benzyl chloride at room temperature to give benzaldoxime (287):

$$PhCH_2Cl + NBu_4[Co(NO)(CO)_2I] \rightarrow PhCH_2NO + [Co(CO)_2I] + NBu_4Cl$$

V. CONCLUSIONS

The chemistry of allylic complexes appears extremely versatile—it gives rise to a large number of new reactions that provide an easier way for preparing known compounds, and, in certain cases, these reactions offer a pathway to the synthesis of new classes of compounds.

Most of the reactions described throughout the chapter can be rationalized with the aid of the general Scheme **V**. In this scheme three different stages are distinguished: oxidative addition, insertion, and termination. Oxidative addition consists of addition of the metal to an undissociated molecule (H_2, alkyl halides, etc.) which on receiving electrons from the metal is dissociated into two fragments, one or both of which become bonded to the metal (75–77,135). This reaction has been compared to the general acid–base reaction, in which the metal acts as a base (269–270*b*,*h*). Concerted three-center additions were also proposed (270*c*). Incursions of radical mechanism were detected in many cases (270*d*,*e*,*f*).

The insertion step can be seen as a nucleophilic addition of a σ-coordinated group which migrates on another ligand. Evidence for the generality of this mechanism has still to be provided, however (137). Insertion on nickel complexes possibly involves pentacoordinated intermediates (57). The process can be repeated several times.

Finally, the termination step is the process by which the bond between the metal and the organic group is cleaved by substitution and reduction or by reductive elimination thus giving rise to a metallic species able to start a new cycle or by going to a metallic species in a higher oxidation state not able to act as a catalyst. This latter way can be the consequence of a substitution reaction, which may be regarded as an electrophilic substitution at the metal-bonded atom, or as an elimination reaction.

In Table 6 several examples of addition reactions on allylnickel halides, comprising these three steps, are classified according to the number of new bonds formed.

Scheme **V** is illustrated by the following three syntheses involving three

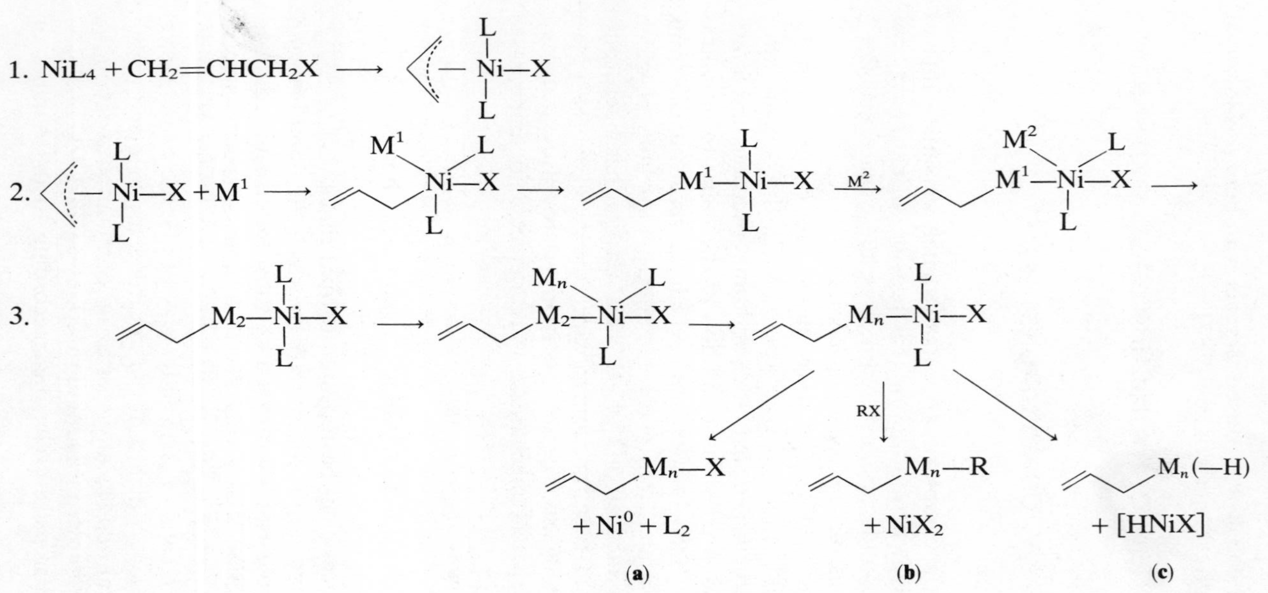

1. $NiL_4 + CH_2{=}CHCH_2X \longrightarrow (\eta^3\text{-allyl})Ni(L)_2X$

2. $(\eta^3\text{-allyl})Ni(L)_2X + M^1 \longrightarrow (CH_2{=}CHCH_2)Ni(M^1)(L)_2X \longrightarrow CH_2{=}CHCH_2{-}M^1{-}Ni(L)_2{-}X \xrightarrow{M^2} CH_2{=}CHCH_2{-}M^1{-}Ni(M^2)(L)_2{-}X \longrightarrow$

3. $CH_2{=}CHCH_2{-}M_2{-}Ni(L)_2{-}X \longrightarrow CH_2{=}CHCH_2{-}M_2{-}Ni(M_n)(L)_2{-}X \longrightarrow CH_2{=}CHCH_2{-}M_n{-}Ni(L)_2{-}X$

(a) $CH_2{=}CHCH_2{-}M_n{-}X + Ni^0 + L_2$

(b) $\xrightarrow{RX}$ $CH_2{=}CHCH_2{-}M_n{-}R + NiX_2$

(c) $CH_2{=}CHCH_2{-}M_n(-H) + [HNiX]$

1. Oxidative addition; 2. stepwise insertion; 3. termination (a) by nucleophiles, (b) by electrophiles, (c) by elimination of HX.

$X = Cl, Br, I, OSO_2Ph, OAc$, etc. $L = CO$ or other ligands

$R = H$ or alkyl $M =$ Molecules to be inserted (L included)

Scheme **V** Some general patterns of stepwise reactions on nickel

TABLE 6

Stepwise Insertions Involving Allylnickel Halides

Substrate for initial reaction	Added compound	Nickel species formed in termination step	Products
	Number of new bonds formed: 1		
$CH_2{=}CHCH_2X$ ($X = Cl$, Br, I, OSO_2Ph OAc, etc.)	$CH_2{=}CHCN$	[H—Ni—X]	*trans*-$CH_2{=}CHCH_2CH{=}CHCN$
	Number of new bonds formed: 2		
$CH_2{=}CHCH_2X$	$HC{\equiv}CCOOCH_3$, HX	NiX_2	*trans*-$CH_2{=}CHCH_2CH{=}CHCOOCH_3$
$CH_2{=}CHCH_2X$	CO, ROH ($R = H$, alkyl)	Ni^0	$CH_2{=}CHCH_2COOR$
$CH_2{=}CHCH_2X$	$CH_2{=}CHCN$, H_2O	Ni(OH)X	$CH_2{=}CHCH_2CH_2CH_2CN$
$H_3COOCCH{=}CHCH_2X$	CH_3COCH_3, H_2O	Ni(OH)X	$CH_2{=}CH{-}C(COOCH_3){-}C(CH_3)_2{-}OH$
	Number of new bonds formed: 3		
$CH_2{=}CHCH_2X$	$HC{\equiv}CH$, CO, ROH	Ni^0	*cis*-$CH_2{=}CHCH_2CH{=}CHCOOR$
$CH_2{=}CHCH_2X$	$H_2C{=}CH_2$, CO, ROH	Ni^0	$CH_2{=}CHCH_2CH_2CH_2COOR$
$CH_2{=}C(CH_3){-}CH_2X$	$PhC{\equiv}CH$, CO, cyclization	[H—Ni—X]	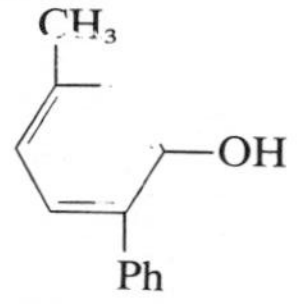 CH_3, OH, Ph
$CH_2{=}CHCH_2X$	$CH_2{=}CHCH_2COOH$, CO, ROH	Ni^0	$CH_2{=}CHCH_2CH_2CH(COOR)CH_2COOH$

TABLE 6 (*cont'd.*)

Substrate for initial reaction	Added compound	Nickel species formed in termination step	Products
Number of new bonds formed: 3 (contd.)			
$CH_2{=}CHCH_2X$	$HC{\equiv}CH$, $HC{\equiv}CH$, $CH_2{=}CHCH_2X$	NiX_2	*cis,cis*-$CH_2{=}CHCH_2$-$CH{=}CH{-}CH{=}CHCH_2CH{=}CH_2$
Number of new bonds formed: 4			
$CH_2{=}C(CH_3){-}CH_2X$	$HC{\equiv}CH$, CO, cyclization, H_2O	Ni(OH)X	CH_3; =O
$H_3COOCCH{=}CHCH_2Br$	$PhC{\equiv}CH$, CO, cyclization, H_2O	Ni(OH)X	Ph; CH_2COOCH_3; O
Number of new bonds formed: 5			
$CH_2{=}CHCH_2X$	$HC{\equiv}CH$, CO, cyclization, CO, H_2O	Ni^0	CH_2COOH; O
$CH_2{=}C(CH_3){-}CH_2X$	$HC{\equiv}CH$, CO, cyclization, cyclization, H_2O	Ni(OH)X	CH_3; O
$CH_2{=}CHCH_2X$	$HC{\equiv}CH$, CO, cyclization, CO, $CH_2{=}CHCH_2X$	NiX_2	$CH_2COCH_2CH{=}CH_2$; O

$CH_3CH{=}CHCH_2X$	$H_2C{=}CH_2$, CO, cyclization, CO, H_2O	Ni^0	CH_3, COOH, O
$CH_3CH{=}CHCH_2X$	, CO, cyclization, CO, H_2O	Ni^0	CH_3, COOH, O ; CH_3, COOH, O

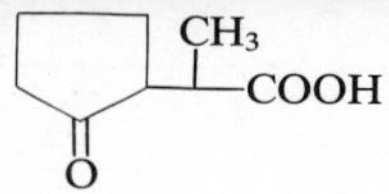

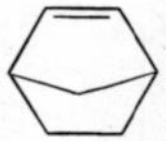

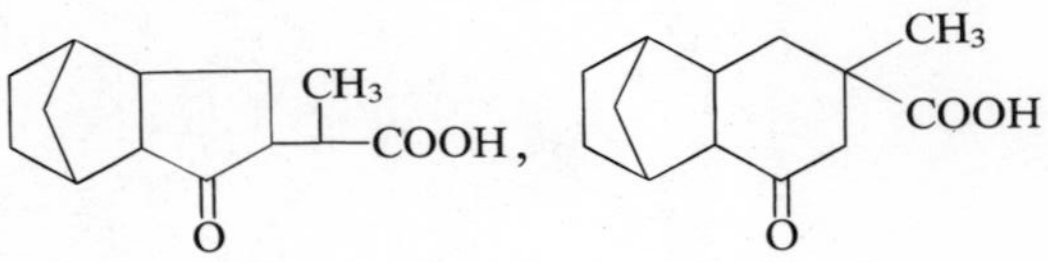

Number of new bonds formed: 6

$CH_2{=}CHCH_2X$	$PhC{\equiv}CH$, CO, cyclization, CO, $PhC{\equiv}CH$, HX	NiX_2	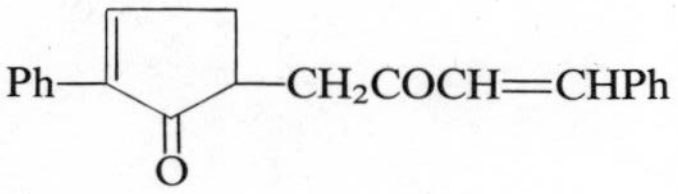 Ph—, $CH_2COCH{=}CHPh$, O
$CH_2{=}CHCH_2X$	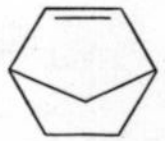 , CO, cyclization, CO, 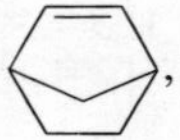 , HX	NiX_2	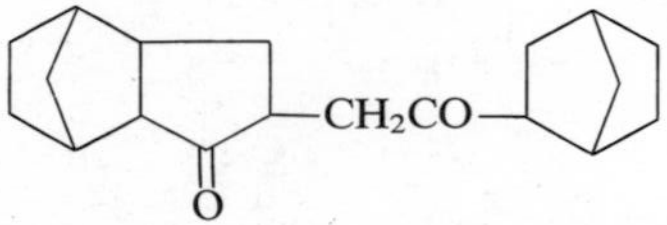 CH_2CO—, O

Number of new bonds formed: 7

$CH_2{=}C(CH_3){-}CH_2X$	$CH_2{=}CHCH_2CH_2CH{=}CHCH_2$-$CH_2CH{=}CH_2$, CO, cyclization, CO, cyclization, CO, H_2O	Ni^0	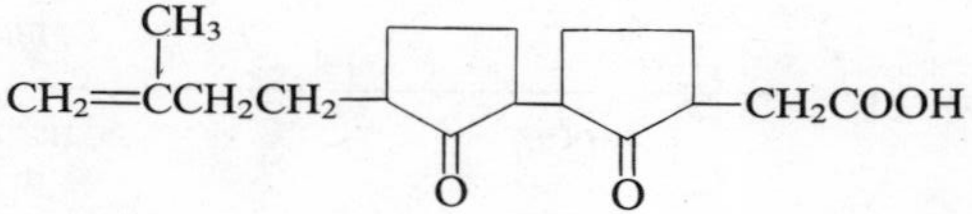 $CH_2{=}C(CH_3)CH_2CH_2$—, O, O, —CH_2COOH

TABLE 6 (*cont'd.*)

Substrate for initial reaction	Added compound	Nickel species formed in termination step	Products
Number of new bonds formed: 7 (contd.)			
$CH_2{=}CHCH_2X$	$H_2C{=}CH_2$, CO, $H_2C{=}CH_2$, CO, $H_2C{=}CH_2$, CO, H_2O	Ni^0	$CH_2{=}CH(CH_2)_3CO(CH_2)_2$-$CO(CH_2)_2COOH$
Number of new bonds formed: 8			
$CH_2{=}CHCH_2X$	$C_4H_9C{\equiv}CH$, CO, cyclization $C_4H_9C{\equiv}CH$, CO, cyclization, H_2O	Ni(OH)X	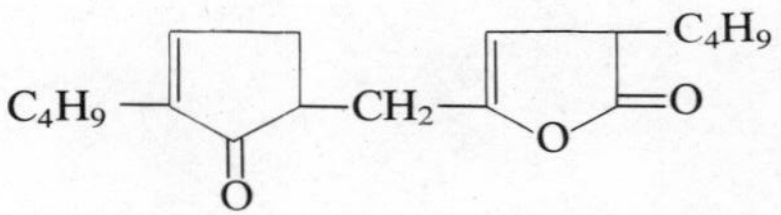 C_4H_9, CH_2, O, O, O, C_4H_9
$CH_3CH{=}CHCH_2X$	$HC{\equiv}CH$, CO, cyclization, CO $HC{\equiv}CH$, CO, cyclization, H_2O	Ni(OH)X	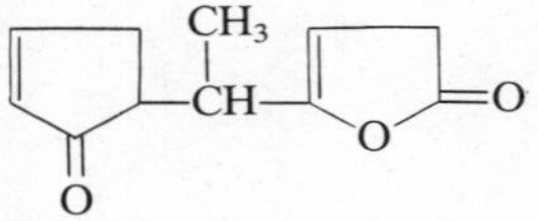 CH_3, CH, O, O, O
Number of new bonds formed: 9			
$CH_2{=}CHCH_2X$	$HC{\equiv}CH$, CO, cyclization, CO, $HC{\equiv}CH$, CO, cyclization, CH_3COCH_3, H_2O	Ni(OH)X	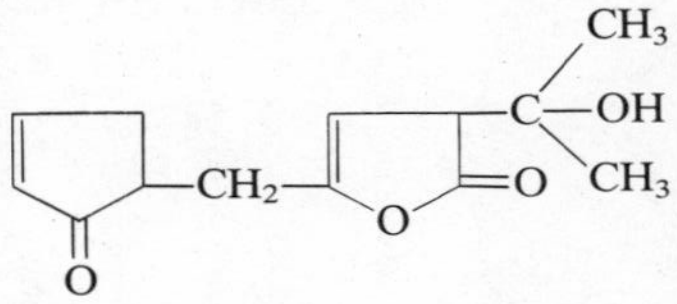 CH_2, O, O, O, C, CH_3, OH, CH_3
$CH_2{=}CHCH_2X$	$HC{\equiv}CH$, CO, cyclization, CO, $HC{\equiv}CH$, $HC{\equiv}CH$, CO, cyclization, H_2O	Ni(OH)X	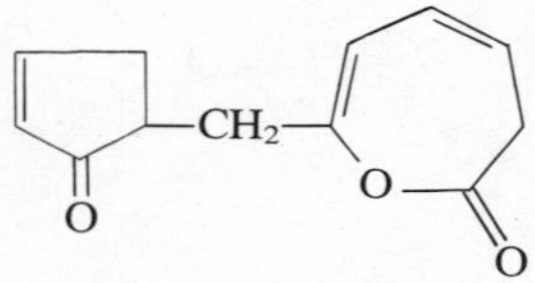 CH_2, O, O, O

different types of termination:

Methyl *cis*-hexa-2,5-dienoate:

1. $Ni(CO)_4 + CH_2{=}CHCH_2X \longrightarrow$ (π-allyl)Ni(CO)(CO)X

2. (π-allyl)Ni(CO)(CO)X $+ HC{\equiv}CH \longrightarrow$ (allyl)(HC≡CH)Ni(CO)(CO)X $\longrightarrow$

   (CH₂=CHCH₂CH=CH)Ni(CO)(CO)X $\xrightarrow{CO}$ (CH₂=CHCH₂CH=CH–C=O)Ni(OC)(CO)X

3. (CH₂=CHCH₂CH=CH–C=O)Ni(OC)(CO)X $\xrightarrow[CH_3OH]{CO}$ $CH_2{=}CHCH_2CH{=}CHC({=}O)OCH_3 + Ni(CO)_4 + HX$

Methyl 3-phenylcyclopenta-3-en-2-on-1-ylacetate:

1. $Ni(CO_4) + H_3COOCCH{=}CHCH_2X \longrightarrow$ (π-allyl–COOCH₃)Ni(CO)(CO)X

2. (π-allyl–COOCH₃)Ni(CO)(CO)X $+ PhC{\equiv}CH \longrightarrow H_3COOCCH{=}CHCH_2CH{=}C(Ph){-}Ni(CO)(CO){-}X \xrightarrow{CO}$

   $H_3COOCCH{=}CHCH{=}C(Ph){-}CO{-}Ni(CO)(CO){-}X \xrightarrow{CO}$ (3-phenylcyclopent-3-en-2-one-1-yl)CH(COOCH₃)–Ni(CO)(CO)X

3. (3-phenylcyclopent-3-en-2-one-1-yl)CH(COOCH₃)–Ni(CO)(CO)X $\xrightarrow{H_2O}$ (3-phenylcyclopent-3-en-2-one-1-yl)CH₂COOCH₃ $+ Ni(OH)X + 2CO$

Methyl *trans*-hexa-2,5-dienoate:

$$1.\quad Ni(CO)_4 + CH_2{=}CHCH_2X \longrightarrow (\pi\text{-}C_3H_5){-}\underset{CO}{\overset{CO}{Ni}}{-}X$$

$$2\quad (\pi\text{-}C_3H_5){-}\underset{CO}{\overset{CO}{Ni}}{-}X + CH_2{=}CHCOOCH_3 \longrightarrow CH_2{=}CHCH_2CH_2\overset{H_3COOC}{CH}{-}\underset{CO}{\overset{CO}{Ni}}{-}X$$

$$3.\quad CH_2{=}CHCH_2CH_2{-}\overset{H_3COOC}{CH}{-}\underset{CO}{\overset{CO}{Ni}}{-}X \longrightarrow$$

$$trans{-}CH_2{=}CHCH_2CH{=}CHCOOCH_3 + [HNiX]$$

Other concepts of some interest have emerged from the review. The first can be expressed as follows: The metal-bonded organic group determines to a large extent the nature of the molecule that inserts into the metal-organic group bond or cleaves it in the termination step.

Table 6 shows that, although several different molecules are present, only certain types are chosen at each stage of the insertion process.

Let us take the nickel–carbon bond in a complex of type

$$-\overset{|}{\underset{|}{C}}-\overset{|}{\underset{|}{Ni}}-X \ (X = \text{halide})$$

and see how it is cleaved by different molecules according to the nature of the nickel-bonded carbon. If the latter is part of an acyl group, nucleophilic agents such as water, alcohols, and various other anions are best suited to cleave the C–Ni bond:

$$RCO-\overset{|}{\underset{|}{Ni}}-X + H_2O \rightarrow RCOOH + Ni + HX$$

Alkyl, vinyl, or aryl groups generally react with electrophiles:

$$R\text{–}Ni\text{–}X + HX \rightarrow RH + NiX_2$$

Allyl–metal bonds are cleaved both by electrophilic and nucleophilic agents depending on the substituent present on the allyl group. For example, in reaction 119 the allylic ligand is attacked by the methoxyl group, whereas in reaction 120, in which an electron-withdrawing substituent is present on the allylic carbon, it is the proton of methanol that attacks the ligand.

$$(\eta^3\text{-}C_3H_5)Ni(CO)(L)X + CH_3OH \xrightarrow{CO} CH_2{=}CHCH_2OCH_3 + Ni(CO)_3L + HX \qquad (119)$$

$$(\eta^3\text{-}CNC_3H_4)Ni(CO)(L)X + CH_3OH \longrightarrow CH_2{=}CHCH_2CN + Ni(OCH_3)X \qquad (120)$$

Alkyl-nickel bonds are mainly cleaved by hydrogen elimination or by coupling. Hydrogen elimination is favored when R and R′ are electron-withdrawing substituents or when R″ and R‴ sterically hinder the insertion of a new molecule:

$$RCHR'\text{–}CR''R'''\text{–}Ni\text{–}X \rightarrow RCR'=CR''R''' + [HNiX]$$

Terminations by coupling of the following type are also known for all the organic groups coordinated to nickel:

$$R\text{–}Ni\text{–}R' \rightarrow RR' + Ni$$
$$R\text{–}Ni\text{–}X + R'X \rightarrow RR' + NiX_2$$

The type of termination and the ease with which it occurs is profoundly influenced by the nature of the ligands on the metal.

As a consequence of the selective behavior of the molecules able to cleave the metal–organic group bond, another concept ensues: The course of a termination or of an insertion step can be changed by appropriate choice of the metal-bonded organic group and of the molecules able to compete for termination or for insertion with those already present in the reaction medium.

Coupling-type reactions also offer a useful example of competition between displacement and coupling of the allyl group, coupling being favored by placing electron-withdrawing substituents on the allyl group. Very effective means for causing preferential insertion of molecules and groups are offered by variation of the effective charge on the metal and by use of appropriate ligands.

An increasing number of new reactions is being obtained by applying the criterion of "assistance to insertion": Chelating olefins of appropriate geometry are able to insert more easily than simple olefins.

Another concept that deserves mention refers to stereoselectivity: By appropriate choice of substituents, ligands and solvents, it is possible to give to the products the desired spatial arrangement. This offers wide perspectives in the field of the synthesis of natural structures.

NOTE ADDED IN PROOF

The concepts outlined previously have been applied to other olefinic substrates. The most interesting results are summarized here. 3-Vinyl-1,5-hexadiene, crotyl chloride, and CO (with $Ni(CO)_4$ as catalyst) gave a 15% yield of products having two fused rings (with and without a carboxylic group), the other products being monocyclopentanone derivatives (288):

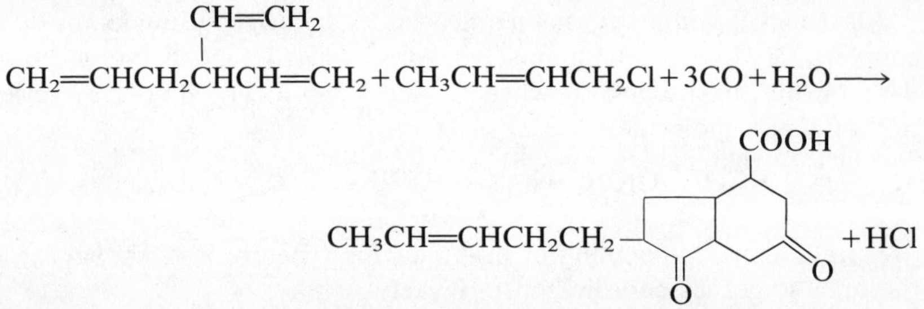

$$CH_2{=}CHCH_2\underset{\displaystyle CH{=}CH_2}{\overset{}{CH}}CH{=}CH_2 + CH_3CH{=}CHCH_2Cl + 3CO + H_2O \longrightarrow$$

$$CH_3CH{=}CHCH_2CH_2\text{–[bicyclic diketone bearing COOH]} + HCl$$

The fact that for the first time a 1,4-pentadiene system has been bridged by CO should be ascribed to the assistance to insertion given by the other two 1,5-hexadiene double bonds. Formation of fused rings is important in view of possible syntheses of analogs of various natural products.

Trans, trans-1,5,9,13-tetradecatetraene with methallyl chloride and CO ($Ni(CO)_4$ as catalyst) was found to give rise selectively to the following compound:

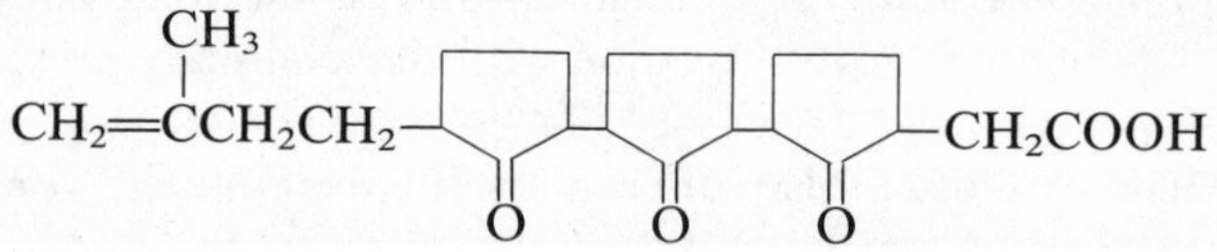

$$CH_2{=}\overset{\displaystyle CH_3}{C}CH_2CH_2\text{–[three linked cyclopentanone rings]–}CH_2COOH$$

The X-ray structure indicates that a stereospecific cyclocooligomerization took place, leading to a regular arrangement of the hydrogen atoms α to the carbonyl functions at opposite sides of the $C{-}\overset{\overset{\displaystyle O}{\|}}{C}{-}C$ planes. The latter are inclined by 71.9° and 72° in respect to each other so that a helicoidal structure results (289).

Endo and *exo*-dicyclopentadiene selectively react only at the strained double bond giving rise to different products, as expected for a stereospecific reaction. The structure of the product obtained from the *endo* form has been determined by X-ray methods. It turns out that a *cis, exo* attack of crotyl group and CO took place at the strained double bond. The cyclopentene double bond appears to be localized, at least to the extent of 70%, in the position that is farthest from the carbonyl group [double-bond length: 1.376(9); single-bond length: 1.439(8)] (290):

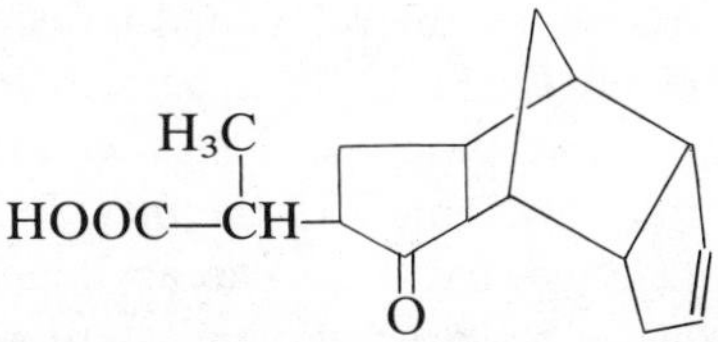

H_3C
HOOC—CH—
O

Strain, chelating power, and minimal steric hindrance on the double bond appear to be the factors necessary, but not sufficient, for insertion. These factors are also important for coordination to several metal ions, as shown, for example, by the formation constants of silver nitrate-olefin complexes (291). Steric hindrance, however, affects insertion much more than coordination ability. What causes insertion to take place easily, once efficient coordination is achieved, is the presence of CO, which has the effect of favoring the first insertion, then the insertion of CO itself, and finally the hydrolytic splitting of the complex, thus pushing the reaction sequence towards an irreversible step. The reaction of an allylnickel bromide with norbornene to give an allylnorbornanylnickel bromide complex does not take place in the absence of CO unless the halide is replaced with acetate (292). The inverse reaction also occurs by replacing acetate with halide, thus suggesting that reversibility can be an important feature of insertion reactions. Reversibility of the carbonylation process was found to be quite general (254).

The above-mentioned characteristic aspects of reactivity of allylnickel complexes containing CO help explain the uniqueness of certain reactions. Palladium is the metal that best approaches nickel, but its behavior is nevertheless very different owing to its lower back-donation ability and to the consequent lower carbanionic character of palladium-bonded organic groups. With transition metals of the second and third rows, insertion of unsaturated species into allylmetal bonds is rather difficult and requires the use of electron-donating ligands like phosphines. For example, allyl chloride adds reversibly in benzene solution to the complex *trans*–[RhCl(CO) L_2] (L = [PMe_2(2–$MeOC_6H_4$)]; in polar solvents, however, insertion of CO occurs to give [$RhCl_2(COCH_2CH{=}CH_2)L_2$] (293). Later work (294) describes insertion of hexafluoro-2-butyne into a methallyliridium bond containing a carbonyl group and PPh_3. The reaction leads to *cis* and *trans* insertion products (as iridium complexes), but CO is not inserted.

Discussion of this work is beyond the scope of this review, and in any event it is premature. It is reported only to point out the complexity of insertion reactions and the need for further work to understand the diverse behavior of transition metal ions. Other studies on insertion reaction of olefins and alkynes into alkyl-metal and hydride-metal bonds in the absence of CO are no doubt relevant in this connection. Some general references are given (295a–e) that pertain to a debate on the possible existence of different types of insertion mechanisms. In this connection, a thorough study was made of the mechanism for insertion of the double bond and carbon monoxide into an acylmanganese bond (296).

For newer developments in regiospecific and stereospecific reactions of allylpalladium complexes with nucleophiles, see refs. 298–300.

REFERENCES

1. R. H. Dewolfe and W. G. Young, *Chem. Rev.*, **56,** 878 (1956).
2. I. G. Farbenind, A.-G., Belg. Pat. 448,884 (1943); *Chem. Abstr.* **41,** 6576 (1947).

2*a*. I. D. Webb and G. T. Borcherdt, *J. Am. Chem. Soc.*, **73,** 2654 (1951).

3. G. P. Chiusoli, Paper presented at the VIII Congress of the Italian Chemical Society, Turin 1958; *Gazz. Chim. Ital.*, **89,** 1332 (1959); *Chim. Ind.* (*Milan*), **41,** 503 (1959).
4. G. P. Chiusoli, *Chim. Ind.* (*Milan*), **41,** 506 (1959); *Angew. Chem.*, **72,** 74 (1960).
5. G. Natta, R. Ercoli, S. Castellano, and P. H. Barbieri, *J. Am. Chem. Soc.*, **76,** 4049 (1954).
6. H. Greenfield, S. Metlin, and I. Wender, *Abstracts, 126th Meeting Am. Chem. Soc.*, New York, September 1954; I. Wender, S. Metlin, S. Ergun, H. W. Sternberg, and H. Greenfield, *J. Am. Chem. Soc.*, **78,** 5401 (1956).
7. A. R. Martın, *Chem. Ind.* (*London*), **1954,** 1536.
8. P. Pino, in Ullmann's *Encyclopädie der Technischen Chemie*, Vol. XIII, Urban & Schwarzenberg-München, Berlin, 1962, p. 60.
9. W. W. Pritchard, U. S. Pat. 2,600,571 (1952); *Chem. Abstr.* **46,** 10188 (1952); U. S. Pat. Reissue 24,653 (1959); *Chem. Abstr.*, **53,** 13051 (1959).
10. F. M. Chaudhari, G. R. Knox, and P. L. Pauson, *J. Chem. Soc. C*, **1967,** 2255.
11. R. F. Heck and D. S. Breslow, *J. Am. Chem. Soc.*, **82,** 750 (1960).
12. H. B. Jonassen, R. I. Stearns, J. Kenttamaa, D. W. Moore, and A. G. Wittaker, *J. Am. Chem. Soc.*, **80,** 2586 (1958).
13. C. L. Aldridge, H. B. Jonassen, and E. Pulkkinen, *Chem. Ind.* (*London*), **1960,** 374.
14. D. W. Moore, H. B. Jonassen, T. B. Joyner, and A. J. Bertrand, *Chem. Ind.* (*London*), **1960,** 1304.
15. J. Smidt and W. Hafner, *Angew. Chem.*, **71,** 284 (1959).
16. I. I. Moiseev, E. A. Fedorovskaya, and Ya. K. Syrkin, *Zhur. Neorg. Khim.*, **4,** 2641 (1959); *Chem. Abstr.*, **54,** 13933 (1960).
17. F. A. Cotton, J. W. Faller, and A. Musco, *Inorg. Chem.*, **6,** 179 (1967).
18. B. L. Shaw, *Proc. Chem. Soc.*, **1960,** 247.
19. H. D. Kaesz, H. B. King, and F. G. A. Stone, *Z. Naturforsch.*, **15b,** 682 (1960).
20. G. Wilke, B. Bogdanovic, P. Hardt, P. Heimbach, W. Keim, M. Kröner, W. Ober-

kirch, K. Tanaka, E. Steinrücke, D. Walter and H. Zimmermann, *Angew. Chem.*, **78,** 157 (1966); *Angew. Chem. Int.*, **5,** 171 (1966).
21. W. R. McClellan, H. H. Hoehn, H. N. Cripps, E. L. Muetterties, and B. W. Howk, *J. Am. Chem. Soc.*, **83,** 1601 (1961), and references cited therein.
21*a*. C. Sourisseau, J. Guillermet, and B. Pasquier, *Chem. Phys. Lett.* **26,** 564 (1974).
22. M. L. H. Green and P. L. I. Nagy, in F. G. A. Stone and R. West, Ed., *Advances in Organometallic Chemistry*, Vol. II, Academic Press, New York, 1964, p. 325.
23. P. Corradini, G. Maglio, A. Musco, and G. Paiaro, *Chem. Commun.* (*London*), **1966,** 618.
24. K. Vrieze, C. McClean, P. Cossee, and C. W. Hilbers, *Rec. Trav. Chim.*, **85,** 1077 (1966); *Chem. Abstr.*, **66,** 24250 (1967).
25. W. E. Smith, *Acta Cryst.*, **18,** 331 (1965).
25*a*. G. E. Coates, M. L. H. Green, P. Powell, and K. Wade, *Principles of Organometallic Chemistry*, Methuen, London, 1968, p. 191.
26. R. Mason and D. R. Russell, *Chem. Commun.* (*London*), **1966,** 26.
27. W. E. Oberhansli and L. F. Dahl, *J. Organometal. Chem.*, **3,** 43 (1965).
28. M. R. Churchill and R. Mason, *Nature*, **204,** 777 (1964).
29. R. Mason and A. G. Wheeler, *Nature*, **217,** 1253 (1968).
30. E. O. Fischer and G. Bürger, *Z. Naturforsch.*, **16b,** 77 (1961).
31. G. Wilke, *Angew. Chem. Intern. Ed.*, **2,** 105 (1963).
32. H. Bönnemann, B. Bogdanovic, and G. Wilke, *Angew. Chem.*, **79,** 817 (1967).
33. J. C. W. Chien and H. C. Dhem, *Chem. Ind.* (*London*), **1961,** 745.
34. J. K. Becconsall and S. O'Brien, *Chem. Commun.* (*London*), **1966,** 302.
35. G. L. Statton and K. C. Ramey, *J. Am. Chem. Soc.*, **88,** 1327 (1966).
36. J. Powell, S. D. Robinson, and B. L. Shaw, *Chem. Commun.* (*London*), **1965,** 78.
37. H. C. Volger and K. Vrieze, *J. Organometal. Chem.*, **13,** 479; 495 (1968).
38. K. Vrieze, A. P. Praat, and P. Cossee, *J. Organometal. Chem.*, **12,** 533 (1968).
39. K. Vrieze, P. Cossee, A. P. Praat, and C. W. Hilbers, *J. Organometal. Chem.*, **11,** 353 (1968).
40. H. Felkin, Y. Gault, and G. Roussi, *Tetrahedron*, **26,** 3761 (1970).
41. P. W. M. N. Van Leeuwen, K. Vrieze, and A. P. Praat, *J. Organometal. Chem.*, **20,** 219 (1969), and references cited therein.
42. P. W. N. M. Van Leeuwen and P. Praat, *J. Organometal. Chem.*, **21,** 501 (1970).
43. S. F. A. Kettle and R. Mason, *J. Organometal. Chem.*, **5,** 573 (1966).
44. P. Ganis, G. Maglio, A. Musco, and A. L. Segre, *Inorg. Chim. Acta*, **3,** 266 (1969).
45. J. W. Faller, M. J. Incorvia, and M. E. Thomsen, *J. Am. Chem. Soc.*, **91,** 518 (1969).
46. J. W. Faller and M. E. Thomsen, *J. Am. Chem. Soc.*, **91,** 6871 (1969).
47. P. Corradini, G. Maglio, A. Musco, and G. Paiaro, *Chem. Commun.* (*London*), **1966,** 618.
48. L. A. Fedorov, *Russ. Chem. Rev.*, **39,** 655 (1970).
49. F. De Candia, G. Maglio, A. Musco, and G. Paiaro, *Inorg. Chim. Acta*, **2,** 333 (1968).
49*a*. J. W. Faller and M. T. Tully, *J. Am. Chem. Soc.* **94,** 2676 (1972).
49*b*. M. Oslinger and J. Powell, *Can. J. Chem.* **51,** 274 (1973).
49*c*. K. Vrieze, H. C. Volger, and P. W. N. M. van Leeuwen, *Inorg. Chim. Acta Rev.* **3,** 109 (1969).
49*d*. A. Musco and P. F. Swinton, *J. Organometal. Chem.* **50,** 333 (1973).
50. F. R. Hartley, *J. Organometal. Chem.*, **21,** 227 (1970).
51. E. J. Corey and M. Semmelhack, *J. Am. Chem. Soc.*, **89,** 2755 (1967).
52. E. J. Corey, M. F. Semmelhack, and L. S. Hegedus, *J. Am. Chem. Soc.*, **90,** 2416 (1968).
53. J. Powell and B. L. Shaw, *J. Chem. Soc.*, **1967,** 1839.

53*b*. F. R. Hartley, *Angew. Chem.* **84,** 657 (1972).
54. M. R. Churchill and T. A. O'Brien, *Chem. Commun.* (*London*), **1968,** 246.
55. A. Sirigu, *Chem. Commun.* (*London*), **1969,** 256; *Inorg. Chem.*, **9,** 2245 (1970).
56. D. Walter and G. Wilke, *Angew. Chem.*, **78,** 941 (1966).
57. F. Guerrieri and G. P. Chiusoli, *J. Organometal. Chem.*, **15,** 209 (1968).
58. K. Noak and F. Calderazzo, *J. Organometal. Chem.*, **10,** 101 (1967).
59. R. F. Heck and D. S. Breslow, *J. Am. Chem. Soc.*, **83,** 1097 (1961).
59*a*. J. J. Eisch and G. A. Damasevilz, *J. Organometal. Chem.* **96,** C19 (1975) and literature cited therein.
60. R. F. Heck, *Advances Chemistry Series*, No. 49, American Chemical Society, Washington, D.C., 1965, p. 181.
61. P. Cossee, *Rec. Trav. Chim.*, **85,** 1151 (1966).
61*a*. A. Wojcicki in F. G. A. Stone and R. West, Ed. *Advances in Organometallic Chemistry*, Vol. 11, Academic Press, New York, 1973 p. 87.
61*b*. A. Wojcicki, ibid. vol. 12, 1974, p. 31.
62. M. Dubini, F. Montino, and G. P. Chiusoli, *Chim. Ind.* (*Milan*), **47,** 839 (1965).
63. M. Dubini, G. P. Chiusoli, and F. Montino, *Tetrahedron Lett.*, **1963,** 1591.
64. E. O. Fischer and G. Bürger, *Z. Naturforsch.*, **17b,** 484 (1962).
65. G. P. Chiusoli and S. Merzoni, *Z. Naturforsch.*, **17b,** 850 (1962).
66. R. F. Heck, *J. Am. Chem. Soc.*, **85,** 2013 (1963).
67. F. Guerrieri and G. P. Chiusoli, *Chem. Commun.* (*London*), **1967,** 781.
68. F. Guerrieri, *Chem. Commun.* (*London*), **1968,** 983.
69. L. Cassar, G. P. Chiusoli, and M. Foa, *Tetrahedron Lett.*, **1967,** 285.
70. R. F. Heck, J. C. W. Chien, and D. S. Breslow, *Chem. Ind.* (*London*), **1961,** 986.
71. R. F. Heck, D. S. Breslow, *J. Am. Chem. Soc.*, **85,** 2779 (1963).
71*a*. L. Cassar, M. Foà and G. P. Chiusoli, Ital. Pat. 868, 994 (1970).
71*b*. G. P. Chiusoli and G. Mondelli, *Chim. Ind.* (Milan) **49,** 856 (1967).
72. D. Medema, R. van Helden, and C. F. Kohll, *Inorg. Chim. Acta*, **3,** 255 (1969).
73. M. C. Gallazzi, L. Porri, and G. Vitulli, *Communication at 10th Congress of Italian Chemical Society*, Padova, 17–21 June 1968.
74. M. Dubini and F. Montino, *J. Organometal. Chem.*, **6,** 188 (1966).
75. P. Collman, *Acc. Chem. Res.*, **1,** 136 (1968).
76. R. Ugo, *Chim. Ind.* (*Milan*), **51,** 1319 (1969).
76*a*. R. Ugo, A. Pasini, A. Fusi, and S. Cenini, *J. Am. Chem. Soc.*, **94,** 7364 (1972).
77. A. J. Deeming and B. L. Shaw, *J. Chem. Soc.* A, **1969,** 1562, and references cited therein.
78. H. Felkin and C. Frajerman, *Tetrahedron Lett.*, **1970,** 1045.
79. Nguyen Trong Anh, *Chem. Commun.* (*London*), **1968,** 1089.
79*a*. J. Halpern, private communication.
80. Y. Takahashi, K. Tsukiyama, S. Sakai, and Y. Ishii, *Tetrahedron Lett.*, **1970,** 1913.
81. G. P. Chiusoli, in R. Ugo, Ed., *Advances in Homogeneous Catalysis*, Manfredi, Milan, 1970, p. 77.
82. F. Guerrieri and G. P. Chiusoli, *Chim. Ind.* (*Milan*), **51,** 1252 (1969).
82*a*. K. Sato, S. Inoue, S. Ota, and J. Fujita, *J. Org. Chem.*, **37,** 462 (1972).
83. F. Guerrieri and G. P. Chiusoli, *Proc., Inorg. Chim. Acta Symposium on Reactivity and Bonding in Transition Organometallic Compounds*, Venice, 8–10 September 1970, C5.
84. G. P. Chiusoli and G. Cometti, *Chim. Ind.* (*Milan*), **45,** 404 (1963).
84*a*. F. Guerrieri, G. P. Chiusoli, and S. Merzoni, *Gazz. Chim. Ital.* **104,** 557 (1974).
85. T. H. Coffield, Y. Kozikowski, and R. D. Closson, *Abstracts, International Conference on Coordination Chemistry*, London 1959.

86. F. A. Hartman, P. J. Pollick, R. L. Downs, and A. Woycicki, *J. Am. Chem. Soc.*, **89,** 2493 (1967).
87. J. E. Thomasson and A. Wojcicki, *J. Am. Chem. Soc.*, **90,** 2709 (1968).
88. R. Pettit, H. Sugahara, J. Wristers, and W. Merck, *Disc. Faraday Soc.*, **47,** 71 (1969).
89. T. H. Whitesides, *J. Am. Chem. Soc.*, **91,** 2395 (1969).
90. D. R. Eaton, *J. Am. Chem. Soc.*, **90,** 4272 (1968).
91. F. D. Mango, *Advances in Catalysis*, Vol. XX, Academic Press, New York, 1969, p. 291.
92. M. I. Lobach, B. D. Babitskii, and V. A. Kormer, *Russ. Chem. Rev.*, **36,** 477 (1967); *Chem. Abstr.*, **68,** 13019 (1968).
93. P. Heimbach, P. W. Jolly, and G. Wilke, in F. G. A. Stone and R. West, Eds., *Advances in Organometallic Chemistry*, Vol. VIII, Academic Press, New York, 1970, p. 29.
94. G. N. Schrauzer, in F. G. A. Stone and R. West, Ed., *Advances in Organometallic Chemistry*, Vol II, Academic Press, New York, 1964, p. 325.
95. C. W. Bird, *J. Chem. Soc. C*, **1967,** 1265.
96. J. Falbe, *Synthesen mit Kohlenmonoxyd*, Springer, Berlin, 1967.
97. R. Hüttel, *Synthesis*, **2,** 225 (1970).
98. J. Tsuji, in E. C. Taylor and H. Wynberg, Ed., *Advances in Organic Chemistry*, Vol. VI, Wiley, New York, 1969, 109.
99. M. Ryang, *Organometal. Chem. Rev. A*, **5,** 67 (1970).
100. R. F. Heck, *Acc. Chem. Res.*, **2,** 10 (1969).
101. R. B. King, *Transition-Metal Organometallic Chemistry*, Academic Press, New York, 1969.
102. M. Orchin, in *Advances in Catalysis*, Vol, XVI, Academic Press, New York (1966) p. 1.
102*a*. H. L. Clarke, *J. Organometal. Chem.*, **80,** 155, 369 (1974).
102*b*. M. Green, *Organometal. Chem.* **1,** 344 (1972); **2,** 407 (1973).
102*c*. M. F. Semmelhack, *Org. React.* **19,** 115 (1972).
102*d*. R. Baker, *Chem. Rev.* **73,** 487 (1973).
102*e*. M. Cooke, *Organometal. Chem.* **3,** 243 (1975).
103. W. Reppe, *Liebigs Ann. Chem.*, **582,** 1 (1953).
104. G. P. Chiusoli, G. Bottaccio, and A. Cameroni, *Chim. Ind.* (*Milan*), **44,** 131 (1962).
105. L. Cassar and G. P. Chiusoli, *Tetrahedron Lett.*, **1965,** 3295; *Chim. Ind.* (*Milan*), **48,** 323 (1966).
106. F. Montino, Ital. Pat. 846,027 (1969); Ger. Offen. 1,936,725 (1969), *Chem. Abstr.*, **73,** 77847 (1970).
107. E. Yoshisato and S. Tsutsumi, *J. Org. Chem.*, **33,** 869 (1968).
108. L. Cassar and M. Foà, *Inorg. Nucl. Chem. Lett.*, **6,** 91 (1970).
109. E. R. H. Jones, G. H. Whitham, and M. C. Whiting, *J. Chem. Soc.*, **1957,** 4628.
110. P. J. Ashworth, G. H. Whitham, and M. C. Whiting, *J. Chem. Soc.*, **1957,** 4633.
111. J. Tsuji, J. Kiji, S. Imamura, and M. Morikawa, *J. Am. Chem. Soc.*, **86,** 4350 (1964).
112. S. Imamura and J. Tsuji, *Tetrahedron*, **25,** 4187 (1969).
113. J. Tsuji, M. Morikawa, and J. Kiji, *Tetrahedron Lett.*, **1963,** 1061.
113*a*. J. Tsuji, S. Imamura, and J. Kiji, *J. Am. Chem. Soc.*, **86,** 4491 (1964).
114. W. T. Dent, R. Long, and G. H. Whitfield, *J. Chem. Soc.*, **1964,** 1588.
115. R. Long and G. H. Whitfield, *J. Chem. Soc.*, **1964,** 1852.
116. R. D. Closson, K. G. Ihrman, and A. H. Filbey U.S. Pat. 3,338,961 (1967); *Chem. Abstr.*, **68,** 2589 (1968).
117. J. Tsuji, J. Kiji, and S. Hosaka, *Tetrahedron Lett.* **1964,** 605; J. Tsuji and S. Hosaka, *J. Am. Chem. Soc.*, **87,** 4075 (1965).

118. C. Bordenca and W. E. Marsico, *Tetrahedron Lett.*, **1967,** 1541.
119. S. Brewis and P. R. Hughes, *Chem. Commun.* (*London*), **1965,** 157.
119*a*. A. Matsuda Jap. Kokai 73 10,020 (1973); Chem. Abstr. **78,** 158973 (1973).
120. G. W. Parshall, *Z. Naturforsch.*, **18b,** 772 (1963).
121. J. Blum, *Tetrahedron Lett.*, **1966,** 1605.
122. M. C. Baird, D. N. Lawson, J. T. Mague, J. A. Osborn, and G. Wilkinson, *Chem. Commun.* (*London*), **1966,** 129.
123. J. Tsuji, *Nippon Kagaku Zasshi*, **88,** 687 (1967); *Chem. Abstr.*, **69,** 77290 (1968).
124. R. G. Schulz, *Tetrahedron Lett.*, **1964,** 301; *Tetrahedron*, **20,** 2809 (1964).
125. M. S. Lupin and B. L. Shaw, *Tetrahedron Lett.*, **1964,** 883.
126. J. Tsuji and T. Susuki, *Tetrahedron Lett.*, **1965,** 3027.
127. T. Susuki and J. Tsuji, *Bull. Chem. Soc. Japan*, **41,** 1954 (1968); *Chem. Abstr.*, **69,** 96861 (1968).
128. A. J. Chalk, *Tetrahedron Lett.*, **1964,** 2627.
129. G. P. Chiusoli and S. Merzoni, *Chim. Ind.* (*Milan*), **43,** 259 (1961).
130. F. Basolo and R. G. Pearson, *Mechanism of Inorganic Reactions*, Wiley, New York, 1958.
131. F. Basolo and A. Woicicki, *J. Am. Chem. Soc.*, **83,** 520 (1961).
132. J. P. Day, F. Basolo, and R. G. Pearson, *J. Am. Chem. Soc.*, **90,** 6927, 6333 (1968).
133. F. Basolo, *Chem. in Britain*, **1969,** 505.
134. J. Halpern, in B. I. Luberoff, Ed., *Advances in Chemistry Series*, No. 70, American Chemical Society, Washington, 1968, p. 1.
135. J. Halpern, *Acc. Chem. Res.*, **3,** 386 (1970).
135*a*. J. Chatt, R. S. Coffey, A. Gough, and D. T. Thompson, *J. Chem. Soc. A*, **1968,** 190.
136. R. D. Closson, J. Kozikowski, and T. H. Coffield, *J. Org. Chem.* **22,** 598 (1957).
137. F. Calderazzo and K. Noack, *Coord. Chem. Rev.*, **1,** 118 (1965).
138. G. P. Chiusoli and A. Cameroni, *Chim. Ind.* (*Milan*), **46,** 1063 (1964).
139. W. T. Dent, R. Long, and A. J. Wilkinson, *J. Chem. Soc.* **1964,** 1585.
140. J. K. Nicholson, J. Powell, and B. L. Shaw, *Chem. Commun.* (*London*), **1966,** 174.
141. J. Tsuji and N. Iwamoto, *Chem. Commun.* (*London*), **1966,** 828.
142. A. N. Nesmeyanov, I. I. Kritskaya, R. V. Kudryavtsev, and Yu. I. Lyakhovetskii, *Izv. Akad. Nauk SSSR, Ser. Khim.*, **1967,** (2) 418; *Chem. Abstr.*, **67,** 43200 (1967).
143. G. P. Chiusoli and L. Cassar, *Angew. Chem. Intern. Ed.*, **6,** 124 (1967).
144. E. R. H. Jones, T. Y. Shen, and M. C. Whiting, *J. Chem. Soc.*, **1951,** 48.
145. G. P. Chiusoli and S. Merzoni, *Chim. Ind.* (*Milan*), **43,** 255 (1961).
146. G. P. Chiusoli, S. Merzoni, and G. Cometti, Ital. Pat. 719,380 (1966); *Chem. Abstr.*, **69,** 51638 (1968).
147. M. Dubini, M. Ferraris, S. Merzoni, and G. P. Chiusoli, Fr. Pat. 1,570,823 (1969); *Chem. Abstr.*, **72,** 89796 (1970).
148. G. P. Chiusoli and G. Agnès, *Z. Naturforsch.*, **17b,** 852 (1962).
149. G. P. Chiusoli and G. Agnès, *Chim. Ind.* (*Milan*), **46,** 548 (1964).
150. G. Agnès and G. P. Chiusoli, *Chim. Ind.* (*Milan*), **49,** 465 (1967).
151. G. P. Chiusoli and F. Minisci, *Gazz. Chim. Ital.*, **88,** 43 (1958).
152. H. Koch and W. Haaf, *Ann.*, **618,** 251 (1958).
153. G. P. Chiusoli, S. Merzoni, and G. Mondelli, *Chim. Ind.* (*Milan*) **46,** 743 (1964).
154. J. S. Sallo and R. D. Fisher, *J. Electrochem. Soc.*, **107,** 277 (1960).
155. R. D. De Mars, *J. Electrochem. Soc.*, **108,** 779 (1961).
156. G. P. Chiusoli and G. Mondelli, Belg. Pat. 619,276 (1962); *Chem. Abstr.*, **58,** 12210 (1963); Brit. Pat. 983,221 (1965).
157. W. Hieber and E. O. Fischer, *Z. Anorg. Allgem. Chem.*, **169,** 292 (1952).

158. E. O. Fischer and W. Hieber, *Z. Anorg. Allgem. Chem.*, **271,** 229 (1953).
159. G. P. Chiusoli, *Atti Accad. Naz. Lincei, Mem. Classe Sci. Fis. Mat. Nat. Sez. IIa*, **26,** 790 (1959); *Chem. Abstr.*, **54,** 8709 (1960).
160. G. P. Chiusoli, *Chim. Ind.* (*Milan*), **41,** 762 (1959).
161. G. P. Chiusoli and S. Merzoni, *Chim. Ind.* (*Milan*), **45,** 6 (1963).
162. G. P. Chiusoli and S. Merzoni, Fr. Pat. 1,342,549 (1963); *Chem. Abstr.*, **60,** 13146 (1964); Brit. Pat. 1,006,008 (1965).
163. E. A. Youngman and F. F. Rust, U.S. Pat. 2,991,306 (1961); *Chem. Abstr.*, **55,** 25759 (1961).
164. G. P. Chiusoli and S. Merzoni, unpublished results.
165. H. Fernholz and L. Schlaefer, Ger. Pat. 1,280,850 (1968); *Chem. Abstr.*, **70,** 19604 (1969).
166. N. L. Bauld, *Tetrahedron Lett.*, **1962,** 859.
167. G. P. Chiusoli and G. Bottaccio, Belg. Pat. 630,619 (1963); *Chem. Abstr.*, **61,** 577 (1964); Brit. Pat. 967,299 (1964).
168. J. B. Mettalia, Jr. and E. H. Specht, *J. Org. Chem.*, **32,** 3941 (1968).
169. G. P. Chiusoli, M. Ferraris, and S. Merzoni, *Proc., XII International Conference on Coordination Compounds,* Sydney, 21–28 Aug. 1969.
170. G. P. Chiusoli and S. Merzoni, Fr. Pat. 1,358,900 (1964); *Chem. Abstr.*, **61,** 14535 (1964); U.S. Pat. 3,238,246 (1966).
171. J. Chatt and B. L. Shaw, *Chem. Ind.* (*London*), **1960,** 931; **1961,** 290.
172. F. Guerrieri and G. P. Chiusoli, *J. Organometal. Chem.*, **19,** 453 (1969).
173. W. Brenner, P. Heimbach, and G. Wilke, *Liebigs Ann. Chem.*, **727,** 194 (1969).
174. R. F. Heck, *J. Am. Chem. Soc.*, **86,** 2819 (1964).
175. M. Tsutsui and H. Zeiss, *J. Am. Chem. Soc.*, **81,** 6090 (1959).
176. R. P. A. Sneeden and H. H. Zeiss, *J. Organometal. Chem.*, **20,** 153 (1969).
177. J. Trocha Grimshaw, and H. B. Henbest, *Chem. Commun.*, **1968,** 757.
178. G. P. Chiusoli, unpublished results.
179. G. P. Chiusoli and G. Mondelli, unpublished results.
180. F. R. Hartley, *Chem. Rev.*, **69,** 799 (1969).
181. F. L. Bowden and A. B. P. Lever, *Organometal. Chem. Rev.*, **3,** 227 (1968).
181*a*. R. Mason and K. M. Thomas, *Annals New York Acad. Sci.* **239,** 225 (1974).
182. N. L. Bauld, *Tetrahedron Lett.*, **1963,** 1841.
183. G. P. Chiusoli, S. Merzoni, and G. Mondelli, *Tetrahedron Lett.*, **1964,** 2777.
184. G. P. Chiusoli, F. Minisci, and A. Quilico, *Gazz. Chim. Ital.*, **87,** 100 (1957).
185. M. Foà, L. Cassar, and M. Tacchi Venturi, *Tetrahedron Lett.*, **1968,** 1357.
186. L. Cassar, G. P. Chiusoli, and M. Foà, *Chim. Ind.* (*Milan*), **50,** 515 (1968).
186*a*. M. Foà and L. Cassar, *Gazz. Chim. Ital.*, **103,** 805 (1973).
187. G. P. Chiusoli and G. Bottaccio, *Chim. Ind.* (*Milan*), **47,** 165 (1965).
188. G. P. Chiusoli, G. Bottaccio, and C. Venturello, *Tetrahedron Lett.*, **1965,** 2875; *Chim. Ind.* (*Milan*), **48,** 107 (1966).
189. G. P. Chiusoli, Lecture at II Intern. Conf. on Metal Carbonyls, Donovaly (High Tatras), 26–28 May 1970.
190. G. P. Chiusoli and S. Merzoni, unpublished results.
191. C. W. Bird, *Transition-Metal Intermediates in Organic Syntheses,* Logos Press, London 1967.
192. P. Pino, A. Miglierina, and E. Pietra, *Gazz. Chim, Ital.*, **84,** 443 (1954); P. Pino and A. Miglierina, *J. Am. Chem. Soc.*, **74,** 5551 (1952).
193. P. Pino, E. Pietra, and B. Mondello, *Gazz. Chim. Ital.*, **84,** 453 (1954).
194. G. P. Mueller and F. L. Mac Artor, *J. Am. Chem.*, **76,** 4621 (1964).

195. E. Weiss and W. Hübel, *J. Inorg. Nucl. Chem.*, **11,** 42 (1959).
196. R. F. Heck, *J. Am. Chem. Soc.*, **85,** 3383 (1963).
197. J. Liebig and F. Wöhler, *Ann.*, **3,** 266 (1832).
198. R. Filler and F. N. Miller, *Can. J. Chem.*, **44,** 838 (1966).
199. H. Simonis, *Ber.*, **45,** 1584 (1912).
200. S. Gould, *Mechanism and Structure in Organic Chemistry*, Holt, New York, 1959, p. 452.
201. G. P. Chiusoli, *Bull. Soc. Chim. France,* **1969,** 1139; *Chem Abstr.*, **71,** 48883 (1969).
202. G. N. Schrauzer, S. Eichler, and D. A. Brown, *Chem. Ber.*, **95,** 2755 (1962).
203. F. Minisci and R. Galli, *Tetrahedron Lett.*, **1962,** 533; *Chim. Ind.* (*Milan*), **45,** 448 (1963).
204. R. L. Shriner in R. Adams, Eds., *Organic Reactions* Vol. I. Wiley, New York, 1942, p. 1.
205. L. Cassar and G. P. Chiusoli, unpublished results.
206. L. Cassar and G. P. Chiusoli, *Tetrahedron Lett.*, **1966,** 2805.
207. R. F. Heck, *J. Am. Chem. Soc.*, **87,** 4727 (1965).
208. R. F. Heck, *J. Am. Chem. Soc.*, **85,** 3387 (1963).
209. G. Albanesi and M. Tovaglieri, *Chim. Ind.* (*Milan*), **41,** 198 (1959); G. Albanesi, *ibid.*, **46,** 1169 (1964).
210. H. W. Sternberg, J. G. Shukys, C. D. Donne, R. Markby, R. A. Friedel, and I. Wender, *J. Am. Chem. Soc.*, **81,** 2339 (1959).
211. J. C. Sauer, R. D. Cramer, V. A. Engelhardt, T. A. Ford, H. E. Holmquist, and B. W. Howk, *J. Am. Chem. Soc.*, **81,** 3677 (1959).
212 Y. Sawa, I. Hashimoto, M. Ryang, and S. Tsutsumi, *J. Org. Chem.*, **33,** 2159 (1968).
213. E. Yoshisato, M. Ryang, and S. Tsutsumi, *J. Org. Chem.*, **34,** 1500 (1969).
214. E. J. Corey and L. S. Hegedus, *J. Am. Chem. Soc.*, **91,** 4926 (1969).
215. L. Cassar and M. Foà, *Chim. Ind.* (*Milan*), **51,** 673 (1969).
215*a*. R. F. Heck, *J. Am. Chem. Soc.* **90,** 5518 (1968); **91,** 6707 (1969); **93,** 6896 (1971); **94,** 2712 (1972).
215*b*. R. C. Larock, *J. Org. Chem.* **40,** 3237 (1975).
216. H. Kröper, in *Houben-Weil Methoden der Organischen Chemie*, Vol. IV, Part 2, Thieme Verlag, Stuttgart, 1955, p. 415.
217. M. Orchin, in *Advances in Catalysis*, Vol. XVI. Academic Press, New York, 1966, p. 1.
218. Y. Takegami, Y. Watanabe, T. Mitsudo, and T. Okajima, *Bull. Chem. Soc. Japan*, **42,** 1992 (1969).
219. M. L. H. Green and P. L. I. Nagy, *J. Chem. Soc.* **1963,** 189.
220. R. P. Hughes and J. Powell, *J. Am. Chem. Soc.*, **94,** 7723 (1972).
221. H. Bönnemann, *Angew. Chem.*, **82,** 699 (1970).
221*a*. J. W. Byrne, H. V. Blaser, and J. A. Osborne, *J. Am. Chem. Soc.* **97,** 3871 (1975).
222. J. Powell and B. L. Shaw, *J. Chem. Soc. A*, **1968,** 159.
223. E. J. Corey and M. F. Semmelhack, *Tetrahedron Lett.*, **1966,** 6237.
224. E. J. Corey and E. Hamanaka, *J. Am. Chem. Soc.*, **86,** 1641 (1964).
225. E. J. Corey and E. K. Wat, *J. Am. Chem. Soc.*, **89,** 2756 (1967).
226. E. J. Corey and E. Hamanaka, *J. Am. Chem. Soc.*, **89,** 2758 (1967).
226*a*. L. S. Hegedus and L. L. Miller, *J. Am. Chem. Soc.* **97,** 459 (1975).
226*b*. E. J. Corey and E. A. Broger, *Tetrahedron Lett.* **1969,** 1779.
226*c*. E. J. Corey and H. A. Kirst, *J. Am. Chem. Soc.* **94,** 667 (1972).
227. R. Traunmüller, O. E. Polansky, P. Heimbach, and G. Wilke, *Chem. Phys. Lett.*, **3,** 300 (1969).

228. E. R. H. Jones, D. G. O'Sullivan, and M. C. Whiting, *J. Am. Chem. Soc.*, **73,** 615 (1951).
229. J. English, Jr., I. D. Gregory, and J. R. Trowbridge, *J. Am. Chem. Soc.*, **73,** 615 (1951).
230. A. S. Dreiding and R. J. Pratt, *J. Am. Chem. Soc.*, **75,** 3717 (1953).
231. L. Mangoni and M. Belardini, *Gazz. Chim. Ital.*, **91,** 390 (1961).
231*a*. L. S. Hegedus, E. L. Waterman, and J. Catlin, *J. Am. Chem. Soc.* **94,** 7155 (1972).
231*b*. L. S. Hegedus and R. K. Stiverson, *J. Am. Chem. Soc.*, **96,** 3250 (1974).
231*c*. K. Sato, S. Inoue and K. Saito, *J. Chem. Soc. Perkin I*, **1973,** 2289.
231*d*. L. S. Hegedus, S. D. Wagner, E. L. Waterman, and K. Siirata Hansen, *J. Org. Chem.* **40,** 593 (1975).
232. G. Wilke, *J. Appl. Chem.*, **17,** 179 (1968); *Chem. Abstr.*, **71,** 70663 (1969).
232*a*. G. Agnés, G. P. Chiusoli, and A. Marraccini, *J. Organometal. Chem.* **49,** 239 (1973).
233. G. P. Chiusoli, *Chim. Ind.* (*Milan*), **43,** 365 (1961).
234. I. Rhee, N. Mizuta, M. Ryang, and S. Tsutsumi, *Bull. Chem. Soc. Japan*, **41,** 1417 (1968); *Chem. Abstr.*, **69,** 85899 (1968).
235. L. Ehrlich and G. F. Emerson, *Chem. Commun.* (*London*), **1969,** 59.
236. F. J. Impastato and K. J. Ihrman, *J. Am. Chem. Soc.*, **83,** 3726 (1961).
237. L. Porri, G. Natta, and M. C. Gallazzi, *Chim. Ind.* (*Milan*), **46**, 428 (1964); *J. Polym. Sci. C*, **1967,** 2525; *Chem. Abstr.*, **68,** 40717 (1968).
238. B. Bogdanovic, P. Heimbach, M. Kröner, and G. Wilke, *Liebigs Ann. Chem.*, **727,** 143 (1969).
239. W. Brenner, P. Heimbach, H. Hey, F. W. Müller, and G. Wilke, *Liebigs Ann. Chem.*, **727,** 161 (1969).
240. P. Heimbach and G. Wilke, *Liebigs Ann. Chem.*, **727,** 183 (1969).
240*a*. A. Misono, Y. Uchida, T. Saito, and K. Uchida, *Bull. Chem. Soc. Japan* **40,** 1889 (1967).
240*b*. H. Bonnemann, G. Grard, W. Kopp, and G. Wilke, *XXIII Int. Congr. Pure Appl. Chem., Special Lectures*, Boston 1971, Vol. 6, Butterworths, London, p. 265.
241. B. D. Babitskii, V. A. Kormer, I. M. Lapuk, and V. I. Skoblikova, *J. Polym. Sci. C*, **1965,** 3219; *Chem. Abstr.*, **70,** 4659 (1969).
242. G. Lugli, W. Marconi, A. Mazzei, and N. Palladino, *Inorg. Chim. Acta*, **3,** 151 (1969).
243. R. Pettit, *Ann. N. Y. Acad. Sci.*, **125,** Art. 1, 89 (1965).
244. F. Asinger and O. Berg, *Chem. Rev.*, **88,** 445 (1955).
245. T. A. Manuel, *J. Org. Chem.*, **27,** 3941 (1962).
246. J. E. Arnet and R. Pettit, *J. Am. Chem. Soc.*, **83,** 2955 (1961).
247. F. Piacenti, M. Cioni, and P. Pino, *Chem. Ind.* (*London*), **1960,** 1240.
248. G. F. Emerson and R. Pettit, *J. Am. Chem. Soc.*, **84,** 4591 (1962).
249. G. F. Emerson, Dissertation, Univ. Texas, Austin 1964.
250. W. T. Hexdrix, F. G. Cowherd, and J. L. von Rosenberg, *Chem. Commun.* (*London*), **1968,** 97.
251. W. D. Bannister, M. Green, and R. N. Haszeldine, *J. Chem. Soc. A*, **1966,** 194.
252. M. Green and R. I. Hancock, *J. Chem. Soc. A*, **1968,** 109.
253. H. D. Murdoch and E. Weiss, *Helv. Chim. Acta*, **45,** 1927 (1962); *Chem. Abstr.*, **58,** 6857 (1963).
254. J. Tsuji, *Acc. Chem. Res.*, **2,** 144 (1969).
255. R. Huttel, *Brennstoff-Chemie*, **9,** 17, 39 (1969).
255*a*. P. M. Maitlis, *The Organic Chemistry of Palladium*, Vols. I and II, Academic Press, New York, 1971.

256. Y. Takahashi, S. Sakai, and Y. Ishii, *Chem. Commun (London)*, **1967,** 1092; *J. Organometal. Chem.*, **16,** 177 (1969).
257. E. I. Smutny, *J. Am. Chem. Soc.*, **89,** 6793 (1967).
258. J. Tsuji, H. Takahashi, and M. Morikawa, *Tetrahedron Lett.*, **1965,** 4387.
259. R. Huttel, *Brennstoff-Chemie*, **9,** 17, 39 (1969).
260. J. Tsuji, S. Hosaka, J. Kiji, and T. Susuki, *Bull. Chem. Soc. Japan* **39,** 141 (1966); *Chem. Abstr.*, **64,** 12563 (1966).
261. G. Agnes and G. P. Chiusoli, *Chim. Ind. (Milan)*, **50,** 194 (1968).
262. M. Ferraris, Ital. Pat. 848,909, 1969; *Chem. Abstr.*, **75,** 88128 (1971).
263. G. P. Chiusoli, G. Agnes, C. A. Ceselli, and S. Merzoni, *Chim. Ind. (Milan)*, **46,** 21 (1964).
264. G. Boffa and G. Battisti, Ital. Pat. 885,309 (1971).
265. G. P. Chiusoli, M. Dubini, M. Ferraris, F. Guerrieri, S. Merzoni, and G. Mondelli, *J. Chem. Soc.*, **1968,** 2889.
266. M. Dubini and F. Montino, Ital. Pat. 845,995 (1969); Ger. Offen. 1,916,533 (1969); *Chem. Abstr.*, **72,** 42834 (1970).
267. M. Dubini and G. Mondelli, Ital. Pat. 814,816 (1968); *Chem. Abstr.*, **71,** 60760 (1969).
268. R. B. King and A. Fronzaglia, *J. Am. Chem. Soc.*, **88,** 709 (1966).
269. A. J. Deeming and B. L. Shaw, *J. Chem. Soc. A*, **1969,** 1802.
270*a*. L. Vaska, *Inorg. Chim. Acta*, **5,** 295 (1971), and references cited therein.
270*b*. P. B. Chock and J. Halpern, *J. Am. Chem. Soc.*, **88,** 354 (1966).
270*c*. R. G. Pearson and W. R. Muir, *J. Am. Chem. Soc.*, **92,** 5519 (1970).
270*d*. J. C. Bradley, D. E. Connor, D. Dolphin, J. A. Labinger, and J. A. Osborn, *J. Am. Chem. Soc.*, **94,** 4043 (1972).
270*e*. A. V. Kramer, J. A. Labinger, J. S. Bradley, and J. A. Osborn, *J. Am. Chem. Soc.* **96,** 7145 (1974).
270*f*. I. H. Elson, D. G. Morrell, and J. K. Kochi, *J. Organometal. Chem.* **84,** C7 (1975).
270*g*. L. S. Hegedus and E. L. Waterman, *J. Am. Chem. Soc.* **96,** 6789 (1974).
270*h*. D. F. Shriver, *Acc. Chem.* **3,** 231 (1970); *J. Organometal Chem.* **94,** 259 (1975).
271. H. C. Volger, K. Vrieze, J. W. F. M. Lemmers, A. P. Pratt, and P. W. N. M. van Leeuwen, *Inorg. Chim. Acta*, **4,** 435 (1970).
272. R. G. Pearson, *Acc. Chem. Res.*, **4,** 152 (1971).
273. L. Cassar, M. Foà, and G. P. Chiusoli, *Organometal. Chem. Syn.*, **1,** 302 (1971).
274. G. P. Chiusoli and G. Cometti, unpublished results.
275. J. A. Scheben, U.S. Pat. 3,627,827 (1971); *Chem. Abstr.*, **76,** 58379 (1972).
275*a*. L. Cassar and A. Giarrusso, *Gazz. Chim. Ital.*, **103,** 793 (1973).
276. M. C. Gallazzi, T. L. Hanlon, G. Vitulli and L. Porri, *J. Organometal. Chem.*, **33,** C45, (1971).
277. G. P. Chiusoli and G. Cometti, *J. Chem. Soc. Chem. Commun.*, **1972,** 1016; G. P. Chiusoli, G. Cometti, and V. Bellotti, *Gazz. Chim. Ital.*, **103,** 569 (1973); *ibid*; **104,** 259 (1974); G. P. Chiusoli, *Acc. Chem. Res.*, **6,** 422 (1973).
277*a*. H. Tolman, *J. Am. Chem. Soc.* **92,** 6777 (1970).
278. G. D. Andretti, G. Bocelli, and P. Sgarabotto, *Cryst. Struct. Commun.*, **3,** 271 (1973); *ibid.*, **3,** 495 (1973).
278*a*. B. Fell, W. Seide and F. Asinger, *Tetrahedron Lett.* **1968,** 1003.
278*b*. S. Brewis and P. R. Hughes, *Chem. Commun.* **1967,** 71.
279. G. P. Chiusoli and S. Merzoni, *J. Chem. Soc. D* **1971,** 522; G. P. Chiusoli, G. Cometti, and S. Merzoni, *Organometal. Chem. Syn.*, **1,** 439 (1972).
280. G. P. Chiusoli, *XXIII Int. Cong. Pure Appl. Chem., Special Lectures*, Boston 1971, Butterworth, London.

280*a*. G. P. Chiusoli and G. Cometti, Ital. Pat. Appl. 25965A/71.
281. L. A. Fedorov, *Russ. Chem. Rev.*, **39,** 655 (1970).
282. W. E. Billups, W. E. Walker, and T. C. Shields, *J. Chem. Soc. D*, **1971,** 1067.
283. J. Tsuji, Y. Mori, and M. Hara, *Tetrahedron*, **28,** 3721 (1972).
284. V. P. Nechiporenko, A. D. Treboganov, V. P. Chernova, G. I. Myagkova, and N. A. Preobrazhenskii, *Zh. Org. Khim.*, **9,** 238 (1973); *Chem. Abstr.*, **78,** 123961 (1973); V. P. Nechiporenko, A. A. Glazkov, G. I. Myagkova and R. P. Evstigneeva, *Zh. Org. Khim.* **9,** 243 (1973); *Chem. Abstr.*, **78,** 123960 (1973).
285. L. Cassar, G. P. Chiusoli, and F. Guerrieri, *Synthesis*, **1973,** 509.
286. M. Foà and M. Tacchi Venturi, *Gazz. Chim. Ital.* **105,** 1199 (1975).
287. M. Foà and L. Cassar, *J. Organometal. Chem.* **30,** 123 (1971).
288. G. P. Chiusoli, G. Cometti, G. Sacchelli, and V. Bellotti; to be published.
289. G. P. Chiusoli, G. Cometti, G. Sacchelli, V. Bellotti, G. D. Andretti, G. Bocelli, and P. Sgarabotto, *Trans. Met. Chem.* **1,** 147 (1976).
290. G. P. Chiusoli, G. Cometti, G. Sacchelli, V. Bellotti, G. D. Andretti, G. Boulli, and P. Sgarabotto, in press.
291. M. A. Muhs and F. T. Weiss, *J. Am. Chem. Soc.* **84,** 4697 (1962).
292. M. C. Gallazzi, L. Porri, and G. Vitulli, *J. Organometal. Chem.* **97,** 131 (1975).
293. H. D. Empsall, E. M. Hyde, C. E. Jones, and B. L. Shaw, *J. Chem. Soc. Dalton* **1974,** 1980.
294. M. Green and S. H. Taylor, *J. Chem. Soc., Dalton* **1975,** 1142.
295. M. H. Chisholm and H. C. Clark, *Acc. Chem. Res.* **6,** 202 (1973).
295a. H. C. Clark, C. R. Jablonski, and C. S. Wong, *Inorg. Chem.*, **14,** 1332 (1975).
295*b*. H. C. Clark and C. S. Wong, *J. Organometal. Chem.* **92,** C31 (1974).
295*c*. A. Deeming, B. F. G. Johnson, and J. Lewis, *J. Chem. Soc. Dalton*, **1973** 1848.
295*d*. G. Bracher, P. S. Pregosin, and L. M. Venanzi, *Angew. Chem.* **87,** 547 (1975).
295*e*. R. Romeo, P. Uguaglisti and U. Belluco, *J. Mol. Cat.* **1,** 325 (1976).
296. B. B. Booth, M. Gardner, and R. N. Haszeldine, *J. Chem. Soc. Dalton* **1975,** 1856, 1863.
297. R. Baker, A. H. Cooke and J. Crimmin, *J. Chem. Soc. Chem. Commun.* **1975,** 727.
298. B. M. Trost and L. Weber, *J. Am. Chem. Soc.* **97,** 1611 (1975).
299. B. M. Trost and P. E. Strege, *J. Am. Chem. Soc.* **97,** 2534 (1975).
300. B. M. Trost and J. R. Verhoeven, *J. Am. Chem. Soc.* **98,** 632 (1976).

BIBLIOGRAPHY

P. W. Jolly and G. Wilke, *The Organic Chemistry of Nickel*, Vols. 1 and 2, Academic Press, New York, 1974, 1975.

R. F. Heck, *Organotransition Metal Chemistry*, Academic Press, New York, 1974.

J. Tsuji, *Organic Syntheses by Means of Transition Metal Complexes*, Springer, Berlin, 1975.

Y. Ishii and M. Tsutsui, *Organotransition-Metal Chemistry*, Plenum Press, New York, 1975.

L. Malatesta and S. Cenini, *Zerovalent Compounds of Metals*, Academic Press, New York, 1974.

H. Alper, *Transition Metal Organometallics in Organic Syntheses*, Vol. 1, Academic Press, New York, 1976.

D. Seyferth, *New Applications of Organometallic Reagents in Organic Syntheses*, Elsevier, Amsterdam, 1976.

Carbon Monoxide Addition to Acetylenic Substrates: Synthesis of Quinones, Acrylic and Succinic Acids, and Their Derivatives

P. PINO, *Swiss Federal Institute of Technology, Department of Industrial and Engineering Chemistry, 8006 Zurich, Universitatstrasse 6, Switzerland*

and

G. BRACA, *Institute of Industrial Organic Chemistry, University of Pisa, Via Risorgimento 35, Italy*

I. INTRODUCTION

The first research on addition of CO to acetylene in the presence of transition metal compounds or metal carbonyls was carried out in Germany between 1938 and 1940 by two different industrial research groups working under the supervision of Reppe and Roelen, respectively. Reppe, who was investigating new aspects of acetylene chemistry (1), reacted acetylene with $Ni(CO)_4$ and water and obtained acrylic acid (2,3) instead of the expected aldehydes. Roelen was investigating the hydroformylation of olefins in the presence of cobalt catalysts (4,5) and as a logical extension tried to carry out the hydroformylation of acetylene. He obtained, however, high-molecular-weight products containing only traces of acrolein, the expected product (6).

Due to the high reactivity of the acetylenic compounds and CO in the presence of transition metals, particularly of group VIII, and to the great industrial importance of some of the products obtained, much scientific and industrial activity has been taking place in this field since these discoveries. As a consequence, the scope of the reaction was extended and it was shown that it is possible to obtain derivatives of mono- and dicarboxylic acids and keto acids, esters of aldehydo-acids, cyclic ketones, and hydroquinones in one step from relatively simple acetylenic compounds and CO. Two related processes, the synthesis of acrylic acid and acrylic esters have been and are being operated in large industrial plants. The synthesis of succinic acid reached the pilot-plant stage; interesting industrial perspectives exist for other processes, for instance, the hydroquinone synthesis.

The present chapter is intended to show the most interesting possibilities offered by reacting acetylenic substrates and CO in the presence of transition metals. We shall not give a complete survey of the literature existing in the field; instead only papers and patents relevant either for the synthesis of new products or for the clarification of reaction mechanisms will be cited. In general, only syntheses giving rise to organic compounds from acetylenic substrates and CO or metal carbonyls will be discussed. The syntheses of organometallic compounds from metal carbonyls and acetylenic substrates will not be treated; this aspect was reviewed (7) and is discussed, at least in part, in Vol. 1 in this series (8).

Disregarding the historical development, the subject of this chapter is organized according to the reactants taking part in the reactions:

(*i*) reactions between acetylenic compounds and CO,
(*ii*) addition of CO and hydrogen to acetylenic compounds (hydrocarbonylation),
(*iii*) addition of a carboxyl group and hydrogen to acetylenic compounds (hydrocarboxylation).

II. REACTIONS BETWEEN ACETYLENIC SUBSTRATES AND CARBON MONOXIDE IN APROTIC SOLVENTS

Acetylenic substrates, unlike olefins, react with CO in the presence of metal carbonyls even in the absence of hydrogen or compounds containing reactive hydrogen atoms. These reactions are related to those of acetylenic substrates with metal carbonyls which give rise to a large number of organometallic carbonyl complexes containing organic ligands (7).

As discussed in Vol. 1 of this series (8), the metal carbonyls are able to organize two or more molecules of the acetylenic substrates forming organic groups which may include CO and which are bound to the metal atom either through a σ- or through a π-bond.

Unfortunately, in most cases, the bonds to the metal are very stable and the organic groups cannot easily be displaced by CO or acetylenic substrates. Therefore, true catalytic syntheses from acetylenic substrates and CO occur in only relatively few instances.

As discussed previously (8), the most common structures that can be synthesized by the metal carbonyls (starting with acetylenic substrates and using CO already present in the metal complex) are the doubly unsaturated keto structure **1**, and the lactone structure **2**. Other reactions in

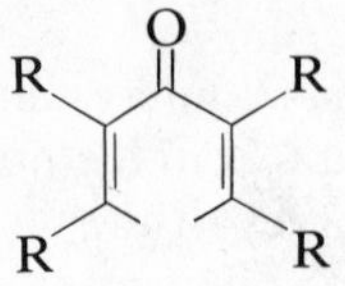

(1)
R = H, alkyl, aryl

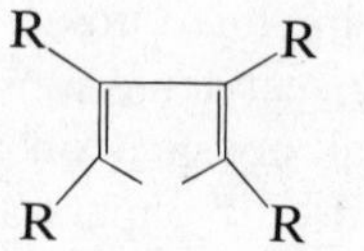

(1a)

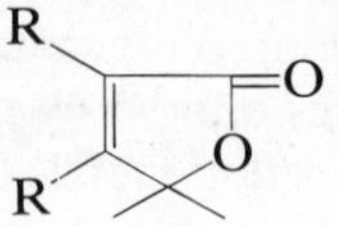

(2)

aprotic solvents involving acetylenic compounds, metal carbonyls as $Fe(CO)_5$, $Fe_3(CO)_{12}$, and $Ni(CO)_4$ and other substances like lithium alkyls (9), diphenylketene (10), and acyl halides (11), producing diketones or lactones will not be discussed here in detail.

A. Synthesis of Cyclopentadienones and *p*-Benzoquinones

Substituted cyclopentadienones and *p*-benzoquinones can be formed from structure **1** (Table 1); the former seems to be much more strongly bound to the metal than the latter. In fact, in the stoichiometric reaction between diphenylacetylene and $Fe(CO)_5$ or π-cyclopentadienyldicarbonyl cobalt, only traces of free tetraphenylcyclopentadienone are formed (12,13); a large amount of the dienone is present in some of the iron and cobalt complexes formed and can be recovered by destroying the complex (12,12*a*).

Alternatively cyclopentadienones might arise from a structure such as **1a** by displacement of a C_4 fragment by CO (8, **13a**).

The organization of acetylenic substrate molecules and CO might proceed via a cyclobutadiene intermediate. Indeed, the tetraphenylcyclobutadiene derivative of $PdCl_2$ reacts with CO under pressure to give tetraphenylcyclopentadienone (15).

From $Ni(CO)_4$ and acetylene, α-hydrindone (**3**) was obtained (17), while from $Ni(CO)_4$ or $(CH_3CN)_3Cr(CO)_3$ and diphenylacetylene, tetraphenylcyclopentadienone was synthesized (16,18).

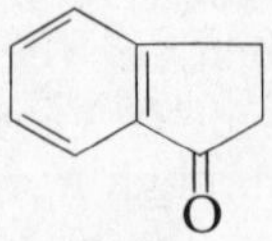

(3)

TABLE 1

Stoichiometric and Catalytic Synthesis of Cyclopentadienones and p-*Benzoquinones from Alkynes and Carbon Monoxide*

Alkynes	Type of synthesis	Catalyst	CO source	Product	Yield (%)	Ref.
HC≡CH	Stoichiometric	—	$Fe(CO)_5$	Quinhydrone	n.d.[a]	14
PhC≡CPh	Stoichiometric	—	CO, 200 atm	Tetraphenylcyclopentadienone	Low	15
PhC≡CPh	Stoichiometric	—	$Ni(CO)_4$	Tetraphenylcyclopentadienone	n.d.	16
HC≡CH	Stoichiometric	—	$Ni(CO)_4$	α-Hydrindone	n.d.	17
PhC≡CPh	Stoichiometric	—	$(CH_3CN)_3Cr(CO)_3$	Tetraphenylcyclopentadienone	Low	18
PhC≡CPh	Stoichiometric	—	$C_5H_5Co(CO)_2$	Tetraphenylcyclopentadienone	Low	13
EtC≡CEt	Stoichiometric	—	$[Rh(CO)_2Cl]_2$	Tetraethylquinone	8	19,20
MeC≡CMe	Stoichiometric	—	$[Rh(CO)_2Cl]_2$	Tetramethylquinone	Low	21
EtC≡CEt	n.d.		$R_2C_2Fe(CO)_4$[b]	Tetraethylquinone	33	8
$F_3CC{\equiv}CCF_3$	Catalytic	$[Rh(CO)_2Cl]_2$	CO, 100 atm	Tetrakis(trifluoromethyl)-cyclopentadienone	Low	22
HC≡CH	Catalytic	$Ru_3(CO)_{12}$	CO, 700 atm	Quinhydrone	14	23

[a] Not determined.
[b] $R = (CH_3)_3C$-.

The synthesis of other substituted cyclopentadienones using cobalt complexes is discussed in Vol. 1 of this series (8).

Quinhydrone and other non-ketonic products were obtained from acetylene and $Fe(CO)_5$ under high pressures (9000 atm) at 110°C (14).

Finally, by reacting $[Rh(CO)_2Cl]_2$ with diethyl or dimethylacetylene (19–21), no free cyclopentadienone derivatives were recovered, but free tetraethyl or tetramethyl-1,4-benzoquinone was formed.

A cyclopentenone derivative of type **4** was obtained from an anomalous stoichiometric carboxylation reaction of 1-bromo-2-butyne with $Ni(CO)_4$ (24).

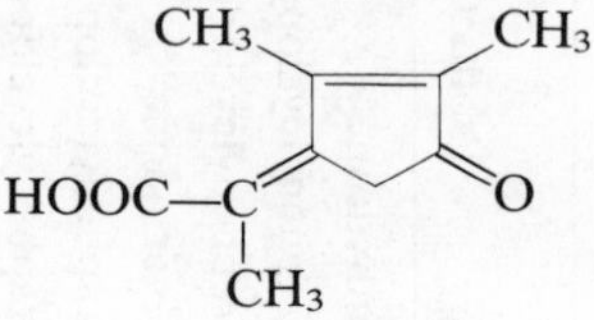

CH_3 CH_3

HOOC—C O

CH_3

(**4**)

The only catalytic synthesis of cyclopentadienone derivatives described to date is the synthesis of tetrakis(trifluormethyl)cyclopentadienone (22) by starting with perfluoro-2-butyne in the presence of $[Rh(CO)_2Cl]_2$ as catalyst. As for the *p*-benzoquinone derivatives, tetraethyl-*p*-benzoquinone was obtained in 33% yield from diethylacetylene in the presence of the complex $(Me_3CC_2CMe_3)Fe(CO)_4$ (8).

However, the true catalytic character of this synthesis was not clearly proved. From acetylene and CO, quinone is formed at high pressure in the presence of ruthenium catalysts, as shown by the presence of quinhydrone in the reaction products (23); probably the hydroquinone is formed because of the presence in the reaction medium of small amounts of water.

The path by which the synthesis of cyclopentadienones and *p*-benzoquinones takes place is still obscure. Both monometallic and cluster complexes, in which one or two metal atoms are bound to the entity **1**, may be intermediate compounds. However, cyclobutadiene metal complexes as well as maleoyl metal complexes may take part in these reactions. From starting with different metal complexes, different types of complexes may well be reaction intermediates. In fact it is known that derivatives of cyclopentadienones are formed in the presence of iron, nickel, rhodium, and chromium complexes (Table 1), whereas quinones are formed mainly in the presence of iron, ruthenium, and rhodium complexes.

For monometallic complexes the paths in Scheme **I** should be considered.

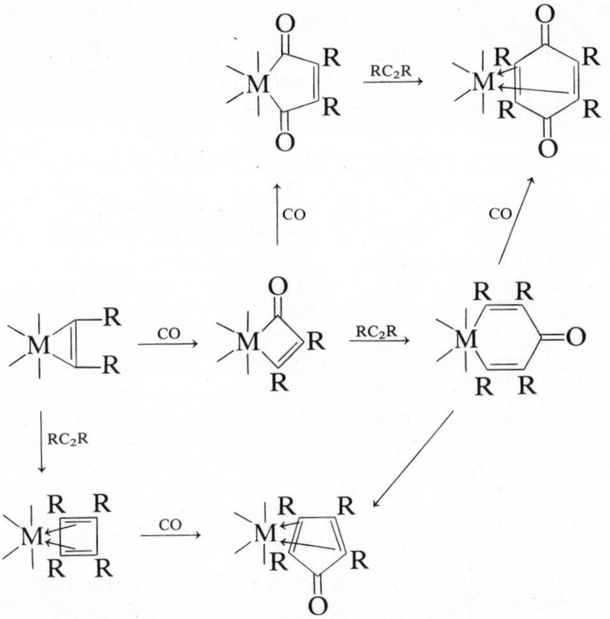

Scheme I

B. Synthesis of Lactones

The catalytic synthesis of unsaturated lactones (Table 2) appears more successful; the organic group is displaced more easily from the intermediate complex by CO and (or) acetylene.

The first hints of this type of reaction arose from an investigation of the influence of the H_2O/C_2H_2 ratio on the synthesis of succinic acid catalyzed by $Co_2(CO)_8$. It was observed that by decreasing the water concentration solid by-products were formed (25). The structure of these products was studied independently by Albanesi (26,27) and by Sauer and his group (28); these authors concluded that the solid products formed from acetylene and CO in aprotic solvents were a mixture of the two lactones **5** and **6**.

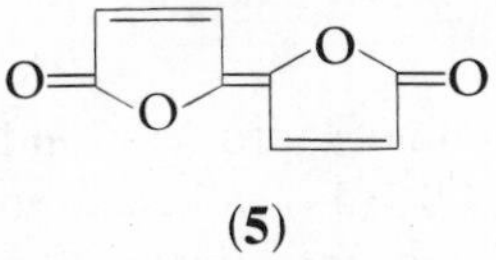

(**5**) 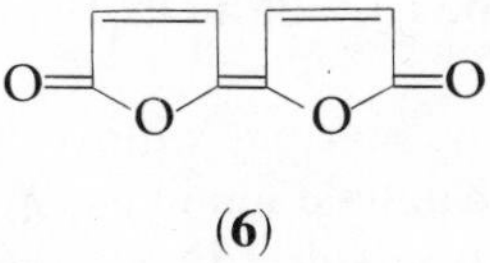 (**6**)

At about the same time, Sternberg, Wender, and others (29) discovered that complexes of the type $RC_2RCo_2(CO)_6$ reacted with CO to give

TABLE 2
Synthesis of Unsaturated Dilactones from Alkynes and Carbon Monoxide Catalyzed by Cobalt Compounds[a]

Alkynes	Solvent	Pressure[b] (atm)	Temp (°C)	Initial CO/C_2H_2 ratio	Products	Yield[c] (%)	Ref.
HC_2H	Acetone[d]	250	90–100	6	**5**+**6**	20–30	26
HC_2H	Tetramethylurea	275–315	90	—	**6**	—	28
HC_2H	Benzene	60	110	1	**7**+**8**	—	33
MeC_2H	Acetone[d]	235	100–110	4	—[e,f]	15–20	27
MeC_2H	Acetone	800–1000	100–110	—	—[e]	41	28
BuC_2H	Acetone	800–1000	100–110	—	—[e]	58	28
PhC_2H	Acetone	800–1000	100–110	—	—[e,g]	6–21	28
EtC_2Et	Acetone	800–1000	100–110	—	—[h]	6	28

[a] $Co_2(CO)_8$ or $Co(OAc)_2$.
[b] Measured at reaction temp.
[c] Based on starting alkyne.
[d] In the presence of small amounts of Ac_2O.
[e] Mixture of disubstituted dilactones.
[f] **9**, **10**, **11** isolated and identified.
[g] Three isomeric products isolated but not identified.
[h] Mixture of tetrasubstituted dilactones.

$RC_2RCo_2(CO)_9$ compounds; here, however, only seven of the nine CO groups present are directly bound to cobalt as terminal carbonyl groups. X-ray examination (30) showed that the complex contained one entity (**2**) bound to two cobalt atoms.

The same type of lactone synthesis was carried out by Reppe (31,32) who, using cobalt catalysts, obtained from acetylene and CO 15–40% yields of lactones **5** and **6**; coumarin was formed as a secondary product.

The synthesis of dilactones such as **5** and **6** (Table 2) can be carried out in various aprotic solvents as acetone or acetonitrile; the presence of tetramethylurea seems to favor the formation of **6**. Temperatures of 90–120°C were used for the synthesis which, although the yield varies, can also be carried out with mono and disubstituted alkynes. Monosubstituted acetylenes give a complicated mixture of isomers difficult to analyze. In the case of methylacetylene (27), isomers **9**, **10**, and **11** were isolated.

Interestingly, starting with 3-diethylamino-1-propyne and CO in the presence of catalytic amounts of $Co_2(CO)_8$ (34), alkylamino-substituted unsaturated furamides **12** are formed in a complicated reaction (44% yield).

To obtain a reasonable yield of the lactones, a CO/C_2H_2 ratio of at least four should be used; with ratios lower than one, compounds **7** and **8** (formally arising from insertion of an acetylene molecule between two lactone rings) were obtained in low yield together with acetylenic oligomers and polymers probably having a linear structure (28).

(**7**) (**8**)

(**9**) (**10**)

(**11**) (**12**)

On the basis of these results, the first step in the lactone synthesis likely involves the synthesis of **13** which was found in the products of the lactone synthesis; this complex can be used as a catalyst instead of $Co_2(CO)_8$ (26).

A double CO insertion into the acetylene–cobalt bonds probably takes place, which is followed by a rearrangement of the —CO—C(R)=C(R)—CO— group (bound to one or two cobalt atoms) to complex **14** (reactions 1, 2).

$$Co_2(CO)_8 + RC{\equiv}CR \longrightarrow (CO)_3Co(\mu\text{-}RC{\equiv}CR)Co(CO)_3 \quad (1)$$

(**13**)

$$\mathbf{13} + 3CO \longrightarrow (CO)_3Co(\mu\text{-}C_4H_2O_2)(\mu\text{-}CO)Co(CO)_3 \quad (2)$$

(**14**)

From the synthesis of **5** and **6** as well as of **7** and **8** and acetylenic polymers, it seems that the entity **2** is readily displaced from **14** both by acetylene and by CO; the way in which the double bond forms between two entities of **2** or between **2** and acetylene is obscure. Either an intramolecular (Eq. 3) or an intermolecular (Eq. 4) path might be postulated for the synthesis of **5** and **6** at high CO/C_2H_2 ratios.

$$2(\mathbf{14}) \rightleftharpoons Co_2(CO)_8 + (CO)_3Co\overset{C(O)}{\underset{C(O)}{—}}Co(CO)_3 \qquad (3)$$

(15)

$$\mathbf{15} + 2CO \rightarrow Co_2(CO)_8 + x\mathbf{5} + y\mathbf{6} \qquad (x + y = 1)$$
$$2(\mathbf{14}) \rightarrow 2Co_2(CO)_8 + m\mathbf{5} + n\mathbf{6} \qquad (m + n = 2) \qquad (4)$$

Both paths can explain non-stereospecific formation of the lactones and formation of many isomers when monosubstituted acetylenes are used.

At low CO/C_2H_2 ratios an insertion reaction of acetylene might take place between the cobalt atoms and the **2** entities present in **14** or **15**, but the available data are not sufficient to explain the formation of **7** and **8**.

In conclusion, the catalytic activity of cobalt in aprotic solvents is strongly dependent, at least qualitatively, on CO pressure; the main products range from the dilactones at high CO pressure to the trimers of substituted acetylenes in the absence of CO (35). The formation of **7** and **8** and of linear polymers which occurs at low CO pressure can be considered as an interesting intermediate case.

III. SYNTHESIS OF ALDEHYDES AND KETONES FROM ACETYLENIC SUBSTRATES

The hydroformylation of acetylenic substrates has not been studied as much as the hydroformylation of olefins. With acetylenic substrates, the hydrocarboxylation reaction has given more interesting results. There are several reasons for this difference. In the first place, acetylenic substrates form stable complexes with metal carbonyls and their derivatives (36), which, therefore, do not act as catalysts at low temperatures. Furthermore, acetylene insertion into the acyl-metal bond, which should be formed during hydroformylation (37,38), is apparently favored over hydrogenolysis of the same bond. Such preference is shown, for instance, in the formation of unsaturated polyketones (39,40).

As with olefins, the reaction of acetylenic substrates with CO and H_2 can give rise both to aldehydes and ketones. Because of the high reactivity of the intermediates in the synthesis of ketones, cyclic ketones containing mostly five- or six-membered rings are often formed instead of aliphatic ketones. Again as with olefins, ketones can form even in the absence of hydrogen if compounds capable of donating hydrogen are present.

We shall treat separately the synthesis of aldehydes and ketones from acetylenic substrates, CO, and H_2 as well as the synthesis of carbonyl compounds by using hydrogen donors instead of hydrogen gas.

The synthesis of hydroquinone from acetylene, CO, and H_2 or hydrogen donors will be considered in this chapter as it occurs under conditions similar to those used for the hydroformylation of acetylenic substrates. Furthermore, hydroquinone can be considered formally as the enolic form of cyclohexen-3,6-dione (**16**).

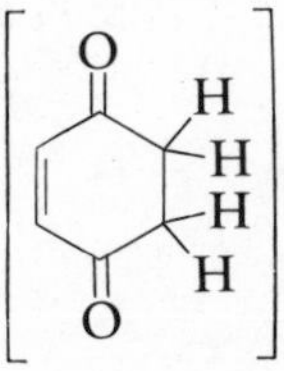

(16)

A. Aldehydes and Ketones from Acetylenic Substrates, Carbon Monoxide, and Hydrogen

The first hydroformylation of acetylene was described by Roelen (6). By using metallic cobalt as catalyst and working at 140–150°C and 1–10 atm he succeeded in transforming acetylene into liquid products containing acrolein. These experiments were never described in detail.

Other authors found that hydroformylation of alkynes occurred more slowly than hydroformylation of olefins (41), and that acetylenes acted as poisons for the hydroformylation of olefins (42). Adkins and Krsek (41) did not succeed in hydroformylating diphenylacetylene but later Greenfield, Wotiz, and Wender (42) converted this compound to diphenylethane under oxo conditions (Table 3). The same authors, however, succeeded in obtaining small quantities of propionaldehyde by reacting acetylene with an excess of cobalt hydrocarbonyl (42). They obtained essentially equivalent amounts of hexanal and 2-methylpentanal

TABLE 3
Hydroformylation of Alkynes

Alkynes	Catalyst	Temp (°C)	Pressure (atm)	CO/H_2	Products	Yield (%)	Ref.
Acetylene[a]	$HCo(CO)_4$	25	1	2	Propanal	Low	42
					High-boiling products		
Acetylene[b]	$Co_2(CO)_8$	100	200	1	Propionaldehyde dimethylacetal	7	47
					Cyclopentanone	—	
					Methyl β-formylpropionate	2	
					Methyl γ-ketopimelate	—	
Acetylene[c]	$Co_2(CO)_8$	110	100	2	Propionaldehyde diethylacetal	Traces	48,49
					Diethyl succinate	34	
					Ethyl γ,γ-diethoxybutyrate	35	
Acetylene[d]	$Co_2(CO)_8$	100	260	1	Cyclopentanone	24	50
1-Butyne[e]	Rh_2O_3	130	200	1	Pentanal	68	51
					2-Methylbutanal	22	
2-Butyne[f]	Rh_2O_3	130	200	1	2-Methylbutanal	65	51
1-Pentyne	$Co_2(CO)_8$	130	300	2	Hexanal	6	42
					2-Methylpentanal	5.5	
1-Hexyne[g]	$RhCl(PPh_3)_3$	110	120	4	Heptanal	15	43
					2-Methylhexanal	15	
1-Hexyne[h]	$RhHCO(PPh_3)_3$	100	50	1	Heptanal	80	46
					2-Methylhexanal	20	
Acetylene	$Ru_3(CO)_{12}$	200	200	12	Hydroquinone	58.5	59
Acetylene[b]	$Rh_2O_3 \cdot 5H_2O$	130	300	2	Methyl β-formylpropionate	8.0	53
					Methyl α-formylpropionate	1.1	
					Hydroquinone	62	

[a] Stoichiometric reaction.
[b] In the presence of MeOH.
[c] In the presence of EtOH; similar results reported operating with isopropanol.
[d] In the presence of water.
[e] $Rh/PPh_3 = 1/30$.
[f] $Rh/PPh_3 = 1/5$.
[g] Hydroformylation of 1-hexene formed by hydrogenation of 1-hexyne.
[h] In molten PPh_3.

by catalytic hydroformylation of 1-pentyne in the presence of $Co_2(CO)_8$. More recently, similar results were obtained with 1-hexyne by Wilkinson and co-workers (43–46) using $(PPh_3)_3RhCl$ as catalyst at 110°C and 120 atm. Better yields were obtained by using an excess of PPh_3; according to Fell and Beutler (51), an Rh/PPh_3 ratio of 1:3 gives a 75% yield of 2-methylbutanal from 2-butyne, but a ratio of at least 1:30 was necessary to obtain 68% of pentanal and 22% of 2-methylbutanal from 1-butyne. In the absence of CO, rapid hydrogenation of the acetylenes took place (43–45).

The formation of aldehydes from acetylenic substrates can be considered as occurring either through a preliminary hydrogenation of the triple bond to a double bond followed by hydroformylation, or by a preliminary hydroformylation followed by hydrogenation of the unsaturated aldehyde thus formed (reaction 5).

Path *a* is in keeping with the facile hydrogenation of the triple bond under the reaction conditions; however, in the case of 1-pentyne, the 1-pentene thus formed, when in the presence of $Co_2(CO)_8$, is known to give rise by hydroformylation to an excess of hexanal (52) and not to approximately equivalent amounts of linear and branched aldehydes as found experimentally with 1-pentyne (43,44). Similarly, with 1-hexyne, hydroformylation in the presence of $(PPh_3)_3RhCl$ as catalyst yielded 50% of heptanal; but when starting with 1-hexene in the presence of a similar catalytic system, a large excess of the linear aldehyde (72% yield) is obtained (44). Therefore path *b* cannot be discarded, although no traces of unsaturated aldehydes were found in the hydroformylation of alkynes with the possible exception of preliminary experiments described by Roelen (6).

$$RC{\equiv}CH \xrightarrow[(a)]{H_2} [RCH{=}CH_2] \xrightarrow[(a)]{CO+H_2} RCH(CHO)CH_3 + RCH_2CH_2CHO$$

$$RC{\equiv}CH \xrightarrow[(b)]{CO+H_2} \left[RC(CHO){=}CH_2 + RCH{=}CHCHO \right] \xrightarrow[(b)]{H_2} RCH(CHO)CH_3 + RCH_2CH_2CHO \qquad (5)$$

The situation has been partially clarified by experiments in which 2-butyne was hydroformylated in the presence of $HRh(CO)(PPh_3)_3$ and (-)-DIOP (44*a*). In this case, (*E*)-2-methyl-2-butenal was isolated in 32% yield. As no (*Z*) isomer of the unsaturated aldehyde was found, hydroformylation of the triple bond seems to occur with the same type of stereochemistry (*cis* addition) as with olefin hydroformylation. The corresponding saturated aldehyde formed (2-methylbutanal) is optically active, the prevailing chirality (*S*) corresponding to that expected for the asymmetric hydroformylation of 2-butene formed according to path *a*. However, the enantiomeric excess (0.2%) is much lower than that expected from hydroformylation of *cis*-2-butene under the same conditions. Furthermore, hydrogenation of (*E*)-2-methylbutenal under the same conditions used for hydroformylation yielded (*S*)-2-methylbutanal with 0.6% optical purity.

From these experiments, it appears that hydroformylation of triple bonds really proceeds according to path *b*. But the experiments do not completely exclude the possibility that formation of the saturated aldehyde, the major product, could also occur according to path *a*.

A completely different picture is revealed when the reaction is carried out in the presence of water or with alcohols as solvents. In methanol with $Co_2(CO)_8$ as catalyst, the reaction between acetylene, CO and H_2 proceeded smoothly even at 100°C and at 200 atm ($CO/H_2 = 1$). Cyclopentanone and β-formylpropionate were the main carbonyl compounds; small amounts of propionaldehyde dimethylacetal and dimethyl γ-ketopimelate also occurred. Beside the carbonyl compounds, methyl esters of mono- and polycarboxylic acids were present (47).

These results were confirmed by Crowe (48,49) who, employing ethanol or isopropanol, found that the main carbonyl compound formed was the ester of β-formylpropionic acid and its acetal (reaction 6). Similar products were obtained in low yields by using methanol as solvent in the presence of soluble rhodium catalysts (53), but the main product was hydroquinone. When aqueous acetone was used, the main product was succinic acid; a 24% yield of cyclopentanone was also obtained (50).

The formation of cyclopentanone and esters of α-ketopimelic acid was also observed in the absence of hydrogen and will be discussed in the following section. However, the formation of propionaldehyde and esters of β-formylpropionic acid occurs only in the presence of hydrogen. Propionaldehyde is probably formed according to reaction 5, path *b* being the more probable one. The esters of β-formylpropionic acid (reaction 6) probably develop from the hydrocarboxylation of acetylene to the corresponding acrylate (path 6*c*); this is known to occur under the same conditions in the absence of hydrogen (54). This hydrocarboxylation is followed by the hydroformylation of the acrylates which is also known to occur under the conditions used (55,56). This route is more likely than path 6*d*; here the postulated acrolein intermediate would be hydrogenated (41,57) rather than hydrocarboxylated under reaction conditions. Another point against the acrolein pathway is that hydrocarboxylation of olefins occurs at much higher temperature (37,38,58).

According to the above hypothesis, acetylenic substrates, unlike olefins, should be more readily hydrocarboxylated than hydroformylated.

$$\begin{array}{ccccc} & & CH_2{=}CHCOOR & & \\ & \overset{CO+ROH}{\underset{(c)}{\nearrow}} & & \overset{CO+H_2}{\underset{(c)}{\searrow}} & \\ CH{\equiv}CH & & & & OHCCH_2CH_2COOR \\ & \overset{CO+H_2}{\underset{(d)}{\searrow}} & & \overset{CO+ROH}{\underset{(d)}{\nearrow}} & \\ & & CH_2{=}CHCHO & & \end{array} \quad (6)$$

The picture changes completely when $Ru_3(CO)_{12}$ is used as catalyst; under the proper conditions (Table 4), hydroquinone is the major product (59–61) (reaction 7).

$$2HC{\equiv}CH + 2CO + H_2 \xrightarrow{Ru_3(CO)_{12}} p\text{-}HOC_6H_4OH \quad (7)$$

The reaction takes place at 180–200°C, a little higher temperature than is used in the hydroformylation of olefins with a ruthenium catalyst (62). To achieve yields of hydroquinone above 50%, the H_2 partial pressure must be kept very low, that is, between 1 and 10 atm. On increasing the H_2 pressure, high-boiling or solid polymeric products of unknown structure are formed.

TABLE 4

Reaction of Acetylene with Carbon Monoxide and Hydrogen in the Presence of $Ru_3(CO)_{12}$[a] (59)

C_2H_2 (moles)	P_{CO} (atm)	P_{H_2} (atm)	Temp (°C)	Time (min)	Hydroquinone Moles	Hydroquinone Yield[b]
0.248	120	10	200	268	0.0725	59
0.240	120	40	200	121	0.0416	37
0.238	75	75	200	128	0.0406	35
0.242	125	7	220	86	0.055	45
0.234	120	15	220	70	0.0495	42
0.238	120	40	220	57	0.0453	38

[a] $Ru_3(CO)_{12}$, 0.1 g; THF, 177 g.

[b] (Moles hydroquinone obtained/0.5 moles acetylene reacted) × 100.

B. Ketones from Acetylene, Carbon Monoxide, and Hydrogen Donors

1. Stoichiometric and Semicatalytic Hydrocarbonylation

The reaction of acetylene with CO in the presence of hydrogen donors (such as water, alcohols, and amines) yields chiefly the corresponding acids, esters, and amides. But, as with olefins, considerable quantities of ketones are formed when these hydrogen donors are used instead of H_2.

The stoichiometric hydrocarbonylation of acetylenic substrates in the presence of hydrogen donors was first carried out by Reppe and his co-workers (17,63). The carbonyls $Fe(CO)_5$, $H_2Fe(CO)_4$, and $HCo(CO)_4$ were used and hydroquinone or substituted hydroquinones were obtained in rather poor yields. The best results were obtained with $Fe(CO)_5$ in an alkaline medium in the presence of water, alcohols, and ethanolamine (Table 5).

It is not clear which of the three compounds functions as the hydrogen donor. The presence in the reaction gases of CO_2, even if occasionally in less than the stoichiometric amount, indicates that water might be the source of the H_2 needed for the hydroquinone synthesis (reaction 8). The stoichiometry of reaction 8 is not well defined at present.

$$4HC{\equiv}CH + Fe(CO)_5 + 2H_2O \longrightarrow 2\,C_6H_4(OH)_2 + [FeCO_3] \qquad (8)$$

Following this work a semicatalytic process was discovered by Reppe, who obtained hydroquinone from acetylene by using $[Fe(NH_3)_6]$-$[Co(CO)_4]_2$ (64,65), CO and water or $Fe(CO)_5$, CO, and hydrated salts (32,66). Here water acts as a hydrogen donor and a stoichiometric amount of CO_2 is formed. High CO pressure (600–700 atm) and high acetylene pressure (up to 80 atm) are necessary to obtain satisfactory yields of hydroquinone, but the reaction is very slow and not more than 1.8 g of hydroquinone is produced per gram of CO donor (Table 6).

The semicatalytic hydroquinone synthesis with $Fe(CO)_5$ takes place with a number of substituted acetylenes (Table 6). Methylacetylene and dimethylacetylene react more rapidly and under milder conditions than does acetylene itself (32).

From the point of view of the reaction mechanism, it is interesting to note that from phenylacetylene or methylacetylene, both 2,5- and 2,6-disubstituted hydroquinones are obtained in a molar ratio of about 1:2 (66). By starting with a 1:1 mixture of propyne and 2-butyne at 700 atm and 190°C, hydroquinone mixtures (containing about 55% of trimethylhydroquinone) in 95% yield (based on the alkynes) are obtained (32).

TABLE 5

Stoichiometric Hydrocarbonylation of Acetylenic Substrates to Hydroquinones

Reaction medium									
Solvent	g	Water (g)	Alkyne (g)	$Fe(CO)_5$ (g)	Temp (°C)	Hydroquinone derivative (g)	$\frac{HQ^{a}\ (g)}{Fe(CO)_5\ (g)}$	Yield[b] (%)	Ref.
Ethanol	80	—	Acetylene						
Monoethanolamine	20	—	20–25 atm	20	80	Hydroquinone (4.33)	0.11	—	17
Monoethanolamine	20	80	Acetylene 14.5 atm	19.6	60–75	Hydroquinone (4.0)	0.20	—	63
Monoethanolamine	20	80	Propyne (16)	20	80–90	2,5-Dimethylhydroquinone (6)	0.30	21.7	63
Monoethanolamine	20	80	2-Butyne (16)	20	110	2,3,5,6-Tetramethylhydroquinone (6.5)	0.33	31.4	63
Monoethanolamine	15	60	Phenylacetylene (34)	15	60–70	2,5-Diphenylhydroquinone (8.9)	0.60	21.6	63
Monoethanolamine	15	60	Hex-3-yn-2,5-diol dimethyl ether (33)	15	70–84	2,3,5,6-Tetrakis(α-methoxyethyl)-hydroquinone (9.3)	0.62	28.5	63
—	—	80	3-Dimethylamino-1-propyne (33)	20	60–70	2,5-bis(Dimethylaminomethyl)-hydroquinone (1.8)	0.09	4.9	63

[a] HQ means hydroquinone derivative.
[b] Moles HQ/(moles alkyne reacted/2) $\times$ 100.

TABLE 6

Semicatalytic Hydrocarbonylation of Alkynes in the Presence of Metal Carbonyls

Reaction medium: Solvent (g)	Reaction medium: Water g	CO donor and catalyst (g)	Alkyne (g)	P_{CO}[a] (atm)	Temp (°C)	Hydroquinone derivative (g)	HQ/catalyst[b] (g/g)	Hydroquinone yield[c] (%)	Ref.
Ethanol (60)	—	$[Fe(NH_3)_6][Co(CO)_4]_2$ (19)	Acetylene 15 atm	15	95	Hydroquinone (17)	0.9	—	64,65
Ethanol (60)	—	$[Fe(NH_3)_6][Co(CO)_4]_2$ (6)	Acetylene 12.5 atm	12.5	90–100	Hydroquinone (10)	1.67	—	64,65
Dioxane (20,000)	520	$Fe(CO)_5$ (1600)	Acetylene 70 atm	680–700	160–165	Hydroquinone (1730)	1.08	69	66
Acetone (200)	6.1	$Fe(CO)_5$ (20) $CaCl_2$ (6.1)	Acetylene 20 atm	680–700	160–165	Hydroquinone (16)	0.8	70	66
Dioxane (320)	5.2	$Fe_2(CO)_9$ (11.9)	Acetylene 80 atm	680–700	160–165	Hydroquinone (16)	1.34	—	66
Dioxane (310)	2.6	$Fe_3(CO)_{12}$ (5.5)	Acetylene 80 atm	680–700	160–165	Hydroquinone (10.1)	1.84	—	66
Dioxane (320)	5.0	$Fe(CO)_5$ (12.8)	Propyne (20)	680–700	140–145	2,5-Dimethylhydroquinone (4.8) 2,6-Dimethylhydroquinone (8.7) Dimethylquinones and quinhydrones (6.6)	1.57	76	66
Dioxane (320)	4.6	$Fe(CO)_5$ (12.8)	2-Butyne (37)	680–700	150–155	2,3,5,6-Tetramethylhydroquinone (31.7) 2,3,5,6-Tetramethylquinone (7.0)	3.03	92	66
Dioxane (320)	8.0	$Fe(CO)_5$ (12.8)	Phenylacetylene (51)	700–720	160–165	2,6-Diphenylhydroquinone (24.4) 2,5-Diphenylhydroquinone (9.5)	2.65	56	66
Dioxane	—[d]	$Fe(CO)_5$	{Propyne (10% by weight) 2-Butyne (10% by weight)}	600–700	170–200	Trimethylhydroquinone 55% Tetramethylhydroquinone 22% 2,5-Dimethylhydroquinone 12% 2,6-Dimethylhydroquinone 11%	—[e]	95	32

[a] Measured at reaction temp.
[b] HQ means hydroquinone derivative.
[c] Moles HQ/(moles alkyne reacted/2) × 100.
[d] Water present, quantity not specified.
[e] 8–15% of CO necessary for carbonylation came from Fe(CO)[illegible]

From 85 to 92% of the CO necessary for the reaction comes from gaseous CO.

An interesting case of ketone formation in stoichiometric carbonylation was reported by Mueller and McArtor (67); they found that diphenylacetylene reacts with $Ni(CO)_4$, HCl, and ethanol in dioxane solution to yield 2,3,4,5-tetraphenylcyclopenta-2-en-1-one (**17**) (reaction 9).

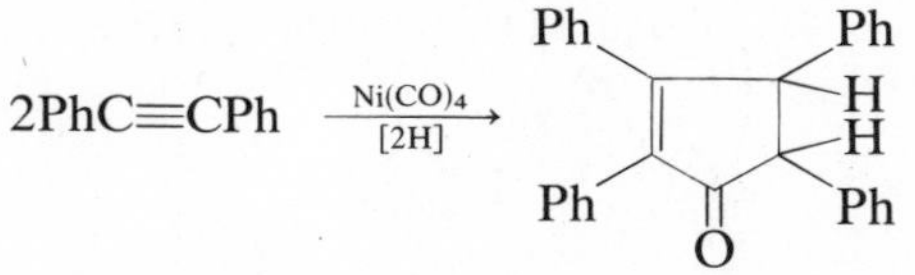

$$2PhC{\equiv}CPh \xrightarrow[{[2H]}]{Ni(CO)_4} \text{2,3,4,5-tetraphenylcyclopent-2-en-1-one} \qquad (9)$$

(**17**)

The authors hypothesize that the hydrogen necessary for the formation of **17** arises from hydrocarboxylation of diphenylacetylene by $Ni(CO)_4$ and HCl and that it is formed by hydrogenation of the corresponding cyclopentadienone.

A peculiar hydrocarbonylation occurs on reacting diphenylacetylene in the presence of $PdCl_2$ and HCl in ethanol. Here, in addition to diethyl diphenylmaleate and ethyl 2,3-diphenylacrylate, there was formed 2,3-diphenyl-γ-crotonolactone in 35–65% yield (68) (reaction 10).

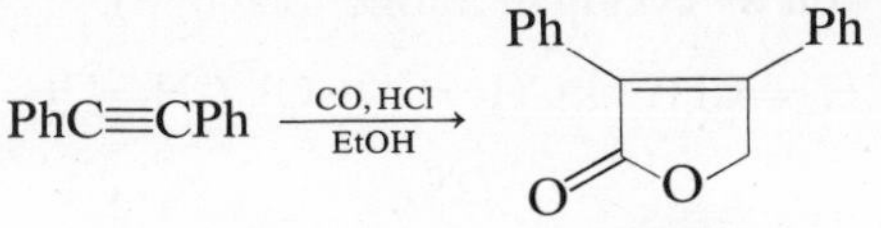

$$PhC{\equiv}CPh \xrightarrow[EtOH]{CO,\,HCl} \text{2,3-diphenyl-}\gamma\text{-crotonolactone} \qquad (10)$$

(**18**)

The origin of the hydrogen necessary for the reaction is not clear, as at least two compounds containing active hydrogen atoms (HCl and EtOH) are present. Two reaction paths seem possible: one involves the following steps: (**a**) hydrocarboxylation of the alkyne; (**b**) hydrocarbonylation of the unsaturated ester; (**c**) cyclization of the β-formyl ester or of the corresponding acid to an oxylactone (68*a*); (**d**) elimination of water; and finally (**e**) isomerization of the resulting unsaturated lactone to **18** (Scheme II).

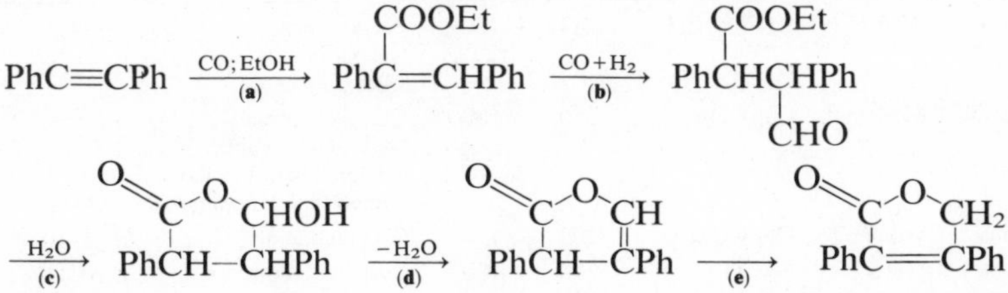

$$PhC{\equiv}CPh \xrightarrow[(a)]{CO;\,EtOH} PhC(COOEt){=}CHPh \xrightarrow[(b)]{CO+H_2} PhCH(COOEt)CH(CHO)Ph$$

$$\xrightarrow[(c)]{H_2O} \underset{\text{PhCH—CHPh}}{\text{O=C—O—CHOH}} \xrightarrow[(d)]{-H_2O} \underset{\text{PhCH—CPh}}{\text{O=C—O—CH}} \xrightarrow[(e)]{} \underset{\text{PhC=CPh}}{\text{O=C—O—CH}_2}$$

(**18**)

Scheme **II**

According to the second path (68), a lactone group bound to two palladium atoms is formed as with cobalt (29); then by hydrogenolysis this gives the lactone **18** (reaction 11). However, no palladium complexes of this type have yet been isolated.

$$\text{[Ph}_2\text{-lactone bridged by two Pd]} \xrightarrow{H_2} \mathbf{18} \qquad (11)$$

2. *Catalytic Hydrocarbonylation*

Interesting results were obtained in the hydrocarbonylation of acetylene when using nickel, cobalt, ruthenium, or rhodium compounds as catalysts. The catalytic hydrocarbonylation yields only ketones but in many cases a simultaneous hydrocarboxylation occurs.

The first work in this field was carried out by Reppe and Magin (39,40). These authors reacted acetylene with CO and water in the presence of $K_2Ni(CN)_3$ or $K_2Ni(CN)_4$ and obtained, besides the expected monocarboxylic acids, polyketones of general formula **19** and ketocarboxylic acids of formula **20** as well as cyclopentanone (Table 7).

$$CH_2{=}CHCO(CH{=}CHCO)_nCH{=}CH_2$$

(19)

$$CH_2{=}CHCO(CH{=}CHCO)_nCH{=}CHCOOH$$

(20)

TABLE 7

Catalytic Synthesis of Ketones from Acetylene, Carbon Monoxide, and Hydrogen Donors

Reaction media							
Solvent	Hydrogen donor	Catalyst	P_{CO} (atm)	Temp (°C)	Ketonic products	Yield[a] (%)	Ref.
—	Water	$K_2Ni(CN)_4$	12	100	Polyketones	—[b]	39,40
					Ketocarboxylic acids	—[b]	
					Cyclopentanone	—[b]	
Methanol	Methanol	$Co_2(CO)_8$	250	100	Cyclopentanone	Traces	69,70
					Cyclopent-2-enone	2-10	
					Methyl γ-ketohexanoate	1	
					Dimethyl γ-ketopimelate	7–15	
Acetone	Water	$Co_2(CO)_8$	200	110	Cyclopentanone	5–10	71
Acetone	Water	$Fe(CO)_5$	200	180	Cyclopentanone	10–15	71

[a] Moles ketone/0.5 moles $C_2H_2 \times 100$.
[b] Not determined.

A large number of different hydrocarbonylation reactions take place when acetylene, CO, and methanol are treated in the presence of cobalt catalysts with relatively high acetylene concentration at 90–110°C and at a CO partial pressure of 200–300 atm. In fact the expected hydrocarboxylation (reaction 12) to methyl acrylate takes place but many other compounds including cyclopentanone, cyclopent-2-en-1-one, methyl γ-ketohexanoate, and dimethyl γ-ketopimelate (69,70) are isolated (Table 7) (Scheme **III**).

$$C_2H_2 + CO + CH_3OH \rightarrow CH_2{=}CHCOOCH_3 \qquad (12)$$

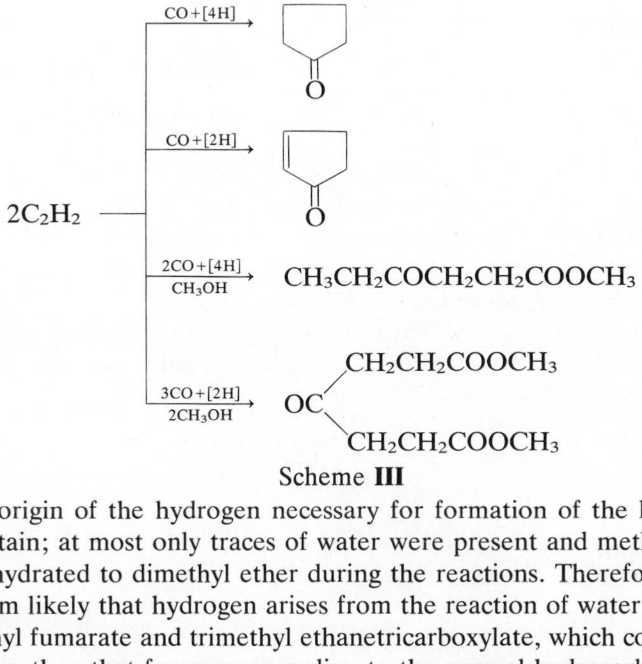

$$2C_2H_2 \xrightarrow{CO+[4H]} \text{cyclopentanone}$$

$$2C_2H_2 \xrightarrow{CO+[2H]} \text{cyclopent-2-en-1-one}$$

$$2C_2H_2 \xrightarrow[CH_3OH]{2CO+[4H]} CH_3CH_2COCH_2CH_2COOCH_3$$

$$2C_2H_2 \xrightarrow[2CH_3OH]{3CO+[2H]} OC(CH_2CH_2COOCH_3)_2$$

Scheme **III**

The origin of the hydrogen necessary for formation of the ketones is not certain; at most only traces of water were present and methanol was not dehydrated to dimethyl ether during the reactions. Therefore it does not seem likely that hydrogen arises from the reaction of water with CO. Dimethyl fumarate and trimethyl ethanetricarboxylate, which contain less hydrogen than that foreseen according to the normal hydrocarboxylation scheme, were present in rather large amounts in the reaction products. The hydrogen necessary for the reactions was thought to originate from the dimethyl succinate (69,70) which may be the obvious intermediate in formation of dimethyl fumarate and of methyl ethanetricarboxylate (reaction 13).

$$C_2H_2 + CO + 2CH_3OH \longrightarrow \left[\begin{array}{c} CH_2COOCH_3 \\ | \\ CH_2COOCH_3 \end{array}\right] \longrightarrow$$

$$\begin{array}{c} HCCOOCH_3 \\ \| \\ H_3COOCCH \end{array} + [2H] \xrightarrow[CH_3OH]{CO} \begin{array}{c} CH(COOCH_3)_2 \\ | \\ CH_2COOCH_3 \end{array} \qquad (13)$$

Although this hypothesis seems likely, the hydrogen necessary for the synthesis of the ketones may have different origins. It may, for example, arise from the decomposition of methanol into CO and H_2, which although improbable cannot in principle be excluded. Finally, since complexes containing a —COCH═CHCO— group bound to cobalt may be intermediates in these syntheses, their reaction with methanol may give rise (reaction 13*a*) to maleic or fumaric esters and to cobalt hydrocarbonyls; the latter may supply the hydrogen necessary for ketone formation.

$$(COCH{=}CHCO)Co_2(CO)_7 + 2CH_3OH \rightarrow HCo(CO)_4 + HCo(CO)_3 + \underset{\displaystyle H_3COOCCH}{\overset{}{HCCOOCH_3}} \quad (13a)$$

(with a double bond between the two CH carbons: $HCCOOCH_3$ ‖ $H_3COOCCH$)

Apart from the origin of hydrogen, the mechanism of the synthesis of the ketonic compounds is far from being elucidated. Another possible pathway to the ketones might be through the intermediate formation of divinyl ketone from acetylene, CO, and hydrogen according to Scheme **IV** (69); this unsaturated ketone may well exist only as part of a cobalt carbonyl complex. The cyclization of divinyl ketone to cyclopent-2-enone was investigated by Nazarov and Nagibina (72), although under different conditions.

$$2H_2C{=}CHCOOCH_3 \xrightarrow{CO+[2H]} OC(CH_2CH_2COOCH_3)_2$$

$$2HC{\equiv}CH + 2CO \xrightarrow{2CH_3OH} 2H_2C{=}CHCOOCH_3$$

$$2HC{\equiv}CH + 2CO \xrightarrow{[2H]} [OC(HC{=}CH_2)_2]$$

$$[OC(HC{=}CH_2)_2] \xrightarrow{2CO,\ 2CH_3OH} OC(CH_2CH_2COOCH_3)_2$$

$$[OC(HC{=}CH_2)_2] \xrightarrow[CH_3OH]{CO+[2H]} OC(CH_2CH_2COOCH_3)(CH_2CH_3)$$

$$[OC(HC{=}CH_2)_2] \rightarrow \text{cyclopent-2-enone} \xrightarrow{[2H]} \text{cyclopentanone}$$

Scheme **IV**

Methyl γ-ketopimelate might arise from methyl acrylate; in fact, this ester was obtained by Matsuda (73) by reacting methyl acrylate with CO and H_2 (20 : 1) at 120°C in the presence of $Co_2(CO)_8$ and pyridine.

Ketonic compounds were also found in the reaction between acetylenic substrates and CO in the presence of water. For instance, in the presence of $Fe(CO)_5$ (71) and $Co_2(CO)_8$ (71) in acid medium or $K_2Ni(CN)_4$ in alkaline medium (40), cyclopentanone was obtained in 15% yield based on acetylene. Furthermore, tetraphenylcyclopentenone was obtained (74) from diphenylacetylene in the presence of $Co_2(CO)_8$.

A rather different catalytic behavior was demonstrated for $RhCl_3$ and $RuCl_3$ (23,53) and for $Ru_3(CO)_{12}$ (60,61,75) (Table 8).

The behavior of rhodium compounds has been described only in patents (23,53). At temperatures of about 150°C and at very high pressures (up to 500 atm), Howk and Sauer (23) obtained relatively low yields of quinhydrone, when the reaction was carried out in the absence of water; but in the presence of water, hydroquinone was the major product. The same results were found by Japanese workers who reported yields of up to 60% of hydroquinone when water was present (53).

The activity of ruthenium catalysts, reported as poor under Sauer's conditions, was more extensively studied by Pino, Braca, and Sbrana (59,61,75,76). These workers obtained hydroquinone in yields of up to 73% (61,75,76), by using $Ru_3(CO)_{12}$, $Ru(acac)_3$, or $K_2Ru(OH)Cl_5$ as

TABLE 8

Synthesis of Hydroquinone from Acetylene, Carbon Monoxide, and Hydrogen Donors with Ruthenium and Rhodium Catalysts[a]

Catalyst						
Compound	g atoms/l	Solvent	Temp (°C)	P_{CO}[b] (atm)	Yield[c] (%)	Ref.
$Ru_3(CO)_{12}$	0.12	Methanol	137–146	750–950	18	23
$Ru_3(CO)_{12}$	0.095	Water	126	890–960	11	23
$Ru(acac)_3$	0.01	Isopropanol	174–178	775–950	19	23
$Ru_3(CO)_{12}$	0.0024	THF	220	250	63	59
$Ru(acac)_3$	0.0024	THF	220	248	61	76
$K_2Ru(OH)Cl_5$	0.0024	THF	230	368	57	76
$Ru_3(CO)_{12}$	0.003	Butanone	250	200	73	76
$RhCl_3$	0.016	Isopropanol	90–104	210–300	26	23
$[Rh(CO)_2Cl]_2$	0.011	Isopropanol	145–200	690–950	33	23
$Rh(acac)_3$	0.01	Isopropanol	140–213	555–920	28	23
$Rh_2O_3 \cdot 5H_2O$	0.0048	Methanol	130	250–350	69	53

[a] Methanol is the hydrogen donor in the first experiment; water and(or) isopropanol is the donor in all other cases.

[b] Measured at reaction temp.

[c] Moles hydroquinone/0.5 moles reacted $C_2H_2 \times 100$.

catalysts, and temperatures between 200 and 250°C with CO at 100–200 atm. During the reaction, about 1 mole of CO_2 was produced for each mole of hydroquinone formed (reaction 14).

$$2C_2H_2 + 3CO + H_2O \longrightarrow p\text{-}HO{-}C_6H_4{-}OH + CO_2 \qquad (14)$$

A study of the influence of different variables on yields and on reaction rates prompted the following conclusions:

(*i*) Acetylene concentration and catalyst concentration affect the initial reaction rate but not the yields.

(*ii*) The water to acetylene ratio, as expected, is very important; the induction period decreases and the initial reaction rate increases on increasing this ratio, whereas the hydroquinone yield decreases when the ratio is below the stoichiometric value (Fig. 1); a similar trend was observed by Reppe (32) in the semicatalytic synthesis of methyl-substituted hydroquinones with $Fe(CO)_5$.

(*iii*) The hydroquinone yield is not significantly changed when the CO partial pressure is varied between 100 and 400 atm; small yields were obtained at lower pressure.

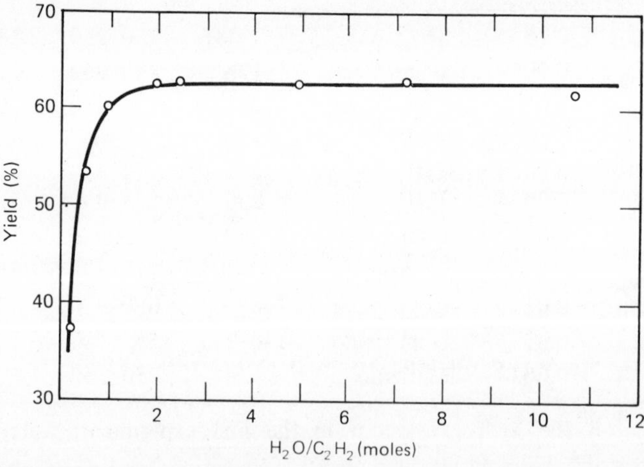

FIG. 1. Effect of H_2O/C_2H_2 molar ratio on the hydroquinone synthesis. Temp 190°C, P_{CO} 200 atm, $Ru_3(CO)_{12}$ 0.1 g, THF 65 g.

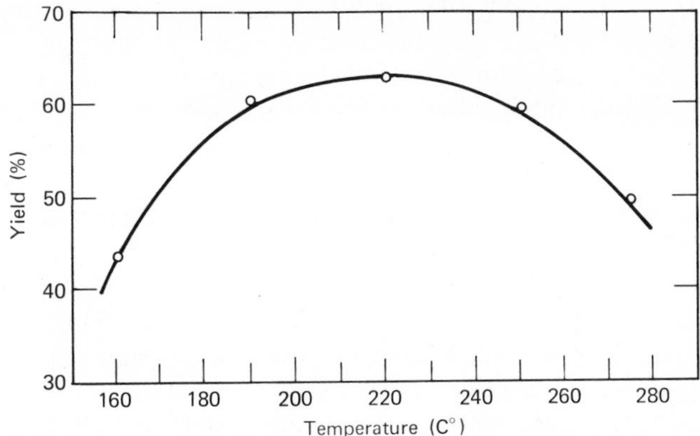

FIG. 2. Effect of temperature on the hydroquinone synthesis, P_{CO} 200 atm, $Ru_3(CO)_{12}$ 0.1 g, $H_2O/C_2H_2 = 2$, THF 65 g.

(*iv*) Temperature has a great effect on the hydroquinone yield; it reaches a maximum at about 220°C (Fig. 2), under the conditions used.

(*v*) The influence of solvent on yields is significant; best results are obtained with ethers or ketones.

(*vi*) The pH of the reaction medium is also important for the hydroquinone synthesis. The reaction does not proceed in an alkaline medium (pH 7.5–8); it proceeds with low yield at pH 6–6.5, but gives best yields at pH 3.5–5. However, when the reaction is carried out in the presence of $Fe(CO)_5$, an alkaline medium seems necessary (17,63).

(*vii*) Hydrogen donors other than water are far less efficient in bringing about the hydroquinone synthesis.

(*viii*) The by-products, which consist of low molecular-weight acids and high molecular-weight liquids and solids of unknown structure, are formed by concurrent reactions, but might also arise at least in part by decomposition of the synthesized hydroquinone; in fact, hydroquinone itself slowly decomposes under reaction conditions in the presence of water to yield CO_2 and solid by-products.

C. Remarks on Hydrocarbonylation of Acetylenic Substrates

If we compare the reactions between acetylenic substrates and CO in the absence and in the presence of H_2 or hydrogen donors, it appears that the carbonylation reaction occurs much more rapidly in their presence to give higher yields, although the selectivity is rather low.

The catalytic formation of cyclic ketones and hydroquinones parallels the formation of cyclopentadienones and *p*-benzoquinones. These products complexed to the transition metals might be the reactive intermediates; then the organic groups in the presence of H_2 or hydrogen donors are hydrogenated, after which weakly π-bonded organic ligands are displaced easily by acetylene or CO. In fact, a facile catalytic hydrogenation of quinone to hydroquinone was observed by reacting *p*-benzoquinone with CO (200 atm) and water in the presence of $Ru_3(CO)_{12}$ at 200°C; these are usual conditions for synthesis of hydroquinone (76).

The formation of β-formylpropionic esters and ketones might occur in the following manner: An entity such as **21**, proposed (77) as a precursor in the synthesis of cyclopentadienone derivatives and isolated in the reaction between hexafluoro-2-butyne and $Ru(CO)_3[P(OMe)_3]_2$ (78), can react with a second molecule of an acetylenic substrate to give a divinylketone skeleton (**22**), which could be the precursor of all the cycloketones and ketoacids (reaction 15). Alternatively, **21** may react with a hydrogen donor to give a molecule of an acrylic derivative which then is hydroformylated to a derivative of β-formylpropionic acid.

$$\mathbf{(21)}\ \text{HC}{=}\text{CH–C(=O)–M (metallacycle)} \begin{cases} \xrightarrow{\text{ROH}} \text{M} + CH_2{=}CHCOOR \\ \xrightarrow{C_2H_2} \text{O}{=}\text{C}_4\text{H}_4\text{M}\ \mathbf{(22)} \end{cases} \qquad (15)$$

Of course, complexes of this type having more than one metal atom could be involved as intermediates, but experimental proof of their existence is lacking at present.

An alternative path (scheme **V**) involving insertion of acetylene into a metal–hydrogen bond (scheme **Va**) cannot be discarded in view of the examples in the literature concerning insertion of acetylenic compounds into H–Ru (79,80), H–Rh (81), H–Ir (82,83), H–Mn (84), and H–Re (84) bonds. A second possibility for the first step may be formation of a cyclic acetylene–ruthenium complex, followed by protonation to vinyl derivative as shown, for instance, in the reaction of $RuC(CF_3){=}C(CF_3)(Cl)(NO)$-$(PMePh_2)_2$ with trifluoroacetic acid (85). The successive CO insertion (scheme **Vb**) is also known for some group VIII metal complexes (86,87) including ruthenium complexes (88).

The resulting acryloyl–metal complex may undergo cleavage by alcohols or water yielding the corresponding acrylic derivatives or an acetylene (scheme **Vc**) and CO (**Vd**) insertion with formation of the γ-keto-1,4-hexadienoyl metal complex.

A less-known way of achieving the synthesis of cyclopentadienone, *p*-benzoquinone, and hydroquinone is ring closure with regeneration of the M–H bond. It might involve insertion of the vinyl group into the M–CO bond followed by elimination of a metal hydride. A double-bond insertion into the metal–acyl bond as shown in scheme **Ve** and **Ve′** was postulated by Heck in the formation of a mixture of cyclohexanone and 2-methyl-2-cyclopentenone from 5-hexenoylcobalt tetracarbonyl (89). A further possibility for the above cyclization is an oxidative addition involving the terminal CH_2 group, as postulated for the codimerization of butadiene and ethylene with iron catalysts (90) and for the dimerization of ethylene with palladium catalysts (91). In this case, the cyclic complex including a metal atom in a seven-membered ring might yield quinone and a metal hydride by a reductive elimination.

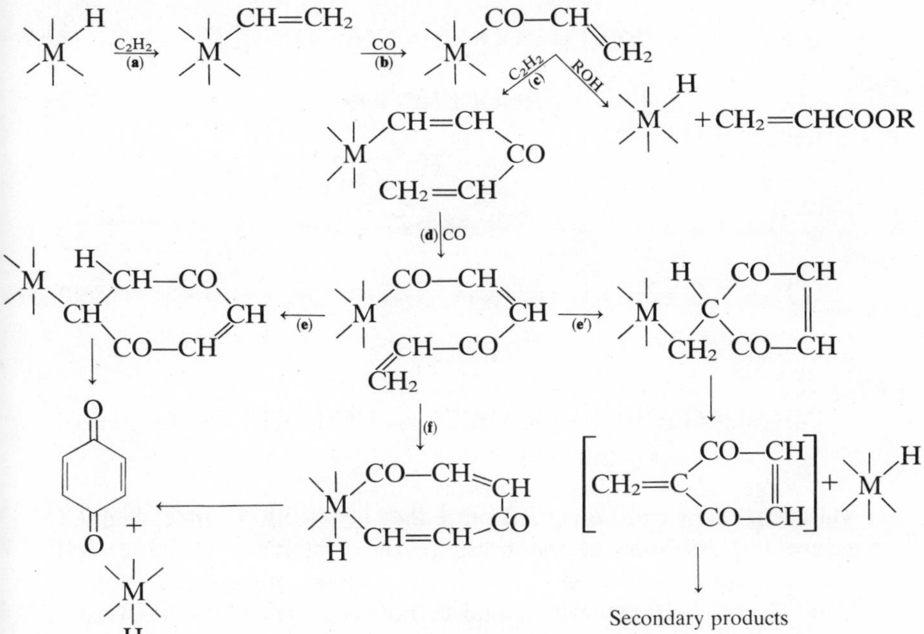

Scheme **V**

IV. ACIDS AND ESTERS FROM ACETYLENIC SUBSTRATES AND CARBON MONOXIDE

The synthesis of acids and esters is by far the most important of the acetylene carbonylation reactions and the only one which has found practical application not only in the laboratory but also on an industrial scale.

Unsubstituted and substituted acrylic and succinic acids and their esters are the main products that can be prepared from acetylenic substrates, CO, and water or alcohols. Moreover, other mono- and polycarboxylic acids and esters were obtained but in smaller yields.

Reactions 16–21 are the most important reactions observed when using acetylene itself as substrate.

$$C_2H_2 + CO + ROH \rightarrow CH_2{=}CHCOOR \qquad R = H, \text{alkyl, aryl} \tag{16}$$

$$C_2H_2 + 2CO + 2ROH \rightarrow \begin{array}{l} CH_2COOR \\ | \\ CH_2COOR \end{array} \tag{17}$$

$$C_2H_2 + 2CO + 2ROH \rightarrow ROOCCH{=}CHCOOR + [2H] \tag{18}$$

$$C_2H_2 + 3CO + 3ROH \rightarrow \begin{array}{c} ROOCCHCOOR \\ | \\ CH_2COOR \end{array} + [2H] \tag{19}$$

$$C_2H_2 + 2CO + 2ROH \rightarrow \begin{array}{c} COOR \\ | \\ CH_3CH \\ | \\ COOR \end{array} \tag{20}$$

$$2C_2H_2 + 2CO + 2ROH \rightarrow ROOCCH{=}CHCH{=}CHCOOR + [2H] \tag{21}$$

These reactions can be carried out either at 90–200°C under high CO pressure (30–200 atm) in the presence of compounds of group VIII metals as catalysts (catalytic procedure) or under milder conditions at 25–100°C and 1–10 atm with metal carbonyls (particularly $Ni(CO)_4$) in quantities stoichiometric with the acetylenic compounds. The type of acids and esters obtained depends not only on the metal used as catalyst but also on the particular metal complex used; the products may also depend on the pH of the reaction medium.

Our survey of the large amount of experimental data on the synthesis of acids and esters from acetylenic substrates will be based on the metal derivatives used as catalyst or reactant. Nickel and cobalt, which have been most widely used, will first be considered separately, then the other group VIII metals will be discussed together.

A. Acids and Esters from Acetylenic Substrates and Carbon Monoxide with $Ni(CO)_4$ as Reactant or Nickel Compounds as Catalysts

1. Stoichiometric Reaction between Acetylenic Substrates and $Ni(CO)_4$

That aldehydes are formed in fair yield by reacting aromatic compounds with $Ni(CO)_4$ in the presence of $AlCl_3$ has been known since 1904 (92). Reppe and co-workers (2,3) observed that acetylenic substrates dissolved in alcohols react rapidly with $Ni(CO)_4$ at 30–70°C when inorganic acids are present. Instead of aldehydes the reaction products consisted of unsaturated esters or, when water was present, unsaturated acids arising from addition of a hydrogen atom and a carboxyl group across the triple bond (hydrocarboxylation). When an excess of acetylenic compound was used, $Ni(CO)_4$ was transformed completely into the nickel salt of the acid present.

For the case of acetylene, the stoichiometry shown in reaction 22 was assumed by Reppe.

$$4C_2H_2 + Ni(CO)_4 + 2HCl + 4H_2O \rightarrow 4CH_2{=}CHCOOH + NiCl_2 + [2H] \quad (22)$$

Hydrogen was not evolved, and according to Reppe's results (2), later confirmed by Ohashi (93–95), the acrylic derivatives were always contaminated by the corresponding propionic acid derivatives and by high-boiling products. These by-products formed in only minute amounts in the synthesis of acrylic acid from acetylene, $Ni(CO)_4$, and water when using acrylic acid for formation of the nickel salt. However, when ethanol was the reaction medium, 12–16% of the ethyl acrylate formed was hydrogenated to the corresponding propionate. The same reaction gave a mixture of high-boiling products consisting mainly of ethyl esters of allylacetic and tetrahydrobenzoic acids. As shown by the overall reactions 23 and 24, the synthesis of these products from acetylene requires hydrogen, besides CO and ethanol.

$$2C_2H_2 + CO + EtOH + H_2 \rightarrow CH_2{=}CHCH_2CH_2COOEt \quad (23)$$

$$3C_2H_2 + CO + EtOH + H_2 \longrightarrow \text{(cyclohexene ring bearing COOEt)} \quad (24)$$

Moreover, in the synthesis of methyl acrylate, the unreacted acetylene was contaminated with ethylene and ethane. The formation of these by-products requires about 4% of the potential hydrogen liberated in the synthesis. The data of Table 9 suggest that the type of acid used has a large effect on the composition of reaction products. In the absence of water, more propionate is obtained with acetic acid than when HCl is present.

The effect of water on the synthesis of acrylic derivatives was confirmed by Jones and co-workers (96,97), particularly when the synthesis of acrylic derivatives from substituted acetylenes was carried out in the presence of weak acids such as glacial acetic acid (2).

The effect of the type of acid used was also investigated by Jones and co-workers (97) in the stoichiometric hydrocarboxylation of phenylacetylene and butylacetylene. They found that the yield of acrylic derivatives was not related to the p*K* of the acid used and that some of the acids used (oxalic, trichloroacetic and α-phenylacrylic acid) did not react. They concluded that the hydrocarboxylation of substituted acetylenes has a stoichiometry much more complicated than that postulated by Reppe for acetylene. Indeed, the effect of the water present in the reaction medium seems to be much more important than the type of acid used.

The role of pH in the stoichiometric hydrocarboxylation has not been completely clarified as yet. It was shown (97) that an acid medium is not essential for starting the reaction between phenylacetylene and $Ni(CO)_4$. In neutral medium, however, only a small amount of 1,3,5-triphenylbenzene and much polymeric material were recovered but no acrylic derivative was found. It was then found (98) that diphenylacetylene can be hydrocarboxylated to α-phenylcinnamic acid in alkaline medium. In this case, the presence of the anion $[Ni_3(CO)_8]^{2-}$ was demonstrated in the reaction medium; this anion is believed to act as a CO donor.

Starting with acetylene in the absence of water, and using less than two equivalents of acid per mole of $Ni(CO)_4$ in the presence of complexing agents (*N*-vinylpyrrolidone, trisodium salt of nitrilotriacetic acid, etc.), a

TABLE 9
Stoichiometric Synthesis of Methyl Acrylate (2)

Reaction medium (wt %)					Products (wt %)				Yield (%)	
MeOH	H_2O	Acid	$Ni(CO)_4$	C_2H_2 (moles)	Methyl acrylate	Acrylic acid	Methyl propionate	Propionic acid	Total[a]	Acrylic derivatives
57	17	HCl 11	15	2.1	85	3	—	12	97	82
67	—	HCl 13.5	19.5	2.0	77	—	23	—	92	71
65.5	—	HCl 13	21.5	2.5	95	—	5	—	82	79
44.5	22.3	AcOH 22.2	11	1.9	70	17	—	14	75	52
60	—	AcOH 30	10	1.1	32	—	68	—	71	23

[a] Based on reacted $Ni(CO)_4$.

mixture of esters of butadiene dicarboxylic acid and of α-ω-dicarboxylic unsaturated acid esters were obtained besides acrylic derivatives (99–101).

Solvents such as ethers, esters, ketones, and pyridine have little effect on the synthesis of α-phenylacrylic acid from phenylacetylene (97). The type of solvent occasionally influences the duration of the induction period usually observed in the reaction.

The type of alcohol used also does not greatly influence the hydrocarboxylation of acetylenic substrates. In such reactions, a large number of acrylic esters of alcohols and glycols was obtained (2,102). This fact might be explained by postulating that, in general, the primary reaction product at low water concentrations is always an acid or an anhydride; these would react with alcohols to give the esters.

As expected, according to this hypothesis, the ester/acid ratio in the hydrocarboxylation of phenylacetylene in alcohols increases with increasing strength of the acid used; the stronger acid should be a better esterification catalyst (97). Moreover, the yield of esters decreases with increasing molecular weight of the alcohol used (103).

Very few limitations have been found in the use of acetylenic substrates in the stoichiometric synthesis of acids and esters. Acetylenic hydrocarbons (Table 10), halogenoacetylenes (Table 11), and acetylenic alcohols, esters, and acids (Table 12) react rapidly but the yields are seldom higher than 60%; nitrogen-containing acetylenic compounds also undergo hydrocarboxylation (104).

With monosubstituted alkylacetylenes, the reaction rate decreases with increasing size of the alkyl group (103). The hydrocarboxylation of acetylenic hydrocarbons (Table 10) can give two types of products according to reaction 25:

$$RC{\equiv}CR' \xrightarrow[H_2O]{Ni(CO)_4} \begin{cases} RC(COOH){=}CHR' \\ RCH{=}CR'(COOH) \end{cases} \qquad (25)$$

In general, the monosubstituted acetylenic hydrocarbons give mostly 2-alkyl derivatives of acrylic acid contrary to what is known for the hydrocarboxylation of terminal olefins (116). An attempt to investigate systematically the stoichiometric hydrocarboxylation of octynes was made by Tetteroo (107). On operating at 50°C in an acetone–acetic acid medium, yields of 30–50% were obtained, based on the acetylenic

TABLE 10
Stoichiometric Hydrocarboxylation of Acetylenic Hydrocarbons with $Ni(CO)_4$

Alkyne	Reaction medium	Temp (°C)	Products	Yield (%)	Ref.
Propyne	Methanol HCl Ethyl ether	45–50	Methyl methacrylate	50	103
1-Butyne	Methanol HCl Ethyl ether	58–60	Methyl 2-ethylacrylate	45	103
1-Hexyne	Ethanol HCl (conc.)	50–70	Ethyl 2-butylacrylate	35–52	96,103
1-Heptyne	Methanol Acetic acid Water	55–60	Methyl 2-pentylacrylate Methyl octa-2-enoate	54 3	106
1-Octyne	Acetone Ethanol Acetic acid	35–40	Ethyl 2-hexylacrylate	63	2
1-Octyne	Acetone Water Acetic acid	35–40	2-Hexylacrylic acid	20	2
1-Octyne	Acetone Water Acetic acid Water	50–55	3-Hexylacrylic acid 2-Hexylacrylic acid	3 28	107

2-Octyne	Acetic acid Acetone	60	2-Methyl-3-pentylacrylic acid	n.d.	107
2-Octyne	Acetone Water Acetic acid	50–55	2-Methyl-3-pentylacrylic acid 2-Pentyl-3-methylacrylic acid	26 21	107
3-Octyne	Acetone Water Acetic acid	50–55	2-Ethyl-3-butylacrylic acid 2-Butyl-3-ethylacrylic acid	25 22	107
4-Octyne	Acetone Water Acetic acid	50–55	2,3-Di-*n*-propylacrylic acid	50	107
2-Nonyne	Acetic acid Water	50–60	2-Hexyl-3-methylacrylic acid 2-Methyl-3-hexylacrylic acid	33	2
5-Decyne	Ethanol Water Acetic acid	75	2,3-Dibutylacrylic acid and ethyl ester	52[a]	109
Phenylacetylene	Ethanol Water Acetic acid	70	2-Phenylacrylic acid	48	2,96
Phenylacetylene	Ethanol HCl (conc.)	55	Ethyl 2-phenylacrylate	33	2,96
1-Phenyl-1-propyne	Acetone Water Acetic acid	30–35	α-Methylcinnamic acid α-Phenylcrotonic acid	54	2,108

TABLE 10 (continued)

Alkyne	Reaction medium	Temp (°C)	Products	Yield (%)	Ref.
Diphenylacetylene	Ethanol Water Acetic acid	70	α-Phenyl-*trans*-cinnamic acid and ethyl ester	48[a]	108,109
Diphenylacetylene	Ethanol Benzene HCl (conc.)	40	Ethyl 2-phenylcinnamate[b]	34	67
Vinylacetylene	Ethanol, Ether, HCl (conc.)	20–40	Diethyl 1-vinylcyclohex-3-ene-1,4-dicarboxylate	n.d.	103,105
Isopropenylacetylene	Methanol Acetic acid	60	1-Methyl-5-isopropenyl-cyclohexene-3,4-dicarboxylic acid	n.d.	110

[a] Combined yields of acid and ester.
[b] Principal reaction product was 2,3,4,5-tetraphenylcyclopenta-2-en-1-one.

TABLE 11

Stoichiometric Hydrocarboxylation of Halogenoacetylenes with $Ni(CO)_4$

Acetylenic compound	Reaction medium	Temp (°C)	Products	Yield, (%)	Ref.
3-Chloro-1-propyne	*a*	65	2,3-Butadienoic acid	6	111–113
3-Bromo-1-propyne	*b*	70	Ethyl 3-bromo-3-butenoate	24	112
3-Chloro-1-butyne	*a*	40	2,3-Pentadienoic acid	13	111
1-Chloro-2-butyne	*a*	45	2-Methyl-2,3-butadienoic acid	15	24
1-Bromo-2-butyne	*a*	45	4-1′-Carboxyethylidene-2,3-dimethylcyclopent-2-enone	5	24
			Dimethylmaleic anhydride	3	
1-Iodo-2-butyne	*a*	45	Dimethylmaleic anhydride	14	24
			4-1′-Carboxyethylidene-2,3-dimethylcyclopent-2-enone	1	
3-Chloro-3-methyl-1-butyne	*a*	40	4-Methyl-2,3-pentadienoic acid and ethyl ester	45[c]	111
5-Bromo-1-pentyne	*d*	70	2-Methylene-5-bromopentanoic acid and ethyl ester	40[c]	114
3-Chloro-1-hexyne	*a*	40	2,3-Heptadienoic acid	10	111
1-Chloro-2-heptyne	*a*	40	2-Butyl-2,3-butadienoic acid	51[c]	111
1-Bromo-2-heptyne	*a*	45	2-Butyl-2,3-butadienoic acid	12	24
			2-Butyl-4-1′-carboxypentylidene-3-methylcyclopent-2-enone	10	
1-Iodo-2-heptyne	*a*	40	2-Butyl-2,3-butadienoic acid	22[c]	24
			2-Butyl-4-1′-carboxypentylidene-3-methylcyclopent-2-enone	15	
2-Chloro-2-methyl-3-octyne	*a*	40	2-Butyl-4-methyl-2,3-pentadienoic acid	13	111
3-Chloro-3-phenyl-1-propyne	*a*	40	4-Phenyl-2,3-butadienoic acid	12	111

[a] Ethanol, acetic acid, sodium acetate and water.
[b] Ethanol and conc. HCl.
[c] Combined yield of acid and ester.
[d] Ethanol, water, and acetic acid.

TABLE 12

Stoichiometric Hydrocarboxylation of Oxygenated Acetylenes with $Ni(CO)_4$[a]

Acetylenic substrate	Temp (°C)	Products	Yield (%)	Ref.
Propargyl alcohol	40	2-Hydroxymethyl acrylic acid	18	115
Propargyl alcohol[b]	45	Ethyl 2-Hydroxymethylacrylate	58	115
		Ethyl *trans*-4-hydroxycrotonate	11	
1-Butyn-3-ol[c]	40–65	2-Methylene-3-hydroxybutanoic acid	n.d.	2
3-Butyn-1-ol	80	α-Methylene-4-butyrolactone	23	96
4-Pentyn-1-ol	80	α-Methylene-5-valerolactone	21	96
4-Pentyn-2-ol	80	α-Methylene-4-methyl-4-butyrolactone	30	96
3-Pentyn-2-ol	75	2-Methyl-4-hydroxy-2-pentenoic acid	60	109
3-Acetoxy-1-hexyne	65	2-Methylene-3-acetoxyhexanoic acid	48	96
2-Methyl-3-butyn-2-ol[d]	65	4-Hydroxy-4-methyl-2-pentenoic acid lactone	17	111
		4-Methyl-2,3-pentadienoic acid	8	
3-Methyl-1-pentyn-3-ol[d]	65	4-Hydroxy-4-methyl-2-hexenoic acid lactone	16	111
		4-Methyl-2,3-hexadienoic acid	18	
2-Methyl-4-pentyn-2-ol	80	2-Methylene-4-methyl-4-hydroxypentanoic acid lactone	7	96
2-Ethynylcyclohexyl acetate[b]	75	2-(1-Acetoxycyclohexyl)acrylic acid	3	96
1-Phenyl-2-propyn-1-ol	80	2-Methylene-3-hydroxy-3-phenylpropanoic acid	10	96
Propargyl acetate	75	2-Methylene-3-acetoxypropanoic acid	40	96
1-Acetoxy-3-butyne	75	2-Methylene-4-acetoxybutanoic acid	47	96
3-Butyn-1-yl-2-tetrahydro-pyranyl ether	75	2-Methylene 4-(2-tetrahydropyranyloxy)butanoic acid	20	96

2-Butyne-1,4-diol-diacetate	65	1,4-Diacetoxy-2-carboxy-2-butene	58	109
4-Pentyn-1-yl-*p*-toluene-sulphonate[e]	60	*p*-Toluenesulphonate of 2-methylene-5-hydroxypentanoic acid	2	114
2-Heptyn-1-yl-p-toluene-sulphonate[f]	40	2-Butyl-2,3-butadienoic acid	31	2,24
		2-Butyl-4,1′-carboxypentylidene-3-methylcyclopent-2-enone	2	
1-Phenyl-1-acetoxy-3-propyne	75	2-Methylene-3-acetoxy-3-phenylpropanoic acid	50	96
3-Pentyn-2-one[g]	65	2-Methyl-4-oxo-2-pentenoic acid	30	109
3-Octyn-2-one	65	2-Butyl-4-oxo-2-pentenoic acid	40	109
Sodium propiolate	70	*Trans,trans*-muconic acid	1	104
Ethyl-3-butynoate	70	Ethyl 3-carboxy-3-butenoate	28	104
Ethyl 4-pentynoate	70	Ethyl 4-carboxy-4-pentenoate	46	104
Ethyl-5-hexynoate	70	Ethyl 5-carboxy-5-hexenoate	40	104
Ethyl-6-heptynoate	70	Ethyl 6-carboxy-6-heptenoate	37	104,109
2-Heptynoic acid	75	Butylfumaric acid	28	109
Diethyl acetylenedicarboxylate	70	Hexaethyl mellitate	18	109

[a] Unless otherwise specified, the reaction medium consisted of ethanol, acetic acid, and water.
[b] With ethanol and acetic acid.
[c] With acetic acid and water.
[d] With butanol and water.
[e] With acetone, water, and acetic acid.
[f] With ethanol, acetic acid, sodium acetate, and water.
[g] With methanol, water, and acetic acid.

substrate. From 1-octyne the prevailing product was 2-methyleneoctanoic acid; from 2- and 3-octyne about a 1:1 ratio for the expected acids was obtained. From 4-octyne only 2,3-di-*n*-propylacrylic acid was obtained, thus showing that no shift of the triple bond occurs.

Among the disubstituted acetylenic hydrocarbons, methylphenylacetylene yields approximately equal amounts of the two isomeric products. In the hydrocarboxylation of substituted diphenylacetylenes, electron-donor substituents (methoxy, methyl, etc.) in the *para* position favor carboxylation of the acetylenic carbon atom adjacent to the *para*-substituted phenyl group; the opposite occurs with electron-withdrawing substituents (Cl and NO_2) (117) (Table 13). Any group in the *ortho* position induces carboxylation of the carbon atom adjacent to the substituted phenyl group.

Diacetylene has not been hydrocarboxylated. A low yield of an unsaturated acid was obtained from 1,5-hexadiyne (103); only polymeric products were obtained from other diacetylenic hydrocarbons (109).

An anomalous reaction occurs when vinylacetylene is treated with $Ni(CO)_4$ and alcohols (103,105). The expected primary product, 2-vinylacrylic acid, dimerizes to the ester of 1-vinylcyclohex-3-ene-1,4-dicarboxylic acid (**23**) and not the isomeric diester **24** suggested by Reppe (reaction 26). With substituted vinylacetylenes, the reaction is favored by the presence of pyridine (114).

$$CH_2{=}CHC{\equiv}CH \xrightarrow[ROH]{Ni(CO)_4} CH_2{=}CH\underset{}{C}(COOR){=}CH_2$$

$$2CH_2{=}CHC(COOR){=}CH_2 \longrightarrow \text{(23)} \quad \text{not} \quad \text{(24)} \qquad (26)$$

(23): ROOC, CH=CH₂ on C-1 of cyclohexene ring; COOR on C-4

(24): ROOC–(cyclohexenyl)–C(COOR)=CH₂

The stoichiometric syntheses of acids and esters from halogenoacetylenes are summarized in Table 11. When the triple bond is far from the halogen atom, as in 1-bromo-4-pentyne, the normal reaction occurs (114). Anomalous results occur when the halogen atom is substituted for an acetylenic hydrogen and with propargylic halides (24,111,112,114). Both 1-iodo-1-hexyne and 1-iodo-2-phenylacetylene are transformed

TABLE 13
Hydrocarboxylation of Monosubstituted Diphenylacetylenes, ArC≡CPh (117)

	Products (wt %)	
Ar group	ArCH=C(Ph)COOH	PhCH=C(Ar)COOH
p-$CH_3OC_6H_4$	37	63
p-$CH_3C_6H_4$	46.5	53.5
C_6H_5	50	50
m-$CH_3OC_6H_4$	53	47
p-ClC_6H_4	62	38
m-ClC_6H_4	66	34
p-$NO_2C_6H_4$	85	15
o-$CH_3OC_6H_4$	32	68
o-$CH_3C_6H_4$	25	75
o-ClC_6H_4	27.5	72.5

into the corresponding hydrocarbons (114) (reaction 27).

$$2RC{\equiv}CI + 2Ni(CO)_4 + 2AcOH \rightarrow 2RC{\equiv}CH + NiI_2 + Ni(OAc)_2 + 8CO \quad (27)$$

Surprisingly, the acetylenic hydrocarbons thus formed are hydrocarboxylated only in the presence of a very large excess of $Ni(CO)_4$ and after long heating times.

When the halogen atom is on the carbon atom adjacent to the triple bond, as in propargyl halides, allenic acids are formed (24,111,113) (reaction 28).

$$HC{\equiv}CCH_2Cl \xrightarrow[H_2O(-HCl)]{Ni(CO)_4} CH_2{=}C{=}CHCOOH \quad (28)$$

This reaction goes also in the absence of added acids, and it is quite general as shown in Table 11. Yields, however, are not too high.

With 1-bromo- or 1-iodo-2-heptyne, large amounts of the ketoacid **25**, as well as some allenic acid, are formed (24). The same type of product is obtained from 1-bromo-2-butyne, whereas 1-iodo-2-butyne gives dimethylmaleic anhydride (**26**) as the main product. The high reactivity of this substrate results in secondary products and accounts for the low yield of allenic acids.

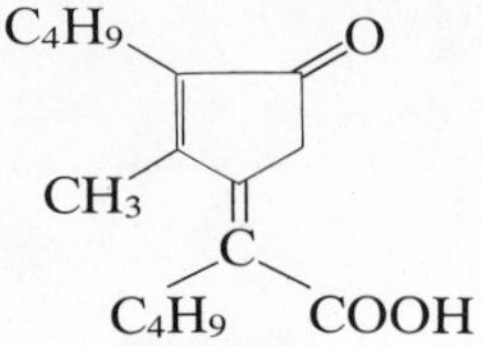

$$C_4H_9,\ CH_3,\ O,\ C,\ C_4H_9,\ COOH \quad \textbf{(25)}$$

$$CH_3,\ CH_3,\ O,\ O,\ O \quad \textbf{(26)}$$

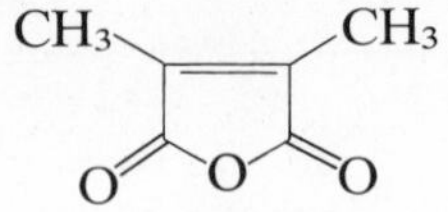

Oxygenated acetylenic substrates (including carbinols, esters, and acetylenic acids) also undergo hydrocarboxylation (Table 12). Some groups such as $—CH_2OH$, $—CH_2OCOCH_3$, $—CH_2CH_2OH$, $—CH_2CH_2OCOCH_3$ facilitate the hydrocarboxylation, while others such as $—COCH_3$, $—COOH$, $—COOEt$, $—CR_2OH$, $—CR_2OCOCH_3$ seem to hinder the reaction (96,104,109).

Jones, Shen, and Whiting (96) investigated the hydrocarboxylation of propargyl alcohol using $Ni(CO)_4$, acetic acid, and water but did not succeed in isolating the reaction products. Under similar conditions, Rosenthal and co-workers (115) obtained α-(hydroxymethyl)acrylic acid in low yield (18%). But after esterification of the reaction products with CH_2N_2, 34% of **27** and 42% of the acetoxy derivative **28** were obtained. These products could also be obtained by direct hydrocarboxylation of propargyl acetate (30% yield) (reaction 28*a*).

A small amount of ethyl *trans*-γ-hydroxycrotonate was isolated in this reaction. This finding is rather surprising since acetylenes having a terminal triple bond are mainly hydrocarboxylated in accord with the Markownikoff rule.

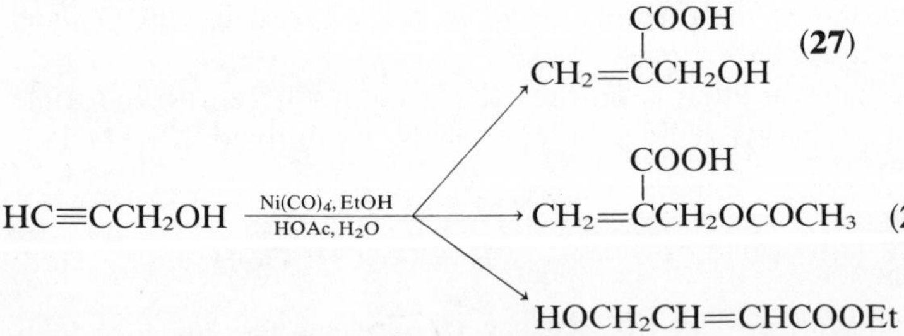

$$HC{\equiv}CCH_2OH \xrightarrow[HOAc,\ H_2O]{Ni(CO)_4,\ EtOH} \begin{cases} CH_2{=}C(COOH)CH_2OH & \textbf{(27)} \\ CH_2{=}C(COOH)CH_2OCOCH_3 & \textbf{(28)} \\ HOCH_2CH{=}CHCOOEt \end{cases} \quad (28a)$$

When the hydrocarboxylation of 2-methyl-3-butyn-2-ol and 3-methyl-1-pentyn-3-ol was carried out in the presence of HCl, the corresponding allenic acids were found in the reaction products (111). The reaction probably involves preliminary formation of the acetylenic chloride which then reacts as a propargyl halide (reaction 29).

$$CH_3\overset{\overset{OH}{|}}{\underset{\underset{CH_3}{|}}{C}}C{\equiv}CH \xrightarrow{HCl} CH_3\overset{\overset{Cl}{|}}{\underset{\underset{CH_3}{|}}{C}}C{\equiv}CH \xrightarrow{Ni(CO)_4,\,BuOH} CH_3\underset{\underset{CH_3}{|}}{C}{=}C{=}CHCOOBu \quad (29)$$

(29) $\xrightarrow[H_2]{Pd}$ **(30)**

In the presence of acids, the lactone **29** was also found in the reaction products. The reaction path outlined seems more likely than that postulated by Bergmann and Zimkin (110) who based their hypothesis on the structure of the hydrogenated product **30**; they proposed that the hydrocarboxylation of the acetylenic alcohol occurred at the acetylenic group.

The hydrocarboxylation of acetylenic esters of general formula $HC{\equiv}C(CH_2)_nCOOR$ occurs readily, except that the first member of the series ($n = 0$) does not react. The corresponding esters of α-methylenedicarboxylic acids are formed (104).

Unlike free propiolic acid, its sodium salt in stoichiometric hydrocarboxylation gives *trans, trans*-muconic acid in small yield (104) (reaction 30).

$$2HC{\equiv}CCOO^- \xrightarrow{Ni(CO)_4} {}^-OOCCH{=}CHCH{=}CHCOO^- \quad (30)$$

Among the nitrogen-containing acetylenic compounds, propiolamide does not undergo hydrocarboxylation but hex-5-ynonitrile gives the expected half-nitrile (104). As primary and secondary amines react easily with CO and acetylene to form acrylamides (2), a complicated reaction mixture would be expected in the hydrocarboxylation of acetylenic amines.

2. *Semicatalytic Hydrocarboxylation*

The stoichiometric hydrocarboxylation of acetylene is not suitable for industrial application chiefly because of the large $Ni(CO)_4$ consumption and because of the formation of propionic acid and its esters which are difficult to separate from the acrylic derivatives. In endeavoring to decrease $Ni(CO)_4$ consumption, it was found that, if CO and acetylene were fed into a solution in which the stoichiometric hydrocarboxylation had already started, rapid absorption of both gases took place. The quantity of acrylic derivatives formed was much higher than that calculated on the basis of $Ni(CO)_4$ present (118,119).

Details of the synthesis of acrylates by this method were investigated by several industrial laboratories (118–121). The pressure is generally a little higher than the vapor pressure of the reaction mixture and the optimum temperature is between 40 and 50°C (122). At this temperature, more than 20 moles of CO react per mole of $Ni(CO)_4$ used. HCl is the most suitable acid; the molar ratio of nickel carbonyl to HCl is maintained between 1 and 1.2 to avoid formation of β-chloropropionate (123), while the acetylene to CO ratio is maintained between 1.4 and 1.6 (120). A higher ratio is advisable for starting the reaction (120).

Different types of activators (such as small quantities of nitrogen- or phosphorus-containing compounds, or $HgCl_2$) seem to stabilize the catalytic agent (121).

The type of alcohol used does not greatly influence the reaction rates or yields (Table 14). The most important by-products were esters of allylacetic and of propionic acid. A large amount of the acrylates used in the last 20 years has been produced by using this method, particularly in the United States.

The semicatalytic hydrocarboxylation of methylacetylene was investigated (124–126,126*a*,126*b*). Higher temperatures (130–160°C) and higher pressures (10–20 atm) than used for acetylene were necessary. The best yields of methyl methacrylate (80–95%) were obtained by using methacrylic acid as the acid component; the by-products were methyl crotonate (methyl crotonate/methyl methacrylate was 0.08 at 130°C and 0.13 at 190°C), isobutyric acid, and acetone; the formation of acetone could be largely avoided by using methacrylic acid instead of HCl as the acid component.

3. *Catalytic Synthesis of Acids and Esters*

When stoichiometric hydrocarboxylation of acetylene was discovered, the industrial importance of the products prompted Reppe and his group to concentrate their research on finding a catalytic process for synthesizing acrylic acid and its esters. Many industrial laboratories, mainly in the United States, followed the same lines of research and secured a large number of patents on the subject, mostly between 1950 and 1965. Reppe and his co-workers were successful in finding an industrial process for the synthesis of acrylic acid and butyl acrylate (127–129); they also developed nickel catalysts for the synthesis of dicarboxylic and ketocarboxylic acids and their esters.

TABLE 14

Semicatalytic Synthesis of Acrylates from Acetylene, Carbon Monoxide, Alcohols, $Ni(CO)_4$*, and Hydrogen Chloride* (118,119)

		Reactants, molar ratio				
Alcohol	Temp (°C)	C_2H_2/ CO	$Ni(CO)_4$/ CO	$Ni(CO)_4$/ acid equivalent × 0.5	Moles gaseous CO reacted/ Total CO reacted[a] (%)	Yield[b] (%)
Methanol	45	1.39	0.082	1.15	76	85
Ethanol	40	1.39	0.1	1.2	72	83
Propanol	42	1.38	0.11	1.17	72	80
Isopropanol	40–43	1.38	0.1	1.16	72	87
Butanol	40–42	1.38	0.1	1.16	72	70
sec-Butanol	40–41	1.38	0.1	1.16	72	88–90
tert-Butanol	55	1.38	0.087	1.13	74	88–90
Decanol	43	1.34	0.123	0.99	67	79–80
2-Ethylhexanol[c]	40	1.1	—	1.2	—	—
tert-Butanol	70–74	1.75	0.17	2.7[d]	60	71
Allyl alcohol	69–70	1.75	0.17	29.4[d]	61	79

[a] Moles gaseous CO used plus 4 moles $Ni(CO)_4$.

[b] Based on moles of C_2H_2 charged.

[c] Ref. 123.

[d] Acrylic acid used instead of HCl.

Despite the large amount of work done in this field, only few papers have been published dealing with the fundamental aspects of the catalytic hydrocarboxylation of acetylene; therefore the relationships between catalyst structure and catalytic activity as well as the reaction mechanism are still little understood.

a. Acrylic Acid Synthesis with Soluble Catalysts. The main problem in the acrylic acid synthesis is the low reaction rate. For this reason high temperature (180–200°C) and high CO and acetylene concentrations in the liquid phase must be used; however, too high CO pressures (P_{CO}) must be avoided. For the catalytic hydrocarboxylation of propargyl alcohol with $Ni(CO)_4$ and water at 85–145°C, a value of P_{CO} was found for each temperature, corresponding to a maximum in the reaction rate. A large drop in reaction rate was observed above these P_{CO} values (115). The stoichiometry of the reaction needs approximately a 1:1 molar ratio of acetylene to CO. Therefore these gases would require compression at elevated pressures to obtain sufficiently high concentrations in the liquid phase. But the compression of gaseous acetylene can be avoided by introducing it into the reactor dissolved in a suitable solvent (54,128) such as THF.

According to a patent by Reppe and Stadler (130), yields of acrylic acid reaching 87% (based on acetylene) can be obtained. The best conditions are: an acetylene to CO molar ratio of 0.9; THF as solvent; a mixture of $Ni(CO)_4$ $CuBr_2$, and HBr as catalyst; operation at 170–220°C; and a total pressure of 40–50 atm. Under these conditions about 2 moles of acrylic acid are produced per gram of $Ni(CO)_4$ per hour.

According to Toepel (128), the reaction is carried out industrially at pressures above 60 atm, at about 200°C in THF, and in the presence of an activated $NiBr_2$ catalyst. The yield is 90% based on acetylene and 85% based on CO. The most important by-products (obtained mainly when water is used as solvent) are the unsaturated compounds $CH_2{=}CHCOOCH_2CH_2COOH$ and $CH_2{=}CHCOOCH_2CH_2COOCH_2$-$CH_2COOH$ formed from two or three molecules, respectively, of acrylic acid. Acrylic acid can be recovered from these by-products by pyrolysis.

The ways of activating nickel catalyst used industrially has been disclosed only in patents. According to patents by Reppe and others (130–132), Cu(II) complexes or salts can be used as activators of $NiBr_2$ or $Ni(CO)_4$.

The synthesis of the sodium or potassium salts of acrylic acid can be carried out in aqueous alkaline medium at 100–180°C by using $K_2Ni(CN)_3$ as catalyst and an acetylene and CO pressure of 30–60 atm (40).

Other than the previously cited paper on the catalytic hydrocarboxylation of propargyl alcohol (115) very little is known about the catalytic synthesis of substituted acrylic acids from acetylenic compounds. A systematic investigation of the hydrocarboxylation of octynes was carried out by Tetteroo (107) whose results are summarized in Table 15. The acids obtained from 2-, 3-, and 4-octyne show that hydrocarboxylation occurs stereospecifically through a *cis* addition to the triple bond. The isomeric composition of the reaction products is very similar to that obtained in the stoichiometric reaction.

b. Acrylate Synthesis with Soluble Catalysts. A detailed description of the catalytic synthesis of ethyl and butyl acrylate (2,133–135) from acetylene and that of butyl 2-butylacrylate from 1-hexyne (136) was reported. For acetylene, the reactions were carried out at 30–40 atm ($C_2H_2/CO = 1$) and at 150–180°C. Nickel halides and their complexes with PPh_3 of the type $(Ph_3P)_2NiX_2$ were generally used as catalysts (137). Results were reported for the synthesis of methyl acrylate by using various nickel compounds and in the presence of activators.

Reppe (2) and Yamamoto (133) showed that the catalytic activity of the nickel halides decreases with increasing ionic character of the nickel–halogen bond ($NiI_2 > NiBr_2 > NiCl_2$). But because NiI_2 favors polymerization of the acrylates, better results were obtained with $NiBr_2$. Unfortunately even $NiBr_2$ is not very stable under reaction conditions and it is hydrolyzed to $Ni(OH)_2$ by alcohols. Best results were obtained in the

TABLE 15

Hydrocarboxylation of Octynes with Catalytic Amounts of Nickel Carbonyl and Nickel Chloride[a] (107)

Alkyne	Conversion (%)	Acid yield[b] (%)	Product composition (%) −(1)	−(2)	−(3)	−(4)[c]
1-Octyne	90	80	19	81	—	—
2-Octyne	93	84	—	57	43	—
3-Octyne	92	84	—	—	53	47
4-Octyne	92	85	—	—	—	100

[a] Temp 200°C, pressure 200 atm; alkyne 0.25 mole, water 0.25 mole, $Ni(CO)_4$ 0.03 mole, $NiCl_2{\cdot}6H_2O$ 9 mmoles.

[b] Based on reacted alkyne.

[c] -( ) — Positions of the carboxylic group along the C_8 chain.

presence of allyl halides which hinder this hydrolysis. Other disadvantages of the use of $NiBr_2$ are its low solubility in alcohols and its transformation into the volatile $Ni(CO)_4$ during the reaction. For these reasons the use of PPh_3 complexes, activated by alkyl halides, was proposed; these are more soluble and are not transformed into the volatile and less active $Ni(CO)_4$.

According to Yamamoto (138), the PPh_3 groups present in the complexes help activate CO by decreasing the ionic character of the nickel–halogen bond and by acting as halogen acceptors. But the use of $NiBr_2(PPh_3)_2$ was not completely satisfactory, chiefly because of the relatively low reaction rate. To overcome this difficulty, many activators were proposed (139–143) but their action and efficiency are difficult to deduce from the patent literature.

As by-products of the catalytic hydrocarboxylation of acetylene in the presence of butanol, Reppe obtained the butyl esters of vinylpropionic, valeric, butoxypropionic, fumaric, and tetrahydrophthalic acids (2). Yamamoto (133), on running the reaction in ethanol, isolated acetaldehyde, diethyl ether, ethyl fumarate, and ethyl succinate.

Esters of hepta-2,4,6-trienoic acid are the main products when the synthesis of acrylates is carried out in the presence of triphenylphosphine nickel halides at high acetylene partial pressures (15 atm) (144).

The synthesis of acrylates in the liquid phase while using supported catalysts was investigated by Bhattacharyya (145–148). Best results were obtained by using SiO_2 as the support, NiI_2 or nickel acetate as catalysts, temperatures between 170 and 270°C, and pressures between 30 and 60 atm. The effect of reaction variables on the conversion to acrylic acid, acrylates, and secondary products was studied. The yields were generally not very high, since a large amount of acetylene was converted into by-products and part of the acrylates was polymerized.

The catalytic synthesis of 2-butylacrylates from 1-hexyne was investigated by Noble (149) and by Sakakibara and Okamoto (136) who used different complexes of the type $[(C_6H_4Y)_3P]_2NiBr_2$ as catalysts ($Y = CH_3$, OCH_3, etc.). The hydrocarboxylation was carried out in butanol at 180–200°C and at a CO pressure of 2–15 atm. The relative initial reaction rate, when using nickel complexes containing, respectively, PPh_3, tris(*m*-tolyl)phosphine, and tris(*p*-methoxyphenyl)phosphine at 190°C and a CO pressure of 5 atm, ranged from 1 to 1.7 to 3.3. For the tris(*p*-methoxyphenyl)phosphine complexes, the reaction rate seems to reach a maximum at a CO pressure of 5 atm and it decreases with increasing CO pressure. The yield based on reacted 1-hexyne is not high. The ester of 2-butylacrylic acid is always predominant; the molar ratio of the two isomeric esters is between 3 and 5. No clear effect of P_{CO}, type of catalyst, and temperature on this ratio has been ascertained.

c. Dicarboxylic Acids From Acetylene, Carbon Monoxide, and Water. Small quantities of esters of dicarboxylic acids such as fumaric, succinic, and tetrahydrophthalic were found among the by-products of the acrylate synthesis. Following this lead, attempts were made to synthesize dicarboxylic acids by semicatalytic or catalytic methods using nickel catalysts. No industrially interesting process grew out of this research, chiefly due to the low selectivity of the catalysts investigated. However, the possibility of synthesizing dicarboxylic acids from acetylene and CO in the presence of nickel catalysts is interesting and will be discussed shortly.

According to Reppe and Magin (40), acetylene and CO reacted in an aqueous solution of K_2CO_3 containing $K_3Ni(CN)_3$ or $K_2Ni(CN)_4$ at 90–100°C and 25 atm of total pressure ($CO/C_2H_2 = 1$) to yield a mixture of products among which cyclopentanone was identified. After hydrogenation of the mixture, succinic acid was isolated. These experiments were repeated by others (150) to determine whether succinic acid was present before hydrogenation and to ascertain the yields based on acetylene. Succinic acid was present before hydrogenation of the reaction products (Table 16) and hence must have been formed during carbonylation. The yield based on reacted acetylene is rather low (10–20%); the P_{CO} greatly influences the reaction rate, but not the yield of succinic acid. Monocarboxylic acids (mainly acrylic and propionic) and ketones (mainly cyclopentanone) occur as by-products.

TABLE 16

Reaction between Acetylene, Carbon Monoxide, and Water in the Presence of Aqueous Alkaline $K_2Ni(CN)_4$[a]

				Products			
$K_2Ni(CN)_4$ (moles/l)	Temp (°C)	Pressure (atm)	Reaction time (min)	Ketones (g[b])	Volatile acids (g[c])	Nonvolatile acids (g[d])	Ref.
0.50	90–100	25	1620	73	177	169	40
0.48	95–100	25	314	25	11.5	17	150
0.50	95	25	50	n.d.[e]	11	15	150
0.46	100–120	150[f]	1140	5.9	13.3	15.2	150

[a] pH at the end of the reaction > 7.
[b] As grams of cyclopentanone.
[c] As grams of acrylic acid.
[d] As grams of succinic acid.
[e] Cyclopentanone was present but not determined quantitatively.
[f] P_{CO} 140 atm.

The synthesis of esters in the presence of nickel catalysts was demonstrated by Reppe and Magin (99–101). This synthesis, either stoichiometric or semicatalytic, was carried out essentially in the absence of water and with $HCl/Ni(CO)_4$ ratios smaller than 2. When this ratio was decreased, the ratio of acrylates to diesters decreased. When ethanol was used, unsaturated esters having the formula $C_{14}H_{20}O_4$ were isolated. After hydrogenation of the high-boiling esters, a mixture of the esters of adipic, suberic, and sebacic acids were obtained. These results led to the stoichiometry shown in reaction 31.

$$2nC_2H_2 + Ni(CO)_4 + 4ROH + 2HCl \rightarrow$$
$$2ROOCCH_2(CH{=}CH)_{n-1}CH_2COOR + NiCl_2 + H_2 \quad (31)$$

4. *Mechanism of Hydrocarboxylation of Acetylenic Substrates*

a. Proposed Mechanisms. The mechanism of the hydrocarboxylation of acetylenic substrates is not well understood, and it is difficult to accommodate all the experimental data in a single path.

The first mechanism proposed by Reppe (2) has only historical interest (reaction 32). This was discarded by Reppe as soon as the hydrocarboxylation of internal acetylenic bonds was discovered.

$$HC{\equiv}CH \longrightarrow [CH_2{=}C{<}] \xrightarrow{CO} [CH_2{=}C{=}C{=}O] \xrightarrow{ROH} CH_2{=}CHCOOR \quad (32)$$

A second hypothesis, proposed by Reppe, involves formation of a cyclopropenone derivative, which then is cleaved to the corresponding acrylic derivatives (reaction 33). The existence of cyclopropenone complexed to a transition metal atom was found possible from a theoretical point of view (151). Furthermore it was shown (16,117) that

$$HC{\equiv}CH + CO \longrightarrow \left[\begin{array}{c} CH{=}CH \\ \diagdown\ \diagup \\ C \\ \| \\ O \end{array}\right] \xrightarrow{ROH} CH_2{=}CHCOOR \quad (33)$$

substituted cyclopropenones in the presence of $Ni(CO)_4$ in an aqueous acid medium give rise to a mixture of carboxylic acids with a composition analogous to that obtained by hydrocarboxylation of the corresponding acetylenic compounds. It was also shown (152) when using bicyclo[5.1.0]-oct-1-3-en-2-one as the substrate that opening of the cyclopropenone ring does not take place by decarbonylation to the corresponding acetylenic derivative as initially supposed (16). However, the fact that cyclopropenones can yield acrylic derivatives, as foreseen by Reppe, does not prove that these compounds are intermediates in the hydrocarboxylation of the acetylenic compounds. In fact the first step of reaction 33 does not seem very likely.

On the basis of present knowledge including formation of a metallacyclobutenone structure in the reaction of $Pt(PPh_3)_4$ and diphenylcyclopropenone (153) and reaction of cyclopropenones with iron and cobalt carbonyls (154), formation from acetylenic compounds or from cyclopropenones of a mixture of carboxylic derivatives having the same isomeric composition was interpreted as shown in Scheme **VI** (152).

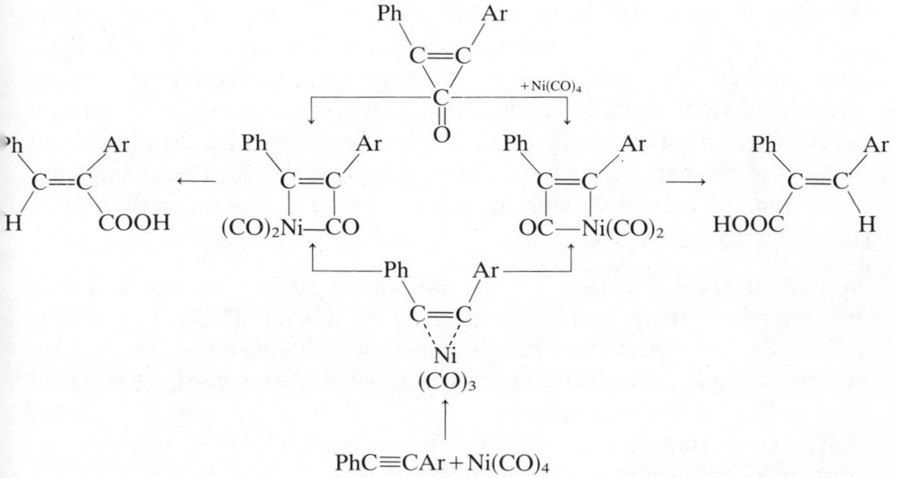

Ph(Ar)C=C (cyclopropenone, C=O) $\xrightarrow{+Ni(CO)_4}$ Ph(Ar)C=C–CO–Ni(CO)$_2$ (two isomeric nickelacyclobutenones)

Ph(H)C=C(Ar)COOH ⟵ (CO)$_2$Ni—CO metallacycle ; OC—Ni(CO)$_2$ metallacycle ⟶ Ph(HOOC)C=C(Ar)H

$PhC{\equiv}CAr + Ni(CO)_4 \longrightarrow$ Ph(Ar)C=C·Ni(CO)$_3$ ⟶ both metallacycles

Scheme **VI**

After Reppe's pioneering work, three main ways of looking at acetylene hydrocarboxylation were developed. According to the first way (155,156), acetylene would be activated by addition of a molecule of hydrogen halide and this reaction would be followed by oxidative addition of the vinyl halogenide to a nickel atom (Scheme **VII**).

$$HC{\equiv}CH + HBr \longrightarrow CH_2{=}CHBr$$

$$CH_2{=}CHBr + Ni(CO)_4 \longrightarrow [CH_2{=}CHCONi(CO)_3]Br$$

$$[CH_2{=}CHCONi(CO)_3]Br \xrightarrow[H_2O]{CO} Ni(CO)_4 + CH_2{=}CHCOOH + HBr$$

Scheme **VII**

The second way of looking at this mechanism (126*b*,157,158) is an oxidative addition of a hydrogen halide to $Ni(CO)_4$ followed by insertion of acetylene into a hydrogen–nickel bond (Scheme **VIII**).

$$Ni(CO)_4 + HCl \rightarrow HNi(CO)_2Cl + 2CO \quad \textbf{(a)}$$

$$HNi(CO)_2Cl + HC{\equiv}CH \rightarrow CH_2{=}CHNi(CO)_2Cl \quad \textbf{(b)}$$

$$CH_2{=}CHNi(CO)_2Cl + CO \rightarrow CH_2{=}CHCONi(CO)_2Cl \quad \textbf{(c)}$$

$$\left.\begin{array}{r} CH_2{=}CHCONi(CO)_2Cl + 2CO \rightarrow Ni(CO)_4 + CH_2{=}CHCOCl \\ CH_2{=}CHCOCl + ROH \rightarrow CH_2{=}CHCOOR + HCl \end{array}\right\} \quad \textbf{(d)}$$

$$CH_2{=}CHCONi(CO)_2Cl + ROH \rightarrow CH_2{=}CHCOOR + HNi(CO)_2Cl \quad \textbf{(d')}$$

Scheme **VIII**

Finally, a third mechanism (117,159) postulates complexation of acetylene to the nickel atom as the first step (Scheme **IX**). The complex thus formed can undergo CO insertion with formation of a four-membered ring which is finally cleaved by a hydrogen halide molecule with formation of an acryloyl nickel derivative (path **IXa**). But if water is present in the complex, an intramolecular protonation of the complexed acetylene can take place with formation of a vinyl–nickel complex which then undergoes a CO insertion.

b. Experimental Evidence for the Postulated Steps. (*1*) *Activation of the Acetylenic Bond for Complexation to the Metal.* There is increasing evidence (7) that acetylenic compounds in the absence of hydrogen form reactive complexes with transition metals which then give rise to cyclization or polymerization products.

It was reported that phenylacetylene reacts with $Ni(CO)_4$ to give *sym*-triphenylbenzene and polymers (97) and that alkynes react with nickel complexes containing carbonyl groups to yield relatively stable complexes (reactions 34 and 35) (160,161).

$$4Ni(CO)_4 + 3CF_3C{\equiv}CCF_3 \rightarrow \underset{\textbf{(31)}}{Ni_4(CO)_3(CF_3C{\equiv}CCF_3)_3} + 13CO \quad (34)$$

$$[(C_5H_5)Ni(CO)_2]_2 + C_2Ph_2 \rightarrow \underset{\textbf{(32)}}{(C_2H_5)_2Ni_2(C_2Ph_2)} + 4CO \quad (35)$$

Also complexes not containing carbonyls of nickel with alkynes are known (reactions 36 and 37) (162,163).

$$Ni(PPh_3)_2C_2H_4 + CH_3C{\equiv}CCH_3 \rightarrow \underset{\textbf{(33)}}{(Ph_3P)_2Ni(CCH_3{\equiv}CCH_3)} + C_2H_4 \quad (36)$$

$$NiBr_2 + 2C_2H_2 \rightarrow \underset{\textbf{(34)}}{NiBr_2(C_2H_2)_2} \quad (37)$$

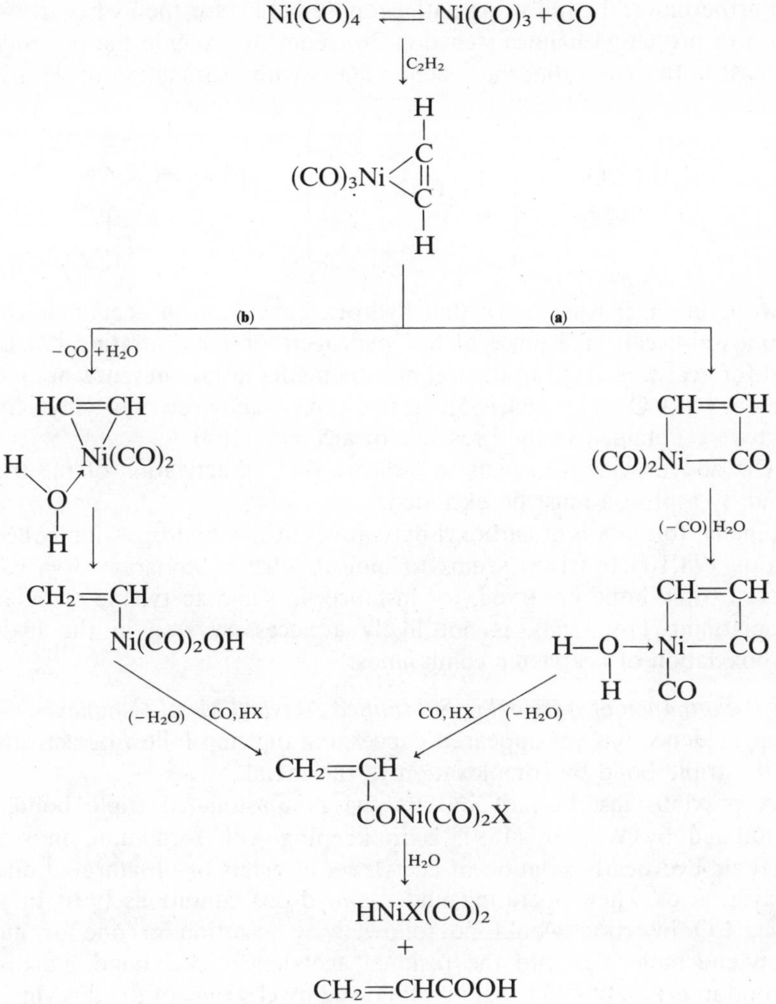

$$Ni(CO)_4 \rightleftharpoons Ni(CO)_3 + CO$$

$\downarrow C_2H_2$

$(CO)_3Ni$–(π-HC≡CH)

(b): $-CO$ | $+H_2O$ → HC=CH / $Ni(CO)_2$ ← H_2O → CH_2=CH–$Ni(CO)_2OH$

(a): CH=CH–CO–$Ni(CO)_2$ (ring) → $(-CO)$ H_2O → CH=CH–CO–Ni(CO)(←O(H)H) (ring)

$(-H_2O)$ CO, HX → CH_2=CH–$CONi(CO)_2X$ ← CO, HX $(-H_2O)$

$\downarrow H_2O$

$HNiX(CO)_2$ + CH_2=CHCOOH

Scheme **IX**

(*2*) *Activation of the Acetylenic Bond by Addition of a Proton.* This type of activation (Scheme **VII**), first postulated by Kröper (156) and developed by Almasi and co-workers (155), seemed to find some support in the distribution of isomeric reaction products from stoichiometric hydrocarboxylation of monosubstituted (106) and disubstituted acetylenes (117). However, a rather large percentage of anti-Markownikoff addition was observed in the case of methylacetylene, 1-hexyne (123,124,136) and propargyl alcohol (115).

Furthermore, the following path proposed (112) for the hydrocarboxylation of propargyl halides (reaction 38) seems to exclude the possibility, at least in this case, that the reaction starts with protonation of the triple bond.

$$ClR_2CC{\equiv}CH + CO \longrightarrow \left[Cl{-}\underset{R}{\overset{R}{C}}{-}\underset{\overset{\|\|\|}{^{\delta+}O}}{\overset{}{C}}{\equiv}CH \right] \longrightarrow [R_2C{=}C{=}\underset{H}{C}{-}\overset{+}{C}O] + Cl^- \quad (38)$$

In addition, it was shown that hydrocarboxylation of acetylenic compounds also can take place at low hydrogen-ion concentration (98,104) and for acetylene itself in neutral aprotic media in the presence of anions such as $[Ni(CO)_3X]^-$ (164,165). In this case, mainly esters of vinylacrylic acid were obtained in the presence of alcohols (165).

The above evidence seems to indicate that an activation of the triple bond by a proton must be excluded.

Finally, formation of carboxyl derivatives at low hydrogen-ion concentration (98,104,164,165) seems to indicate that protonation of a complexed triple bond observed, for instance, in some acetylenic complexes of platinum (166–168), is not likely a necessary step in the hydrocarboxylation of acetylenic compounds.

(3) *Formation of Acryloyl or Substituted Acryloyl Metal Complexes.* No clear evidence has yet appeared concerning the step following activation of the triple bond by complexation to the metal.

A previous insertion of CO on the nonprotonated triple bond, as postulated by Whiting (159), is in keeping with formation, in semicatalytic hydrocarboxylation of acetylene, of esters of unsaturated dicarboxylic acids when operating under anhydrous conditions (99). In this case, CO insertion would be followed by insertion of one or more acetylene molecules into the residual acetylene–nickel bond. Finally a second insertion of CO should be followed by cleavage of the diacylnickel bonds (reaction 39) and by a partial hydrogenation.

Moreover hydrocarboxylation of diphenylacetylene (98) and sodium propiolate (104) at low hydrogen-ion concentrations is best accounted for by postulating a preliminary CO addition to a nonprotonated double bond.

An analogous CO insertion was also suggested on the basis of product composition by Kinugasa and Agawa (10) from the reaction of diphenylacetylene with diketene and $Ni(CO)_4$. Finally, isolation of the first metallacyclobutenone complex (153) corroborates the soundness of the formula postulated for the intermediate **35**.

$$\text{Ni}(\text{HC}{\equiv}\text{CH}) \xrightarrow{\text{CO}} \underset{(\mathbf{35})}{\text{Ni}(\text{CO})(\text{CH}{=}\text{CH})} \xrightarrow{n\text{C}_2\text{H}_2} \text{Ni}(\text{CO})(\text{CH}{=}\text{CH})_n(\text{HC}{=}\text{CH}) \xrightarrow{\text{CO}} \tag{39}$$

$$\text{Ni}(\text{C}{=}\text{O})(\text{CH}{=}\text{CH})_{n+1}(\text{C}{=}\text{O}) \xrightarrow{2\text{ROH}} \text{ROOCCH}_2(\text{CH}{=}\text{CH})_n\,\text{CH}_2\text{COOR}$$

No conclusive experimental evidence exists on the postulated oxidative addition of acids to nickel carbonyl complexes and on the insertion of complexed or uncomplexed triple bonds into a nickel–hydrogen bond. However, very unstable nickel hydrides containing CO have been known for some time (169–171).

It is also doubtful that a nickel carbonyl hydride can be an intermediate in the stoichiometric synthesis of acrylates at low temperatures, since $Ni(CO)_4$ reacts very slowly with HCl under these conditions (172). But such a complex may function as an intermediate in the catalytic synthesis of acrylates at higher temperatures. Indeed, the presence of hydrides helps to explain the formation of hydrogenated products formed during the reaction.

Admitting that a possible step following complexation of the acetylenic bond to nickel is insertion of the complexed triple bond into a nickel–hydrogen bond, a σ-vinyl or a substituted vinyl complex should be formed. Insertions of acetylenic substrates bearing electron-withdrawing substituents into metal-hydrogen bonds are well known (79–84). In the case of rhodium (173) and ruthenium (88), it was shown that σ-vinyl derivatives readily undergo CO insertion.

The thermal instability of acylnickel species (174–176) indicates that CO insertion into the alkyl–nickel bond is not favored; in fact, isolation of σ-acylnickel complexes was achieved in the presence of bulky phosphite or phosphine ligands that apparently prevent expansion of coordination around Ni(II) and decarbonylation of the acyl group (174). This experimental finding, however, cannot be taken as evidence against the existence of an acylnickel complex intermediate that readily undergoes nucleophilic attack by water or alcohols with formation of the corresponding acids or esters.

(4) *Decomposition of Acyl Nickel Derivatives to Carboxylic Derivatives.* There is no significant controversy concerning this step,

which is believed to occur by nucleophilic attack of water or alcohols on the acyl group bound to the metal. An interesting observation was made by Chiusoli and Cameroni (177) about this nucleophilic attack. After reacting 1-octyne with $Ni(CO)_4$ in the presence of a little methanol and acetic acid (reaction 40), the mixed anhydride was detected by infrared.

$$C_6H_{13}C{\equiv}CH \xrightarrow[CH_3COOH]{Ni(CO)_4} C_6H_{13}\underset{\underset{CH_2}{\|}}{C}COOCOCH_3 \qquad (40)$$

The anhydride reacts further with acetic acid to give α-methyleneoctanoic acid and acetic anhydride. This is a possible path by which carboxylic acids can be formed even in the presence of alcohols, as found by Jones and co-workers (96,97,109,114).

c. Final Remarks. The experimental evidence to now indicates that the Heck mechanism (Scheme **VIII**) and the two mechanisms proposed by Bird (Scheme **IXa** and **IXb**) are the most probable. An initial activation of the triple bond by the transition metal atom seems in both cases very likely. But the experimental evidence is not sufficient to decide whether CO insertion precedes or follows addition of the hydrogen atom to the acetylenic compound. It might very well be that both mechanisms are operative, depending on reaction conditions.

Some important aspects of the nature of the intermediate organometallic complexes, however, are still unknown or controversial. Perhaps the most obscure aspect of the hydrocarboxylation mechanism is the number of nickel atoms present in the catalytic complexes. It is known that nickel clusters can be formed by reacting $Ni(CO)_4$ with acetylenic compounds (7); the deep brown color that develops before the start of the stoichiometric reaction suggests the presence of complexes containing more than one nickel atom. The molecular complexity of the catalytic species may be important in directing the reaction toward different reaction products. In the carbonylation of diphenylacetylene in aqueous alkaline medium, for example, the anion $[Ni_3(CO)_8]^{2-}$ yields mostly tetraphenylbutadiene (98). On the other hand, $K_2Ni(CN)_3$, which in alkaline solution in the presence of CO forms the dimeric anion **36**, mainly gives rise to succinic acid and cyclopentanone when starting with acetylene (40).

A further important aspect that needs clarification is the oxidation number of nickel during these reactions. According to the postulations of Heck (157) and of Bird (117), Ni(II) should be present in the catalytic species. But according to Ehrreich, Nickerson, and Ziegler (122), deactivation of the catalyst observed in the semicatalytic acrylate synthesis is

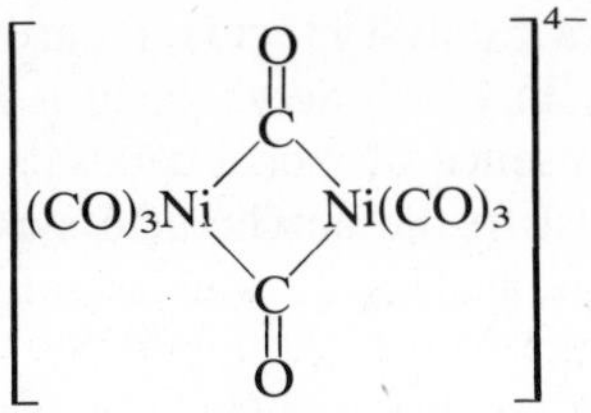

$$\left[\begin{array}{ccc} & O & \\ & \| & \\ & C & \\ (CO)_3Ni & & Ni(CO)_3 \\ & C & \\ & \| & \\ & O & \end{array}\right]^{4-}$$

(36)

related to an oxidation of nickel present in the complex to Ni(II), which cannot be reduced under reaction conditions.

A sound knowledge of the path for the oxidation of Ni(0) to Ni(II), which probably is connected with formation of propionic and vinylacrylic esters, would probably yield useful information on the catalytic species involved in the process.

*5. Industrial Aspects of the Synthesis of Acrylic Acid**

Reppe discovered the stoichiometric synthesis of acrylic acid from acetylene in 1939, during his pioneering research on acetylene chemistry. The catalytic synthesis of acrylic acid from acetylene, CO, and water was successfully achieved in the same year (2,127–129,178). This work became generally known in 1945. In view of the great industrial importance of acrylic acid and its derivatives for the synthesis of high polymers, many industrial laboratories started to work in this new field of acetylene chemistry. This research resulted in a large number of patents between 1950 and 1965 in which the stoichiometric process was modified and the stoichiometric–catalytic or semicatalytic process was developed (118,119,179–181).

The production capacity of acrylic acid and its esters in the United States and Western Europe in 1973 was about 570,000 tons per year. It is estimated that in the same year about 65% of these plants used processess based on acetylene—about half used the catalytic process and half the stoichiometric–catalytic or semicatalytic process. The amount of acrylic acid production based on acetylene demonstrates that the process is technically well mastered. On the other hand, despite the large number of papers dealing with the experimental synthetic aspects of acetylene chemistry, only a few deal with the fundamental aspects of acetylene hydrocarboxylation. The relationships between structure and activity of

* By W. Himmele, F. J. Müller, A. Magin, and N. v. Kutepow. The editors are particularly grateful to these authors for the text of this subsection and to BASF for permission to publish it.

the catalysts, as well as of catalyst precursors, are still largely unclarified. For this reason, the way in which acrylic acid is formed from acetylene, CO, and water in the presence of nickel catalysts is still largely a matter of speculation. In contrast to the acrylic acid synthesis, the syntheses of mono- and polyunsaturated acids as well as saturated dicarboxylic and ketocarboxylic acids (which were developed by Reppe's groups starting with acetylene and using particular nickel catalysts) have never reached industrial significance.

a. Acrylic Acid Synthesis in the Presence of Soluble Catalysts. Equation 41 not only indicates relationships between starting material and the reaction product but also suggests reaction conditions for the catalytic synthesis of acrylic acid.

$$HC{\equiv}CH + CO + H_2O \rightarrow H_2C{=}CHCOOH \qquad \Delta H = -56 \text{ kcal/mole} \quad (41)$$

Two of the reagents, acetylene and CO, are gases at normal temperature and pressure; the third reaction component and the reaction product are liquids under these conditions. Critical data for the four compounds show that, over a very large interval of pressure and temperature, a system containing these four components must be a two-phase system in which a liquid and a gas phase coexist. It is expected therefore, according to the Le Chatelier–Braun principle, that the formation of acrylic acid would be favored by a pressure increase.

According to Eq. 41, best results should be obtained by using a gaseous mixture containing acetylene and CO in a 1:1 ratio. But a property characteristic of acetylene prohibits increasing the pressure of such a gas mixture at will. Gas mixtures containing acetylene spontaneously decompose above certain pressure limits; the limit for a 1:1 mixture of acetylene and CO is 6 atm. Beyond this limit, an explosion takes place and the final pressure reaches about 10 times higher than the starting pressure. It is possible to overcome this risk by using suitable reactors (182). A much greater risk, however, arises when the reaction with acetylene is carried out in the gas phase because it appears that above 60 atm the explosion becomes a detonation and gives a pressure increase of 50–100 times. In this case the highest pressure reached cannot be precisely defined because the change from an explosion to a detonation depends on the form and the size of the space occupied by the gas.

A large amount of research on the decomposition of acetylene–carbon monoxide mixtures, in which the decomposition was initiated by melting a platinum wire, resulted in determination of the pressures at which C_2H_2–CO mixtures of different composition decompose. It appears that gaseous mixtures containing less than 20% of acetylene do not decompose up to a total pressure of 200 atm (Fig. 3).

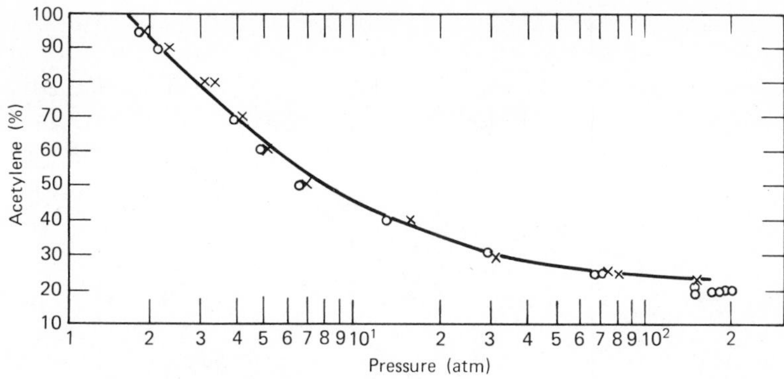

FIG. 3. Pressure of acetylene–carbon monoxide mixtures at which decomposition takes place.

An industrial synthesis of acrylic acid from a water and C_2H_2–CO mixture containing 17–20% of acetylene can be carried out without danger under normal safety rules for working with gaseous acetylene at 200 atm and temperatures up to 300°C (183). But when using this process, it is impossible to reach the optimum C_2H_2–CO ratio (1:1). This ratio can be reached in a high-pressure reactor by dissolving acetylene in a proper solvent (i.e., ketones, esters, lactones, substituted amides, cyclic ethers, nitriles, or hydrocarbons) and compressing the solution to the pressure chosen for the synthesis (54,128,184). However, to guarantee a safe operating procedure, it is necessary in these cases to know the tendency of acetylene solutions to decompose depending on acetylene concentration, temperature, and pressure.

To use the above procedure for the synthesis of acrylic acid, it is only necessary to avoid formation of too high an acetylene concentration in that part of the reactor in which a gas phase is present. This risk can also be avoided when the synthesis takes place under limiting conditions by carrying out the reaction in a reactor completely filled with the liquid reaction mixture, that is, carrying out the chemical reaction under hydraulic conditions. By using these conditions, a continuous increase in reaction rate was observed between 45 and 100 atm of total pressure (Fig. 4).

The catalytic synthesis of acrylic acid is a reaction occurring in the presence of homogeneous catalysts, the catalysts being dissolved in the reaction medium. As catalysts, the group VIII metals able to form metal carbonyls can be used under conditions necessary for the synthesis of acrylic acid. Particularly high catalytic activity was shown by nickel compounds in acid media. Addition of activators increases the activity of

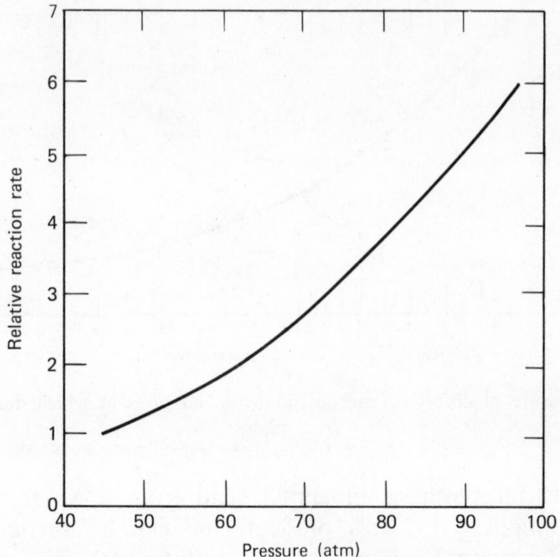

FIG. 4. Dependence of reaction rate on total pressure in the Reppe catalytic synthesis of acrylic acid. The reaction rate at 40 atm was taken as reference.

nickel. Completely different classes of compounds can act as activators: namely, cyclic and open chain amides (185), organic compounds containing >N–C–S–, >N–C–Se, >N–C–Te groups, selenium or selenium-containing organic compounds (185–187), compounds of the type $X(CH_2)_n$-Y (X and Y = 0, S, or N-containing groups) (188–189), metals or compounds of metals of groups II–V, as well as of the first subgroups of the periodic system (190–196).

The dependence of the hydrocarboxylation rate on CO pressure was shown, for example, by using propargyl alcohol as substrate (115). In the presence of the above-mentioned substances and using proper CO pressures, the catalytic activity of $NiBr_2$, used as the main catalyst component, can be increased to such an extent that, despite low nickel concentrations, yields of acrylic acid as high as 90% with respect to acetylene and of 85% with respect to CO can be obtained. The most important secondary products are diacrylic and polyacrylic acids from which acrylic acid can be recovered by pyrolysis.

Potassium or sodium salts of acrylic acid are formed in aqueous alkaline solutions at a temperature of 100–180°C at acetylene and CO pressures of 30–60 atm when using $K_2Ni(CN)_3$ as catalyst (40).

B. Acids and Esters from Acetylenic Substrates with Cobalt Catalysts

Hydrocarboxylation of acetylenic substrates can also be carried out in the presence of cobalt compounds, but reaction conditions and products obtained are much different from those reported for hydrocarboxylation conducted in the presence of nickel compounds.

First of all, stoichiometric hydrocarboxylations using $HCo(CO)_4$ or $Co_2(CO)_8$ as the metal carbonyl component have been rather unsuccessful. Acetylene was reported to react with $HCo(CO)_4$ at room temperature to give the complex $C_2H_2Co_2(CO)_6$ [isolated as $C_2H_2Co_2(CO)_5PPh_3$ or $C_2H_2Co_2(CO)_4(PPh_3)_2$] in yields of about 50%. Decomposition of $C_2H_2Co_2(CO)_6$ in ethanol at 100°C under CO pressure gave a mixture of ethyl succinate, ethyl propionate, and ethyl acrylate (197). When $Co_2(CO)_8$ was reacted with an equimolar amount of acetylene in the presence of CO and water in THF, the reaction started at 70°C to give succinic acid in 43% yield based on reacted acetylene (199). Stable complexes $(CF_3C{\equiv}CH)Co_2(CO)_6$, $CF_3CH_2C[Co(CO)_3]_3$ and $(CF_3C{\equiv}CH)_3Co_2(CO)_4$ were isolated by reacting trifluoropropyne with $Co_2(CO)_8$ in pentane at 100°C (198).

The stoichiometric hydrocarboxylation of substituted methynyl tricobaltenneacarbonyl complexes **37**, reported by Tominaga and coworkers (200), took place above 70°C under CO pressure in methanol in the presence of an organic base. Methyl propionate, dimethyl methylmalonate, and dimethyl succinate were obtained in yields of 14–69%, 5–61%, and 3–21%, respectively, depending on the type of base and the CO partial pressure.

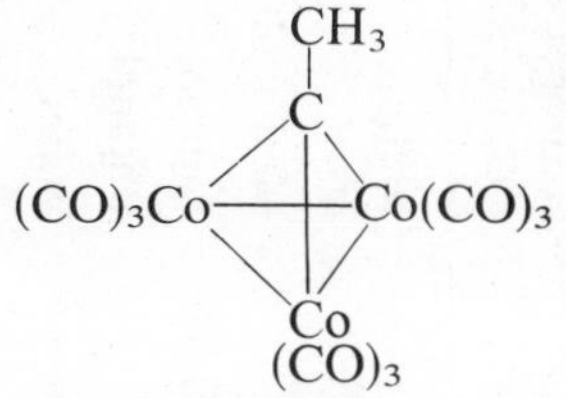

CH_3
C
$(CO)_3Co$ — $Co(CO)_3$
Co
$(CO)_3$

(**37**)

The stoichiometric hydrocarboxylation of acetylenic substrates with cobalt complexes occurs only above 70°C in low yields, probably because of the high stability of the intermediate cobalt carbonyl acetylene complexes. But the catalytic activity of the cobalt compounds seems much higher than that of the nickel compounds.

With cobalt, catalytic hydrocarboxylation occurs in the absence of hydrogen halides, alkyl halides, or halide anions even at 80–100°C under

TABLE 17
Hydrocarboxylation of Acetylene with Certain Nickel or Cobalt Catalysts

CO/C_2H_2	Catalyst	Hydrogen donor	Temp (°C)	Pressure[a] (atm)	Hydrocarboxylation product: Compound	Hydrocarboxylation product: Moles product/g atom Ni or Co per hr	Yield[b] (%)	Ref.
1	$NiBr_2$:BuBr pyridine	Methanol	160	30	Methyl acrylate	0.9	n.d.	202
30	$Co_2(CO)_8$	Methanol	110	210	Methyl acrylate	16.5	58	203
1	$(PPh_3)_2NiBr_2$:BuBr	Butanol	170	30	Butyl acrylate	3.1	n.d.	202
17.5	$Co_2(CO)_8$	Butanol	130	160	Butyl acrylate	25.3	87	203
1	$NiBr_2:CuBr_2$ Et_4NBr	Water	190	45	Acrylic acid	8.0	n.d.	204
15	$Co_2(CO)_8$	Water	110	200	Succinic acid	8.0	80	199

[a] Measured at reaction temp.
[b] Based on reacted C_2H_2.

fairly high CO pressure (100–200 atm). The catalytic activities of cobalt and nickel are compared in Table 17.

At first glance, the selectivity with cobalt catalysts seems much lower (Table 17). In the first example, carried out in methanol, relatively large amount of ketones, ketoesters, and diesters in addition to acrylic derivatives were found (54,201); with nickel only small amounts of mono- and diesters were formed (2). With cobalt, however, conditions were found whereby the hydrocarboxylation of acetylene could be directed to yield more than 90% of methyl acrylate (203,205) or succinic acid (25,50,71).

The hydrocarboxylation of acetylenic compounds in the presence of cobalt catalysts will be discussed in two parts. The first will deal with the synthesis of esters and the second with the synthesis of acids. The use of compounds containing active hydrogen atoms in the hydrocarboxylation will be mentioned in another section.

1. Esters from Acetylenic Substrates, Carbon Monoxide, and Alcohols

Natta and Pino (54,70,201,206) endeavored to find a suitable catalytic synthesis of methyl acrylate by reacting acetylene with CO in methanol using Raney cobalt or $Co_2(CO)_8$ as catalyst. As shown in Table 18, the yields of methyl acrylate under these conditions were rather low; the products were a mixture of esters of mono-, di-, and tricarboxylic acids, in addition to ketoesters and ketones (see Section III-B).

The formation of these products can be depicted as given in reactions 42–47.

$$C_2H_2 + CO + CH_3OH \rightarrow CH_2{=}CHCOOCH_3 \tag{42}$$

$$C_2H_2 + 2CO + 2CH_3OH \rightarrow \begin{array}{l} CH_2COOCH_3 \\ | \\ CH_2COOCH_3 \end{array} \tag{43}$$

$$C_2H_2 + 2CO + 2CH_3OH \rightarrow \begin{array}{r} CHCOOCH_3 \\ \| \\ CH_3OCOCH \end{array} \quad + [2H] \tag{44}$$

$$C_2H_2 + 3CO + 3CH_3OH \rightarrow \begin{array}{l} CH_2COOCH_3 \\ | \\ CH(COOCH_3)_2 \end{array} \quad + [2H] \tag{45}$$

$$2C_2H_2 + 2CO + 2CH_3OH \rightarrow \begin{array}{l} CH{=}CHCOOCH_3 \\ | \\ CH{=}CHCOOCH_3 \end{array} \quad + [2H] \tag{46}$$

$$2C_2H_2 + 2CO + 2CH_3OH \rightarrow \begin{array}{l} CH_2CH_2COOCH_3 \\ | \\ CH{=}CHCOOCH_3 \end{array} \tag{47}$$

TABLE 18
Reaction of Acetylene, Carbon Monoxide, and Methanol with $Co_2(CO)_8$ *as Catalyst*[a] (206)

Products	Yield[b] (mole %)
Methyl acrylate	3–20
Dimethyl succinate	10–30
Dimethyl fumarate	0.5–6
Trimethyl ethanetricarboxylate	2–5
Dimethyl *trans-trans*-muconate	Traces
Dimethyl 2-hexenedioate	1
Ketones and ketoesters	10–26

[a] Temp 95°C; P_{CO} 250 atm.
[b] Based on C_2H_2 charged.

The syntheses of fumaric, ethanetricarboxylic and muconic esters are dicarboxylations rather than hydrocarboxylations.

A systematic investigation was carried out on the effect of different reaction variables on the cobalt-catalyzed hydrocarboxylation of acetylene in the presence of alcohols (54,201,203,205).

The yield of methyl acrylate rapidly increases with decreasing initial acetylene concentration (203,205) (Fig. 5). These results agree qualitatively with previously published data (54,201) and have general validity independent of the alcohol used (Table 19) (203).

Carrying out the reaction between acetylene, CO, and methanol in the presence of preformed methyl acrylate, the yields obtained, even when using low acetylene concentrations, were low and decreased with increasing preformed methyl acrylate concentration. These results show that the low yields obtained at high acetylene concentration are at least, in part, due to successive reactions of methyl acrylate, the rates of which are obviously dependent on the ester concentration. A further decrease in the ester yield may be due to other side reactions favored by high acetylene concentrations.

A smaller but significant increase in methyl and butyl acrylate yield at constant acetylene concentration was observed by increasing CO pressure from 50 to 260 atm. On operating at 110°C and using an acetylene concentration of about 0.13 mol/l in the presence of $Co_2(CO)_8$ (3.3 g/l), the yields of methyl acrylate were 70 and 93.5% at CO pressures of 50 and 260 atm, respectively.

Temperature has only a small effect on methyl acrylate yield, but it has

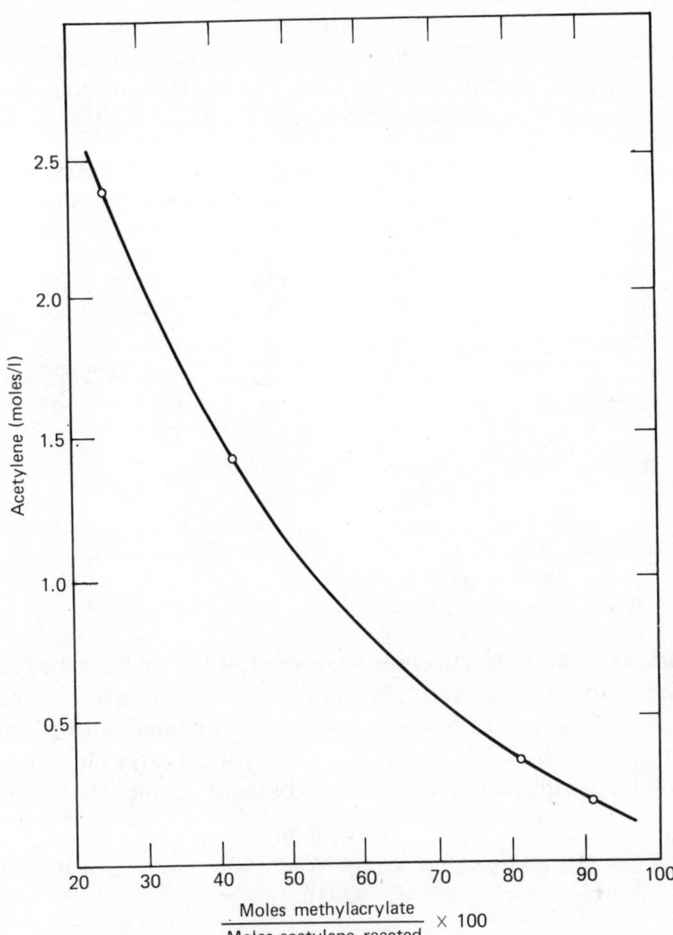

FIG. 5. Effect of acetylene concentration on yield of methyl acrylate (203). Temp. 110°C, P_{CO} 200 atm, $Co_2(CO)_8$ 3.3 g/1, solvent methanol.

a large influence on the composition and yields of diesters (201,206). At a total pressure of 300 atm, raising the temperature from 90 to 170°C increases the methyl acrylate yield from 9 to 12.4%, but the yield of diesters decreases from 50 to 21%. The yield of methyl fumarate increases from 0.8 to 6% when the temperature is raised from 80 to 120°C; at the same time the dimethyl succinate yield decreases from 31 to 17%.

The effect of changing catalyst concentration on the synthesis of acrylates is shown in Table 20. Both in the methyl acrylate and butyl acrylate syntheses, the yield increases with decreasing catalyst concentration. The effect is larger for the methyl than for the butyl ester.

TABLE 19
Effect of Acetylene Concentration on Acrylate Yields in the Reaction of Acetylene, Carbon Monoxide, and Various Alcohols in the Presence of $Co_2(CO)_8$[a] (203)

Alcohol	Acetylene (moles/l) (initial)	P_{CO} (atm) (initial)	Time (hr)	Acrylate yield[b] (%)
Ethanol	0.95	147	1.5	33
	1.57	145	2	26
Propanol	0.75	141	2	41
	1.25	139	2	35
Isopropanol	0.8	155	1	41
	1.39	161	1.5	36
Butanol	0.32	140	1	61
	0.53	159	1	58
	0.87	161	1	49

[a] Temp 130°C; $Co_2(CO)_8$ 3.3 g/l.
[b] Based on reacted acetylene.

A number of aliphatic alcohols were used in the hydrocarboxylation of acetylene using similar reaction conditions; the yields obtained with *t*-butyl alcohol were much lower than those obtained with primary and secondary alcohols. With primary alcohols, the best yields of acrylates were obtained with methanol and with butanol (Table 21) (203).

TABLE 20
Effect of $Co_2(CO)_8$ *Concentration on the Syntheses of Methyl and Butyl Acrylates* (203)

Alcohol	C_2H_2 (moles/l) (initial)	$Co_2(CO)_8$ (g/l)	P_{CO} (atm)	Temp (°C)	Acrylate yield[a] (%)
Methanol	0.14	8.58	218	92	72
	0.13	5.72	214	96	79
	0.13	2.88	223	99	86
	0.37	3.43	211	108	73
	0.36	1.43	209	104	82
Butanol	0.84	20.0	127	112	47
	0.84	3.34	127	132	60
	0.84	1.33	131	131	59

[a] Based on C_2H_2 reacted.

TABLE 21
Synthesis of Acrylates from Acetylene, Carbon Monoxide, and Various Alcohols with $Co_2(CO)_8$ (203)[a]

Alcohol	C_2H_2 (moles/l) (initial)	Acrylate yield[b] (%)
Methanol	0.90	56
Ethanol	0.95	33
Propanol	0.75	41
Butanol	0.87	54
Dodecanol	0.49	42
Isopropanol	0.8	41
t-Butanol	0.4	19

[a] Temp 130°C; P_{CO} 150 atm; $Co_2(CO)_8$ 3.3 g/l.
[b] Based on C_2H_2 reacted.

2. *Acids from Acetylenic Substrates, Carbon Monoxide, and Water*

Reppe and Stadler (207) obtained acrylic acid by treating equimolar amounts of acetylene and CO at 200°C and 50 atm in THF as solvent in the presence of $CoBr_2$ and $CuBr_2$. At higher CO pressure (200 atm) in the presence of $Co_2(CO)_8$, the reaction started at lower temperature (80–100°C) and, working in neutral to acid media, conditions were found in which more than 80% of the reacted acetylene is transformed into succinic acid. Operating in an alkaline medium and using $K_3Co(CN)_6$ as catalyst (40,150), only traces of succinic acid were obtained at 120–125°C and a total pressure of 22 atm. No reaction takes place in an alkaline medium when using $Co_2(CO)_8$ as catalyst (150). If the reaction is carried out in water with no other added solvent at 120°C and 200–300 atm of CO, only small quantities of succinic and propionic acids are found (25); part of the CO is transformed into CO_2 and the $Co_2(CO)_8$ reacts to form cobalt carbonate and cobalt succinate. With solvents such as acetone or dioxane, the reaction occurs in homogeneous phase and practically no acetylene is left after a few hours under the same conditions.

Succinic acid is the principal product but small quantities of propionic and acrylic acids and cyclopentanone are also formed (reactions 48–52). Operation at very low ratios of water to acetylene gives a solid product containing the *cis*- and *trans*-2,4,6-octatrien-1:4, 5:8-diolides (5,6,26).

$$C_2H_2 + 2CO + 2H_2O \rightarrow HOOCCH_2CH_2COOH \qquad (48)$$

$$C_2H_2 + CO + H_2O \rightarrow CH_2{=}CHCOOH \qquad (49)$$

$$C_2H_2 + CO + H_2O + H_2 \rightarrow CH_3CH_2COOH \qquad (50)$$

$$2C_2H_2 + CO + 2H_2 \longrightarrow \text{cyclopentanone} \qquad (51)$$

$$CO + H_2O \rightleftharpoons CO_2 + H_2 \qquad (52)$$

The hydrogen necessary for these reactions can be supplied by the water gas shift reaction (52). The amount of CO_2 found in the gas phase at the end of the reaction corresponds to the amounts of propionic acid and cyclopentanone formed according to reactions 50 and 51 (50).

The effects of certain variables on the synthesis of succinic acid (50,199) were investigated in some detail and are reported below.

As mentioned earlier, the solvent has a great effect on the yield of succinic acid (Table 22); best yields are obtained in THF or acetone, whereas low yields result when using toluene, ethyl acetate, or water (25,50,199). The highest yield of cyclopentanone based on reacted acetylene was obtained when using ethyl acetate; the solvent is partially hydrolyzed under reaction conditions.

The effect of temperature was investigated by Natta and Albanesi (50) between 70 and 160°C (Fig. 6). The cyclopentanone yield is higher below 100°C (22% at 70°C) and the succinic acid yield is lower below this temperature (69%). Between 100 and 160°C no great change in the yield of these products is noted.

TABLE 22

Effect of Solvent on the Synthesis of Succinic Acid from Acetylene, Carbon Monoxide, and Water with $Co_2(CO)_8$[a]

Solvent	Yield[b] (%)			Ref.
	Succinic acid	Acrylic and propionic acids	Cyclopentanone	
Water[c]	17	Traces	Traces	71
Acetone	85	4	12	50
Tetrahydrofuran	79	7	7	199
Dioxane	65	13	17	71
Ethyl acetate	25	—[d]	25	199
Toluene	8	0	13	199
Cyclopentanone	82	5	n.d.	71

[a] Initial P_{CO} 180–200 atm; $Co_2(CO)_8$ 6.65×10^{-2} moles/1; temp 110°C; molar ratio $H_2O/C_2H_2 = 10/1$.

[b] Based on reacted C_2H_2.

[c] Reaction carried out at 140°C.

[d] Ethyl acetate partially hydrolyzed.

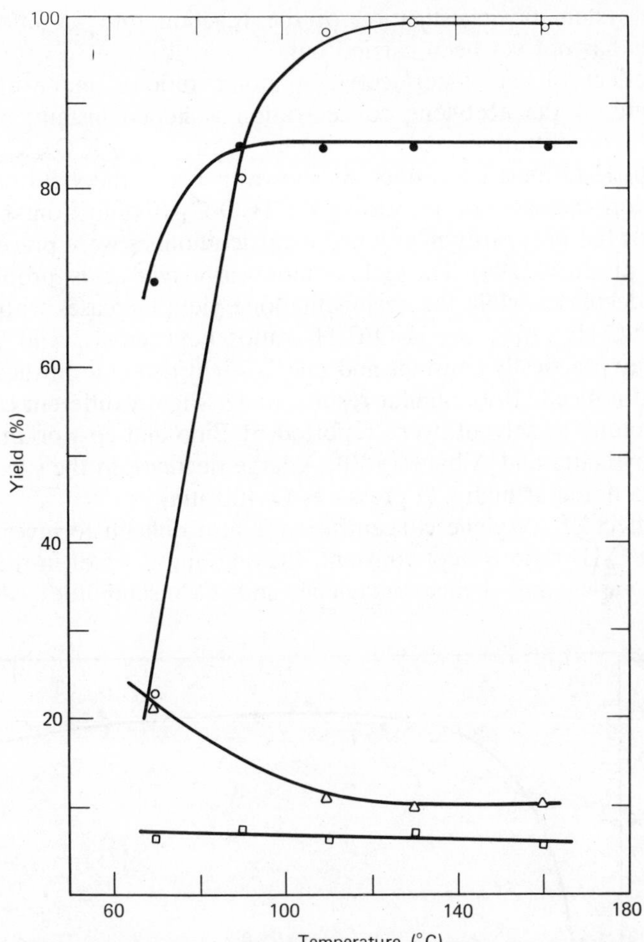

FIG. 6. Effect of temperature on the synthesis of succinic acid (50). Solvent acetone, pressure 261 atm, C_2H_2/H_2O 0.09–0.11, $C_2H_2/Co_2(CO)_8$ 15.5–20; ○: total conversion, ●: succinic acid, △: cyclopentanone; □: monocarboxylic acids.

The yield of monocarboxylic acids is not affected by temperature change in the range investigated.

The effect of CO pressure on the yield of succinic acid was investigated in acetone (25,50) and in THF (199). In acetone the maximum yield was found at 260 atm and 100°C when using a C_2H_2/H_2O ratio of 0.06; at the same temperature in THF the maximum occurred at an initial CO pressure of 240 atm when using a C_2H_2/H_2O ratio of 0.115.

In THF, the yields of cyclopentanone and that of monocarboxylic acids are not as greatly affected by the CO pressure as the yield of succinic acid. A systematic investigation of the reaction rate at different CO pressures has not yet been carried out.

The effect of the water/acetylene molar ratio is more difficult to investigate; if the acetylene concentration is kept constant, either the catalyst concentration or the molar ratios $Co_2(CO)_8/C_2H_2$ and $Co_2(CO)_8/H_2O$ must be varied. As shown in Fig. 7 the yield of succinic acid rapidly increases on increasing the H_2O/C_2H_2 ratio from 1 to 6. At low ratios the previously mentioned octatriendiolides were present in the reaction products (199). The yield of monocarboxylic acids (propionic and acrylic) decreases while the cyclopentanone yield increases with increasing H_2O/C_2H_2 ratios. At H_2O/C_2H_2 ratios between 8.7 and 22.8, the yields stay practically constant and only a slight decrease in succinic acid yield was noticed (199). Similar results, under slightly different conditions using acetone as solvent, were reported by Pino and co-workers (25) as well as by Natta and Albanesi (50). A large decrease in the succinic acid yield was noted at high CO pressures (>400 atm).

The effect of acetylene concentration is also difficult to investigate; if the H_2O/C_2H_2 ratio is kept constant, the amount of water in the organic solvent varies and hence acetylene and CO solubilities also vary.

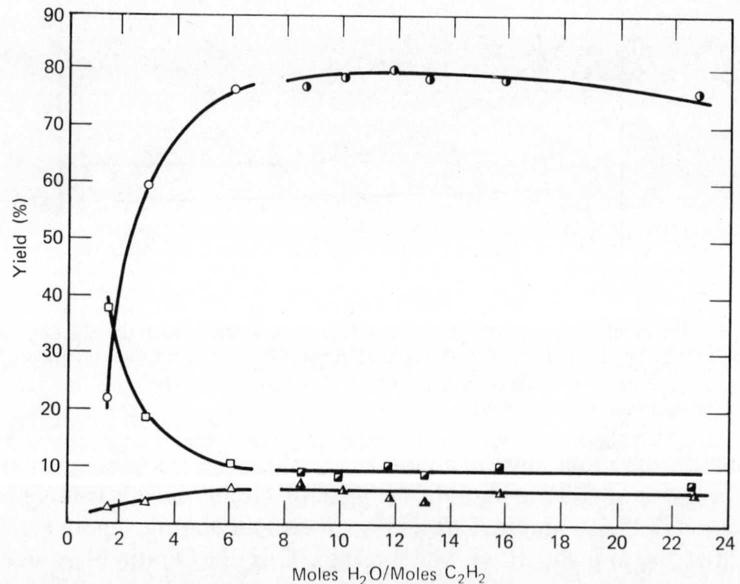

FIG, 7. Effect of H_2O/C_2H_2 molar ratio on succinic acid synthesis in THF (199). Temp. 110°C, P_{CO} 200 atm, $Co_2(CO)_8$ 6.6×10^{-2} moles/1. ○: succinic acid, □: monocarboxylic acids, △: cyclopentanone at C_2H_2 concentration of 1–1.3 moles/1. ◐: succinic acid, ◪: monocarboxylic acids, ▲: cyclopentanone at C_2H_2 concentration of 1.6–1.9 moles/1.

TABLE 23

Effect of Acetylene Concentration on Synthesis of Esters and Acids from Acetylene, Carbon Monoxide, and Methanol or Water with $Co_2(CO)_8$

C_2H_2 (initial) (moles/l)	Yield (%) Esters: Methyl acrylate	Esters: Polycarboxylic acid esters	Acids: Monocarboxylic	Acids: Succinic
0.256	79	18	29	62
0.713	66	27	14	80
1.14	56	29	7	79

Furthermore, if catalyst concentration is kept constant, not only the $C_2H_2/Co_2(CO)_8$ ratio but the C_2H_2/H_2O ratio varies in different experiments. Even with these limitations, the results obtained in two series of experiments (one in THF and the other in acetone) are interesting. At constant catalyst concentration and at a H_2O/C_2H_2 molar ratio of 10 in THF, the total yield of distillable monocarboxylic acids, succinic acid, and cyclopentanone decreases slightly with increasing C_2H_2 concentration. The yield of succinic acid reaches a maximum at an acetylene concentration of about 0.7 moles/l of solution.

The yields of monocarboxylic acids are in keeping with the previously given results in the acrylate synthesis (Table 23) (199).

It is interesting to note that, according to Natta and Albanesi (50), when the reaction is carried out in acetone, the percentage of acrylic acid in the volatile acids increases up to 80% with decreasing acetylene concentration.

3. *Acids and Esters from Various Acetylenic Substrates*

The results of hydrocarboxylation of a number of acetylenic hydrocarbons with CO and water in the presence of $Co_2(CO)_8$ in acetone are summarized in Table 24. Reactivity decreases going from acetylene itself to monosubstituted and finally to disubstituted acetylenes. The phenyl group has a larger deactivating effect than the methyl or ethyl group. Owing to the high temperature used in the hydrocarboxylation of diphenylacetylene, large amounts of secondary products were obtained, such as diphenylethane and 2,3,4,5-tetraphenylcyclopentenone, arising from the addition of hydrogen or CO and hydrogen to diphenylacetylene (208).

TABLE 24
Hydrocarboxylation of Acetylenic Hydrocarbons with Carbon Monoxide and Water with $Co_2(CO)_8$

Alkyne	Temp (°C)	P_{CO} (initial) (atm)	Products	Yield[a] (%)	Ref.
$C_2H_5C{\equiv}CH$	125	200	Ethylsuccinic acid	n.d.[b]	208
$CH_3C{\equiv}CCH_3$	140	200	*meso*-2-3-Dimethylsuccinic acid	n.d.[b]	208
$PhC{\equiv}CH$	140	200	racemic Phenylsuccinic acid	44	71,208
$PhC{\equiv}CPh$	215	150	*meso*-Diphenylsuccinic acid 84%[c] racemic Diphenylsuccinic acid 16%	9	208

[a] Based on charged acetylenic hydrocarbon.
[b] Not determined.
[c] Secondary products are 1,2-diphenylethane and 2,3,4,5-tetraphenylcyclopentenone.

In the hydrocarboxylation of disubstituted alkynes, the *meso*-dicarboxylic acid is the main acid formed. No racemic 2,3-dimethylsuccinic acid was found in the case of dimethylacetylene. With diphenylacetylene, 16% of the racemic acid was found in the mixture of dicarboxylic acids; in this case, however, the racemic products could arise from a successive transformation of the *meso* compound which might be the primary reaction product. In fact, it is known that *meso*-2,3,-diphenylsuccinic acid is partially transformed into the corresponding racemic compound above 200°C (209): about the temperature used in the present synthesis. A similar transformation occurs with 2,3-dimethylsuccinic acid at 180°C, but this is well above the temperature used for its synthesis (210).

4. Remarks on Possible Reaction Mechanisms

Natta and Pino (54) as well as Reppe and Magin (40) attempted to explain the formation of succinic acid derivatives by postulating preliminary formation of acrylic derivatives which are further hydrocarboxylated to succinic derivatives (reactions 53 and 54).

$$C_2H_2 + CO + ROH \rightarrow CH_2{=}CHCOOR \qquad (53)$$

$$CH_2{=}CHCOOR + CO + ROH \rightarrow \begin{array}{l} CH_2COOR \\ | \\ CH_2COOR \end{array} \qquad (54)$$

This hypothesis was not plausible since it was found that hydrocarboxylation of the acrylates in the presence of cobalt catalysts occurs at much higher temperatures than those for the acrylate synthesis. Wender and co-workers (29) demonstrated that two CO molecules could add to acetylene; on isolating the complex $C_4H_2O_2Co_2(CO)_7$ (**38**), they proposed that this or similar complexes might be intermediates in the synthesis of succinic acid and its derivatives.

The interaction of $HCo(CO)_4$ with acetylene in the gas phase was first investigated by Greenfield, Wotiz, and Wender (42) who obtained $C_2H_2Co_2(CO)_6$ (**39**) and propionaldehyde. More recently, Iwashita, Tamura, and Wakamatsu (197) clearly showed that, in the presence of methanol (in which $HCo(CO)_4$ does not decompose very rapidly), acetylene and presumably acetylenic derivatives do not react with $HCo(CO)_4$ to give the corresponding vinyl derivatives (reaction 55) but only to form the known complex **39** (reaction 56).

$$C_2H_2 + HCo(CO)_4 \rightarrow CH_2{=}CHCo(CO)_4 \qquad (55)$$

$$C_2H_2 + 2HCo(CO)_4 \rightarrow \underset{(\mathbf{39})}{C_2H_2Co_2(CO)_6} + 2CO + H_2 \qquad (56)$$

Furthermore, the complex **40**, obtained from 1-pentyne and $Co_2(CO)_8$ reacts with methanol and HCl to give hexanoic, 2-methylpentanoic, and 2-methylenepentanoic acids (211,212) (reaction 57).

$$(CO)_3Co\text{–}\underset{\underset{CH_2CH_2CH_3}{|}}{\overset{H}{C}{\equiv}C}\text{–}Co(CO)_3 \ (\mathbf{40}) \xrightarrow[HCl]{CH_3OH} \begin{cases} CH_3(CH_2)_4COOCH_3 & 10\% \\ CH_3(CH_2)_2CH(CH_3)COOCH_3 & 61\% \\ CH_2{=}C(C_3H_7)COOCH_3 & 29\% \end{cases} \qquad (57)$$

The corresponding acetylene complex (**39**) reacts with ethanol under CO pressure at 100°C (197) to give diethyl succinate, ethyl acrylate, and ethyl propionate. In the absence of CO, only ethyl acrylate and propionate are formed (reaction 58).

According to these experiments, a possible path for the hydrocarboxylation of acetylene in the presence of $Co_2(CO)_8$ is shown in Scheme **X**.

$$\mathbf{39} \begin{cases} \xrightarrow[EtOH,\ 100°C]{CO\ (100\ atm)} \begin{cases} CH_2{=}CHCOOEt & 34\% \\ CH_3CH_2COOEt & 15\% \\ CH_2COOEt\text{–}CH_2COOEt & 46\% \end{cases} \\ \xrightarrow[100°C]{EtOH} \begin{cases} CH_2{=}CHCOOEt & 6\text{–}12\% \\ CH_3CH_2COOEt & 1\text{–}15\% \end{cases} \end{cases} \qquad (58)$$

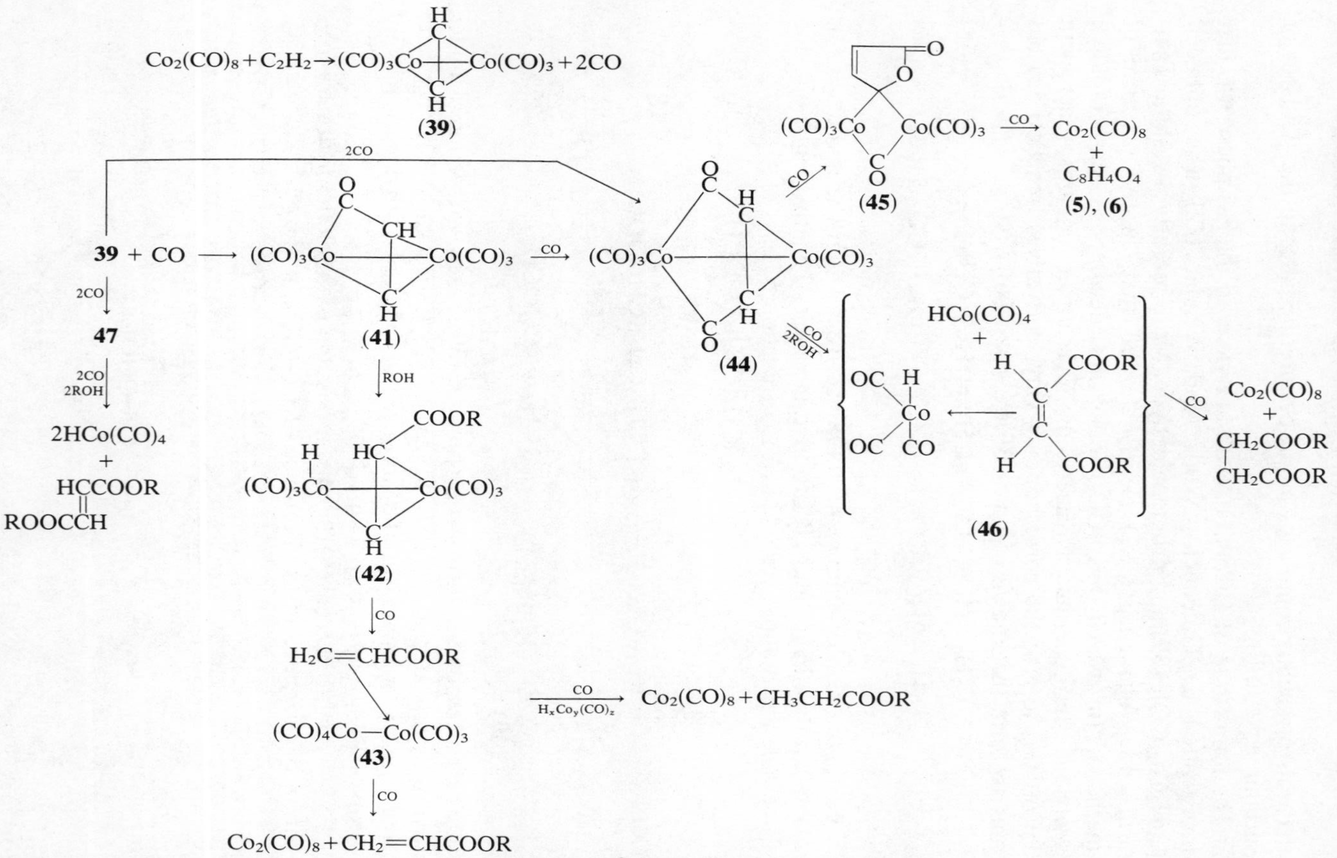

$$Co_2(CO)_8 + C_2H_2 \rightarrow (CO)_3Co(HC{\equiv}CH)Co(CO)_3 + 2CO \quad (\mathbf{39})$$

$$\mathbf{39} + CO \longrightarrow (\mathbf{41}) \xrightarrow{CO} (\mathbf{44})$$

$$\mathbf{39} \xrightarrow{2CO} (\mathbf{44})$$

$$(\mathbf{44}) \xrightarrow{CO} (\mathbf{45}) \xrightarrow{CO} Co_2(CO)_8 + C_8H_4O_4 \; (\mathbf{5}), (\mathbf{6})$$

$$(\mathbf{44}) \xrightarrow[2ROH]{CO} \left\{ HCo(CO)_4 + \text{H(COOR)C=C(H)COOR} \right\} (\mathbf{46}) \xrightarrow{CO} Co_2(CO)_8 + \text{CH}_2\text{COOR}\text{–CH}_2\text{COOR}$$

$$\mathbf{39} \xrightarrow{2CO} \mathbf{47} \xrightarrow[2ROH]{2CO} 2HCo(CO)_4 + \text{HCCOOR}{=}\text{ROOCCH}$$

$$(\mathbf{41}) \xrightarrow{ROH} (\mathbf{42}) \xrightarrow{CO} H_2C{=}CHCOOR + (CO)_4Co{-}Co(CO)_3 \; (\mathbf{43})$$

$$(\mathbf{43}) \xrightarrow{CO} Co_2(CO)_8 + CH_2{=}CHCOOR$$

$$H_2C{=}CHCOOR \xrightarrow[H_xCo_y(CO)_z]{CO} Co_2(CO)_8 + CH_3CH_2COOR$$

Scheme **X**

The kinetics of formation of complex **39** was investigated by several authors (213,213*a*,214) and discussed by Poe (214*a*); preliminary dissociation of $Co_2(CO)_8$ into $Co_2(CO)_7$ and CO seems likely.

Complex **39** can add one molecule of CO to form **41**, which, however, has never been isolated. At low CO pressure and low complex concentration in the presence of large excess of water or alcohol, a nucleophilic attack of the Lewis base on the acyl group occurs to give **42**; this may give **43**, which contains a molecule of an acrylic derivative still bound to cobalt. The acrylic molecule can either be displaced by CO or hydrogenated by a cobalt hydride species likely present in solution. Albanesi, for instance (215), reports the hydrogenation of **48** to complex **49** under conditions similar to those used for the hydrocarboxylation of acetylene (reaction 59).

$$\underset{\textbf{(48)}}{ROOC\text{—}CH\text{=}CH\text{—}C[Co(CO)_3]_3} \xrightarrow{H_xCo_y(CO)_z} \underset{\textbf{(49)}}{ROOC\text{—}CH_2\text{—}CH_2\text{—}C[Co(CO)_3]_3} \tag{59}$$

In the presence of high CO pressure and (or) low alcohol or water concentration, **41** can add a second molecule of CO to give **44**. This complex then can either add a third molecule of CO and isomerize to **45** or react with alcohol to give complex **46**. The presence of **45** was in fact observed during the succinic acid synthesis (71); furthermore, bifurandiones (**5**) and (**6**) were isolated when the reaction was carried out at low water/acetylene ratios.

The existence of maleoyl cobalt compounds that may be intermediates between **44** and **46** was demonstrated at least in mononuclear complexes (215*a*).

As expected, complex **46** has never been isolated; it undoubtedly reacts very rapidly in the presence of CO to give succinic acid derivatives and $Co_2(CO)_8$.

The addition of two CO molecules to **39** could take place in two steps according to the path **39** → **41** → **44** or in one step according to the path **39** → **44**. The structure of **39** is known (216); two molecules of CO might insert between each of two acetylenic carbon atoms and the same cobalt atom (*cis* addition) to give **44**. The addition could also take place between the acetylenic carbon atoms and the two different cobalt atoms (*trans* addition) to give **47** (reaction 60). In the latter case, the unstable **47** could

(CO)$_3$Co—(HC≡CH)—Co(CO)$_3$ (39) $\xrightarrow{+2CO}$ *cis* addition → (44); *trans* addition → (47) (60)

react with two molecules of CO and two alcohol molecules to give fumarates and two molecules of $HCo(CO)_4$. At present, there is no indication that complex **49**, isolated during the succinic acid synthesis, is a catalytic intermediate. Since **45** is transformed into **48** in the presence of mineral acids (215), this transformation may also occur in the presence of a large concentration of succinic acid at the temperature used for the succinic acid synthesis (100–120°C).

This mechanism explains the stereospecificity observed in the hydrocarboxylation of dimethylacetylene (208). The two acyl groups in **44**, which yield **46**, must be *cis*; *cis* addition of H_2 to the complexed double bond in **46**, as is usual in homogeneous hydrogenation (217), would then give *meso*-dimethylsuccinic acid.

Fumarates (but not maleates), believed to arise by dehydrogenation of succinates (69), probably are formed by the route **39** → **47**, followed by nucleophilic attack of water or alcohol on **47**. In general, internal olefins with *trans* substituents form weaker π-bonds with transition metals than the corresponding *cis* compounds (218); therefore, a displacement of the fumaric derivative, before double-bond hydrogenation, can be expected. Cobalt hydrocarbonyls are also formed in this step and may be a source of hydrogen for the formation of ketones and ketoesters. However, the possibility that dimethyl fumarate is formed by isomerization of the corresponding maleate cannot be disregarded since this reaction takes place in the presence of $HCo(CO)_4$ (219).

Another source of hydrogen during the synthesis of acids may be the reaction of $Co_2(CO)_8$ with water (50,220) to give $HCo(CO)_4$ and CO_2

which was found in the gases from the succinic acid synthesis; it is interesting to note that no fumaric acid was found in this synthesis.

Ethanetricarboxylic acid esters might result from the hydrocarboxylation of either **44** or **46**; the hydrocarboxylation of fumaric or maleic acid esters does not take place under the mild conditions of acetylene carboxylation (221,222).

The manner of formation of muconic acid esters, which on hydrogenation would give dihydromuconic acid derivatives, is still obscure. Only cobalt complexes containing one (36) or three (223) acetylene molecules were isolated, but we can envisage formation of a muconic acid complex with two cobalt atoms bridged by two acetylene molecules. The small quantities of muconic and dihydromuconic derivatives found may be formed in the type of dimerization reaction that occurs in the stoichiometric hydrocarboxylation of sodium propiolate with $Ni(CO)_4$ (104) or in some palladium-catalyzed carboxylations (224). It is interesting that a dimethyl muconate(tricarbonylmanganese) complex is formed by heating *cis*-$CH_3OOCCH{=}CHMn(CO)_5$ at 75°C; the complex obtained is binuclear with structure **50** (84).

CH_3O CH
C CH
O CO CO
OC—Mn —Mn—CO
OC CO O
CH C
C OCH_3
H

(50)

C. Acids and Esters from Acetylenic Substrates and Carbon Monoxide with Transition Metals Other Than Nickel and Cobalt as Catalysts

Transition metals (mainly of group VIII) other than nickel and cobalt were used in the synthesis of acids and esters from acetylenic substrates, CO, and water or alcohols. In this section we will briefly discuss the use of iron, ruthenium, rhodium, and palladium compounds.

1. *Iron Compounds*

The readily available $Fe(CO)_5$ was first used by Reppe (2) in the stoichiometric hydrocarboxylation of acetylene, but low yields of ethyl acrylate were reported for the reaction in ethanol in the presence of HCl. In alkaline medium or under conditions where $H_2Fe(CO)_4$ could be present in ethanolic solution, ethyl acrylate, besides hydroquinone, was formed at 80–100°C in 20–25% yield based on the $Fe(CO)_5$ used (2). Acrylic derivatives were also formed as secondary products in the semi-catalytic synthesis of hydroquinones from alkynes, CO, and water in the presence of $Fe(CO)_5$ at 170–200°C and a total pressure of 700-800 atm (32).

Attempts to carry out the catalytic synthesis of methyl acrylate with $Fe(CO)_5$ (225) in methanol at 200°C resulted in a mixture of methyl acrylate, methyl propionate, and methyl formate; the yield of methyl acrylate (based on reacted acetylene) was about 50% when using an acetylene concentration of 0.2 mole/l of solution and an initial CO pressure of 150 atm. The yield decreased to 25% at an initial CO pressure of 50 atm. Dimethyl fumarate was isolated as one of the by-products. Similar results were obtained in the synthesis of acids. By using the system $Fe(CO)_5/CuBr_2/Br_2$ as catalyst in THF as solvent (207) at 208°C, a reaction product containing 80% of acrylic acid was obtained.

On using $Fe(CO)_5$ and acetone as solvent at 180°C and an initial CO pressure of 200 atm (71), yields (based on reacted acetylene) of 34% of succinic acid were obtained. In the same experiment, a yield of 11% of a mixture of acrylic and propionic acids was also obtained.

Hydrocarboxylation of terminal alkynes, $RCH_2C{\equiv}CH$ ($R = CH_3$, C_3H_7, C_4H_9) with CO and water in the presence of $Fe(CO)_5$ and triethylamine in methanol gave the corresponding esters $R(CH_2)_3COOCH_3$ or $RCH_2CH(CH_3)COOCH_3$ as low-boiling products and $RCH_2CH(COOCH_3)CH_2COOCH_3$ and γ-lactones as a high-boiling mixture (226).

The catalytic reaction of propargyl alcohol as well as its acetate or methyl ether with CO and water and methanol in basic medium in the presence of $Fe(CO)_5$ yields mainly dimethyl methylsuccinate arising from the dicarboxylation of the triple bond and elimination of the functional group at C-3 (227). According to the proposed mechanism the catalytically active species would be $HFe(CO)_4^-$ and the reaction starts with a nucleophilic attack of the anion on the terminal acetylenic carbon atom followed by elimination of the oxy or acetoxy group analogously to the mechanism shown in reaction 38. 3-Methyl-1-butyn-3-ol and 1-ethynylcyclohexanol reacted in a similar manner but gave mixtures of unsaturated monoesters, γ-lactones, and both saturated and unsaturated diesters (227).

It thus appears that the behavior of iron is more similar to that of cobalt than to that of nickel. Due to the low practical importance of these reactions, no systematic investigation was carried out on the effect of reaction variables on the yield of acids and esters.

2. *Ruthenium Compounds*

As already mentioned, the most thoroughly investigated reaction in which ruthenium compounds (Ru acetylacetonate, $Ru_3(CO)_{12}$, $K_2Ru(OH)Cl_5$) were used as catalysts is the reaction between C_2H_2, CO, and H_2O in the presence of ethers or ketones as solvents.

The most important reaction product here is hydroquinone, but monocarboxylic acids (mainly propionic acid) were formed as by-products. The yield of these acids varies between 3 and 15%, based on reacted acetylene, reaching the higher values in the presence of halogenated ruthenium compounds. Solid and high-boiling carboxylic derivatives of unknown nature are also present in the reaction products (228). With $Ru_3(CO)_{12}$ in anhydrous alcohols, the reaction between acetylene and CO gives lower yields of hydroquinone (20–30%) than those obtained in the presence of water; the quantity of esters of monocarboxylic acids was also lower (228).

The greater selectivity of ruthenium catalysts for the synthesis of hydroquinone than for hydrocarboxylation is surprising. But the reaction of acetylene with $Ru_3(CO)_{12}$ has not been investigated to any great extent (229–231), and no sound explanation for this selectivity is at hand.

3. *Rhodium Compounds*

No systematic investigation has been reported on the use of rhodium compounds as catalysts for hydrocarboxylation of acetylenic substrates. In the presence of $Rh_2O_3 \cdot 5H_2O$ (53), acetylene, and CO in methanol at 130°C and at initial CO pressures of 100–300 atm give mainly hydroquinone, but dimethyl fumarate (10–20% based on reacted C_2H_2) and smaller quantities of dimethyl succinate (1–10%) and methyl acrylate (2–6%) were also obtained.

Propargyl alcohols were also carboxylated in the presence of rhodium compounds like $RhCl_3$ (232); dimethylpropargyl alcohol in benzene solution gives 31% of teraconic anhydride (**51**).

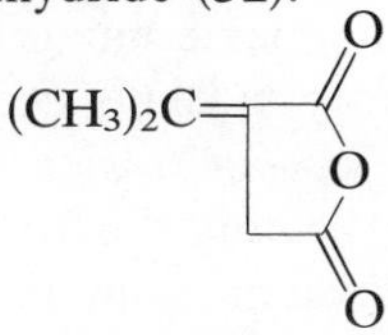

$(CH_3)_2C=$ O O O

(**51**)

Vinyl and acryloyl rhodium complexes were investigated by Wilkinson's group (173) and rhodium complexes containing acetylenic ligands were studied by Maitlis and co-workers (20,21). Insertions of acetylenic compounds into rhodium–hydrogen bonds were also investigated (81). This field seems very promising from the point of view of investigating the mechanism of the hydrocarboxylation reaction.

4. Palladium Compounds

The use of palladium catalysts for the hydrocarboxylation of acetylenic substrates was first mentioned in 1962 (233). Rather large quantities of catalyst were often used; since CO is not supplied by the metal carbonyl as with $Ni(CO)_4$, the reaction should better be classified as semicatalytic. Acetylene was reacted in butanol in the presence of palladium dichloride, iodine, and water at 65°C with equimolar mixtures of acetylene and CO under a pressure of 10–20 atm (233). The relative yields of the butyl esters were: acrylate 1, propionate 0.41, fumarate+maleate 1.8, and succinate 0.04. By increasing the quantities of water, the amount of succinic acid increases, but the sum of the dicarboxylic acids esters remains constant; the yields of maleate and fumarate esters are increased by adding Fe^{3+} ions and a small percentage of air (234). The same reaction was investigated later by Lines and Long (235). They found that the highest catalytic activity was shown by a mixture of PdI_2 (1 mole) and NaI (2 moles) when operating at 75°C and with an acetylene/CO ratio of 1. The formation of butyl fumarate and succinate increased steadily with time, while that of butyl acrylate reached a maximum at a reaction time of about 5 hr and then it decreased. The amount of butyl acrylate and the reaction rate increased in the presence of hydrogen halides (236). The number of moles of CO absorbed per mole of C_2H_2 increased when the C_2H_2 to CO molar ratio fell below 1.

Palladium black plus HI was used as catalyst in hydrocarboxylation of acetylene, phenylacetylene, and methylacetylene. In the case of phenylacetylene, addition of PPh_3 or $AsPh_3$ increases the yield of linear esters (236*a*).

According to Chiusoli and co-workers (224), dimethyl maleate and dimethyl muconate can be obtained as major products from acetylene at room temperature and atmospheric pressure when using a methanolic solution of $PdCl_2$ and thiourea and adding a small percentage of air (4%) to the acetylene–carbon monoxide mixture. By increasing the C_2H_2/CO ratio, the yield of dimethyl maleate decreased and diesters of other unsaturated acids such as octa-2,4,6-trien-1,8-dicarboxylic acid as well as deca-2,4,6,8-tetraen-1,10-dicarboxylic acid formed.

Triphenylphosphine-palladium catalytic systems under mild conditions yield acrylates from acetylene (236*b*) and methacrylic derivatives from methylacetylene (236*c*). Catalytic systems of the same type activated by $SnCl_2 \cdot 2H_2O$ yield prevailingly linear α, β-unsaturated esters from monosubstituted acetylenes (236*d*).

The effect of the type of acetylenic substrate on hydrocarboxylation in the presence of palladium catalysts was investigated mainly by Tsuji and Nogi (68,237,238) and by Heck (239). Alkynes and acetylenic substrates containing carboxyl, hydroxyl, and halogen substituents were hydrocarboxylated mainly with $PdCl_2$ in the presence or absence of HCl and with alcohols or benzene as reaction media. Two carboxyl groups are introduced; as with cobalt, both a carboxylation and a hydrocarboxylation takes place (reactions 61, 62) with preference for reaction 61 (Table 25).

$$RC{\equiv}CR + 2CO + 2R'OH \xrightarrow{PdCl_2} R'OOCRC{=}CRCOOR' + [2H] \quad (61)$$

$$RC{\equiv}CR + 2CO + R'OH \xrightarrow{PdCl_2} R'OOCHRCCHRCOOR' \quad (62)$$

The main by-products of these reactions were chloroesters **53**, **56**, **61** (Fig. 8) observed in the dicarboxylation of some alkynes (Table 25) (239). Presumably the chloroesters are formed by reduction of $PdCl_2$ to Pd(0). Metallic palladium is always separated, and only by using oxygen and $CuCl_2$ as reoxidants is it possible to make these reactions catalytic at room temperature.

Some anomalous reactions were found with diphenylacetylene, propargyl alcohol, and propargyl halides. They resemble reactions observed in the stoichiometric hydrocarboxylation of the corresponding acetylenic substrates with $Ni(CO)_4$ and acids or in catalytic hydrocarboxylation in the presence of $Co_2(CO)_8$.

The formation of lactones of type **62** (Fig. 8) was noted earlier by Whiting's group (111) in the hydrocarboxylation of the 2-methylbut-3-yn-2-ol. Allenic acids and their carbonylation products were obtained by several authors (24,110–113) mostly in the hydrocarboxylation of halogenated acetylenic substrates (Tables 11 and 12); muconic acid esters and esters of ethanetricarboxylic acid were found in the hydrocarboxylation of acetylene with $Co_2(CO)_8$ (54,70,201,206).

The reactions of propargyl alcohol in the presence of palladium catalyst are somewhat unusual; the hydroxyl group is either methylated and the triple bond is hydrocarboxylated, or three carboxylic groups are added with formation of aconitic acid esters as **68** that is formed in conditions under which **66** is not hydrocarboxylated. The formation of **68** is believed to occur according to reaction 63.

$$\begin{array}{l} CH \\ \| \! | \\ C \\ | \\ CH_2OH \end{array} + 2CH_3OH + 2CO \longrightarrow \begin{array}{l} CHCOOCH_3 \\ \| \\ CCOOCH_3 \\ | \\ CH_2OH \end{array} + [2H]$$

$$\xrightarrow[CH_3OH]{CO} \begin{array}{l} CHCOOCH_3 \\ \| \\ CCOOCH_3 \\ | \\ CH_2COOCH_3 \end{array} + H_2O \quad (63)$$

(68)

TABLE 25
Hydrocarboxylation of Acetylenic Substrates with Palladium Dichloride

Substrate	Reaction	Temp (°C)	Products: Structure[a]	Products: Yield (%)	Ref.
	Initial pressure of CO = 2 *atm*[b]				
PhC≡CH	CH_3OH	25	**52**	134	239
			53	8	
PhC≡CH	CH_3COCH_3, H_2O	25	**54**	86	239
PhC≡CCH_3	CH_3OH	25	**55**	156	239
			56	63	
$CH_3(CH_2)_4C$≡CH	CH_3OH	25	**57**	44	239
			58	183	
EtC≡CEt	CH_3OH	25	**59**	134	239
$(CH_3)_3CC$≡CH	CH_3OH	25	**60**	180	239
			61	20	
	Initial pressure of CO = 100 *atm*[c]				
PhC≡CPh	EtOH, HCl	100	**62**	66	68
			63	26	
			64	n.d.	
HC≡CCH_2Cl	CH_3OH	20	**65**	17	238,240
			66	30–70	
HC≡CCH_2OH	CH_3OH	100	**66**	26	238
			67	24	
			68	5	
HC≡CCH_2OH	CH_3OH, HCl	100	**66**	12	238
			67	36	
			68	3	
HC≡C—C(CH_3)(CH_3)—OH	C_6H_6	100	**69**	42	238
HC≡C—C(CH_3)(CH_3)—OH	CH_3OH	100	**70**	n.d.	238
			71	n.d.	
			72	n.d.	
			73	n.d.	
HO—C(CH_3)(CH_3)—C≡C—C(CH_3)(CH_3)—OH	CH_3OH	100	**74**	49	238
			75	14	
HC≡CCOOEt	EtOH, HCl	25	**76**	19	237
			77	{40	
			78		
			79	9.7	
EtOOCC≡CCOOEt	EtOH	25	**77**	{42.5	237
			78		
			80	6.5	
EtOOCC≡CCOOEt	EtOH, HCl	25	**77**	{24	237
			78		
			80	51	

[a] Structures given in Fig. 8.
[b] Yield based on $PdCl_2$ used.
[c] Yields based on reacted acetylenic compound.

(*Z*)—Ph(COOMe)C═CH(COOMe) (**52**)

(*Z*)— and (*E*)—Ph(Cl)C═CH(COOMe) (**53**)

Ph O O O

(**54**)

(*Z*)—Ph(COOMe)C═CMe(COOMe) (**55**)

(*Z*)— and (*E*)—Ph(Cl)C═CMe(COOMe) (**56**)

(*Z*)—(COOMe)CH═C(*n*—Bu)(COOMe) (**57**)

(*E*)—(COOMe)CH═C(*n*—Bu)(COOMe) (**58**)

(*Z*)—Et(COOMe)C═CEt(COOMe) (**59**)

(*Z*)—(*t*—Bu)(COOMe)C═CH(COOMe) (**60**)

(*Z*)— and (*E*)—(*t*—Bu)ClC═CH(COOMe) (**61**)

Ph Ph O O

(**62**)

(*Z*)—Ph(COOEt)C═CPh(COOEt) (**63**)

(*E*)—PhCH═CPh(COOEt) (**64**)

CH_2═CCl(CH_2COOMe) (**65**)

CH_2═C(COOMe)(CH_2COOMe) (**66**)

CH_2═C(COOMe)(CH_2OMe) (**67**)

(*Z*)—(COOMe)CH═C(COOMe)(CH_2COOMe) (**68**)

═C(CH_3)$_2$ O O O

(**69**) (continued)

(*Z*)—[Me_2(OMe)C]CH=CH(COOMe) (**70**)

Me_2C=C(COOMe)(CH_2COOMe) (**71**)

(*Z*)—[Me_2(OMe)C](COOMe)C=CH(COOMe) (**72**)

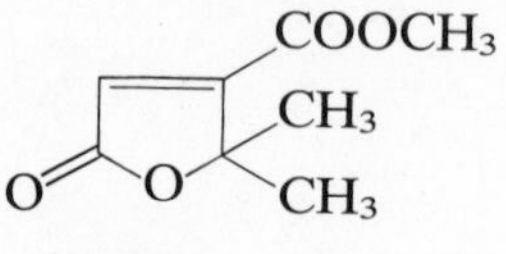

(**73**) 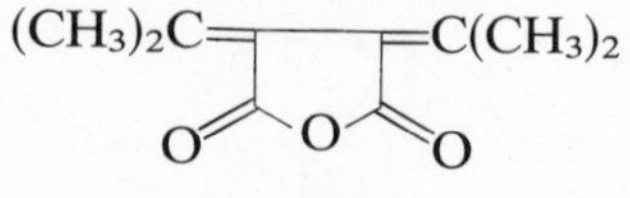 (**74**) 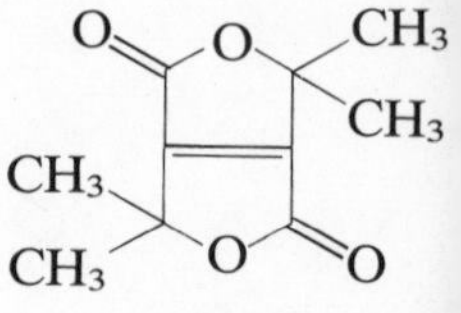 (**75**)

(*E*)—(COOEt)CH=CH(COOEt) (**76**)

$(COOEt)_2$C=CH(COOEt) (**77**)

[$(COOEt)_2$CH]CH_2(COOEt) (**78**)

$(COOEt)_2$C=CH—CH=C$(COOEt)_2$ (**79**)

(COOEA$_2$CHCH$(COOEt)_2$ (**80**)

FIG. 8. Structures of compounds from hydrocarboxylation of acetylenic substrates with $PdCl_2$. See Table 25 for details.

Considerable difficulties exist in explaining the mechanism of the hydrocarboxylation of acetylenic substrates in the presence of palladium catalysts. At least in some of the experiments, it is not certain that the reaction occurs in homogeneous phase, and experiments are reported in which heterogeneous catalysts were used (68,241).

Although complexes of the type $(PdCOCl_2)_2$ are known (242), no stable complexes of palladium containing CO and acetylenic compounds have been isolated, and the few known Pd complexes containing acetylenic ligands (243-245) do not allow formulation of a sound reaction mechanism. In fact, the products obtained by reacting acetylenic substrates with $PdCl_2$ and subsequently adding CO are different from those obtained by reacting acetylenic substrates with palladium compounds in the presence of CO (68).

According to Heck (239), the dicarboxylation of alkynes may be accounted for as shown in Scheme **XI**. The fact that small amounts of methyl phenylpropiolate were formed in the phenylacetylene carboxylation suggests that elimination of [HPdCl] occurs from an intermediate monocarboxylated vinyl palladium compound. The chloroesters, present as by-products, may be formed either by a reductive elimination of Pd(0) from the monocarboalkoxylated intermediate or by $PdCl_2$ addition to the acetylenic substrate followed by carbonylation and alcoholysis.

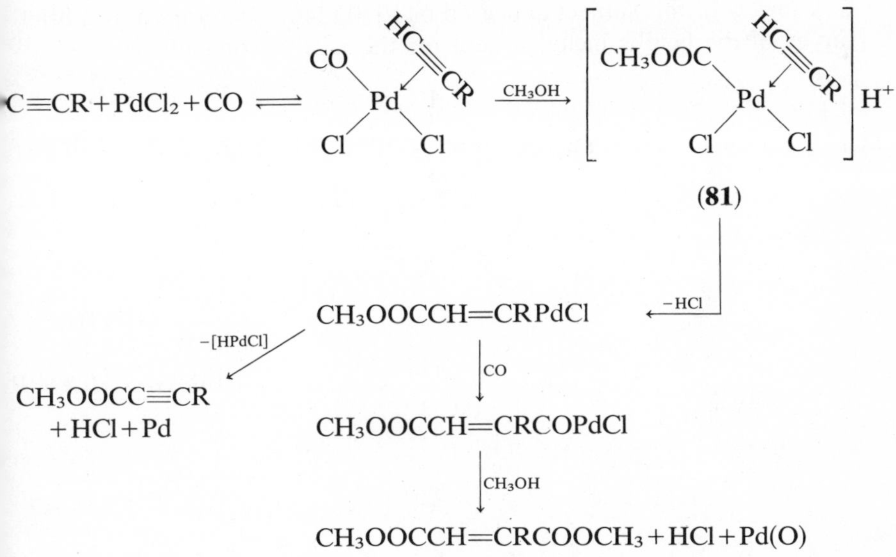

$$\text{RC}\equiv\text{CR} + \text{PdCl}_2 + \text{CO} \rightleftharpoons \text{Pd(CO)(Cl)}_2(\text{HC}\equiv\text{CR}) \xrightarrow{\text{CH}_3\text{OH}} [\text{Pd(CH}_3\text{OOC)(Cl)}_2(\text{HC}\equiv\text{CR})]\,\text{H}^+ \; (\mathbf{81})$$

$$(\mathbf{81}) \xrightarrow{-\text{HCl}} \text{CH}_3\text{OOCCH}{=}\text{CRPdCl}$$

$$\text{CH}_3\text{OOCCH}{=}\text{CRPdCl} \xrightarrow{-[\text{HPdCl}]} \text{CH}_3\text{OOCC}\equiv\text{CR} + \text{HCl} + \text{Pd}$$

$$\text{CH}_3\text{OOCCH}{=}\text{CRPdCl} \xrightarrow{\text{CO}} \text{CH}_3\text{OOCCH}{=}\text{CRCOPdCl} \xrightarrow{\text{CH}_3\text{OH}} \text{CH}_3\text{OOCCH}{=}\text{CRCOOCH}_3 + \text{HCl} + \text{Pd(O)}$$

Scheme **XI**

An analogous interpretation is suggested by Kaliya and co-workers (236) on the basis of kinetic evidence obtained in the hydrocarboxylation of acetylene to acrylates in the presence of a $PdBr_2[P(OPh)_3]_2$–HBr catalytic system. They suggest that acetylene inserts into the palladium–carbon bond of the alkoxycarbonyl intermediate **81** formed by carbonylation of a palladium alcoholatė originating from initial interaction of the palladium compound with alcohol.

V. HYDROCARBOXYLATION OF ACETYLENIC SUBSTRATES IN THE PRESENCE OF HYDROGEN DONORS OTHER THAN WATER OR ALCOHOLS

As discussed in the previous sections, the hydrocarboxylation of acetylenic compounds likely takes place according to the general Scheme **XII**.

The hydrocarboxylation of acetylenes in the presence of thiols, acids and ammonia or amines has not been extensively investigated since Reppe and co-workers (2) carried out their early experiments mainly with acetylene.

A review of this subject appeared in 1970 (246); therefore only a brief survey of the field is included here for the sake of completeness.

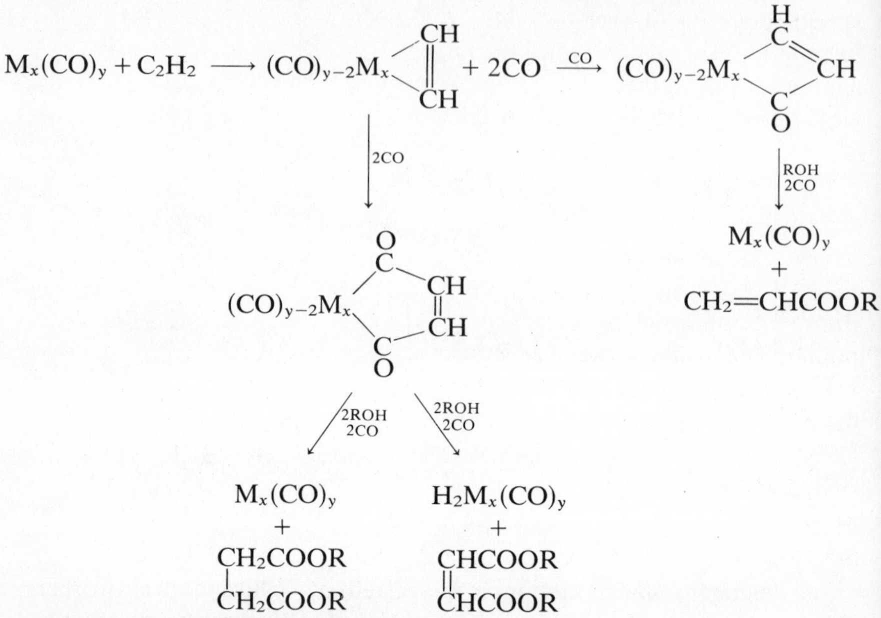

$$M_x(CO)_y + C_2H_2 \longrightarrow (CO)_{y-2}M_x(\text{CH}{=}\text{CH}) + 2CO \xrightarrow{CO} (CO)_{y-2}M_x(\text{CH}{=}\text{CH}\text{C}{=}\text{O}) \xrightarrow[2CO]{ROH} M_x(CO)_y + CH_2{=}CHCOOR$$

$$(CO)_{y-2}M_x(\text{CH}{=}\text{CH}) \xrightarrow{2CO} (CO)_{y-2}M_x(\text{C(O)CH}{=}\text{CHC(O)})$$

$$(CO)_{y-2}M_x(\text{C(O)CH}{=}\text{CHC(O)}) \xrightarrow[2CO]{2ROH} M_x(CO)_y + \begin{array}{l} CH_2COOR \\ | \\ CH_2COOR \end{array}$$

$$(CO)_{y-2}M_x(\text{C(O)CH}{=}\text{CHC(O)}) \xrightarrow[2CO]{2ROH} H_2M_x(CO)_y + \begin{array}{l} CHCOOR \\ \| \\ CHCOOR \end{array}$$

ROH could also be RSH, RCOOH′, RNH_2, RNHR′, etc.; R′ = alkyl

Scheme **XII**

A. Synthesis of Acid Chlorides

The possibility of obtaining acyl chlorides from acetylene, CO, and HCl was first shown by Sauer (247). He obtained acryloyl chloride as the main product at 120–200°C, a high CO pressure (500–1000 atm), and using rhodium or ruthenium compounds as catalysts in the absence of solvents or in the presence of aprotic solvents. In toluene, β-chloropropionyl chloride and succinoyl chloride were also obtained, according to the following stoichiometry (reaction 64).

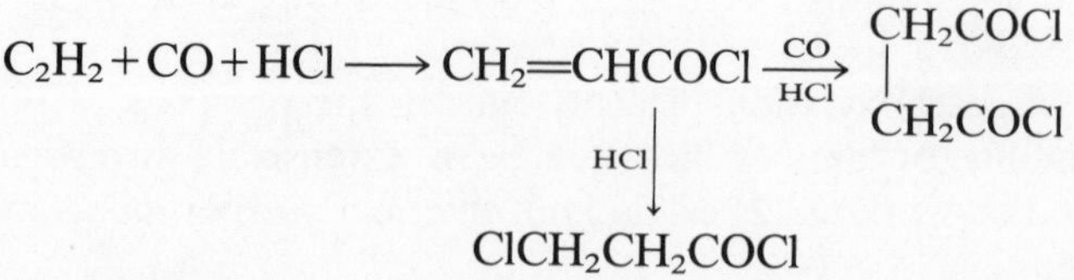

$$C_2H_2 + CO + HCl \longrightarrow CH_2{=}CHCOCl \xrightarrow[HCl]{CO} \begin{array}{l} CH_2COCl \\ | \\ CH_2COCl \end{array} \tag{64}$$

$$CH_2{=}CHCOCl \xrightarrow{HCl} ClCH_2CH_2COCl$$

The same type of reaction occurred with HF in place of HCl; both reactions are typical hydrocarboxylations.

An interesting stoichiometric reaction of the same type, but corresponding to a carboxylation, was discovered by Tsuji, Morikawa, and Iwamoto (248). By reacting acetylene and CO with a stoichiometric amount of $PdCl_2$, they obtained fumaroyl and maleoyl chlorides in 31.7% yield based on $PdCl_2$ (reaction 65).

$$C_2H_2 + 2CO \xrightarrow{PdCl_2} \begin{matrix} CHCOCl \\ \| \\ CHCOCl \end{matrix} + \begin{matrix} CHCOCl \\ \| \\ ClOCCH \end{matrix} \tag{65}$$

Surprisingly, the main reaction product (38.5% based on $PdCl_2$) was *trans,trans*-muconyl chloride accompanied by traces of the *cis,cis* compound. The synthesis was carried out in two steps: the reaction between $PdCl_2$ and acetylene, and then the carbonylation. Since it is known that acetylenic compounds are dimerized by $PdCl_2$ to the corresponding cyclobutadiene compounds (249), the mechanism shown in reaction 66 seems likely.

$$2C_2H_2 + PdCl_2 \longrightarrow Cl_2Pd \leftarrow \begin{matrix} CH{=}CH \\ | \quad | \\ CH{=}CH \end{matrix} \xrightarrow{CO} Cl_2Pd \begin{matrix} COCH{=}CH \\ \quad | \\ COCH{=}CH \end{matrix} \longrightarrow \begin{matrix} CH{=}CHCOCl \\ | \\ CH{=}CHCOCl \end{matrix} + Pd \longrightarrow \begin{matrix} CHCOCl \\ \| \\ CH{-}CH \\ \| \\ ClOCCH \end{matrix} \tag{66}$$

The same authors showed that the *cis-cis*-muconic acid chloride isomerizes to the all *trans* compound under conditions of this reaction (248).

The same type of acetylene carbonylation can be carried out with a mixture of CO and phosgene in the presence of $(PPh_3)_2PdCl_2$ (250).

B. Synthesis of Thioesters

A stoichiometric synthesis of a series of thioesters from acetylene at 40–45°C, in toluene with $Ni(CO)_4$ as CO donor, was described by Reppe (2). The best yield (77%) was obtained by using $PhCH_2SH$ as the active hydrogen compound. The reaction seems analogous to the synthesis of esters by the same route, but the data are not sufficient for a rigorous comparison of the behavior of thiols and alcohols as Lewis bases.

C. Synthesis of Anhydrides

Little is known about the synthesis of anhydrides from acetylenic substrates, CO, and acids in the presence of transition metal catalysts. In general, mixed anhydrides should be formed according to reaction 67 (2).

$$C_2H_2 + CO + RCOOH \rightarrow \begin{matrix} CH_2{=}CHCO \\ \quad\quad\quad\quad \diagdown \\ \quad\quad\quad\quad\quad O \\ \quad\quad\quad\quad \diagup \\ RCO \end{matrix} \quad (67)$$

However, according to Chiusoli and Cameroni (177), who detected anhydrides by infrared in the products of the reaction of 1-octyne with $Ni(CO)_4$ in the presence of a large excess of acid, a secondary reaction takes place, as shown in reaction 68.

$$\begin{matrix} CH_2{=}CHCO \\ \quad\quad\quad\quad \diagdown \\ \quad\quad\quad\quad\quad O \\ \quad\quad\quad\quad \diagup \\ RCO \end{matrix} + RCOOH \rightarrow \begin{matrix} RCO \\ \quad\quad \diagdown \\ \quad\quad\quad O \\ \quad\quad \diagup \\ RCO \end{matrix} + CH_2{=}CHCOOH \quad (68)$$

From acrylic acid is obtained acrylic anhydride in 88% yield when using acetylene and $Ni(CO)_4$ in an inert solvent at 40–50°C (251).

An especially interesting reaction is what seems to be a one-step synthesis of acrylic anhydride from C_2H_2, CO, and a stoichiometric amount of water (2), but the yield of anhydride is low even with large amounts of catalyst.

D. Synthesis of Amides

Reppe (2) was the first to synthesize substituted acrylamides from acetylene and various amines in the presence of $Ni(CO)_4$ or of nickel catalysts and CO (Table 26). Free acids were formed during the reactions. Ammonia itself reacts very slowly. A semicatalytic synthesis of acrylamide was achieved by Neher, Specht, and Newman (252). They reacted acetylene and $Ni(CO)_4$ in acrylic acid and then successively added CO, acetylene, $Ni(CO)_4$, HCl, and an excess of ammonia at 50–90°C (252). In general, best yields were obtained when using aniline and secondary amines; with primary amines secondary reactions sometimes occur (reaction 69); NiI_2 or $Ni(CN)_2$ were used as catalysts. Acrylamide can further react with acetylene and CO to yield bisacryloylamines (reaction 70). Acetamide was successfully used in place of amines (2),

TABLE 26

Synthesis of Amides from Acetylene, Carbon Monoxide, and Amines or Related Nitrogen Compounds

Nitrogen compound	Catalyst	Temp (°C)	Pressure (atm) ($CO/C_2H_2 = 1$)	Products	Yield[a] (%)	Ref.
Ammonia	$Ni(CO)_4$	50–90	1	Acrylamide	—	252
Ethylamine	NiS_2	100–140	30	*N*-ethylacrylamide	20–30	253
Ethylamine	NiI_2	168	28	*N*-ethylacrylamide (dimer)	48	2
Butylamine	$K_2Ni(CN)_4$	130	20	*N*-butylacrylamide(dimer)	—	2
Aniline	NiI_2	100–170	35	Acrylanilide	43	2,253
Aniline	$Co_2(CO)_8$	185	320[b]	Succinic acid dianilide	—	254
Diethylamine	$Ni(CN)_2$	130	25	Acrylic acid *N,N*-diethylamide	66	2
				Succinic acid bis(*N,N*-diethylamide)	12	
Pyrrolidine	$NiBr_2$	190	30	*N*-acryloylpyrrolidine	35	253
Piperidine	NiI_2	170–190	30	*N*-acryloylpiperidine	—	253
Dicyclohexylamine	$Ni(CO)_4$[c]	40–45	1	*N,N*-dicyclohexylacrylamide	—	2
Acetamide	$Ni(CO)_4$[c]	40–45	1	*N*-acetylacrylamide	—	2
Diphenylamine	NiI_2	150–180	25	*N,N*-diphenylacrylamide (polymer)	—	2,253
Urea	$Ni(CO)_4$	40–45	1	Polymer	—	2
Aniline[d]	$NiBr_2$	150–180	20–30	2-Phenylacrylic acid anilide	—	253

[a] Based on the nitrogen compound.
[b] $CO/C_2H_2 = 10$.
[c] Stoichiometric synthesis.
[d] Phenylacetylene used as reactant.

whereas urea gave only polymers (2).

$$C_2H_2+CO+RNH_2 \rightarrow CH_2{=}CHCONHR \qquad (69)$$

$$CH_2{=}CHCONHR+C_2H_2+CO \rightarrow \begin{matrix} CH_2{=}CHCO \\ \quad\quad\quad\quad \diagdown \\ \quad\quad\quad\quad\quad NR \\ \quad\quad\quad\quad \diagup \\ CH_2{=}CHCO \end{matrix} \qquad (70)$$

With $Co_2(CO)_8$ as catalyst, the reaction between acetylene, CO, and aniline gave succinic acid dianilide (254). Diphenylurea was isolated as a secondary product (reaction 71). The fate of the hydrogen resulting from this secondary reaction is unknown.

$$2PhNH_2+CO \xrightarrow{Co_2(CO)_8} PhNHCONHPh+[2H] \qquad (71)$$

With $Ni(CN)_2$ as catalyst and diethylamine, the diethylamide of succinic acid and diethyl acrylamide were obtained in yields of 12 and 66% respectively (2).

VI. CONCLUDING REMARKS

An attempt has been made to point out the most important facts about the carbonylation of acetylenic substrates. The discussion was limited to the use of group VIII metal compounds. These were the most thoroughly investigated, although other metal derivatives were used. Ketones, for example, were obtained in the presence of chromium compounds (18,255), and rhenium derivatives were used in carbonylation reactions (256). The decomposition by HCl of the adducts obtained from alkyl- and acylmanganesepentacarbonyl with acetylenic compounds (257,258) also yielded carbonylation products.

Acetylenic substrates are more reactive with transition metal complexes than are olefins. Their reactions with CO give rise to a larger variety of products than do the corresponding olefinic substrates. In general, acetylenes give both mono- and dicarboxylic derivatives.

There are two types of reactions in the synthesis of carboxylic acids and their derivatives: The more general one corresponds to a true hydrocarboxylation (reaction 72); the second, catalyzed mainly by palladium and cobalt compounds, corresponds to a dicarboxylation (reaction 73).

In dicarboxylation, the two hydrogen atoms liberated are probably responsible for the secondary products formed; for palladium complexes, reduction to the metal may occur in this way. ROR′, RNH_2, RNR'_2, RSR′,

$$C_2H_2 \xrightarrow{CO,\,ROH} \begin{cases} \rightarrow CH_2{=}CHCOOR \\ \rightarrow \begin{array}{l} CH_2COOR \\ | \\ CH_2COOR \end{array} \end{cases} \tag{72}$$

$$C_2H_2 \xrightarrow{CO,\,ROH} ROOCCH{=}CHCOOR + [2H] \tag{73}$$

and so on may be used instead of ROH to give the corresponding derivatives of carboxylic acids.

Hydrocarbonylation, formation of carbonyl compounds, is the most important reaction with olefinic substrates, but it does not give good yields with acetylenic substrates. The dissimilar behavior in the hydrocarbonylation of olefins and of acetylenic substrates with the same catalysts may be taken as an indication that metal carbonyl hydrides play a less important role in the reaction of acetylenic substrates. The fact that insertion reactions of acetylenic substrates into metal–hydrogen bonds are known (79–84,258) does not necessarily mean that the alkenyl–metal complexes formed are important intermediates in hydrocarboxylation.

As formation of π-complexes between the unsaturated substrates and the metal is largely accepted as the first reaction step, the main differences should exist in subsequent steps. Despite the fact that the insertion of acetylenic substrates into metal–hydrogen bonds was observed (79–84,258), the second step often seems to be an insertion of one or two molecules of CO between the complexed acetylenic substrates and the metal. But, according to the most widely accepted mechanism, the second step with an olefinic substrate would be formation of a metal alkyl intermediate.

The two different intermediate complexes might have dissimilar reactivity with alcohols, water, or hydrogen (Scheme **XIII**). In fact, cyclic organometallic derivatives of type **82** might not be cleaved easily by hydrogen. Finally, the metal carbonyl hydride formed in the reaction with alcohols or water could give rise, in the case of acetylenic substrates, to saturated dicarboxylic acids or to various secondary products; however, the hydride should not be necessary for the catalytic cycle.

This path is not necessarily true for the synthesis of acrylic derivatives from acetylenic substrates which can give rise to intermediates of type **83**. This complex, as already mentioned for **82**, might react more easily with alcohols or water than with hydrogen.

The path hypothesized above does not take into account significant differences observed for various metal compounds. The typical behavior of the different group VIII metal compounds can be briefly summarized

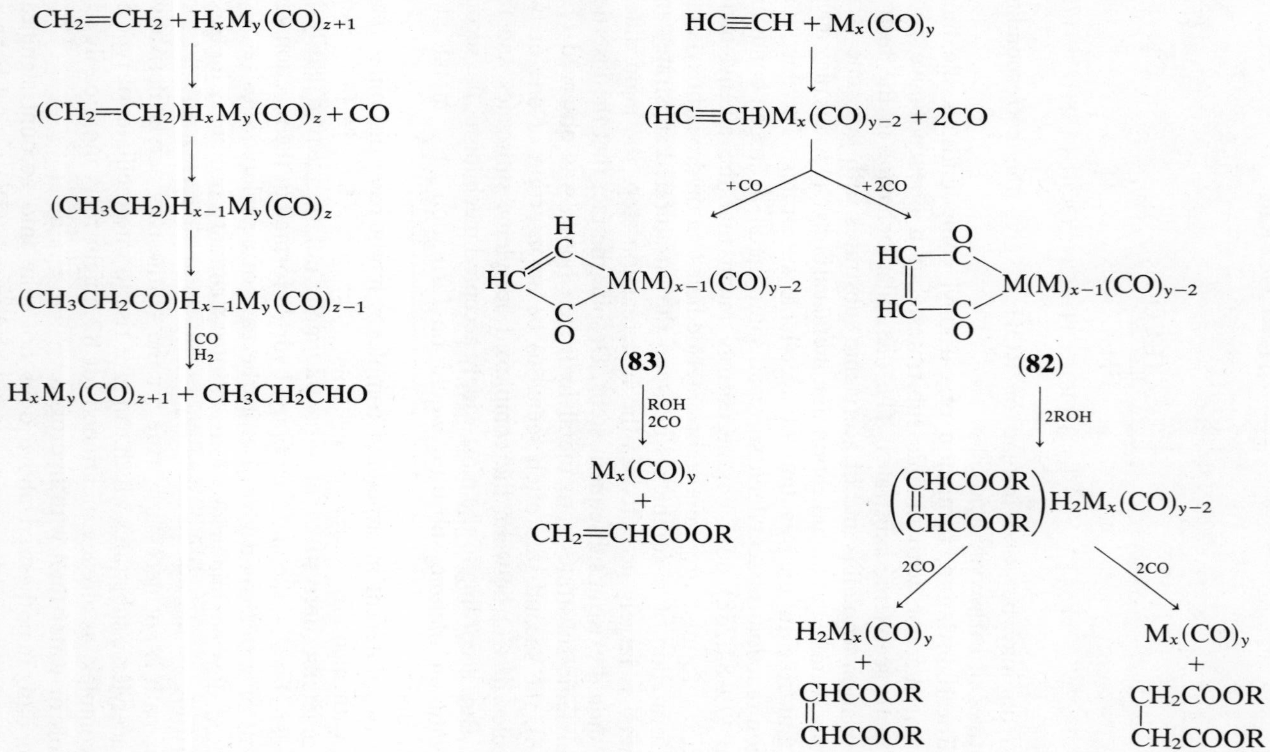

$$CH_2{=}CH_2 + H_xM_y(CO)_{z+1} \longrightarrow (CH_2{=}CH_2)H_xM_y(CO)_z + CO \longrightarrow (CH_3CH_2)H_{x-1}M_y(CO)_z \longrightarrow (CH_3CH_2CO)H_{x-1}M_y(CO)_{z-1} \xrightarrow{CO,\ H_2} H_xM_y(CO)_{z+1} + CH_3CH_2CHO$$

$$HC{\equiv}CH + M_x(CO)_y \longrightarrow (HC{\equiv}CH)M_x(CO)_{y-2} + 2CO$$

$$(HC{\equiv}CH)M_x(CO)_{y-2} \xrightarrow{+CO} \mathbf{(83)}\ [HC{=}CH{-}C(O)]M(M)_{x-1}(CO)_{y-2}$$

$$(HC{\equiv}CH)M_x(CO)_{y-2} \xrightarrow{+2CO} \mathbf{(82)}\ [C(O){-}HC{=}CH{-}C(O)]M(M)_{x-1}(CO)_{y-2}$$

$$\mathbf{(83)} \xrightarrow{ROH,\ 2CO} M_x(CO)_y + CH_2{=}CHCOOR$$

$$\mathbf{(82)} \xrightarrow{2ROH} (CHCOOR{=}CHCOOR)H_2M_x(CO)_{y-2}$$

$$(CHCOOR{=}CHCOOR)H_2M_x(CO)_{y-2} \xrightarrow{2CO} H_2M_x(CO)_y + CHCOOR{=}CHCOOR$$

$$(CHCOOR{=}CHCOOR)H_2M_x(CO)_{y-2} \xrightarrow{2CO} M_x(CO)_y + CH_2COOR{-}CH_2COOR$$

Scheme **XIII**

as follows: Catalytic activity seems greater for Ru, Rh, and Pd than for Fe, Co, and Ni. In each triad, catalytic activity reaches a maximum and selectivity a minimum for the second member. Cobalt and palladium show maximum selectivity toward formation of dicarboxylic acids; in particular, palladium seems to favor dicarboxylation.

No generalization can be attempted for the detailed mechanisms of the reactions mentioned in this chapter, since they are largely unknown. Research in this field has proceeded in two steps: investigation of the composition of the reaction products accompanied by an empirical search for the most suitable reaction conditions, then investigation of the synthesis and structure of those metal complexes that seem logically connected with the reaction intermediates. Only after enough knowledge has been acquired in both fields can attempts be made to understand the reaction mechanisms in detail. At present, our most advanced knowledge exists in the field of cobalt-catalyzed reactions. Information acquired from investigating cobalt complexes, have, for instance, completely changed the views on the mechanism of the succinic acid synthesis formerly proposed on the basis of the nature of the reaction products.

As for nickel, a better knowledge of nickel–acetylene complexes is needed, whereas in the case of palladium, ruthenium, and rhodium, the first step, identification of the reaction products formed under different conditions, is yet to be completed.

The most important aspect still to be investigated is the nature of reaction variables influencing catalytic selectivity. This field is certainly open to further progress, as shown, for instance, by cobalt-catalyzed carbonylations. Here, despite the originally low selectivity, conditions were found that gave yields of more than 85% of methyl acrylate or succinic acid. Despite difficulties in elucidating details of the catalytic action exerted by transition metals, research on the carbonylation of acetylenic substrates revealed the surprising ability of transition metal complexes to selectively arrange many small unsaturated molecules in the catalytic complexes, this ability is strongly affected by the number of the metal atoms and type of ligands present in the catalytic complexes.

Acknowledgments

We thank Prof. G. Albanesi of the University of Parma for careful reading of the manuscript and for useful suggestions and Dr. G. Sbrana of the University of Pisa for continuous discussions and readings of the text.

REFERENCES

1. W. Reppe, *Ann.*, **560,** 1 (1948).
2. W. Reppe, *Ann.*, **582,** 1 (1953).
3. W. Reppe, Ger. Pat. 855,110 (1939).
4. O. Roelen, Ger. Pat. 849,548 (1938); U.S. Pat. 2,327,066 (1943); Belg. Pat. 436,625 (1939).
5. O. Roelen, *Angew. Chem.*, **A.60** (3), 213 (1948).
6. O. Roelen, *Naturforschung und Medizin in Deutschland*, Bd. 36, *Präp. Org. Chem.*, Vol. I, 1948, p. 157.
7. F. L. Bowden and A. B. Lever, *Organometal. Chem. Rev.*, **3,** 227 (1968).
8. W. Hübel, in I. Wender and P. Pino, Ed., *Organic Syntheses via Metal Carbonyls*, Vol. I, Interscience, New York, 1968, p. 273–342.
9. Y. Sawa, I. Hashimoto, M. Ryang, and S. Tsutsumi, *J. Org. Chem.*, **33,** 2159 (1968).
10. K. Kinugasa and T. Agawa, *Organometal. Chem. Syn.*, **1,** 427 (1972).
11. M. Foà and L. Cassar, *Gazz. Chim. Ital.*, **103,** 805 (1973).
12. W. Hübel and E. H. Braye, *J. Inorg. Nucl. Chem.*, **10,** 250 (1959).

12*a*. R. S. Dickson and L. J. Michel, *Aust. J. Chem.*, **26,** 1791 (1973).

13. M. D. Rausch and R. A. Genetti, *J. Org. Chem.*, **35,** 3888 (1970).

13*a*. F. W. Grevels, D. Schulz, and E. Koerner von Gustorf, *Angew. Chem.*, **86,** 558 (1974).

14. R. Bühler, R. Geist, R. Mündich, and H. Plieninger, *Tetrahedron Lett.*, **1973,** 1919.
15. L. Vallarino and G. Santarella, *Gazz. Chim. Ital.*, **94,** 252 (1964).
16. C. W. Bird and J. Hudec, *Chem. Ind.* (*London*), **1959,** 570.
17. W. Reppe and H. Vetter, *Ann.*, **582,** 143 (1953).
18. D. P. Tate, J. M. Augl, W. M. Ritchey, B. L. Ross, and J. G. Grasselli, *J. Am. Chem. Soc.*, **86,** 3261 (1964).
19. P. M. Maitlis and S. McVey, *J. Organometal. Chem.*, **4,** 254 (1965).
20. S. McVey and P. M. Maitlis, *J. Organometal. Chem.*, **19,** 169 (1969).
21. J. W. Kang, S. McVey, and P. M. Maitlis, *Can. J. Chem.*, **46,** 3189 (1968).
22. R. S. Dickson and G. Wilkinson, *J. Chem. Soc.*, **1964,** 2699.
23. B. W. Howk and J. C. Sauer, U.S. Pat. 3,055,949 (1962).
24. P. J. Ashworth, G. H. Whitham, and M. C. Whiting, *J. Chem. Soc.*, **1957,** 4633.
25. P. Pino, L. Giuffré, and G. Albanesi, *Sintesi di acido succinico da acetilene, ossido di carbonio ed acqua*, Tamburini Ed., Milan, 1955.
26. G. Albanesi and M. Tovaglieri, *Chim. Ind.* (*Milan*), **41,** 189 (1959).
27. G. Albanesi, *Chim. Ind.* (*Milan*), **46,** 1169 (1964).
28. J. C. Sauer, R. D. Cramer, V. A. Engelhardt, T. A. Ford, H. E. Holmquist, and B. W. Howk, *J. Am. Chem. Soc.*, **81,** 3677 (1959).
29. H. W. Sternberg, J. G. Shukys, C. Delle Donne, R. Markby, R. A. Friedel, and I. Wender, *J. Am. Chem. Soc.*, **81,** 2339 (1959).
30. O. S. Mills and G. Robinson, *Proc. Chem. Soc.*, **1959,** 156.
31. W. Reppe and A. Magin, Ger. Pat. 1,071,077 (1965).
32. W. Reppe, N. v. Kutepow, and A. Magin, *Angew. Chem. Intern. Ed.*, **8,** 727 (1969).
33. G. Albanesi, R. Farina, and A. Taccioli, *Chim. Ind.* (*Milan*), **48,** 1151 (1966).
34. J. C. Sauer, B. W. Howk, and R. T. Stiehl, *J. Am. Chem. Soc.*, **81,** 693 (1959).
35. C. Hoogzand and W. Hübel, in I. Wender and P. Pino, Ed., *Organic Syntheses via Metal Carbonyls*, Vol. I, Interscience, New York, 1968, pp. 343–371.
36. H. W. Sternberg, H. Greenfield, R. A. Friedel, J. Wotiz, R. Markby, and I. Wender, *J. Am. Chem. Soc.*, **76,** 1457 (1954).

37. D. S. Breslow and R. F. Heck, *Chem. Ind.* (*London*), **1960,** 467.
38. R. F. Heck and D. S. Breslow, *J. Am. Chem. Soc.*, **83,** 4023 (1961).
39. W. Reppe and A. Magin, U.S. Pat. 2,562,393 (1950).
40. W. Reppe and A. Magin, U.S. Pat. 2,604,490 (1951).
41. H. Adkins and G. Krsek, *J. Am. Chem. Soc.*, **71,** 3051 (1949).
42. H. Greenfield, J. Wotiz, and I. Wender, *J. Org. Chem.*, **22,** 542 (1957).
43. F. H. Jardine, J. A. Osborn, G. Wilkinson, and J. F. Young, *Chem. Ind.* (*London*), **1965,** 560.
44. D. Evans, J. A. Osborn, and G. Wilkinson, *J. Chem. Soc. A*, **1968,** 3133.
44*a*. C. Botteghi and C. Salomon, *Tetrahedron Lett.*, **1974,** 4285.
45. G. Wilkinson, Fr. Pat. 1,459,643 (1965).
46. C. K. Brown, D. Georgiou, and G. Wilkinson, *J. Chem. Soc. A*, **1971,** 3120.
47. G. Natta, *Brennstoff Chem.*, **36,** 176 (1955).
48. B. F. Crowe, *Chem. Ind.* (*London*), **1960,** 1000.
49. B. F. Crowe, *Chem. Ind.* (*London*), **1960,** 1506.
50. G. Natta and G. Albanesi, *Chim. Ind.* (*Milan*), **48,** 1157 (1966).
51. B. Fell and M. Beutler, *Tetrahedron Lett.*, **1972,** 3455.
52. F. Piacenti, P. Pino, R. Lazzaroni, and M. Bianchi, *J. Chem. Soc. C*, **1966,** 488.
53. Ajinomoto Co., Inc., Fr. Pat. 1,486,666 (1966).
54. G. Natta and P. Pino, *Chim. Ind.* (*Milan*), **31,** 245 (1949).
55. F. Piacenti, P. Pino, and P. L. Bertolaccini, *Chim. Ind.* (*Milan*), **44,** 600 (1962).
56. Y. Takegami, C. Yokokawa, and Y. Watanabe, *Bull. Chem. Soc. Japan*, **39,** 2430 (1966).
57. R. W. Goetz and M. Orchin, *J. Am. Chem. Soc.*, **85,** 2782 (1963).
58. G. Natta, P. Pino, and E. Mantica, *Gazz. Chim. Ital.*, **80,** 680 (1950).
59. P. Pino, G. Braca, G. Sbrana, and A. Cuccuru, *Chem. Ind.* (*London*), **1968,** 1732.
60. P. Pino, G. Braca, and G. Sbrana, Swiss Pat. 442,346 (1963).
61. P. Pino, G. Braca, and G. Sbrana, Swiss Pat. 489,450 (1968).
62. G. Braca, G. Sbrana, F. Piacenti, and P. Pino, *Chim. Ind.* (*Milan*), **52,** 1091 (1970).
63. W. Reppe and A. Magin, Ger. Pat. 870,698 (1944).
64. W. Reppe, Ger. Pat. 874,910 (1951).
65. W. Reppe, U.S. Pat. 2,702,304 (1952).
66. W. Reppe and A. Magin, Ger. Pat. 1,232,974 (1965).
67. G. P. Mueller and F. L. McArtor, *J. Am. Chem. Soc.*, **76,** 4621 (1954).
68. J. Tsuji and T. Nogi, *J. Am. Chem. Soc.*, **88,** 1289 (1966).
68*a*. P. Pino, G. Gaudiano, M. Cecchetti, and F. Piacenti, *Ann. Chim.* (*Rome*), **51,** 786 (1961).
69. P. Pino and A. Miglierina, *J. Am. Chem. Soc.*, **74,** 5551 (1952).
70. P. Pino, E. Pietra, and B. Mondello, *Gazz. Chim. Ital.*, **84,** 453 (1954).
71. G. Natta and P. Pino, U.S. Pat. 2,851,486 (1955).
72. I. N. Nazarov and T. D. Nagibina, *Bull. Acad. Sci. S.S.S.R. Classe Sci. Chim.*, **1946,** 641; *Chem. Abstr.*, **42,** 7733 (1948).
73. A. Matzuda, *Bull. Chem. Soc. Japan*, **42,** 571 (1969).
74. P. L. Bertolaccini, Thesis, University of Pisa, 1957.
75. P. Pino, G. Braca, and G. Sbrana, Swiss Pat. 473,076 (1966).
76. P. Pino, G. Braca, and G. Sbrana, unpublished results.
77. G. N. Schrauzer, *Chem. Ind.* (*London*), **1958,** 1403.
78. R. Burt, M. Cooke, and M. Green, *J. Chem. Soc. A*, **1970,** 2981.
79. T. Blackmore, M. I. Bruce, F. G. A. Stone, R. E. Davis, and A. Garza, *Chem. Commun.*, **1971,** 852.

80. T. Blackmore, M. I. Bruce, and F. G. A. Stone, *J. Chem. Soc. Dalton*, **1974,** 106.
81. B. L. Booth and A. D. Lloyd, *J. Organometal. Chem.*, **35,** 195 (1972).
82. W. H. Baddley and M. S. Fraser, *J. Am. Chem. Soc.*, **91,** 3661 (1969).
83. H. C. Clark and R. K. Mittal, *Can. J. Chem.*, **51,** 1511 (1973).
84. B. L. Booth and R. G. Hargreaves, *J. Chem. Soc. A*, **1969,** 2766.
85. J. Clemens, M. Green, and F. G. A. Stone, *J. Chem. Soc. Dalton*, **1973,** 375.
86. G. P. Chiusoli and S. Merzoni, *Chim. Ind.* (*Milan*), **43,** 259 (1961).
87. R. F. Heck, *J. Am. Chem. Soc.*, **86,** 2819 (1964).
88. G. Sbrana, G. Braca, and E. Benedetti, *J. Chem. Soc. Dalton*, **1975,** 754.
89. R. F. Heck, *J. Am. Chem. Soc.*, **85,** 3116 (1963).
90. A. Carbonaro, A. Greco, and G. Dall'Asta, *J. Organometal. Chem.*, **20,** 177 (1969).
91. A. D. Ketley, L. P. Fisher, A. J. Berlin, C. R. Morgan, E. H. Gorman, and T. R. Steadman, *Inorg. Chem.*, **6,** 657 (1967).
92. J. Dewar and H. O. Jones, *J. Chem. Soc.*, **1904,** 212.
93. K. Ohashi, S. Suzuki, and H. Ito, *J. Chem. Soc. Japan, Ind. Chem. Sect.*, **55,** 120 (1952).
94. K. Ohashi, S. Suzuki, and H. Ito, *J. Chem. Soc. Japan, Ind. Chem. Sect.*, **55,** 607 (1952).
95. S. Suzuki, K. Uno, H. Ionezawa, and H. Ito, *J. Chem. Soc. Japan, Ind. Chem. Sect.*, **55,** 718 (1952).
96. E. R. H. Jones, T. Y. Shen, and M. C. Whiting, *J. Chem. Soc.*, **1950,** 230.
97. E. R. H. Jones, T. Y. Shen, and M. C. Whiting, *J. Chem. Soc.*, **1951,** 766.
98. H. W. Sternberg, R. Markby, and I. Wender, *J. Am. Chem. Soc.*, **82,** 3638 (1960).
99. W. Reppe and A. Magin, Brit. Pat. 943,721 (1961).
100. W. Reppe and A. Magin, Ger. Pat. 1,173,458 (1961).
101. W. Reppe and A. Magin, Ger. Pat. 1,215,139 (1963).
102. M. Maki, *J. Fuel Soc. Japan*, **32,** 410 (1953); *Chem. Abstr.*, **49,** 851 (1955).
103. A. Yakubovic and E. Volkova, *Dokl. Akad. Nauk S.S.S.R* **84,** 1183 (1952).
104. E. R. H. Jones, G. H. Whitham, and M. C. Whiting, *J. Chem. Soc.*, **1954,** 1865.
105. A. Yakubovich and E. Volkova, *Zhur. Obsch. Khim.*, **30,** 3972 (1960); *Chem. Abstr.*, **55,** 25798 (1961).
106. G. P. Chiusoli and S. Merzoni, *Chim. Ind.* (*Milan*), **51,** 612 (1969).
107. J. M. J. Tetteroo, Thesis, Aachen Rheinisch-Westfalischen Technischen Hochschule, July 1965.
108. W. Reppe and A. Simon, Ger. Pat. 857,635 (1952).
109. E. R. H. Jones, T. Y. Shen, and M. C. Whiting, *J. Chem. Soc.*, **1951,** 48.
110. E. D. Bergmann and E. Zimkin, *J. Chem. Soc.*, **1950,** 3455.
111. E. R. H. Jones, G. H. Whitham, and M. C. Whiting, *J. Chem. Soc.*, **1957,** 4628.
111*a*. K. Ohkuma, A. Ueda, and S. Hashimoto, *Japan Kokai*, **75,** 25524 (1975); *Chem. Abstr.*, **83,** 58155 (1975).
112. R. W. Rosenthal and L. H. Schwartzman, *J. Org. Chem.*, **24,** 836 (1959).
113. G. P. Chiusoli, *Chim. Ind.* (*Milan*), **41,** 513 (1959).
114. E. R. H. Jones, T. Y. Shen, and M. C. Whiting, *J. Chem. Soc.*, **1951,** 763.
115. R. W. Rosenthal, L. H. Schwartzman, N. P. Greco, and R. Proper, *J. Org. Chem.*, **28,** 2835 (1963).
116. P. Pino, F. Piacenti, M. Bianchi, and R. Lazzaroni, *Chim. Ind.* (*Milan*), **50,** 106 (1968).
117. C. W. Bird and E. M. Briggs, *J. Chem. Soc. C*, **1967,** 1265.
118. H. T. Neher, E. H. Specht, and A. Neuman, U.S. Pat. 2,582,911 (1950).
119. A. Neuman, H. T. Neher, and E. H. Specht, U.S. Pat. 2,778,848 (1953).

120. I. Dakli and L. Corsi, U.S. Pat. 2,881,205 (1956).
121. G. A. Elliot, and S. A. Furbusch, U.S. Pat. 3,002,016 (1959).
122. J. E. Ehrreich, R. G. Nickerson, and C. E. Ziegler, *Ind. Eng. Chem., Process Design Develop.*, **4,** 77 (1965).
123. S. Suzuki, H. Ito, H. Shoda, and S. Tsuchiya, *Yuki Gosei Kagaku Kyokai Shi*, **26,** 59 (1968); *Chem. Abstr.*, **68,** 68428 (1968).
124. Y. Sakakibara, *Bull. Chem. Soc. Japan*, **37,** 1601 (1964).
125. S. Kunichika, Y. Sakakibara, and T. Nakamura, *Bull. Chem. Soc. Japan*, **41,** 390 (1968).
126. J. Happel and C. J. Marsel, Belg. Pat. 639,260; *Chem. Abstr.*, **62,** 9019 (1965).
126*a*. J. Happel and S. Umemura, U.S. Pat., 3,812,175 (1974); *Chem. Abstr.*, **81,** 92193 (1974).
126*b*. J. Happel, S. Umemura, Y. Sakakibara, H. Blanck, and S. Kunichika, *Ind. Eng. Chem., Process Des. Develop.*, **14,** 44 (1975).
127. T. P. Forbath, *Chem. Eng.*, **1962,** 96.
128. R. Toepel, *Chim. Ind.* (*Paris*), **91,** 139 (1964).
129. BASF, *Hydrocarbon Processing*, **46** (11), 140 (1967).
130. W. Reppe and R. Stadler, U.S. Pat. 3,023,237 (1953).
131. W. Reppe, N. v. Kutepow, P. Raff, E. Henkel, H. Friederich, and W. Himmele, Ger. Pat. 965,323 (1954).
132. W. Reppe, H. Friederich, E. Henkel, N. v. Kutepow, and W. Himmele, Ger. Pat. 1,003,721 (1954).
133. K. Yamamoto, *Bull. Chem. Soc. Japan*, **27,** 491 (1954).
134. K. Yamamoto and M. Oku, *Bull. Chem. Soc. Japan*, **27,** 505 (1954).
135. S. Shoda, N. Ito, and T. Miki, U.S. Pat. 3,670,012 (1972).
136. Y. Sakakibara and T. Okamoto, *Bull. Inst. Chem. Res. Kyoto Univ.*, **45,** 175 (1967).
137. W. Reppe, H. Kröper, and H. J. Pistor, Ger. Pat. 763,693 (1941).
138. K. Yamamoto, *Bull. Chem. Soc. Japan*, **27,** 516 (1954).
139. W. Reppe and co-workers, (*a*) Ger. Pat. 854,948 (1939); (*b*) Ger. Pat. 805,641 (1948); (*c*) Ger. Pat. 872,205 (1941); (*d*) U.S. Pat. 2,738,364 (1951); (*e*) Ger. Pat. 965,323 (1954); (*f*) Ger. Pat. 1,003,721 (1954); (*g*) Ger. Pat. 1,011,415 (1953); (*h*) Brit. Pat. 779,277 (1953); (*i*) Brit. Pat. 790,930 (1954); (*j*) U.S. Pat. 2,806,040 (1955); (*k*) Ger. Pat. 949,654 (1956); (*l*) U.S. Pat. 2,845,451 (1955); (*m*) Brit. Pat. 802,544 (1956); (*n*) Ger. Pat. 1,042,571 (1955); (*o*) U.S. Pat. 2,886,591 (1956); (*p*) Brit. Pat. 824,520 (1956); (*q*) Brit. Pat. 822,731 (1955); (*r*) Ger. Pat. 1,060,382 (1956); (*s*) Ger. Pat. 1,076,672 (1956); (*t*) U.S. Pat. 3,023,237 (1953); (*u*) U.S. Pat. 2,925,436 (1957); (*v*) Ger. Pat. 1,009,620 (1955); (*w*) Ger. Pat. 942,809 (1953).
140. J. T. Dunn and W. R. Proops, (*a*) Fr. Pat. 1,229,843 (1958); (*b*) U.S. Pat. 2,966,510 (1958); (*c*) U.S. Pat. 2,966,511 (1958); (*d*) U.S. Pat. 2,967,882 (1958); (*e*) U.S. Pat. 2,967,883 (1958); (*f*) U.S. Pat. 2,967,884 (1958); (*g*) U.S. Pat. 2,992,270 (1959); (*h*) U.S. Pat. 3,013,067 (1959); (*i*) U.S. Pat. 3,019,256 (1959).
141. B. J. Luberoff and co-workers, (*a*) U.S. Pat. 2,882,297 (1956); (*b*) U.S. Pat. 2,882,298 (1956); (*c*) U.S. Pat. 2,882,299 (1956); (*d*) U.S. Pat. 3,025,319 (1958); (*e*) U.S. Pat. 3,025,322 (1958).
142. H. H. Mathews, U.S. Pat. 2,903,479 (1956).
143. J. M. Leathers and W. P. Coker, U.S. Pat. 2,964,558 (1958).
144. H. Friederich and H. Hoffman, Ger. Pat. 1,042,572 (1958).
145. S. Bhattacharyya and A. Sen, *J. Appl. Chem.*, **1963,** 498.
146. S. Bhattacharyya and A. Sen, *Ind. Eng. Chem., Process Design Develop.*, **3** (2), 169 (1964).

147. S. Bhattacharyya and D. Bhattacharyya, *J. Appl. Chem.*, **1966,** 18.
148. S. Bhattacharyya and D. Bhattacharyya, *J. Appl. Chem.*, **1966,** 202.
149. M. L. Noble, Brit. Pat. 713,325 (1954).
150. P. Pino and A. Bargellini, unpublished results.
151. L. E. Orgel, *Intern. Conf. on Coordination Chem., Chem. Soc. Special Publ No.* 13, London, 1959, p. 93.
152. G. Ayrey, C. W. Bird, E. M. Briggs, and A. F. Harmer, *Organometal. Chem., Syn.*, **1,** 187 (1970/71).
153. W. Wong, S. J. Singer, W. D. Pitts, S. F. Watkins, and W. H. Baddley, *Chem. Commun.*, **1972,** 672.
154. C. W. Bird, E. M. Briggs, and J. Hudec, *J. Chem. Soc. C*, **1967,** 1862.
155. M. Almasi, L. Szabo, S. Farkas, F. Kacso, O. Vegh, and I. Muresan, *Studi Cercetari Chim. Ac. Rep. Pop. Romine*, **1960,** 509.
156. H. Kröper, *Allgemeine Chemische Methoden* (Houben Weil), Vol. IV, Part 2, Georg Thieme, Stuttgart, 1955, p. 414.
157. R. F. Heck, *J. Am. Chem. Soc.*, **85,** 2013 (1963).
158. P. L. Pauson, *Proc. Chem. Soc.*, **1960,** 297.
159. M. C. Whiting, *Proc. Chem. Soc.*, **1958,** 50.
160. R. B. King, M. I. Bruce, J. R. Phillips, and F. G. A. Stone, *Inorg. Chem.*, **5,** 684 (1966).
161. O. S. Mills and B. W. Shaw, *Acta Crystal.*, **18,** 562 (1965).
162. G. Wilke and G. Herrmann, *Angew. Chem.*, **74,** 693 (1962).
163. R. J. Tedeschi and G. Moore, U.S. Pat. 3,474,120 (1967).
164. L. Cassar and M. Foà, *Inorg. Nucl. Chem. Lett.*, **6,** 291 (1970).
165. M. Foà and L. Cassar, *Gazz. Chim. Ital.*, **102,** 85 (1972).
166. B. E. Mann, B. L. Shaw, and N. I. Tucker, *Chem. Commun.*, **1970,** 1333.
167. B. E. Mann, B. L. Shaw, and N. I. Tucker, *J. Chem. Soc. A*, **1971,** 2667.
168. P. B. Tripathy and D. M. Roundhill, *J. Am. Chem. Soc.*, **92,** 3825 (1970).
169. H. Behrens and F. Lohofer, *Chem. Ber.*, **94,** 1391 (1961).
170. H. Behrens, H. Zizlsperger, and R. Rausch, *Chem. Ber.*, **94,** 1497 (1961).
171. W. Hieber and J. Ellermann, *Z. Naturforschung,* **18b,** 595 (1963).
172. F. Calderazzo, R. Ercoli, and G. Natta, in I. Wender and P. Pino, Ed., *Organic Syntheses via Metal Carbonyls*, Vol. I, Interscience, New York, 1968, p. 67.
173. M. C. Baird, J. T. Mague, J. A. Osborn, and G. Wilkinson, *J. Chem. Soc. A*, **1967,** 1347.
174. S. Otsuka, A. Nakamura, T. Yoshida, M. Naruto, and K. Ataka, *J. Am. Chem. Soc.*, **95,** 3180 (1973).
175. H. F. Klein, *Angew. Chem.*, **85,** 403 (1973).
176. K. Fischer, K. Jonas, P. Misbach, R. Stabba, and G. Wilke, *Angew. Chem.*, **85,** 1002 (1973).
177. G. P. Chiusoli and A. Cameroni, *Chim. Ind. (Milan)*, **46,** 1063 (1964).
178. N. v. Kutepow, *Ullmanns Encyklopädie der techn. Chemie*, in press.
179. M. Salkind, E. H. Riddle, and R. W. Keefer, *Ind. Eng. Chem.*, **51,** 1328 (1959).
180. H. Lautenschlager and H. Friederich, Ger. Pat. 1,046,030 (1956).
181. H. Lautenschlager and H. Friederich, Ger. Pat. 1,058,048 (1956).
182. W. Reppe, *Chemie und Technik der Acetylen-Druck-Reaktionen* Verlag Chemie, Weinheim/Bergstrasse, 1952, pp. 1–19.
183. W. Reppe, F. Reicheneder, G. Stengel, and A. Ziegler, Ger. Pat. 1,076,672 (1956).
184. W. Büche and A. Ziegler, Ger. Pat. 1,018,396 (1955).
185. W. Reppe, H. Friederich, H. Lautenschlager, and H. Laib, Ger. Pat. 949,654 (1954).

186. W. Reppe, H. Friederich, E. Henkel, and H. Lautenschlager, Ger. Pat. 1,000,806 (1954).
187. H. Friederich and E. Henkel, Ger. Pat. 1,042,571 (1955).
188. H. Lautenschlager, H. Friederich, E. Henkel, N. v. Kutepow, W. Himmele, and P. Raff, Ger. Pat. 1,009,620 (1955).
189. H. Lautenschlager and H. Friederich, Ger. Pat. 1,029,365 (1956).
190. W. Reppe, O. Hecht, and E. Reindl, Ger. Pat. 881,650 (1953).
191. W. Reppe and R. Stadler, Ger. Pat. 942,809 (1953).
192. W. Reppe and R. Stadler, Ger. Pat. 944,789 (1953).
193. R. Stadler and F. Becke, Ger. Pat. 1,060,382 (1956).
194. R. Stadler and H. Lautenschlager, Ger. Pat. 1,064,503 (1957).
195. W. Reppe, N. v. Kutepow, P. Raff, E. Henkel, H. Friederich, and W. Himmele, Ger. Pat. 965,323 (1954).
196. W. Reppe, H. Friederich, E. Henkel, N. v. Kutepow, and W. Himmele, 1,003,721 (1954).
197. Y. Iwashita, F. Tamura, and H. Wakamatsu, *Bull. Chem. Soc. Japan*, **43,** 1520 (1970).
198. D. A. Harbourne and F. G. A. Stone, *J. Chem. Soc. A*, **1968,**1765.
199. P. Pino, G. Braca, G. Sbrana, and P. Cerrai, unpublished results.
200. K. Tominaga, N. Yamagami, and H. Wakamatsu, *Tetrahedron Lett.*, **1970,** 2217.
201. G. Natta and P. Pino, *Chim. Ind.* (*Milan*), **34,** 449 (1952).
202. W. Reppe, W. Schweckendiek, and H. Friederich, U.S. Pat. 2,738,364 (1951).
203. G. Braca, Thesis, Pisa University, Feb. 1961.
204. W. Reppe, H. Friederich, E. Henkel, N. v. Kutepow, and W. Himmele, Ger. Pat. 1,003,721 (1954).
205. P. Pino, Swiss Pat. 387,017 (1961).
206. P. Pino, A. Miglierina, and E. Pietra, *Gazz. Chim. Ital.*, **84,** 443 (1954).
207. W. Reppe and R. Stadler, Ger. Pat. 942,809 (1953).
208. P. Pino and P. L. Bertolaccini, unpublished results.
209. C. I. Reimer, *Chem. Ber.*, **14,** 1802 (1881).
210. W. A. Bone and H. W. Perkin, *J. Chem. Soc.*, **1880,** 267.
211. R. Markby, I. Wender, R. A. Friedel, F. A. Cotton, and H. W. Sternberg, *J. Am. Chem. Soc.*, **80,** 6529 (1958).
212. H. W. Sternberg, R. Markby, and I. Wender, *Chim. Ind.* (*Milan*), **42,** 41 (1960).
213. P. C. Ellgen, *Inorg. Chem.*, **11,** 691 (1972).
213*a*. M. R. Tirpak, J. H. Wotiz, and C. A. Hollingsworth, *J. Am. Chem. Soc.*, **80,** 4265 (1958).
214. F. Ungvary and L. Markó, *Proc. Symposium Chemistry of Hydroformylation and Related Reactions*, Veszprem (Hungary), 1972, p. 53.
214*a*. A. J. Poe, *J. Organometal. Chem.*, **94,** 235 (1975).
215. G. Albanesi and E. Gavezzotti, *Chim. Ind.* (*Milan*), **47,** 1322 (1965).
215*a*. R. S. Dickson and H. P. Kirsch, *Aust. J. Chem.*, **27,** 61 (1974).
216. W. G. Sly, *J. Am. Chem. Soc.*, **81,** 18 (1959).
217. J. A. Osborn, F. H. Jardine, J. F. Young, and G. Wilkinson, *J. Chem. Soc. A*, **1966,** 1711.
218. J. R. Joy and M. Orchin, *J. Am. Chem. Soc.*, **81,** 310 (1959).
219. P. Taylor and M. Orchin, *J. Organometal. Chem.*, **26,** 389 (1971).
220. W. Hieber, J. Sedlmeier, and W. Abeck, *Chem. Ber.*, **86,** 700 (1953).
221. M. Stolle and P. Balle, *Helv. Chim. Acta*, **21,** 1551 (1938).
222. H. J. Hagemeyer and D. C. Hull, U.S. Pat. 2,610,203 (1952).
223. U. Kruerke and W. Hübel, *Chem. Ber.*, **94,** 2829 (1961).

224. G. P. Chiusoli, C. Venturello, and S. Merzoni, *Chem. Ind.* (*London*), **1968,** 977.
225. P. Pino and R. Di Ciolo, unpublished results.
226. T. Matsuda, H. Kondu, and T. Nakamura, *Kogyo Kagaku Zasshi*, **74,** 1135 (1971).
227. F. Wada and T. Matsuda, *Bull. Chem. Soc. Japan*, **46,** 510 (1973).
228. P. Pino, G. Braca, and G. Sbrana, unpublished results.
229. C. T. Sears and F. G. A. Stone, *J. Organometal. Chem.*, **11,** 644 (1968).
230. O. Gambino, G. Cetini, E. Sappa, and M. Valle, *J. Organometal. Chem.*, **20,** 195 (1969).
231. E. Sappa, G. Cetini, O. Gambino, and M. Valle, *J. Organometal. Chem.*, **20,** 201 (1969).
232. J. Tsuji and T. Nogi, Japan. Pat. 68 0946 (1964).
233. G. Jacobsen and H. Spathe, Ger. Pat. 1,138,760 (1960).
234. H. Michihisa and T. Shimada, Japan. Pat. 73 11088; *Chem. Abstr.*, **79,** 18132 (1973).
235. C. Lines and R. B. Long, *Symposium, Homogeneous Catalytic Reactions Involving Palladium*, Minneapolis Meet. Am. Chem. Soc. Inorg. Chem. Div., 1969, p. 159.
236. O. L. Kaliya, O. N. Temkin, N. G. Mekhryakova, and R. M. Feld, *Dokl. Akad. Nauk SSSR*, **199,** 1321 (1971).
236*a*. K. Mori, T. Mizoroki, and A. Ozaki, *Chem. Lett.*, **1,** 39 (1975).
236*b*. O. L. Kaliya, O. N. Temkin, G. S. Kirschenkova, and R. M. Flid, *Kinetics and Catalysts*, **10,** 979 (1969).
236*c*. D. M. Fenton, U.S. Pat., 3,641,137 (1972); *Chem. Abstr.*, **76,** 141495 (1972).
236*d*. J. F. Knifton, U.S. Pat., 3,904,672 (1975); *Chem. Abstr.*, **83,** 205782 (1975).
237. J. Tsuji and T. Nogi, *J. Org. Chem.*, **31,** 2641 (1966).
238. T. Nogi and J. Tsuji, *Tetrahedron*, **25,** 4099 (1969).
239. R. F. Heck, *J. Am. Chem. Soc.*, **94,** 2712 (1972).
240. J. Tsuji, T. Nogi, and M. Morikawa, Japan. Pat. 73 12726; *Chem. Abstr.*, **79,** 31507 (1973).
241. J. Tsuji and T. Nogi, *Tetrahedron Lett.*, **1966,** 1801.
242. W. Manchot and J. König, *Chem. Ber.*, **59,** 883 (1926).
243. E. O. Greaves and P. M. Maitlis, *J. Organometal. Chem.*, **6,** 104 (1966).
244. E. O. Greaves, C. J. Lock and P. M. Maitlis, *Can. J. Chem.*, **46,** 3879 (1968).
245. A. Babaeva and T. Beresneva, *Russ. J. Inorg. Chem.*, **11,** 1048 (1966).
246. J. Falbe, *Carbon Monoxide in Organic Syntheses*, Springer-Verlag, Berlin, 1970, p. 97.
247. J. C. Sauer, U.S. Pat. 3,097,237 (1960).
248. J. Tsuji, M. Morikawa, and N. Iwamoto, *J. Am. Chem. Soc.*, **86,** 2095 (1964).
249. A. T. Blomquist and P. M. Maitlis, *J. Am. Chem. Soc.*, **84,** 2329 (1962).
250. N. v. Kutepow, K. Bittler, D. Neubauer, and H. Reis, Ger. Pat. 1,236,752 (1964).
251. W. A. Raczynski, U.S. Pat. 2,738,368 (1956); *Chem. Abstr.*, **50,** 15577 (1956).
252. H. T. Neher, E. H. Specht, and A. Newman, U.S. Pat. 2,773,063 (1956).
253. *BIOS Final Report No. 1811*, 1947, p. 37.
254. P. Pino and C. Paleari, *Gazz. Chim. Ital.*, **81,** 646 (1951).
255. R. P. A. Sneeden, T. F. Burger, and H. H. Zeiss, *J. Organometal. Chem.*, **4,** 397 (1965).
256. L. H. Slaugh and R. D. Mullineaux, U.S. Pat. 3,239,571 (1966).
257. B. L. Booth and R. G. Hargreaves, *J. Chem. Soc. A*, **1970,** 308.
258. B. L. Booth and R. G. Hargreaves, *J. Organometal. Chem.*, **33,** 365 (1971).

Carbonylation of Organic Halides

THOMAS A. WEIL, *Amoco Chemicals Corporation, Naperville, Illinois 60540*

and

L. CASSAR AND M. FOÁ, *Istituto Ricerche G. Donegani-Montedison-, Via Del Lavoro 4, 28100, Novara, Italy*

I. INTRODUCTION

In this section, carbonylation reactions of organic halides in the presence of metal carbonyls and other transition metal complexes and CO will be discussed. Alkyl and aryl halides will be considered; allyl halides have been covered in the fourth chapter of this volume. Reactions leading to new complexes are not covered, unless they are necessary in describing the mechanism of a particular reaction.

Fundamental studies of elementary processes of organometallic chemistry allow us to indicate the probable mechanistic scheme for carbonylation of an organic halide. In the catalytic synthesis of derivatives of organic acids, this scheme would consist of the following steps for a nonanionic metal carbonyl catalyst:

1. Oxidative addition (1):

$$RX + M^n(CO)_m \rightarrow R\text{—}\underset{\displaystyle X}{\underset{|}{M}}{}^{n+2}(CO)_{m-y} + yCO$$

2. Carbon monoxide insertion (2):

$$\underset{\displaystyle \mathrm{X}}{\mathrm{R{-}M^{n+2}}}(\mathrm{CO})_{m-y} \rightarrow$$

$$\mathrm{R{-}\underset{\displaystyle O}{\overset{}{C}}{-}\underset{\displaystyle X}{M^{n+2}}(CO)_{(m-y)-1}} \xrightarrow{\mathrm{CO}} \mathrm{R{-}\underset{\displaystyle O}{C}{-}\underset{\displaystyle X}{M^{n+2}}(CO)_{m-y}}$$

3. Reductive elimination (3):

$$\mathrm{R{-}\underset{\displaystyle O}{C}{-}\underset{\displaystyle X}{M^{n+2}}(CO)_{m-y} + yCO \rightarrow R{-}\underset{\displaystyle O}{C}{-}X + M^{n}(CO)_{m}}$$

At the end of these three steps the metal complex returns to the low valence state, ready to start a new cycle; thus, a catalytic process is achieved.

For anionic metal carbonyls, the intermediate alkyl and acyl metal carbonyls do not contain the halogen, X (4), so reductive elimination occurs via attack by a nucleophile on the acyl group.

Besides acids and their functional derivatives, several kinds of organic products, such as ketones and aldehydes, can also be obtained from organic halides and a metal carbonyl. In these cases, reductive elimination is replaced by other processes, such as coupling or hydrolysis. For example, ketones, diketones, and aldehydes can be obtained according to the equation:

$$\mathrm{R{-}\underset{\displaystyle O}{C}{-}\underset{\displaystyle L}{M^{n+2}}(CO)_{n} + R'Y \rightarrow R{-}\underset{\displaystyle O}{C}{-}R' + M^{n+2} + \mathit{n}CO + Y^{-} + L}$$

R = alkyl, aryl

L = halogen, CO, phosphine, etc.

R′ = alkyl, aryl, hydrogen

II. CARBONYLATION OF ARYL HALIDES

The carbonylation of aryl halides with CO in the presence of a transition metal (Table 1) was extensively studied. Heterogeneous catalysts such as $NiCl_2$ on pumice (5) and NiI_2 on SiO_2 (6) were found active in carbonylating chlorobenzene to benzoic acid in high yield:

$$\mathrm{PhCl + CO + H_2O \xrightarrow[\mathrm{NiI_2}]{>290^{\circ}\mathrm{C}} PhCOOH + HCl}$$

With NiI_2 on SiO_2, the effects of the quantity of catalyst, CO pressure, reaction time, and reaction temperature on the yield of benzoic acid were studied (6). Optimum yields (76%) are obtained when operating at

295°C, under 450 atm of CO for 3 hr. Interestingly, even in the absence of the nickel salt, carbonylation products can be obtained, but higher temperature (360°C) and pressure (640 atm) of CO are needed.

As for homogeneous catalysts, $Ni(CO)_4$, its derivatives, and precursors were used most frequently. The carbonylation products obtained are: aromatic acids when the reaction occurs in the presence of water; esters, in the presence of alcohols; and amides or nitriles, in the presence of amines, formamides, or oxamides. Acyl fluorides are also obtained when the reaction occurs in the presence of NaF (Table 1).

Homogeneous catalytic carbonylation also occurs in the presence of basic compounds (25–31). With stoichiometric amounts of alkali metal carbonates or orthophosphates, aryl halides are readily transformed to phthalic anhydrides at 250–375°C, under CO pressure in the presence of $Ni(CO)_4$ (32*a–c*). Precursors of $Ni(CO)_4$, such as Raney nickel or $NiCl_2$, work equally well; however, no phthalic anhydride was obtained with either iron or cobalt carbonyl as catalyst.

Benzene and benzoic anhydride are also formed in the reaction; the amount of benzene formed is roughly equivalent to that of the phthalic anhydride. A reaction scheme in which benzoic anhydride is the precursor of phthalic anhydride is envisioned thus:

$$2PhBr + CO \xrightarrow[Ni(CO)_4]{Na_2CO_3} PhC(=O)\text{—}O\text{—}C(=O)\text{—}Ph \longrightarrow$$

$$\text{(phthalic anhydride: } C_6H_4(CO)_2O) + PhH$$

The conversion of aryl halides to carboxylic acids and their derivatives in the presence of sodium formate or sodium formate and $Ni(CO)_4$ was also reported (Table 1) (21,22). The reaction occurs under a pressure of CO in the presence of nickel or nickel compounds at 290–360°C. In the absence of nickel, the reaction will proceed at a much slower rate, if the CO pressure is sufficiently high. The reaction is not applicable to *o*-dihalogen compounds or alkyl halides. An ionic mechanism is proposed in which the halogen is displaced by the formyl anion:

$$C_6H_5\text{—}Cl \xrightarrow{HCOONa} \left[C_6H_5 \cdots (Cl)(H)\,C(=O)\text{—}ONa \right] \longrightarrow C_6H_5\text{—}CO_2Na + HCl$$

TABLE 1

Conversion of Aryl Halides to Aromatic Acids and Derivatives under High Pressure of Carbon Monoxide

Aryl halide	Other reagents or solvents	Catalyst	Product	Yield (%)	Ref.
Chlorobenzene	Et_2O, I_2	$Ni(CO)_4$	Ethyl benzoate	20	17
Chlorobenzene	—	$Ni(CO)_4$	Benzoic acid	21	9
Chlorobenzene	H_2O	NiI_2 on SiO_2	Benzoic acid	80	6
Chlorobenzene	H_2O, NaOH	Ni naphthenate	Benzoic acid	—	31
Chlorobenzene	H_2O	$NiCl_2$ on pumice	Benzoic acid	96	5
Chlorobenzene	H_2O, BuOAc	Ni and NiI_2	Benzoic acid	95	12
Chlorobenzene	NaF	$Ni(CO)_4$	Benzoyl fluoride	27	32*b*
Chlorobenzene	—	$PdCl_2$	Benzoyl chloride	80	14
Chlorobenzene	$(NH_2)_2CO$	$Ni(CO)_4$	Benzamide Benzonitrile	40 19	20
Chlorobenzene	$(NH_2)_2CO$	$Ni(CO)_4$	Benzamide Benzonitrile	15 60	20
p-Chloroanisole	HCO_2Na, CH_3OH, HCO_2CH_3	$Ni(CO)_4$	*p*-Anisic acid Methyl *p*-anisoate	—[a]	21,22
o-Chlorotoluene	BuOAc, H_2O	NiI_2, Ni	*o*-Toluic acid	61	12
p-Chlorotoluene	CH_3OH	$Ni(CO)_4$	Methyl *p*-toluate *p*-Toluic acid	50 5	17
p-Chlorotoluene	CH_3OH	$Fe(CO)_5$	Methyl *p*-toluate	—	17
p-Chlorotoluene	NaOH, H_2O	Ni naphthenate	*p*-Toluic acid	90	31
p-Chlorotoluene	PhOH	$Co(OAc)_2$ or $Ni(CO)_4$	Phenyl *p*-toluate	10	17

p-Chlorotoluene	HCO_2CH_3	$Ni(CO)_4$	Methyl *p*-toluate	70	15
			p-Toluic acid	9	
p-Chlorotoluene	$CO(OEt)_2$	$Ni(CO)_4$	Ethyl *p*-toluate	80	15
p-Chlorotoluene	$HC(OEt)_3$	$Ni(CO)_4$	Ethyl *p*-toluate	80	15
p-Chlorotoluene	HOAc	$Ni(CO)_4$	*p*-Toluic acid	87	19
p-Chlorotoluene	aq. HCl, C_6H_6	$Co_2(CO)_8$	*p*-Toluic acid	15	9
p-Chlorotoluene	aq. HCl, C_6H_6	$Fe(CO)_5$	*p*-Toluic acid	35	9
p-Chlorotoluene	Hexyl alcohol	$Ni(CO)_4$	*p*-Toluic acid	—	17
			Hexyl *p*-toluate		
			Dihexyl ether		
p-Chlorotoluene	NH_3	$Ni(CO)_4$	*p*-Toluamide	43	20
			p-Tolunitrile	15	
p-Chlorotoluene	$HCONH_2$	$Ni(CO)_4$	*p*-Toluamide	40	20
			p-Tolunitrile	40	
p-Chlorotoluene	Cyclohexylamine	$Ni(CO)_4$	N-Cyclohexyl-*p*-toluamide	—	20
o-Dichlorobenzene	I_2, H_2O	$Ni(CO)_4$	Phthalic acid	90	18
o-Dichlorobenzene	aq. HCl	$Ni(CO)_4$	*o*-Chlorobenzoic acid	2	9
o-Dichlorobenzene	NaF	$Ni(CO)_4$	*o*-Chlorobenzoyl fluoride	—	32*b*
p-Dichlorobenzene	BuOAc, H_2O	Ni, NiI_2	Terephthalic acid	31	12
p-Dichlorobenzene	H_2O	$Ni(CO)_4$	Terephthalic acid	50	8,16
p-Dichlorobenzene	H_2O	$(Ph_3P)_2NiBr_2 \cdot BuBr$	Terephthalic acid	20	10
			p-Chlorobenzoic acid	80	
p-Dichlorobenzene	CH_3OH	$Ni(CO)_4$	Terephthalic acid / Dimethyl terephthalate	55	17

[a] Yield reported to be good.

TABLE 1 (cont'd.)

Aryl halide	Other reagents or solvents	Catalyst	Product	Yield (%)	Ref.
p-Dichlorobenzene	HCO_2Na, HCO_2CH_3	$Ni(CO)_4$	Terephthalic acid	8	21,22
			Dimethyl terephthalate	80	
p-Dichlorobenzene	CH_3OAc, HCO_2CH_3	$Ni(CO)_4$	Dimethyl terephthalate	35	15
			Methyl *p*-chlorobenzoate	30	
p-Dichlorobenzene	Et_2O	$Ni(CO)_4$	Ethyl *p*-chlorobenzoate	48	17
p-Dichlorobenzene	CH_3OH, CH_3OAc	$(PPh_3)_2Ni(CO)_2$	Methyl *p*-chlorobenzoate	30	11
			Dimethyl terephthalate	10	
p-Dichlorobenzene	$(CH_3)_2O$	Cu–Ni–Zn	Methyl *p*-chlorobenzoate	11	23,24
			Dimethyl terephthalate	80	
p-Dichlorobenzene	NaOH, CH_3OH	$Ni(OAc)_2$	Dimethyl terephthalate	—	25
p-Dichlorobenzene	CH_3ONa, CH_3OH	$Ni(CO)_4$	Terephthalic acid	4	26,27
			Dimethyl terephthalate	85	
p-Dichlorobenzene	HCO_2Na	Cu, Ag, Fe, Co, or Ni	Terephthalic acid	65	28
			p-Chlorobenzoic acid	11	
p-Dichlorobenzene	$(CH_3)_2O$, C_6H_6	Cu–Zn	Methyl *p*-chlorobenzoate	20	29
			Dimethyl terephthalate	21	
p-Dichlorobenzene	NaF	$Ni(CO)_4$	*p*-Chlorobenzoyl fluoride	—	32*b*
p-Dichlorobenzene	$HCONH_2$	$Ni(CO)_4$	*p*-Chlorobenzonitrile	—	20
			p-Chlorobenzamide		
			Terephthalonitrile		
Sodium *p*-chlorobenzoate	HCO_2CH_3, CH_3OH	$Ni(CO)_3(PPh_3)$	Dimethyl terephthalate } Terephthalic acid }	87	22

m-Dichlorobenene	HCO_2Na	$Ni(CO)_4$	Isophthalic acid	—[a]	21
p-Chlorobenzoic acid	CH_3OH	$(Ph_3P)_2NiBr_2\cdot BuBr$ or NiI_2	Dimethyl terephthalate	67	10
p-Chlorobenzonitrile	$(NH_2)_2CO$	$Ni(CO)_4$	Terephthalonitrile	30	20
1-Chloronaphthalene	H_2O	NiI_2	1-Naphthoic acid	52	18
2-Chloronaphthalene	H_2O, BuOAc	NiI_2, Ni	2-Naphthoic acid	52	12
Bromobenzene	KOAc, H_2O	$Ni(OAc)_2\cdot 4H_2O$	Benzoic acid	94	13
Bromobenzene		$PdBr_2$ or $PtCl_2$	Benzoyl bromide	—	14*a*
Bromobenzene	HCO_2CH_3	$Ni(CO)_4$	Methyl benzoate	—	15
Bromobenzene	HCO_2K, EtOH, HCO_2Et	$Ni(CO)_4$	Ethyl benzoate Benzoic acid	—	22
Bromobenzene	NaF	$Ni(CO)_4$	Benzoyl fluoride	18	32*b*
o-Bromotoluene	KOAc, H_2O	$Ni(OAc)_2\cdot 4H_2O$	*o*-Toluic acid	86	7
m-Bromotoluene	KOAc, H_2O	$Ni(OAc)_2\cdot 4H_4O$	*m*-Toluic acid	99	7
p-Bromotoluene	KOAc, H_2O	$Ni(OAc)_2\cdot 4H_2O$	*p*-Toluic acid	94	7
1-Bromo-3,4-dimethylbenzene	KOAc, H_2O	$Ni(OAc)_2\cdot 4H_2O$	3,4-Dimethylbenzoic acid	99	7
1-Bromo-2,4-dimethylbenzene	KOAc, H_2O	$Ni(OAc)_2\cdot 4H_2O$	2,4-Dimethylbenzoic acid	95	7
1-Bromo-2,5-dimethylbenzene	KOAc, H_2O	$Ni(OAc)_2\cdot 4H_2O$	2,5-Dimethylbenzoic acid	100	7
1-Bromo-2,4,5-trimethylbenzene	KOAc, H_2O	$Ni(OAc)_2\cdot 4H_2O$	2,4,5-Trimethylbenzoic acid	95	7
Bromomesitylene	KOAc, H_2O	$Ni(OAc)_2\cdot 4H_2O$	2,4,6-Trimethylbenzoic acid	67	7

TABLE 1 (cont'd.)

Aryl halide	Other reagents or solvents	Catalyst	Product	Yield (%)	Ref.
1-Bromonaphthalene	KOAc, H_2O	$Ni(OAc)_2$	1-Naphthoic acid	83	13,30
p-Dibromobenzene	Isobutyric acid	$Ni(CO)_4$	Terephthalic acid	—	19
p-Dibromobenzene	KOAc, H_2O	$Ni(OAc)_2 \cdot 4H_2O$	Terephthalic acid	73	13
p-Dibromobenzene	CH_3OH	$Ni(CO)_4$	Dimethyl terephthalate	—	17
p,p'-Dibromodiphenyl	CH_3ONa, xylene, CO_2	Pd on carbon	Dimethyl *p-p'*-diphenyl-dicarboxylate	65	14*b*
Iodobenzene		Ni	Benzoic acid	80	18
Iodobenzene	BuOAc, H_2O	NiI_2, Ni	Benzoic acid	100	12

Recently, the importance of the presence of bases, such as potassium acetate or benzoate, in the carbonylation was pointed out by Nakayama and Mizoroki (7,13,30). In the presence of KOAc, using nickel acetate as catalyst at 250°C and 200 atm of CO, aryl bromides are transformed into aromatic acids with yields of over 90%. The amount of KOAc strongly influences carbonylation of the aryl halides. With bromobenzene, the yield of benzoic acid rises from 1% in the absence of KOAc to 90% in the presence of a stoichiometric amount of this salt. The conversion of bromobenzene to benzoic acid is inhibited by the HBr formed; removal of HBr by KOAc probably accounts for the increased rate of reaction and increased yield of benzoic acid.

A kinetic study of the carbonylation of bromobenzene, when the amount of KOAc is greater than that of aryl bromide, has shown that the rate is proportional to the concentrations of $Ni(CO)_4$, bromobenzene, and water, and that the rate is inversely proportional to the partial pressure of CO. According to Nakayama and Mizoroki the following equations can explain the rate expression, with Eqs. 1 and 2 being rate determining.

$$PhBr + Ni(CO)_4 \rightleftarrows (PhBr)Ni(CO)_3 + CO \qquad (1)$$

$$(PhBr)Ni(CO)_3 + H_2O \rightarrow PhCOOH + (HBr)Ni(CO)_2 \qquad (2)$$

$$(HBr)Ni(CO)_2 + KOAc \rightarrow Ni(CO)_2 + KBr + HOAc \qquad (3)$$

$$Ni(CO)_2 + 2CO \rightarrow Ni(CO)_4 \qquad (4)$$

In the carbonylation of 1-bromonaphthalene at 100 atm of CO, the rate expression changes. The rate of this reaction depends only on $Ni(CO)_4$ and KOAc. To explain these results, steps 3 and 4 are said to be rate determining.

Other catalysts, including $Co_2(CO)_8$ (9,17–19), iron carbonyl (9,17,19), Pd(II) and Pt(II) halides (14*a*), and Pd on carbon (14*b*) were used as catalysts in the carbonylation of aryl halides. In Pd(II)-catalyzed carbonylations, high yields of the corresponding aroyl halides are obtained when operating at 160°C and 80 atm CO. Patents also claim, as catalysts, compounds of Rh, Ir, Os, and Ru (for example $RhCl_3$, $[RhCl(CO)_2]_2$, and $IrCl_3$) (33,34).

So far we have considered only the carbonylation of aromatic halides under high CO pressures. In 1963, Bauld (35) reported that $Ni(CO)_4$ reacts with aryl iodides in alcoholic solvents to give aroate esters (Table 2, column one). In aprotic solvents, benzils are formed (Table 2, column two). Enediol diesters in addition to the benzils are formed when hindered iodides are used (Table 2, column three). A scheme in which aroylnickel carbonyl complexes are intermediates was proposed

TABLE 2

Conversion of Aryl Iodides to Methyl Aroate Esters, Benzils, and Enediol Diesters (35)

$ArI \xrightarrow{Ni(CO)_4} ArCO_2R\ (R = Me)$		$ArCOCOAr$	+ $ArC(OCOAr){=}C(OCOAr)Ar$
Ar	A	B	C
	Conversion (%) to A (in MeOH)	Conversion (%) to B (in THF)	Conversion (%) to C (in THF)
Phenyl	60	80	0
p-Chlorophenyl	—	47	0
m-Tolyl	—	50	0
m-Chlorophenyl	70	—	—
1-Naphthyl	80	35	29
o-Tolyl	—	10	26
Mesityl	0	0	0

(Scheme **I**). Aryl chlorides and bromides and alkyl halides do not undergo this reaction.

$$ArI + Ni(CO)_4 \longrightarrow [ArCONiI(CO)_n]$$

$$[ArCONiI(CO)_n] \xrightarrow[B]{THF} NiI_2 + ArCOCOAr + CO$$

$$[ArCONiI(CO)_n] \xrightarrow[A]{ROH} ArCO_2R + HI + Ni(CO)_n$$

$$NiI_2 + ArCOCOAr + CO \longrightarrow ArC(OCOAr){=}C(OCOAr)Ar$$

Scheme **I**

By reacting $Ni(CO)_4$ (36) with pentafluoroiodobenzene and pentafluorobromobenzene at atmospheric pressure, decafluorobiphenyl and decafluorobenzophenone are formed. A radical mechanism has been suggested for this reaction.

In the presence of $Fe_3(CO)_{12}$, benzophenone has been obtained from iodobenzene (37) in 53% yield at 110°C; a phenyliron carbonyl halide is proposed as the intermediate in this reaction.

Corey and Hegedus (38) and Nakayama and Mizoroki (39) pointed out the importance of the presence of a base in the carbonylation of aromatic iodides at atmospheric pressure. The latter workers proposed that the function of the base is to neutralize the HI formed during the carbonylation, and that $Ni(CO)_3$ is the species that catalyzes the reaction.

Recently the carbonylation of aromatic bromides and chlorides under 1 atm of CO at about 100°C was achieved (40). This carbonylation is possible by using aprotic solvents such as DMF, DMSO, or HMPA. (Table 3) and by using a base such as $Ca(OH)_2$. A variety of aromatic acids, in high yields, are obtained (Table 4). The presence of chloride or bromide ion plays an important role in activating this carbonylation. The carbonylation of aromatic chlorides and bromides exhibits an autocatalytic behavior due to formation of chloride and bromide ions during the reaction. The rate is, in fact, strongly enhanced by adding chloride and bromide ions ($Cl^- > Br^-$). The presence of iodide ions has a deactivating effect, whereas other salts such as benzoate, perchlorate, and fluoride have very little effect on the rate (Fig. 1). The reaction is facilitated by electron-withdrawing groups on the aromatic nucleus. The relative rates correlated

TABLE 3
Synthesis of Calcium Salt of 1-Naphthoic Acid from 1-Chloronaphthalene by Carbonylation with Nickel Carbonyl[a] (40)

Solvent	Naphthoic acid calcium salt yield (%)
Dimethyl sulfoxide	90
N,N-Dimethylformamide	95
N-Methylformamide	12[b]
Formamide	—[c]
N,N-Dimethylacetamide	90
N-Methylpyrrolidone	90
Hexamethylphosphoramide	71
Propylene carbonate	—[c]

[a] Solvent 40 cc; $Ca(OH)_2$ 3.0 g; $Ni(CO)_4$ 1.5 cc; 1-chloronaphthalene 4.8 g. All reactions at 110–112°C for 6–8 hr.

[b] 31% of naphthoic acid *N*-methylamide also formed.

[c] $Ni(CO)_4$ decomposed.

TABLE 4
Carbonylation of Aromatic Halides in Dimethylformamide[a] (40)

Substrate	Time (hr)	Product	Yield (%)
1-Chloronaphthalene	6	1-Naphthoic acid	95[b]
2-Chloronaphthalene	8	2-Naphthoic acid	97[c]
Bromobenzene	8	Benzoic acid	88
o-Bromotoluene	8	*o*-Toluic acid	94
p-Bromotoluene	14	*p*-Toluic acid	97
o-Bromoanisole	8	*o*-Anisic acid	80
m-Bromoanisole	8	*m*-Anisic acid	83
p-Bromoanisole	13	*p*-Anisic acid	85
p-Bromobenzonitrile	3	*p*-Cyanobenzoic acid	80[d]
o-Chlorobromobenzene	8	*o*-Chlorobenzoic acid	87[e]
m-Chlorobromobenzene	5	*m*-Chlorobenzoic acid	88
p-Chlorobromobenzene	5	*p*-Chlorobenzoic acid	85

[a] Ratio $Ni(CO)_4$:halide = 1 : 4. All reactions at 100–102°C except 1- and 2-chloronaphthalene at 110–112°C.
[b] Naphthalene 5%.
[c] Naphthalene 3%.
[d] Benzonitrile 16%.
[e] Chlorobenzene 10%.

with the σ of Hammett ($\rho = 2.7$), and the reactivity of different aromatic halides increases in the order Br > Cl > F. Aromatic iodides react even in the absence of the base.

In the presence of a strong base, such as NaOH, the carbonylation product is not formed. Under these conditions the hydrocarbon (ArH) is obtained in high yield via substitutive hydrogenation of the halide (ArX). Comparison of the relative rates of carbonylation of some aromatic halides with the $Ni(CO)_4$–$Ca(OH)_2$ system and the $[Ni_3(CO)_8]^{2-}$ complex (41) led us to postulate an anionic nickel carbonyl as the catalyst (40).

Corey and Hegedus (38) showed that carbonylation of vinyl halides and aromatic iodides is facilitated by the presence of sodium methoxide in methanol or potassium *t*-butoxide in butanol; they suggested that anionic species are responsible for this carbonylation. However, they found that aromatic bromo derivatives are not reactive under the conditions used for iodobenzene. This result strongly indicates the important role of the dipolar aprotic solvent in carbonylation of aromatic halides.

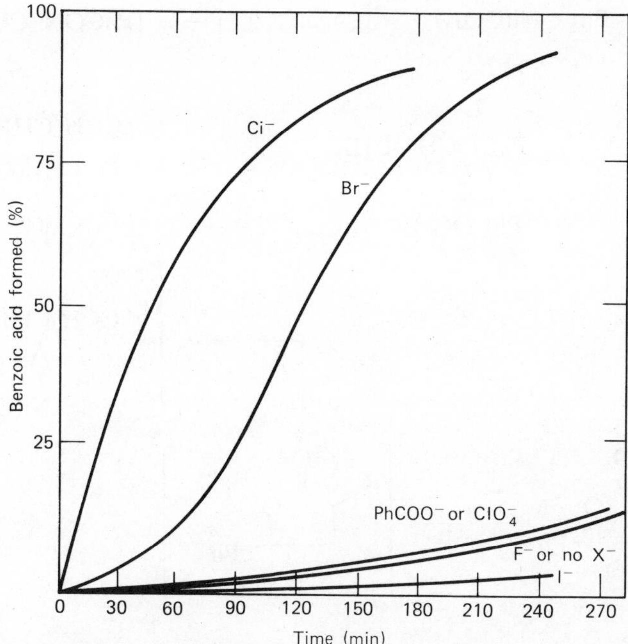

FIG. 1. Carbonylation of bromobenzene in the presence of anions. Temperature 96°C; solvent DMSO : H_2O (95 : 5); X^-, 0.23 mole/1.; $Ni(CO)_4$, 0.17 mole/1.; $Ca(OH)_2$, 0.4 mole/1.; PhBr, 0.20 mole/1.

III. CARBONYLATION OF ARYL HALIDES WITH INSERTION OF UNSATURATED COMPOUNDS

Besides functioning as a catalyst for the direct carbonylation of aryl iodides, $Ni(CO)_4$ will also catalyze carbonylation with insertion of unsaturated compounds. Insertion of acetylene was reported (42) and is covered further in the fourth chapter by Chiusoli and Cassar. When iodobenzene is reacted with $Ni(CO)_4$ in the presence of olefins, a variety of compounds including lactones and unsaturated ketones are formed (43). The proposed mechanism of the reaction is outlined in Scheme **II** and the results listed in Table 5.

When *N*-benzylidene alkylamines (43a) are reacted with iodobenzene and $Ni(CO)_4$ in DMF, 1-alkyl-2-phenylindolin-3-one and benzoylamides are formed; see page 531.

When benzene is used as solvent the main product is *N*-alkyl-*N*-(α-phenylphenacyl)benzamide:

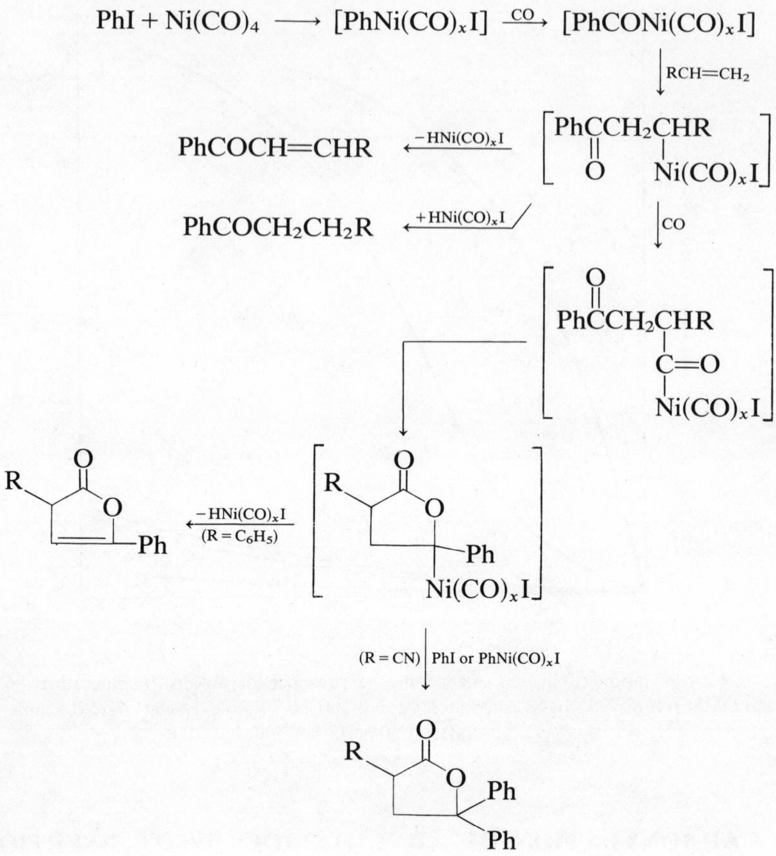

$$PhI + Ni(CO)_4 \longrightarrow [PhNi(CO)_xI] \xrightarrow{CO} [PhCONi(CO)_xI]$$

$$[PhCONi(CO)_xI] \xrightarrow{RCH{=}CH_2} [PhC(=O)CH_2CHR(Ni(CO)_xI)]$$

$$[PhC(=O)CH_2CHR(Ni(CO)_xI)] \xrightarrow{-HNi(CO)_xI} PhCOCH{=}CHR$$

$$[PhC(=O)CH_2CHR(Ni(CO)_xI)] \xrightarrow{+HNi(CO)_xI} PhCOCH_2CH_2R$$

$$[PhC(=O)CH_2CHR(Ni(CO)_xI)] \xrightarrow{CO} [PhC(=O)CH_2CHR(C{=}O)Ni(CO)_xI]$$

R, O, O, Ph, $Ni(CO)_xI$

$\xleftarrow[(R = C_6H_5)]{-HNi(CO)_xI}$ R, O, O, Ph

$\xrightarrow[(R = CN)]{PhI \text{ or } PhNi(CO)_xI}$ R, O, O, Ph, Ph

Scheme **II**

TABLE 5
Reaction of Iodobenzene with $Ni(CO)_4$ *in the Presence of Olefinic Compounds* (43)

Substrate	Products	Yield[a] (%)
$CH_2{=}CHPh$[b]	$PhCOCH{=}CHPh$	5
	$PhCH{=}CHPh$	4
	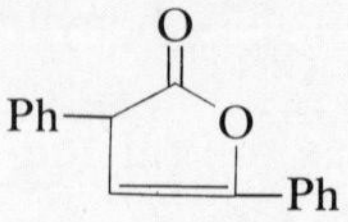 Ph, O, O, Ph	19

TABLE 5 (*Cont'd.*)

Substrate	Products	Yield[a] (%)
CH_2=CHPh	PhCOCH=CHPh	1
	$PhCOCH_2CH_2Ph$	43
	[lactone structure: Ph-substituted furanone with Ph]	25
CH_2=CHCN	PhCOCH=CHCN	1
	$PhCOCH_2CH_2CN$	29
	[lactone structure: NC-substituted γ-lactone with two Ph]	30
CH_2=CHCOOEt	$PhCOCH_2CH_2COOEt$	24

[a] Based on iodobenzene consumed.
[b] THF as solvent; benzene used in all other cases.

$$PhI + PhCH{=}NCH_3 \xrightarrow[\text{DMF}]{Ni(CO)_4} \text{[N-methyl-2-phenylindolin-3-one]} + (PhC({=}O){-}N(CH_3){-}CH(Ph){-})_2$$

$$PhI + PhCH{=}NR \xrightarrow[\text{benzene}]{Ni(CO)_4} PhC({=}O){-}N(R){-}CH(Ph){-}C({=}O)Ph$$

IV. CARBONYLATION OF BENZYL HALIDES AND SOME OTHER ALKYL HALIDES

The reaction of benzyl halides with $Ni(CO)_4$ in various solvents was reported by Tsutsumi and co-workers (Scheme **III**, (44)). In polar non-aromatic solvents CO insertion occurs, while in aromatic solvents almost no insertion takes place (Table 6). The rate of carbonylation for the benzyl halides is in the following order: I > Br > Cl. Increased solvent polarity also increases the reactivity.

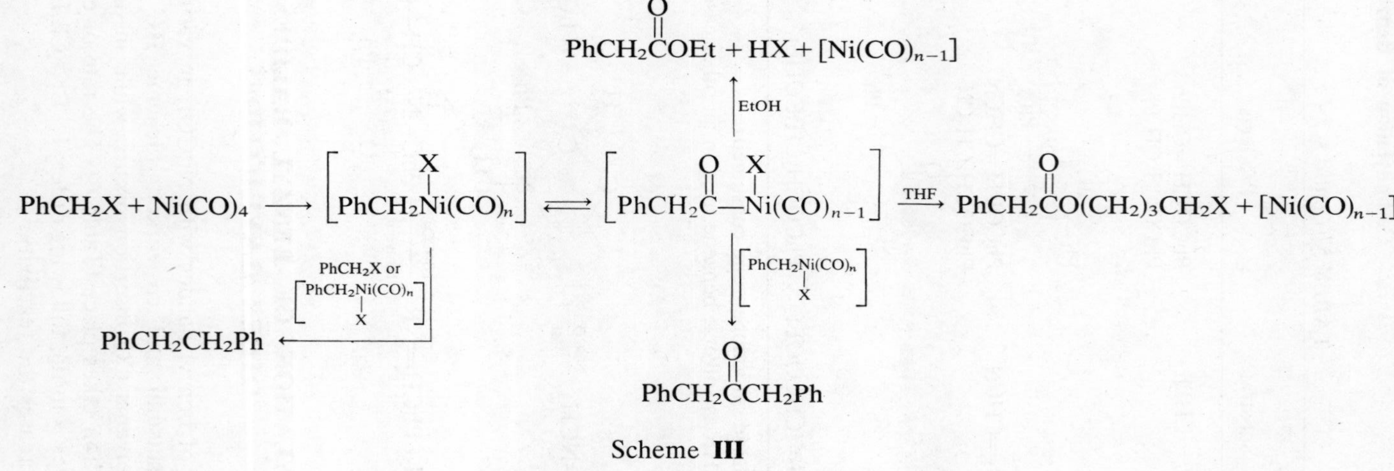
PhCH₂X + Ni(CO)₄
PhCH₂Ni(CO)ₙ
X
PhCH₂C—Ni(CO)ₙ₋₁
O
THF
PhCH₂CO(CH₂)₃CH₂X + [Ni(CO)ₙ₋₁]
EtOH
PhCH₂COEt + HX + [Ni(CO)ₙ₋₁]
PhCH₂X or
PhCH₂CH₂Ph
PhCH₂CCH₂Ph

Scheme **III**

TABLE 6

Effect of Solvent on Products of Carbonylation of Benzyl Halides with $Ni(CO)_4$ (44)

Solvent	Halide	Products	Yield[a] (%)
EtOH	$PhCH_2Cl$	No reaction	
	$PhCH_2Br$	Bibenzyl	9
		Dibenzyl ketone	2
		Ethyl phenylacetate	71
	$PhCH_2I$	Bibenzyl	4
		Dibenzyl ketone	32
		Ethyl phenylacetate	57
Cyclohexane	$PhCH_2Cl$	No reaction	
	$PhCH_2Br$	No reaction	
	$PhCH_2I$	Bibenzyl	31
		Dibenzyl ketone	4
THF	$PhCH_2Cl$	No reaction	
	$PhCH_2Br$	Bibenzyl	18
		Dibenzyl ketone	18
		4-Bromobutylphenylacetate	2
	$PhCH_2I$	Bibenzyl	4
		Dibenzyl ketone	15
		4-Iodobutylphenylacetate	62
DMF	$PhCH_2Cl$	Bibenzyl	6
		Dibenzyl ketone	51
	$PhCH_2Br$	Bibenzyl	4
		Dibenzyl ketone	93
	$PhCH_2I$	Dibenzyl ketone	95
Benzene	$PhCH_2Cl$	Diphenylmethane	36
		Dibenzylbenzenes	11
	$PhCH_2Br$	Diphenylmethane	37
		Dibenzylbenzenes	11
	$PhCH_2I$	Diphenylmethane	11
		Bibenzyl	33
		Dibenzyl ketone	3

[a] Based on benzyl halide.

TABLE 7

Carbonylation of Benzyl Chlorides to Phenylacetic Acids and Esters in DMF *with* $Ni(CO)_4$[a] (45)

Substrate	Solvent	Base	$RCl/Ni(CO)_4$	Products	Yield (%)
$PhCH_2Cl$	H_2O	CaO[b]	23	$PhCH_2COOH$	95
$PhCH_2Cl$	CH_3OH	Et_3N	12	$PhCH_2COOCH_3$	32
$PhC_6H_4CH_2Cl$	EtOH	Et_3N	12	$PhC_6H_4CH_2COOEt$	92

[a] Temp = 75–81°C; molar ratio of X^- to $Ni(CO)_4 = 0.3$.

[b] CaO added to keep pH between 6.3 and 6.7.

In benzene as solvent, diphenylmethane and higher benzylated compounds are produced. The mechanism of the reaction is said to involve abstraction of halogen from the benzyl halide by nickel to form the benzyl cation, which in turn attacks the aromatic ring (44).

The use of a mixture of aprotic and hydroxylic solvents, in the presence of halide ions and $Ni(CO)_4$, enabled the catalytic carbonylation of benzyl chloride at 1 atm of CO (45) (Table 7):

$$PhCH_2Cl + CO + ROH + B: \xrightarrow[\text{DMF}]{Ni(CO)_4,\,X^-} PhCH_2COOR + BHCl$$

$$R = H, \text{ alkyl or aryl} \qquad B{:} = \text{base}$$

The base is added to neutralize the acid formed during the reaction.

Halotricarbonylnickel anions (46) seem to be the catalysts for this carbonylation; the anions are formed thus:

$$Ni(CO)_4 + X^- \rightleftarrows [Ni(CO)_3X]^- + CO$$

Phenacyl bromide reacts with $Ni(CO)_4$; the reaction shows a dramatic solvent effect (47).

$$XC_6H_4\overset{O}{\overset{\|}{C}}CH_2Br + Ni(CO)_4 \xrightarrow{\text{THF}} XC_6H_4\overset{O}{\overset{\|}{C}}CH_2CH_2\overset{O}{\overset{\|}{C}}C_6H_4X$$

$$XC_6H_4\overset{O}{\overset{\|}{C}}CH_2Br + Ni(CO)_4 \xrightarrow{\text{DMF}} \text{2-}XC_6H_4\text{-4-}C_6H_4X\text{-furan}$$

In a similar reaction between an α-bromoketone and $Ni(CO)_4$ in DMF, β-epoxyketones were isolated as precursors to furan formation (48).

$$t\text{-BuC(=O)CH}_2\text{Br} \xrightarrow{Ni(CO)_4} t\text{-BuC(=O)CH}_2\text{C}(t\text{-Bu})(\text{—O—})\text{CH}_2 \longrightarrow \text{2-}t\text{-Bu-4-}t\text{-Bu-furan}$$

Potassium hexacyanodinickelate reacts with benzyl halide in the presence of CO to yield the corresponding symmetrical ketones (49, 50) (Table 8).

$$K_4Ni_2(CN)_6 + RBr \rightarrow K_2[Ni(CN)_3Br] + [K_2RNi(CN)_3] \xrightarrow{CO} R\text{—}\underset{\underset{O}{\|}}{C}\text{—}R + K_2[Ni(CN)_3Br]$$

Under the same conditions, vinyl halides give esters and aldehydes, alkyl iodides are less reactive, and iodobenzene is unaffected. Nickel complexes containing a carbon–nickel σ-bond were postulated as intermediates, but none were isolated.

As with $Ni(CO)_4$, $Fe_3(CO)_{12}$ reacts with benzyl halides to give the corresponding symmetrical ketones (37,51). The order of reactivity is benzyl chloride < benzyl bromide < benzyl iodide. When $Fe(CO)_5$ is used instead of $Fe_3(CO)_{12}$, benzyl chloride reacts to give a hydrocarbon polymer, $(\text{—}CH_2C_6H_4)_n$, $n \sim 62$, with no CO insertion products formed. The mechanism of the reaction between organohalides and $Fe_3(CO)_{12}$ is thought to involve formation of a benzyl carbonyl halide intermediate, whose decomposition produces the symmetrical ketone.

TABLE 8

Reaction of Some Halides with $K_4Ni_2(CN)_6$ *and Carbon Monoxide* (49,50)

Halide	Product	Yield[a] (%)
$PhCH_2Br$	Dibenzyl ketone	90
$p\text{-}CH_3C_6H_4CH_2Br$	Bis(4-methylbenzyl) ketone	67
n-BuI	Dibutyl ketone	4
trans-PhCH═CHBr	Methyl *trans*-cinnamate	57
	Cinnamaldehyde	10

[a] Yields based on amount of $K_4Ni_2(CN)_6$ used.

$$PhCH_2X + Fe_3(CO)_{12} \rightarrow PhCH_2\overset{X}{\overset{|}{Fe}}(CO)_n \rightarrow PhCH_2\underset{\underset{O}{\|}}{\overset{\overset{X}{|}}{C}}Fe(CO)_{n-1}$$

$$PhCH_2\underset{\underset{O}{\|}}{\overset{\overset{X}{|}}{C}}Fe(CO)_{m-1} + PhCH_2X \longrightarrow PhCH_2\underset{\underset{O}{\|}}{C}CH_2Ph$$

The reaction of benzyl halides with $Fe_3(CO)_{12}$ was also carried out in the presence of various olefins (37,51) resulting in benzylation of the olefin. The reaction is assumed to involve addition of the σ-bonded intermediate (from reaction of the polynuclear carbonyl and benzyl chloride) to the carbon–carbon double bond (insertion of olefin into the C–M bond).

$$R{-}FeX(CO)_n + CH_2{=}CHY \longrightarrow R{-}\overset{\overset{CH_2=CHY}{|}}{Fe}X(CO)_n \longrightarrow$$

$$RCH_2{-}\underset{\underset{FeX(CO)_n}{|}}{C}HY \xrightarrow{H^+} RCH_2CH_2Y \quad (\text{or} \rightarrow RCH{=}CHY)$$

$$RCH_2{-}\underset{\underset{FeX(CO)_n}{|}}{C}HY \xrightarrow{CH_2=CHY} RCH_2\underset{\underset{Y}{|}}{C}HCH_2\underset{\underset{FeX(CO)_n}{|}}{C}HY \xrightarrow{H^+} RCH_2\underset{\underset{Y}{|}}{C}HCH_2CH_2Y$$

$X = Cl$; $Y = CN$, COOEt

Alkyl bromides can be converted into the corresponding aldehydes by reaction with $Na_2Fe(CO)_4$ in the presence of PPh_3, after quenching with acetic acid (52). $Na_2Fe(CO)_4$ is prepared by reducing $Fe(CO)_5$ with sodium amalgam in THF. A number of different primary alkyl bromides were used (Table 9); secondary alkyl bromides and benzyl bromides gave poorer yields of the corresponding aldehydes. Several iron-containing

$$Fe(CO)_5 \xrightarrow[THF]{Na\text{-}Hg} Na_2Fe(CO)_4 \xrightarrow{RBr,\ PPh_3}$$

$$\left[OC{-}\underset{\underset{RCO}{|}}{\overset{\overset{PPh_3}{|}}{Fe}}(CO)_2 \right]^- Na^+ \xrightarrow{HOAc} RCHO$$

TABLE 9
Reaction of $Na_2Fe(CO)_4$ *with Alkyl Bromides* (52)

Alkyl bromide	Product	Yield (%)
$CH_3(CH_2)_4Br$	$CH_3(CH_2)_4CHO$	99
$CH_3(CH_2)_8Br$	$CH_3(CH_2)_8CHO$	91
$PhCH_2CH_2Br$	$PhCH_2CH_2CHO$	86
$C_6H_{11}CH_2Br$	$C_6H_{11}CH_2CHO$	96
$(CH_3)_2C{=}CH(CH_2)_2Br$	$(CH_3)_2C{=}CH(CH_2)_2CHO$	81
$CH_3(CH_2)_5CH(CH_3)Br$	$CH_3(CH_2)_5CH(CH_3)CHO$	50

complexes were also in the product, but specific identification was not made. In the absence of PPh_3, the major product was alkane, with only small amounts of aldehyde formed.

Potassium iron carbonylates react with alkyl halides and CO at 1 atm and 30°C to give the corresponding aldehyde or ester (52,53). Thus, *n*-propyl iodide reacts with $KHFe(CO)_4$ and CO in the presence of EtOH and I_2 to give *n*-butyraldehyde, as the major product and ethyl *n*-butyrate as the minor product, according to the following reactions (54):

$$KHFe(CO)_4 + CH_3CH_2CH_2I \xrightarrow{CO} n\text{-}C_3H_7C(=O)Fe(H)(CO)_4 \longrightarrow n\text{-}C_3H_7CHO$$

$$n\text{-}C_3H_7C(=O)Fe(H)(CO)_4 \xrightarrow{ROH,\ I_2} n\text{-}C_3H_7COOR$$

The *n*-butyl iron hydrocarbonyl intermediate can either undergo reduction to give the aldehyde or react with EtOH and I_2 to give the ester. *n*-Propyl iodide also reacts with $K_2Fe(CO)_4$ and CO in the presence of EtOH and I_2 to give ethyl *n*-butyrate as the major product, with some aldehyde also formed. In this case, dibutyrl iron carbonyl is proposed as the intermediate and reaction of this with I_2 and EtOH gives the ester.

Ethyl α- and β-bromopropionate each reacts with $K_2Fe(CO)_4$ and 1 atm of CO at 30–50°C to give a mixture of diethyl methylmalonate and diethyl succinate (54).

The percentage of isomerization of α-bromopropionate to the β-product was small (2–6%), however, for β-bromopropionate, the percentage of isomerization to the α-product was as high as 95%. Scheme **IV** was proposed to explain the results.

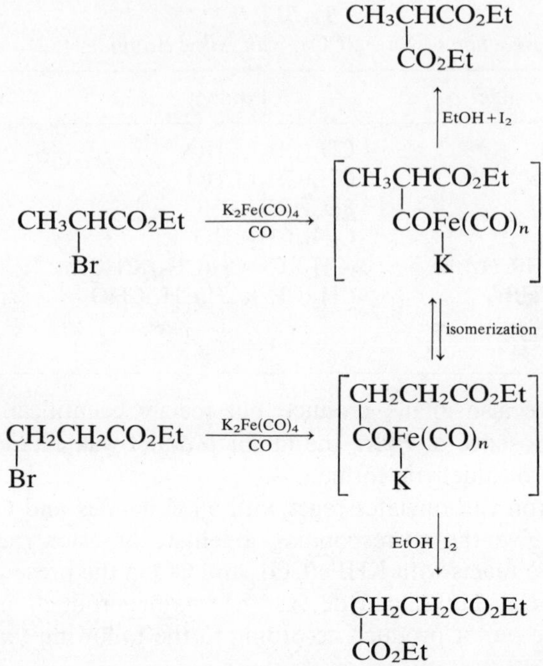

$$CH_3CH(Br)CO_2Et \xrightarrow[CO]{K_2Fe(CO)_4} \left[CH_3CH(CO_2Et)COFe(CO)_nK \right] \xrightarrow{EtOH + I_2} CH_3CH(CO_2Et)CO_2Et$$

$$\left[CH_3CH(CO_2Et)COFe(CO)_nK \right] \underset{\text{isomerization}}{\rightleftharpoons} \left[CH_2(COFe(CO)_nK)CH_2CO_2Et \right]$$

$$CH_2(Br)CH_2CO_2Et \xrightarrow[CO]{K_2Fe(CO)_4} \left[CH_2(COFe(CO)_nK)CH_2CO_2Et \right] \xrightarrow{EtOH \; I_2} CH_2(CO_2Et)CH_2CO_2Et$$

Scheme **IV**

$K_2Fe(CO)_4$ was also reacted with α- and with β-bromoethylbenzene to ascertain the isomerization of the resulting complexes. The aldehyde products retained the geometry of the starting substrate (maximum isomerization observed was 3%) indicating very little interconversion between the two iron complexes.

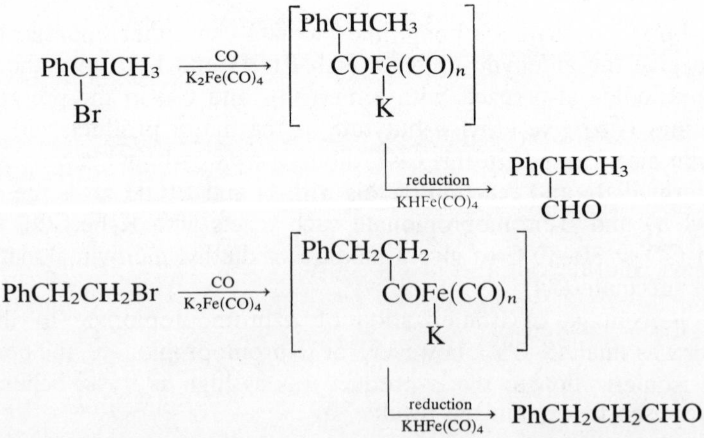

$$PhCH(Br)CH_3 \xrightarrow[K_2Fe(CO)_4]{CO} \left[PhCH(CH_3)COFe(CO)_nK \right] \xrightarrow[KHFe(CO)_4]{\text{reduction}} PhCH(CH_3)CHO$$

$$PhCH_2CH_2Br \xrightarrow[K_2Fe(CO)_4]{CO} \left[PhCH_2CH_2COFe(CO)_nK \right] \xrightarrow[KHFe(CO)_4]{\text{reduction}} PhCH_2CH_2CHO$$

Alkyl bromides, iodides, and tosylates have been converted to ketones in high yields (Table 10) by use of $Na_2Fe(CO)_4$ (55*a*, *b*). Four different paths were used:

$$Na_2Fe(CO)_4 + RX \rightarrow [RFe(CO)_4]^- \xrightarrow{R'X} \underset{\displaystyle O}{\overset{}{R\underset{\|}{C}R'}} \qquad (a)$$

$$Na_2Fe(CO)_4 \xrightarrow[PPh_3]{RX} [R{-}\underset{\underset{\displaystyle O}{\|}}{C}{-}Fe(CO)_3PPh_3]^- \xrightarrow{R'X} R\underset{\underset{\displaystyle O}{\|}}{C}R' \qquad (b)$$

$$Na_2Fe(CO)_4 \xrightarrow[CO]{RX} [R{-}\underset{\underset{\displaystyle O}{\|}}{C}{-}Fe(CO)_4]^- \xrightarrow{R'X} R\underset{\underset{\displaystyle O}{\|}}{C}R' \qquad (c)$$

$$Na_2Fe(CO)_4 \xrightarrow{RCOCl} [R{-}\underset{\underset{\displaystyle O}{\|}}{C}{-}Fe(CO)_4]^- \xrightarrow{R'X} R\underset{\underset{\displaystyle O}{\|}}{C}R' \qquad (d)$$

Some of the ketones prepared are reported in Table 10. By using $Na_2Fe(CO)_4$, γ-ethylenic bromides and tosylates are converted to cyclohexanones and β-allenyl bromides to cyclopentanones (55*c*). The intermediate alkyl and acyl iron carbonyl anions have been separated, characterized (56*a*) and transformed to acid derivatives (56*b*). Alkyl–acyl reversible transformations (57) in these anionic iron carbonyl complexes depends strongly on the nature of the cation. The rate of alkyl migration decreases in the series $Li^+ > Na^+ \gg (Ph_3P)_2N^+$.

These results are consistent with a mechanism involving ion pairing phenomena, which may be important in the chemical reactivity of metal carbonyl anions. This ion pairing hypothesis is fortified by other observations such as the effects of solvents and of complexing molecules on the cation.

Tetracarbonyl cobalt anion was also found effective in the carbonylation of benzyl and alkyl halides (4). The use of this anionic carbonyl has been found particularly useful for the synthesis of phenylacetic salts (58) from benzyl chlorides and CO at atmospheric pressure in the presence of a base such as $Ca(OH)_2$. Thus, phenylacetic salts are obtained in high yield (>90%) and with good catalytic activity (>150 moles per mole of $[Co(CO)_4]^-$).

Benzyl chloride can also be transformed into phenylacetyl chloride in moderate yields in the presence of $RhClCO(PPh_3)_2$ at 180°C and 150 atm of CO (59).

Patents claim the carbonylation of alkyl halides in the presence of various transition metals (Pd on carbon (60), $RhCl_3$, $[RhCl(CO)_2]_2$, $IrCl_3$, $OsCl_3$, etc.) under CO (33,34).

TABLE 10

Synthesis of Aliphatic Ketones Using Sodium Tetracarbonylferrate (55)

RX[a]	R′X[a]	Method[b]	Ligand	Solvent[c]	Product	Yield (%)
n-$C_8H_{17}Br$	C_2H_5I	(a)		MP	n-$C_8H_{17}COC_2H_5$	99
n-$C_8H_{17}I$	C_2H_5I	(a)		MP	n-$C_8H_{17}COC_2H_5$	98
C_2H_5I	n-$C_8H_{17}I$	(a)		MP	n-$C_8H_{17}COC_2H_5$	91
n-$C_5H_{11}Br$	C_2H_5I	(b)	PPh_3	MP	n-$C_5H_{11}COC_2H_5$	95
sec-$C_8H_{17}OTs$	CH_3I	(c)	CO	THF-HMPA	*sec*-$C_8H_{17}COCH_3$	79
n-$C_8H_{17}Br$	CH_3OCH_2Cl	(b)	PPh_3	MP	n-$C_8H_{17}COCH_2OCH_3$	31
n-$C_5H_{11}COCl$	C_2H_5I	(d)		THF-HMPA	n-$C_5H_{11}COC_2H_5$	99
$Br(CH_2)_4CO_2C_2H_5$	C_2H_5I	(a)		MP	$C_2H_5CO(CH_2)_4CO_2C_2H_5$	80

[a] The amount of $Na_2Fe(CO)_4$ used corresponded to that of RX or to a slight excess of the former.

[b] See text, which shows the four methods.

[c] MP = *N*-methyl-2-pyrrolidone; THF in first step, HMPA added with the iodide.

NOTE ADDED IN PROOF

The carbonylation of aryl (61,66–68), benzyl (61,65,67,68), heterocyclic (61), and vinyl halides (61,68) was shown to occur with palladium-triphenylphosphine complexes as catalysts in the presence of a base under mild conditions:

$$RX + CO + R'OH + B \xrightarrow{\text{Pd complex}} RCOOR' + BHX$$

The reaction tolerates a variety of functional groups and shows appreciable stereospecificity with cis and trans vinyl halides to produce esters with retained configuration.

When the reaction is carried out with primary or secondary amines instead of an alcohol, amides are obtained in high yields (62).

Aryl, heterocyclic, and vinyl halides react under about 70 atm of synthesis gas ($1H_2:1CO$) in the presence of basic tertiary amines and a dihalobis(triphenylphosphine)palladium(II) catalyst at 80–150°C to form aldehydes in good yields (69):

$$RX + CO + H_2 \xrightarrow[R'_3N]{PdX_2(PPh_3)_2} RCHO + HX$$

An important feature of the palladium-catalyzed carbonylation of alkyl halides is the observed stereospecificity of the CO insertion step (3–5).

Cooke and Parlman (70) reported a facile conversion of alkyl halides and tosylates to ethyl ketones through the alkylation of $Na_2Fe(CO)_4$ in the presence of ethylene:

$$RX \xrightarrow[C_2H_4]{Na_2Fe(CO)_4} \xrightarrow{H^+} RCOCH_2CH_3$$

REFERENCES

1*a*. J. P. Collman and W. R. Roper, in F. G. A. Stone, and R. West, Eds., *Advances in Organometal. Chem.*, Academic Press, New York–London, Vol. VII, 1968, p. 53.
1*b*. J. P. Collman, *Acc. Chem. Res.*, **1,** 136 (1968).
1*c*. R. Ugo, *Coord. Chem. Rev.*, **3,** 319 (1968).
1*d*. J. Halpern, *Acc. Chem. Res.*, **3,** 386 (1970).
1*e*. R. Cramer, *Acc. Chem. Res.*, **1,** 186 (1968).
2*a*. R. F. Heck, *Adv. Chem. Ser.*, **49,** 181 (1965).
2*b*. K. Noack and F. Calderazzo, *J. Organometal. Chem.*, **10,** 101 (1967).
2*c*. A. Wojcicki, in F. G. A. Stone and R. West, Eds., *Advances in Organometal. Chem.*, Academic Press, New York–London Vol XI, 1973, p. 87.
3*a*. G. P. Chiusoli and A. Cameroni, *Chim. Ind.* (*Milan*), **46,** 1063 (1964).
3*b*. G. P. Chiusoli, G. Cometti, and S. Merzoni, *J. Organometal. Chem. Syn.*, **1,** 439 (1972).

4. R. F. Heck, in I. Wender and P. Pino, Eds., *Organic Synthesis via Metal Carbonyls*, Interscience, New York–London, Vol. I, 1968, p. 373.
5*a*. H. Dieterle and W. Eschenbach, Ger. Pat. 537,610 (1927); through *Chem. Abstr.*, **26,** 1300 (1932).
5*b*. H. Dieterle and W. Eschenbach, *Arch. Pharm.*, **265,** 187 (1927); through *Chem. Abstr.*, **21,** 1975 (1927).
6. S. K. Palit and N. Tripathy, *J. Appl. Chem.*, **19,** 301 (1969).
7. M. Nakayama and T. Mizoroki, *Bull. Chem. Soc. Japan*, **42,** 1124 (1969).
8. V. I. Romanovskii and A. A. Artem'ev, *Zhur. Vsesoyuz. Khim. Obshchestva im. D. I. Mendeleeva*, **5,** 472 (1960); through *Chem. Abstr.*, **55,** 3511 (1961).
9. H. Bliss and R. W. Southworth, U.S. Pat. 2,565,461 (1951).
10. S. E. Yakushkina and N. V. Kislyakova, *Izvest. Akad. Nauk SSSR, Otdel. Khim. Nauk*, **1958,** 1119; through *Chem. Abstr.*, **53,** 3134 (1959).
11. H. J. Leibeu, U.S. Pat. 2,640,071 (1953).
12. K. Yamamoto and K. Sato, *Bull. Chem. Soc. Japan*, **27,** 389 (1954).
13. T. Mizoroki and M. Nakayama, *Bull. Chem. Soc. Japan*, **40,** 2203 (1967).
14*a*. I. L. Mador and J. A. Scheban, U.S. Pat. 3,452,090 (1969).
14*b*. R. N. Knowles, U.S. Pat. 3,636,082 (1969).
15. G. E. Tabet, U.S. Pat. 2,565,464 (1951).
16. V. I. Romanovskii and A. A. Artem'ev, *Zhur. Vsesoyuz. Khim. Obshchestva im. D. I. Mendeleeva*, **5,** 476 (1960); through *Chem. Abstr.*, **55,** 1506 (1961).
17. W. W. Prichard and G. E. Tabet, U.S. Pat. 2,565,462 (1951).
18. K. Yamamoto and K. Sato, Japan Pat. 2424 (1952); through *Chem. Abstr.*, **48,** 2105 (1954).
19. G. E. Tabet, U.S. Pat. 2,565,463 (1951).
20. G. E. Tabet, U.S. Pat. 2,691,670 (1954).
21. H. Kroeper, F. Wirth, and O. Huchler, *Angew. Chem.*, **72,** 867 (1960).
22. H. Kroeper, F. Wirth, and O. H. Huchler, Ger. Pat. 1,074,028 (1960); through *Chem. Abstr.*, **55,** 12364 (1961).
23. H. J. Leibeu, U.S. Pat. 2,734,912 (1956).
24. H. J. Leibeu, U.S. Pat. 2,773,090 (1956).
25. H. Kroeper, R. Sinn, and F. Wirth, Ger. Pat. 1,112,971 (1960); through *Chem. Abstr.*, **56,** 1399 (1962).
26. H. Kroeper, F. Wirth, and O. Huchler, U.S. Pat. 2,914,554 (1959); through *Chem. Abstr.*, **54,** 3322 (1960).
27. H. Kroeper, F. Wirth, and O. Huchler, Ger. Pat. 1,066,574 (1959); through *Chem. Abstr.*, **55,** 12362 (1961).
28. Badische Anilin- und Soda-Fabrik Akt. Ges., Brit. Pat. 815,835 (1959); through *Chem. Abstr.*, **54,** 1449 (1960).
29. H. J. Leibeu, U.S. Pat. 2,691,671 (1954); through *Chem. Abstr.*, **49,** 14807 (1955).
30. M. Nakayama and T. Mizoroki, *Bull. Chem. Soc. Japan*, **43,** 569 (1970).
31. M. Suzuki and T. Kondo, Japan. Pat. 22064 (1965); through *Chem. Abstr.*, **64,** 5002 (1966).
32*a*. W. W. Prichard, *J. Am. Chem. Soc.*, **78,** 6137 (1956).
32*b*. W. W. Prichard, U.S. Pat. 2,696,503 (1954).
32*c*. W. W. Prichard, U.S. Pat. 2,680,751 (1954).
33. F. E. Paulik, A. Hershman, J. F. Roth, J. H. Craddock, and D. Forster, French Pat. 1,573,130 (1969); through *Chem. Abstr.*, **72,** 100054 (1970).
34. F. E. Paulik, A. Hershman, J. F. Roth, J. H. Craddock, W. R. Knox, and R. E. Schultz, French Pat. 1,573,131 (1969).

35. N. L. Bauld, *Tetrahedron Lett.*, **1963,** 1841.
36. W. F. Beckert and T. U. Lowe, Jr., *J. Org. Chem.*, **32,** 1215 (1967).
37. I. Rhee, M. Ryang, and S. Tsutsumi, *J. Organometal. Chem.*, **9,** 361 (1967).
38. E. J. Corey and L. S. Hegedus, *J. Am. Chem. Soc.*, **91,** 1233 (1969).
39. M. Nakayama and T. Mizoroki, *Bull. Chem. Soc. Japan*, **44,** 508 (1971).
40. L. Cassar and M. Foà, *J. Organometal. Chem.* **51,** 381 (1973).
41. H. W. Sternberg, R. Markby, and I. Wender, *J. Am. Chem. Soc.*, **82,** 3638 (1960).
42. G. P. Chiusoli, S. Merzoni, and G. Mondelli, *Tetrahedron Lett.*, **1964,** 2777.
43. E. Yoshisato, M. Ryang, and S. Tsutsumi, *J. Org. Chem.*, **34,** 1500 (1969).
43*a*. M. Ryang, Y. Toyoda, S. Murai, N. Sonoda and S. Tsutsumi, *J. Org. Chem.*, **38,** 62 (1973).
44. E. Yoshisato and S. Tsutsumi, *J. Org. Chem.*, **33,** 869 (1968).
45. L. Cassar and M. Foà, Ital. Pat. 845,993 (1969).
46. L. Cassar and M. Foà, *Inorg. Nucl. Chem. Lett.*, **6,** 291 (1970).
47. E. Yoshisato and S. Tsutsumi, *Chem. Commun.*, **1968,** 33.
48. E. Yoshisato, and S. Tsutsumi, *J. Am. Chem. Soc.*, **90,** 4488 (1968).
49. I. Hashimoto, M. Ryang, and S. Tsutsumi, *Tetrahedron Lett.*, **1969,** 3291.
50. I. Hashimoto, H. Tsuruta, M. Ryang, and S. Tsutsumi, *J. Org. Chem.*, **35,** 3748 (1970).
51. I. Rhee, N. Mizuta, M. Ryang, and S. Tsutsumi, *Bull. Chem. Soc. Japan*, **41,** 1471 (1968).
52. M. P., Cooke, Jr., *J. Am. Chem. Soc.*, **92,** 6080 (1970).
53. Y. Takegami, Y. Watanabe, H. Masada, and I. Kanaya, *Bull. Chem. Soc. Japan*, **40,** 1456 (1967).
54. H. Masada, M. Mizuno, S. Suga, Y. Watanabe and Y. Takegami, *Bull. Chem. Soc. Japan*, **43,** 3824 (1970).
55*a*. J. P. Collman, S. R. Winter, and D. R. Clark, *J. Am. Chem. Soc.*, **94,** 1788 (1972).
55*b*. J. P. Collman, and N. W. Hoffman, *J. Am. Chem. Soc.*, **95,** 249 (1973).
55*c*. J. Y. Merour, J. L. Roustán, C. Charrier, J. Collin, and J. Benaim, *J. Organometal. Chem.*, **51,** C24 (1973).
56*a*. W. O. Siegl and J. P. Collman, *J. Am. Chem. Soc.*, **94,** 2516 (1972).
56*b*. J. P. Collman, S. R. Winter, R. G. Komoto, *J. Am. Chem. Soc.*, **95,** 249 (1973).
57. J. P. Collman, T. N. Cawse, and J. I. Brauman, *J. Am. Chem. Soc.*, **94,** 5905 (1972).
58. L. Cassar, M. Foà, and G. P. Chiusoli, Ital. Pat. 868,994 (1970).
59. K. Ohno and J. Tsuji, *J. Am. Chem. Soc.*, **90,** 99 (1968).
60. R. N. Knowles, U.S. Pat. 3,772,384 (1971).
61. A. Schoenberg, I. Bartoletti, and R. F. Heck, *J. Org. Chem.*, **39,** 3318 (1974).
62. A. Schoenberg and R. F. Heck, *J. Org. Chem.*, **39,** 3327 (1974).
63. K. S. Y. Lau, R. W. Friess, and J. K. Stille, *J. Am. Chem. Soc.*, **96,** 4983 (1974).
64. P. K. Wong, K. S. Y. Lau, and J. K. Stille, *ibid.*, 5957.
65. J. K. Stille, L. F. Hines, R. W. Friess, P. K. Wong, D. E. James, and K. Lau in *Advances in Chemistry Series* No. 132 (1974), Am. Chem. Soc. p. 90.
66. M. Hidai and Y. Uchida in Y. Ishii and M. Tsutsui, *Organotransition-Metal Chemistry*, Plenum Press, New York–London, 1975, p. 265.
67. J. K. Stille and P. W. Wong, *J. Org. Chem.* **40,** 532 (1975).
68. M. Hidai, T. Hikita, Y. Wada, Y. Fujikura, and Y. Uchida, *Bull. Chem. Soc. Japan*, **48,** 2075 (1975).
69. A. Schoenberg and R. F. Heck, *J. Am. Chem. Soc.* **96,** 7761 (1974).
70. M. P. Cooke Jr, and R. M. Parlman, *J. Am. Chem. Soc.*, **97,** 6863 (1975).

Organic Syntheses with Iron Pentacarbonyl

HOWARD ALPER, *University of Ottawa, Department of Chemistry, Ottawa, Ontario, Canada KIN 6N5*

The catalytic utility of iron pentacarbonyl (and other metal carbonyls) has been amply demonstrated for a variety of reactions described in Vol. 1 (123*a*). The past few years witnessed an increased awareness of the importance of $Fe(CO)_5$ as a *reagent* for organic synthesis. This chapter, which discusses the reagent capabilities of $Fe(CO)_5$, is generally organized according to the type of reaction effected by the metal carbonyl towards a particular functional group. Some general mechanistic pathways have been proposed for many of the described reactions, the author's intention being to stimulate more investigations into this aspect of metal carbonyl chemistry.

I. ISOMERIZATION

Iron pentacarbonyl is capable of effecting hydrogen migration reactions with certain dienes and vinylcyclopropanes, the products usually being isolated as diene–iron tricarbonyl complexes. The iron tricarbonyl group must be removed so that the reactions can be of synthetic utility in organic chemistry. A number of reagents are available for liberating the organic ligand; the choice of reagents depends on the nature and reactivity of the complex as well as of the free diene. Although numerous diene–iron tricarbonyl complexes have been obtained via isomerization of a reactant with $Fe(CO)_5$, usually only those cases will be considered where the isomerized free diene was isolated from the complex. An excellent review of diene–iron tricarbonyl complexes exists in the literature (1).

A. Conjugated Dienes

Treatment of *cis*-1,3-pentadiene with $Fe(CO)_5$ at 150–160°C for 80 hr gives *trans*-1,3-pentadiene-iron tricarbonyl in low yield (2). The free diene can be obtained by reacting the complex with ferric chloride in ethanol. There is less steric hindrance in the *trans* complex than in the *cis;* consequently, isomerization to give the least hindered complex occurs for *cis*-1,3-pentadiene as well as for many other cisoid dienes having *cis* substituents (1).

A diene must be in the cisoid conformation in order to form a diene–iron tricarbonyl complex (3). Thus reaction of transoid steroid dienes, for example, cholesta-3,5-dienes (**1**), with $Fe(CO)_5$ in refluxing dibutyl ether followed by decomplexation with $FeCl_3$ in ethanol gives the cisoid isomer **3** in good yield (4). The driving force for isomerization is the necessity of the diene to possess a cisoid conformation in order for complex formation to take place. This reaction sequence has provided for

$Fe(CO)_5$, Bu_2O → ; $FeCl_3$ →

R ; R ; $Fe(CO)_3$; R

(**1**) (**2**) (**3**)

R = H, CH_3, OCH_3, C_6H_5

the synthesis of compounds not readily available by other means (e.g., **3**, R = OCH_3). In addition, these reactions illustrate the utility of $Fe(CO)_5$ as a reagent for converting heteroannular dienes into their thermodynamically *less* stable homoannular cisoid isomers (5). Cais and Maoz (6) have reported a similar isomerization of *retro*-ionylidene acetate (**4**) but neither product yield nor experimental details was given for the decomplexation step.

CH_3 $CH_2COOCH_2CH_3$ CH_3 (**4**) $\xrightarrow[Bu_2O]{Fe(CO)_5}$ CH_3 $-Fe(CO)_3$ $CH_2COOCH_2CH_3$ $\downarrow$ $FeCl_3/CH_3CH_2OH$ CH_3 $CH_2COOCH_2CH_3$

B. Nonconjugated Dienes

The two double bonds of a diene should be conjugated, or in a suitable orientation if nonconjugated, for complexation to occur with the iron tricarbonyl group. Many nonconjugated dienes form conjugated diene–iron tricarbonyl complexes on reaction with $Fe(CO)_5$. For example, thermal (7) or photochemical (8) reaction of 1,4-cyclohexadiene with $Fe(CO)_5$ gives 1,3-cyclohexadiene-iron tricarbonyl (**5**) from which the free diene **6** may be obtained by treatment with $FeCl_3$ (8,9). Photolysis of a mixture of the complex and PPh_3 in benzene results in CO (10) rather than diene ligand displacement, the latter process reported to occur with several other complexes (11,12).

$\xrightarrow[\Delta \text{ or } h\nu]{Fe(CO)_5}$ $-Fe(CO)_3$ (**5**) $\xrightarrow{FeCl_3}$ (**6**)

$\downarrow$ $PPh_3, h\nu$

$-Fe(CO)_2(PPh_3)$ + $-Fe(CO)(PPh_3)_2$

Birch and co-workers (8) studied the reaction of a number of substituted 1,4-cyclohexadienes with $Fe(CO)_5$. Reaction of 1-methoxy-1,4-cyclohexadiene with the metal carbonyl under ultraviolet irradiation gives complexes **7** and **8**. Treatment of the product mixture with MnO_2 in methylene chloride gave pure 1-methoxy-1,3-cyclohexadiene (**9**), along with unreacted **8**. MnO_2 is a milder oxidizing agent than $FeCl_3$ or ceric

OCH3 | Fe(CO)5, hν → OCH3 —Fe(CO)3 (7) + OCH3 Fe(CO)3 (8); (7) —MnO2→ OCH3 (9) + **8**

ammonium nitrate (to be discussed later) and hence can effect liberation of the diene ligand from the complex without appreciable hydrolysis of the formed enol ether or secondary oxidation.

1,4-Pentadiene can be converted to *trans*-1,3-pentadiene in the manner described above (9). Kiji and Iwamoto (13) reported that reaction of *cis*-1,4-hexadiene (**10**) with $Fe(CO)_5$ followed by decomplexation with $FeCl_3$ gives *trans*-1,3-hexadiene (**13**) as the major product with *trans,trans*-2,4-hexadiene (**14**) formed in small amounts. Treatment of

H H CH3 (10) —Fe(CO)5→ C2H5 Fe(CO)3 (11) + CH3— —CH3 Fe(CO)3 (12); (15) —Fe(CO)5→ (11); (11) —FeCl3→ C2H5 (13) + CH3— —CH3 (14)

1,5-hexadiene (**15**) with $Fe(CO)_5$ also gives **11** as the major product with minor quantities of **12** (2,7). Another 1,5-diene, 4-vinylcyclohexene, reacted with $Fe(CO)_5$ and then $FeCl_3$ to give 1-ethyl-1,3-cyclohexadiene (14). Frankel and co-workers (12) have shown that treatment of equimolar quantities of $Fe(CO)_5$ and methyl linoleate gives a mixture of conjugated diene–iron tricarbonyl complexes from which a mixture of *trans,trans*-conjugated dienes can be obtained by reaction with $FeCl_3$ or

PPh_3. Extended photolysis of 1,5-cyclooctadiene gives the corresponding iron tricarbonyl complex **16** which can then be converted to the conjugated complex **17** on further irradiation (15,16). 1,3-Cyclooctadiene could be generated by reaction of **17** with ceric ion. However, 1,5-cyclooctadiene can be converted to the 1,3-isomer with catalytic amounts of $Fe(CO)_5$ (7).

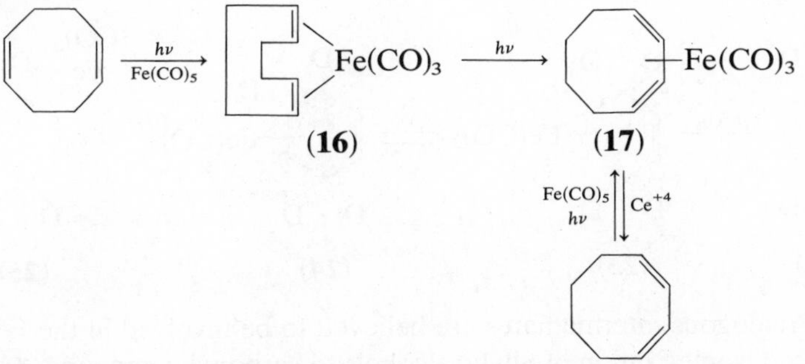

$hν$, $Fe(CO)_5$ → $Fe(CO)_3$ (**16**) $hν$ → $Fe(CO)_3$ (**17**)

$Fe(CO)_5$, $hν$ ⇅ Ce^{+4}

Experiments with *d*-limonene have shown (1,14) that optically active nonconjugated dienes gave inactive complexes when treated with $Fe(CO)_5$ and thus the products formed on liberation of the ligand from the complex were also inactive.

A limitation of this isomerization reaction is encountered with nonconjugated dienes having either blocking groups [e.g., **18** (17), **19** (18)] or a suitable geometry for complexation [e.g., **20** (19,20); **21** (21)]; complexation of $Fe(CO)_3$ to the nonconjugated diene will occur here rather than isomerization.

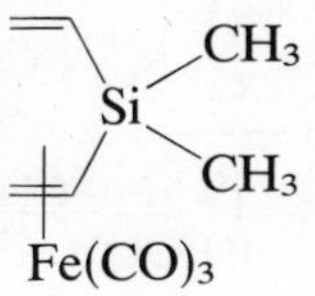

CH_3 Si CH_3 $Fe(CO)_3$ (**18**)

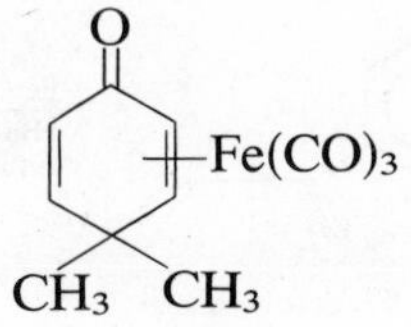

O $Fe(CO)_3$ CH_3 CH_3 (**19**)

Fe $(CO)_3$ (**20**)

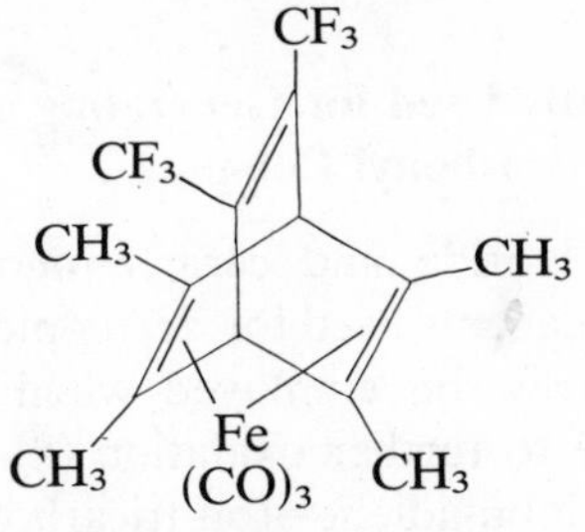

CF_3 CF_3 CH_3 CH_3 CH_3 Fe $(CO)_3$ CH_3 (**21**)

A simple approach to the mechanism for the formation of conjugated diene–iron tricarbonyl complexes by reaction of nonconjugated dienes with $Fe(CO)_5$ involved deuterium-labeling of the methylene groups of 1,4-cyclohexadiene and locating the deuterium in the product (22). Photolysis of a mixture of **22** and $Fe(CO)_5$ gave **25**, clearly indicating that the isomerization occurs via a π-allylhydroiron tricarbonyl intermediate

D D D D D D $(CO)_3$ Fe D D
$Fe(CO)_5$ → $Fe(CO)_3$ ⇌ $Fe(CO)_3$ → H D
D D D D D D D D

(22) **(23)** **(24)** **(25)**

(**24**). Analogous intermediates are believed to be involved in the $Fe(CO)_5$ catalyzed isomerization of allylic alcohols to carbonyl compounds (23,24).

C. Vinylcyclopropanes

Sarel and co-workers (25) have found that reaction of vinylcyclopropanes, activated by the presence of a 2-aryl group, with $Fe(CO)_5$ gives the iron tricarbonyl complex of the highly reactive 2-aryl-1,3-*trans*-pentadiene. The workers did not report any attempt to generate the free diene from the complex.

X X
$Fe(CO)_5$ / Bu_2O / 140° →
CH_3
$Fe(CO)_3$

X = H, CL, OCH_3

D. Reagents Used for Generating a Free Diene from the Corresponding Iron Tricarbonyl Complex

Ferric chloride and ceric ammonium nitrate (26–28) are the most common reagents used for decomplexation. These oxidizing agents should not, however, be employed when dienes to be generated are readily susceptible to further oxidation (8) or isomerization (29). A novel reaction of cyclobutadiene–iron tricarbonyl with $FeCl_3$ or ceric ion in acetone saturated with LiCl was reported by Pettit and co-workers (30).

Triphenylphosphine gives either the desired diene (11,12) or unwanted CO ligand displacement reactions (10,30). Nesmayanov and co-workers (31) have shown that esters of phenylphosphinous acid displace CO rather than the diene ligand from an iron tricarbonyl complex. MnO_2 is a mild oxidizing agent ideally suited to the generation of reactive dienes (8).

II. DEHYDRATION AND DEHYDROSULFURATION

Dehydration and dehydrosulfuration of primary amides and thionamides, respectively, by $Fe(CO)_5$ in refluxing dibutyl ether results in the formation of nitriles (32). The yield of nitrile (60–66%) from a thionamide was generally twice that from the corresponding amide.

$$\mathrm{PhC(=O)NH_2} \xrightarrow[32\%]{Fe(CO)_5} \mathrm{PhCN}$$

$$\mathrm{PhC(=S)NCH_2} \xrightarrow[64\%]{Fe(CO)_5} \mathrm{PhCN}$$

N-(*p*-methoxybenzyl)-*p*-methoxythionbenzamide (**27**) was unexpectedly obtained as a by-product from the reaction of *p*-methoxythionbenzamide with $Fe(CO)_5$.

$$p\text{-}CH_3OC_6H_4CSNH_2 + Fe(CO)_5 \xrightarrow{Bu_2O} p\text{-}CH_3OC_6H_4CN + p\text{-}CH_3OC_6H_4CSNHCH_2C_6H_4OCH_3\text{-}p$$

(**26**) (61%) (**27**) (5.4%)

The nitrile may be formed via initial attack of the sulfur (or oxygen) of the reactant on the metal atom to form an imine intermediate **28**. Dimerization of **28** gives **29** [see ref, 123*a*, pp. 123–126, and (33) for analogous structures]; collapse of this complex in dibutyl ether produces the nitrile. Complex **29**, $R = CH_3OC_6H_4–$, was isolated from the reaction of *p*-methoxythionbenzamide with $Fe_2(CO)_9$ in THF at room temperature (**30** and **31** were also formed under these conditions) and when heated in dibutyl ether, gave the nitrile in 77% yield.

In complex **28** sulfur has *d* orbitals available for bonding to iron; this situation does not exist for the corresponding oxygen complex. The greater yield of nitrile from a thionamide than from an amide may be due

$$2R{-}\overset{S}{\overset{\|}{C}}{-}\ddot{N}H_2 + 2Fe(CO)_4 \xrightarrow{-H_2} 2R{-}C\begin{smallmatrix}SFe(CO)_4\\ NH\end{smallmatrix} + 2CO$$

(28)

$$\xrightarrow{-2CO} (CO)_3Fe{\leftarrow}S{-}Fe(CO)_3 \quad \textbf{(29)}$$

$$\textbf{(30)}\ [Fe(CO)_3]_2(\mu\text{-}S)_2Fe(CO)_3 \; + 2RC{\equiv}N \xleftarrow[+Fe(CO)_3]{-H_2} \textbf{(29)}$$

(30) **(31)** **(29)**

to *d*-orbital stabilization of **28** providing the driving force for reaction.

N-substituted thionamides and amides, which cannot form nitriles, react with $Fe(CO)_5$ in dibutyl ether to give aldimines. This reaction is particularly useful for preparing cyclic Schiff bases such as **32** (34). The reaction may involve complex **33** (analogous to **28**) as an intermediate

$$\begin{matrix}PhCONHPh & \xrightarrow{15\%} \\ PhCSNHPh & \xrightarrow{47\%}\end{matrix} \xrightarrow{Fe(CO)_5} PhCH{=}NPh$$

$$\overset{S}{\overset{\|}{C}}{-}NH\ \text{[ring with } (CH_2)_6] \xrightarrow[52\%]{Fe(CO)_5} \overset{H}{C}{=}N\ \text{[ring with } (CH_2)_6]$$

(32)

$$R{-}\underset{X-Fe(CO)_4}{\underset{|}{C}}{=}N{-}R' \qquad X = O,S$$

(33)

which can undergo hydrogenolysis of the C–O or C–S bond. Unlabeled *N*-benzylideneaniline was obtained from the reaction of *N*-deuteriothionbenzanilide with $Fe(CO)_5$, indicating that the hydrogen attached to the aldimine carbon comes from the solvent. The mechanism for conversion of **33** to the product is not clear.

Treatment of several primary and secondary amides with $Fe(CO)_5$, $Fe_2(CO)_9$, or $Fe_3(CO)_{12}$, in refluxing chloroform results in formation of iron halide complexes (35).

N-methylacetanilide, a fully substituted amide, does not react with $Fe(CO)_5$ in refluxing dibutyl ether.

Some preliminary studies have been carried out on the reactions of thioureas with $Fe(CO)_5$. Many thioureas decompose in refluxing dibutyl ether, but 1,1-diphenyl-2-thiourea, stable under these reaction conditions, gives diphenylamine in 72% yield on reaction with $Fe(CO)_5$ (36). No diphenylcyanamide was obtained as would have been expected had the reaction followed the pathway described for the primary thionamide.

$$Ph_2NCSNH_2 + Fe(CO)_5 \xrightarrow{Bu_2O} Ph_2NH$$

III. DEOXIMATION

$Fe(CO)_5$ is a useful reagent for regenerating carbonyl compounds from oximes under anhydrous aprotic conditions (37). Several aromatic ketones have been regenerated from their ketoximes using lead tetraacetate in methylene chloride(38). Treatment of equimolar quantities of oxime and $Fe(CO)_5$ in the presence of a catalytic amount of boron trifluoride etherate gives carbonyl compounds in 55–81% yield. The reaction is applicable to a wide variety of oximes of aldehydes and ketones including saturated, α,β-unsaturated, aromatic, and hindered oximes (Table 1). The role of boron trifluoride etherate is uncertain.

TABLE 1
Yields of Carbonyl Compounds from Oximes

Parent carbonyl compound	Reflux time (hr)	Yield (%)	Ref.
Cyclohexanone	20	81	37
4-Methyl-4-trichloromethyl-cyclohexadien-1-one	10	72	37,18
Fluorenone	16	69	37
Cholest-4-en-3-one	16	67	37
Santonin	18	67	37
p-Bromobenzaldehyde	12	60	34
O-Methylpodocarpinal	17	55	37

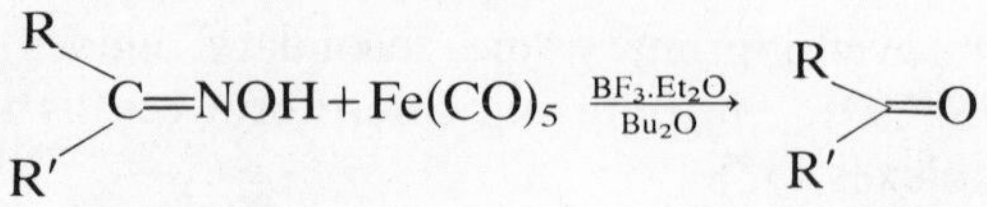

$$RR'C{=}NOH + Fe(CO)_5 \xrightarrow[Bu_2O]{BF_3.Et_2O} RR'C{=}O$$

Furthermore, the catalyst is not required in some cases (e.g, 4-methyl-4-trichloromethylcyclohexadien-1-one (18)). Carbonyl compounds were not obtained from reaction of fluorenone phenylhydrazone (**34**), *N*-2,6-trichloro-*p*-benzoquinoneimine (**35**), or acetophenone semicarbazone [**36** (39)].

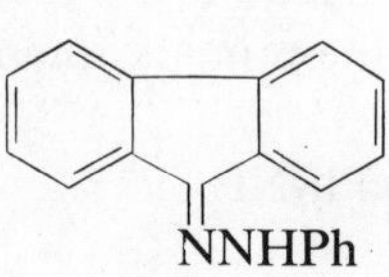

NNHPh Cl Cl N Cl O Ph—C—CH$_3$ NNHCONH$_2$

(**34**) 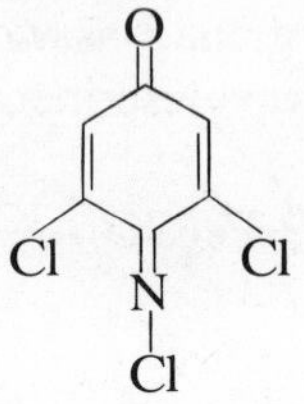 (**35**) 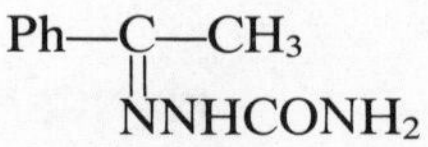 (**36**)

IV. REACTIONS OF $Fe(CO)_5$ WITH HALIDES—COUPLING AND DEHALOGENATION

A. Alkyl, Allyl, Vinyl, and Aryl Halides

In 1961, Coffey reported a convenient synthesis of tetraarylethylenes in good yields by reacting certain *gem*-dihalides with $Fe(CO)_5$ in refluxing benzene (40):

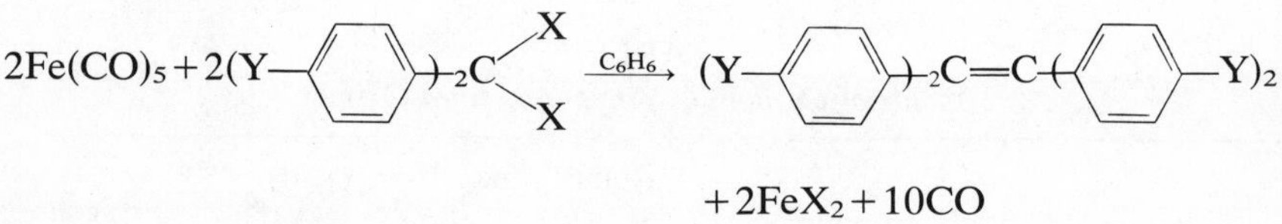

$$2Fe(CO)_5 + 2(Y{-}C_6H_4{-})_2CX_2 \xrightarrow{C_6H_6} (Y{-}C_6H_4{-})_2C{=}C({-}C_6H_4{-}Y)_2 + 2FeX_2 + 10CO$$

X = Cl, Y = H, CH_3, $(CH_3)_3C$, Cl
X = Br, Y = NO_2

Coffey noted that in order for reaction to occur, activating groups such as aryl, cyano, carbalkoxy, or halo must be attached to the carbon bearing the halogen atoms. Halides such as diethyl dibromomalonate and dibromomalononitrile do in fact react with the metal carbonyl but give iron-containing products of no synthetic utility.

The reaction of $Fe(CO)_5$ and CCl_4 under the conditions described gives some hexachloroethane but mostly tar (40,41). Carbon tetrabromide, ethyl dibromoacetate, and benzotrichloride undergo similar reactions with the metal carbonyl. Coffey observed no reaction between Fe(CO)$_5$ and any monohalide. A carbene mechanism was proposed for the reaction of Fe(CO)$_5$ and dihalodiarylmethanes, but little evidence was presented to support this suggestion (40). A more likely mechanism involves iron insertion into one of the C–X bonds to give the σ-bonded complex **37** which can react with additional *gem*-dihalide, possibly via a four-centered transition state, to form the *vic*-dihalide **38**. The latter, in the presence of more Fe(CO)$_5$, dehalogenates to the olefin. Complex **37** may lose CO prior to reaction with additional *gem*-dihalide in which case ferrous halide instead of dihalotetracarbonyliron will be produced along with **38**.

$$Ar_2CX_2 + Fe(CO)_5 \longrightarrow Ar_2C(X){-}Fe(CO)_4{-}X \ (\mathbf{37}) + CO$$

$$\mathbf{37} \xrightarrow{Ar_2CX_2} \left[\begin{array}{c} X \\ | \\ Ar_2{-}C\cdots X \\ \vdots \qquad \vdots \\ Ar_2{-}C\cdots Fe(CO)_4 \\ | \qquad | \\ X \qquad X \end{array}\right] \longrightarrow Ar_2C(X){-}C(X)Ar_2 \ (\mathbf{38}) + Fe(CO)_4X_2$$

$$5CO + FeX_2 + Ar_2C{=}CAr_2 \xleftarrow{Fe(CO)_5} Ar_2C(X){-}C(X)Ar_2 \ (\mathbf{38})$$

$$Fe(CO)_4X_2 \xrightarrow{-4CO} FeX_2$$

Several complexes analogous to **37** have been isolated in other reactions and will be discussed below. The formation of a *vic*-dihalide was observed by Coffey in the reaction of 9,9-dibromofluorene with Fe(CO)$_5$: By using benzene as solvent, bisfluorenylidene (**39**) was obtained; however, reaction in dioxane results in the formation of 9,9′-dibromobisfluorenyl (**40**) along with the olefin. Coffey has also shown that

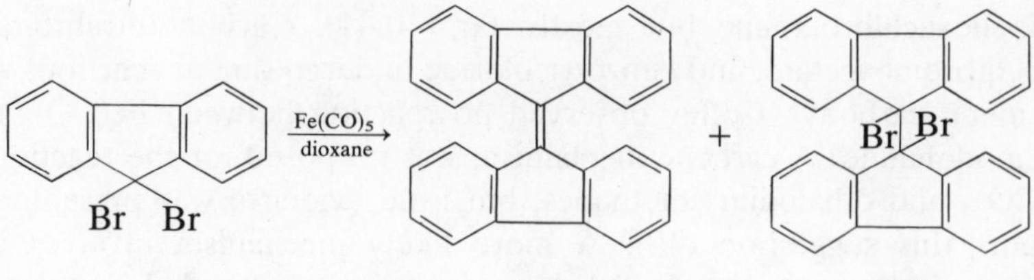

$\xrightarrow[\text{dioxane}]{Fe(CO)_5}$ **(39)** + **(40)**

$Fe(CO)_5$ reacts with the *vic*-dihalide 1,2-dichloro-1,1,2,2-tetraphenylethane to form tetraphenylethylene (40). The more reactive $Fe_2(CO)_9$ effects analogous dehalogenation reactions (30,42,43).

Bruce (44) was able to effect reaction of decafluorobenzhydryl bromide with $Fe(CO)_5$. The expected product, 1,1,2,2-tetrakis (pentafluorophenyl)-ethane, was obtained in 37% yield along with small amounts of bis-(pentafluorophenyl)methane. Perfluoroalkyliodides react with $Fe(CO)_5$ in benzene (45) or in the absence of solvent (46) to give complexes analogous to **37**.

$$R_fI + Fe(CO)_5 \rightarrow R_fFe(CO)_4(I) + CO$$

$$R_f = CF_3,\ C_2F_5,\ n\text{-}C_3F_7$$

The thermal reaction of several simple halides not having strongly electron-withdrawing groups with $Fe(CO)_5$ was investigated. Treatment of benzyl chloride with $Fe(CO)_5$ in THF at 30°C for 30 hr gives polymeric material $[(PhCH_2)_n,\ n\sim 62]$ (47). The dimeric product, bibenzyl, is obtained on reaction of the halide with $Ni(CO)_4$ in refluxing benzene. A patent (48) has claimed the formation of polymeric material of structure **41** by heating α,α-dichloro-*p*-xylene with $Fe(CO)_5$ in xylene at 100–110°C. 1,4-Dichloro-2-butene (stereochemistry not given) reacts in a similar manner with $Fe(CO)_5$ to give **42**.

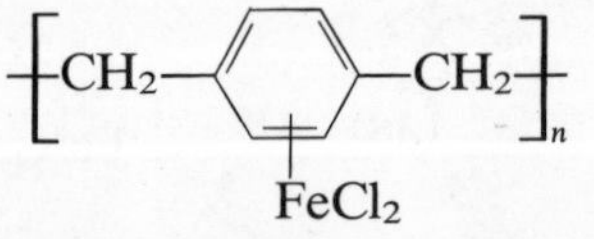

$[CH_2-C_6H_4(FeCl_2)-CH_2]_n$ **(41)**

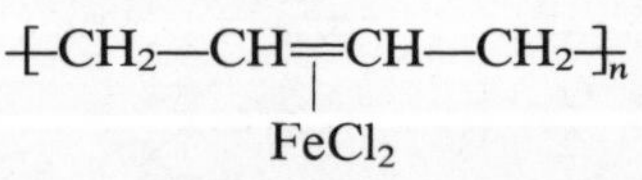

$[CH_2-CH{=}CH(FeCl_2)-CH_2]_n$ **(42)**

Coffey (40) reported that allyl chloride did not react with $Fe(CO)_5$ in refluxing benzene. Plowman and Stone (49) similarly noted that both allyl chloride and allyl bromide reacted very slowly with the metal carbonyl at 40°C. However, allyl iodide did react with $Fe(CO)_5$ in a reasonable length of time to give a σ-bonded organoiron complex. The three allyl halides, along with several substituted ones, usually give π-allyliron-tricarbonyl halides (50,51) or allyl halide–iron tetracarbonyl complexes (52). 1,5-Dienes have not been isolated from these reactions. Vinyl halides, on irradiation, give the corresponding iron tetracarbonyl complexes (53–55).

E. K. von Gustorf and co-workers (16,55,56) studied the irradiation of a number of simple saturated halides with $Fe(CO)_5$, but few details were reported. Cobalt-60 irradiation of a mixture of $Fe(CO)_5$ and CCl_4 at 20°C gives hexachloroethane, among other products (56). The investigators noted similar reactions of $Fe(CO)_5$ with benzyl chloride, acetyl and benzoyl chlorides, trichloroacetic acid, and $CHCl_3$. Evidence was obtained for the formation of **43** on irradiation of a mixture of bromoform and $Fe(CO)_5$ (16,56), methylene bromide being formed on treating **43**

$$Fe(CO)_5 + CHBr_3 \xrightarrow{h\nu} (CO)_4Fe\begin{matrix} \diagup CHBr_2 \\ \diagdown Br \end{matrix} + CO$$

(**43**)

with acid. Moreover, **43** is a good polymerization catalyst possibly acting

$$\mathbf{43} + CHBr_3 \rightarrow 2Br_2CH\cdot + 4CO + FeBr_2$$

as a radical generator (16). Such a radical process could also be invoked as a rationale for the Coffey reaction (40). Irradiation of a solution of the metal carbonyl in dibromodifluoromethane gives **44** (57).

$$\begin{matrix} Br—CF_2 \diagdown \\ \qquad Fe(CO)_4 \\ Br—CF_2 \diagup \end{matrix}$$

(**44**)

Photolysis of a mixture of $Fe(CO)_5$ and a halide bearing no easily abstractable β-hydrogens results in coupling reactions (16,55) analogous to those reported by Coffey (40) and Bruce (44). Elimination reactions

$$2RX + Fe(CO)_5 \xrightarrow{h\nu} R—R + 5CO + FeX_2$$

R = e.g., cycloheptatrien-7-yl

have been observed for those halides having readily abstractable β-hydrogens; for example, *t*-butyl bromide and $Fe(CO)_5$ form isobutylene, which reacts further to give mainly the trimers **45** and **46**. The proposed mechanism, involving initial iron incorporation into the C–Br bond, is outlined in Scheme **I** (16).

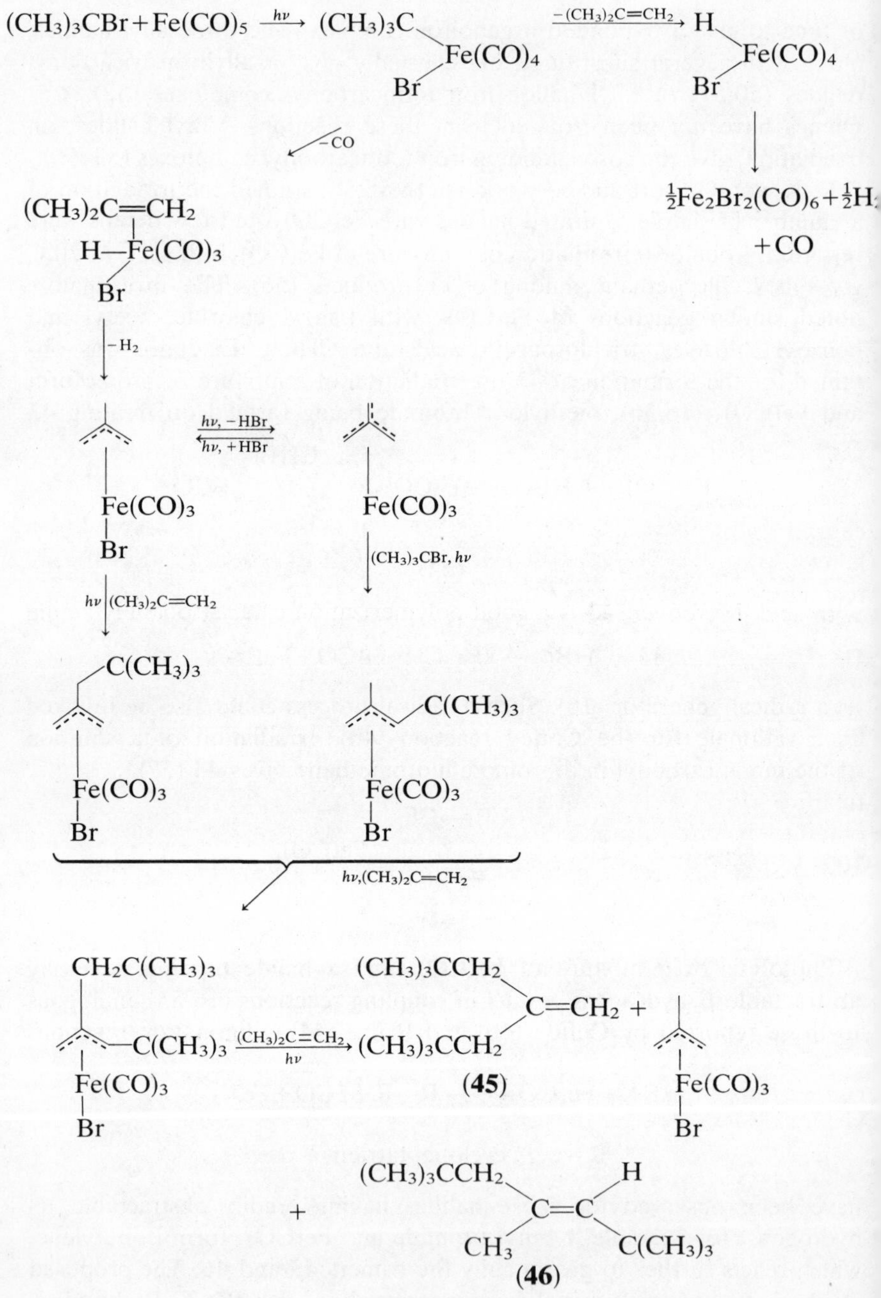

$(CH_3)_3CBr + Fe(CO)_5 \xrightarrow{h\nu} (CH_3)_3C(Br)Fe(CO)_4 \xrightarrow{-(CH_3)_2C{=}CH_2} H(Br)Fe(CO)_4 \longrightarrow \tfrac{1}{2}Fe_2Br_2(CO)_6 + \tfrac{1}{2}H_2 + CO$

$(CH_3)_3C(Br)Fe(CO)_4 \xrightarrow{-CO} (CH_3)_2C{=}CH_2 \cdot H{-}Fe(CO)_3{-}Br \xrightarrow{-H_2} (\eta^3\text{-}C_4H_7)Fe(CO)_3Br$

$(\eta^3\text{-}C_4H_7)Fe(CO)_3Br \underset{h\nu,\ +HBr}{\overset{h\nu,\ -HBr}{\rightleftharpoons}} (\text{trimethylenemethane})Fe(CO)_3$

$(\eta^3\text{-}C_4H_7)Fe(CO)_3Br \xrightarrow{h\nu,\ (CH_3)_2C{=}CH_2}$ (allyl bearing $CH_2C(CH_3)_3$ on the central carbon)$Fe(CO)_3Br$

$(\text{trimethylenemethane})Fe(CO)_3 \xrightarrow{(CH_3)_3CBr,\ h\nu}$ (methylallyl bearing $C(CH_3)_3$ on a terminal carbon)$Fe(CO)_3Br$

both $\xrightarrow{h\nu,(CH_3)_2C{=}CH_2}$

$(\eta^3\text{-allyl})(CH_2C(CH_3)_3)(C(CH_3)_3)Fe(CO)_3Br \xrightarrow[h\nu]{(CH_3)_2C{=}CH_2} ((CH_3)_3CCH_2)_2C{=}CH_2$ **(45)** $+ (\eta^3\text{-}C_4H_7)Fe(CO)_3Br$

$+ (CH_3)_3CCH_2(CH_3)C{=}CH(C(CH_3)_3)$ **(46)**

Scheme **I**

Aryl halides are generally quite unreactive with $Fe(CO)_5$. For example, iodobenzene does not react with the metal carbonyl at temperatures between 30 and 60° (46). Many reactions of $Fe(CO)_5$ with various functional groups can be carried out in the presence of an aryl halide functionality without affecting the latter group.

B. α-Haloketones

α-Haloketones react with $Fe(CO)_5$ in refluxing 1,2-dimethoxyethane to give 1,4-diketones and monoketones, products expected on the basis of some of the reactions described above (58). The much more toxic reagent, $Ni(CO)_4$, in THF, can also be used to effect this transformation (59).

C. Tin Halides

Stone and co-workers (60) reported that reaction of tri-*n*-butyltin chloride with $Fe(CO)_5$ gives the carbonyl insertion product, dibutyl ketone (among other products), in low yield. The mechanism of this reaction is not known.

D. Nitrogen Halides

A number of papers have appeared concerning the dehalogenation of fluorinated nitrogen compounds with $Fe(CO)_5$. Englin and Filatov (61) reported that reaction of tetrafluorohydrazine and $Fe(CO)_5$ in a sealed tube at −10°C gives difluorodiimide in 90% yield. The same product is obtained by reaction of nitrogen trifluoride with the metal carbonyl at 100–200°C. These reactions have been postulated to occur via the

$$CF_3{-}N(X)(Cl) + Fe(CO)_5 \xrightarrow[20°C]{CCl_4} CF_2{=}N{-}X \qquad (62)$$

X = F, Cl — X = F, 30%; X = Cl, 40%

$$XNFC(Y)FCF_2NF_2 + Fe(CO)_5 \xrightarrow{n\text{-}C_7H_{16}} NF_2CF_2CN + FN{=}CFCN + CNCN \qquad (63)$$

	NF_2CF_2CN	$FN{=}CFCN$	CNCN
X=Y=F	25.5%	20.4%	47.4%
X=Cl, Y=F	25%	17.5%	42.3%
X=F, Y=H	15%	17%	45%

$$(CF_3)_2N{-}X + Fe(CO)_5 \xrightarrow[<0°]{} CF_3N{=}CF_2 \qquad (64)$$

X = Br, Cl — X = Br, 96.4%; X = Cl, 97%

difluoronitrogen radical or fluoronitrene. Dehalogenation was also observed for halogenated primary and secondary amines, several examples of which are given below. Carbon monoxide, ferrous and ferric halides, halogen, and, in some cases, carbonyl fluoride, are by-products of these reactions. The three principal products of the halogenated ethylenediamine–metal carbonyl reaction result from successive dehalogenation processes (63). No studies have been reported on the reaction of monohaloamines with $Fe(CO)_5$, but treatment of $Ni(CO)_4$ with these compounds in ether gives ureas (carbamoyl halides were sometimes obtained in trace amounts) in low yields (65).

E. Sulfonyl and Sulfenyl Halides

A rapid and convenient method for converting sulfonyl chlorides to thiolsulfonate esters involves treating the reactant with a 1:1 mixture of $Fe(CO)_5$ and boron trifluoride etherate in *N,N*-dimethylacetamide (DMAC) or tetramethylurea (TMU) (66). The author has found dipolar

$$RSO_2Cl + Fe(CO)_5 \xrightarrow[\text{DMAC or TMU}]{BF_3 \cdot Et_2O} RSSO_2R$$

aprotic solvents such as these to be highly effective in promoting many $Fe(CO)_5$ reactions and in giving improved yields when compared to the same reactions carried out in relatively non-polar solvents such as dibutyl ether. Boron trifluoride etherate is not required for this particular reaction, but lower yields of thiolsulfonate ester result in its absence. Moderate to good yields of ester have been obtained from a variety of sulfonyl chlorides (Table 2). An intriguing aspect of this reaction is the wide variation in the temperature at which a vigorous exothermic reaction is noted; the presence of *meta* or *para* electron-attracting substituents (e.g., *m*-nitro-) on benzenesulfonyl chloride allows the reaction to take place at a much lower temperature than the reaction of parent sulfonyl chloride, while with electron-releasing *para* substituents (e.g., *p*-CH_3-), the initial exothermic reaction occurs at higher temperatures. Similar results were observed with aliphatic sulfonyl chlorides.

The proposed mechanism for this reaction is similar to that outlined for Coffey's reaction, that is, iron insertion into the sulfur–chlorine bond of the sulfonyl chloride to give **47** which can then react with additional sulfonyl chloride to form the disulfone **48** and the relatively unstable dichlorotetracarbonyliron. Alternatively, **47** may first eliminate CO to give **49** which could be converted to **48** with more reactant. Deoxygenation of **48** with more $Fe(CO)_5$ gives the α-disulfoxide which exists as the

TABLE 2
Reaction of $Fe(CO)_5$ *and* $BF_3{\cdot}Et_2O$ *with Sulfonyl Chlorides*

Sulfonyl chloride	Temp (°C).	Yield of thiolsulfonate (%)
Benzene[a]	50	71
p-Toluene	58	48
p-Methoxybenzene	63–65	41
p-Bromobenzene	30	50
m-Nitrobenzene	0	66
2-Naphthalene	60	45
Methane	63	36
n-Hexadecane	73–75	42
Trichloromethane	−20	—

[a] Diphenyl disulfide was obtained as a by-product in 4–6% yield.

thiolsulfonate ester. Several results support the proposed mechanism: (*a*)

$$RSO_2Cl \xrightarrow[-CO]{Fe(CO)_5} RSO_2Fe(Cl)(CO)_4\ (\mathbf{47}) \xrightarrow[-Fe(CO)_4Cl_2]{RSO_2Cl} RSO_2SO_2R\ (\mathbf{48})$$

$$\mathbf{47} \xrightarrow{-4CO} RSO_2FeCl\ (\mathbf{49}) \xrightarrow[-FeCl_2]{RSO_2Cl} RSO_2SO_2R\ (\mathbf{48})$$

$$\mathbf{48} \xrightarrow{Fe(CO)_5} RSOSOR \longrightarrow RSSO_2R$$

Lindner and co-workers (67) reported the preparation of **47**, $R = CF_3$, in 80% yield by reaction of the corresponding sulfonyl chloride with $Fe(CO)_5$ in *n*-heptane at temperatures below −20°C. On warming to room temperature, **47** was converted to **49**, $R = CF_3$. Solvated **49**, $R = CF_3$, could be directly obtained in 98% yield by reacting the sulfonyl chloride and $Fe(CO)_5$ in THF at −20°C; (*b*) In agreement with Alper's results, *p*-toluenesulfonyl chloride is much less reactive towards $Fe(CO)_5$ than trifluoro- or trichloromethanesulfonyl chloride and reaction in THF at room temperature for 4 days gives **50** and **48**, $R = p\text{-}CH_3C_6H_4$-(68). Lindner and co-workers do not discuss the formation of **48** ($R = p\text{-}CH_3C_6H_4$–), give its yield, or mention its separation from **50** in the

$$4\,p\text{-}CH_3C_6H_4SO_2Cl + 2Fe(CO)_5 \xrightarrow{THF} 2\,p\text{-}CH_3C_6H_4SO_2FeCl_2(THF) + 10CO + \mathbf{48},\quad R = p\text{-}CH_3C_6H_4$$

93%

(50)

experimental section of their paper; (*c*) Diphenyldisulfone (**48,** R = C_6H_5–) reacts with $Fe(CO)_5$ in hot DMAC to give the thiolsulfonate ester in 65% yield together with 7% of diphenyl disulfide (a further deoxygenated product) (66).

Group VI metal carbonyls react with sulfonyl chlorides at a slower rate than $Fe(CO)_5$. For example, it requires 90 days to obtain the tungsten analog of **47**, $CF_3SO_2W(CO)_5Cl$, in 2% yield by reaction of $W(CO)_6$ with trifluoromethanesulfonyl chloride. Reaction of the group VI carbonyls with CF_3SO_2Cl at 120°C for 12 hr gave CF_3SO_2MCl (M = Cr, Mo, W), similar to **49** (68). In the dipolar aprotic solvent, tetramethylurea, reaction of sulfonyl chlorides with $M(CO)_6$ (M = Cr, Mo, W) occurs in just 2.0–2.5 hr and gives disulfides in good yields (particularly when M = Mo), thereby providing a convenient synthesis of these compounds (69).

Disulfides may also be obtained by reaction of sulfenyl halides with metal carbonyls. Alkyl, aryl, and perfluoroalkylsulfenyl chlorides react with $Fe(CO)_5$, $Cr(CO)_6$, or $Ni(CO)_4$ in THF to give disulfides in high yield

$$M(CO)_n + 2RSCl \xrightarrow[THF]{<0°C} RSSR + MCl_2 + nCO$$

$$M = Cr, Fe, Ni\ (n = 6,5,4, \text{respectively})$$

(70). The same conversion can be effected using $Fe(CO)_5$ or $Mo(CO)_6$ in tetramethylurea at higher temperatures (34). For example, perchloro- or

$$2\,(2,4\text{-}(NO_2)_2C_6H_3SCl) + Mo(CO)_6 \xrightarrow[65\text{–}70°C,\ 4\,hr]{TMU} (2,4\text{-}(NO_2)_2C_6H_3S)_2 + MoCl_2 + 6CO$$

95%

perfluoroalkanesulfenyl chlorides react with $Fe(CO)_5$ in THF to form perhalodisulfides which then undergo partial dehalogenation, again expected on the basis of previously described reactions (70). The mechanism of the sulfenyl halide–disulfide reaction is, most likely, similar to the pathways proposed for the other halide reactions.

Treatment of 2-naphthalenesulfenylthiocyanate with $Fe(CO)_5$ in THF at low temperature gave 2-naphthyl disulfide (71).

F. Acid Halides

Filatov and Englin (72) investigated the reaction of $Fe(CO)_5$ with acid halides under quite drastic conditions. Acetyl chloride behaves like many other halides and on treatment with $Fe(CO)_5$ at 180°C, gives biacetyl in 86.5% yield. Monofluoroacetyl chlorides react with $Fe(CO)_5$ to form the corresponding acetyl fluorides in addition to the expected α-diketone.

$$\mathrm{FCHXCOCl} + \mathrm{Fe(CO)_5} \xrightarrow{180\text{–}200^\circ\mathrm{C}} \mathrm{(FCHXCO)_2} + \mathrm{FCHXCOF}$$

X = H, Cl

X = H,	42.0%	16.6%
X = Cl,	58.4%	17.1%

Only the acid fluoride is formed on reaction of highly fluorinated acid chlorides with $Fe(CO)_5$. Trifluoroacetyl fluoride decomposes on treatment with $Fe(CO)_5$ at 200°C for 6 hr to give hexafluoroethane, among other products.

$$\mathrm{Fe(CO)_5} + \mathrm{F_2CXCOCl} \xrightarrow{180\text{–}190^\circ\mathrm{C}} \mathrm{F_2CXCOF}$$

X = H, F $\qquad$ X = H, 35%; X = F, 61%

By-products of these reactions include ferrous and ferric chlorides and fluorides. The change in product distribution with increased fluorine content of the reactant is probably due to a corresponding increase in the formation of iron fluorides (as outlined below), which can then react with the acid chloride to produce the fluoride. An acyl radical intermediate, **51**, has been proposed for this reaction (72), but a mechanism involving initial iron insertion into the C–Cl bond of the acid chloride appears equally probable.

$$\mathrm{(Y)(X)(F)C{-}COCl} + \mathrm{Fe(CO)_5} \longrightarrow \mathrm{(Y)(X)(F)C{-}\overset{O}{\overset{\|}{C}}\cdot} + \mathrm{FeCl_2} + \mathrm{FeCl_2} + \mathrm{CO}$$

(51)

$$2\mathrm{Y{-}C(X)(F){-}\overset{O}{\overset{\|}{C}}\cdot} \longrightarrow \mathrm{(Y)(X)(F)C{-}\overset{O}{\overset{\|}{C}}{-}\overset{O}{\overset{\|}{C}}{-}C(X)(Y)(F)}$$

$$\mathrm{CXYF{-}COCl} + Fe(CO)_5 \longrightarrow CO + C + FeCl_2 + FeCl_3 + FeF_2 + FeF_3 + HCl(HF) + COF_2(COFCl)$$

$$\mathrm{CXYF{-}COCl} \xrightarrow{FeF_2,\ FeF_3} \mathrm{CXYF{-}COF} + FeCl_2,\ FeCl_3$$

X = Y = H; X = H, Y = Cl
X = H, Y = F; X = Y = F

Acid chlorides react with straight-chain and cyclic ethers in the presence of $Fe(CO)_5$ to form alkyl esters and alkyl chlorides (73). The metal carbonyl here functions as a Lewis acid catalyst.

V. CARBONYL INSERTION

A number of functional groups react with $Fe(CO)_5$ to give carbonyl insertion products.

A. Diazonium Salts

Schrauzer (74) showed that reaction of $Fe(CO)_5$ with greater than an equimolar quantity of aryldiazonium halide in aqueous acidic acetone at temperatures of 5–10°C gives the corresponding acid in yields ranging from 11 to 41%. Aryl halides were frequently obtained as by-products in

$$p\text{-}X\text{-}C_6H_4N_2^{\oplus}Cl^{\ominus} + Fe(CO)_5 \xrightarrow[\text{aq. acetone}]{H^+} p\text{-}X\text{-}C_6H_4COOH$$

X(Yield, %) = Cl(41), NO_2(25), CH_3(12), OCH_3(17), H(11)

low yield. Using an excess of $Fe(CO)_5$, the corresponding symmetrical ketone was obtained as the major product. Ryang (75) has noted that benzenediazonium tetrafluoroborate reacts with $Fe(CO)_5$ [or $Ni(CO)_4$] in

$$p\text{-}Cl\text{-}C_6H_4N_2^{\oplus}Cl^{\ominus} + \text{excess } Fe(CO)_5 \xrightarrow[\text{aq. acetone}]{+5°C} \underset{9\%}{p\text{-}Cl\text{-}C_6H_4COOH} + \underset{44\%}{(Cl\text{-}C_6H_4)_2CO}$$

aqueous acetone to give not only the carboxylic acid, but also biphenyl and azobenzene. Any differences in the experimental conditions to those used by Schrauzer were not stated; Scheme **II** involving ionic intermediates can account for all of the products formed in these reactions. Alternatively (or in addition), radicals may play an important role here.

$$ArN_2^{\oplus}X^{\ominus} \xrightarrow{Fe(CO)_5} ArN{=}N{-}\overset{\oplus}{Fe}(CO)_5$$

$$\downarrow -N_2$$

$$ArN{=}NAr \xleftarrow{ArN{=}N\overset{\oplus}{Fe}(CO)_5} Ar{-}\overset{\oplus}{Fe}(CO)_5 \xrightarrow{X^{\ominus}} ArX$$

$$Ar{-}\overset{\oplus}{Fe}(CO)_5 \longrightarrow ArAr$$

$$\downarrow$$

$$ArCAr\,(C{=}O) \xleftarrow{ArN_2^{\oplus}} Ar{-}\underset{\|}{\overset{}{C}}{-}\overset{\oplus}{Fe}(CO)_4\,(C{=}O) \xrightarrow{H_2O} ArCOOH$$

Scheme **II**

B. Amines

Simple primary or secondary saturated aliphatic and heterocyclic amines react with $Fe(CO)_5$ to produce formamides and the presence of water results in the formation of $HFe(CO)_4^-$ and carbamates [(76–78); see (123*a*), pp. 447–448, for a discussion of the mechanism of these reactions]. Ionic complexes were obtained by reaction of $Fe(CO)_5$ with aromatic heterocyclic amines such as pyrazole, imidazole (79), pyridine, and related compounds [(123*a*), pp. 104–107]. Reaction of $Fe(CO)_5$ and triethylamine followed by acid treatment gives $Fe_3(CO)_{12}$ (80).

C. Hydrazine

Hieber and co-workers (81) showed that hydrazine hydrate reacts with the metal carbonyl to form the carbonyl insertion product, semicarbazide. A systematic investigation of the reaction of $Fe(CO)_5$ with substituted hydrazines is currently in progress in our laboratories.

D. Organolithium Compounds

Treatment of organolithium compounds with $Fe(CO)_5$ in diethyl ether results in the formation of a variety of products depending on reaction conditions and the nature of the organolithium (82). Alkyllithiums (e.g., RLi, $R = n\text{-}C_4H_9$, $n\text{-}C_5H_{11}$) react at −60°C to give ketones. Benzhydrol and benzoin derivatives as well as aldehydes and ketones have been obtained by reaction of fairly concentrated solutions of organolithium with $Fe(CO)_5$. Some examples include:

$$n\text{-}C_5H_{11}Li + Fe(CO)_5 \xrightarrow[-60°C]{Et_2O} (n\text{-}C_5H_{11})_2CO \quad 14\%$$

$$m\text{-}CH_3C_6H_4Li + Fe(CO)_5 \xrightarrow[-60°C]{Et_2O} m\text{-}CH_3C_6H_4CHO \quad 34.2\%$$

$$o\text{-}CH_3C_6H_4Li + Fe(CO)_5 \xrightarrow[-60°C]{Et_2O} (o\text{-}CH_3C_6H_4)_2CHOH + (o\text{-}CH_3C_6H_4)_2CO$$

$$PhLi + Fe(CO)_5 \xrightarrow[-60°]{Et_2O} PhCOCH(OH)Ph + Ph_3CH + Ph\text{—}Ph$$

These reactions most likely occur via lithium acyltetracarbonylferrates formed by nucleophilic attack of the organolithium compound on one of the carbonyl carbons. Fischer and Kiener (83) isolated several of these adducts as their tetramethylammonium salts. The nucleophilic attack may

$$Li^{\oplus}R^{\ominus} + (OC)Fe(CO)_4 \longrightarrow Li[RC(=O)Fe(CO)_4]$$

$R = CH_3, Ph$

$$Li[RC(=O)Fe(CO)_4] \xrightarrow{(CH_3)_4N^{\oplus}Br^{\ominus}} (CH_3)_4N[RC(=O)Fe(CO)_4] + LiBr$$

occur specifically at one of the axial carbonyls since triphenylphosphine–iron tetracarbonyl gave a *trans* adduct on reaction with the organolithium while *trans*-bis(triphenylphosphine)iron tricarbonyl failed to react with the nucleophile (84).

Rapid addition of $Fe(CO)_5$ to a *dilute* ether solution of an aryllithium at −60°C followed by acidic workup gives aromatic aldehydes in 24.2–65.4% yields (85). Under these reaction conditions, dimerization of the

$$\text{2,5-}(CH_3)_2C_6H_3Li \xrightarrow{Fe(CO)_5} \text{2,5-}(CH_3)_2C_6H_3CHO \quad 65.4\%$$

acyltetracarbonylferrate or reaction of the complex with additional organolithium does not readily occur. To complement this reaction, *aliphatic* aldehydes were obtained in good yields by reaction of sodium tetracarbonylferrate(-II) with preferably a primary (but not benzylic) bromide and PPh_3 in THF (86).

Instead of protonating the lithium acyltetracarbonylferrate, one could treat it with an alkyl or aryl halide and obtain an unsymmetrical ketone (87). For example, benzyl phenyl ketone was obtained in 57% yield by

$$Li[RC(=O)Fe(CO)_4] \xrightarrow{R'X} R\text{—}C(=O)\text{—}R'$$

$$Li[RC(=O)Fe(CO)_4] \xrightarrow{R'C(=O)X} RC(=O)R' + RC(=O)C(=O)R'$$

the reaction of phenyllithium with $Fe(CO)_5$ followed by treatment with benzyl bromide. Unsymmetrical ketones were also obtained in modest yields, along with small amounts of α-diketones, by using an acyl rather than an alkyl or aryl halide. Carbonyl insertion products were also obtained from the reaction of lithium phenylacetylide with $Fe(CO)_5$ at $-15°C$ followed by treatment with iodine in methanol (88).

E. Grignard Reagents

There have been no reports on reactions of Grignard reagents with $Fe(CO)_5$, but Darensbourg and Darensbourg recently noted that benzyl and cyclohexylmagnesium chlorides react more slowly with chromium and tungsten hexacarbonyls than do the corresponding organolithium compounds (84,89,90).

$$M(CO)_6 + RMgX \xrightarrow{THF} (CO)_5M\text{—}C(O^{\ominus}\ \overset{\oplus}{M}gX)(R)$$

$$M = Cr, W$$

VI. DEOXYGENATION

A. *N*-Oxides, Azoxy compounds, Nitrones

Hieber and Lipp (91), in a paper concerned with the reaction of oxides of organic nitrogen, sulfur, phosphorus, and arsenic compounds with iron carbonyls to give ionic complexes, noted, without experimental details, that pyridine-*N*-oxide reacts with $Fe(CO)_5$ at high temperature to form pyridine and iron oxide. Alper and Edward (92) showed that $Fe(CO)_5$ is indeed a useful reagent for deoxygenating a variety of aliphatic,

$$R_3N \rightarrow O + Fe(CO)_5 \xrightarrow[\substack{\text{or} \\ \text{Diglyme} \\ 130\text{–}142°\text{C}}]{Bu_2O} R_3N$$

aromatic, and heterocyclic *N*-oxides. Diglyme and dibutyl ether have been used as solvents for the reaction, and the presence of lactam, olefin, and alkoxy functional groups as well as water (as a hydrate) apparently do not interfere with the reaction. The procedure involved is very simple and good yields of tertiary amine were obtained. This technique of deoxygenation is competitive with, or superior to, many others reported in the literature [e.g., sulfur dioxide (93)].

Application of this reaction to several azoxybenzenes gave azobenzenes

$$\underset{\substack{\downarrow \\ O}}{ArN{=}NAr} + Fe(CO)_5 \xrightarrow[Bu_2O]{140°\text{C}} ArN{=}NAr$$

in 65–77% yield (92). Similarly, hexafluoroazomethane was obtained in

$$\underset{\substack{\downarrow \\ O}}{CF_3N{=}NCF_3} + Fe(CO)_5 \xrightarrow{150°\text{C}} CF_3N{=}NCF_3 + C_2F_6 + CO + CO_2 + N_2 + FeF_2$$

50% yield by reaction of hexafluoroazoxymethane and $Fe(CO)_5$ under more drastic conditions (94).

α-Phenyl-*N*-phenylnitrone reacts in an analogous manner to give N-benzylideaniline (92).

$$\overset{\substack{O \\ \uparrow}}{PhCH{=}NPh} + Fe(CO)_5 \xrightarrow{Bu_2O} PhCH{=}NPh$$

A possible mechanism for these reactions involves attack of the oxide oxygen on a terminal carbon of $Fe(CO)_5$ (analogous to the organolithium reactions) followed by elimination of CO_2 and iron tetracarbonyl (which can either trimerize to $Fe_3(CO)_{12}$ or decompose to iron oxide).

$$R_3\overset{\oplus}{N}-\overset{\ominus}{O} + Fe(CO)_4 \longrightarrow R_3\overset{\oplus}{N}-O-C(=O)-\overset{\ominus}{Fe}(CO)_4 \longrightarrow R_3N + CO_2 + Fe(CO)_4$$

B. Sulfoxides

Iron pentacarbonyl is the reagent of choice for effecting deoxygenation of sulfoxides to sulfides (95). Reaction of $Fe(CO)_5$ and a dialkyl, diaryl,

$$R_2S \rightarrow O + Fe(CO)_5 \xrightarrow[\text{Diglyme or } Bu_2O]{130\text{–}135°C} R_2S$$

$$\text{2–8 hr}$$

or heterocyclic sulfoxide (or disulfoxide) in hot diglyme or dibutyl ether gives sulfides in generally excellent yields. By comparison, PPh_3 (96), PCl_3 (97), and trichlorosilane (98) are principally effective only for aromatic sulfoxides. Hexachlorodisilane can give substantial quantities of chlorosulfide as a by-product (99), whereas $LiAlH_4$ is a sluggish reagent for sulfoxide deoxygenation.

The $Fe(CO)_5$ reaction may proceed via an analogous pathway to that described for the *N*-oxide reaction. Alternatively, sulfur–iron bonded (100) or disproportionation complexes (91) may be involved.

C. Phosphine Oxides

Reaction of trialkyl or triarylphosphine oxides with $Fe(CO)_5$ in nonpolar or polar solvents at 60–140°C gives disproportionation complexes analogous to that obtained by photolysis of a mixture of $Fe(CO)_5$ and triphenylphosphine oxide at 80°C in benzene (36,58,91).

D. Nitroso Compounds

1. *C-Nitroso Compounds*

Refluxing an equimolar mixture of nitrosobenzene and $Fe(CO)_5$ in dibutyl ether gives azobenzene in 75% yield (92). Trifluoronitrosomethane reacts with $Fe(CO)_5$ [1.0/0.37 mole ratio of $CF_3NO/Fe(CO)_5$]

under pressure to give the azoxy and azo compounds. Carbon tet-

$$PhNO + Fe(CO)_5 \xrightarrow{Bu_2O} PhN{=}NPh$$

$$CF_3NO + Fe(CO)_5 \longrightarrow CF_3N{=}\overset{\overset{O}{\uparrow}}{N}CF_3 + CF_3N{=}NCF_3 + \text{others}$$

rafluoride, among other products, was obtained by using a 1.0/0.7 mole ratio of nitroso compound to metal carbonyl (94).

The mechanism proposed for the nitrosobenzene reaction is similar to that described earlier for the deoxygenations of various *N*-oxides, that is, deoxygenation, in this case, to a nitrene intermediate [probably complexed to iron (101)—see discussion on nitro compounds].

$$Ar{-}N{=}O \quad Fe(CO)_4{-}C{\equiv}O \longrightarrow Ar{=}\overset{\oplus}{N}{-}O{-}\underset{\underset{O}{\|}}{C}{-}\overset{\ominus}{Fe}(CO)_4$$

$$ArN{=}NAr \xleftarrow{Fe(CO)_5} \underset{\underset{O}{\downarrow}}{ArN{=}NAr} \xleftarrow{ArNO} Ar{-}\ddot{N}{:} + CO_2 + Fe(CO)_4$$

(ArNO also converts the intermediate directly to the azoxy compound.)

Rather interesting behavior was exhibited by nitrosophenols towards $Fe(CO)_5$ (92). Although *p*-nitrosophenol exists predominantly (83–86% in dioxane solution) in the quinone–monoxime tautomeric form, it reacts with $Fe(CO)_5$ in refluxing dioxane to give 4,4′-dihydroxyazobenzene as the only product in 41% yield. No *p*-benzoquinone, which would be formed by deoximation of the major tautomer, was detected even when the reaction was conducted in the presence of catalytic quantities of boron trifluoride etherate (37). 5-Isopropyl-2-methyl-1,4-benzoquinone (**55**) was obtained in low yield (11%) from the reaction of 5-methyl-4-nitroso-2-isopropylphenol with $Fe(CO)_5$ in dibutyl ether. This nitrosophenol exists almost completely in the quinone-monoxime form (**53**) and yet gives **54** as the major product (18% yield). The quinone was the major product when a catalytic amount of boron trifluoride etherate was present in the reaction mixture. The formation of amine rather than azo compound may be due to the low concentration of the nitrosophenol tautomer making hydrogen abstraction from the solvent by the nitrene [free or complexed], or its precursor, the predominant process [i.e., azoxy compound formation cannot compete]. These results seem to indicate that the rate of deoxygenation of the nitrosophenol is much greater than the rate of deoximation of the oxime. The appropriate experiments to confirm or refute this idea are in progress.

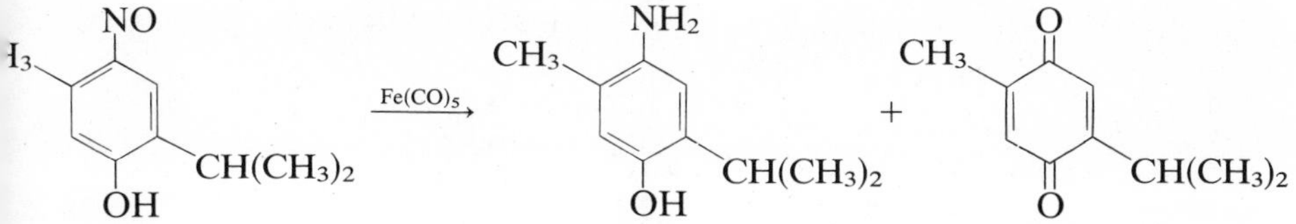

2. *N-Nitroso Compounds*

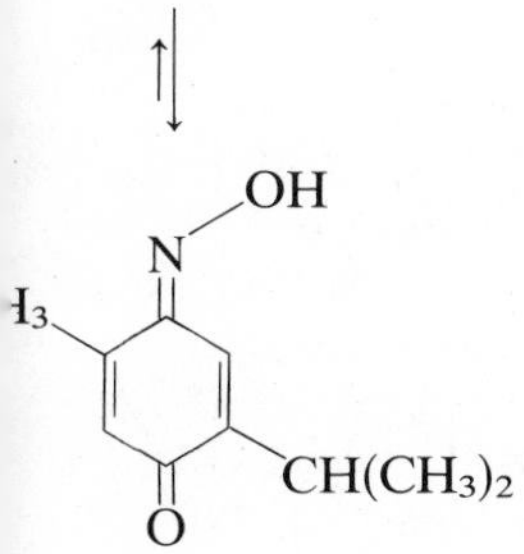

Treatment of $Fe(CO)_5$, in dibutyl ether, with an approximately equimolar quantity of a nitrosamine having at least one aryl group gave high yields of secondary amines (92). The initial steps for the proposed mechanism of this reaction, azamine formation (**56**, free or complexed), are analogous to the proposed mechanisms for the other deoxygenations. Dimerization of **56** gives **57** which can then decompose at the reaction temperature to nitrogen and secondary amino radicals. Hydrogen abstraction from the solvent by the radical gives the amine. Group VI metal carbonyls behaved similarly, but no CO_2 was evolved in these reactions

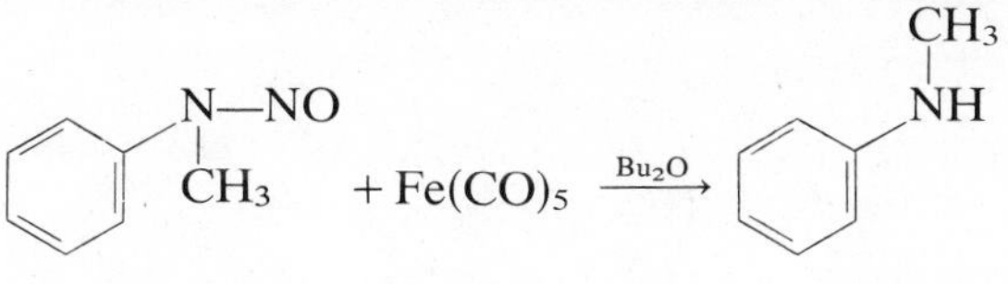

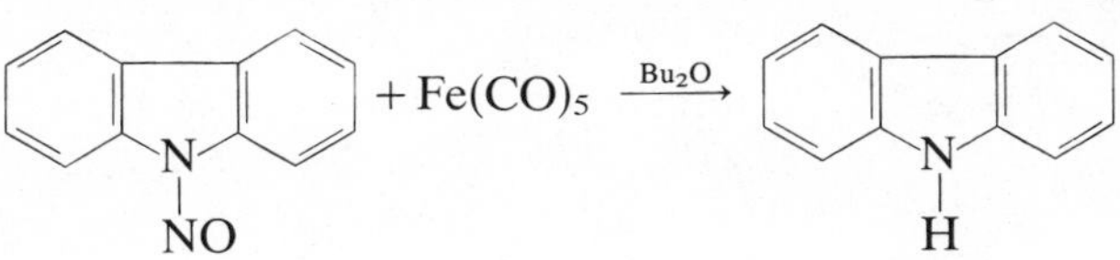

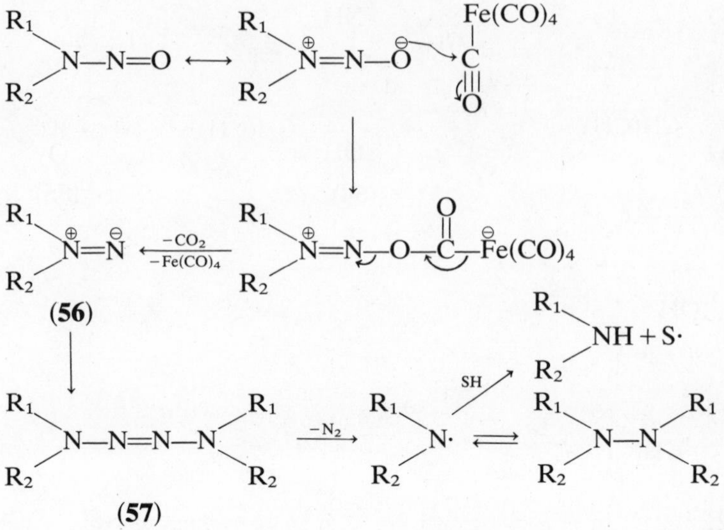

(56)

(57)

(102). Using isooctane as the solvent for the $Fe(CO)_5$ reaction gave the hydrazine as the major product.

Piperidine (6.5%) and its *N*-formyl derivative (10%) were obtained in low yields by treatment of *N*-nitrosopiperidine with $Fe(CO)_5$ in dibutyl ether (102). The major product (48%) was the tetrasubstituted urea, carbonyl dipiperidine. The corresponding carbonyl insertion product was formed in 57% yield from *N*-nitrosomorpholine thereby demonstrating the usefulness of $Fe(CO)_5$ as a reagent for preparing tetrasubstituted ureas from nonaromatic nitrosamines. Group VI metal carbonyls, by contrast, react with nonaromatic nitrosamines to give amine-metal carbonyl complexes.

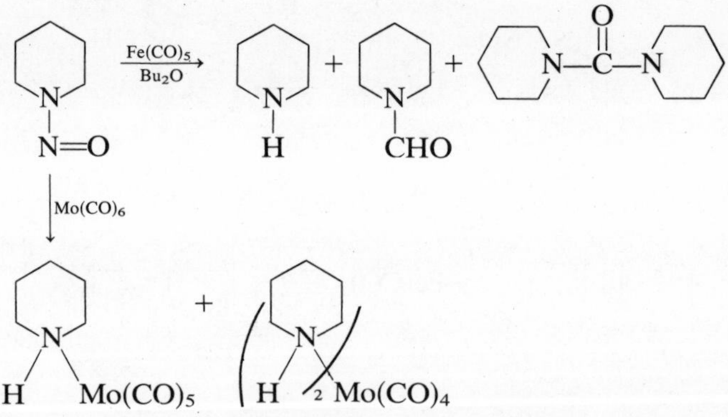

E. Nitro Compounds

1. *Aromatic Nitro Compounds*

Ultraviolet or ^{60}Co γ-radiation of various nitrobenzenes and $Fe(CO)_5$ gave the partially deoxygenated monomeric or dimeric nitrosobenzene–iron complexes (103–105). A crystal structure study of one of the dimeric complexes indicated the novel structure **58** (106). No reports have

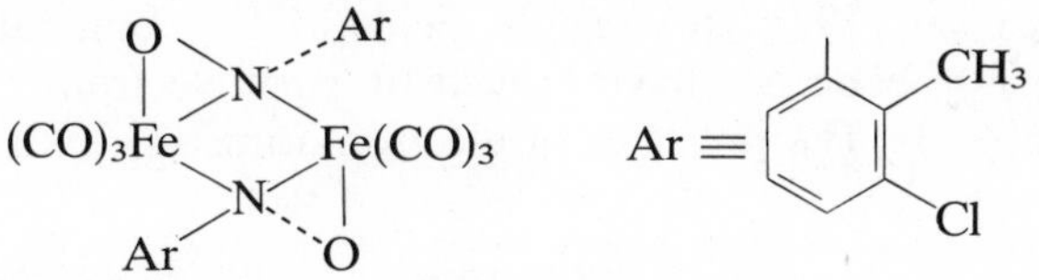

(58)

appeared on the successful generation of the free nitroso compound from the complex—a reaction that would make the deoxygenation procedure useful to the organic chemist. It should be noted that azobenzene was obtained as a low-yield by-product from ultraviolet irradiation of a mixture of $Fe(CO)_5$ and nitrobenzene (105).

Thermal reaction of $Fe(CO)_5$ and aromatic nitro compounds produces azo and azoxy compounds, amines, or complexes, depending both on the reagent concentration and on the nature of the nitro compound (92). When equimolar quantities of nitrobenzene and $Fe(CO)_5$ were refluxed in dibutyl ether, azoxybenzene was obtained in 64% yield and azobenzene was formed in small amounts (3%). Azobenzene was obtained in 76% yield by using a 1.5/1.0 mole ratio of metal carbonyl to nitrobenzene.

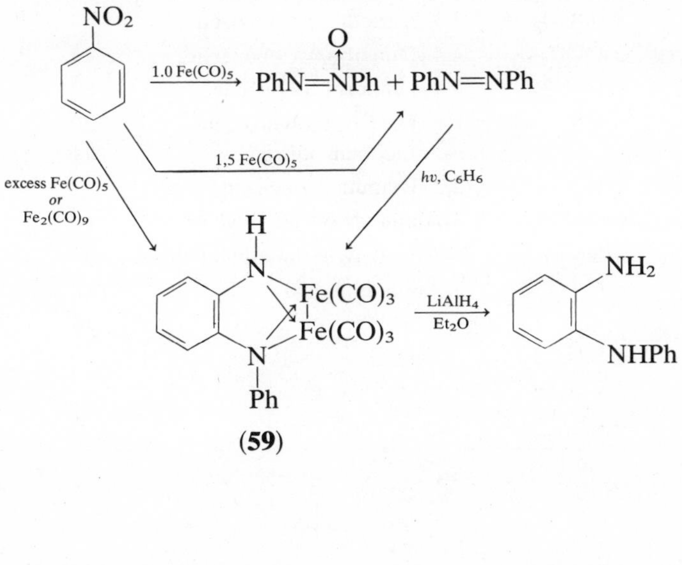

(59)

Treatment of nitrobenzene with excess $Fe(CO)_5$ [alternatively, $Fe_2(CO)_9$ could be used] gave **59** in 54–56% yield. The same complex was prepared by Pauson and co-workers (107) by photolysis of a benzene solution containing azobenzene and $Fe(CO)_5$ (108). Decomplexation of **59** with $LiAlH_4$ gives *o*-semidine (107), thus enabling reduction of the nitro group and *ortho* substitution to occur by a simple two-step process.

The reactions of a series of substituted nitrobenzenes with $Fe(CO)_5$ [1.5/1.0 mole ratio of $Fe(CO)_5$/nitro compound] were investigated by Alper and Edward (92). Reasonable yields of azo compound were obtained from nitrobenzenes having *meta* or *para* electron-releasing substituents (Table 3). The presence of electron-attracting *para* substituents

TABLE 3
Products from the Reaction of Substituted Nitrobenzenes (1.0 mole) with $Fe(CO)_5$ *(1.5 mole) in Dibutyl Ether*

Substituent	Product	Yield (%)
H	Azobenzene	76
o-CH_3	2,2′-Dimethylazobenzene	67
	o-Toluidine	13
m-CH_3	3,3′-Dimethylazobenzene	48
	m-Toluidine	6
p-CH_3	4,4′-Dimethylazobenzene	67
o-C_2H_5	2,2′-Diethylazobenzene	39
	o-Ethylaniline	35
o-OCH_3	*o*-Methoxyaniline	74
m-OCH_3	3,3′-Dimethoxyazobenzene	37
p-OCH_3	4,4′-Dimethoxyazobenzene	43
p-F	4,4′-Difluoroazobenzene	67
p-CN	4,4′-Dicyanoazobenzene	20
	p-Aminobenzonitrile	11
o-NO_2	2,2′-Diaminoazobenzene	11
m-$N(CH_3)_2$	3-Amino-*N,N*-dimethylaniline	60
p-$N(CH_3)_2$	4-Amino-*N,N*-dimethylaniline	77
o-C_6H_5	2-Aminobiphenyl	58
	Carbazole	15
p-OH	—	—
p-COOH	—	—
p-CHO	—	—

substantially reduced the yield of azo compound and gave considerable amounts of amine (e.g., *p*-CN). In several instances, such substituents inhibited reaction completely (*p*-OH, *p*-COOH). The metal carbonyl can act as a Lewis acid towards the dimethylamino group (*meta* or *para*) and thus alter its electronic properties; hence the aniline was obtained in these cases. The presence of *ortho* substituents (CH_3, C_2H_5, OCH_3, NO_2, C_6H_5) usually resulted in formation of amines or mixtures of amines and azo compounds. Treatment of 2-nitrobiphenyl with $Fe(CO)_5$ gave some insertion product, carbazole, with 2-aminobiphenyl being the major reaction product. Another cyclization product, benzo[*c*]cinnoline, was obtained in good yield from reaction of 2,2′-dinitrobiphenyl and $Fe(CO)_5$.

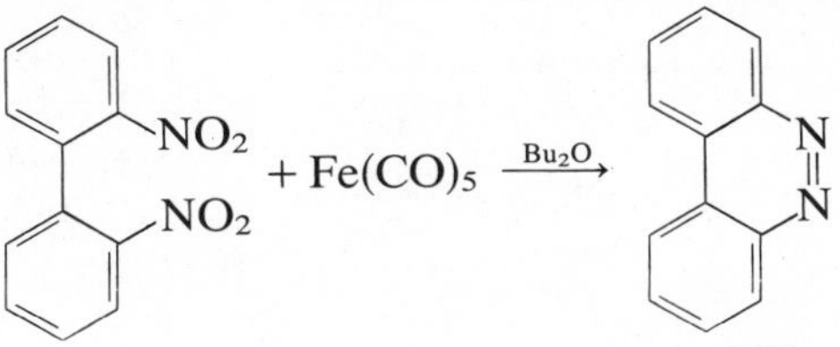

NO_2 / NO_2 + $Fe(CO)_5$ $\xrightarrow{Bu_2O}$ N=N

65%

1-Nitronaphthalene gave the corresponding amine in 70% yield by reaction with $Fe(CO)_5$ but only traces of 1,8-diaminonaphthalene were obtained from 1,8-dinitronaphthalene. The metal carbonyl is not selective towards the nitro or amine oxide functionalities since 4-nitropyridine-1-oxide gave 4,4′-azoxypyridine (59) in 22% yield along with traces of **60** and **61**. However, as described earlier, $Fe(CO)_5$ reacts with the sulfonyl chloride group in the presence of, and without affecting, a nitro group (66).

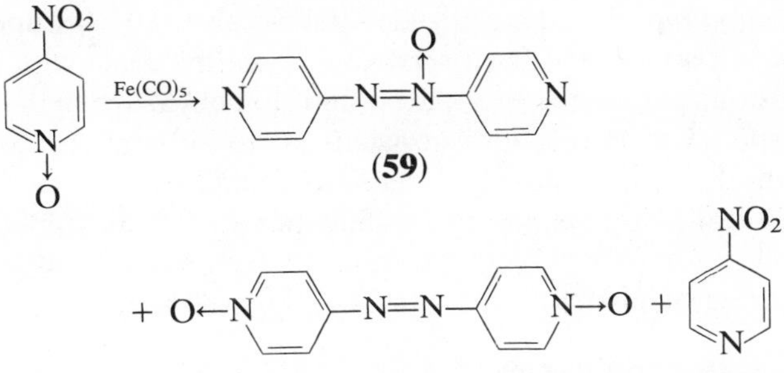

NO_2 … N→O $\xrightarrow{Fe(CO)_5}$ N … N=N(→O) … N

(**59**)

+ O←N … N=N … N→O + NO_2 … N

(**60**) (**61**)

The azoxy and azo compounds obtained in these reactions may arise via the pathways outlined for the deoxygenation of nitrosobenzene, assuming initial deoxygenation of the nitro to the nitroso function [It is conceivable that the azoxybenzenes may also be formed from complexes of the type obtained by von Gustorf et al. (104,106)]. The nitrene may be free or complexed as **62** or **63**. Loss of $Fe(CO)_3$ from **62** gives **63**. Collapse of **63** under the reaction conditions would give the azo compound. If an *ortho* substituent is present on the benzene ring, then Ar–carbonyl repulsion in **62** and Ar–Ar repulsion in **63** would be relieved by formation of the sterically more favorable **64** which can then be converted to the amine as outlined. Should the nitrene be free, the presence of electron-attracting groups on the benzene ring would destabilize the electron-deficient nitrene so that it abstracts hydrogen from solvent more rapidly (amine formation) and hence is less available for coupling reactions.

(CO)$_3$Fe, Fe(CO)$_3$, Ar—N, N—Ar, Fe(CO)$_3$ **(62)** $\xrightarrow{-Fe(CO)_3}$ Ar, Ar, N—N, (CO)$_3$Fe—Fe(CO)$_3$ **(63)**

(63) $\xrightarrow{\Delta}$ ArN=NAr

(63) $\xrightarrow{\text{solvent}}$ Ar, H, Ar, H, N, N, (CO)$_3$Fe- - - -Fe(CO)$_3$ **(64)** $\longrightarrow$ ArNHFe(CO)$_3$ $\xrightarrow{\text{solvent},\ -Fe(CO)_3}$ ArNH$_2$

Complexes of types **62–64** have been obtained from reaction of aliphatic nitro compounds with $Fe(CO)_5$ [see below, also (101)]. Complexes of type **62** were also isolated from reactions of methyl azide with $Fe_2(CO)_9$ (101) and from photolysis of $Fe(CO)_5$ and diphenyldiazomethane (108); those of type **63** from reactions of azides (101,110) and azo compounds (111) with $Fe_2(CO)_9$ or rigid *cis*-azo compounds with $Fe(CO)_5$ [(112,113)—note discrepancy in melting points for the reported benzo[*c*]cinnoline complex (110,112)]; those of type **64** isolated from reactions of methyl or phenyl azide with $Fe_2(CO)_9$ (101).

2. Nonaromatic Nitro Compounds

Aliphatic and cycloalkyl nitro compounds react with $Fe(CO)_5$ in either diglyme or di-*n*-butyl ether to give the carbonyl insertion products, for-

$$RNO_2 \xrightarrow{Fe(CO)_5} RNHCHO + RNHC(=O)NHR$$

mamides or ureas (34). Sterically small alkyl groups such as n-propyl give formamides as the major product while the presence of bulky groups (e.g., $R = (CH_3)_3C$, 1-adamantyl) results in predominant urea formation. Scheme **III** can satisfactorily account for the products and for the observed steric effects. Of the four polynuclear iron complexes, **67** is the

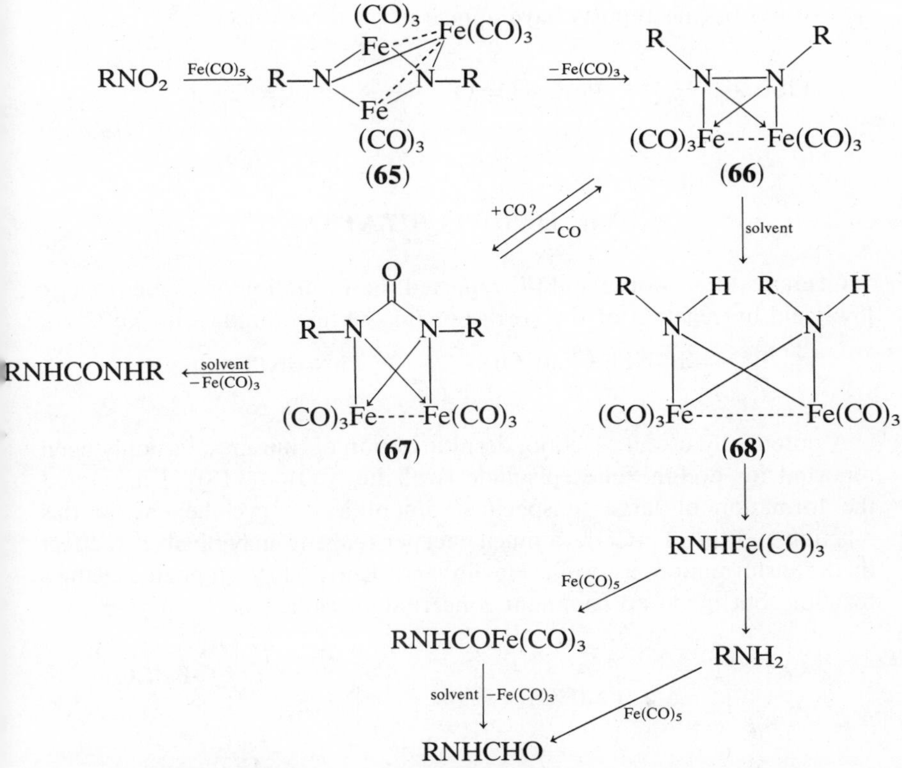

$RNO_2 \xrightarrow{Fe(CO)_5}$ (**65**) $\xrightarrow{-Fe(CO)_3}$ (**66**) $\xrightarrow{solvent}$ (**68**) $\rightarrow RNHFe(CO)_3$

(**66**) $\underset{+CO?}{\overset{-CO}{\rightleftharpoons}}$ (**67**) $\xrightarrow[-Fe(CO)_3]{solvent}$ RNHCONHR

$RNHFe(CO)_3 \xrightarrow{Fe(CO)_5} RNHCOFe(CO)_3 \xrightarrow[-Fe(CO)_3]{solvent} RNHCHO$

$RNHFe(CO)_3 \rightarrow RNH_2 \xrightarrow{Fe(CO)_5} RNHCHO$

Scheme **III**

most sterically favored and thus when R is bulky, the urea complex will be the major constituent [shown with $Fe_2(CO)_9$] and will give urea [also some formamide] on decomplexation. Complexes of type **67** were also obtained by reaction of $Fe_2(CO)_9$ with methyl azide (101), nitromethane (101), and isocyanates (114). When R is relatively small, reaction with $Fe_2(CO)_9$ in benzene at room temperature gives **65** as the predominant product. The formamide was produced by heating **65** in diglyme or dibutyl ether [**66** and **68** are intermediates in this process]. Aliphatic amines are much more basic than aromatic amines and thus will react more readily with the Lewis acid, $Fe(CO)_5$, to give the formamide (See carbonyl insertion: amines).

F. Ketenes

Ketenes, which themselves are preparable by carbonyl insertion in reaction of diazo compounds with $Fe(CO)_5$ or preferably $Ni(CO)_4$ (115), deoxygenate when photolyzed with $Fe(CO)_5$ in benzene to give the complex **69** (116). Interesting products may be formed on decomplexation of **69**, but no reports have appeared on this subject.

$$Ph_2CN_2 \xrightarrow[Fe(CO)_5]{Ni(CO)_4\ or} Ph_2C{=}C{=}O \xrightarrow[h\nu]{Fe(CO)_5} Ph_2C{=}C\begin{matrix}Fe(CO)_4\\ |\\ Fe(CO)_4\end{matrix}$$

(69)

VII. DESULFURIZATION

Otsuka and co-workers (117) reported the formation of azobenzene in low yield by reaction of the corresponding sulfur diimide with $Fe(CO)_5$.

$$PhN{=}S{=}NPh + Fe(CO)_5 \xrightarrow[20\ hr\ reflux]{cyclohexene} \underset{19\%}{PhN{=}NPh} + \underset{trace}{PhNH_2}$$

A potentially useful reaction, desulfurization of thiiranes, has only been reported for cyclohexene episulfide [with $Fe_3(CO)_{12}$ (118)]. King noted the formation of large (unspecified) quantities of cyclohexene in this reaction and thus $Fe(CO)_5$, a much cheaper reagent, may be able to effect this transformation as well. Havlin and Knox (119) repeated King's reaction, but made no comment concerning cyclohexene formation.

$$C_6H_{10}S + Fe_3(CO)_{12} \longrightarrow C_6H_{10} + S_2Fe_3(CO)_9$$

(CO)$_3$Fe---Fe(CO)$_3$, S, S, Fe(CO)$_3$

(**30**) 10%

Desulfurization of thiophene occurs on using $Fe(CO)_5$ or $Fe_3(CO)_{12}$ to give a ferracyclopentadiene complex (120).

VIII. MISCELLANEOUS REACTIONS

A. Acetylenes and Nitriles

Chapters by Hubel and by Hoogzand and Hubel in Vol. 1 (123*a*) describe the cyclization reactions of acetylenes and benzonitrile with $Fe(CO)_5$ and other carbonyls.

B. Peroxides

Iron pentacarbonyl reacts with di-*t*-butyl peroxide to give the iron(III) alkoxide (121) and thus reactions of other functional groups with $Fe(CO)_5$ probably cannot be carried out in the presence of the peroxide group.

C. Deoxygenation of a Carbonyl Group of $Fe(CO)_5$

Wannagat and Seyffert (122) reported that treatment of the sodium salt of 1,1,1,3,3,3-hexamethyldisilazane with $Fe(CO)_5$ gives hexamethyldisiloxane and an iron carbonyl complex. This is the first example of a

$$Fe(CO)_5 + NaN[Si(CH_3)_3]_2 \rightarrow [(CH_3)_3Si]_2O + [Fe(CO)_4(CN)]Na$$

reaction of $Fe(CO)_5$ in which substitution of the carbonyl oxygen occurs without Fe–C bond cleavage.

D. Carbon Dioxide Extrusion

Rosenblum and Gatsonis (123) reported an elegant preparation of cyclobutadiene–iron tricarbonyl, a precursor to cyclobutadiene, by photolysis of α-pyrone and $Fe(CO)_5$ in ether. Intriguing possibilities arise for reaction of substituted pyrones and other similar systems with $Fe(CO)_5$.

IX. SUPPLEMENT

This supplement furnishes more recent information on the correspondingly numbered and entitled sections and subsections in the main text. In addition, a few new topics are discussed.

I. ISOMERIZATION

A. Conjugated Dienes

Reaction of monoterpenes with $Fe(CO)_5$ gives diene–iron tricarbonyl complexes. The latter, in several instances, undergo double-bond and skeletal rearrangement (e.g., by thermolysis) to afford other complexes, from which the organic ligand can be liberated by using ceric ion in ether (124). All *trans*-α-sinesal was synthesized in 51% yield by reacting the readily available Δ^9-*cis*-tetraene isomer first with $Fe(CO)_5$, followed by decomplexation of the resulting complex with $FeCl_3$ (125).

Additional examples of the $Fe(CO)_5$ effected isomerization of substituted 1,4-cyclohexadienes (e.g., ethers, acids, esters) to conjugated 1,3-diene-iron tricarbonyl complexes have been published by Birch and co-workers (126,127). Ferric chloride in ethanolic HCl, among other reagents, was used to generate the conjugated diene from the complex.

The prebullvalene **70** reacts with $Fe(CO)_5$ in hot dibutyl ether to give the *cis*-9,10-dihydronaphthalene diiron hexacarbonyl complex **71** as the major product (36.5% yield). Treatment of **71** with three equivalents of ceric ion afforded the iron tricarbonyl complex **72** in 54% yield and **73** in 17% yield. An essentially quantitative yield of the naphthalenic ether (**73**) was realized by using a 6:1 ratio of Ce^{4+} to **71**. Thus, ceric ion treatment here results not only in decomplexation but in further oxidation to the aromatic system (128).

$Fe(CO)_5$ → **(70)** → **(71)** [Fe(CO)₃ complexes]; Ce^{+4} → **(72)** + **(73)**

A π-allylhydro iron tricarbonyl intermediate (e.g., **24**) occurs in the isomerization of non-conjugated dienes to conjugated diene–iron tricarbonyl complexes (22). Such species are also involved in the $Fe(CO)_5$ catalyzed isomerization of allyl to vinyl ethers (129), and in the $Fe_3(CO)_{12}$ catalyzed isomerization of 3-ethyl-1-pentene (130). The $Fe(CO)_5$ catalyzed photoisomerization of *N*-allylamides to 2-propenyl derivatives may possibly occur via this intermediate as well (131).

B. Nonconjugated Dienes

Corey and Moinet (131*a*) utilized the isomerization of nonconjugated to conjugated dienes, via a diene-$Fe(CO)_3$ complex, in a prostaglandin synthesis. 1,4-Dihydropyridines readily isomerize to 1,3-dihydropyridines by treatment first with iron carbonyls, and then with trimethylamine oxide (131*b*).

C. Vinylcyclopropanes

Irradiation of equimolar quantities of vinylcyclopropanes and $Fe(CO)_5$ in benzene or hexane solution results in the formation of cyclohexenones (in low to moderate yields) via a 1,5-carbonyl insertion process (132*a*). In one instance, an intermediate diene–iron tricarbonyl complex was tentatively identified and subsequently oxidatively degraded to the cyclohexenone. However, the reaction may occur via a π-allyl complex of the type obtained by Aumann (133) by low temperature irradiation of vinylcyclopropane and $Fe(CO)_5$.

D. Reagents Used for Generating a Free Diene from the Corresponding Iron Tricarbonyl Complex

Ferric chloride in ethanolic HCl has been reported to be generally superior to $FeCl_3$ in ethanol for liberation of the organic ligand from a diene–iron carbonyl complex (126,134). Several papers, in addition to those noted above, recorded the use of ceric ammonium nitrate for decomplexation (135,136). Employment of lead tetraacetate or thallium trifluoroacetate as decomplexating reagents results in further oxidation (126). Shvo and Hazum (137) showed that trimethylamine oxide (in benzene or acetone) can liberate dienes, including oxidation-sensitive organic ligands, in reasonable yields. Deoxygenation of the reagent to trimethylamine occurs, and thus this reaction is related to the previously reported deoxygenation of amine oxides by $Fe(CO)_5$ (92).

II. DEHYDRATION AND DEHYDROSULFURATION

A series of novel bi- and trinuclear complexes have been isolated from the reaction of thioureas and thionamides with $Fe_2(CO)_9$ (138). It is conceivable that one or more of these types of complexes are intermediates in the reactions of thioamides and 1,1-diphenyl-2-thiourea with $Fe(CO)_5$.

III. DEOXIMATION

Treatment of amide oximes with $Fe(CO)_5$ in THF affords amidines in high yields. The deoximation of ketoximes by $Fe(CO)_5$ (37) proceeds via an imine intermediate (138*a*).

IV. REACTIONS OF $Fe(CO)_5$ WITH HALIDES—COUPLING AND DEHALOGENATION

A. Alkyl, Allyl, Vinyl, and Aryl Halides

Methyl iodide does not react with $Fe(CO)_5$ at room temperature (139), although ethane is formed at elevated temperatures (140). In addition, ethylene was obtained from $Fe(CO)_5$ and 1,2-dichloroethane (140). Irradiation of dibromodifluoromethane with $Fe(CO)_5$ gives tetrafluoroethylene and **44**, among other products (141).

Support for the proposed mechanism for conversion of *gem*-dihalides to olefins (with decarbonylation of **37** occurring prior to reaction with additional *gem*-dihalide) was obtained by investigation of the reactions of dichlorodiphenylmethane and 1,2-dichloro-1,1,2,2-tetraphenylethane with $Fe(CO)_5$ at 5°C (142).

1,2-Oxazet-3-enes are formed by $Fe(CO)_5$ induced partial dehalogenation of perhalo-1,2-oxazetidines (143). Treatment of perfluorodihydrodiazete with $Fe(CO)_5$ at 150–180°C affords tetrafluoroethylene (41.5–50.0% yield) and dicyanogen (32–50%) (144).

B. α-Haloketones

A detailed study of the reactions of primary, secondary, and tertiary aryl and alkyl α-bromoketones with $Fe(CO)_5$ was published (142). Use of 1,2-dimethoxyethane as reaction solvent, and subsequent aqueous work-up, generally gives the 1,4-diketone as the major product and a monoketone as the by-product (β-epoxy ketones were also isolated in several instances). α-Deuteriomonoketones could be obtained by using D_2O for reaction workup. The reaction apparently occurs via organoiron

$$4\text{-PhC}_6\text{H}_4\text{C(O)CH}_2\text{Br} + Fe(CO)_5 \xrightarrow[\text{then } H_2O]{(CH_3OCH_2)_2} [4\text{-PhC}_6\text{H}_4\text{C(O)CH}_2]_2 + 4\text{-PhC}_6\text{H}_4\text{C(O)CH}_3$$

63%   30%

tetracarbonyl halide and organoiron halide complexes structurally similar to the intermediates of the *gem*-dihalide and sulfonyl halide–$Fe(CO)_5$ reactions. It is possible to apply the reaction to the preparation of unsymmetrical 1,4-diketones by first isolating the intermediate organoiron bromide [using $Fe_2(CO)_9$].

Iron pentacarbonyl, or better $Fe_2(CO)_9$, reacts with α,α'-dibromoketones and either 1,3-dienes (145) or enamines (146) to give good yields of 4-cycloheptenones and 2-cyclopentenones, respectively. The metal carbonyl can also effect coupling of α,α'-dibromoketones with aryl substituted olefins (147), amides (148), and five-membered ring heterocycles (149–151). These reactions represent novel approaches to the synthesis of tropane alkaloids, furanones, and cyclopentanones. The mechanism proposed for these cyclization processes is initially similar to that found for the α-bromo ketone–$Fe(CO)_5$ reaction (152).

Stereospecific dehalogenation of organic halides can be effected by $Fe(CO)_5$ and KOH (152*a*).

D. Nitrogen Halides

Dehalogenation of bis(trifluoromethyl)amine oxide by $Fe(CO)_5$ leads to perfluoroazapropene (153).

E. Sulfonyl and Sulfenyl Halides

Further evidence was obtained for the intermediacy of **47** and **49** in the RSO_2Cl-$Fe(CO)_5$ reactions (142).

F. Acid Halides

Biacetyls are produced, among other products, from the high-temperature (190–200°C) reactions of acetic, trifluoroacetic, and difluoroacetic anhydrides with $Fe(CO)_5$ (154). These results are similar to those reported for using acid halides as the substrate (72).

G. Benzohydroxamoyl Chlorides

Iron pentacarbonyl is a useful reagent for converting benzohydroxamoyl chlorides to nitriles (155). The reaction probably occurs by initial oxidative addition [directly or via a π-complex] to give an iron tetracarbonyl complex, collapse of which affords a nitrile oxide. The nitrile then results by deoxygenation of the nitrile oxide in the presence of more $Fe(CO)_5$.

V. CARBONYL INSERTION

D. Organolithium Compounds—Acyltetracarbonylferrates (155*a*)

Sodium tetracarbonylferrate, generated from $Fe(CO)_5$ and sodium amalgam in THF, can act as a source of acyltetracarbonylferrates. Collman and co-workers found that stereospecific carbonylation of bromides, iodides, or tosylates with *in situ* generated $Fe(CO)_4^{2-}$ gives ketones in reasonable yields (156). These transformations can be effected in a variety of ways (157, 158, 158*a*) including carbonylation of the alkyltetracarbonylferrate and exposure of the resultant acyl tetracarbonylferrate to a different halide [procedure of ref. 87]. γ-Ethylenic bromides or tosylates can be converted into cyclohexenones by using $Fe(CO)_5$ and Na/Hg, and a β-allenic bromide was reported to be transformed into a 2-cyclopentenone in 30% yield by using the same reagents (159).

Halogenation, in ethanol, of an acyltetracarbonylferrate gives ethyl esters (160,161). In certain instances, isomerization of the acyltetracarbonylferrate intermediate can be important. Halogenation (or oxidation) in aqueous media of either alkyl or acyltetracarbonylferrates (derived from primary or secondary halides and tosylates) affords aliphatic carboxylic acids in good yields (161). Hemifluorinated ketones were synthesized in variable yields by treatment of perfluoroacid chlorides with alkyl or acyltetracarbonylferrates (162), a reaction which only occurs in low–moderate yields for nonfluorinated acid chlorides (87)

Acid chlorides react with $Fe(CO)_4^{2-}$ to again give acyltetracarbonylferrates, which upon acidification with acetic acid, affords aliphatic or aromatic aldehydes in good yields (163). Use of phthaloyl dichloride as the reactant acid halide results in formation of biphthalidylidene in 23% yield (164). Fairly good yields of aldehydes and aldehydic acids can be realized by treating acid anhydrides first with $Fe(CO)_4^{-2}$, and then with HCl or acetic acid (165).

E. Grignard Reagents

Darensbourg and co-workers (166) reported the formation of an acyltetracarbonylferrate from treatment of benzylmagnesium chloride with $Fe(CO)_5$.

F. Strained Olefins

Iron pentacarbonyl can induce dimerization and carbonyl insertion of strained olefins under thermal or photolytic conditions (167,168). For instance, Mantzaris and Weissberger (167) found that thermal reaction of benzonorbornadiene (**74**) with $Fe(CO)_5$ gave **75** in 57% yield and **76** in 27% yield. Using very similar reaction conditions, Wege and co-workers (168) also observed the formation of **75** in 57% yield, but dimer **76** was isolated in only 11% yield. However, only the carbonyl insertion product was obtained by irradiation of **74** and $Fe(CO)_5$. A metallocycle intermediate (168,169) was proposed for this highly stereospecific carbonyl insertion process (170).

(**74**) $\xrightarrow[\Delta]{Fe(CO)_5}$ (**75**) + (**76**)

VI. DEOXYGENATION

A. *N*-oxides, Azoxy Compounds, Nitrones, Nitroxyl Radicals, and Hydroxylamines.

Reaction of stable nitroxyl radicals with $Fe(CO)_5$ in refluxing benzene affords air and moisture sensitive nitroxyl–iron compounds lacking carbonyl groups. Deoxygenation of nitroxyl radicals to amines can be attained with the *in situ* generated hydridoundecacarbonyltriferrate anion (171).

Bis(trifluoromethyl)amine was obtained by $Fe(CO)_5$ induced deoxygenation of bis(trifluoromethyl)hydroxylamine (153).

B. Sulfoxides, Sulfines

Treatment of diaryl sulfines with $Fe(CO)_5$ gives thiobenzophenones in good yield (171*a*).

D. Nitroso Compounds

2. *N*-Nitroso Compounds

A study of the reaction of $Fe(CO)_5$ with several benzylnitrosamines indicates that tetrazenes (**57**) are not involved in the conversion of nitrosamines to amines (172).

E. Nitro Compounds

1. Aromatic Nitro Compounds

Landesberg and co-workers (173) showed that nitroarenes can be reduced to anilines by $Fe_3(CO)_{12}$ and methanol $[HFe_3(CO)_{11}^-]$. A variety of functionalities [e.g., ketone, ester] can be tolerated and this useful reaction is believed to proceed initially via the same pathway as that described for the deoxygenation of *N*-oxides (92).

Although complexes of structural type **63** were obtained by reaction of iron carbonyls with azo compounds (111–113) under mild conditions, azodicarboxylic acid diisopropyl ester reacts with $Fe(CO)_5$ at elevated temperatures to give 2,3-dimethylbutane in 93% yield (174).

2. Nonaromatic Nitro Compounds

A paper has appeared regarding the reactions of nonaromatic nitro compounds with iron carbonyls (175, also see text).

VII. DESULFURIZATION

Desulfurization of thiiranes by $Fe_2(CO)_9$ at 80°C is a stereospecific process (176). This reaction can also be effected at low temperatures [15°C, $Fe_2(CO)_9$] or with $Fe(CO)_5$ (177).

Olefins can be prepared by desulfurization of thionocarbonates while using $Fe(CO)_5$ in hot xylene. The reaction is not stereospecific (178). Carbene complexes were also isolated from several of these reactions (179).

Good yields of hydrocarbons and amines result by reaction of thioketones and thioamides, respectively, with $Fe(CO)_5$ and KOH (179*a*).

VIII. MISCELLANEOUS REACTIONS

A. Acetylenes and Nitriles

Irradiation of cyclooctyne in $Fe(CO)_5$ or with $Fe(CO)_5$ in benzene affords modest yields of carbonyl insertion products [e.g., benzoquinone derivatives]. Cyclotrimerization occurred when thermal conditions were used (180).

Treatment of diaryl acetylenes with $Fe(CO)_5$ and diphenylketene in hot mesitylene gives reasonable yields of 3-cyclopentene-1,2-diones (181). Reaction of diynes and carbodiimides with $Fe(CO)_5$ affords small amounts of pyrrolidone derivatives, among other products (182).

A fluorinated iron phthalocyanine was readily synthesized by exposure of *o*-dicyanotetrafluorobenzene to $Fe(CO)_5$ in 1-methylnaphthalene (183).

C. Deoxygenation of a Carbonyl Group of $Fe(CO)_5$.

Phosphineimines react with $Fe(CO)_5$ in THF to give isocyanide complexes and triphenylphosphine oxide, via a novel deoxygenation of a carbonyl group of the metal carbonyl (184).

E. Reductive Alkylation and Hydrogenation

Reductive methylation of an aldehyde or a ketone [structural types $RCOCH_3$, $RCOCH_2R'$] can be accomplished in high yield using formaldehyde and $Fe(CO)_5$-KOH [1:3 ratio, i.e., *in situ* generation of $HFe(CO)_4^-$] (185). This reaction probably occurs via an α,β-unsaturated carbonyl compound which is known to undergo hydrogenation, at the carbon–carbon double bond, in the presence of $Fe(CO)_5$ and base in moist solvents (186). The reductive alkylation process has been applied to the synthesis of 3-substituted indoles (187).

Schiff bases and diazines such as phthalazine can be conveniently reduced to amines and dihydroazines, respectively, by $HFe_3(CO)_{11}^-$ or $HFe(CO)_4^-$ (34,188,189). The latter can also effect reductive amination (190).

F. Ylids

Benzylidene and alkylidene phosphoranes (phosphorus ylids) react with $Fe(CO)_5$ in THF to give stilbenes in low to moderate yields, and 1,4-diketones were obtained from phenacylidene nitrogen, sulfur, or phosphorus ylids (191). An ylide carbene adduct was isolated from the reaction of methylenetriphenylphosphorane with $Fe(CO)_5$ (192).

G. Thioketones

Iron pentacarbonyl reacts with thiobenzophenones in benzene to afford sulfur-donor ligand *ortho*-metalated complexes (e.g., **77**). Superior yields of these complexes result when $Fe_2(CO)_9$ is used as the reagent. Photolytic or oxidative cleavage of these complexes affords thiolactones (e.g., **78**) in good yields (193). This transformation can also be effected in the presence of Lewis bases such as amines or phosphines, i.e., nonoxidatively (194). *ortho*-Mercuration, accompanied by ether or ester formation, results on exposure of the *ortho*-metalated complexes to mercuric acetate in alcohol (e.g., **79**) or ether solvents. Since the complexes need only be generated *in situ*, this reaction provides a simple entry into aromatics difficult or impossible to prepare by other means.

Ph
H—C S $Fe(CO)_3$
Fe
$(CO)_3$
(77)

$Ce^{+4}/(CH_3)_2CO$
$h\nu$, C_2H_5OH

Ph
H—C S
C
O
(78)

$Hg(OAc)_2$
CH_3OH

Ph
H—C—OCH_3
—HgOAc
(79)

REFERENCES

1. R. Pettit and G. F. Emerson, in F. G. A. Stone and R. West, Eds., *Advances in Organometallic Chemistry*, Vol. 1, Academic Press, New York, **1964,** p. 1.
2. G. F. Emerson, J. E. Mahler, R. Kochhar, and R. Pettit, *J. Org. Chem.*, **29,** 3620 (1964).
3. B. F. Hallam and P. L. Pauson, *J. Chem. Soc.*, **1958,** 642.
4. H. Alper and J. T. Edward, *J. Organometal. Chem.*, **14,** 411 (1968).
5. L. F. Fieser and M. Fieser, *Steroids*, Reinhold, New York, **1959,** p. 264.
6. M. Cais and N. Maoz, *J. Organometal. Chem.*, **5,** 370 (1966).
7. J. E. Arnet and R. Pettit, *J. Am. Chem. Soc.*, **83,** 2954 (1961).
8. A. J. Birch, P. E. Cross, J. Lewis, D. A. White, and S. B. Wild, *J. Chem. Soc., A*, **1968,** 332.
9. R. Pettit, G. F. Emerson, and J. Mahler, *J. Chem. Educ.*, **40,** 175 (1963).
10. F. M. Chaudhari and P. L. Pauson, *J. Organometal. Chem.*, **5,** 73 (1966).
11. T. A. Manuel and F. G. A. Stone, *J. Am. Chem. Soc.*, **82,** 366 (1960).
12. E. N. Frankel, E. A. Emken, H. M. Peters, V. L. Davison, and R. O. Butterfield, *J. Org. Chem.*, **29,** 3292 (1964).
13. J. Kiji and M. Iwamoto, *Bull. Chem. Soc. Japan*, **41,** 1483 (1968).
14. G. F. Emerson, Ph.D. Thesis, The University of Texas, Austin, Texas, 1964.
15. E. K. von Gustorf and J. C. Hogan, *Tetrahedron Lett.*, **1968,** 3191.
16. E. K. von Gustorf and F.-W. Grevels, *Fortschr. Chem. Forsch.*, **13,** 366 (1969).
17. J. W. Fitch and H. E. Herbold, *Inorg. Chem.*, **8,** 1926 (1970).
18. H. Alper and J. T. Edward, *J. Organometal. Chem.*, **16,** 342 (1969).
19. R. Pettit, *J. Am. Chem. Soc.*, **81,** 1266 (1959).
20. M. L. H. Green, L. Pratt, and G. Wilkinson, *J. Chem. Soc.*, **1960,** 989.
21. R. B. King, *J. Am. Chem. Soc.*, **84,** 4705 (1962).
22. H. Alper, P. C. LePort, and S. Wolfe, *J. Am. Chem. Soc.*, **91,** 7553 (1969).
23. F. G. Cowherd and J. L. von Rosenberg, *J. Am. Chem. Soc.*, **91,** 2157 (1969).
24. W. T. Hendrix, F. G. Cowherd, and J. L. von Rosenberg, *Chem. Commun.*, **1968,** 97.

25. S. Sarel, R. Ben-Shoshan, and B. Kirson, *Israel J. Chem.*, **10,** 787 (1972), and references cited therein.
26. L. Watts, J. D. Fitzpatrick, and R. Pettit, *J. Am. Chem. Soc.*, **87,** 3253 (1965).
27. J. D. Holmes and R. Pettit, *J. Am. Chem. Soc.*, **85,** 2531 (1963).
28. H. Kappeler and J. Wild, Ger. Pat., 1,801,661 (1969); *Chem. Abstr.*, **71,** 61600 (1969).
29. D. Dell, N. Maoz, and M. Cais, *Israel J. Chem.*, **7,** 783 (1969).
30. G. F. Emerson, L. Watts, and R. Pettit, *J. Am. Chem. Soc.*, **87,** 131 (1965).
31. A. N. Nesmayanov, K. N. Anisimov, and N. E. Kolonova, *Izv. Akad. Nauk SSSR, Otd. Khim. Nauk*, **1962,** 772.
32. H. Alper and J. T. Edward, *Can. J. Chem.*, **46,** 3112 (1968).
33. G. Bor, *J. Organometal. Chem.*, **11,** 195 (1968).
34. H. Alper, unpublished results.
35. P. P. Singh and R. Rivest, *Can. J. Chem.*, **46,** 1773 (1968).
36. H. Alper, Ph.D. Thesis, McGill University, 1967.
37. H. Alper and J. T. Edward *J. Org. Chem.*, **32,** 2938 (1967).
38. M. M. Frojmovic and G. Just, *Can. J. Chem.*, **46,** 3719 (1968).
39. H. Alper and R. A. Partis, unpublished results.
40. C. E. Coffey, *J. Am. Chem. Soc.*, **83,** 1623 (1961).
41. A. Mittasch, *Z. Angew. Chem.*, **41,** 827 (1928).
42. G. Berens, F. Kaplan, R. Rimerman, B. W. Roberts, and A. Wissner, *J. Am. Chem. Soc.*, **97,** 7076 (1975).
43. W. R. Roth and J. D. Meier, *Tetrahedron Lett.*, **1967,** 2053.
44. M. I. Bruce, *J. Organometal. Chem.*, **10,** 495 (1967).
45. T. A. Manuel, S. L. Stafford, P. M. Treichel, and F. G. A. Stone, *J. Am. Chem. Soc.*, **83,** 249 (1961).
46. R. B. King, S. L. Stafford, P. M. Treichel, and F. G. A. Stone, *J. Am. Chem. Soc.*, **83,** 3604 (1961).
47. I. Rhee, M. Ryang, and S. Tsutsumi, *J. Organometal. Chem.*, **9,** 361 (1967).
48. M. Asfazadourian and M. Prilleux, Fr. Pat. 1,418,851 (1965); *Chem. Abstr.*, **65,** 10747*ef* (1966).
49. R. A. Plowman and F. G. A. Stone, *Z. Naturforsch.*, **17b,** 575 (1962).
50. R. F. Heck and C. R. Boss, *J. Am. Chem. Soc.*, **86,** 2580 (1964).
51. R. F. Heck, U.S. Pat. 3,338,936 (1967); *Chem. Abstr.*, **68,** 49788 (1968).
52. E. K. von Gustorf, M. C. Henry, and C. DiPietro, *Z. Naturforsch.*, **21b,** 42 (1966).
53. R. Fields, M. M. Germain, R. N. Haszeldine, and P. W. Wiggans, *J. Chem. Soc., A*, **1970,** 1969.
54. E. K. von Gustorf, M. C. Henry, and D. J. McAdoo, *Ann. Chem.*, **707,** 190 (1967).
55. E. K. von Gustorf, F.-W. Grevels, and J. C. Hogan, *Angew. Chem. Interm. Ed.*, **8,** 899 (1969); J. C. Hogan, Ph.D. Thesis, Boston College, Boston, Mass. 1969.
56. E. K. von Gustorf, M.-J. Jun, H. Hun, and G. O. Schenck, *Angew. Chem.* **75,** 1120 (1963).
57. F. Seel and G.-V. Roschenthaler, *Angew. Chem.*, **82,** 182 (1970); *Angew. Chem. Intern. Ed.*, **9,** 166 (1970).
58. H. Alper and E. C. H. Keung, unpublished results.
59. E. Yoshisato and S. Tsutsumi, *Chem. Commun.*, **1968,** 33.
60. J. D. Cotton, S. A. R. Knox, I. Paul, and F. G. A. Stone, *J. Chem. Soc., A*, **1967,** 264.
61. M. A. Englin and A. S. Filatov, *Zh. Obshch. Khim.*, **39,** 1214 (1969); *J. Gen. Chem. USSR*, **39,** 1184 (1969).
62. V. A. Ginsburg and K. N. Smirnov, *Zh. Obshch. Khim.*, **37,** 1413 (1967); *J. Gen. Chem. USSR*, **37,** 1343 (1967).

63. A. S. Filatov, M. A. Englin, and V. I. Yakutin, *Zh. Obshch. Khim*, **39,** 1325 (1969); *J. Gen. Chem. USSR*, **39,** 1295 (1969).
64. M. Green and A. E. Tipping, *J. Chem. Soc.*, **1965,** 5774.
65. S. Fukuoka, M. Ryang, and S. Tsutsumi, *Tetrahedron Lett.*, **1970,** 2553.
66. H. Alper, *Tetrahedron Lett.*, **1969,** 1239.
67. E. Lindner, H. Weber, and G. Vitzthum, *J. Organometal. Chem.*, **13,** 431 (1968).
68. E. Lindner, G. Vitzthum, and H. Weber, *Z. Anorg. Allgem. Chem.*, **373,** 122 (1970).
69. H. Alper, *Angew. Chem.*, **81,** 706 (1969); *Angew. Chem. Intern. Ed.*, **8,** 677 (1969).
70. E. Lindner and G. Vitzthum, *Angew. Chem.*, **81,** 532 (1969); *Angew. Chem. Intern. Ed.*, **8,** 518 (1969).
71. J. Roy, *Z. Naturforsch.*, **25b,** 1063 (1970).
72. A. S. Filatov and M. A. Englin, *Zh. Obshch. Khim.*, **39,** 533 (1969); *J. Gen. Chem. USSR*, **39,** 502 (1969).
73. H. Alper and J. T. Edward, *Can. J. Chem.*, **48,** 1623 (1970).
74. G. N. Schrauzer, *Chem. Ber.*, **94,** 1891 (1961).
75. M. Ryang, *Organometal. Chem. Rev., A*, **5,** 67 (1970).
76. W. F. Edgell, M. T. Yang, B. J. Bulkin, R. Bayer, and N. Koizumi, *J. Am. Chem. Soc.*, **87,** 3080 (1965).
77. W. F. Edgell and B. J. Bulkin, *J. Am. Chem. Soc.*, **88,** 4839 (1966).
78. B. J. Bulkin and J. A. Lynch, *Inorg. Chem.*, **7,** 2654 (1968).
79. F. Seel and V. Sperber, *Angew. Chem.*, **80,** 38 (1968); *Angew. Chem. Intern. Ed.*, **7,** 70 (1968).
80. M. Heintzeler, Ger. Pat. 928,044 (1955); *Chem. Abstr.*, **52,** 3284 (1958).
81. W. Hieber, F. Sonnekalb, and E. Becker, *Chem. Ber.*, **63,** 973 (1930).
82. M. Ryang, Y. Sawa, H. Masada, and S. Tsutsumi, *Kogyo Kagaku Zasshi*, **66,** 1086 (1963); *Chem. Abstr.*, **62,** 7670 (1965).
83. E. O. Fischer and V. Kiener, *J. Organometal. Chem.*, **23,** 215 (1970); E. O. Fischer, V. Kiener, D. St. P. Bunbury, E. Frank, P. F. Lindley, and O. S. Mills, *Chem. Commun.*, **1968,** 1378.
84. D. J. Darensbourg and M. Y. Darensbourg, *Inorg. Chem.*, **9,** 1691 (1970).
85. M. Ryang, I. Rhee, and S. Tsutsumi, *Bull. Chem. Soc. Japan*, **37,** 341 (1964).
86. M. P. Cooke, Jr., *J. Am. Chem. Soc.*, **92,** 6080 (1970).
87. Y. Sawa, M. Ryang, and S. Tsutsumi, *Tetrahedron Lett.*, **1969,** 5189; *J. Org. Chem.*, **35,** 4183 (1970).
88. I. Rhee, M. Ryang, and S. Tsutsumi, *Tetrahedron Lett.*, **1969,** 4593.
89. D. J. Darensbourg and M. Y. Darensbourg, *Inorganica Chimica Acta, Proc. Third International Symposium on Reactivity and Bonding in Transition Organometallic Compounds*, Venice, **1970,** D5.
90. M. Y. Darensbourg and D. J. Darensbourg, *Abstracts 160th Am. Chem. Soc. National Meeting*, Chicago, Illinois, September, **1970,** Inorg. 114.
91. W. Hieber and A. Lipp, *Chem. Ber.*, **92,** 2085 (1959).
92. H. Alper and J. T. Edward, *Can. J. Chem.*, **48,** 1543 (1970).
93. F. A. Daniker and B. E. Hackley, *J. Org. Chem.*, **31,** 4267 (1966).
94. A. S. Filatov and M. A. Englin, *Zh. Obshch. Khim.*, **39,** 783 (1969); *J. Gen. Chem. USSR*, **39,** 743 (1969).
95. H. Alper and E. C. H. Keung, *Tetrahedron Lett.*, **1970,** 53.
96. J. P. A. Castrillon and H. H. Szmant, *J. Org. Chem.*, **30,** 1338 (1965).
97. I. Granoth, A. Kalir, and Z. Pelah, *J. Chem. Soc., C*, **1969,** 2424.
98. T. H. Chan, A. Melnyk, and D. N. Harpp, *Tetrahedron Lett.*, **1969,** 201.
99. K. Naumann, G. Zon, and K. Mislow, *J. Am. Chem. Soc.*, **91,** 7012 (1969).

100. W. Strohmeier, J. F. Guttenberger, and G. Popp, *Chem. Ber.*, **98,** 2248 (1965).
101. M. Dekker and G. R. Knox. *Chem. Commun.*, **1967,** 1243.
102. H. Alper, *Organometal. Chem. Syn.*, **1,** 69 (1970/1971).
103. E. K. von Gustorf, *Angew. Chem. Intern. Ed.*, **5,** 739 (1966).
104. E. K. von Gustorf, M. C. Henry, R. E. Sacher, and C. DiPietro, *Z. Naturforsch.*, **21b,** 1152 (1966).
105. E. K. von Gustorf and M.-J. Jun, *Z. Naturforsch.*, **20b,** 521 (1965).
106. M. J. Barrow and O. S. Mills, *Angew. Chem.*, **81,** 898 (1969); *Angew. Chem. Intern. Ed.*, **8,** 879 (1969).
107. M. M. Bagga, W. T. Flannigan, G. R. Knox, and P. L. Pauson, *J. Chem. Soc., C,* **1969,** 1534; M. M. Bagga, P. L. Pauson, F. J. Preston, and R. I. Reed, *Chem. Commun.*, **1965,** 543.
108. P. E. Baikie and O. S. Mills, *Chem. Commun.*, **1966,** 707.
109. M. M. Bagga, P. E. Baikie, O. Saikie, O. S. Mills, and P. L. Pauson, *Chem. Commun.*, **1967,** 1106.
110. C. D. Campbell and C. W. Rees, *Chem. Commun.*, **1969,** 537.
111. S. Otsuka, T. Yoshida, and A. Nakamura, *Inorg. Chem.*, **8,** 2514 (1969).
112. R. P. Bennett, *Inorg. Chem.*, **9,** 2184 (1970).
113. R. J. Doedens, *Inorg. Chem.*, **9,** 429 (1970).
114. W. T. Flannigan, G. R. Knox, and P. L. Pauson, *Chem. Ind. (London)*, **1967,** 1094.
115. C. Ruchardt and G. N. Schrauzer, *Chem. Ber.*, **93,** 1840 (1960).
116. O. S. Mills and A. D. Redhouse, *J. Chem. Soc., A,* **1968,** 1282; O. S. Mills and A. D. Redhouse, *Chem. Commun.*, **1966,** 444.
117. S. Otsuka, T. Yoshida, and A. Nakamura, *Inorg. Chem.*, **7,** 1833 (1968).
118. R. B. King, *Inorg. Chem.* **2,** 326 (1963).
119. R. Havlin and G. R. Knox, *J. Organometal. Chem.*, **4,** 247 (1965).
120. H. D. Kaesz, R. B. King, T. A. Manuel, L. D. Nichols, and F. G. A. Stone, *J. Am. Chem. Soc.*, **82,** 4749 (1960).
121. H. Schott and G. Wilke, *Angew. Chem.*, **81,** 896 (1969); *Angew. Chem. Intern. Ed.*, **8,** 877 (1969).
122. U. Wannagat and D. Seyffert, *Angew. Chem.*, **77,** 457 (1965); *Angew. Chem. Intern. Ed.*, **4,** 438 (1965).
123. M. Rosenblum and C. Gatsonis, *J. Am. Chem. Soc.*, **89,** 5074 (1967).
123*a*. I. Wender and P. Pino, Eds., *Organic Syntheses via Metal Carbonyls,* Vol. 1, Interscience, New York, 1968.
124. D. V. Banthorpe, H. Fitton, and J. Lewis, *J. Chem. Soc., Perkin Trans. 1,* **1973,** 2051.
125. E. Bertele and P. Schudel, *Helv. Chim. Acta,* **50,** 2445 (1967).
126. A. J. Birch, K. B. Chamberlain, M. A. Haas, and D. J. Thompson, *J. Chem. Soc., Perkin Trans. 1,* **1973,** 1882.
127. A. J. Birch and D. H. Williamson, *J. Chem. Soc., Perkin Trans. 1,* **1973,** 1892.
128. J. Altman and D. Ginsburg, *Tetrahedron,* **27,** 93 (1971).
129. A. J. Hubert, A. Georis, R. Warin, and P. Teyssié, *J. Chem. Soc., Perkin Trans. 2,* **1972,** 366.
130. C. P. Casey and C. R. Cyr, *J. Am. Chem. Soc.*, **95,** 2248 (1973).
131. A. J. Hubert, P. Moniotte, G. Goebbels, R. Warin, and P. Teyssié, *J. Chem. Soc., Perkin Trans. 2,* **1973,** 1954.
131*a* E. J. Corey and G. Moinet, *J. Am. Chem. Soc.*, **95,** 7185 (1973).
131*b* H. Alper, *J. Organometal. Chem.*, **96,** 95 (1975).
132. R. Victor, R. Ben-Shoshan, and S. Sarel, *Tetrahedron Lett.*, **1970,** 4253.
132*a*. S. Sarel, A. Felzenstein, R. Victor, and J. Yovell, *Chem. Commun.*, **1974,** 1025.

133. R. Aumann, *J. Am. Chem. Soc.*, **96,** 2631 (1974).
134. A. J. Birch, K. B. Chamberlain, and D. J. Thompson, *J. Chem. Soc., Perkin Trans. 1*, **1973,** 1900.
135. M. Korat, D. Tatarsky, and D. Ginsburg, *Tetrahedron*, **28,** 2315 (1972).
136. R. E. Ireland, G. G. Brown, Jr., R. H. Stanford, Jr., and T. C. McKenzie, *J. Org. Chem.*, **39,** 51 (1974).
137. Y. Shvo and E. Hazum, *Chem. Commun.*, **1974,** 336.
138. H. Alper and A. S. K. Chan, *Inorg. Chem.*, **13,** 225 (1974).
138*a*. A. Dondoni and G. Barbaro, *Chem. Commun.*, **1975,** 761.
139. M. Pankowski and M. Bigorgne, *J. Organometal Chem.*, **30,** 227 (1971).
140. M. D. Vorob'ev, V. A. Shpanskii, and M. A. Englin, *Zh. Vses. Khim. Obshchest.*, **17,** 692 (1972); *Chem. Abstr.*, **78,** 71286 (1973).
141. F. Seel and G. V. Roschenthaler, *Z. Anorg. Allgem. Chem.*, **396,** 297 (1971).
142. H. Alper and E. C. H. Keung, *J. Org. Chem.*, **37,** 2566 (1972).
143. I. V. Ermakova, V. A. Chimishkyan, and M. A. Englin, *Zh. Org. Khim.*, **8,** 186 (1972).
144. M. A. Englin, A. S. Filatov, and N. F. Alekseeva, *Zh. Org. Khim.*, **7,** 2611 (1971); *J. Org. Chem.* (*USSR*), **7,** 2710 (1971).
145. R. Noyori, S. Makino, and H. Takaya, *J. Am. Chem. Soc.*, **93,** 1272 (1971).
146. R. Noyori, K. Yokoyama, S. Makino, and Y. Hayakawa, *J. Am. Chem. Soc.*, **94,** 1772 (1972).
147. R. Noyori, K. Yokoyama, and Y. Hayakawa, *J. Am. Chem. Soc.*, **95,** 2722 (1973).
148. R. Noyori, Y. Hayakawa, S. Makino, N. Hayakawa, and H. Takaya, *J. Am. Chem. Soc.*, **95,** 4103 (1973).
149. R. Noyori, Y. Baba, S. Makino, and H. Takaya, *Tetrahedron Lett.*, **1973,** 1741.
150. R. Noyori, S. Makino, Y. Baba, and Y. Hayakawa, *Tetrahedron Lett.*, **1974,** 1049.
151. R. Noyori, Y. Baba, and Y. Hayakawa, *J. Am. Chem. Soc.*, **96,** 3336 (1974).
152. R. Noyori, Y. Hayakawa, M. Funakura, H. Takaya, S. Murai, R. Kobayashi, and S. Tsutsumi, *J. Am. Chem. Soc.*, **94,** 7202 (1972).
152*a*. H. Alper, *Tetrahedron Lett.*, **1975** (2257).
153. A. F. Videiko and M. A. Englin, *Zh. Org. Khim.*, **8,** 2049 (1972); *J. Org. Chem.* (*USSR*), **8,** 2095 (1972).
154. A. S. Filatov and M. A. Englin, *Zh. Org. Khim*, **7,** 2316 (1971); *J. Org. Chem.* (*USSR*), **7,** 2407 (1971).
155. N. A. Genco, R. A. Partis, and H. Alper, *J. Org. Chem.*, **38,** 4365 (1973).
155*a*. J. P. Collman, Accounts Chem. Res., **8,** 342 (1975).
156. J. P. Collman, S. R. Winter, and D. R. Clark, *J. Am. Chem. Soc.*, **94,** 1788 (1972).
157. W. O. Siegl and J. P. Collman, *J. Am. Chem. Soc.*, **94,** 2516 (1972).
158. J. P. Collman, J. W. Cawse, and J. I. Brauman, *J. Am. Chem. Soc.*, **94,** 5905 (1972).
158*a*. M. P. Cooke, Jr. and R. M. Parlman, *J. Am. Chem. Soc.*, **97,** 6863 (1975).
159. J. Y. Merour, J. L. Roustan, C. Charrier, J. Collin, and J. Benaim, *J. Organometal. Chem.*, **51,** C24 (1973).
160. H. Masada, M. Mizuno, S. Suga, Y. Watanabe, and Y. Takegami, *Bull. Chem. Soc. Japan*, **43,** 3824 (1970).
161. J. P. Collman, S. R. Winter, and R. G. Komoto, *J. Am. Chem. Soc.*, **95,** 249 (1973).
162. J. P. Collman and N. W. Hoffman, *J. Am. Chem. Soc.*, **95,** 2689 (1973).
163. Y. Watanabe, T. Mitsudo, M. Tanaka, K. Yamamoto, T. Okajima, and Y. Takegami, *Bull. Chem. Soc. Japan*, **44,** 2569 (1971).
164. T. Mitsudo, Y. Watanabe, M. Tanaka, K. Yamamoto, and Y. Takami, *Bull. Chem. Soc. Japan*, **45,** 305 (1972).

165. Y. Watanabe, M. Yamashita, T. Mitsudo, M. Tanaka, and Y. Takegami, *Tetrahedron Lett.*, **1973,** 3535.
166. M. Y. Darensbourg, H. C. Condor, D. J. Darensbourg, and C. Hasday, *J. Am. Chem. Soc.*, **95,** 5919 (1973).
167. J. Mantzaris and E. Weissberger, *J. Am. Chem. Soc.*, **96,** 1873 (1974), and references cited therein.
168. L. Lombardo, D. Wege, and S. P. Wilkinson, *Aust. J. Chem.*, **27,** 143 (1974).
169. J. Mantzaris and E. Weissberger, *J. Am. Chem. Soc.*, **96,** 1880 (1974).
170. J. Grandjean, P. Laszlo, and A. Stockis, *J. Am. Chem. Soc.*, **96,** 1622 (1974).
171. H. Alper, *J. Org. Chem.*, **38,** 1417 (1973).
171*a*. H. Alper, *J. Organometal. Chem.*, **84,** 347 (1975).
172. A. Tanaka and J. P. Anselme, *Tetrahedron Lett.*, **1971,** 3567.
173. J. M. Landesberg, L. Katz, and C. Olsen, *J. Org. Chem.*, **37,** 930 (1972).
174. M. A. Englin, A. S. Filatov, and N. F. Alekseeva, *Zh. Org. Khim.*, **7,** 2319 (1971); *J. Org. Chem.* (*USSR*), **7,** 2410 (1971).
175. H. Alper, *Inorg. Chem.*, **11,** 976 (1972).
176. B. M. Trost and S. D. Ziman, *J. Org. Chem.*, **38,** 932 (1973).
177. H. Alper and B. D. Sash, unpublished results.
178. J. Daub, V. Trautz, and U. Erhardt, *Tetrahedron Lett.*, **1972,** 4435.
179. J. Daub, U. Erhardt, J. Kappler, and V. Trautz, *J. Organometal. Chem.*, **69,** 423 (1974).
179*a*. H. Alper, *J. Org. Chem.*, **40,** 2694 (1975).
180. H. Kolshorn, H. Meier, and Eu. Muller, *Tetrahedron Lett.*, **1971,** 1469.
181. K. Kinugasa and T. Agawa, *Organometal. Chem. Syn.*, **1,** 427 (1972).
182. K. Kinugasa and T. Agawa, *J. Organometal Chem.*, **51,** 329 (1973).
183. J. G. Jones and M. V. Twigg, *Inorg. Chem.*, **8,** 2018 (1969).
184. H. Alper and R. A. Partis, *J. Organometal. Chem.*, **35,** C40 (1972).
185. G. Cainelli, M. Panunzio, and A. Umani-Ronchi, *Tetrahedron Lett.*, **1973,** 2491.
186. R. Noyori, I. Umeda, and T. Ishigami, *J. Org. Chem.*, **37,** 1542 (1972).
187. G. P. Boldini, M. Panunzio, and A. Umani-Ronchi, *Chem. Commun.*, **1974,** 359.
188. H. Alper, *J. Org. Chem.*, **37,** 3972 (1972).
189. H. Alper, *J. Organometal. Chem.*, **50,** 209 (1973).
190. Y. Watanabe, M. Yamashita, T. Mitsudo, M. Tanaka, and Y. Takegami, *Tetrahedron Lett.*, **1974,** 1879.
191. H. Alper and R. A. Partis, *J. Organometal Chem.*, **44,** 371 (1972).
192. W. C. Kaska, D. K. Mitchell, R. F. Reichederfer, and W. D. Korte, *J. Am. Chem. Soc.*, **96,** 2847 (1974).
193. H. Alper and A. S. K. Chan, *J. Am. Chem. Soc.*, **95,** 4905 (1973).
194. H. Alper and W. G. Root, *J. Am. Chem. Soc.*, **97,** 4251 (1975).

Decarbonylation Reactions Using Transition Metal Compounds

J. TSUJI, *Tokyo Institute of Technology, Ōokayama, Meguro-ku, Tokyo, Japan,* 152

I. INTRODUCTION

Carbonylations of olefinic compounds catalyzed by transition metal complexes to form various carbonyl compounds are useful reactions and have been widely studied. On the other hand, the reverse reaction of carbonylation, namely the decarbonylation of carbonyl compounds with similar catalysts, has received less attention and only relatively recently.

Systematic studies of the decarbonylation reactions were prompted several years ago by the discovery of several efficient catalysts. The decarbonylation reaction has become important from a mechanistic viewpoint based on coordination chemistry, and for organic synthesis. In organic synthesis, one carbon unit can be removed by the decarbonylation.

In this review, progress in the study of decarbonylation reactions using transition metal complexes is surveyed. The decarbonylations can be discussed from several standpoints. Reviews from the standpoint of organic synthesis, given by Tsuji and Ohno (1,2), survey all examples and applications of the decarbonylation reactions using transition metal complexes in organic synthesis. In the present review, the decarbonylation reactions are discussed with emphasis on reaction mechanisms based on coordination chemistry and organometallic chemistry.

II. CARBONYLATION AND DECARBONYLATION AS REVERSIBLE PROCESSES

Several transition metal complexes are active catalysts for the carbonylation of olefinic and acetylenic compounds to form various carbonyl compounds (3.4). The mechanisms of the carbonylation reactions of olefinic and acetylenic compounds catalyzed by transition metal complexes have been studied extensively. Especially, the mechanism of the oxo reaction has received much attention (see second chapter), and the overall mechanism of the oxo reaction was presented as follows (5).

$$RCH{=}CH_2 + HCo(CO)_3$$

$$\rightleftharpoons Co(CO)_3CH_2CH_2R \overset{CO}{\rightleftharpoons} Co(CO)_3COCH_2CH_2R$$

$$\xrightarrow{H_2} HCo(CO)_3 + RCH_2CH_2CHO$$

It is well established that an essential step in the mechanism of carbonylation reactions is a reversible CO insertion as shown above; namely, CO is inserted into a metal alkyl σ-bond to form an acyl complex in the presence of CO or of other ligands (6,7). On the other hand, the acyl complex is reconverted into the alkyl complex under certain conditions. The reversible CO insertion is also called acyl–alkyl rearrangement.

$$R{-}\overset{|}{\underset{|}{M}}{-} \underset{-CO}{\overset{CO}{\rightleftharpoons}} R{-}\overset{O}{\overset{\|}{C}}{-}\overset{|}{\underset{|}{M}}{-}$$

A typical and well-established example of this process is the following reversible CO insertion reaction of manganese carbonyl.

$$\mathrm{(CH_3CO)Mn(CO)_4(PPh_3)} \underset{+\mathrm{CO}}{\overset{-\mathrm{CO}}{\rightleftharpoons}} \mathrm{(CH_3)Mn(CO)_4(PPh_3)}$$

Temperature, CO pressure, and structure of the complex are decisive factors for the equilibrium of the acyl–alkyl rearrangement. A large number of kinetic and stereochemical studies have been carried out on this reaction, and there are ample other examples supporting reversible CO insertion reactions with various transition metal complexes (8–14).

So far extensive studies have been carried out mostly on the forward carbonylation reactions. However, if the reversibility of the CO insertion can be utilized efficiently, it should be possible to carry out decarbonylations of carbonyl compounds. The decarbonylation reaction, if carried out smoothly, is useful in organic chemistry for removing a functional group with loss of one carbon unit. By using the above assumption, several good catalysts have been discovered and efficient decarbonylations of certain carbonyl compounds have become possible. For example, acyl halides and aldehydes are decarbonylated by several transition metal complexes, both catalytically and stoichiometrically.

First, the several requirements or limitations of the decarbonylation reactions are considered from the standpoints of both carbonyl compounds and transition metal complexes. In these decarbonylations, a carbon–carbon bond is cleaved. What is the driving force of the carbon–carbon bond cleavage? For a decarbonylation to be possible, the essential first step is coordination of the carbonyl compound with the transition metal complex. The carbonyl compound is activated in one sense by the coordination, and an acyl complex is usually formed in the first step. Then follows the acyl–alkyl rearrangement, with carbon–carbon bond cleavage, to form an alkyl complex. Once the acyl complex is formed, the rearrangement to the alkyl complex is rather easy, when the reaction is carried out by heating in the absence of CO.

$$\mathrm{R{-}C({=}O){-}X} + {-}\mathrm{M}{-} \rightleftharpoons \mathrm{R{-}C({=}O){-}M{-}X} \rightleftharpoons \mathrm{R{-}M(CO){-}X}$$

For formation of the acyl complex as the essential step in decarbonylation, the general reaction is an oxidative addition of the carbonyl compound. Oxidative addition reactions are well-established common reactions of certain transition metal complexes, especially of noble metal complexes (15,16). Many simple covalent molecules, such as H_2, hydrogen halides, alkyl halides, halogens, and acyl halides, can add oxidatively to coordinatively unsaturated complexes. Thus coordinatively unsaturated complexes that can add carbonyl compounds oxidatively are possible candidates as catalysts for decarbonylation. In other words, the existence of available sites on the metal complexes, including latent or solvent occupied sites, is important in the reaction.

In order to determine whether a compound is coordinatively unsaturated, it is necessary to know the maximum coordination numbers for complexes with different configurations of the metal d-electrons, d^n. For complexes of d^6 spin-paired configuration, six coordination is regarded as saturated; for d^8 and d^{10} complexes, five and four coordinations are considered to be saturated. Hence coordination of fewer ligands than these numbers constitutes coordinative unsaturation. Thus, unsaturated noble metal complexes are possible catalysts for decarbonylation reactions. A typical example of oxidative addition related to decarbonylation is formation of an acyl complex by reaction of $Pd(PPh_3)_4$ and $Pt(PPh_3)_4$ with acyl halides (17,18).

$$RCOCl + Pd(PPh_3)_4 \rightarrow RCOPdCl(PPh_3)_2 + 2PPh_3$$

The above four-coordinated palladium and platinum complexes (d^{10}) are considered to be saturated. However, in solution two coordinated molecules of PPh_3 dissociate easily, and the complex becomes unsaturated (19,20). Thus they undergo oxidative addition. In this sense, the initial palladium and platinum complexes are potentially unsaturated.

However, the oxidative addition reaction is not a sufficient condition for decarbonylation. It should be noted that the reactions stop without completing the decarbonylation if the product of the oxidative addition is too stable.

The final step of the decarbonylation is reductive elimination of the alkyl complexes to form alkanes or alkenes by ligand coupling or by β-elimination. This step is rather easy, because the alkyl–metal σ-bonds are usually not stable.

$$RCH_2CH_2{-}\overset{|}{\underset{|}{M}}{-}X \rightarrow RCH_2CH_2{-}X \text{ (or } RCH{=}CH_2 + HX) + {-}\overset{|}{\underset{|}{M}}{-}$$

There are two kinds of decarbonylation reactions, namely, stoichiometric and catalytic. Of course catalytic decarbonylation is most useful because transition metal complexes are usually rather expensive. If the metal carbonyl complexes formed by decarbonylation are too stable, and CO is not liberated from them, the reaction stops at this stage. In such cases the reaction is stoichiometric with respect to the complex. Thus, some complexes are used only for the stoichiometric reaction. If the original complexes are regenerated with evolution of CO, then they can be used as efficient catalysts. In general, high temperature is necessary for the acyl–alkyl rearrangement and reductive elimination. It is apparent that smooth oxidative addition, acyl–alkyl rearrangement, reductive elimination, and CO evolution are important steps for efficient catalytic decarbonylation. So far, the most efficient catalysts for decarbonylation are found among the complexes of rhodium and palladium.

Although not widely studied, nickel, cobalt, and iron complexes are used for decarbonylation of some carbonyl compounds, such as isocyanates or ketenes. Oxidative addition is not so common for these base metal complexes, and the driving force of the initial step may be somewhat different from that of the noble metal complexes. For this reaction, coordination of CO and some other species (such as carbenes or nitrenes) plays a decisive role.

What kinds of carbonyl compounds can be decarbonylated? For reversible CO insertion and oxidative addition reactions, acyl halides and aldehydes are the most promising candidates. It should be pointed out that these compounds can be prepared by carbonylation of olefinic compounds catalyzed by transition metal complexes. It is now established that acyl complexes are formed by the oxidative addition reactions of acyl halides to the metal complexes (17,18,21). From these considerations, it is reasonably expected that acyl halides can be decarbonylated.

Aldehydes, the other promising candidates for decarbonylation, are easily prepared by the oxo reaction from olefins, CO, and H_2. The utility of aldehydes rests on the weak C–H bond. However, so far only one example of the oxidative addition of aldehydes has been reported. Harvie and Kemmitt (22) reported that aldehydes and potentially unsaturated $Pt(PPh_3)_4$ reacted to give the following diacyl complex, probably with evolution of H_2.

$$2RCHO + Pt(PPh_3)_4 \rightarrow RCO{-}\overset{\displaystyle PPh_3}{\underset{\displaystyle PPh_3}{\overset{|}{\underset{|}{Pt}}}}{-}COR + 2PPh_3 + H_2$$

$$2RCHO + Pt(PPh_3)_4 \rightarrow Pt(COOR)_2(PPh_3)_2$$

However, Tripathy and Roundhill (23) proposed that the product of the reaction of aldehydes with the platinum complex was not the diacyl compound, but a dicarboxylate. Further studies are necessary to confirm the nature of the reaction. Still it is highly probable that aldehydes undergo oxidative addition during decarbonylation.

It is understandable that esters and ketones, other common carbonyl compounds, cannot be decarbonylated with transition metal complexes because they are unable to coordinate readily with metals. Acid anhydrides and carboxylic acids are somewhat more promising, because they have weak C–O and/or O–H bonds, which might undergo oxidative addition reactions. Actually there are a few examples of the decarbonylation of these carbonyl compounds. Some other carbonyl compounds that can coordinate with the complexes in other ways can be decarbonylated. For example, carbonyl groups that have cumulative double bonds, such as in ketene and isocyanates, are possible candidates for decarbonylation.

In the following sections, various decarbonylation reactions are surveyed according to the metal species used.

III. DECARBONYLATIONS WITH RHODIUM COMPLEXES

A. Introduction

The most efficient and hence useful compounds for decarbonylation so far discovered are rhodium complexes, especially chlorotris(triphenylphosphine)rhodium, $RhCl(PPh_3)_3$. The synthesis of this useful complex was reported by Wilkinson and co-workers (24), and by Bennett and Longstaff (25). This complex is well known as an efficient catalyst for homogeneous hydrogenation as a result of extensive studies by Wilkinson and co-workers (26). Tsuji and Ohno (21,27–29) found that the complex can be used for both stoichiometric and catalytic decarbonylations of acyl halides and aldehydes under homogeneous conditions. Stoichiometric decarbonylation of aldehydes was studied also by Wilkinson and co-workers (30). Further, Blum (31) found that the complex is active for catalytic decarbonylation of aroyl halides.

Chlorocarbonylbis(triphenylphosphine)rhodium, $RhCl(CO)(PPh_3)_2$, which was used by Tsuji and Ohno for catalytic decarbonylation at high temperature, is another good catalyst. The decarbonylation reactions using these rhodium complexes are important and useful for organic synthesis as well as for mechanistic studies and associated coordination chemistry.

First, the stoichiometric decarbonylation reaction with $RhCl(PPh_3)_3$ will be considered. One of the interesting and important properties of this complex is its ability to react easily with CO to form $RhCl(CO)(PPh_3)_2$. The $RhCl(PPh_3)_3$ complex can acquire the CO moiety not only from molecular CO but also from various carbonyl groups and some types of oxygenated compounds, such as aldehydes and allyl alcohol, to form

$RhCl(CO)(PPh_3)_2$. In solution, $RhCl(PPh_3)_3$, which has the d^8 configuration, dissociates to form a three-coordinate, probably solvated, unsaturated complex with liberation of 1 mole of PPh_3 (30,32). This solvated species is highly unsaturated and undergoes the oxidative addition reaction easily to form a five-coordinate d^6 complex. This is, however, still coordinatively unsaturated and can undergo further reactions such as the acyl–alkyl rearrangement in order to reach coordinative saturation. Thus, it is understandable that this complex is useful for decarbonylation reactions.

$$\text{Cl}-\underset{\underset{\text{PPh}_3}{|}}{\overset{\overset{\text{PPh}_3}{|}}{\text{Rh}}}-\text{PPh}_3 \xrightarrow{S} \text{Cl}-\underset{\underset{S}{|}}{\overset{\overset{\text{PPh}_3}{|}}{\text{Rh}}}-\text{PPh}_3 + \text{PPh}_3$$

(S = solvent)

B. Decarbonylation of Aldehydes

Stoichiometric decarbonylation with $RhCl(PPh_3)_3$ will be discussed first (27,33). When this complex and aldehydes are mixed in solution, decarbonylation of the aldehydes proceeds smoothly and $RhCl(CO)(PPh_3)_2$ and alkanes are formed.

$$\text{RCHO} + \text{RhCl(PPh}_3)_3 \rightarrow \text{RH} + \text{PPh}_3 + \text{RhCl(CO)(PPh}_3)_2$$

This reaction affords the easiest way to decarbonylate aldehydes under mild, neutral conditions. The reaction proceeds homogeneously in organic solvents such as benzene or dichloromethane, in which the complex is moderately soluble. The progress of the decarbonylation can be followed easily by a color change from dark red to yellow. When the color of the solution turns to yellow, decarbonylation is complete.

Although no quantitative studies were carried out, it can be said that steric effects have a great influence on the decarbonylation reaction. Primary aldehydes are decarbonylated most easily at room temperature, or on warming for a short time. When the decarbonylation is very slow at room temperature, it can be carried out more rapidly in boiling benzene, toluene, or xylene. Under these conditions, most secondary aldehydes are decarbonylated. Aldehydes with large steric hindrance, however, cannot be decarbonylated in boiling benzene.

Furthermore, when aldehydes and $RhCl(PPh_3)_3$ are heated in toluene or xylene for a long period of time, the complex is converted into a brick red, dimeric complex shown below, which is insoluble in these solvents and precipitates, stopping the decarbonylation.

$$RhCl(PPh_3)_3 \text{ or } RhCl(PPh_3)_2S \rightarrow [(Ph_3P)_2Rh(\mu\text{-}Cl)_2Rh(PPh_3)_2] + PPh_3 \text{ or } S$$

This difficulty can be overcome by using nitriles as the solvent. Benzonitrile or acetonitrile seems to solvate the three-coordinate species strongly, and prevents formation of the dimeric complex. Thus, it is possible to decarbonylate highly hindered aldehydes in benzonitrile. For example, 9,10-dihydro-9,10-ethanoanthracene-11-carboxaldehyde and 1-methyl-3-cyclohexene-1-carboxaldehyde are decarbonylated rapidly in

$$\text{(9,10-dihydro-9,10-ethanoanthracene-11-CHO)} + RhCl(PPh_3)_3 \longrightarrow \text{(9,10-dihydro-9,10-ethanoanthracene)} + RhCl(CO)(PPh_3)_2 + PPh_3$$

$$\text{(1-}CH_3\text{-1-CHO-3-cyclohexene)} + RhCl(PPh_3)_3 \longrightarrow \text{(4-}CH_3\text{-cyclohexene)} + RhCl(CO)(PPh_3)_2 + PPh_3$$

benzonitrile in a few minutes. Some examples of stoichiometric decarbonylations using $RhCl(PPh_3)_3$ are shown in Table 1.

In decarbonylation with $RhCl(PPh_3)_3$, the products are mostly saturated alkanes. However, in some cases, olefins are formed as a minor product. For example, when heptanal was decarbonylated, the formation of 1-hexene as a minor product was confirmed.

$$CH_3(CH_2)_5CHO \xrightarrow{RhCl(PPh_3)_3} \text{hexane } (86\%) + \text{1-hexene } (14\%)$$

The decarbonylation of aldehydes with $RhCl(PPh_3)_3$ proceeds under very mild conditions in high yields without change of other functional groups. Thus the method has wide application in organic synthesis, allowing the removal of one carbon unit via the carbonyl group. Several examples of the application of this decarbonylation method have been reported. In one example, in order to introduce a double bond at Δ^1 of a 3-ketosteroid, a formyl group was introduced at the carbon adjacent to the ketone, and the double bond was formed by reaction with 2,3-dichloro-5,6-dicyano-1,4-benzoquinone(DDQ). Finally the formyl group was removed smoothly by using $RhCl(PPh_3)_3$, and an α,β-unsaturated ketone was formed (34).

Aldehydes[a]	Solvent	Temp (°C)	Time, min(hr)	Yield of $RhCl(CO)(PPh_3)_2$ (%)	Product	Yield[b] (%)	Ref.
Benzaldehyde	Toluene	Reflux	(1)	92	Benzene	83	27
Butyraldehyde	Benzene	Room temp	(2)	65			27
Cinnamaldehyde	Benzene	Room temp	(12)	65	Styrene		27
Cinnamaldehyde	Benzene	Reflux	15	93	Styrene	70	27
3-Phenylpropionaldehyde	Benzene	Reflux	10	90	Ethylbenzene	67	27
p-Chlorobenzaldehyde	Benzene	Reflux	(3)	90	Chlorobenzene	85	27
Salicylaldehyde	Toluene	Reflux	20	76	Phenol	70	27
Heptaldehyde	CH_2Cl_2	Room temp	(24)	90	Hexane (86%) 1-Hexene (14%)	78	27
9,10-Dihydro-9,10-ethanoanthracene-11-carboxaldehyde	PhCN	160	3	100	9,10-Dihydro-9,10-ethanoanthracene	67	27
1-Methyl-3-cyclohexene-1-carbaldehyde	PhCN	160	1	90	3-Methylcyclohexene	71	27
trans-α-Methylcinnam aldehyde	Toluene	Reflux	5	90	β-Methylstyrene *cis*(91), *trans*(9)	88	27
trans-α-Methylcinnam-aldehyde	PhCN	160	1	91	β-Methylstyrene *cis*(96), *trans*(4)	86	27
trans-α-Ethylcinnamaldehyde	PhCN	160	1	88	β-Ethylstyrene *cis*(94), *trans*(6)	82	27
(*R*)-(+)-2,2-Diphenyl-1-methyl-cyclopropanecarbaldehyde (0.72 g)	Xylene	Reflux	(12)		(*S*)-(+)-1-Methyl-2,2-diphenylcyclopropane (0.29 g)		45
(*R*)-(−)-α-Phenyl-α-methyl-butyraldehyde (0.66 g)	PhCN	110	(2)		(*S*)-(+)-2-Phenylbutane (0.14 g)		45

[a] In all cases, 1–3 g of $RhCl(PPh_3)_3$ and an excess of aldehydes were used.
[b] Yields based on $RhCl(PPh_3)_3$.

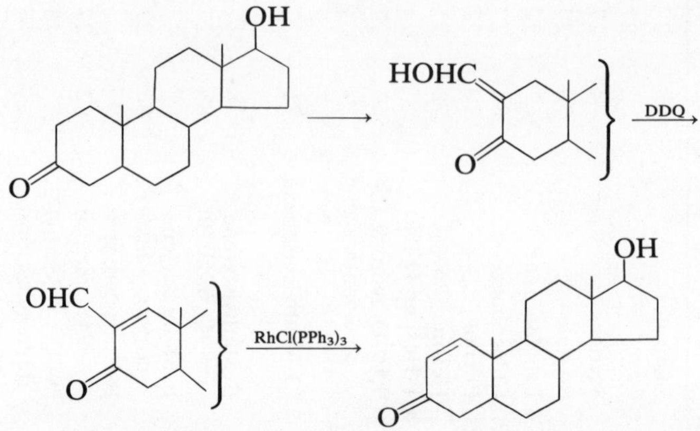

OH

HOHC

DDQ

O

OHC

RhCl(PPh$_3$)$_3$

O

OH

O

Also the side chain of 25,26,27-trisnorlanosterol-24-carboxylic acid was shortened via several steps, the final one being the removal of the formyl group by decarbonylation (35).

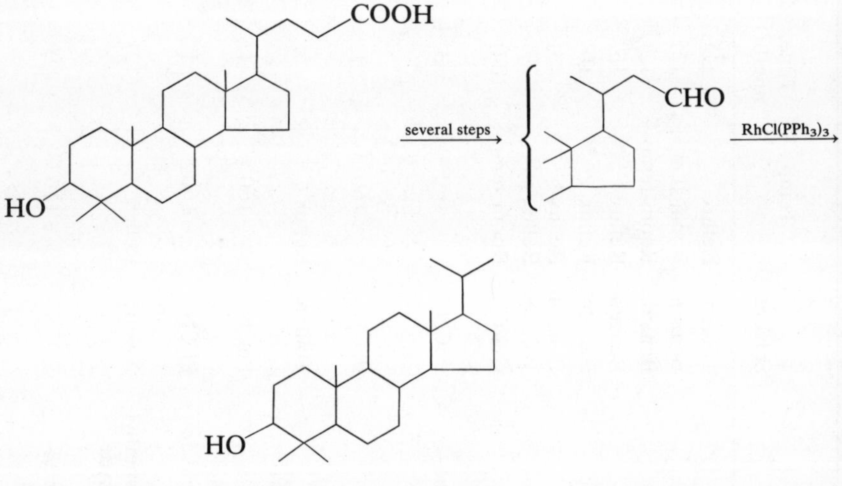

COOH

several steps

CHO

RhCl(PPh$_3$)$_3$

HO

HO

A method of stereoselective introduction of an angular methyl group was devised by Dawson and Ireland (36) using a combination of the Claisen rearrangement and decarbonylation as shown below. At first, the —CH_2CHO group was introduced stereoselectively at the angular position of the α,β-unsaturated ketone via the Claisen rearrangement and then the carbonyl group was removed with the rhodium complex. This procedure was applied to the conversion of 4-cholesten-3-one into 5-methyl-3-coprostene (37).

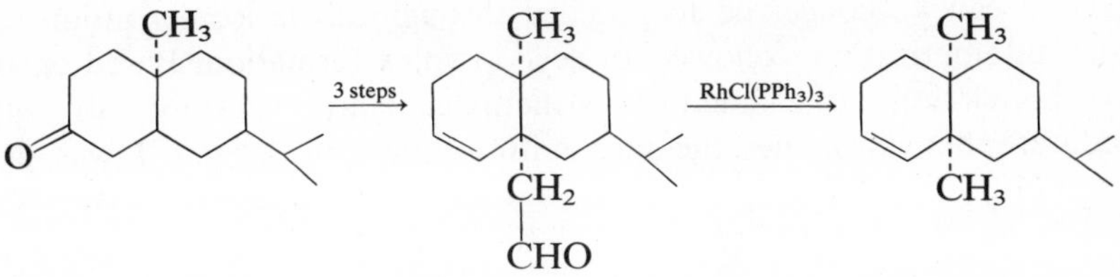

Synthesis of 9,11-secoprogesterone from 9-dehydroprogesterone was accomplished by ozonization, followed by the decarbonylation of the generated aldehyde with $[(PPh_3)_2RhCl]_2$ (38). Similarly 15,16-secoprogesterone was produced via decarbonylation (39).

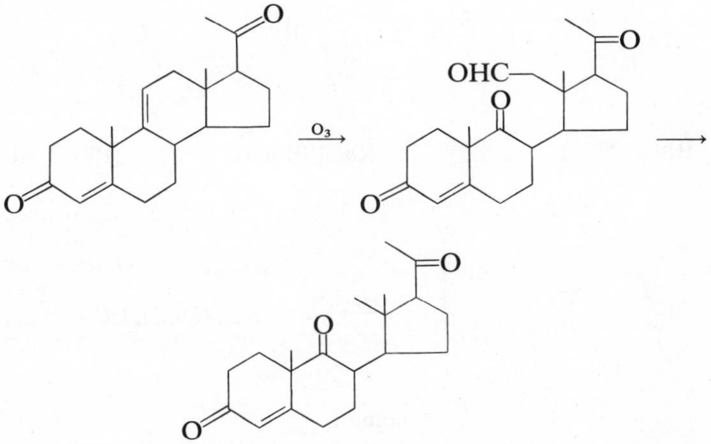

The transformation of allylic alcohols into aldehydes, a reaction catalyzed by noble metal complexes (40), suggests an interesting route to the decarbonylation of this type of alcohol. Bennett and Longstaff (25) found that $RhCl(PPh_3)_3$ is converted into $RhCl(CO)(PPh_3)_2$ by treatment with allyl alcohol. Also geraniol is decarbonylated (41). Detailed studies of the decarbonylation of allylic alcohols were carried out by Emery, Oehlschlager, and Unrau (42); these alcohols can be decarbonylated by heating with a stoichiometric amount of $RhCl(PPh_3)_3$ in acetonitrile or benzonitrile as solvent. The reaction can be expressed as follows:

$$RhCl(PPh_3)_3 + RCH{=}CHCH_2OH \rightarrow$$

$$\underset{\text{(major)}}{RCH_2CH_3} + \underset{\text{(minor)}}{RCH{=}CH_2} + PPh_3 + RhCl(CO)(PPh_3)_2$$

For example, cinnamyl alcohol was converted into ethylbenzene (76%) and styrene (4%) by heating with $RhCl(PPh_3)_3$ at 110–150°C. The reaction can be conceived to proceed through allylic isomerization and enol tautomerization, followed by acyl complex formation. Based on the decarbonylation of dideutero-α-methylcinnamyl alcohol to give dideuterophenylpropane, the mechanism shown in Scheme **I** was proposed (42).

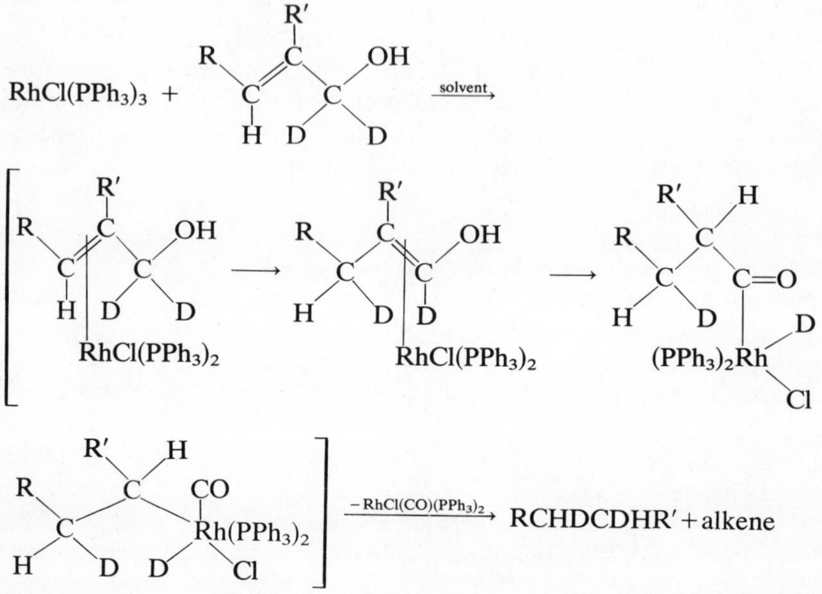

$RhCl(PPh_3)_3$ + R′ R C OH C C H D D $\xrightarrow{\text{solvent}}$

[R′ R C OH C C H D D $RhCl(PPh_3)_2$ ⟶ R′ R C OH C C H D D $RhCl(PPh_3)_2$ ⟶ R′ H R C C C=O H D D $(PPh_3)_2Rh$ Cl

R′ H R C CO C $Rh(PPh_3)_2$ H D D Cl] $\xrightarrow{-RhCl(CO)(PPh_3)_2}$ RCHDCDHR′ + alkene

Scheme **I**

The stoichiometric decarbonylation proceeds under very mild conditions, and 1 mole of $RhCl(PPh_3)_3$ is converted into $RhCl(CO)(PPh_3)_2$. It would be desirable to reconvert $RhCl(CO)(PPh_3)_2$ into $RhCl(PPh_3)_3$, so that the rather expensive complex could be used repeatedly. However, the reconversion seems to be impossible so far under any conditions. Fortunately, however, the catalytic decarbonylation of aldehydes is possible with $RhCl(CO)(PPh_3)_2$ at high temperature. The catalytic decarbonylation of aldehydes at 200°C with either $RhCl(PPh_3)_3$ or $RhCl(CO)(PPh_3)_2$ proceeds smoothly and is especially effective for aromatic aldehydes. Some side reactions, such as the aldol condensation, occur with aliphatic aldehydes. Some results of the catalytic decarbonylation of aldehydes are shown in Table 2 and Table 8 (in part). Metallic rhodium and platinum, supported on alumina or silica, are active catalysts for decarbonylation at high temperature to give olefins. An important industrial application of the decarbonylation of aldehydes catalyzed by metallic rhodium or

TABLE 2
Catalytic Decarbonylation of Aldehydes Using $RhCl(PPh_3)_3$ *or* $RhCl(CO)(PPh_3)_2$

Aldehyde	Amt (g)	Catalyst (g)	Temp (°C)	Time (hr)	Product	Yield (%)	Ref.
p-Chlorobenzaldehyde	8	0.2[a]	220	9	Chlorobenzene	77[b]	27
Salicylaldehyde	5	0.1[a]	210	8.5	Phenol	80[b]	27
Cinnamaldehyde	8	0.1[a]	230	4	Styrene	76[b]	27
1-Naphthaldehyde		—[c]	—	2	Naphthalene	88	60
2-Naphthaldehyde		—[c]	—	2.5	Naphthalene	77	60

[a] Using $RhCl(CO)(PPh_3)_2$.
[b] Yield based on aldehyde.
[c] Using $RhCl(PPh_3)_3$.

platinum was found with isobutyraldehyde (43). In the cobalt carbonyl-catalyzed oxo reaction of propylene, *n*-butyraldehyde is the main product, accompanied by small amounts of isobutyraldehyde, a less useful product. The reconversion of isobutyraldehyde into propylene, CO, and H_2 with rhodium or platinum metal contained in a fixed bed catalyst at 250–350°C was reported. The generated propylene, CO, and H_2 are recycled to the oxo reaction system. Some results of the gas-phase decarbonylation of aldehydes are shown in Table 3.

The mechanism of the stoichiometric decarbonylation of aldehydes seems to be the following, although no intermediate complex has been isolated that would directly support the proposal (27,33). The first step is the oxidative addition (or electrophilic addition, from another point of view) of aldehydes to the highly unsaturated d^8 complex. The oxidative addition is a nucleophilic attack on the carbon atom of the aldehyde with concomitant hydrogen transfer to the rhodium. In this step, Rh(I) is

TABLE 3
Catalytic Decarbonylation of Aldehydes in the Gas Phase with Metallic Rhodium and Platinum Catalysts (43)

Aldehyde	Product	Conversion (%)	Selectivity (%)		
			Olefin	CO	H_2
Isobutyraldehyde	Propylene	82	93	100	84
Butyraldehyde	Propylene	76	87	100	83
3-Methylbutyraldehyde	Isobutylene	97	83	97	81
2-Ethylhexaldehyde	Heptene	85	89	98	82
Benzaldehyde	Benzene	67	100	100	

oxidized to Rh(III) in what is probably the rate-determining step in the decarbonylation reaction. The acyl complex thus formed is a five-coordinate d^6 complex, which is still unsaturated. This complex then undergoes the acyl–alkyl rearrangement with alkyl migration, followed by hydrogen transfer to the alkyl group to give the alkane and $RhCl(CO)(PPh_3)_2$. This last step is the reductive elimination. From the above discussion, it would be expected that factors making the aldehydes a better electrophile, or alternatively, making the rhodium more nucleophilic by changing ligands, would lead to an increase in the decarbonylation rate. The formation of a small amount of olefins in the decarbonylation can be ascribed to hydrogen transfer from the bound alkyl group in the intermediate.

$$H{-}\overset{\displaystyle L}{\overset{|}{Rh}}{-}CH_2CH_2R \rightarrow H{-}\underset{\displaystyle H}{\underset{|}{\overset{\displaystyle L}{\overset{|}{Rh}}}}{-}\begin{matrix} CHR \\ \| \\ CH_2 \end{matrix} \rightarrow LRh + \text{olefin} + H_2$$

Kinetic studies of the aldehyde decarbonylation were carried out by Baird, Nyman, and Wilkinson (33). The observed rate for the decarbonylation of valeraldehyde in dichloromethane at 20°C is given by

$$\text{Rate} = k_{obs}[RhCl(PPh_3)_3]$$

$$\text{where } k_{obs} = k_1 + k_2(C_4H_9CHO) \text{ with } k_1 = 5.68 \times 10^{-5}\ \text{sec}^{-1}$$

$$\text{and } k_2 = 1.81 \times 10^{-4}\ 1 \cdot \text{mole}^{-1}\ \text{sec}^{-1}$$

Based on these studies, the decarbonylation can be depicted as shown in Scheme **II**.

The mechanism of the catalytic decarbonylation with $RhCl(CO)(PPh_3)_2$ may be the following. The oxidative addition of aldehyde to $RhCl(CO)(PPh_3)_2$ gives a six-coordinate saturated complex. At high temperature, CO is removed to give a five-coordinate intermediate, making the acyl–alkyl rearrangement possible to give a six-coordinate alkyl complex. Another possibility is that $RhCl(CO)(PPh_3)_2$ first loses CO to give an unsaturated intermediate, which then undergoes oxidative addition of aldehyde to give the five-coordinate complex. The last step is the reductive elimination with the coupling of the alkyl group and hydrogen to give an alkane and the regeneration of the original complex.

According to these mechanisms, it would be expected that decarbonylation should proceed without changing the stereochemistry of the carbon atom directly connected to the aldehyde group. It has been shown that *cis*-substituted styrenes are formed predominantly by decarbonylation of

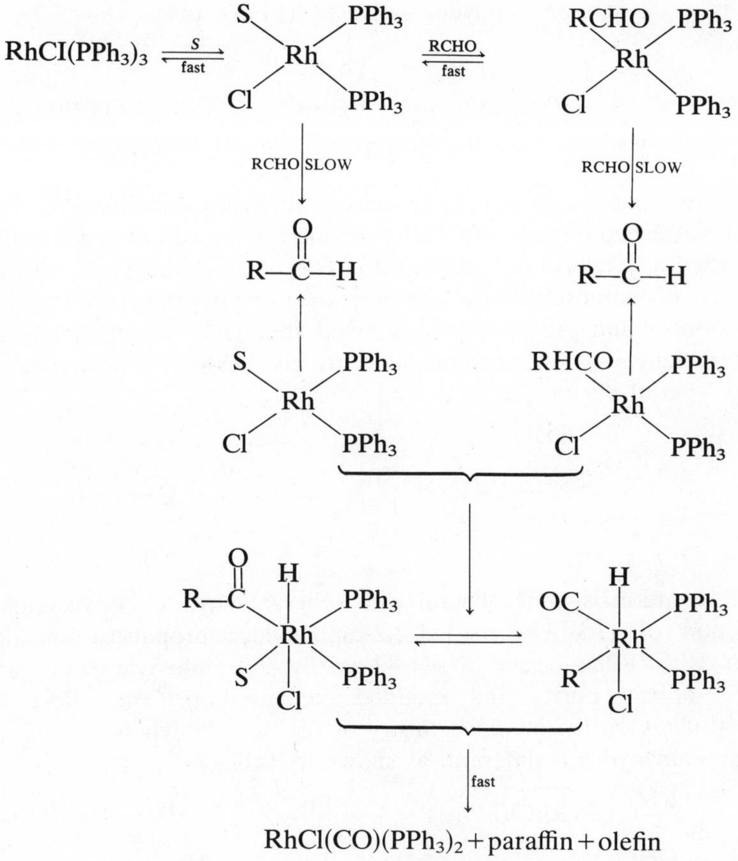

$$RhCl(PPh_3)_3 \underset{\text{fast}}{\overset{S}{\rightleftharpoons}} RhCl(S)(PPh_3)_2 \underset{\text{fast}}{\overset{RCHO}{\rightleftharpoons}} RhCl(RCHO)(PPh_3)_2$$

RCHO | SLOW

$$R{-}C(=O){-}H$$

$$RhCl(S)(PPh_3)_2 \qquad RhCl(RHCO)(PPh_3)_2$$

$$RC(O)RhH(S)Cl(PPh_3)_2 \rightleftharpoons RRhH(CO)Cl(PPh_3)_2$$

fast

$$RhCl(CO)(PPh_3)_2 + \text{paraffin} + \text{olefin}$$

Scheme **II**

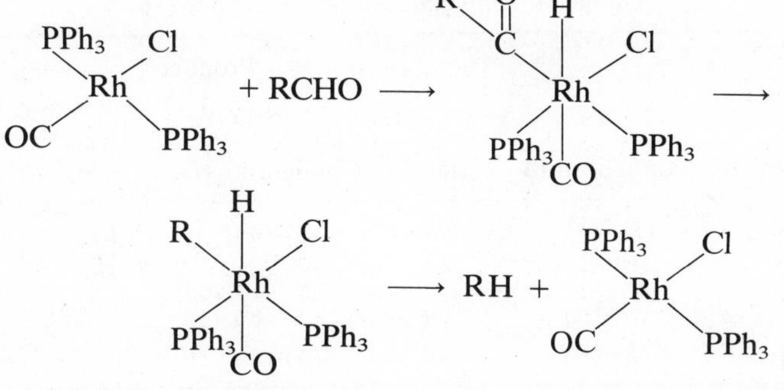

$$RhCl(CO)(PPh_3)_2 + RCHO \longrightarrow RC(O)RhHCl(CO)(PPh_3)_2 \longrightarrow$$

$$RRhHCl(CO)(PPh_3)_2 \longrightarrow RH + RhCl(CO)(PPh_3)_2$$

trans-α-alkylcinnamaldehydes with $RhCl(PPh_3)_3$ (44).

$$\text{PhHC=C(R)CHO} + RhCl(PPh_3)_3 \rightarrow \text{PhHC=C(R)H} + RhCl(CO)(PPh_3)_2 + PPh_3$$

More convincingly, optically active aldehydes containing the formyl group attached directly to a cyclopropane ring as well as to trigonally and tetrahedrally hybridized carbon atoms were decarbonylated with a high degree of stereoselectivity and with overall retention of configuration. Walborsky and Allen (45,46) showed that (*R*)-(-)-2-methyl-2-phenylbutyraldehyde was decarbonylated to give (*S*)-(+)-2-phenylbutane of 81% optical purity.

$$\text{(CHO)(CH}_3\text{)(C}_2\text{H}_5\text{)(Ph)C} \xrightarrow[\text{benzonitrile}]{RhCl(PPh_3)_3,\ 110^\circ C} \text{(H)(CH}_3\text{)(C}_2\text{H}_5\text{)(Ph)C}$$

The retention is quite general with other aldehydes. For example, the reaction of (*R*)-(+)-1-methyl-2,2-diphenylcyclopropanecarboxaldehyde with $RhCl(PPh_3)_3$ gave (*S*)-(+)-1-methyl-2,2-diphenylcyclopropane of 94% optical purity and retained configuration. The effect of α-substituents on the stereochemistry of the decarbonylation of the cyclopropyl aldehydes is different, as shown in Table 4.

$$\text{Ph}_2\text{C}_3\text{H}_2(\text{CH}_3)(\text{CHO}) + RhCl(PPh_3)_3 \longrightarrow \text{Ph}_2\text{C}_3\text{H}_2(\text{CH}_3)(\text{H}) + RhCl(CO)(PPh_3)_2 + PPh_3$$

TABLE 4
Decarbonylation of Optically Active 1-*Substituted* 2,2-*Diphenylcyclopropanecarboxaldehydes* (46)

			Product		
1-Substituent	Configuration	$[\alpha]_{Hg}$ (deg)	Configuration	$[\alpha]_{Hg}$ (deg)	Optical purity (%)
CH_3	(*R*)-(+)	99	(*S*)-(+)	141	94
Cl	(*S*)-(+)	153	(*S*)-(+)	168	83
F	(*S*)-(−)	164	(*S*)-(−)	22	73
OCH_3	(*S*)-(−)	49	(*S*)-(+)	75	6

That the decarbonylation is clearly intramolecular, that is, that the aldehyde hydrogen is the one that becomes bonded to the carbon atom from which the carbonyl is removed, is shown by the results of the decarbonylation of 1-methyl-2,2-diphenylcyclopropanecarboxaldehyde-*d*. When the aldehyde containing 96±3% deuterium at the aldehyde carbon is decarbonylated, the product contains 93±3% deuterium in the C-1 position. Furthermore, since C-1 deuterated aldehydes can be prepared readily and cheaply, this mild procedure provides a good method of preparing hitherto inaccessible optically active deuterated hydrocarbons.

$$\mathrm{RCDO + RhCl(PPh_3)_3 \rightarrow RD + PPh_3 + RhCl(CO)(PPh_3)_2}$$

Based on these experimental results, Walborsky and Allen proposed a radical pair mechanism for the decarbonylation reactions.

Aldehydes are decarbonylated with radical initiators such as di-*tert*-butyl peroxide. In the radical-initiated decarbonylation, skeletal rearrangement takes place during decarbonylation to give a different carbon chain. Thus, the following results of comparative studies of decarbonylation are important in the discussion of the mechanism of decarbonylation with rhodium complexes. The decarbonylation of β-phenylisovaleraldehyde, with $RhCl(PPh_3)_3$ at 160°C for 5 min (stoichiometric), or a catalytic amount of either $RhCl(CO)(PPh_3)_2$ or $RhCl(PPh_3)_3$ at 210°C for 3 hr afforded only *tert*-butylbenzene, and no rearranged product was obtained (47).

$$\mathrm{Ph{-}C(CH_3)_2{-}CH_2CHO \xrightarrow{\text{Rh catalyst}} Ph{-}C(CH_3)_2{-}CH_3}$$

It is well known that β-phenylisovaleraldehyde undergoes a neophyl rearrangement when subjected to radical decarbonylation induced by di-*tert*-butyl peroxide. Depending on the reaction conditions, the rearrangement takes place to an extent of 90% (48).

$$\mathrm{Ph{-}C(CH_3)_2{-}CH_2CHO \xrightarrow{\text{peroxide}} PhCH_2{-}CH(CH_3){-}CH_3 + Ph{-}C(CH_3)_2{-}CH_3 + PhCH_2{-}C(CH_3){=}CH_2 + PhCH{=}C(CH_3)_2}$$

These results indicate that the mechanism of rhodium-catalyzed decarbonylation is clearly different from that of radical-initiated decarbonylation.

The oxo reaction, the formation of aldehydes from olefins, CO, and H_2, is essentially the reverse of decarbonylation. Since rhodium is a decarbonylation catalyst, it is reasonable to expect that rhodium complexes should catalyze the carbonylation of olefins to form aldehydes. Indeed, both $RhCl(PPh_3)_3$ (49–52) and $RhCl(CO)(PPh_3)_2$ (53), a number of other rhodium complexes (54,55) and even rhodium oxide combined with a phosphine (56) were found to be very active catalysts for the oxo reaction.

$$RCH{=}CH_2 + CO + H_2 \rightarrow RCH_2CH_2CHO + \underset{\displaystyle CH_3}{\underset{|}{RCHCHO}}$$

In studying the mechanism of the hydroformylation reaction catalyzed by rhodium complexes, Wilkinson and co-workers (57,58) carried out the following experiment. Treatment of $RhH(CO)(PPh_3)_3$ with a mixture of ethylene and CO gave an acyl complex, $Rh(COC_2H_5)(CO)(PPh_3)_2$, by insertion of ethylene and CO. This complex was then converted into propionaldehyde and $RhH(CO)(PPh_3)_2$ by the action of H_2. Also propionaldehyde and $RhCl(CO)(PPh_3)_2$ were obtained by the action of HCl.

$$CH_2{=}CH_2 + RhH(CO)(PPh_3)_3 \longrightarrow Rh(C_2H_5)(CO)(PPh_3)_2 \xrightarrow{CO}$$

$$Rh(COC_2H_5)(CO)(PPh_3)_2 \begin{cases} \xrightarrow{H_2} C_2H_5CHO + RhH(CO)(PPh_3)_2 \\ \xrightarrow{HCl} C_2H_5CHO + RhCl(CO)(PPh_3)_2 \end{cases}$$

C. Decarbonylation of Acyl Halides

Acyl halides are decarbonylated with rhodium complexes both catalytically and stoichiometrically. In the stoichiometric decarbonylation, intermediate complexes are isolated and hence the mechanism of the reaction is more definitely established. Thus by the reaction of various acyl halides with $RhCl(PPh_3)_3$ under mild conditions, a five-coordinate acyl complex can be isolated as a crystalline compound (26,27). This complex is converted into $RhCl(CO)(PPh_3)_2$ and an olefin by a reductive elimination reaction. For example, palmitoyl chloride formed the acyl complex, which was converted into pentadecene and $RhCl(CO)(PPh_3)_2$ by heating. In this way, the stoichiometric decarbonylation of acyl halides can be carried out under mild conditions with or without isolation of an acyl complex. Phenylpropionyl chloride was decarbonylated to styrene by refluxing in benzene with $RhCl(PPh_3)_3$. When there is no hydrogen at the

$$RCH_2CH_2COX + RhCl(PPh_3)_3 \rightarrow RCH{=}CH_2 + RhCl(CO)(PPh_3)_2 + PPh_3 + HX$$

$$R'COX + RhCl(PPh_3)_3 \rightarrow R'X + RhCl(CO)(PPh_3)_2 + PPh_3$$

$$R' {=} PhCH_2, Ph$$

β-position, the corresponding halides are formed. Examples of the stoichiometric decarbonylation of acyl halides are shown in Table 5.

The decarbonylation of acyl halides is most interesting and useful when it is catalytic rather than stoichiometric. In catalytic decarbonylation, either $RhCl(CO)(PPh_3)_2$ or $RhCl(PPh_3)_3$ can be used, and aliphatic and aromatic acyl halides are decarbonylated. Also $RhCl_3$ and metallic rhodium can be used (21,27,31,59). In contrast to stoichiometric decarbonylation, which is carried out at rather low temperature, catalytic decarbonylation proceeds at about 200°C under homogeneous conditions. The decarbonylation has wide application and is especially useful for aromatic acyl halides. Results are shown in Table 6.

Benzoyl chloride and bromide are decarbonylated smoothly to give chlorobenzene and bromobenzene by heating at 200°C in the presence of a catalytic amount of rhodium complex. The reaction proceeds so smoothly that when a catalytic amount of the rhodium complex is added to benzoyl chloride and subjected to slow distillation from a Claisen flask, chlorobenzene rather than benzoyl chloride is collected.

$$PhCOCl \rightarrow PhCl + CO$$

In decarbonylation reactions of relatively low-boiling acyl halides, it is essential to distil off the decarbonylation product as it is formed; otherwise the reaction comes to a standstill after a certain period, which varies

TABLE 5

Stoichiometric Decarbonylation of Acid Halides with $RhCl(PPh_3)_3$

Acyl halides[a]	Solvent	Temp (°C)	Time, min (hr)	Product	Yield[b]	Ref.
$PhCH_2COCl$	Benzene	Reflux	10	$PhCH_2Cl$	81	21
$PhCH_2CH_2COCl$	Toluene	Reflux	20	$PhCH{=}CH_2$	71	21
$PhCH_2COCl$	CH_2Cl_2	Room	(48)	$PhCH_2Cl$	86	21
PhCOCl	none	180	2	PhCl	61	27

[a] An excess of acyl halide used.

[b] Yields based on the rhodium complex.

TABLE 6

Catalytic Decarbonylation of Acyl Halides Using Rhodium Complexes: $RhCl(CO)(PPh_3)_2 = A$; $RhCl(PPh_3)_3 = B$

Acyl halide	Amt (g)	Catalyst (g)	Temp (°C)	Time hr (min)	Product (%)	Yield[a]	Ref.
Palmitoyl chloride	8	A(0.1)	230	0.5	Pentadecene	54	27
Phenylacetyl chloride	8	A(0.1)	220	2	Benzyl chloride	60	27
Heptanoyl bromide	4	A(0.1)	200	0.5	1-Hexene (61), 2- and 3-hexenes (39)	85	27
Octanoyl chloride	8	A(0.1)	190–200	1	Heptenes	91	27
Octanoyl chloride	8	A(0.1)	200	1	1-Heptene (71) *trans*-2-heptene (24) *cis*-2-heptene (5)	90	27
Sebacoyl chloride	35	A(0.3)	210–220	3.5	1,7-Octadiene (25) 1,6-octadiene (34) 1,5-octadiene (13) other octadienes (28)	85	27
Benzoyl chloride	8	A(0.2)	200	4	Chlorobenzene	85	27
Benzoyl bromide	8	A(0.2)	220	1.5	Bromobenzene	87	27
p-Toluoyl chloride	8	A(0.25)	230	8.5	*p*-Chlorotoluene	66	27
p-Toluoyl bromide	30	A(0.6)	250	11	*p*-Bromotoluene	43	27
p-Methoxybenzoyl chloride	5	A(0.27)	140–250	12	*p*-Chloroanisole	46	27
3,4-Dichlorobenzoyl chloride	10	A(0.8)	250	1.5	1,3,4-Trichlorobenzene	96	27
Isophthaloyl chloride	5	A(0.2)	240–250	4	*m*-Dichlorobenzene (75) *m*-chlorobenzoyl chloride (25)	92	27

p-Iodobenzoyl chloride	3	B	at bp		*p*-Chloroiodobenzene	78	31
2,4-Dichlorobenzoyl chloride	30	B	at bp	0.5	1,2,4-Trichlorobenzene	98	31
2-Naphthoyl chloride	4	B	at bp		2-Chloronaphthalene	94	31
p-Chlorobenzoylchloride	5	B	at bp	8	*p*-Dichlorobenzene	79	31
2,4-Dichlorobenzoyl chloride	30	B	at bp	0.5	1,2,4-Trichlorobenzene	98	31
2-Naphthoyl chloride	4	B	at bp		2-Chloronaphthalene	94	31
p-Chlorobenzoyl chloride	5	B	at bp	8	*p*-Dichlorobenzene	79	31
o-Bromobenzoyl chloride	5	B	at bp	0.5	*o*-Bromochlorobenzene	78	31
2-Methyl-1-naphthoyl chloride	4	B	at bp		1-Chloro-2-methyl-naphthalene	93	31
5-Acenaphthenoyl chloride	—	B	300	(1–3)	5-Chloroacenaphthene	81	60
2-Fluorenoyl chloride	—	B	280	(1–3)	2-Chlorofluorene	60	60
2-Anthracenoyl chloride	—	B	310	(1–3)	2-Chloroanthracene	81	29
6-Chrysenoyl chloride	—	B	280	(1–3)	6-Chlorochrysene	48	29
1-Pyrenoyl chloride	—	B	250	(1–3)	1-Chloropyrene	94	29
1-Pyrenoyl bromide	—	B	290	(1–3)	1-Bromopyrene	89	29
Benzoyl iodide	2–10	B	175–250	—	Iodobenzene	62	62
o-Chlorobenzoyl iodide	2–10	B	175–250	—	*o*-Chlorobenzene	63	62
m-Iodobenzoyl iodide	2–10	B	175–250	—	*m*-Diiodobenzene	64	62
o-Iodobenzoyl iodide	2–10	B	175–250	—	*o*-Diiodobenzene	72	62
Benzoyl fluoride	—	B[c]	110	3	Fluorobenzene	—[d]	61
p-Toluoyl fluoride	—	B[c]	110	3	*p*-Fluorotoluene	—[d]	61
p-Fluorobenzoyl fluoride	—	B[c]	110	3	*p*-Difluorobenzene	—[d]	61

[a] Yield based on the acyl halide.
[b] Unless otherwise stated, the amount of catalyst B used was 50–100 mg.
[c] Amount of catalyst used not stated.
[d] 280–580% product obtained based on B.

from compound to compound. For example, Blum, Oppenheimer, and Bergmann (60) observed that *o*-, *m*-, and *p*-chlorobenzoyl chlorides, heated with a small amount of $RhCl(PPh_3)_3$ at 217°C, cease to react after 300, 40, and 10 min, respectively, and that the yields of each of the three dichlorobenzenes after these periods were 8.0, 15.3, and 1.7%. Benzoyl fluorides are decarbonylated to give fluorobenzenes at 80–120°C, but the catalytic turnover is somewhat low (5–6 times) (61). These decarbonylation reactions of aromatic acyl halides are extremely useful as a method of introducing halogen into aromatic rings.

$$PhCOF \rightarrow PhF$$

Aroyl iodides are decarbonylated more easily than chlorides and bromides, and the reaction is carried out at lower temperature (62). For example, benzoyl iodide was converted into iodobenzene in 62% yield by heating at 210°C for 5 min with a catalytic amount of $RhCl(PPh_3)_3$.

$$PhCOI \rightarrow PhI + CO$$

In the case of *o*-halogenobenzoyl iodides, Blum, Roseman, and Bergmann (62) observed that the catalytic decarbonylation of the *o*-iodinated acid chloride or bromide in the presence of iodine gave 30 and 47% yields, respectively, of *o*-diiodobenzene, accompanied by a small amount of the normal products, *o*-chloro- and *o*-bromoiodobenzene. For this reaction, intermediate formation of a benzyne species was proposed:

$$o\text{-}IC_6H_4COX + RhCl(PPh_3)_3 \longrightarrow [C_6H_4(I)(CO)\text{—}Rh(X)(PPh_3)_2] \longrightarrow$$

$$C_6H_4 \text{ (benzyne)} \xrightarrow{I_2} o\text{-}C_6H_4I_2$$

Besides aroyl halides, aroyl cyanides are decarbonylated to give nitriles, showing that the cyanide group acts as a pseudohalogen (60).

$$p\text{-}ClC_6H_4COCN \xrightarrow{3{\cdot}5\ \text{hr},\ 95\%} p\text{-}ClC_6H_4CN + CO$$

When long-chain acyl halides are decarbonylated and olefins are distilled off as soon as they are formed, it is possible to isolate the terminal olefin as the main product rather than internal olefins. For example, the decarbonylation of octanoyl bromide gave the following results (27):

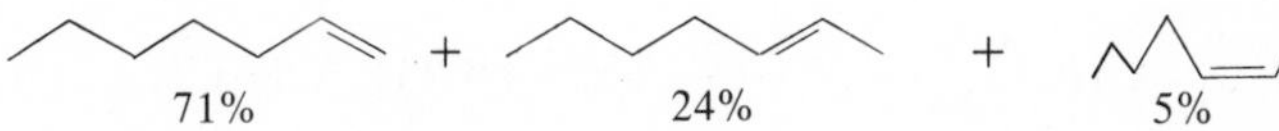

71% + 24% + 5%

Formation of terminal olefins becomes more selective by the addition of PPh_3 (59).

Stille and co-workers (63) carried out decarbonylations of two deuterated norbornanecarboxylic acid chlorides. First, *exo*-3-deutero-*exo*-2-norbornanecarboxylic acid chloride was subjected to decarbonylation with $RhCl(CO)(PPh_3)_2$ to give an 86% overall yield of products as shown below. The norbornene formed contained all the deuterium on the carbon atoms of the double bond. The results indicate that decarbonylation is occurring primarily by abstraction of the *exo*-3-deuterium, but that it does not appear to be concerted. (The norbornyl chloride was formed by addition of the evolved DCl and HCl to the norbornene).

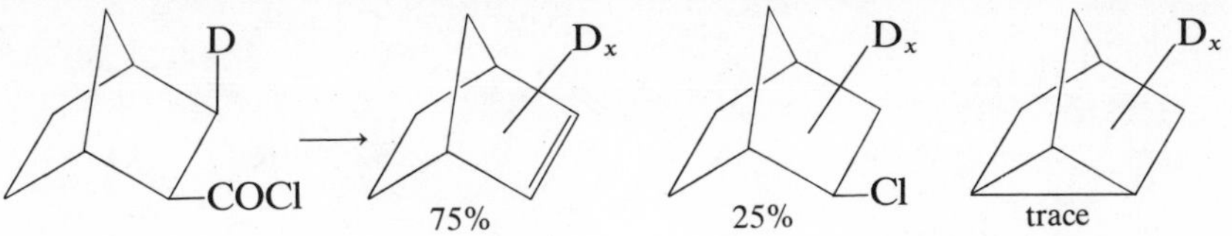

D, COCl → D_x (75%), D_x, Cl (25%), D_x (trace)

	Starting acid chloride	Norbornene (75%)	Norbornyl chloride (25%)	Nortricyclene (trace)
D_0	0%	61.1%	14.2	10.5
D_1	98.7	35.5	67.4	63.5
D_2	1.3	3.3	18.4	26.1

In a similar comparison of the formation of halides versus olefins, Stille and co-workers (47) observed the predominant formation of the chloride in the decarbonylation of 2,2-dimethylvaleroyl chloride with $RhCl(CO)$-$(PPh_3)_2$:

$$CH_3CH_2CH_2C(CH_3)_2{-}COCl \xrightarrow[17\ \text{hrs.}]{190\text{–}205°C}$$

$$\underset{74.4\%}{CH_3CH_2CH_2C(CH_3)_2{-}Cl} + \underset{16.8\%}{CH_3CH_2CH{=}C(CH_3)_2} + \underset{7.3\%}{CH_3CH_2CH_2C(CH_3){=}CH_2}$$

In the formation of olefin, removal of a secondary rather than a primary hydrogen is preferred. Thus when 2-methylvaleroyl chloride was decarbonylated at 190–200°C, the ratio of 2-pentene to 1-pentene was 18:7.

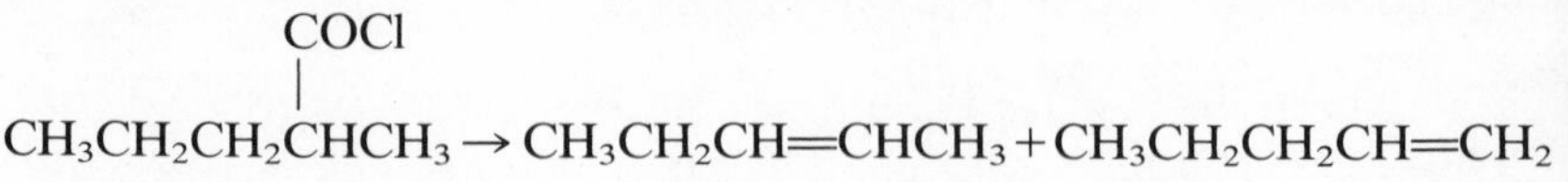

$$CH_3CH_2CH_2CH(COCl)CH_3 \rightarrow CH_3CH_2CH{=}CHCH_3 + CH_3CH_2CH_2CH{=}CH_2$$

The decarbonylation of dicarboxylic acid chlorides with six or less carbon atoms does not proceed satisfactorily. On the other hand, smooth decarbonylation of sebacoyl chloride gave a mixture of octadienes.

The mechanism of the decarbonylation of acyl halides using rhodium complexes may be described in the following way. In stoichiometric decarbonylation with $RhCl(PPh_3)_3$, the first step is oxidative addition of acyl halide to the d^8 complex to form a d^6 trivalent five-coordinate acyl complex. The acyl–alkyl rearrangement then follows to give a six-coordinate alkyl–carbonyl complex. Finally the alkyl–carbonyl complex is converted into olefin and $RhCl(CO)(PPh_3)_2$ by abstraction of hydrogen from a carbon atom β to the carbonyl group. When there is no hydrogen in the β-position, alkyl or aryl halides are formed (Scheme **III**).

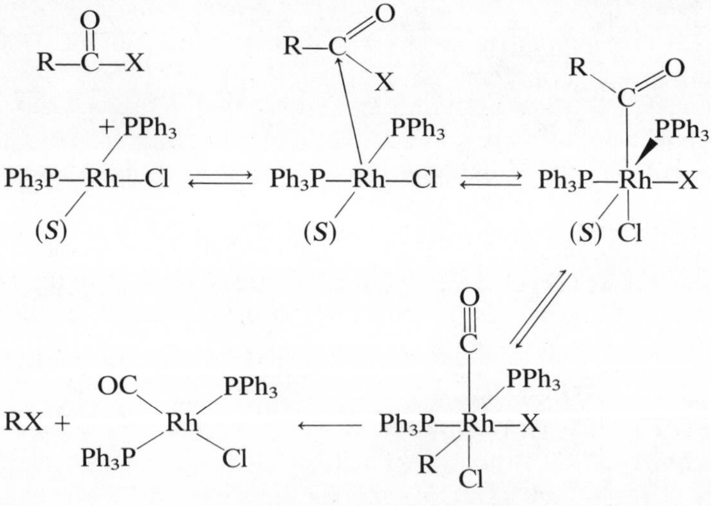

$$R{-}C(=O){-}X + Ph_3P{-}Rh(PPh_3)(S){-}Cl \rightleftharpoons Ph_3P{-}Rh(PPh_3)(S)(\leftarrow C(=O)(X)R){-}Cl \rightleftharpoons Ph_3P{-}Rh(PPh_3)(S)(Cl)(C(=O)R){-}X$$

$$\rightleftharpoons Ph_3P{-}Rh(PPh_3)(R)(Cl)(C{\equiv}O){-}X \longrightarrow RX + Rh(OC)(PPh_3)(Ph_3P)(Cl)$$

Scheme **III**

Catalytic decarbonylation with $RhCl(CO)(PPh_3)_2$ proceeds through oxidative addition of acyl halide to form an acyl complex in the first step.

When the acyl complex is heated in the absence of carbon monoxide, CO is lost to form a five-coordinate unsaturated acyl complex, which undergoes the acyl–alkyl rearrangement to form an alkyl complex. Finally olefins or alkyl halides are formed with regeneration of $RhCl(CO)(PPh_3)_2$. Consequently, the reaction can be carried out catalytically.

$$RhCl(CO)(PPh_3)_2 + RCOX \rightleftharpoons RCORh(CO)Cl(X)(PPh_3)_2$$

$$\downarrow -CO \quad \uparrow CO$$

$$RhCl(PPh_3)_3 + RCOX \longrightarrow RCORhCl(X)(PPh_3)_2$$

$$\updownarrow$$

$$RhCl(CO)(PPh_3)_2 + \underset{\text{(or olefin)}}{RX} \rightleftharpoons RRhCl(X)(CO)(PPh_3)_2$$

Each step in this mechanism is probably reversible and it is therefore expected that the same complexes should function as carbonylation catalysts. Actually, the carbonylation of halides, such as allyl bromide or benzyl halide, was found possible with $RhCl(CO)(PPh_3)_2$ as catalyst (28).

$$CH_2{=}CHCH_2Br + CO \xrightarrow[130°C]{RhCl(CO)(PPh_3)_2} CH_2{=}CHCH_2COBr$$

Here, CO is inserted into the carbon–halogen bond. The mechanism of the carbonylation reaction can be said to be exactly the reverse of decarbonylation. Carbonylation proceeds in the presence of CO, while decarbonylation takes place in the absence of CO, with the same catalyst. Ample experimental results have been reported which support this decarbonylation mechanism. In the following, some of the results are discussed.

The acyl rhodium complex was isolated by the reaction of $RhCl(PPh_3)_3$ with acyl halides in solution at room temperature. The complex showed an acyl absorption at 1715 cm^{-1}. Further transformation of the acyl complexes was carried out. For example, when dichloro(3-phenylpropionyl)bis(triphenylphosphine)rhodium was heated to 170°C under reduced pressure (100 mm), styrene and $RhCl(CO)(PPh_3)_2$ were obtained.

$$PhCH_2CH_2CORhCl_2(PPh_3)_2 \rightarrow \underset{77\%}{PhCH{=}CH_2} + \underset{99\%}{RhCl(CO)(PPh_3)_2} + HCl$$

From dichloropalmitoylbis(triphenylphosphine)rhodium, *trans*-2-pentadecene was formed as a main product. When a chloroform–ether solution of the acyl complex was treated with an equivalent amount of iodine at room temperature, the acyl complex was converted into an iodine-containing complex and a pure terminal olefin. This procedure is useful for the conversion of long-chain acyl halides into terminal olefins with one less carbon atom. By the action of CO under pressure (50 atm) at room temperature, the acyl complex was converted into an acyl halide and $RhCl(CO)(PPh_3)_2$ (27).

$$Rh(RCO)(Cl)_2(PPh_3)_2 + CO \rightarrow RCOCl + RhCl(CO)(PPh_3)_2$$

The reaction of benzoyl chloride with $RhCl(PPh_3)_3$ at 30–35°C gave the acyl complex, which absorbs at 1702 and 1657 cm^{-1}. Recrystallization of this benzoyl complex gave a phenyl complex formed by the acyl–alkyl rearrangement (64).

$$PhCOCl + RhCl(PPh_3)_3 \rightarrow RhCl_2(PhCO)(PPh_3)_2 \rightarrow RhCl_2(Ph)(CO)(PPh_3)_2$$

The reaction of acetyl chloride gave both the acetyl and methyl complexes. These complexes are unstable and easily gave methyl chloride and $RhCl(CO)(PPh_3)_2$. The reaction of acetyl chloride was quite different depending on the solvents used. In chloroform, the freshly prepared acyl complex was converted into $RhCl_2(COCH_3)(CO)(PPh_3)_2$ by the action of CO. This complex was converted into $RhCl(CO)(PPh_3)_2$ and acetyl chloride in a few minutes. The reaction of $RhCl(PPh_3)_3$ with acetyl chloride is summarized in Scheme **IV**.

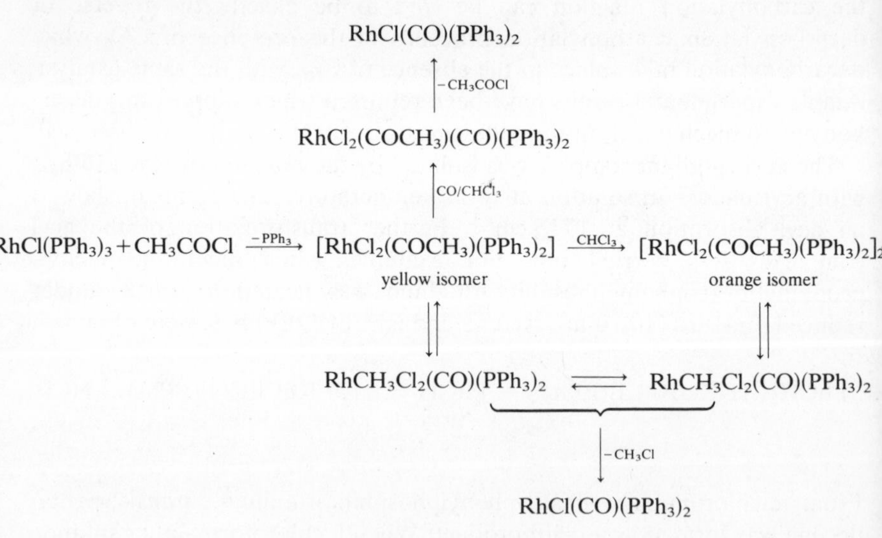

Scheme **IV**

With propionyl chloride, the corresponding acyl complex was isolated more easily. By the action of CO, the acyl complex was converted into $RhCl(CO)(PPh_3)_2$ and propionyl chloride through intermediate formation of $RhCl_2(COC_2H_5)(CO)(PPh_3)_2$.

$$RhCl(PPh_3)_3 + C_2H_5COCl \longrightarrow [RhCl_2(COC_2H_5)(PPh_3)_2] \xrightarrow{CO}$$

$$RhCl_2(COC_2H_5)(CO)(PPh_3)_2 \longrightarrow RhCl(CO)(PPh_3)_2 + C_2H_5COCl$$

In the oxidative addition of acyl halides to the rhodium complex, the ease of addition for a given complex is in the order: $Cl < Br < I$, and depends on the nature of the phosphine. The ease of the acyl–alkyl rearrangement seems to be in the order: $Ph > CH_3 > C_2H_5$; the ease of rearrangement is expected to decrease with the presence of electron-withdrawing substituents in the *para* position of the aromatic rings.

The alkyl and acyl complexes of rhodium were synthesized independently by olefin insertion (64). Thus with ethylene insertion, the ethyl complex was isolated, which then reacted with CO at 25°C and 1 atm to give ethyl chloride and $RhCl(CO)(PPh_3)_2$. CO insertion took place to form the acyl complex at a lower temperature (−60 to −30°C). Also,

$$RhCl(PPh_3)_3 \xrightarrow{HCl} RhCl_2H(PPh_3)_2 \xrightarrow{CH_2=CH_2} RhCl_2Et(PPh_3)_2$$

$$RhCl_2Et(PPh_3)_2 \xrightarrow{CO,\ 25°C} EtCl + RhCl(CO)(PPh_3)_2$$

$$RhCl_2Et(PPh_3)_2 \xrightarrow{CO,\ -30°C} RhCl_2(EtCO)(PPh_3)$$

acryloyl complex was obtained by the action of CO on vinyl complex.

$$RhCl_2(C_2H_3)(PPh_3)_2 + CO \rightarrow RhCl_2(COCH{=}CH_2)(PPh_3)_2$$

From these experimental results, it is apparent that oxidative addition to $RhCl(PPh_3)_3$ is a facile reaction.

On the other hand, attempts to isolate an acyl complex from oxidative addition to $RhCl(CO)(PPh_3)_2$ have not yet been successful. $RhCl_2(COR)(PPh_3)_2(CO)$ seems to be less stable than $RhCl_2(COR)(S)(PPh_3)_2$, due to the effect of coordinated CO on the loss of acyl halide. However, the addition reaction of acetyl bromide to a similar complex, $RhBr(CO)(PEt_2Ph)_2$, was reported (65).

$$RhBr(CO)(PEt_2Ph)_2 + CH_3COBr \rightarrow RhBr_2(CH_3CO)(CO)(PEt_2Ph)_2$$

In another example, the reaction of $Rh(PPh_3)_2C_6H_4PPh_2$ with benzoyl chloride gives the phenyl complex shown by oxidative addition and subsequent acyl–alkyl rearrangement. This is a coordinatively unsaturated d^8 complex. At high temperature with a catalytic amount of the rhodium complex, benzoyl chloride is decarbonylated to benzene (66). With oxalyl chloride the corresponding complex was isolated at room temperature.

Ph3P PPh2 Rh Ph3P + PhCOCl ⟶ Cl Ph3P PPh2 Rh Ph CO

Ph3P PPh2 Rh Ph3P + COCl | COCl ⟶ Cl | CO Cl PPh2 Rh CO PPh3

As a model reaction for the first step in the carbonylation of alkyl halides catalyzed by $RhCl(CO)(PPh_3)_2$, the oxidative addition of methyl iodide to $RhCl(CO)(PPh_3)_2$ or a corresponding alkylphosphine complex gave $RhCl(I)(CH_3)(CO)(PPh_3)_2$. Subsequent isomerization via methyl transfer gave the acetyl complex (67,68).

$$RhCl(CO)(PPh_3)_2 + CH_3I \rightarrow RhCl(I)(CO)(CH_3)(PPh_3)_2 \rightarrow RhCl(I)(CH_3CO)(PPh_3)_2$$

If the mechanism of decarbonylation requires an oxidative addition to form the acyl complex as the first step, the decarbonylation of carboxylic acids should not be easy. For carboxylic acids, if there is any possibility of oxidative addition, it will be at the O–H bond rather than at the C–O bond. However, it was reported that carboxylic acids can be decarbonylated to give olefins by treatment with $RhCl_3(PEt_2Ph)_3$ (69). In this reaction, about 3 moles of olefin are formed per mole of the rhodium complex, which is finally converted into $RhCl(CO)(PEt_2Ph)_2$. The mechanism is not known.

$$CH_3(CH_2)_4COOH \longrightarrow$$

In related reactions, the catalytic decarbonylation of certain aromatic acid anhydrides with rhodium complexes was reported by Blum and Lipshes (70). For example, when benzoic anhydride was heated with a catalytic amount of $RhCl(PPh_3)_3$ or $RhCl(CO)(PPh_3)_2$ at 240°C for 4.5 hrs, CO was evolved, and the following products were obtained: benzoic acid (30%), fluorenone (36%), biphenyl (6%), and benzophenone (3%). At 225°C, the reaction is more selective, and benzoic acid and fluorenone are primary products. At temperatures above 240°C, the formation of biphenyl increases.

$$2PhC(=O){-}O{-}C(=O)Ph \xrightarrow{225°C} \text{fluorenone} + 2PhCOOH + PhC(=O)Ph + Ph{-}Ph + CO$$

From studies with substituted benzoic anhydrides and mixed anhydrides, it was suggested that the reaction proceeds via formation of an arylaroylrhodium complex followed by electrophilic attack of the aroyl group at the *ortho* position of the aryl ligand to give an arylrhodium complex. The complex is decomposed to give fluorenone (71).

RhCl(PPh₃)₃ → Rh—CO · Rh · C=O · Rh—C≡O

Another possibility is aryl migration to give an aroylrhodium complex, which then decomposes either to fluorenone or to biphenyl. For this reaction anhydrous $RhCl_3$ is also active.

D. Decarbonylation of Ketenes

Although less extensively studied than acyl halides and aldehydes, the decarbonylation of diphenylketene using vanadium, titanium, and rhodium complexes has been reported. The reaction is stoichiometric. Here, the stoichiometric decarbonylation of diphenylketene with rhodium, as well as with vanadium and titanium complexes, is surveyed. The catalytic decarbonylation of diphenylketene with a cobalt complex is reviewed later.

Ketene is a highly unsaturated compound: The mechanism via simple oxidative addition and acyl–alkyl rearrangement does not hold for the decarbonylation of ketene. Other types of coordination must be considered. Ketene has a cumulative double bond system and forms complexes with transition metal compounds. Thus, diphenylketene reacts with $(\pi\text{-}C_5H_5)_2V$ and $(\pi\text{-}C_5H_5)Ti(CO)_2$ to form 1:1 complexes. Interestingly, these complexes decompose to tetraphenylethane when heated with protic solvents such as water and alcohol (72). Usually ketene itself reacts

$$\begin{aligned} Ph_2C{=}C{=}O + (\pi\text{-}C_5H_5)_2V &\rightarrow (\pi\text{-}C_5H_5)_2V(Ph_2C{=}C{=}O) \\ &\rightarrow Ph_2CHCHPh_2 \end{aligned}$$

easily with these protic solvents to form carboxylic acid derivatives. But, on coordination with the metal complexes, decarbonylation takes place.

Further studies on the decarbonylation of ketene revealed that $RhCl(PPh_3)_3$ is active for the stoichiometric decarbonylation and its mechanism has been made quite clear (73). When the rhodium complex was heated with diphenylketene in toluene, tetraphenylethylene and tetraphenylallene were obtained together with $RhCl(CO)(PPh_3)_2$ and triphenylphosphine oxide. It should be noted that triphenylphosphine oxide is obtained rather than PPh_3. The first step of the reaction seems to

$$Ph_2C{=}C{=}O + RhCl(PPh_3)_3 \rightarrow Ph_2C{=}CPh_2 + Ph_2C{=}C{=}CPh_2$$

be coordination of the C=C bond to rhodium, followed by C–C bond cleavage to form a complex containing coordinated carbene and CO. The coupling of two coordinated carbenes forms tetraphenylethylene. The formation of the allene proceeds as follows. The carbene species reacts with PPh_3 to form a ylide, which then reacts with another molecule of diphenylketene as in the Wittig reaction to form tetraphenylallene. As supporting evidence for this mechanism of allene formation, tetraphenylallene was obtained as the main product, together with a small amount of tetraphenylethylene (93:7), when the reaction was carried out in the presence of an excess of PPh_3.

$$Ph_2C{=}C{=}O + RhCl(PPh_3)_3 \longrightarrow Ph_2C{=}RhCl(CO)(PPh_3)_2 + PPh_3$$

$$\downarrow PPh_3$$

$$Ph_2C{=}PPh_3 + RhCl(CO)(PPh_3)_2$$

$$\downarrow Ph_2C{=}C{=}O$$

$$\begin{array}{l} \quad Ph_2C{-}PPh_3 \\ \qquad | \qquad | \\ Ph_2C{=}C{-}O \end{array}$$

$$\downarrow$$

$$Ph_2C{=}C{=}CPh_2 + Ph_3PO$$

E. Miscellaneous Decarbonylations

Unlike aldehydes, the decarbonylation of ketones seems to be difficult. So far, no example of the decarbonylation of common ketones has been published. However, as an exceptional example, the stoichiometric decarbonylation reaction of mono- and diacetylenic ketones with $RhCl(PPh_3)_3$ has been reported. For example, 1,5-diphenylpentadiyn-3-one was decarbonylated to give diphenylbutadiyne in 78% yield by heating with the rhodium complex (74). The mechanism of this peculiar reaction is unknown.

$$PhC{\equiv}C{-}\overset{\overset{\displaystyle O}{\|}}{C}{-}C{\equiv}CPh + RhCl(PPh_3)_3 \rightarrow PhC{\equiv}C{-}C{\equiv}CPh + RhCl(CO)(PPh_3)_2$$

In connection with the facile carbon–carbon bond cleavage which takes place in this decarbonylation of diacetylenic ketones, another reaction should be mentioned. Müller and Segnitz (75) found that diacetylenic ketones undergo a rearrangement when treated with $PtCl_4$ as catalyst.

$$PhC{\equiv}C{-}\overset{\overset{\displaystyle O}{\|}}{C}{-}C{\equiv}CPh \xrightarrow{PtCl_4} PhC{\equiv}C{-}C{\equiv}C{-}\overset{\overset{\displaystyle O}{\|}}{C}{-}Ph$$

These results indicate some specific interactions of acetylenic ketones with noble metal complexes.

The many examples of decarbonylation using $RhCl(PPh_3)_3$ make it apparent that this complex is extremely useful for extracting CO from various carbonyl compounds. The complex can also be used for removing

CO coordinated to metals. Thus cyclopentadienyldicarbonyl(α-phenylpropionyl)iron was converted to a corresponding alkyl complex (76). This acyl–alkyl rearrangement of the iron complex is not possible by heating or irradiation. The rhodium complex extracts CO at room temperature very smoothly, making the acyl–alkyl rearrangement possible.

$$C_5H_5Fe(CO)_2\text{—}\overset{\overset{\displaystyle O}{\|}}{C}\text{—}\underset{\underset{\displaystyle CH_3}{|}}{C}HPh + RhCl(PPh_3)_3 \rightarrow$$

$$C_5H_5Fe(CO)_2\text{—}\overset{\overset{\displaystyle CH_3}{|}}{C}HPh + PPh_3 + RhCl(CO)(PPh_3)_2$$

An example of the transfer of CO from an acyl complex of iridium is shown in the following equation (57). The propionyl group is decomposed into ethylene and CO.

$$Ir(COC_2H_5)(CO)_2(PPh_3)_2 + RhCl(PPh_3)_3 \rightarrow$$

$$RhCl(CO)(PPh_3)_2 + IrH(CO)(PPh_3)_2 + CO + C_2H_4$$

IV. DECARBONYLATIONS WITH PALLADIUM COMPLEXES

A. Introduction

Palladium is an efficient catalyst for the carbonylation of olefinic and acetylenic compounds to form various saturated and unsaturated esters, lactones, and acyl halides (77–79). Aldehydes can also be formed, in low yield, from olefins by using a palladium catalyst (80). Metallic palladium, palladium salts, and most efficiently, palladium complexes with phosphine ligands are used for these carbonylation reactions. Later it was found that palladium is an efficient catalyst for both carbonylation and decarbonylation as reversible processes (81). Acyl halides and aldehydes are decarbonylated catalytically with palladium catalysts (82). Metallic palladium, rather than palladium complexes, is used for these decarbonylation reactions.

B. Decarbonylation of Aldehydes

Aldehydes can be decarbonylated catalytically in the presence of metallic palladium at high temperature. For some time, it has been known that certain aldehydes can be decarbonylated when heated above 180°C in the presence of metallic palladium and other hydrogenation catalysts. Considerable work has been reported on the palladium-catalyzed decarbonylation of aldehydes from both mechanistic and synthetic viewpoints. The products of decarbonylation are olefins and the corresponding saturated hydrocarbons.

$$RCH_2CH_2CHO \rightarrow RCH{=}CH_2 + RCH_2CH_3 + CO + H_2$$

Since the decarbonylation of aldehydes proceeds smoothly in high yield, it has been used for synthetic purposes. Examples are shown in Table 7. In the five-step irone synthesis from α-pinene, an aldehyde group was introduced by ozonization in the first step. The *cis*-pinonic aldehyde thus formed was decarbonylated to give pinonone and pinonenone in 80% yield by heating with a palladium catalyst at 220°C (83,84). Another application to synthesis is the two-step preparation of

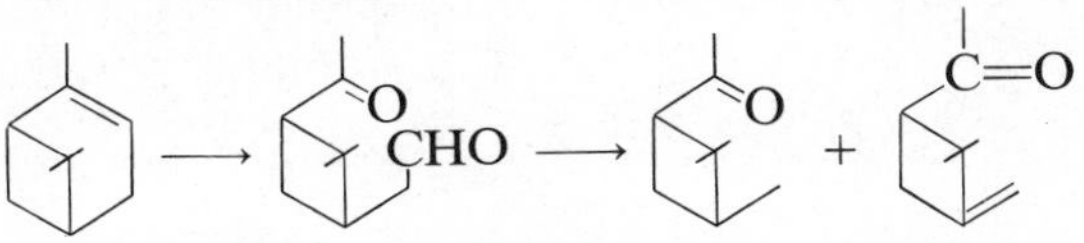

apopinene from the easily available α-pinene. The methyl group of α-pinene is oxidized to a formyl group with selenium oxide to form myrtenal; the compound is then decarbonylated with palladium on $BaSO_4$ at 195°C to give apopinene in an overall yield of 55% (85).

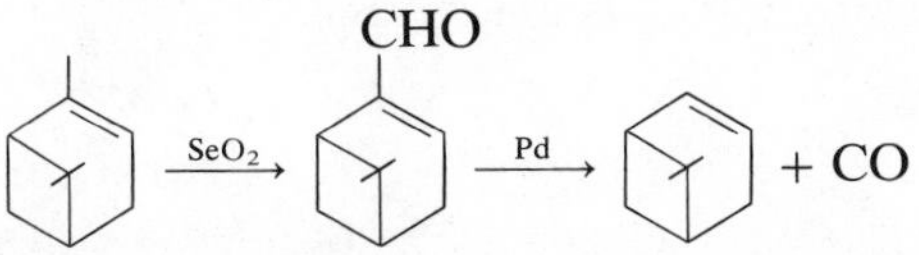

Lower boiling aldehydes in the gas phase can be decarbonylated by using a solid catalyst containing palladium. Kinetic studies were carried out on the decarbonylation of butyraldehyde over a palladium film to give both propylene and propane (86). As mentioned earlier (43), platinum metal contained in a fixed-bed catalyst is used for gas-phase decarbonylation of aldehydes.

Trans-α-substituted cinnamaldehydes can be decarbonylated to give *cis*-styrene derivatives as a main product, when the product is distilled out as formed (Table 8) (87). From this result, it is clear that the initial product is largely if not completely, the *cis*-styrene. This is a mechanistically important observation.

The decarbonylation of aromatic aldehydes over palladium seems to be fairly general (88). However, several exceptions are known. 1-Naphthaldehyde readily lost CO at 210°C, but 2-naphthaldehyde did not (89).

TABLE 7
Decarbonylation of Aldehydes Catalyzed by Metallic Palladium

Aldehydes	Catalyst	Temp (°C)	Time (hr)	Products (%)	Ref.
cis-Pinonic aldehyde	5% Pd/$BaSO_4$	200		1-Acetyl-2,2,3-trimethylcyclobutane (80)	83
Myrtenal	5% Pd/$BaSO_4$	155–195	4	Apopinene (85)	85
Citral (125 ml)	Pd/$BaSO_4$	212	10	Geraniolene (70 ml)	88
Citronellal (80 ml)	Pd/$BaSO_4$	209	5	Citronellene (40–45 ml)	88
trans-α-Methylcinnamaldehyde	10% Pd/C	180	23	*trans*-1-phenylpropene, *cis*-1-phenylpropene (92)	87
Benzaldehyde	5% Pd/C	179	1	Benzene (78)	89
p-Tolualdehyde	5% Pd/C	199	0.5	Toluene (88)	89
o-Methoxybenzaldehyde	5% Pd/C	243	1	Anisole (94)	89
m-Nitrobenzaldehyde	5% Pd/C	205	1.75	Nitrobenzene (86)	89
p-Nitrobenzaldehyde	5% Pd/C	205	0.5	Nitrobenzene (79)	89
Pyridine-2-aldehyde	5% Pd/C	180	1	Pyridine (68)	89
1-Naphthaldehyde	5% Pd/C	220	0.5	Naphthalene (80)	89
2-Naphthaldehyde	5% Pd/C	250–255	0.5	No reaction	89
9-Anthraldehyde	5% Pd/C	240–245	0.5	Anthracene (84)	89
Methyl-2-formyl-2′-biphenylcarboxylate	5% Pd/C	240–250	0.5	2-Biphenylcarboxylic acid (20)	89
2-Formyl-2′-biphenyl carboxylic acid	5% Pd/C	240–250	0.5	No reaction	89
1-Formyl-9-fluorenone	5% Pd/C	240–250	0.25	9-Fluorenone (82)	89

TABLE 8

Decarbonylation of Substituted Cinnamaldehydes

$$\text{Ph(H)C=C(R)CHO} \xrightarrow{\text{catalyst}} \underset{cis}{\text{Ph(H)C=C(R)H}} + \underset{trans}{\text{Ph(H)C=C(H)R}} + \underset{\text{X}}{\text{PhCH}_2\text{CH}_2\text{R}}$$

R	Amt (g)	Catalyst	Amt (g)	Method[a]	Reaction Temp (°C)	Reaction Time (hr)	Yield (%)	Product distribution *cis*	Product distribution *trans*	Product distribution X	Ref.
CH_3	—[b]	Pd/C	—[c]	A	180	23	92	85	16	0	87
CH_3	—[b]	Pd/C	—[c]	B	180	24	48	10	90	0	87
C_2H_5	—[b]	Pd/C	—[c]	A	210	23	85	61	34	4	87
C_2H_5	—[b]	Pd/C	—[c]	B	200	46	49	10	21	67	87
n-C_3H_7	—[b]	Pd/C	—[c]	A	200	18	48	35	41	22	87
n-C_3H_7	—[b]	Pd/C	—[c]	B	210	74	49	8	12	77	87
CH_3	10	$RhCl(CO)(PPh_3)_2$	0.1	A	250–60	7	77	88	12	0	27,28
CH_3	10	—[d]	0.25	A	250–60	4	87	90	10	0	27,28
C_2H_5	10	$RhCl(CO)(PPh_3)_2$	0.3	A	250–60	3	87	80	20	Trace	27,28

[a] In method A, product was distilled as rapidly as possible; in method B the reaction mixture was distilled to remove product only after decarbonylation was complete.

[b] Aldehydes, 40–100 g, were used.

[c] 1 wt % of 10% Pd on charcoal used.

[d] $RhClCO[(p\text{-}CH_3OC_6H_4)_3P)]_2$.

Only a trace of gas was generated from the latter compound even at 250°C. The decarbonylation of high-boiling aliphatic aldehydes was much slower than that of aromatic aldehydes, requiring a reaction time of 9 hr before gas evolution stopped. The gas contained CO and some H_2.

Aldehydes can be decarbonylated with free radical generators with skeletal rearrangement as described earlier (48). Unlike free radical decarbonylation, no carbon skeletal rearrangement was observed in the palladium-catalyzed decarbonylation of tetralin-2-carboxaldehyde, β-phenylisovaleraldehyde, and 3,3-dimethyl-4-phenylbutyraldehyde (90–93).

$$PhCH_2{-}\overset{\displaystyle CH_3}{\underset{\displaystyle CH_3}{\overset{|}{\underset{|}{C}}}}{-}CH_2CHO \xrightarrow{Pd/C} PhCH_2{-}\overset{\displaystyle CH_3}{\underset{\displaystyle CH_3}{\overset{|}{\underset{|}{C}}}}{-}CH_3$$

Wilt and Abegg (92) have reported that *tert*-butylbenzene, contaminated by a small amount of the rearranged products, is the main product of the palladium-catalyzed decarbonylation of β-phenylisovaleraldehyde.

$$Ph{-}\overset{\displaystyle CH_3}{\underset{\displaystyle CH_3}{\overset{|}{\underset{|}{C}}}}{-}CH_2CHO \rightarrow Ph{-}\overset{\displaystyle CH_3}{\underset{\displaystyle CH_3}{\overset{|}{\underset{|}{C}}}}{-}CH_3 + \text{rearranged products (from traces to 6\%)}$$

A mechanism based on free radicals on the catalyst surface was proposed. However, these are the same results observed with rhodium complexes and a mechanism based on coordination chemistry is most likely, as will be discussed later.

Formates can be considered as having an aldehyde group; thus, palladium-catalyzed decarbonylation is expected. Matthews and coworkers (94) found that octyl formate can be decarbonylated to 1-octanol

$$H{-}\underset{\displaystyle O}{\underset{\|}{C}}{-}OC_8H_{17} \rightarrow C_8H_{17}OH + CO$$

in 95.7% yield by heating with palladium on charcoal at 200°C. The decarbonylation is dependent on the structure of the formate. Thus benzyl formate was converted into toluene and CO_2 rather than CO.

The decarbonylation of formamides to form amines catalyzed by $PdCl_2$ was briefly reported (95). This decarbonylation is reversible, and amines can be carbonylated to formamides by the catalytic action of palladium on carbon (96).

$$R_2NCHO \rightleftharpoons R_2NH + CO$$

The removal of a —CH_2OH group catalyzed by palladium is known. In this reaction, primary alcohols are dehydrogenated to aldehydes and then decarbonylated with metallic palladium. For example, 1-hydroxymethyl-5,6,7,8-tetrahydronaphthalene was decarbonylated to give naphthalene by heating to 270–300°C with palladium for 13 hr (90).

$$\text{1-(CH}_2\text{OH)-5,6,7,8-tetrahydronaphthalene} \xrightarrow{Pd/C} \underset{95.6\%}{\text{naphthalene}} + \underset{4\%}{\text{1-methylnaphthalene (CH}_3\text{)}}$$

C. Decarbonylation of Acyl Halides

The carbonylation of olefins to form carboxylic acid derivatives is catalyzed by palladium. As will be discussed later in the section on mechanisms, it was shown by Tsuji, Ohno, and Kajimoto (81) that the reverse reaction, that is, formation of olefins from acyl halides, is also possible when using palladium catalysts. Palladium on carbon is an active catalyst. Palladium salts such as $PdCl_2$, which are readily reduced to black metallic palladium when heated with acyl halides, can conveniently be used for the decarbonylation (82). When heated at 200°C or above, higher aliphatic acyl halides are decarbonylated in high yield to give olefins having one less carbon atom. Examples of the palladium-catalyzed decarbonylation of acyl halides are shown in Table 9. The product is a

$$RCH_2CH_2COCl \rightarrow R'CH{=}CHR'' + CO + HCl$$

mixture of isomeric internal olefins, even when distilled from the reaction mixture as soon as it is formed. For example, when decanoyl chloride is heated with $PdCl_2$ at 200°C in a distilling flask, rapid evolution of CO and HCl stops after 1 hr, during which time a mixture of nonene isomers is distilled off in high yield. The isomerization of the double bond is very fast, and a mixture of isomeric olefins is always obtained. Hexenes obtained by the decarbonylation of heptanoyl bromide consisted of 7.5% of 1-hexene and 92.5% of a mixture of 2- and 3-hexenes.

When double-bond migration of olefins produced by decarbonylation is impossible, other products are formed in addition to olefins. When phenylpropionyl chloride was decarbonylated, the main products were

TABLE 9

Decarbonylation of Acyl Halides Catalyzed by Noble Metal Compounds (82)

Acyl halide	Amt (g)	Catalyst (g)	Time (hr)	Temp (°C)	Products	Yield %
$CH_3(CH_2)_8COCl$	5	$PdCl_2$ (0.1)	1.5	200	Nonenes	90
$CH_3(CH_2)_7COCl$	5	$PdCl_2$ (0.2)	1.5	200	Octenes	83
$CH_3(CH_2)_6COBr$	5	$PdCl_2$ (0.2)	1.5	200	*trans*-2- and 3-Heptenes (76) *cis-2- and* 3-heptenes (24) 1-heptene	80
$CH_3(CH_2)_5COBr$	5	$PdCl_2$ (0.1)	1.0	200	1-Hexene (7.5) 2- and 3-hexenes (92.5)	80
$PhCH_2CH_2COCl$	10	$PdCl_2$ (0.2)	1.0	200	Styrene	53
CH_2CH_2COCl | CH_2CH_2COCl	8	$PdCl_2$ (0.2)	0.5	280	Cyclopentanone	30
$CH_3(CH_2)_8COCl$	5	$Pd(acac)_2$ (0.2)	1.5	200	Nonenes	80
$CH_3(CH_2)_8COCl$	5	$PdCl_2(PhCN)_2$ (0.2)	1.0	200	Nonenes	20
$CH_3(CH_2)_8COCl$	5	PdO (0.2)	2.0	200	Nonenes	86
$CH_3(CH_2)_8COCl$	5	$Pd(NO_3)_2$ (0.2)	1.5	200	Nonenes	76
$Ch_3(CH_2)_8COCl$	5	$RhCl_3$ (0.2)	1.5	200	Nonenes	86
$CH_3(CH_2)_8COCl$	5	Rh (0.1)	0.3	200	Nonenes	85
cyclo-$C_8H_{15}COCl$	5	$PdCl_2$ (0.2)	3	260	Cyclooctene	41
PhCOCl	5	Rh (0.2)	3	200	Chlorobenzene	Trace
$CH_3(CH_2)_8COCl$	20	5% Pd/C (1–2)	10	200–30	Nonenes	58
$CH_3(CH_2)_7COCl$	20	5% Pd/C (1–2)	8	200–20	Octenes	47
$CH_3(CH_2)_6COBr$	20	5% Pd/C (1–2)	5	200–10	Heptenes	66
$Ch_3(CH_2)_{14}COCl$	16	5% Pd/C (1–2)	8	200–20	Pentadecenes	30
$PhCH_2COCl$	20	5% Pd/C (1–2)	3	200–18	Benzyl chloride	42
$PhCH_2CH_2COCl$	20	5% Pd/C (1–2)	6	200–5	Styrene	58
PhCOCl	10	5% Pd/C (1–2)	24	200–10	Chlorobenzene	5

styrene (53%) and 1,5-diphenyl-1-penten-3-one (10%) with a small amount of unidentified products. The mechanism of formation of the

$$PhCH_2CH_2COCl \rightarrow PhCH{=}CH_2 + PhCH_2CH_2COCH{=}CHPh + \text{other products}$$

unsaturated ketone is given in the discussion of the mechanism of palladium-catalyzed decarbonylations. Phenylacetyl chloride was decarbonylated to give benzyl chloride as well as unidentified high molecular weight compounds.

$$PhCH_2COCl \rightarrow PhCH_2Cl + CO$$

In contrast to aliphatic acyl halides which give olefins in good yields, aromatic acyl halides are not decarbonylated satisfactorily. This is quite different from rhodium-catalyzed decarbonylations.

D. Mechanism of Palladium-catalyzed Decarbonylation of Aldehydes and Acyl Halides

The mechanism of decarbonylation of aldehydes with palladium is shown below. The first step is oxidative addition of aldehyde to palladium to form an acyl complex. The acyl–alkyl rearrangement then follows to give an alkyl complex. Elimination of a β-hydrogen then gives olefins as

$$RCH_2CH_2CHO + Pd \rightarrow RCH_2CH_2CO\text{—}Pd\text{—}H \rightarrow$$

$$RCH_2CH_2\text{—}\underset{\displaystyle CO}{\underset{|}{Pd}}\text{—}H \rightarrow RCH{=}CH_2 + Pd + H_2 + CO$$

products. It is known that the alkylpalladium phosphine complex with a β-hydrogen is unstable, and decomposes to an olefin by β-hydrogen elimination to form palladium hydride which further decomposes to give zerovalent palladium (97). Moiseev and Vargaftik (98) studied the de-

$$RCH_2CH_2\text{—}\overset{\displaystyle L}{\overset{|}{\underset{\displaystyle L}{\underset{|}{Pd}}}}\text{—}Cl \rightarrow RCH{=}CH_2 + Pd + HCl + 2L \qquad L = \text{phosphine}$$

composition of the alkyl–palladium bond by reacting ethylmagnesium bromide with $PdCl_2$ in ether. Equal amounts of ethylene and ethane were formed by the following reaction.

$$C_2H_5MgBr + PdCl_2 \rightarrow C_2H_5PdCl + MgBrCl$$

$$C_2H_5PdCl \rightarrow CH_2{=}CH_2 + HCl + Pd$$

$$C_2H_5MgBr + HCl \rightarrow CH_3CH_3 + MgBrCl$$

Similarly the mechanism of the decarbonylation of acyl halides is as shown below.

$$RCH_2CH_2COCl + Pd \rightarrow RCH_2CH_2CO\text{—}Pd\text{—}Cl \rightarrow$$

$$RCH_2CH_2\text{—}\overset{\displaystyle CO}{\overset{|}{Pd}}\text{—}Cl \rightarrow \begin{cases} RCH{=}CH_2 + CO \\ H\text{—}Pd\text{—}Cl \\ \quad\downarrow \\ Pd + HCl \end{cases}$$

This mechanism is similar in essence to those of the decarbonylations catalyzed by rhodium complexes, the only difference being that palladium has no definite ligand, which makes the mechanism somewhat unclear. However, it was observed that part of the palladium catalyst goes into solution during the decarbonylation: It seems likely that the decarbonylation proceeds partly in solution and partly on the surface of the metallic palladium. Isolation of the acyl complex from $PdCl_2$ or metallic palladium and acyl halides is not possible. However, isolation of an acyl complex by reaction of a zerovalent palladium complex, $Pd(PPh_3)_4$, with an acyl halide was reported (17).

$$RCOCl + Pd(PPh_3)_4 \rightarrow \begin{array}{ccc} PPh_3 & & Cl \\ & \diagdown \; \diagup & \\ & Pd & \\ & \diagup \; \diagdown & \\ RCO & & PPh_3 \end{array} + 2PPh_3$$

Although there is no direct evidence for the mechanism of decarbonylation of acyl halides given above, the following results supported formation of an acyl complex as an intermediate in the decarbonylation. When β-phenylpropionyl chloride was heated with $PdCl_2$ at 200°C, styrene was formed as a main product of the decarbonylation, accompanied by a small amount of 1,5-diphenyl-1-penten-3-one. The formation of the ketone can be rationalized as occurring by insertion of styrene formed by decarbonylation into the intermediate acyl–palladium bond.

$$PhCH_2CH_2CO{-}Pd{-}Cl \rightleftharpoons PhCH_2CH_2{-}\overset{\displaystyle CO}{\overset{|}{Pd}}{-}Cl \rightleftharpoons PhCH{=}CH_2 + HPdCl +$$

$$\xrightarrow{PhCH=CH_2} PhCH_2CH_2CO{-}\underset{\displaystyle PhCH=CH_2}{\underset{\uparrow}{Pd}}{-}Cl \longrightarrow PhCH_2CH_2CO{-}\underset{\displaystyle H}{\underset{|}{CH}}{-}\underset{\displaystyle PdCl}{\underset{|}{CH}}{-}Ph$$

$$\swarrow$$

$$PhCH_2CH_2COCH{=}CHPh + [HPdCl] \rightarrow Pd + HCL$$

In connection with the interaction of palladium with acyl halide, it is worthwhile to consider a mechanism for the well-known Rosenmund reduction. This reduction involves the conversion of an acyl halide into an aldehyde by bubbling hydrogen gas through a refluxing solution of the acid halide in toluene or xylene in the presence of metallic palladium catalyst (99). The first step is formation of an acyl–palladium bond by oxidative addition of acyl halide to metallic palladium. The acyl–palladium complex does not decarbonylate at the temperature of toluene or xylene reflux. Instead, lower temperature and the presence of hydrogen gas transform the acyl–palladium complex into an aldehyde and metallic palladium. The mechanism of the Rosenmund reduction can be expressed in the following way. In the main path of this reduction, the

$$RCOCl + Pd \longrightarrow RCOPdCl \begin{cases} \xrightarrow[\text{main path}]{H_2} RCHO + Pd + HCl \\ \xrightarrow{-CO} R\text{—}Pd\text{—}Cl \xrightarrow{H_2} RH + Pd + HCl \\ \qquad\qquad \downarrow \\ \qquad\quad \text{olefin} + Pd + HCl \end{cases}$$

acyl–palladium bond is cleaved by hydrogen to form aldehyde. At the same time, as a minor path, the acyl–alkyl rearrangement takes place to form the alkyl complex, which then gives either olefin or paraffin as the decarbonylation product. With a favorable structure of the acyl complex, the acyl–alkyl rearrangement becomes a main path. Actually, the formation of a decarbonylation product has been reported in several cases as an abnormal reaction in the Rosenmund reduction. For example, a diene was formed via decarbonylation in an attempted Rosenmund reduction of the chloride of acetyloleanolic acid (100). This example shows that even a sterically hindered angular group can be removed by the palladium catalyst. Sterically hindered acyl halides such as diphenyl and triphenylacetyl chloride and naphthoyl chloride undergo decarbonylation in attempted Rosenmund reductions (101). Diphenylacetyl chloride gave tetraphenylethane as a major product, accompanied by diphenylmethane.

$$Ph_3CCOCl \rightarrow Ph_3CH$$

$$Ph_2CHCOCl \rightarrow \underset{47\%}{Ph_2CHCHPh_2} + \underset{13.5\%}{Ph_2CH_2}$$

Also acid chlorides attached to heterocyclic rings gave decarbonylated products (102): Examples are 1-phenylpyrazole-5-carbonyl chloride, 1-phenyl-5-methyl-1,2,3-triazole-4-carbonyl chloride, and coumarin-3-carbonyl chloride. The decisive factor for decarbonylation in the Rosenmund reduction seems to be the structure of the acyl halide and the temperature. For the optimal yield of aldehyde, it is important to keep the temperature near the lowest point at which HCl is evolved.

Oxidative addition of acyl halide to metallic palladium to form the acyl–palladium complex is essential for the mechanism of the Rosenmund reduction proposed above. Concerning the addition of acyl halide to metallic palladium, Chiusoli and Agnes made an interesting observation

(103). On an attempted Rosenmund reduction of 2,5-hexadienoyl chloride, they obtained phenol in high yield. In this facile cyclization reaction, the first step must be addition of the chloride to form an acyl–palladium bond. Coordination of the terminal double bond to palladium, followed by cyclization through insertion of the double bond at the palladium–acyl bond, is the probable mechanism. In this case the formation of a stable product, phenol, is certainly the main driving force of the cyclization.

Cl
C=O
Pd →
Cl Pd
C=O
→
O
ClPd
→
OH
+ Pd + HCl

V. DECARBONYLATIONS WITH NICKEL, COBALT, AND IRON COMPLEXES

Metallic nickel used for catalytic hydrogenation is also active for the decarbonylation of aldehydes at high temperature, although it has not been studied extensively (104). For example, heptanal was decarbonylated to hexene by heating with a nickel catalyst at 250°C (105). In addition, like metallic palladium, nickel is used for removal of the —CH_2OH group. The reaction was discovered by Sabatier and Senderens (106,107), who found that primary alcohols are dehydrogenated to aldehydes and then decarbonylated to alkanes by nickel hydrogenation catalysts. This method is used to convert an alcohol to a hydrocarbon with

$$RCH_2OH \rightarrow RCHO + H_2$$
$$\searrow$$
$$RH + CO$$

one less carbon atom. Thus, 1-decanol was converted into a nonane and nonene mixture at 230°C in 3 hr with Raney nickel catalyst (108). Bird

$$\underset{150\text{ g}}{CH_3(CH_2)_8CH_2OH} \rightarrow \underset{76\text{ g}}{C_9H_{20} + C_9H_{18}}$$

and Hudec (109) reported that diphenylcyclopropenone is decarbonylated to diphenylacetylene by using metal carbonyls such as nickel, cobalt, and iron carbonyls as the catalyst.

Ph Ph
C=C
C
‖
O
→ PhC≡CPh + CO

Aromatic polycarboxylic acids, and their anhydrides, which contain carboxyl groups on adjacent carbon atoms, can be partially decarbonylated by using catalytic amounts of $Co_2(CO)_8$ under CO and H_2 pressure (110,111). One of each ortho pair of carboxyl groups is selectively removed. For example, pyromellitic anhydride is converted in 90% yield into isophthalic acid in the presence of $Co_2(CO)_8$ at 200°C in 5 hr under 3000 psi pressure of synthesis gas. Replacement of one or two CO groups

$$\text{pyromellitic anhydride} + H_2 \xrightarrow[CO/H_2]{Co_2(CO)_8} \text{HO}_2\text{C–}C_6H_4\text{–CO}_2\text{H (isophthalic acid)} + 2CO$$

in $Co_2(CO)_8$ by any of several trialkylphosphine ligands increases catalyst stability and activity; PPh_3 destroys the decarbonylative activity of $Co_2(CO)_8$. The results are shown in Table 10.

The reaction can be understood by the following mechanism.

$$\text{phthalic anhydride} + HCo(CO)_4 \rightleftharpoons o\text{-}(CO_2H)C_6H_4\text{–C(=O)–Co(CO)}_4 \underset{+CO}{\overset{-CO}{\rightleftharpoons}}$$

$$o\text{-}(CO_2H)C_6H_4\text{–Co(CO)}_4 \xrightarrow{H_2} C_6H_5CO_2H + HCo(CO)_4$$

TABLE 10

Decarbonylation of Benzenepolycarboxylic Acid Anhydrides with Synthesis Gas and $Co_2(CO)_8$[a] (111)

Anhydride	Product acid	Yield (%)
Phthalic[b]	Benzoic	91
Hemimellitic(1,2,3)	Benzoic	65
	Isophthalic	15
Trimellitic(1,2,4)	Isophthalic	100
Pyromellitic(1,2,4,5)	Isophthalic	90
Mellitic(1,2,3,4,5,6)	Mellitic[c]	80

[a] 85 ml of toluene, 3500 psig of $1H_2:1CO$, 5 hrs at 200°.

[b] 175 ml of toluene as solvent.

[c] Recovered as the anhydride.

Conversion of bis-carboxylic acid anhydrides to olefins was achieved by using $(Ph_3P)_2Ni(CO)_2$ (112). For example, treatment of *endo*-bicyclo(2.2.1)heptane-2,3-dicarboxylic acid anhydride with $(Ph_3P)_2$-$Ni(CO)_2$ in triglyme gave bicyclo(2.2.1)heptene in 53% yield. The reaction can be explained by the following mechanism.

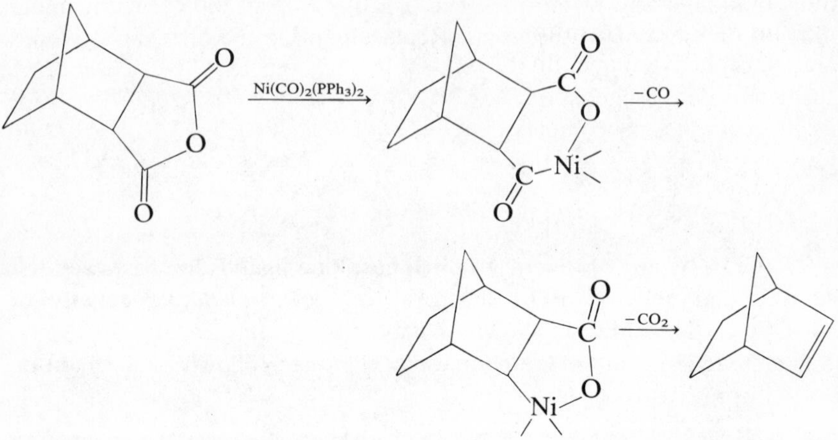

$\xrightarrow{Ni(CO)_2(PPh_3)_2}$ $\xrightarrow{-CO}$ $\xrightarrow{-CO_2}$

Corresponding thioanhydrides gave higher yields. For the decarbonylation of thioanhydride. $(PPh_3)_3RhCl$ and $Fe_2(CO)_9$ can also be used.

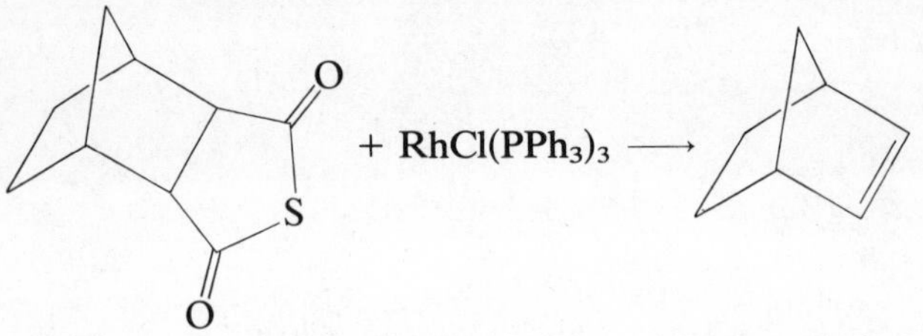

$+ RhCl(PPh_3)_3 \longrightarrow$

$+ CO + Ph_3PS + RhCl(CO)(PPh_3)_2$

Decarbonylation of acyl halides with these metal carbonyls is not common. 3-Butenoyl chloride is decarbonylated with a stoichiometric amount of $Ni(CO)_4$ and the product is 1,5-hexadiene (113). This reaction probably involves formation of a π-allynickel complex, the allylic ligand of which couples to form 1,5-hexadiene. Reductive decarbonylation of

$$CH_2{=}CHCH_2COCl + Ni(CO)_4 \rightarrow$$

$$CH_2{=}CHCH_2CH_2CH{=}CH_2 + NiCl_2 + CO$$

acyl halides with a hydridonitrogen complex of cobalt was reported to give a mixture of products as shown below (114).

$$Co(PPh_3)_3N_2H + RCOCl : Et_2O \rightarrow \begin{cases} RH + CoCl_2(PPh_3)_2 & \text{major} \\ RR + N_2 + C_2H_4 + C_2H_6 + CO & \\ \quad + Co(CO)_2(PPh_3)_2 & \text{minor} \end{cases}$$

The decarbonylation of specific acyl halides is known. Thus α-chlorodiphenylacetyl chloride was decarbonylated by refluxing with a catalytic amount of $Co_2(CO)_8$ in toluene. The products were tetraphenylethylene and tetraphenylethane. In this reaction it is assumed that ketene was formed by dehydrogenation, then decarbonylation of the ketene took place to form a carbene complex. Finally, coupling of the carbene gave the products (115). This reaction depends on the structure of the acyl halide.

$$Ph{-}\underset{Ph}{\overset{Cl}{C}}{-}COCl \xrightarrow[-Cl_2]{Co_2(CO)_8} \left[Ph_2C{=}C{=}O \right] \longrightarrow Ph_2C{=}CPh_2 + Ph_2CHCHPh_2$$

For example, α-chloro-α-phenylpropionyl chloride was converted into a trinuclear cobalt complex by decarbonylation and rearrangement. The

$$CH_3{-}\underset{Ph}{\overset{Cl}{C}}{-}COCl \xrightarrow{Co_2(CO)_8} PhCH_2CCo_3(CO)_9$$

stoichiometric decarbonylation of ketene with a rhodium complex was described earlier. Unlike the rhodium complex, $Co_2(CO)_8$ can be used to catalyze the decarbonylation of ketene (73). On decarbonylation of diphenylketene in refluxing toluene at 110°C, tetraphenylethylene was obtained in 68% yield. In this reaction, the formation of a carbene complex is essential in the decarbonylation. In addition to $Co_2(CO)_8$,

$$2Ph_2C{=}C{=}O \xrightarrow{Co_2(CO)_8} Ph_2C{=}CPh_2$$

other complexes such as $Co_4(CO)_{12}$ and π-$CpCo(CO)_2$ are also active catalysts. $Ni(CO)_4$ and $Fe(CO)_5$ are inactive.

The reverse of the decarbonylation of ketene to form a carbene, namely the carbonylation of carbene, is known. Thus diphenyldiazomethane is converted into an ester of diphenylacetic acid by reaction with $Ni(CO)_4$ in alcohol (116). It is understandable that the diazo compound is first converted into a carbene, and then a CO attack on the carbene gives the ketene. The ketene reacts with the alcohol to give an ester.

$$Ph_2CN_2 \xrightarrow{Ni(CO)_4} Ph_2CNi(CO)_{n-1} \longrightarrow Ph_2C{=}C{=}O \xrightarrow{ROH} Ph_2CHCOOR$$

Like ketene, isocyanates have cumulative double bonds, which could interact with transition metal complexes. An interesting reaction of isocyanates catalyzed by metal carbonyls was reported by Ulrich, Tucker, and Sayigh (117); namely, the conversion of isocyanates into carbodiimides, which involves removal of a —C=O unit from R—N=C=O. Thus *o*-tolyl isocyanate was refluxed with $Fe(CO)_5$ and the temperature raised from 180 to 250°C with evolution of CO_2. Vacuum distillation afforded an 85.3% yield of di-*o*-tolycarbodiimide. This reaction involves the formation of an iron isonitrile complex and the decarbonylation of the isocyanate. Some examples of this reaction are shown in Table 11.

$$RN{=}C{=}O + Fe(CO)_5 \longrightarrow \begin{matrix} (CO)_4Fe - C{=}O \\ | \quad\quad | \\ R{-}N{=}C - O \end{matrix} \longrightarrow (CO)_4Fe{=}C{=}N{-}R + CO_2$$

$$(CO)_4Fe{=}C{=}N{-}R \xrightarrow{R-N=C=O} \begin{matrix} (CO)_4Fe - C{=}N{-}R \\ | \quad\quad | \\ O{=}C - N{-}R \end{matrix} \longrightarrow Fe(CO)_5 + R{-}N{=}C{=}N{-}R$$

It is interesting to compare this reaction with the formation of carbodiimides from isocyanates by catalytic action of certain phosphine oxides as shown below (118). The similarity of the two mechanisms is apparent.

$$R{-}N{=}C{=}O + R_3P{=}O \longrightarrow \begin{matrix} R{-}N - C{=}O \\ | \quad\quad | \\ R_3P - O \end{matrix} \longrightarrow R{-}N{=}PR_3 + CO_2$$

$$R{-}N{=}PR_3 \xrightarrow{R-N=C=O} \begin{matrix} R{-}N - PR_3 \\ | \quad\quad | \\ R{-}N{=}C - O \end{matrix} \longrightarrow R{-}N{=}C{=}N{-}R + R_3P{=}O$$

TABLE 11
Formation of o-*Tolylcarbodiimides from* o-*Tolyl Isocyanate* (117)

Metal carbonyl[a]	Time (min)	Yield of diimide (%)
$Fe(CO)_4CNPh$	25	83.0
$Fe(CO)_5$	36	85.3
$Fe_2(CO)_9$	43	82.3
$W(CO)_6$	68	85.6
$Mo(CO)_6$	200	85.0

[a] 1% by wt.

VI. DECARBONYLATIONS USING RUTHENIUM AND IRIDIUM COMPLEXES

Ruthenium complexes are active for carbonylation reactions, and considerable work has been reported on their use for this purpose. On the other hand, few studies of decarbonylations with ruthenium complexes have been reported.

Stoichiometric decarbonylation of aldehydes is possible with trichlorohexakis(diethylphenylphosphine)diruthenium chloride (119). This reaction is slower than that catalyzed by rhodium complexes; it also differs from rhodium in other ways. With the ruthenium complex, the main product of the decarbonylation is an olefin rather than a saturated hydrocarbon. Furthermore, H_2 is not evolved by formation of the olefin; instead, the hydrogen is consumed by reduction of the aldehydes to alcohols. For example, when butyraldehyde was decarbonylated, the products were propylene (80%) and butanol, if an excess of aldehyde was used. The mechanisms of the decarbonylations using rhodium and

$$Ru_2Cl_3(Et_2PhP)_6Cl + 4CH_3CH_2CH_2CHO \rightarrow$$
$$2RuCl_2(CO)(Et_2PhP)_3 + 2CH_3CH{=}CH_2 + 2CH_3CH_2CH_2CH_2OH$$

ruthenium complexes are somewhat different as shown by the following experiment carried out with a deuterated aldehyde. As expected from the

$$CD_3CH_2CHO + RhCl(PPh_3)_3 \rightarrow CD_3CH_3$$

$$CH_3CH_2CDO + Ru_2Cl_3(Et_2PhP)_6Cl \rightarrow \underset{80\text{–}90\%}{CH_3CH_3} + C_2H_5D$$

$$CD_3CH_2CHO + Ru_2Cl_3(Et_2PhP)_6Cl \rightarrow \underset{\text{main product}}{CD_2HCH_2D} +$$

$$D_2C{=}CHD + DHC{=}CHD + HDC{=}CD_2 + D_2C{=}CH_2$$

mechanism mentioned earlier, CD_3CH_3 was obtained by decarbonylation of CD_3CH_2CHO with the rhodium complex. On the other hand, the decarbonylation of CH_3CH_2CDO with the ruthenium complex gave C_2H_6 as the main product (80–90%), the remainder being C_2H_5D. Also decarbonylation of CD_3CH_2CHO resulted in the scrambling of deuterium in the decarbonylated products. Several possible mechanisms were proposed to account for these results. As one way to explain the redistribution of deuterium in the products, the following complex formation was proposed:

$$C_2H_5CHO + (Ru) \rightleftharpoons \begin{array}{c} H \quad H \\ | \\ H_3C-\overset{|}{C}H-C{=}O \\ \downarrow \\ Ru \end{array} \rightleftharpoons$$

H H
C=C
H3C O
Ru
H
⇌
H H
C=C
H—C O
H Ru
H H

Formic acid and formaldehyde were decarbonylated by aqueous $Ru(II)Cl_2$; kinetic studies were carried out on this reaction (120).

$$RuCl_4^{2-} + HCOOH \rightarrow Ru(CO)(H_2O)Cl^{2-}$$

The iridium complex, $IrCl(CO)(PPh_3)_2$ (Vaska's complex) undergoes a large number of oxidative addition reactions. However, because of the stability of the products of such oxidative addition reactions, this complex does not seem to have good catalytic activity. Blum and Kraus (59) found that this complex can be a catalyst for decarbonylation of acyl halides with a β-hydrogen atom to give olefins. Thus 3-phenylpropionyl, 4-phenylbutyryl, and 5-phenylvaleryl chlorides were converted in boiling xylene to mixtures of the corresponding terminal and internal olefins in almost quantitative yield. However, aroyl halides were not decarbonylated with this complex. Thus 4-(*p*-chloroformylphenyl)-butyryl chloride was converted into *p*-allylbenzoyl chloride and its isomerized products.

$$PhCH_2CH_2CH_2COCl \xrightarrow{IrCl(CO)(PPh_3)_2} PhCH_2CH{=}CH_2$$

$$ClC(=O)-C_6H_4-CH_2CH_2CH_2COCl \longrightarrow ClC(=O)-C_6H_4-CH_2CH{=}CH_2$$

Results of the decarbonylation of acyl halides by $IrCl(CO)(PPh_3)_2$ are shown in Table 12. Benzoyl halides are inert because they form very stable complexes which do not decompose into phenyl halides even upon heating at 300°C. Model reactions of CO insertion with iridium com-

$$IrCl(CO)(PPh_3)_2 + PhCOCl \rightarrow PhIrCl_2(CO)(PPh_3)_2 + CO$$

plexes have been carried out by Wilkinson and co-workers (57). Kubota and Blake (121) also studied the oxidative addition of acyl halides to

TABLE 12
Decarbonylation of Acyl Chlorides by $IrCl(CO)(PPh_3)_2$ (59)

Acyl chloride (g/g catalyst)	Solvent (ml)	Temp (°C)	Time (min)	Product (%)
Undecanoyl (4.9/0.1)	Benzene (20)	78	1560	Decenes (18)
Undecanoyl (2.0/0.2)	*p*-Xylene (20)	135	900	Decenes (95)
Undecanoyl (61/0.6)		Reflux	20	1-Decene (41), isomerized decenes (49)
Lauroyl (1.0/0.1)		Reflux	15	1-Undecene (38), isomerized undecenes (46)
3-Phenylpropionyl (4.0/0.05)		Reflux	30	Styrene (100)
4-Phenylbutyryl (4.0/0.05)		Reflux	15	Allylbenzene (75), *cis*-β-methylstyrene (3), *trans*-β-methylstyrene (20)
5-Phenylvaleryl (1.5/0.4)	*p*-Xylene	135	330	4-Phenyl-1-butene (3.2), *cis*-1-phenyl-2-butene (2.6), *trans*-1-phenyl-2-butene (12.5), *trans*-1-phenyl-1-butene (40.8)
5-Phenylvaleryl (4.0/0.06)		Reflux	10	4-Phenyl-1-butene (40), *cis* and *trans*-1-phenyl-2-butene (38), *trans*-1-phenyl-1-butene (12)

iridium complexes, $IrL_2Cl(CO)$, IrL_2ClN_2, to form acyl complexes, and various aspects of acyl–alkyl rearrangement of these acyl complexes.

In connection with the decarbonylation reaction of carbonyl compounds, the interesting results of Chatt and of Vaska should be mentioned. It was found that certain noble metal compounds can be converted into carbonyl complexes by reaction with an alcohol. Detailed studies on this carbon monoxide-abstracting ability of certain noble metal complexes have been made. Chatt, Shaw, and Field (122) observed that $Ru_2Cl_3(PEt_2Ph)_6Cl$ reacts with ethanol in the presence of KOH to give the hydridocarbonylruthenium complex. Hydrogen and CO are derived from the ethanol, and by this reaction the ethanol is converted into

$$Ru_2Cl_3(PEt_2Ph)_6Cl + 2KOH + 2C_2H_5OH \rightarrow 2RuHCl(CO)(PEt_2Ph)_3 + 2CH_4 + 2KCl + 2H_2O$$

methane. This type of reaction, namely the decomposition of ethanol into methane, H_2, and CO, has never been realized in ordinary organic chemistry. This example shows the interesting versatility of transition metal complex-catalyzed reactions. The following mechanism has been proposed. The initial step is formation of a ruthenium ethoxide moiety;

$$CH_3\text{—}\overset{H}{\underset{H}{C}}\text{—O—Ru—} \;(Cl) \longrightarrow CH_3\text{—}\overset{H}{C}\text{=O—}\underset{H}{\text{Ru}}\text{—} \longrightarrow CH_4 + \left[\text{C}\equiv\text{O—}\underset{H}{\text{Ru}}\right] \longrightarrow \text{O}\equiv\text{C—}\underset{H}{\text{Ru}}\text{—}$$

this is followed by hydride transfer to ruthenium to give an acetaldehyde complex, which then breaks down to give methane and the carbonyl complex. Chatt and Shaw (65) also reported the formation of $RhCl(CO)(PEt_2Ph)_2$ by reaction of $RhCl_3(PEt_2Ph)_3$ with ethanol.

A similar reaction observed by Vaska is the following (123). Again,

$$(NH_4)_2OsBr_6 + 4PPh_3 + RCH_2OH \rightarrow RH + HBr + PPh_3Br_2 + 2NH_4Br + OsHBr(CO)(PPh_3)_3$$

RCH_2OH was converted into RH while donating hydrogen and CO to the osmium complex. Also $IrCl_3$ is converted into $IrCl(CO)(PPh_3)_2$ by refluxing with PPh_3 in ethylene glycol. Certainly the CO in the final complex is derived from ethylene glycol, but the mechanism is not clear.

In another example, Rusina and Vlcek (124) reported that oxygen-containing compounds such as THF, DMF, dioxane, and cyclohexanone can donate CO to form a carbonyl complex of rhodium:

$$RhCl_2 \cdot 3H_2O + PPh_3 \xrightarrow{\text{THF, etc.}} RhCl(CO)(PPh_3)_2$$

ACKNOWLEDGEMENTS

The author expresses his sincere appreciation to Dr. H. G. Tennent and Dr. J. Blum for careful reading of the manuscript and for suggestions.

VII. ADDENDUM

The chapter was originally completed at the end of 1970. This addendum briefly summarizes the major achievements reported from 1971 in the field of decarbonylation.

A. Decarbonylations with Rhodium Complexes

A convenient and ingenious method for regeneration of $RhCl(PPh_3)_3$ from $RhCl(CO)(PPh_3)_2$ was discovered by Stille and Fries (125). By this method, the rather expensive $RhCl(PPh_3)_3$ can be reused. The reaction sequence for this method is as follows:

$$RhCl(CO)(PPh_3)_2 + C_6H_5CH_2Cl \longrightarrow$$

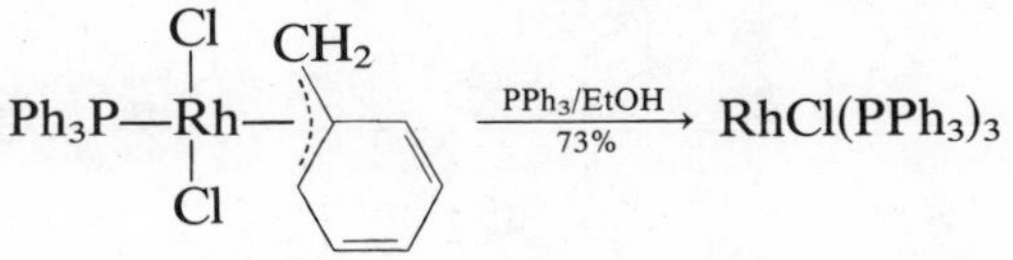

$$Ph_3P\text{—}RhCl_2(\eta^3\text{-}CH_2C_6H_5) \xrightarrow[73\%]{PPh_3/EtOH} RhCl(PPh_3)_3$$

Decarbonylation under mild conditions with $RhCl(PPh_3)_3$ was used for removing aldehyde groups in the total syntheses of natural products such as occidentalol (126), desoxypodocarpate (127), and grandisol (128), respectively:

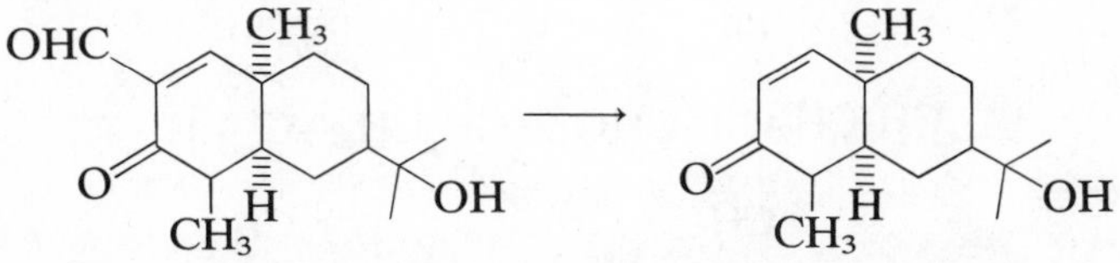

[Structure: OHC-substituted bicyclic enone with CH_3, H, CH_3 and OH groups] $\longrightarrow$ [Structure: corresponding bicyclic enone without CHO, with CH_3, H, CH_3 and OH groups]

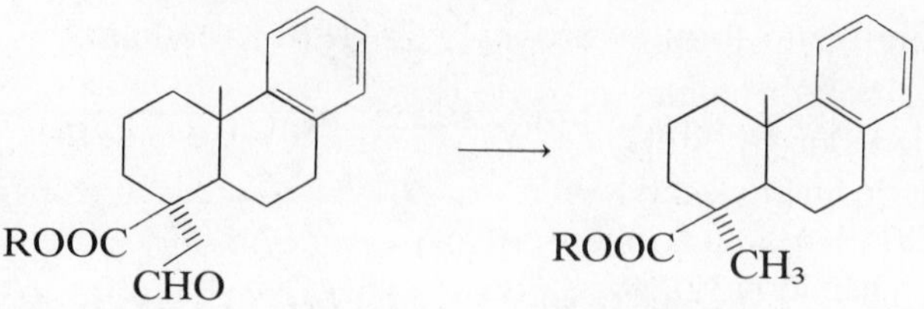

ROOC CHO → ROOC CH_3

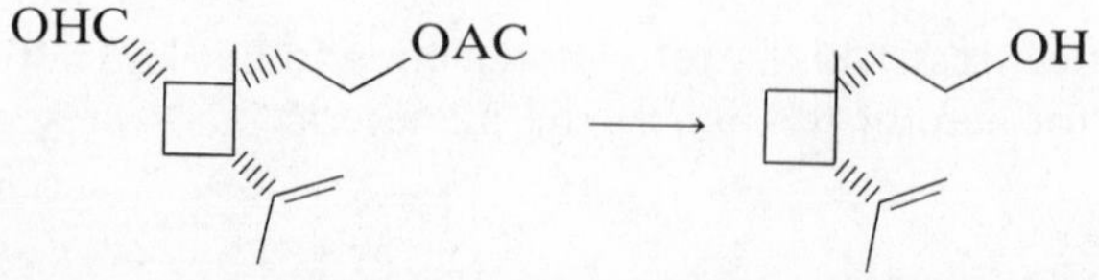

OHC OAC → OH

Aldehydes having olefinic bonds at a suitable position undergo cyclization without decarbonylation when treated with $RhCl(PPh_3)_3$. For example, 1-al-4-ene systems were converted to cyclopentanone derivatives. The cyclization can be explained by oxidative addition of the aldehyde to form an acyl–rhodium bond, followed by intramolecular olefin insertion (129). Unexpectedly, (+)-citronellal was converted to (+)-neoisopulegol

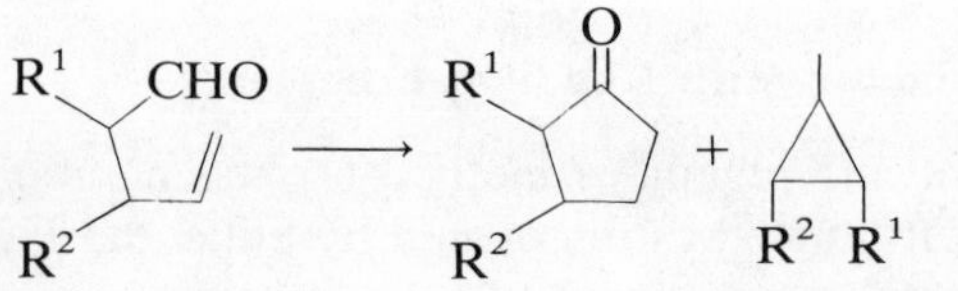

R^1 CHO R^2 → R^1 O R^2 + R^2 R^1

and (−)-isopulegol rather than to the corresponding six-membered ketones (130). Details of the stereochemical studies on the decarbonylation of aldehydes have been given (131). (−)-(*R*)-2-Phenylbutanal and (+)-(*R*)-1-methyl-2,2-diphenylcyclopropanecarboxaldehyde were decarbonylated with $RhCl(PPh_3)_3$ to give (+)-(*S*)-2-phenylbutane and (+)-(*S*)-1-methyl-2,2-diphenylcyclopropane with 81% and 94% retention of optical activity.

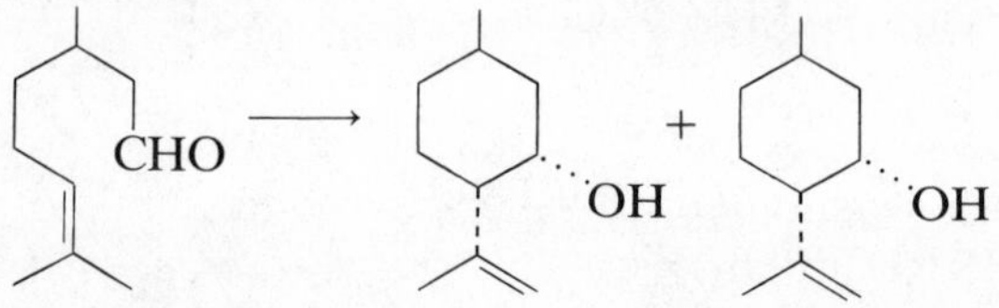

CHO → OH + OH

$$CH_3CH_2\overset{\overset{\large Ph}{|}}{\underset{\underset{\large CH_3}{|}}{C^*}}\text{—}CHO \rightarrow CH_3CH_2\text{—}\overset{\overset{\large Ph}{|}}{\underset{\underset{\large CH_3}{|}}{C^*}}\text{—}H$$

Extensive mechanistic, kinetic, and stereochemical studies on decarbonylation of halides of various acids with the rhodium complexes were carried out by Stille and co-workers. Optically active (*S*)-(−)-α-trifluoromethylphenylacetyl chloride was subjected to decarbonylation with $RhCl(PPh_3)_3$. α-Trifluoromethylbenzyl chloride obtained in 71% yield was racemic (132). An optically active alkylrhodium complex was

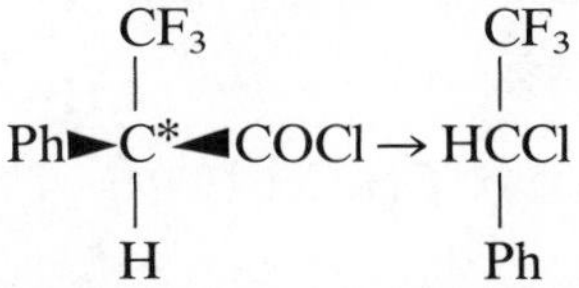

$$\mathrm{Ph}\blacktriangleright\overset{\displaystyle CF_3}{\underset{\displaystyle H}{C^*}}\blacktriangleleft\mathrm{COCl} \rightarrow \mathrm{H}\overset{\displaystyle CF_3}{\underset{\displaystyle Ph}{C}}\mathrm{Cl}$$

prepared by the following oxidative addition reaction of the benzyl chlorosulfite with elimination of SO_2. This step was assumed to be an inversion. The decomposition of this complex afforded the benzyl chloride in 24% yield with 62% net inversion. Thus decomposition of the alkylrhodium complex to the halide should proceed with retention. These results show that racemization in the decarbonylation probably takes place in an acyl–alkyl rearrangement step.

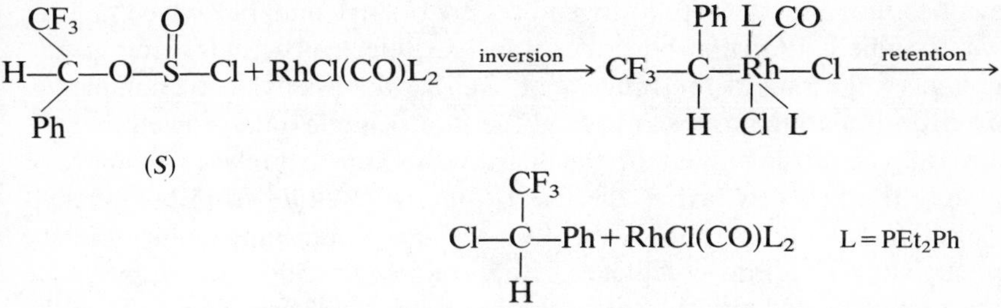

$$\underset{(S)}{\mathrm{H{-}C(CF_3)(Ph){-}O{-}S(=O){-}Cl}} + RhCl(CO)L_2 \xrightarrow{\text{inversion}} CF_3{-}C(Ph)(H){-}Rh(L)(CO)(Cl)(L){-}Cl \xrightarrow{\text{retention}}$$

$$\underset{(R)}{\mathrm{Cl{-}C(CF_3)(H){-}Ph}} + RhCl(CO)L_2 \qquad L = PEt_2Ph$$

The decarbonylation of *erythro*- and *threo*-2,3-diphenylbutanoyl chloride with $RhCl(PPh_3)_3$ gave exclusively *trans*- and *cis*-α-methylstilbene, respectively. These results are best explained either by an acyl-alkyl rearrangement with retention and a *cis*-β-hydride elimination, or concerted *cis* elimination (133).

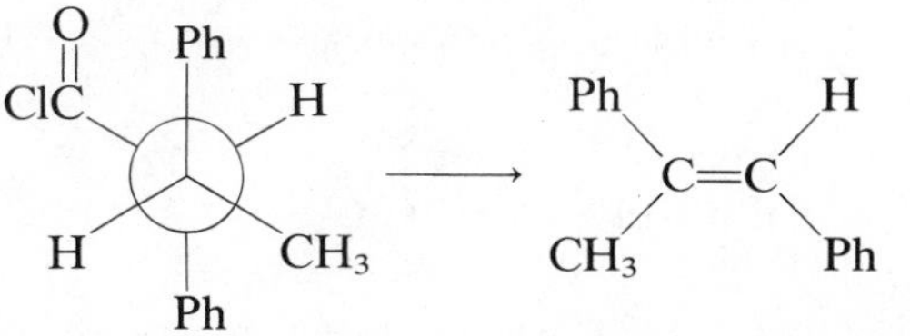

[Newman projection: front carbon Ph (top), ClC(=O) (upper left), H (upper right); back carbon H (lower left), CH$_3$ (lower right), Ph (bottom)] ⟶ (Ph)(CH$_3$)C=C(H)(Ph)

erythro   *trans*-α-methylstilbene

The decarbonylation of several branched acid chlorides having primary, secondary, and tertiary hydrogens was carried out, where two choices of types of hydrogens (primary vs. secondary, primary vs. tertiary, and secondary vs. tertiary) were available. The results of the catalytic decarbonylation of 2-methylpentanoyl chloride, 2,2-dimethylpentanoyl chloride, and 2-ethyl-3-methylbutanoyl chloride with $RhCl(CO)(PPh_3)_2$ in the presence of PPh_3 showed that this olefin-forming reaction preferred the Saytzeff elimination (133).

$$\mathrm{H_3C\overset{CH_3}{\underset{H}{C}}-\overset{H}{\underset{COCl}{C}}CH_2CH_3} \rightarrow \begin{cases} \mathrm{CH_3CH_2CH_2\overset{CH_3}{C}{=}CH_2} & 10\% \\ \mathrm{H_3CCH_2CH{=}\overset{CH_3}{C}CH_3} & 44\% \quad (2) \\ \mathrm{H_3CCH{=}CH\overset{CH_3}{C}HCH_3} & 46\% \quad (1) \end{cases}$$

(the statistically corrected ratio appears in parentheses)

Para-substituted benzoyl and phenylacetyl complexes of rhodium were synthesized and their rearrangements to phenyl and benzyl complexes were studied. With the benzoyl complexes, electron-withdrawing groups enhanced the rate of rearrangement. But for the phenylacetyl complexes, electron-donating groups enhanced the rate. Kinetic data showed that the acyl–alkyl rearrangement of the benzoylrhodium complexes is approximately ten times as fast as the loss of phenyl chloride from the phenylrhodium complexes. However, the acyl–alkyl rearrangement for the phenylacetylrhodium complexes is approximately 20 times slower than the loss of benzyl chloride from the benzylrhodium complexes. Thus the rate determining step is different for the phenyl and benzyl complexes (134).

Blum and Kraus (135) reported that the addition of PPh_3 inhibits migration of a terminal double bond formed by the decarbonylation of acyl halides in the presence of rhodium catalysts to an internal double bond.

Decarbonylation of one of the ketonic groups of α- and β-diketones was reported (136). A simple one-step method for the synthesis of benzo-

$$\mathrm{R\overset{O}{\overset{\|}{C}}(CH_2)_n{-}\overset{O}{\overset{\|}{C}}{-}R' + RhCl(PPh_3)_3 \rightarrow R\overset{O}{\overset{\|}{C}}(CH_2)_n{-}R' + RhCl(CO)(PPh_3)_2}$$
$$n = 0, 1$$

and dibenzofluorenones was provided by the decarbonylation of anhydrides of benzoic and naphthoic acids (137).

$$\text{PhC(=O)–O–C(=O)Ph} \longrightarrow \text{fluorenone}$$

Details of the decarbonylation of diacetylenic ketones in boiling xylene was reported (138). This reaction was explained by the following mechanism.

$$RC{\equiv}C\overset{O}{\overset{\|}{C}}C{\equiv}CR + RhClL_3 \longrightarrow L(L)(Cl)Rh(C{\equiv}CR)(\overset{O}{\overset{\|}{C}}{-}C{\equiv}CR) \longrightarrow (RC{\equiv}C)(C{\equiv}CR)Rh(L)_2(Cl)(CO)$$

$L = PPh_3$

$$\longrightarrow RC{\equiv}CC{\equiv}CR + RhCl(CO)L_2$$

B. Decarbonylation with Palladium Catalysts

Fenton and Olivier (139) reported that carboxylic acids and esters decompose to give olefins and CO around 200–250°C in the presence of a palladium catalyst. Acid anhydrides are converted to olefins at 150°C.

$$CH_3CH_2CH_2COOH \rightarrow Ch_3CH{=}CH_2 + CO + H_2O$$

They found the following equilibrium, which is achieved by heating acids in a closed system. Based on this equilibrium they claimed that branched acids can be converted into linear ones by removing the latter from the reaction system. In this connection it is worthwhile to point out that

$$(CH_3)_2CHCOOH \rightleftarrows CH_3CH_2CH_2COOH$$

straight-chain alkyliridium complexes were always formed by the oxidative addition of acyl halides branched at the α-position to $IrCl(PPh_3)_3$. For example, 2-ethylbutanoyl chloride formed the n-pentyliridium complex. The reaction was explained by rearrangement due to steric hindrance of the branched chains (140). Also, conversion of linear acids to

$$CH_3CH_2CH(CH_2CH_3)\overset{O}{\overset{\|}{C}}Cl + IrCl(PPh_3)_3 \longrightarrow Ir(Cl)_2(PPh_3)_2(CO)(CH_2CH_2CH_2CH_2CH_3)$$

their isomeric branched-chain analogs was carried out by decarbonylation and carbonylation. Thus, stearic acid was converted into the acid chloride, which was decarbonylated to alkenes by $PdCl_2$ or $RhCl_3$. The olefins formed were carbonylated by the Koch method to give branched-chain acids (141).

An elegant application of palladium-catalyzed decarbonylation of aldehydes is the removal of 4-carboxybenzaldehyde in terephthalic acid production. This method has industrial importance for the production of pure terephthalic acid free from 4-carboxylbenzaldehyde, which is an undesired by-product of *p*-xylene oxidation (142). Benzoic acid formed by the decarbonylation can be separated from terephthalic acid easily.

CHO / COOH —(Pd on carbon, 260°C)→ COOH + CO

Decarbonylation catalyzed by $PdCl_2$ was applied to the synthesis of dicyclohepta(*cd,gh*)pentalene from 1,6;8,13-ethanediylidene[14]annulene-15-carboxylic acid (143).

COCl →

C. Miscellaneous Examples

Raney nickel-catalyzed decarbonylation of formate esters was reported (144). $IrCl(CO)(PPh_3)_2$ does not catalyze the decarbonylation of aroyl

$$R_2CHOCH(=O) \xrightarrow[>150°C]{Ni} R_2CHOH + CO$$

halides probably due to the stability of aryliridium complexes. Decarbonylation of acyl halides bearing a β-hydrogen proceeds with this complex to give a mixture of olefins (135).

REFERENCES

1. J. Tsuji and K. Ohno, *Synthesis*, **1969,** 157.
2. J. Tsuji, *Advances in Organic Chemistry*, Vol. 6, Interscience, New York, 1969, p. 109.

3. J. P. Cadlin, K. A. Taylor, and D. T. Thompson, *Reactions of Transition Metal Complexes*, Elsevier, Amsterdam, 1968, p. 119.
4. C. W. Bird, *Transition Metal Intermediates in Organic Synthesis.* Logos Press, London, 1967, pp. 112, 239.
5. F. Calderazzo, R. Ercoli, and G. Natta, *Organic Syntheses via Metal Carbonyls*, Vol. 1, Interscience, New York, 1968, p. 249.
6. G. Booth and J. Chatt, *Proc. Chem. Soc.*, **1961,** 67.
7. B. L. Booth, M. Green, R. N. Haszeldine, and N. P. Woffenden, *J. Chem. Soc. A*, **1970,** 1979; and **1969,** 920.
8. T. H. Coffield, R. P. Closson, and J. Kazoikowski, *J. Org. Chem.*, **22,** 598 (1957).
9. F. Calderazzo and F. A. Cotton, *Inorg. Chem.*, **1,** 30 (1962).
10. F. Calderazzo and K. Noack, *J. Organometal. Chem.*, **4,** 250 (1965).
11. K. Noack and F. Calderazzo, *J. Organometal. Chem.*, **10,** 101 (1967).
12. K. Noack, M. Ruch, and F. Calderazzo, *Inorg. Chem.*, **7,** 345 (1968).
13. R. J. Mawby, F. Basolo, and R. G. Pearson, *J. Am. Chem. Soc.*, **86,** 3994; and 5043 (1964).
14. I. S. Butler, F. Basolo, and R. G. Pearson, *Inorg. Chem.*, **6,** 2074 (1967).
15. J. P. Collman, *Acc. Chem. Res.*, **1,** 136 (1968).
16. J. P. Collman and W. R. Roper, *Advances in Organometal. Chem.*, Vol. 7, 54 (1968).
17. P. Fitton, M. P. Johnson, and J. E. McKeon, *Chem. Commun.*, **1968,** 6.
18. C. D. Cook and G. S. Jauhal, *Can. J. Chem.*, **45,** 301 (1967).
19. L. Malatesta and C. Cariello, *J. Chem. Soc.*, **1958,** 2323.
20. R. Ugo, F. Cariati, and G. LaMonica, *Chem. Commun.*, **1966,** 868.
21. J. Tsuji and K. Ohno, *J. Am. Chem. Soc.*, **88,** 3452 (1966).
22. I. Harvie and R. D. Kemmitt, *Chem. Commun.*, **1970,** 198.
23. P. B. Tripathy and D. M. Roundhill, *J. Am. Chem. Soc.*, **92,** 3825 (1970).
24. J. F. Young, J. A. Osborn, F. H. Jardine, and G. Wilkinson, *Chem. Commun.*, **1965,** 131.
25. M. A. Bennett and P. A. Longstaff, *Chem. Ind.*, **1965,** 846.
26. J. A. Osborn, F. H. Jardine, J. F. Young, and G. Wilkinson, *J. Chem. Soc. A*, **1966,** 1711.
27. K. Ohno and J. Tsuji, *J. Am. Chem. Soc.*, **90,** 99 (1968).
28. J. Tsuji and K. Ohno, *Tetrahedron Lett.*, **1966,** 4713.
29. J. Tsuji and K. Ohno, *Tetrahedron Lett.*, **1965,** 3969.
30. M. C. Baird, D. N. Lawson, J. Mague, J. A. Osborn, and G. Wilkinson, *Chem. Commun.*, **1966,** 129.
31. J. Blum, *Tetrahedron Lett.*, **1966,** 1605 and 3041.
32. R. L. Augustine and J. F. Van Peppen, *Chem. Commun.*, **1970,** 497; H. Arai and J. Halpern, *Chem. Commun.*, **1971,** 1571.
33. M. C. Baird, C. J. Nyman, and G. Wilkinson, *J. Chem. Soc. A*, **1968,** 348.
34. Y. Shimizu, H. Mitsuhashi, and E. Caspi, *Tetrahedron Lett.*, **1966,** 4113.
35. R. J. Anderson, R. P. Hanzlik, K. B. Sharpless, and E. E. van Tamelen, *Chem. Commun.*, **1969,** 53.
36. D. J. Dawson and R. E. Ireland, *Tetrahedron Lett.*, **1968,** 1899.
37. R. E. Ireland and G. Pfister, *Tetrahedron Lett.*, **1969,** 2145.
38. N. S. Crossley and R. Dowell, *J. Chem. Soc. C*, **1971,** 2496.
39. N. S. Crossley, *J. Chem. Soc. C*, **1971,** 2491.
40. J. K. Nicholson and B. L. Shaw, *Proc. Chem. Soc.*, **1963,** 282.
41. A. J. Birch, and K. A. W. Walker, *J. Chem. Soc. C*, **1966,** 1894.
42. A. Emery, A. C. Oehlschlager, and A. M. Unrau, *Tetrahedron Lett.*, **1970,** 4401.

43. J. Falbe, H. Tummes, and H. Hahn, *Angew. Chem.*, **82,** 181 (1970); and Ger. Pat. 1,917,244, *Chem. Abstr.*, **74,** 41915 (1971).
44. J. Tsuji and K. Ohno, *Tetrahedron Lett.*, **1967,** 2173.
45. H. M. Walborsky and L. E. Allen, *Tetrahedron Lett.*, **1970,** 823.
46. H. M. Walborsky and L. E. Allen, *J. Am. Chem. Soc.*, **93,** 5466 (1971).
47. J. K. Stille, D. B. Fox, L. F. Hines, R. W. Fries, and R. D. Hughes, *Symposium on Homogeneous Catalytic Reactions involving Palladium*, Am. Chem. Soc., 1969, p. B149.
48. C. Büchart, *Chem. Ber.*, **94,** 2599 (1961).
49. J. A. Osborn, G. Wilkinson, and J. F. Young, *Chem. Commun.*, **1965,** 17.
50. F. H. Jardine, J. A. Osborn, G. Wilkinson, and J. F. Young, *Chem. Ind.* (*London*), **1965,** 560.
51. D. Evans, J. A. Osborn, and G. Wilkinson, *J. Chem. Soc. A*, **1968,** 3133.
52. C. K. Brown and G. Wilkinson, *Tetrahedron Lett.*, **1969,** 1725.
53. J. H. Craddock, A. Hershman, F. E. Paulik, and J. F. Roth, *Ind. Eng. Chem.*, **8,** 291 (1969).
54. L. H. Slaugh and R. D. Mullineaux, U. S. Pat. 3,239,566 (1966), *Chem. Abstr.*, **64,** 15745 (1966).
55. M. Takesada, and Wakamatsu, *Bull. Chem. Soc. Japan*, **43,** 2192 (1970).
56. R. L. Pruett and J. A. Smith, *J. Org. Chem.*, **34,** 327 (1969).
57. G. Yagupsky, C. K. Brown, and G. Wilkinson, *Chem. Commun.*, **1969,** 1244; *J. Chem. Soc. A*, **1970,** 1392.
58. D. Evans, G. Yagupsky, and G. Wilkinson, *J. Chem. Soc. A*, **1968,** 2660.
59. J. Blum and S. Kraus, *J. Organometal. Chem.*, **33,** 227 (1971); and *Israel J. Chem. Suppl.*, **8,** 3 (1970).
60. J. Blum, E. Oppenheimer, and E. D. Bergmann, *J. Am. Chem. Soc.*, **89,** 2338 (1967).
61. G. A. Olah and P. Kreinenbuhl, *J. Org. Chem.*, **32,** 1614 (1967).
62. J. Blum, H. Roseman, and E. D. Bergmann, *J. Org. Chem.*, **33,** 1928 (1968).
63. J. K. Stille, R. Hughes, L. Hines, and R. Fries, unpublished results; private communication from Prof. Stille.
64. M. C. Baird, J. T. Mague, J. A. Osborn, and G. Wilkinson, *J. Chem. Soc. A*, **1967,** 1347.
65. J. Chatt and B. L. Shaw, *J. Chem. Soc. A*, **1966,** 1437.
66. W. Keim, *J. Organometal. Chem.*, **19,** 161 (1969).
67. I. C. Doucek and G. Wilkinson, *J. Chem. Soc. A*, **1969,** 2604.
68. R. F. Heck, *J. Am. Chem. Soc.*, **86,** 2796 (1964).
69. R. H. Prince and K. A. Raspin, *Chem. Commun.*, **1966,** 156.
70. J. Blum and Z. Lipshes, *J. Org. Chem.*, **34,** 3076 (1969).
71. J. Blum, D. Milstein, and Y. Sasson, *J. Org., Chem.*, **35,** 3233 (1970).
72. P. Hong, K. Sonogashira, and N. Hagihara, *Bull. Chem. Soc. Japan*, **39,** 1821 (1966).
73. P. Hong, K. Sonogashira, and N. Hagihara, *Nippon Kagaku Zasshi*, **89,** 74 (1968), *Chem. Abstr.*, **69,** 2599*a* (1968).
74. E. Müller, A. Segnitz, and E. Langer, *Tetrahedron Lett.*, **1969,** 1129.
75. E. Müller and A. Segnitz, *Synthesis*, **1970,** 147.
76. J. J. Alexander and J. Wojcicki, *J. Organometal. Chem.*, **15,** 23 (1968).
77. J. Tsuji, *Acc. Chem. Res.*, **2,** 144 (1969).
78. K. Bittler, N. v. Kutepow, D. Neubauer, and H. Reis, *Angew. Chem.*, **80,** 352 (1968).
79. R. Hüttel, *Synthesis*, **1970,** 225.
80. J. Tsuji, N. Iwamoto, and M. Morikawa, *Bull. Chem. Soc. Japan*, **38,** 2213 (1965).
81. J. Tsuji, K. Ohno, and T. Kajimoto, *Tetrahedron Lett.*, **1965,** 4565.

82. J. Tsuji and K. Ohno. *J. Am. Chem. Soc.*, **90,** 94 (1968).
83. H. E. Eschinazi, *J. Am. Chem. Soc.*, **81,** 2905 (1959).
84. J. M. Conia and C. Faget, *Bull. Soc. Chim. France*, **1964,** 1963.
85. H. E. Eschinazi and H. Pines, *J. Org. Chem.*, **24,** 1369 (1959).
86. J. F. Hemidy and F. G. Gault, *Bull. Soc. Chim. France*, **1965,** 1710.
87. N. E. Hoffman, A. T. Kanakkanatt, and R. F. Schneider, *J. Org. Chem.*, **27,** 2687 (1962).
88. H. E. Eschinazi, *Bull. Soc. Chim. France*, **1952,** 967.
89. J. O. Hawthorn and M. H. Wilt, *J. Org. Chem.*, **25,** 2215 (1960).
90. M. S. Newman and H. V. Zahm, *J. Am. Chem. Soc.*, **65,** 1097 (1943).
91. M. S. Newman and N. Gill, *J. Org. Chem.*, **31,** 3860 (1966).
92. J. W. Wilt and V. P. Abegg, *J. Org. Chem.*, **33,** 923 (1968).
93. N. E. Hoffman and T. Puthenpurackal, *J. Org. Chem.*, **30,** 420 (1965).
94. J. S. Matthews, D. C. Ketter, and R. F. Hall, *J. Org. Chem.*, **35,** 1694 (1970).
95. T. Saegusa, T. Tsuda, and H. Isegawa, *Abstr. Ann. Meeting Japan Chem. Soc., Tokyo*, **1970,** 1203.
96. J. Tsuji and N. Iwamoto, *Chem. Commun.*, **1966,** 380.
97. G. Calvin and G. E. Coates, *J. Chem. Soc.*, **1960,** 2008.
98. I. I. Moiseev and M. N. Vargaftik, *Dokl. Akad. Nauk SSSR*, **166,** 370 (1966); through *Chem. Abstr.*, **64,** 11,248 (1966).
99. E. Mossetig and R. Mozingo, *Organic reactions*, Vol. 4, Wiley, New York, 1948 p. 362.
100. T. Nozoe and T. Kinugasa, Nippon Kagaku Zasshi, **59,** 772 (1938).
101. J. G. Burr, Jr., *J. Am. Chem. Soc.*, **73,** 3502 (1951).
102. C. A. Rajahn and A. Seitz, *Ann.*, **437,** 297 (1924).
103. G. P. Chiusoli and G. Agnes, *Chim. Ind.* (*Milan*), **46,** 548 (1964).
104. M. S. Newman and F. T. J. O'Leary, *J. Am. Chem. Soc.*, **68,** 258 (1946).
105. T. J. Suen and S. Fan, *J. Am. Chem. Soc.*, **64,** 1460 (1942); **65,** 1243 (1943).
106. P. Sabatier and J. B. Senderens, *Ann. Chem. Phys.*, **4**(8), 474 (1905).
107. P. Sabatier and J. B. Senderens, *Compt. Rend.*, **136,** 738,921 (1903).
108. E. J. Badin, *J. Am. Chem. Soc.*, **65,** 1809 (1943).
109. C. W. Bird and J. Hudec, *Chem. Ind.* (*London*), **1959,** 570.
110. I. Wender, S. Friedman, W. A. Steiner, and R. B. Anderson, *Chem. Ind.* (*London*), **1958,** 1964.
111. S. Friedman, S. R. Harris, and I. Wender, *Ind. Eng. Chem., Prod. Res. Develop.*, **9,** 347 (1970).
112. B. M. Trost and F. Chen, *Tetrahedron Lett.*, **1971,** 2603.
113. G. P. Chiusoli, S. Merzoni, and G. Mondelli, *Tetrahedron Lett.*, **1964,** 2777.
114. S. Tyrlik and H. Stepowska, *Tetrahedron Lett.*, **1969,** 3593.
115. P. Hong and N. Hagihara, *Abstr. Ann. Meeting Japan Chem. Soc.*, **1968,** 1971.
116. C. Ruchardt and G. N. Schrauzer, *Chem. Ber.*, **93,** 1840 (1960).
117. H. Ulrich, B. Tucker, and A. A. R. Sayigh, *Tetrahedron Lett.*, **1967,** 1731.
118. T. W. Campbell, J. J. Monagle, and V. S. Foldi, *J. Am. Chem. Soc.*, **84,** 3673 (1962).
119. R. H. Prince and K. A. Raspin, *J. Chem. Soc. A*, **1969,** 612.
120. J. Halpern and A. L. W. Kemp, *J. Am. Chem. Soc.*, **88,** 5147 (1966).
121. M. Kubota and D. M. Balke, *J. Am. Chem. Soc.*, **93,** 1368 (1971).
122. J. Chatt, B. L. Shaw, and A. E. Field, *J. Chem. Soc.*, **1964,** 3466.
123. L. Vaska, *J. Am. Chem. Soc.*, **86,** 1943 (1964).
124. A. Rusina and A. A. Vlcek, *Nature*, **1965,** 295.
125. R. W. Fries and J. K. Stille, *Synthesis Inorg. Metal-Org. Chem.*, **1,** 295 (1971).

126. M. Sergent, M. Mongrain, and P. Deslongchamps, *Can. J. Chem.*, **50,** 336 (1972).
127. B. M. Trost and M. Preckel, *J. Am. Chem. Soc.*, **95,** 7862 (1973); B. M. Trost and M. J. Bogdanowiz, *ibid.*, **95,** 2038 (1973).
128. P. D. Hobbs and P. D. Magnus, *Chem. Commun.*, **1974,** 856.
129. K. Sakai, J. Ide, O. Oda, and N. Nakamura, *Tetrahedron Lett.*, **1972,** 1287.
130. K. Sakai and O. Oda, *Tetrahedron Lett.*, **1972,** 4375.
131. H. M. Walborsky and L. E. Allen, *J. Am. Chem. Soc.*, **93,** 5465 (1971).
132. J. K. Stille and R. W. Fries, *J. Am. Chem. Soc.*, **96,** 1514 (1974).
133. J. K. Stille, F. Huang, and M. T. Regan, *J. Am. Chem. Soc.*, **96,** 1518 (1974).
134. J. K. Stille and M. T. Regan, *J. Am. Chem. Soc.*, **96,** 1508 (1974).
135. J. Blum and S. Kraus, *J. Organometal. Chem.*, **33,** 227 (1971).
136. K. Kaneda, H. Azuma, M. Wayaku, and S. Teranishi, *Chem. Lett.*, **1974,** 215.
137. J. Blum, M. Ashkenasy, and Y. Pickholtz, *Synthesis*, **1974,** 220.
138. E. Müller and A. Segnitz, *Ann. Chem.*, **9,** 1583 (1973).
139. D. M. Fenton and K. L. Olivier, *Chemtech.*, **1973,** 1583.
140. M. A. Bennett and R. Charles, *J. Am. Chem. Soc.*, **94,** 666 (1972).
141. T. A. Foglia, I. Schmeltz, and P. A. Barr, *Tetrahedron*, **30,** 11 (1974).
142. K. Matsuzawa, T. Kimura, Y. Murao, and H. Hashizume, Ger. Offen. 2232,253 (Japan. Kokai, 73 14643), *Chem. Abstr.*, 78, 98244 (1973); Japan. Kokai, 72 44213, *Chem. Abstr.*, **78,** 84043 (1973).
143. H. Reel and E. Vogel, *Angew. Chem. Intern. Ed.*, **11,** 1013 (1972).
144. M. R. Sandner and D. J. Trecker, *J. Org. Chem.*, **38,** 3954 (1973).

Addition of Hydrogen Cyanide to Olefins

E. S. BROWN, *Union Carbide Corporation, Chemicals and Plastics, Research and Development Department, South Charleston, West Virginia 25303*

Alkyl nitriles can be synthesized by the addition of hydrogen cyanide to olefins (hydrocyanation).

$$\rangle C{=}C\langle \; + HCN \rightarrow \; \rangle CH{-}\overset{|}{\underset{|}{C}}{-}CN$$

Addition of HCN to olefins containing activating substituents, for example, aryl, —COOR, —CN, —COR groups, commonly occurs in the presence of basic catalysts, for instance, cyanide ion (1–3). Use of AlR_3 catalysts was also reported for the hydrocyanation of α,β-unsaturated ketones (4–9).

Hydrocyanation of nonactivated olefins does not occur under these conditions. Certain transition-metal complexes do catalyze this reaction and have provided the impetus for significant interest in its synthetic utility.

Heterogeneous vapor-phase hydrocyanations employing supported transition metals were reported (10–12). Such reactions, however, are outside the scope of this discussion. This chapter will review what is now known about homogeneous, transition-metal catalyzed hydrocyanation of both activated (conjugated) and nonactivated olefins.

I. HYDROCYANATION OF ACTIVATED OLEFINS

Arthur and Pratt were the first to effect hydrocyanation of activated olefins by employing a transition-metal complex catalyst. They reported

TABLE 1
Homogeneous Hydrocyanation of Activated Olefins

Catalyst	Olefin	Product(s)	Yield (%)	Ref.
$Co_2(CO)_8$	$PhCH=CH_2$	$PhCH(CN)CH_3$	52	14–16
$Co_2(CO)_8$	1-cyclohexenyl–$CH=CH_2$	1-cyclohexenyl–$CH(CN)CH_3$	14	14–16
$Co_2(CO)_8$	$CH_2=CHCH=CH_2$	$CH_2=CHCH(CH_3)CN$, $CH_3CH=CHCH_2CN$, $CH_2=CH(CH_2)_2CN$	40	14–16
$Co_2(CO)_8$	$CH_2=C(CH_3)CH=CH_2$	Mixture of hexenenitriles	50	14–16
$CoH[P(OPh)_3]_4$	$CH_2=CHCH=CH_2$	$CH_3CH=CHCH_2CN$, $CH_2=CHCH(CH_3)CN$	—[a]	17
$Ni[P(OR)_3]_4$, $P(OR)_3$, metal salt	$CH_2=CHCH=CH_2$	$CH_3CH=CHCH_2CN$	72	29
$Ni[P(OR)_3]_4$, $P(OR)_3$, metal salt	$PhCH=CH_2$	$PhCH(CN)CH_3$, $PhCH_2CH_2CN$	8	28
NiL_4[b], BH_4^-	$CH_2=CHCH=CH_2$	$CH_3CH=CHCH_2CN$, $CH_2=CH(CH_2)_2CN$	30	18,27
NiL_4[b], BH_4^-	$CH_2=C=CH_2$	$CH_2=CHCH_2CN$, $CH_2=C(CH_3)CN$, $CH_3CH=CHCN$	54	18,27
$K_4[Ni(CN)_4]$	$CH_2=CHCOOEt$	$NCCH_2CH_2COOEt$	79	21
$K_4[Ni(CN)_4]$	$CH_2=CHCN$	$NC(CH_2)_2CN$	93	21
$K_4[Ni(CN)_4]$	$(CH_3)_2C=CHCOCH_3$	$(CH_3)_2C(CN)CH_2COCH_3$	13	21
$K_4[Ni(CN)_4]$	$PhCH=CHNO_2$	$PhCH(CN)CH_2NO_2$	97	21
$K_4[Ni(CN)_4]$	$CH_2=CHCH=CH_2$	Mixture of monocyanobutenes and dicyanobutanes	—[a]	21

$CaK_2[Ni(CN)_4]$	$CH_2{=}CHCN$	$CH_2(CN)CH_2CN$	25	21
$CaK_2[Ni(CN)_4]$	$CH_2{=}C(CH_3)COOCH_3$	$CH_2(CN)CH(CH_3)COOCH_3$	14	21
$Na_4[Ni(CN)_4]$	$CH_2{=}C(CH_3)COOCH_3$	$CH_2(CN)CH(CH_3)COOCH_3$	22	21
$Na_4[Ni(CN)_4]$	$CH_2{=}CHCOOEt$	$CH_2(CN)CH_2COOCH_3$	90	21
$Ni(CO)_4$	$CH_2{=}CHCH{=}CH_2$	$CH_3CH{=}CHCH_2CN$, $CH_2{=}CH(CH_2)_2CN$	10[c]	13
$Ni(CO)_4$, PPh_3	$CH_2{=}CHCH{=}CH_2$	$CH_3CH{=}CHCH_2CN$, $CH_2{=}CH(CH_2)_2CN$	63[c]	13
$Ni(CO_4)$, $AsPh_3$	$CH_2{=}CHCH{=}CH_2$	$CH_3CH{=}CHCH_2CN$, $CH_2{=}CH(CH_2)_2CN$	28[c]	13
$Pd[P(OR)_3]_4$, boron compound, metal salt	$PhCH{=}CH_2$	Hydrocyanated styrene	—[a]	19
Cu_2Cl_2	$CH_2{=}CHCH{=}CH_2$	$CH_3CH{=}CHCH_2CN$	27	23
Cu_2Cl_2	[cyclopentadiene ring]	[cyclopentene ring]–CN	—[a]	22
Cu_2Cl_2	$CH_2{=}CHCH{=}CH_2$	$CH_3CH{=}CHCH_2CN$	21	22
Cu_2Cl_2	$CH_2{=}CHC(CH_3){=}CH_2$	Mixture of hexenenitriles	90	22

[a] Yield not given.
[b] $L = CO$, $P(OR)_3$.
[c] Calculated from reported data.

the effectiveness of $Ni(CO)_4$ for formation of pentenenitriles (13) from HCN and butadiene.

$$CH_2{=}CHCH{=}CH_2 + HCN \xrightarrow[100-140°C]{Ni(CO)_4} CH_3CH{=}CHCH_2CN$$

The presence of either PPh_3 or $AsPh_3$ improved the pentenenitrile yield.

Dicobalt octacarbonyl was subsequently shown to be catalytically active for butadiene hydrocyanation as well as for hydrocyanation of isoprene and styrene (14–16).

A number of other metal complexes were reported to possess catalytic activity, for example, $CoH[P(OPh)_3]_4$ (17), $Ni[P(OR)_3]_4$ (18–20), $Pd[P(OR)_3]_4$ (19), $[Ni(CN)_4]^{4-}$ (21), and Cu_2Cl_2 (22,23).

The olefins successfully hydrocyanated represent a wide variety of structural types. Esters of acrylic and methacrylic acids, acrylonitrile, styrene, α-nitrostyrene, 1-vinylcyclohexene and cyclopentadiene as well as butadiene were derivatized (see Table 1). As with $Ni(CO)_4$ and $Co_2(CO)_8$, butadiene preferentially reacts by conjugate addition of HCN in the presence of other zerovalent nickel complexes (18), cobalt(I) complexes (17), and copper complexes (22,23). Conjugate addition of HCN to cyclopentadiene was reported to occur in the presence of cuprous chloride (22). In the $Co_2(CO)_8$-catalyzed hydrocyanation of 1-vinylcyclohexene, however, 1,2-addition of HCN is apparently favored.

CH=CH$_2$ (cyclohexenyl) + HCN $\xrightarrow{Co_2(CO)_8}$ CH(CN)CH$_3$ (cyclohexenyl)

Of the limited number of catalysts investigated to date, tetracyanonickelate(0) anion has consistently given the best yields of HCN adducts. However, at present there appears to be no compelling reason to prefer transition metals over basic catalysts for hydrocyanation of activated olefins.

II. HYDROCYANATION OF NONACTIVATED OLEFINS

Arthur et al. in 1954 discovered that $Co_2(CO)_8$ and certain of its derivatives were catalysts for hydrocyanation of a number of nonactivated olefins (14). Subsequently, other complexes of cobalt (17,24) and complexes of ruthenium (33–35), rhodium (17), nickel (18,26–33), palladium (19,34,35), molybdenum, and tungsten (36) were shown to possess catalytic activity.

Ethylene reacts with HCN in the presence of $Co_2(CO)_8$ (14–16), $Ni[(PR_3)]_4$ (28), or $Pd[(PR_3)]_4$ (34) to form propionitrile in good yields.

$$CH_2{=}CH_2 + HCN \xrightarrow[130°C]{Co_2(CO)_8} CH_3CH_2CN$$

Propylene is reported to form isobutyronitrile exclusively in the presence of $Co_2(CO)_8$ (14).

$$CH_3CH{=}CH_2 + HCN \xrightarrow[130°C]{Co_2(CO)_8} CH_3CH(CN)CH_3$$

In contrast, $Pd[P(OPh)_3]_4$ affords a mixture of *n*-butyronitrile and isobutyronitrile (24).

Arthur and co-workers (14) observed that terminal olefins were more readily hydrocyanated in the presence of $Co_2(CO)_8$ than olefins having an internal double bond. 1-Butene, for example, affords a good yield (67%) of 2-cyanobutane while 2-butene under the same reaction conditions gives a rather modest yield (9.3%) of 2-cyanobutane. They also report that 5-hexenenitrile reacts with greater facility than does 3-pentenenitrile.

The hydrocyanation of 3-pentenenitrile has received considerable attention since it represents the second step of a synthesis of adiponitrile from butadiene. In the presence of zerovalent nickel (18,26–33) and palladium (19) or complexes of ruthenium (25), rhodium (17), molybdenum, and tungsten (36) isomerization of the 3-pentenenitrile to 4-pentenenitrile occurs. Furthermore, 4-pentenenitrile undergoes hydrocyanation more readily than the 3-isomer and affords adiponitrile.

$$CH_2{=}CHCH{=}CH_2 + HCN \rightarrow CH_3CH{=}CHCH_2CN$$

$$CH_3CH{=}CHCH_2CN \rightarrow CH_2{=}CHCH_2CH_2CN$$

$$CH_2{=}CHCH_2CH_2CN + HCN \rightarrow CH_2(CN)CH_2CH_2CH_2CN$$

Hydrocyanation of cycloalkenes has received limited attention; the addition of HCN to cyclopentene and to 4-cyanocyclohexene in the presence of zerovalent nickel complexes were reported (26,27). Use of a cobalt catalyst prepared *in situ* for HCN addition to 4-cyano-1-cyclohexene was also reported (17).

Cyclopentene + HCN ⟶ cyclopentyl–CN

4-Cyanocyclohexene (CN) + HCN ⟶ 1,4-dicyanocyclohexane (CN, CN)

A number of olefins containing the bicyclo[2.2.1]hept-2-ene structure are reported to react with HCN. Bicyclo[2.2.1]hept-2-ene itself reacts with HCN to give 2-cyanobicyclo[2.2.1]heptane in the presence of $Co_2(CO)_8$ (14,15), $Co_2(CO)_6(PR_3)_2$ (24), and $Pd[P(OPh)_3]_4$ (34). Complexes of the type $Co_2(CO)_6(PR_3)_2$ appear to be more effective as catalysts than the parent $Co_2(CO)_8$ (32). In the presence of $Pd[P(OPh)_3]_4$, exclusive formation of *exo*-2-cyanobicyclo[2.2.1]heptane was reported.

There is no evidence for formation of the *endo* isomer. Whether the reaction proceeds through a *trans* or *cis* addition of HCN was not established, however.

Reaction of 5-cyanobicyclo[2.2.1]hept-2-ene affords dinitrile products in the presence of either $Co_2(CO)_8$ (14) or $Pd[PR_3]_4$ (14–16,35).

Here also use of $Pd[P(OPh)_3]_4$ leads to HCN addition in such a fashion that the entering cyano substituent assumes the *exo*-orientation. This result appears independent of the steric disposition of the initial cyano substituent.

Bicyclo[2.2.1]heptadiene reacts in a stepwise manner with two equivalents of HCN, first to form an 86:14 mixture of *exo*- and *endo*-5-cyanobicyclo[2.2.1]hept-2-ene (35).

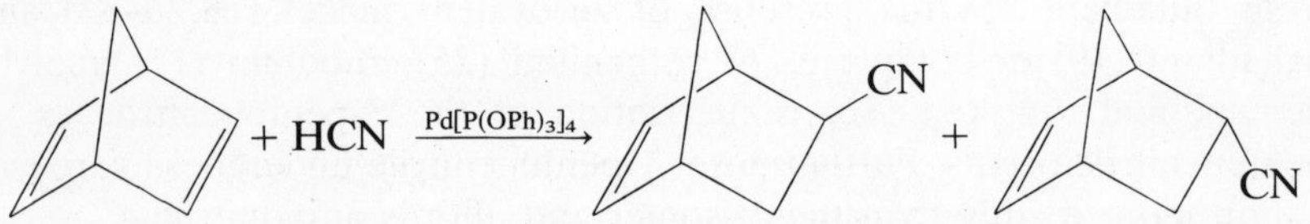

+ HCN $\xrightarrow{Pd[P(OPh)_3]_4}$ CN + CN

In a slower step, the second double bond reacts with HCN to give the four products expected from exclusive *exo* orientation of the second cyano substituent.

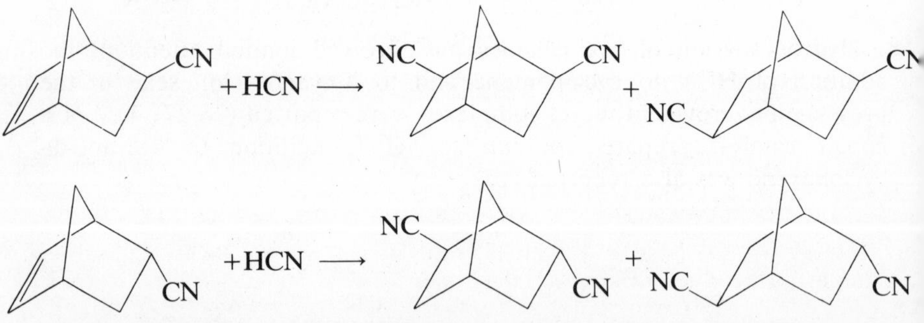

CN + HCN ⟶ NC CN + NC CN

CN + HCN ⟶ NC CN + NC CN

Hydrocyanation of 5-vinylbicyclo[2.2.1]hept-2-ene also provides several products (35). Under the conditions employed, isomerization of the vinyl substituent leads to formation of 5-ethylidenebicyclo[2.2.1]hept-2-ene concurrent with addition of HCN to the strained double bond.

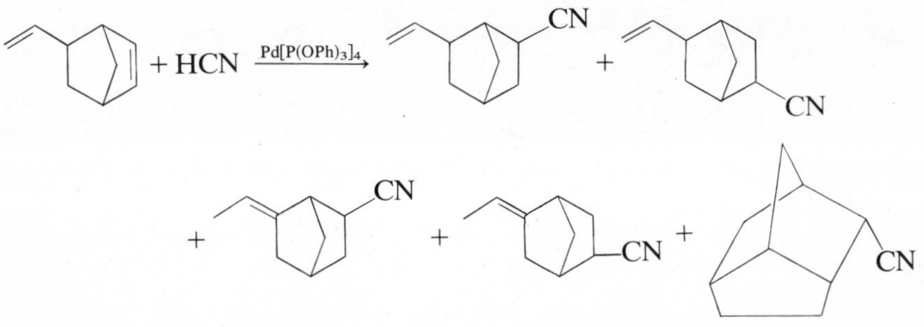

$$\text{5-vinylbicyclo[2.2.1]hept-2-ene} + HCN \xrightarrow{Pd[P(OPh)_3]_4} \text{(CN isomers)} + \ldots + \ldots + \ldots$$

CN CN CN CN CN

Unexpectedly, formation of a tricyclic nitrile, *endo*-2-cyanotricyclo[4.2.1.$0^{3,7}$]nonane (cyanobrendane), also occurred. Formation of the latter from *endo*-5-vinylbicyclo[2.2.1]hept-2-ene is rationalized as follows:

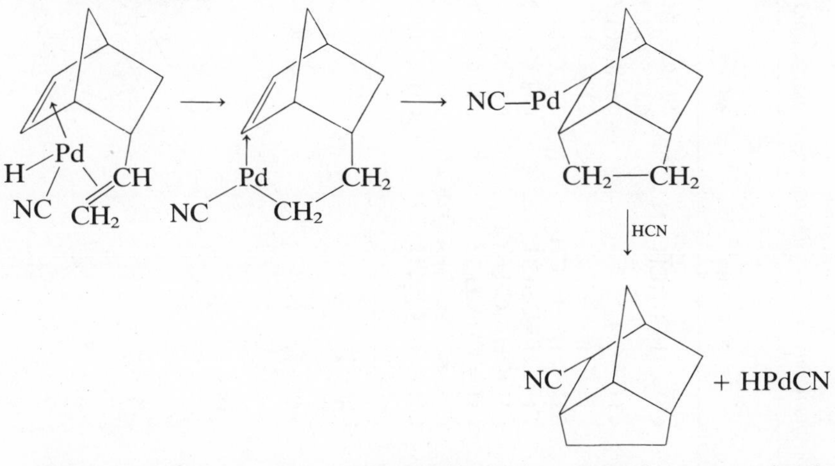

H Pd NC CH CH_2 → Pd NC CH_2 CH_2 → NC—Pd CH_2—CH_2

HCN

NC + HPdCN

Hydrocyanation of vinyl- and allylsilanes is reported to occur in the presence of tetrakis[triphenylphosphite]palladium(0) (37). Vinylsilanes frequently behave in a manner typical of activated olefins, for example, undergo Michael-type additions, but hydrocyanation of these silanes requires the use of transition-metal catalysts.

Vinyltriethoxysilane reacts with HCN to form 2-cyanoethyltriethoxysilane as the predominant product (yield: 73%). Other vinylsilanes react similarly (see Table 2).

$$CH_2{=}CHSi(OEt)_3 + HCN \xrightarrow{Pd[P(OPh)_3]_4} CH_2(CN)CH_2Si(OEt)_3$$

Reaction of vinyltriethoxysilane catalyzed by tetrakis[triphenylphosphite]nickel(0), affords a somewhat lower nitrile yield and a greater proportion of α-cyanoethyltriethoxysilane.

TABLE 2
Homogeneous Hydrocyanation of Nonactivated Olefins

Catalyst	Olefin	Product(s)	Yield (%)	Ref.
$RuCl_2(PPh_3)_3$	$CH_3CH{=}CHCH_2CN$	$NC(CH_2)_3CN$; $CH_3CH(CN)(CH_2)_2CN$	—[a]	25
$Co_2(CO)_8$	$CH_2{=}CH_2$	CH_3CH_2CN	64	14–16
$Co_2(CO)_8$	$CH_3CH{=}CH_2$	$CH_3CH(CN)CH_3$	65	14–16
$Co_2(CO)_8$	$CH_3CH_2CH{=}CH_2$	$CH_3CH_2CH(CN)CH_3$	68	14–16
$Co_2(CO)_8$	$CH_3CH{=}CHCH_3$	$CH_3CH(CN)CH_2CH_3$	9, 43[b]	14–16
$Co_2(CO)_8$	$CH_3(CH_2)_5CH{=}CH_2$	$CH_3(CH_2)_5CH(CN)CH_3$	5, 24[b]	14–16
$Co_2(CO)_8$	$CH_2{=}CH(CH_2)_3CN$	$CH_3CH(CN)(CH_2)_3CN$	36	14–16
$Co_2(CO)_8$	$CH_2{=}CH(CH_2)_3COOCH_3$	$CH_3CH(CN)(CH_2)_3COOCH_3$	19	14–16
$Co_2(CO)_8$	$CH_3CH{=}CHCH_2CN$	$CH_3CH(CN)CH_2CH_2CN$	7,15	14–16
$Co_2(CO)_8$	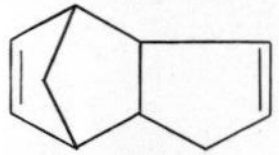	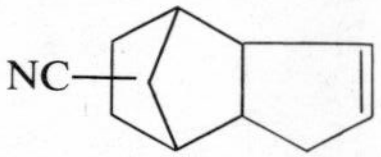 NC—	24[b]	14–16
$Co_2(CO)_8$	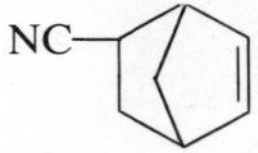 NC—	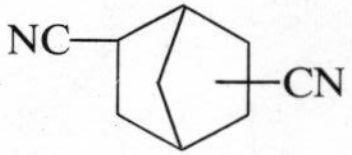 NC— —CN	62[b]	14–16
$Co_2(CO)_8$	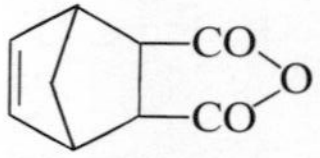 CO, O, CO	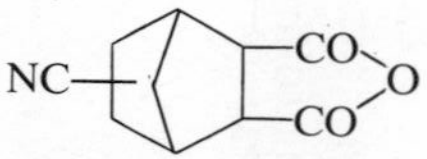 NC— CO, O, CO	28[b]	14–16
$Co_2(CO)_6(PPh_3)_2$	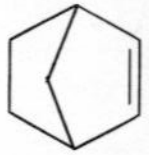	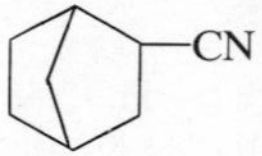 —CN	22	24
$Co_2(CO)_6[PPh(OPh)_2]_2$	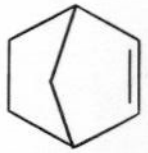	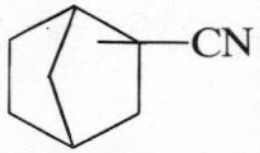 —CN	18	24

$CoH[P(OPh)_3]_4$	$CH_3CH{=}CHCH_2CN$	$NC(CH_2)_4CN$ } $CH_3CH(CN)(CH_2)_2CN$ }	5–30[c]	17
$CoH[P(OPh)_3]_4$	$CH_2{=}CHCH(CH_3)CN$	$NCCH_2CH(CH_3)CH_2CN$	9[c]	17
Co^{2+} salts, reducing agent, $P(OR)_3$	$CH_3CH{=}CHCH_2CN$	$NCCH_2CH_2(CH_2)_2CN$	1–41[c]	17
Co^{2+} salts, reducing agent, $P(OR)_3$	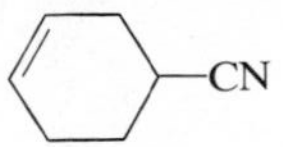 [structure]—CN	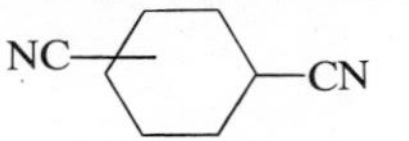 NC—[structure]—CN	—[a]	17
Co^{2+} salts, reducing agent, $P(OR)_3$	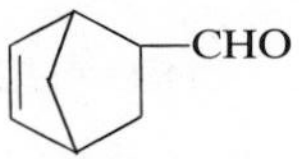 [structure]—CHO	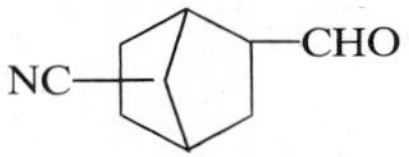 NC—[structure]—CHO	—[a]	17
Co^{2+} salts, reducing agent, $P(OR)_3$	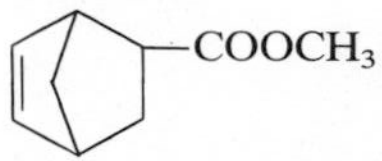 [structure]—$COOCH_3$	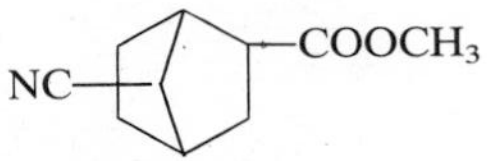 NC—[structure]—$COOCH_3$	—[a]	17
Co^{2+} salts, reducing agent, $P(OR)_3$	$CH_2{=}CHCH_2CN$	$NC(CH_2)_3CN$	—[a]	17
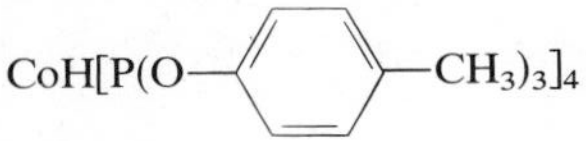 $CoH[P(O{-}C_6H_4{-}CH_3)_3]_4$	$CH_3CH{=}CHCH_2CN$	$NC(CH_2)_4CN$	0.5[c]	17
$RhCl(PPh_3)_3$	$CH_3CH{=}CHCH_2CN$	$NC(CH_2)_4CN$	0.2[c]	17
$Ni[P(OR)_3]_4$, Zn^{2+} or Co^{2+} salts	$CH_3CH{=}CHCH_2CN$	$NC(CH_2)_4CN$	11–64[c]	26
NiL_4[d], BH_4^-	$CH_3CH{=}CHCH_2CN$	$NC(CH_2)_4CN$	0.2–60	18,27
NiL_4[d], BH_4^-	$CH_2{=}CHCH_2CN$	$NC(CH_2)_3CN$ } $NCCH_2CH(CH_3)CN$ }	0.5–7	18,27
$Ni[P(OR)_3]_4$, $P(OR)_3$, metal salts	$CH_3CH{=}CHCH_2CN$	$NC(CH_2)_4CN$	0.3–10	28

[a] Yield not given.
[b] PPh_3 added.
[c] Approximate yield calculated from literature data.
[d] $L = P(OR)_3$, CO, PPh_3, $AsPh_3$, etc.

TABLE 2 (*contd.*)

Catalyst	Olefin	Product(s)	Yield (%)	Ref.
$Ni[P(OR)_3]_4$, $P(OR)_3$, metal salts	$CH_2{=}CHCH_2CN$	$NC(CH_2)_3CN$ $CH_3CH(CN)CH_2CN$	40	28
$Ni[P(OR)_3]_4$, $P(OR)_3$, metal salts	$CH_2{=}CHCH_2CH_2CN$	$NC(CH_2)_4CN$	37	28
$Ni[P(OR)_3]_4$, $P(OR)_3$, metal salts	$CH_2{=}CHCH(CH_3)CN$	$NCCH_2CH_2CH(CH_3)CN$	63	28
$Ni[P(OR)_3]_4$, $P(OR)_3$, metal salts	$CH_2{=}C{-}CH_2$ / $CH_2{-}CHCN$ (ring)	$NCCH_2CH{-}CH_2$ / $CH_2{-}CHCN$ (ring)	60	28
$Ni[P(OR)_3]_4$, $P(OR)_3$, metal salts	$C_4H_9CH{=}CH_2$	$C_4H_9CH_2CH_2CN$ (74%) $C_4H_9CH(CN)CH_3$ (24%)	23	28
$Ni[P(OR)_3]_4$, $P(OR)_3$, metal salts	[cyclopentene]	[cyclopentyl]–CN	20	28
$Ni[P(OR)_3]_4$, $P(OR)_3$, metal salts	[cyclohex-3-enyl]–CN	NC–[cyclohexane-1,4-diyl]–CN	—[a]	28
$Ni[P(OR)_3]_4$, $P(OR)_3$, metal salts	$C_6H_{13}CH{=}CH_2$	$C_6H_{13}CH_2CH_2CN$ (71%) $C_6H_{13}CH(CN)CH_3$ (23%)	40	28
$Ni[P(OR)_3]_4$, $P(OR)_3$, metal salts	$CH_2{=}CH_2$	CH_3CH_2CN	34	28
$Ni[P(OR)_3]_4$, $P(OR)_3$, metal salts	[norbornene]	[norbornyl]–CN	77	28
$Ni[P(OR)_3]_4$, $P(OR)_3$, metal salts	$CH_3CH{=}CHCH(CH_3)_2$	$(CH_3)_2CHCH_2CH(CN)CH_3$ $(CH_3)_2CHCH_2CH_2CH_2CN$	27	28
$Ni[P(OR)_3]_4$, $P(OR)_3$, metal salts	$CH_3(CH_2)_2C(CH_3){=}CH_2$	$CH_3(CH_2)_2CH(CH_3)CH_2CN$ (53%) $(CH_3)_2CH(CH_2)_3CN$ (36%) $(CH_3)_2CHCH_2CH(CN)CH_3$ (7%)	9	28

$Ni[P(OR)_3]_4$, $P(OR)_3$, metal salts	$CH_3(CH_2)_2CH{=}CHCH_3$	$CH_3(CH_2)_5CN$ (75%) $CH_3(CH_2)_3CH(CN)CH_3$ (25%)	36	28
$Ni[P(OR)_3]_4$, $P(OR)_3$, metal salt, or boron compound	$CH_3CH{=}CHCH_2CN$	$NC(CH_2)_4CN$	0.5–80	29
Ni^{2+} salt, Zn or Cd powder, $P(OR)_3$	$CH_3CH{=}CHCH_2CN$	$NC(CH_2)_4CN$	68	30
$Pd[P(OR)_3]_4$, metal salt or boron compound	$CH_3CH{=}CHCH_2CN$	$NC(CH_2)_4CN$	4–20	19
$Pd[P(OR)_3]_4$, metal salt or boron compound	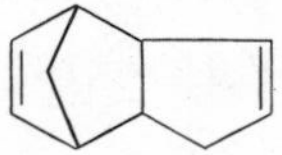	hydrocyanated dicyclopentadiene	—[a]	19
$Pd[P(OPh)_3]_4$, $P(OPh)_3$	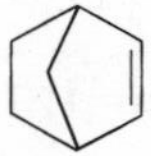	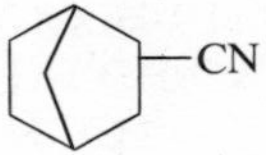 CN	83	34,35
$Pd[P(OPh)_3]_4$, $P(OPh)_3$	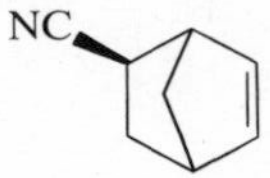 NC	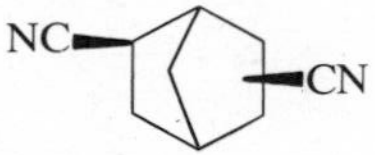 NC CN	58	34
$Pd[P(OPh)_3]_4$, $P(OPh)_3$	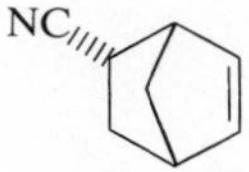 NC	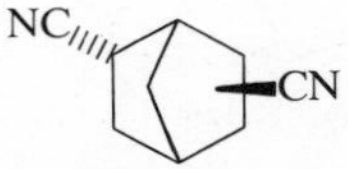 NC CN	57	35
$Pd[P(OPh)_3]_4$, $P(OPh)_3$	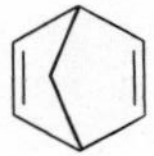	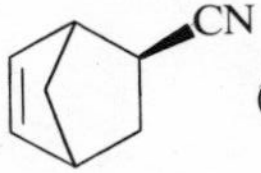  CN (86%), CN (14%)	49	35
$Pd[P(OPh)_3]_4$, $P(OPh)_3$	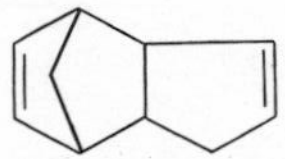	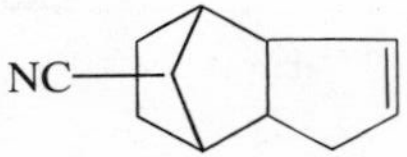 NC	35	35

TABLE 2 (contd.)

Catalyst	Olefin	Product(s)	Yield (%)	Ref.
$Pd[P(OPh)_3]_4$, $P(OPh)_3$		—CN, —CN, CN	26	35
$Pd[P(OR_3)]_4$, $P(OPh)_3$	$CH_2{=}CH_2$	CH_3CH_2CN	—[a]	34
$Pd[P(OPh)_3]_4$	$CH_2{=}CHSi(OEt)_3$	$NCCH_2CH_2Si(OEt)_3$ $CH_3CH(CN)Si(OEt)_3$	73	37
$Pd[P(OPh)_3]_4$	$CH_2{=}CHSi(CH_3)(OEt)_2$	$NCCH_2CH_2Si(CH_3)(OEt)_2$ $CH_3CH(CN)Si(CH_3)(OEt)_2$	31	37
$Pd[P(OPh)_3]_4$	$CH_2{=}CHSi(CH_3)_2(OEt)$	$NCCH_2CH_2Si(CH_3)_2(OEt)$ $CH_3CH(CN)Si(CH_3)_2(OEt)$	31	37
$Pd[P(OPh)_3]_4$	$[CH_2{=}CHSi(CH_3)_2]_2O$	$[NCCH_2CH_2Si(CH_3)_2]_2O$	85	37
$Pd[P(OPh)_3]_4$	$CH_2{=}CHCH_2Si(OEt)_3$	$NC(CH_2)_3Si(OEt)_3$ $CH_3CH(CN)CH_2Si(OEt)_3$	49	37
$Mo[P(O{-}C_6H_4{-}CH_3)_3]_3(CO)_3$ BPh_3, $P(O{-}C_6H_4{-}CH_3)_3$	$CH_3CH{=}CHCH_2CN$	$NC(CH_2)_4CN$	—[a]	36
$W[P(OPh)_3]_3(CO)_3$, BPh_3	$CH_3CH{=}CHCH_2CN$	$NC(CH_2)_4CN$	—[a]	36

Allyltriethoxysilane affords an equimolar mixture of 2-methyl-3-(triethoxysilyl)propionitrile and 4-(triethoxysilyl)butyronitrile.

$$CH_2{=}CHCH_2Si(OEt)_3 + HCN \rightarrow CH_2(CN)CH_2CH_2Si(OEt)_3 + CH_3CH(CN)CH_2Si(OEt)_3$$

III. MECHANISM OF HYDROCYANATION

Most investigations to date have been synthetically oriented; that is, only one kinetic investigation of catalytic hydrocyanation has appeared. Consequently, there is limited experimental basis for mechanistic considerations of the reaction. However, it is useful to discuss several mechanistic possibilities.

There are two stoichiometric reactions of transition-metal complexes that require consideration in discussing the mechanism of hydrocyanation. First, the oxidative addition of HCN was reported to occur with complexes of Rh(I) (38), Ir(I) (38,39), and Pt(0) (40) to yield the corresponding hydridocyanide complexes.

$$RhCl(PPh)_3 + 2HCN \longrightarrow Rh(H)(CN)(Cl)(HCN)(PPh_3)_2$$

(Octahedral Rh: H and Cl axial; PPh_3, CN, (HCN), PPh_3 equatorial.)

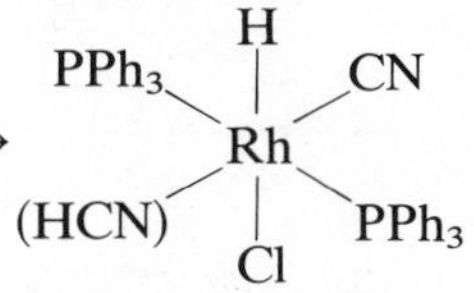

$$IrCl(CO)(PPh_3)_2 + HCN \longrightarrow Ir(CN)(H)(Cl)(CO)(PPh_3)_2$$

(Octahedral Ir: CN and Cl axial; PPh_3, H, OC, PPh_3 equatorial.)

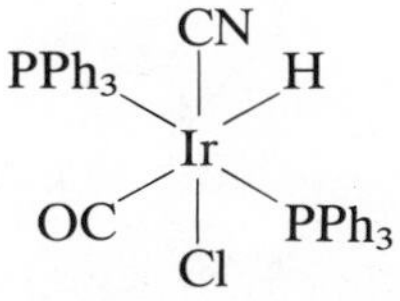

$$Pt(PPh_3)_{3\ \text{or}\ 4} + HCN \rightarrow Pt(H)(CN)(PPh_3)_2$$

In an analogous reaction, tetrakis[triethylphosphite]nickel(0) reacts with HCN to form a mixture of hydrido complexes (41–43).

$$Ni[P(OEt)_3]_4 + HCN \rightarrow Ni(H)[P(OEt)_3]_3CN + [Ni(H)[P(OEt)_3]_4]CN$$

It seems reasonable to suggest that oxidative addition of HCN to the metal-complex catalyst may be involved in the reaction mechanism.

A metal-hydride intermediate in hydrocyanation is given credence by the observed catalytic activity of the cobalt hydride, $CoH[P(OPh)_3]_4$, and also by the reported occurrence of olefin isomerization during some hydrocyanations. Corain demonstrated that bis[1,4-bis(diphenylphosphino)butane]nickel(0) catalyzes isomerization of 1-pentene in the presence of HCN (44).

The second stoichiometric reaction that requires consideration in this context is the formation of alkyl nitriles from alkylpentacyanocobaltate(II) complexes. A mechanism for this reaction was first proposed by Kwiatek and Seyler (46) and has been subsequently substantiated by the kinetic studies of Johnson, Tobe, and Wong (47,48). It is believed that the reaction proceeds as follows:

$$[RCo(CN)_5]^{3-} + H^+ \rightleftarrows [RCo(CN)_4CNH]^{2-}$$

$$\downarrow$$

$$RC{\equiv}N + [Co(CN)_4]^{3-} + BH^+ \xleftarrow{B:} [R\overset{\overset{\displaystyle N{-}H}{\|}}{C}Co(CN)_4]^{2-}$$

Combining the notions of HCN oxidative addition and of hydrogen isocyanide insertion, the following scheme may be written.

$$M + HCN \rightleftharpoons H{-}M{-}CN$$

$$H{-}M{-}CN + \text{olefin} \rightleftharpoons HM(\text{olefin})CN$$

$$H{-}M(\text{olefin})CN \rightarrow \text{alkyl}{-}M{-}CN$$

$$\text{alkyl}{-}M{-}CN + HCN \rightarrow \text{alkyl}{-}M(CNH)CN$$

$$\text{alkyl}{-}M(CNH)CN \rightarrow \text{alkyl}{-}\overset{\overset{\displaystyle NH}{\|}}{C}{-}M{-}CN$$

$$\text{alkyl}{-}\overset{\overset{\displaystyle NH}{\|}}{C}{-}M{-}CN \rightarrow \text{alkyl}{-}C{\equiv}N + HMCN$$

As an alternative, the product-forming step can be visualized as a reductive elimination of alkyl nitrile from the alkyl cyano–metal complex. There is no precedent for this particular elimination (49); however, the reverse reaction (oxidative addition of RCN to M°) was observed (51,52).

In conjunction with a discussion of mechanism, the matter of catalyst deactivation must be considered. Hydrocyanations catalyzed by $Co_2(CO)_8$ require large quantities of the complex (~10 mole %) to achieve satisfactory olefin conversions. This apparently results from the reaction of $Co_2(CO)_8$ with HCN to form compounds containing CN and CO ligands which are less effective than $Co_2(CO)_8$ itself for hydrocyanation (15).

Similar deactivating processes occur in hydrocyanations catalyzed by zerovalent nickel and palladium complexes. In the course of hydrocyanation of 3-pentenenitrile catalyzed by tetrakis[triphenylphosphite]nickel, for example, insoluble nickel salts (corresponding to about 60% of the added nickel) precipitate primarily in the form of $Ni(CN)_2$. (Nickel cyanide is catalytically inactive under the reaction conditions.) When the same reaction is conducted in the presence of excess triphenylphosphite (6 moles/mole of nickel complex), a slower rate of hydrocyanation results, but only about 20% of the $Ni[P(OPh)_3]_4$ is converted to $Ni(CN)_2$ (29).

Similarly, loss of catalytic activity in $Pd[P(OPh)_3]_4$-catalyzed hydrocyanations appears related to the formation of palladium(II) cyanide complexes (34). The deactivation process may be suppressed by employing excess phosphorus-containing ligand, but at the expense of a reduced reaction rate.

The palladium(II) and nickel(II) cyanides may arise during hydrocyanation via further interaction of the postulated hydridocyano–metal complexes with HCN, reaction (a), or of the alkyl metal cyanide complex with HCN, reaction (*b*):

$$\text{H—M—CN} + \text{HCN} \rightarrow [H_2M(CN)_2] \rightarrow M(CN)_2 + H_2 \qquad (a)$$

$$\text{R—M—CN} + \text{HCN} \rightarrow [RM(H)(CN)_2] \rightarrow M(CN)_2 + RH \qquad (b)$$

Reactions similar to (*a*) were reported for zerovalent nickel and palladium complexes with acids (40).

Lewis acids, for example BR_3 compounds and various metal salts, were utilized as promoters (26,27,29), and it would appear that they operate by formally altering the catalyst deactivation reaction since it is known that the stoichiometric reaction of $Ni[P(O(Et)_3]_4$ with HCN is altered by the presence of $ZnCl_2$ (41). Rather than forming a mixture of $[Ni(H)\text{-}P(OEt)_3]_4]^+CN^-$ and $NiH[P(OEt)_3]_3CN$, the complex $Ni[P(OEt)_3]_4$ forms the cationic pentacoordinate hydride $[Ni(H)(P(OEt)_3)_4]^+[ZnCl_2CN]^-$.

Taylor and Swift (45) investigated the effect of Lewis acid promoters on the hydrocyanation of 1-hexene by zerovalent nickel phosphite complex. They found that the rate of hydrocyanation increased with increasing levels of added $ZnCl_2$ until the $ZnCl_2/Ni(0)$ mole ratio reached unity and subsequently remained constant, thus suggesting a 1:1 $ZnCl_2$–NiL_4 adduct. They proposed the following mechanism for the hydrocyanation:

$$NiL_4 + HCN + ZnCl_2 \rightleftarrows (NiHL_3)^+(ZnCl_2CN)^- + L$$

$$(HNiL_3)^+ + RCH{=}CH_2 \rightleftarrows (RCH_2CH_2NiL_3)^+ + (R\underset{}{\overset{CH_3}{\overset{|}{C}}}HNiL_3)^+$$

$$(RCH_2CH_2NiL_3)^+ + (ZnCl_2CN)^- \rightarrow RCH_2CH_2CN + NiL_3^+ + ZnCl_2$$

Catalyst deactivation by HCN was attributed to these reactions:

$$(RCH_2CH_2NiL_3)^+ + HCN \rightarrow RCH_2CH_3 + (L_3NiCN)^+$$

$$(L_3NiCN)^+ + (ZnCl_2CN)^- \rightarrow NiL_2(CN)_2 + ZnCl_2 + L$$

The presence of excess ligand L, would retard catalyst deactivation by HCN in the following way:

$$(HNiL_3)^+ + L \rightleftarrows (HNiL_4)^+$$

They further noted that the exact nature of the Lewis acid cocatalyst also influenced the ratio of normal to branched nitriles formed from hexene. Taylor and Swift also reported that under their experimental conditions, the rate of 1-hexene isomerization was greater than the rate of hydrocyanation.

The influence of excess ligand on rate of catalyst deactivation is reminiscent of ligand effect on the sulfuric acid decomposition of tetrakis[triethylphosphite]nickel(0) (50). Initially, protonation of nickel occurs to form $[Ni(H)(P(OEt)_3]HSO_4$, followed by a decomposition step involving a second H_2SO_4 molecule. The latter reaction requires a prior dissociation of a triethylphosphite ligand, resulting in an inverse relation of the rate of Ni(0) decomposition and ligand concentration.

In closing, we can say that hydrocyanation of olefins by transition-metal catalysts has been demonstrated to be a potentially useful synthetic method. In fact, hydrocyanation of pentenenitrile to adiponitrile may well have commercial potential (53).

REFERENCES

1. D. T. Mowry, *Chem. Rev.*, **42,** 189 (1948).
2. *Cyanides in Organic Chemistry, A Literature Review*, Electrochemicals Department, E. I. DuPont de Nemours and Co., Wilmington, Delaware, 1962.
3. P. Kurtz, *Ann.*, **572,** 23 (1951).
4. W. Nagata and M. Yoshioka, *Yuki Gosei Kagaku Hyokai Shi.*, **26,** 2 (1968); *Chem. Abstr.*, **68,** 77327 (1968).
5. W. Nagata, M. Yoshioka, and S. Hirai, *J. Am. Chem. Soc.*, **94,** 4635 (1972).

6*a*. W. Nagata. M. Yoshioka, and M. Murakami, *J. Am. Chem. Soc.*, **94,** 4644 (1972).

6*b*. *Idem*, p. 4654.

7. W. Nagata, M. Yoshioka, and T. Terasawa, *J. Am. Chem. Soc.*, **94,** 4672 (1972).
8. W. Nagata and M. Yoshioka, *Org. Syn.*, **52,** 100 (1972).
9. J. W. Teter, U.S. Pat. 2,385,741 (1945); *Chem. Abstr.*, **40,** 590 (1946).
10. D. D. Davis and L. Scott, U.S. Pat. 3,278,575 (1966); *Chem. Abstr.*, **66,** 55068 (1967).
11. D. D. Davis, U.S. Pat. 3,278,576 (1966); *Chem. Abstr.*, **65,** 20016 (1966).
12. D. D. Davis, U.S. Pat. 3,282,981 (1966); *Chem. Abstr.*, **66,** 10612 (1967).
13. P. Arthur, Jr. and B. C. Pratt, U.S. Pat. 2,571,099 (1951); *Chem. Abstr.*, **46,** 3068 (1952).
14. P. Arthur, Jr., D. C. England, B. C. Pratt, and G. M. Whitman, *J. Am. Chem. Soc.*, **70,** 5364 (1954).
15. P. Arthur, Jr. and B. C. Pratt, U.S. Pat. 2,666,780 (1954); *Chem. Abstr.*, **49,** 1776 (1955).

16. P. Arthur, Jr. and B. C. Pratt, U.S. Pat. 2,666,748 (1954); *Chem. Abstr.*, **49,** 1774 (1955).
17*a*. W. C. Drinkard, Jr. and B. W. Taylor, Belg. Pat. 727,382.
17*b*. Neth. Pat. 68 15,812.
18*a*. W. C. Drinkard, Jr., U.S. Pat. 3,496,218.
18*b*. Fr. Pat. 1,544,658.
18*c*. Neth. Pat. 67 06,555.
18*d*. Brit. Pat. 1,112,539; *Chem. Abstr.*, **69,** 26810 (1968).
18*e*. Belg. Pat. 698,332.
19*a*. W. C. Drinkard, Jr. and R. V. Lindsey, Jr., Ger. Offen. 1,806,098 (1969); *Chem. Abstr.*, **71,** 49343 (1969).
19*b*. Belg. Pat. 723,128.
19*c*. Neth. Pat. 68 15,487.
20. W. C. Seidel and C. A. Tolman, U.S. Pat. 3,850,973 (1974); *Chem. Abstr.*, **82,** 97704 (1975).
21*a*. G. R. Coraor and W. Z. Heldt, U.S. Pat. 2,904,581 (1959); *Chem. Abstr.*, **54,** 4393 (1960).
21*b*. Brit. Pat. 845,086.
21*c*. Can. Pat. 628,973.
22. W. A. Schulze and J. E. Mahan, U.S. Pat. 2,422,859 (1947); *Chem. Abstr.*, **42,** 205 (1948).
23. D. D. Coffman, L. F. Salisbury, and N. D. Scott, U.S. Pat. 2,509,859 (1950); *Chem. Abstr.*, **44,** 8361 (1950).
24. E. A. Rick and E. S. Brown, unpublished results.
25*a*. W. C. Drinkard, Jr., Ger. Offen. 1,807,088 (1969); *Chem. Abstr.*, **71,** 38383 (1969).
25*b*. Belg. Pat. 723,381.
25*c*. Neth. Pat. 68 15,746.
26*a*. W. C. Drinkard, Jr. and R. J. Kassal, Fr. Pat. 1,529,134 (1968); *Chem. Abstr.*, **71,** 30092 (1969).
26*b*. Belg. Pat. 700,420.
26*c*. Brit. Pat. 1,146,330.
26*d*. Neth. Pat. 57 08,812.
27*a*. W. C. Drinkard, Jr., Ger. Offen. 1,806,096 (1969); *Chem. Abstr.*, **71,** 30093 (1968).
27*b*. Neth. Pat. 68 15,560.
27*c*. Belg. Pat. 723,126.
28*a*. W. C. Drinkard, Jr. and R. V. Lindsey, Brit. Pat. 1,104,140 (1968); *Chem. Abstr.*, **68,** 77795 (1968).
28*b*. U.S. Pat. 3,496,215.
28*c*. Belg. Pat. 698,333.
28*d*. Neth. Pat. 68 05,556.
28*e*. U.S. Pat. 3,496,217.
29*a*. Yuan-Tsan Chia, W. C. Drinkard, Jr., and E. N. Squire, Brit. Pat. 1,178,950 (1970); *Chem. Abstr.*, **72,** 89831 (1971).
29*b*. U.S. Pat. 3,766,237.
29*c*. Belg. Pat. 720,659.
29*d*. Neth. Pat. 68 12950.
30. A. W. Anderson and M. O. Unger, Neth. Pat. 69 07449.
31. Y. L. Mok, U.S. Pat. 3,846,474 (1974); *Chem. Abstr.*, **82,** 124818 (1975).
32. H. E. Shook, Ger. Offen. 2,421,081 (1974); *Chem. Abstr.*, **82,** 155842 (1975).
33. C. M. King and M. T. Musser, U.S. Pat. 3,864,380 (1975); *Chem. Abstr.*, **82,** 155421 (1975).

34. E. A. Rick and E. S. Brown, *Chem. Commun.*, **1969,** 112.
35. E. A. Rick and E. S. Brown, *Am. Chem. Soc. Div. Pet. Chem. Preprints,* **14,** B29 (1969).
36*a*. W. C. Drinkard, Jr. and B. W. Taylor, Ger. Offen. 1,807,089 (1969); *Chem. Abstr.*, **71,** 101350 (1969).
36*b*. Belg. Pat. 723,380.
36*c*. Neth. Pat. 68 15,744.
37. E. S. Brown and E. A. Rick, *Abstracts, 2nd North American Meeting, The Catalysis Society*, Houston Texas, Feb. 24, 1971.
38. H. Singer and G. Wilkinson, *J. Chem. Soc. A*, **1969,** 2516.
39. L. Benzoni, *Chem. Ind.* (*Milan*), **50,** 1227 (1968).
40. F. Cariati, R. Ugo, and F. Bonati, *Inorg. Chem.*, **5,** 1128 (1966).
41. W. C. Drinkard, Jr. and R. V. Lindsey, Jr., Belg. Pat. 723,386.
42. W. C. Drinkard, Jr., D. R. Eaton, J. P. Jesson, and R. V. Lindsey, Jr., *Inorg. Chem.*, **9,** 392 (1970).
43. R. A. Schunn, *Inorg. Chem.*, **9,** 394 (1970).
44. B. Corain, *Chem. Ind.* (*London*), **1971,** 1465.
45*a*. B. W. Taylor and H. E. Swift, *J. Catal.*, **26,** 254 (1972).
45*b*. U.S. Pat. 3,778,462 (1973); *Chem. Abstr.*, **80,** 59520 (1974).
46. J. Kwiatek and J. K. Seyler, *J. Organometal. Chem.*, **3,** 433 (1965).
47. M. D. Johnson, M. L. Tobe, and Lai-Yoong Wong, *Chem. Commun.*, **1967,** 298.
48*a*. M. D. Johnson, M. L. Tobe, and Lai-Yoong Wong, *J. Chem. Soc. A*, **1969,** 2516.
48*b*. *Ibid.*, **1968,** 923.
48*c*. *Ibid.*, **1969,** 929.
49. J. Chatt, R. S. Coffey, A. Gough, and D. T. Thompson, *J. Chem. Soc. A*, **1968,** 190.
50. C. A. Tolman, *J. Am. Chem. Soc.*, **92,** 4217 (1970).
51. J. L. Burmeister and L. M. Edwards, *J. Chem. Soc. A*, **1971,** 1663.
52. D. H. Gerlock, A. R. Kane, G. W. Parshall, J. P. Jesson, and E. L. Muetterties, *J. Am. Chem. Soc.*, **93,** 3543 (1971).
53. *Chem. Eng. News*, **49,** April 5, 1971, p. 31.

Hydrosilation Catalyzed by Group VIII Complexes

J. F. HARROD, *McGill University, Montreal 101, Quebec, Canada*

and

A. J. CHALK, *Givaudan Corporation, Clifton, New Jersey 07014*

I. INTRODUCTION

From a quantitative point of view the "direct reaction" of organohalides with elemental silicon, or alloys thereof, is the most important chemical process for the formation of silicon–carbon bonds (1). This process is used for producing the methyl- and phenylsilicon chlorides upon which the silicones industry has been built. If the "direct reaction" is considered to be the skeleton of silicone chemistry, then the flesh and blood is surely the hydrosilation reaction.* The latter, which may include

* We coined the term "hydrosilation" as an analog of "hydroboration." Other workers have shown a preference for the term "hydrosilylation."

the addition of a silicon–hydrogen bond to any type of unsaturated molecule (e.g., alkene, alkyne, carbonyl, nitrile, cycloalkane), provides a route to a wealth of functionally substituted organosilicon derivatives whose variety and ease of synthesis surpasses that of any element except carbon itself. The archetypal and most intensively studied hydrosilation reaction is the addition of silicon hydrides* to alkenes, thus:

$$\equiv\text{Si—H} + \gt\text{C}=\text{C}\lt \;\rightarrow\; \equiv\text{Si—}\overset{|}{\underset{|}{\text{C}}}\text{—}\overset{|}{\underset{|}{\text{C}}}\text{—H} \qquad (1)$$

Reaction 1 finds two major applications in silicone technology. In the first case, when the unsaturated molecule is a functionally substituted ethylene derivative, the product may be a substituted alkylsilicon compound which can be polymerized, or copolymerized, to a polysiloxane with special properties. For example, the incorporation of β-cyanoethyl residues into a silicone gum via hydrosilation of acrylonitrile provides a route to highly oil-resistant rubbers suitable for lining self-sealing fuel tanks (2). A second major use of reaction 1 is as a crosslinking reaction, either for the curing of gums to elastomers (Eq. 2) or of polyfunctional monomers to resins (Eq. 3) (3).

$$\left(\text{O—}\underset{\text{CH}_3}{\overset{\text{CH}_3}{\text{Si}}}\right)_m\text{O—}\underset{\text{H}}{\overset{\text{CH}_3}{\text{Si}}}\left(\text{O—}\underset{\text{CH}_3}{\overset{\text{CH}_3}{\text{Si}}}\right)_n + \left(\text{O—}\underset{\text{CH}_3}{\overset{\text{CH}_3}{\text{Si}}}\right)_x\text{O—}\underset{\text{CH}_3}{\overset{\text{CH}=\text{CH}_2}{\text{Si}}}\left(\text{O—}\underset{\text{CH}_3}{\overset{\text{CH}_3}{\text{Si}}}\right)_y \longrightarrow \begin{matrix}\left(\text{O—}\underset{\text{CH}_3}{\overset{\text{CH}_3}{\text{Si}}}\right)_m\text{O—}\underset{\text{CH}_2}{\overset{\text{CH}_3}{\text{Si}}}\left(\text{O—}\underset{\text{CH}_3}{\overset{\text{CH}_3}{\text{Si}}}\right) \\ | \\ \left(\text{O—}\underset{\text{CH}_3}{\overset{\text{CH}_3}{\text{Si}}}\right)_x\text{O—}\underset{\text{CH}_3}{\overset{\text{CH}_2}{\text{Si}}}\left(\text{O—}\underset{\text{CH}_3}{\overset{\text{CH}_3}{\text{Si}}}\right)\end{matrix} \qquad (2)$$

$$\left(\underset{\text{H}}{\overset{\text{CH}_3}{\text{Si}}}\text{—O}\right)_x + \left(\underset{\text{CH}=\text{CH}_2}{\overset{\text{CH}_3}{\text{Si}}}\text{—O}\right)_y \longrightarrow \text{Infinite network} \qquad (3)$$

Although certain hydrosilation reactions can be made to proceed thermally, the required forcing conditions make them of little practical value. Fortunately, a large number of different types of catalyst are available that can effect addition of silicon hydrides to unsaturated molecules under mild conditions, with high yields and with a variety of

*We use this expression to indicate a compound containing at least one Si–H bond. The alternative description of such compounds as "silanes" is less explicit and may be interpreted to exclude certain classes of compounds such as "siloxanes."

selectivities. The latter can be a very important factor; for example, most catalysts yield the relatively useless α-cyanoethyl product from a silicon hydride addition to acrylonitrile. The production of the β-cyanoethylsilane requires a special copper chloride/amine catalyst (2).

The general chemistry of the hydrosilation reaction has been excellently reviewed elsewhere (1). The object of this chapter is to describe the catalysis of hydrosilation by group VIII metal complexes from a mechanistic point of view, rather than from the perspective of a synthetic organosilicon chemist. The metal-complex-catalyzed hydrosilation reaction provides an extraordinary variety of reaction types whose chemistries yield insight into many general aspects of coordination catalysis. The chemistries of silicon hydrides and of molecular hydrogen are very similar with respect to the various mechanisms of catalytic activation. It is also to be hoped that a detailed understanding of the catalytic activation of silicon hydrides will be of some value in solving one of the most intractable problems of coordination catalysis: the catalytic activation of paraffins.

II. History and Scope of Platinum-Complex-Catalyzed Hydrosilation

A. SPEIER'S CATALYST

The exceptional activity of chloroplatinic acid as a hydrosilation catalyst was first reported by Speier, Webster, and Barnes (4) and this catalyst, particularly when used as a solution in an alcohol, is often referred to as Speier's catalyst. The minute amounts of catalyst required, 10^{-5}–10^{-8} moles of platinum per mole of reactant, make this reaction one of the few cases where platinum may be regarded as a disposable catalyst.

A veritable plethora of patents describes chloroplatinic acid and other complexes as catalysts. The most common variation is the prereaction of chloroplatinic acid with an alcohol. The chemistry of these prereacted alcohol solutions is complex; the platinum causes dehydration, dehydrogenation, and coupling of allylic alcohols (produced by dehydrogenation of higher aliphatic alcohols) to diallyl ethers. The resulting solutions are usually claimed to possess higher activity and/or better miscibility with the reactants.

Another common and useful procedure is to use platinum–olefin complexes as catalysts (5,6). In this case the best procedure is to react an ethylene complex, such as dichlorobis(ethylene)-μ-dichlorodiplatinum(II) with the unsaturated reactant. Most cases give a rapid replacement of the ethylene by the unsaturated reactant compound and produces a highly soluble and active analog of the ethylene derivative.

Induction periods and the high exothermicity of hydrosilation reactions can make large-scale reactions quite hazardous, especially with the highly active platinum complex catalysts. High exotherms often lead to spontaneous reflux with liquid systems. In polymer-forming reactions, where heat dissipation is much less efficient, spontaneous blowing, combustion, or explosion may result if the catalyst concentration is too high. Another problem with exotherms is the thermal destruction of the catalyst, especially with trialkyl- and triarylsilanes. In spite of the high exotherms, solvents are rarely used in hydrosilation reactions. More commonly, dilution is achieved by use of an excess of one of the reactants, particularly the unsaturated reactant.

B. The Chemistry of Platinum-Complex-Catalyzed Hydrosilation

Because of the extremely low catalyst concentrations, and the resulting extreme sensitivity to impurity effects, kinetic studies have been of little value in elucidating the mechanism of platinum-catalyzed hydrosilation. Most of what is known of the mechanism has been deduced from the reactivity patterns which emerge on varying the composition of the various reactants. Most of the reactivity patterns were established by Speier and his collaborators (4,7) in the period 1955–1965. Some salient features of hydrosilation may be summarized as follows:

(*i*) Addition of most silicon hydrides to alkenes gives exclusively terminal products, even when the double bond is remote from the terminal position (4,7*b*), for example,

$$Cl_3SiH + CH_3CH_2CH_2CH{=}CHCH_2CH_2CH_3 \rightarrow Cl_3Si(CH_2)_7CH_3 \quad (4)$$

There are exceptions to this rule. For example, the addition of *sym*-tetramethyldisiloxane to 3-heptene yields substantial amounts of product with silicon attached to the 3- and 4- positions of the heptyl group, as well as the predominant 1-heptyl product (7*a,d*).

(*ii*) In the course of hydrosilation of an *n*-alkene, the alkene may undergo extensive double-bond migration (7*c*). In the case of chlorosilanes, a terminal olefin may be isomerized to an equilibrium distribution of isomers prior to completion of hydrosilation.

(*iii*) Extensive exchange may occur between a silicon deuteride and an olefin (7*c*). The addition of excess trichlorosilane-*d* to isobutylene resulted in the following stoichiometry:

$$3.5\ Cl_3SiD + CH_2{=}C(CH_3)_2 \rightarrow Cl_3SiC_4H_{6.5}D_{2.5} + 1.5\ Cl_3SiH + Cl_3SiD \quad (5)$$

The deuterium was located at all positions of the isobutyl groups of the product isobutyltrichlorosilane but the secondary position was 70% deuterated. In addition, unreacted isobutylene was found to be partially deuterated.

(*iv*) Certain allylic compounds such as allyl chloride and allyl acetate undergo a catalytic reduction to propylene in the presence of silicon hydrides and platinum complex catalysts (7*c*,*e*).

(*v*) Certain combinations of reactants gave multiple insertion products, for example,

$$CH_3SiCl_2H + 2CF_2{=}CF_2 \rightarrow CH_3SiCl_2(CF_2)_3CF_2H \qquad (5a)$$

On the basis of the these facts, and the demonstration of reversible oxidative addition of silicon hydrides to some d^8 complexes, the hypothetical mechanism shown in Fig. 1 was proposed as a model for catalysis of hydrosilation within the coordination sphere of the metal complex (5). All the steps shown in Fig. 1 find analogies in homogeneous catalytic hydrogenation and hydroformylation. The critical steps of the cycle may be summarized thus:

(*i*) Whatever the starting form of the catalyst, it is rapidly converted to a platinum(II) complex of the unsaturated substrate. The complexed and uncomplexed unsaturated reactants and products are in relatively rapid equilibrium.

(*ii*) Platinum(II) complexes undergo rapid reversible oxidative addition of silicon hydride.

(*iii*) Insertion of unsaturated ligand into a Pt–H bond occurs reversibly and at a rate competitive with reductive elimination of silicon hydride.

(*iv*) Reversal of step (*iii*) may result in β-elimination of a group such as chloride, or acetate, rather than hydride.

(*v*) A slow insertion of a second unsaturated ligand into the alkyl produced by step (*iii*). In general, this step does not compete with step (*vi*).

(*vi*) Reductive elimination of a coordinated alkyl with a coordinated silyl to yield alkylsilicon product and regenerate catalyst.

In Fig. 1 only the ligands derived from reactants are specified, since the detailed nature of the catalyst species is not known. The reasonable assumption that the platinum is either four-coordinate [Pt(II)], or six-coordinate [Pt(IV)], is made. Most, but not all, experimental observation can be rationalized in terms of Fig. 1.

The exchange reactions of Cl_3SiD with olefins require the reversibility of all steps up to the final reductive elimination of hydrosilation product. The irreversibility of the overall reaction therefore requires the irreversibility of the final step.

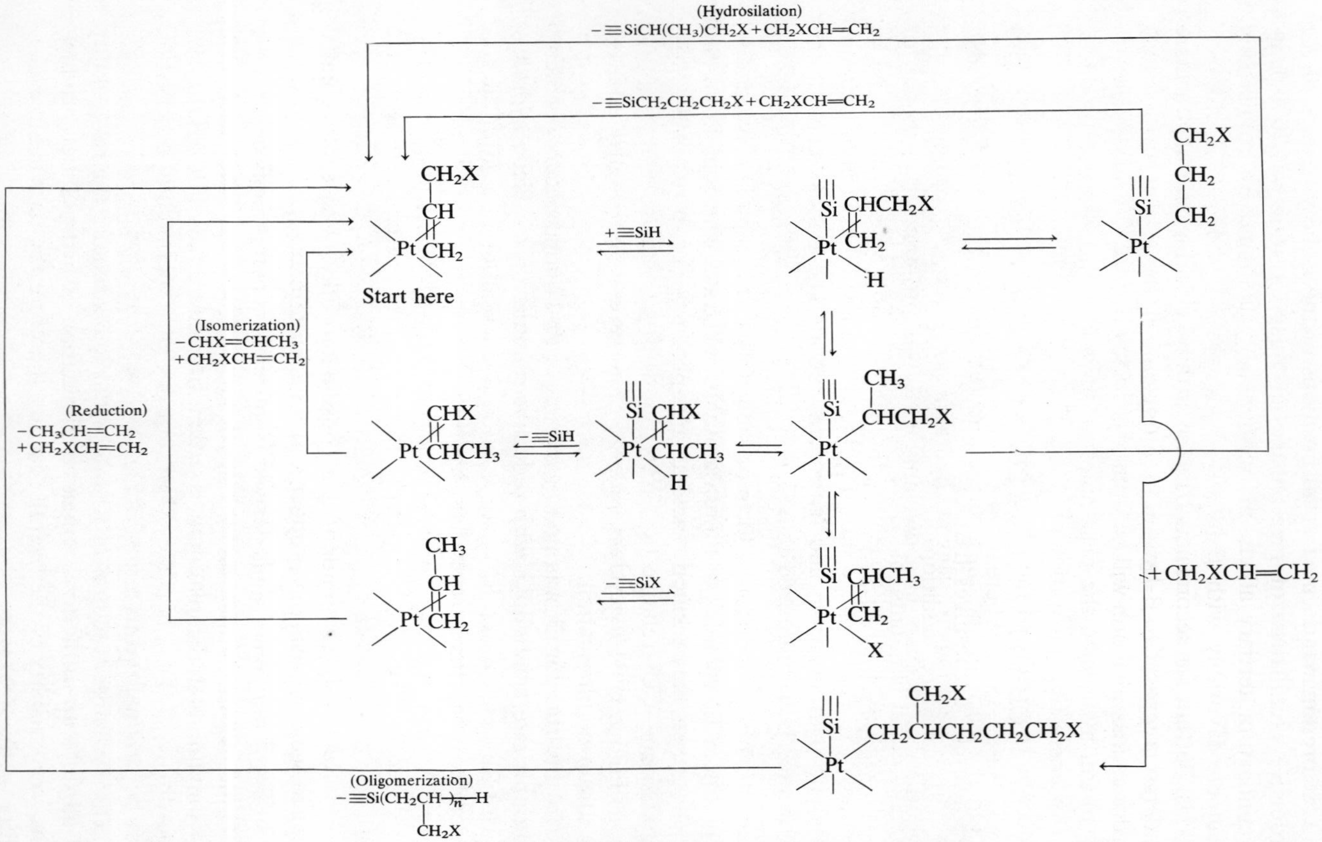

(Hydrosilation)

$-\equiv SiCH(CH_3)CH_2X + CH_2XCH{=}CH_2$

$-\equiv SiCH_2CH_2CH_2X + CH_2XCH{=}CH_2$

Start here

$+\equiv SiH$

(Isomerization)
$-CHX{=}CHCH_3$
$+CH_2XCH{=}CH_2$

(Reduction)
$-CH_3CH{=}CH_2$
$+CH_2XCH{=}CH_2$

$-\equiv SiH$

$-\equiv SiX$

$+CH_2XCH{=}CH_2$

(Oligomerization)
$-\equiv Si(CH_2CH)_n\!-\!H$ with CH_2X side chain

FIG. 1. Mechanism for hydrosilation of olefins

The predominance of terminal products requires that the rate of reductive elimination of primary alkyls be much greater than that of secondary or tertiary alkyls, or, more likely, that reductive elimination compete favorably with β-hydride elimination in the case of primary alkyls, but not for secondary or tertiary alkyls. If reductive elimination is very fast compared to β-hydride elimination, no isomerization of olefin will occur and the reaction will be indistinguishable from a concerted addition of silicon hydride to the unsaturated substrate.

An interesting illustration of how delicate may be the balance of factors affecting the nature of the products in hydrosilation was reported by Selin and West (8). Whereas the addition of trichlorosilane to 1-methylcyclohexene gives trichlorosilylcyclohexylmethane as the only detectable product, addition to 1-methyl-d_3-cyclohexene gave approximately equal amounts of 1-methyl-d_3-2-trichlorosilylcyclohexane and trichlorosilylcyclohexyl-d_1-methane-d_2.

$$Cl_3SiH + \text{(1-methyl-}d_3\text{-cyclohexene)} \longrightarrow \text{(1-D-cyclohexyl)}{-}CD_2SiCl_3 + \text{(2-}SiCl_3\text{-cyclohexyl)}{-}CD_3 \quad (6)$$

It appears that a kinetic isotope effect slows down β-hydride elimination from —CD_3 relative to —CH_3, to the point that reductive elimination of a 1-methyl-d_3-cyclohexyl ligand has a rate roughly equal to that of β-deuteride elimination.

At the time Fig. 1 was first proposed, reaction 7 was the only known example of oxidative addition of a silicon hydride to a d^8 complex (5). Subsequently, numerous examples of such additions to Rh(I) and Ir(I) complexes have been reported and are described in detail in Section III.

$$\underset{\mathbf{(1)}}{(Ph_3P)(Cl)Ir(OC)(PPh_3)} + {\equiv}Si{-}H \rightleftharpoons \underset{\mathbf{(2)}}{(Ph_3P)(\equiv Si)(H)Ir(OC)(Cl)(PPh_3)} \quad (7)$$

To date, no known analog of reaction 7 involving addition of a silicon hydride to platinum(II) has been reported. On the other hand, numerous silyl platinum(II) complexes such as **3** and **4** have been synthesized (9,10).

$$\underset{\mathbf{(3)}}{(PhMe_2P)(Cl)Pt(Ph_2MeSi)(PMe_2Ph)} \qquad \underset{\mathbf{(4)}}{\left[(Ph_2PCH_2CH_2PPh_2)Pt(PEt_3)(Si(Me)_3)\right]^+ Cl^-}$$

A most important observation was reported by Yamamoto et al. (55). Bis(triphenylphosphine)ethyleneplatinum(0) was found to catalyze the reaction of 1-hexene and methyldichlorosilane in almost quantitative yield. When a large amount of catalyst was used, the complex hydrido(methyldichlorosilyl)bis(triphenylphosphine)platinum(II) crystallized from the reaction mixture.

In the absence of olefin the complex $(Ph_3P)_2Pt(SiMeCl_2)_2$ was produced. Trichlorosilane only gave the bis(silyl)platinum(II) complex whether olefin was present or not.

Although the platinum(0) complex catalyzed hydrosilation of 1-hexene with other silanes, for example, $(EtO)_2MeSiH$, Me_2ClSiH, Me_3SiH, Me_2EtSiH, these silanes reacted with the complex to give dark red-brown products, soluble in benzene and analyzing as $[Pt(Ph_3P)C_6H_6]_x$. This behavior is undoubtedly related to the formation of dark brown colors when chloroplatinic acid, or other platinum catalysts are decomposed by the exotherm during hydrosilation. It is also of interest that similar dark brown complexes are produced by reaction of silicon hydrides with chloro(olefin) complexes of Pd(II) and Pt(II), both in the presence and absence of PPh_3 (11).

The observation of catalytic activity related to oxidative addition to a Pt(0) center suggests that the replacement of Pt(II) and Pt(IV) in Fig. 1 by Pt(0) and Pt(II) would perhaps make a more acceptable model for platinum-catalyzed hydrosilation. The appropriate changes in coordination numbers and geometries would of course also have to be made. The suggestion that Pt(0) is the lower oxidation state involved in the homogeneous reaction may also be interpreted to be more in line with the fact that metallic platinum is a moderately good hydrosilation catalyst.

Although the mechanism of Fig. 1 does not require a *cis* addition, the occurrence of *cis* stereoselectivity is readily rationalized. In fact such *cis* addition seems to be a common feature of coordination catalyzed additions to multiple bonds.

The 2-trichlorosilyl-1-methyl-d_3-cyclohexane produced in reaction 6 was uniquely *trans*, the product of a *cis* addition of Pt–H to the double bond of 1-methyl-d_3-cyclohexene (8). This stereospecificity may result from the concerted nature of the insertion requiring *cis* addition.

A quite remarkable example of unusual stereochemistry in hydrosilation is provided by the addition of trimethyl- or trichlorosilane to norbornadiene in the presence of chloroplatinic acid (14). The reaction gave the three products **5**, **6**, and **7**, with **7** predominating. This reaction represents a rare example of addition to norbornadiene to form an *endo*norbornenyl product. It has been suggested that concerted elimination of a chelated norbornenyl ligand forces formation of the sterically unfavorable *endo*

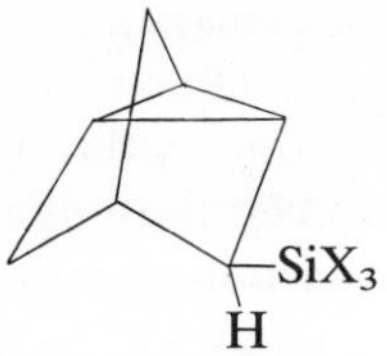

SiX_3 H **(5)** H H SiX_3 H **(6)** H H H SiX_3 **(7)**

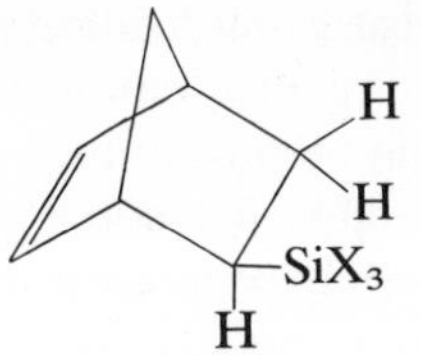

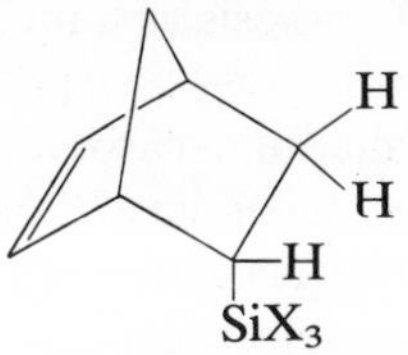

product, or of the nortricyclene **5**. The small amount of the *exo* product may arise through elimination of a monodentate σ-norbornenyl residue.

Several series of experiments using optically active silicon hydrides reported by Sommer et al. indicated that the oxidative addition of such hydrides to metal complexes occurs with retention at silicon, as also does the corresponding reductive elimination (15). The hydrosilation of 1-octene by (+)MePh(1-$C_{10}H_7$)SiH also occurred with retention of configuration. Again, although these results are not required by the proposed mechanism, they are not unexpected consequences of concerted oxidative addition and reductive elimination processes. Even less diagnostic significance may be attributed to these results in view of the demonstration that silicon can undergo free radical reactions with retention of configuration (16).

C. Isomerization as a Side Reaction in Hydrosilation

Double-bond migration is characteristic of most coordination catalytic reactions involving higher alkyl complexes of transition elements as transient intermediates. With a single exception, virtually all of the experimental evidence supports the mechanism for isomerization of olefins schematically illustrated in reaction 8 (17–19).

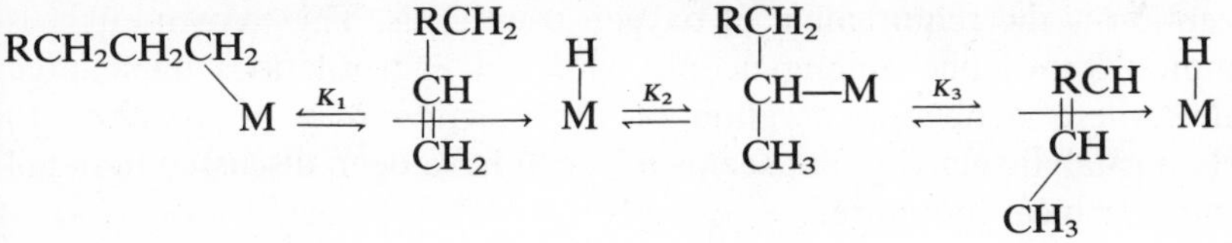

$$RCH_2CH_2CH_2{-}M \underset{}{\overset{K_1}{\rightleftharpoons}} \left(\begin{matrix}RCH_2\\ | \\ CH\\ \| \\ CH_2\end{matrix}\right) \rightarrow \begin{matrix}H\\ |\\ M\end{matrix} \overset{K_2}{\rightleftharpoons} \begin{matrix}RCH_2\\ |\\ CH{-}M\\ |\\ CH_3\end{matrix} \overset{K_3}{\rightleftharpoons} \left(\begin{matrix}RCH\\ \| \\ CH\\ / \\ CH_3\end{matrix}\right) \rightarrow \begin{matrix}H\\ |\\ M\end{matrix} \qquad (8)$$

Studies of hydrogen shifts in specifically deuterated olefins reveal that in some cases isomerization occurs by a $3 \rightarrow 1$ shift, in others by a $2 \rightarrow 1$ shift, and in others by a mixture of $3 \rightarrow 1$ and $2 \rightarrow 1$ shifts. The $3 \rightarrow 1$ shift occurs if $K_2 \gg K_1$, whereas a $2 \rightarrow 1$ shift occurs when $K_1 \gg K_2$. Mixed shifts result when $K_1 \approx K_2$. In our original investigations (17) we had some difficulty accepting the possibility that K_2 could be much greater than K_1 and therefore favored the alternative π-allyl-metal hydride hypothesis as a mechanism for isomerization in those cases where a $3 \rightarrow 1$ shift was observed. Subsequent demonstrations that addition of a palladium–carbon bond to an olefin can indeed proceed uniquely through the secondary alkyl intermediate has convinced us that even those reactions occurring via a $3 \rightarrow 1$ shift proceed according to reaction 8 (20). A possible exception to this rule is the isomerization of conjugated dienes by iron carbonyl where the evidence seems to favor a π-allyl-hydride mechanism (21).

In the hydrosilation reaction there is great variability in the relative rates of hydrosilation and olefin isomerization. In most cases the isomerization is considerably faster than hydrosilation. However, during addition of trialkoxysilanes to 1-alkenes, hydrosilation occurs in the almost complete absence of olefin isomerization (5). The latter presumably represents a case where the reductive elimination of alkylsilicon product is faster than β-hydride elimination from the metal alkyl intermediate. The fact that isomerization is very rapid during addition of trialkyl- and trichlorosilane, but negligible with trialkoxysilanes, is noteworthy. Studies on relative stabilities of oxidative adducts of silicon hydrides to transition metal complexes show the stability sequence to be: $Cl_3SiH > (RO)_3SiH > R_3SiH$ (5,11,22). Why the isomerization behavior of $(RO)_3SiH$ is so different from the other silicon hydrides is not clear. That such variations should occur, however, is not surprising in view of the complexity of the reaction.

D. Objections to the Chalk–Harrod Mechanism

To our knowledge, the mechanism shown in Fig. 1 can satisfactorily account for all of the numerous experimentally observed characteristics of platinum-catalyzed hydrosilation, with a single serious exception. This exception is the requirement of oxygen cocatalysis. The oxygen effect is common knowledge among people who run hydrosilations on a large scale, where deliberate aeration of the reaction may be required to sustain catalytic activity, but seems never to have been discussed in detail in the technical literature.

An excellent example of the oxygen effect is provided by the reaction of $(EtO)_3SiH$ with 1-olefins in the presence of either $H_2PtCl_6 \cdot 6H_2O$ or $[(C_2H_4)PtCl_2]_2$ (11). If the reactants are carefully outgassed under vacuum and then distilled onto the catalyst and allowed to warm to room temperature, the yellow color of the catalyst is discharged as usual but no reaction takes place. These mixtures show no sign of reaction even after months at 100°C. However, if such a mixture, containing equimolar reactants and a catalyst concentration of 10^{-4} M, is exposed to air after several months' storage, the reaction exotherm causes spontaneous boiling within a few seconds.

Olefin complexes of the type $[(\text{olefin})PtCl_2]_2$, prepared by displacement of ethylene from Zeise's dimer, do not manifest the above-mentioned oxygen effect if they are prepared under the usual conditions of synthesis for such compounds. However, if these complexes are prepared from carefully purified olefins on a vacuum line, with careful avoidance of contact with oxygen, they are totally inactive as hydrosilation catalysts with triethoxysilane. Catalytic activity immediately appears if the olefin complexes are exposed to oxygen and are then used as catalysts under completely anaerobic conditions. It is therefore evident that the active catalytic species is *not* a simple platinum–olefin complex, but an oxidation product of such a complex. We have never identified the superactive species, but the following organic species showed no cocatalytic activity: peroxides, hydroperoxides, ketones, aldehydes, carboxylic acids, alcohols, and aliphatic ethers.

Contrary to the clear need for oxygen cocatalysis in the case of triethoxysilane, trichlorosilane and triethylsilane were successfully reacted with 1-olefins under the most stringently anaerobic conditions achievable using normal vacuum line techniques.

At present there is no natural explanation of the oxygen effect in the mechanism shown in Fig. 1. The formation of olefin complex does not require oxygen and the discharge of the typical yellow color of the olefin complex, even under conditions where hydrosilation does not occur, suggests that the oxidative addition of $\equiv$SiH is occurring readily. We also know of no reason why either insertion of olefin into Pt–H, or reductive elimination of alkylsilane should be promoted by O_2. Most likely, the effect of O_2 is to introduce a new oxygen-containing ligand into the coordination sphere of the metal with concurrent loss of a chloride ligand. Both the steric and electronic consequences of replacing Cl^- with an oxygen-containing ligand, such as OH^- or OR^-, are likely to be substantial. It is not all unusual for coordination catalyses to be extremely sensitive to the nature of the ligands on the metal (23).

Cocatalysis of hydrosilation by oxygen has also been observed for rhodium-catalyzed reactions (24,25). In this case the phenomenon is more general in that it is observed with a variety of silicon hydrides and with a variety of catalysts including $RhCl(PPh_3)_3$, $RhCOCl(PPh_3)_2$, $RhHCO(PPh_3)_3$, and $[Rh(CO)_2Cl]_2$. As with the platinum example, no reaction was observed when the reactants were mixed under anaerobic conditions (under nitrogen) for times up to 60 hr. Exposure to air resulted in immediate exothermic reaction. Unlike the platinum case, peroxides were equally as effective as air (25). Rhodium–phosphine complexes also catalyze the hydrogenation of olefins, and here too oxygen has an effect, but here it only results in an enhancement of the rate (26). This has been attributed to an enhanced dissociation of the rhodium complex resulting from oxidation of phosphine (26–28). A similar explanation could explain the effect of oxygen on these complexes when used as hydrosilation catalysts.

Some interesting preliminary results on asymmetric catalytic hydrosilation, by platinum(II) complexes containing chiral phosphines, have been reported by Yamamoto et al. (56). Hydrosilation of α-methylstyrene with methyldichlorosilane in the presence of *cis*-dichloro(ethylene)[(*R*)-benzylmethylphenylphosphine]platinum(II) gave a 5% excess of the *R*-2-phenylpropylmethyldichlorosilane. Optical yield was lower when dichlorobis[(*R*)-methylphenylpropylphosphine]-di-μ-chlorodiplatinum(II) was used as catalyst, or when 2-methyl-1-butene was used as substrate. With *cis*-dichloro(ethylene)[(*S*)-1-phenylethylamine]platinum(II) as catalyst, the reaction products were racemic.

Although the asymmetric induction achieved in these reactions is modest, there will undoubtedly be more activity in this area and the potential synthetic applications of hydrosilation will be of much greater interest with this added dimension.

III. NICKEL AND PALLADIUM COMPLEXES AS HYDROSILATION CATALYSTS

Although platinum complexes show exceptional activity as hydrosilation catalysts, the other members of the nickel group have proven comparatively ineffective. The ineffectiveness of simple palladium(II) complexes as catalysts is due to the great ease with which they are reduced to the metal by silicon hydrides. Reports have described the use of certain Pd(0) complexes as catalysts for hydrosilation, particularly of dienes (29). In the cobalt group, iridium complexes are relatively ineffective, whereas those of rhodium show a high level of catalytic activity (24,30,31).

Nickel(II) halide complexes with chelating phosphines, for example, 1-1′-bis(dimethylphosphino)ferrocene (dmpf); 1,2-bis(dimethylphosphino) ethane (diphos), were found to be moderately active catalysts for the hydrosilation of olefins (120°C; 20 hr; 95% yield of products) (57). When methyldichlorosilane was used as reactant (2 moles per mole of olefin) the predominant product was alkylmethylchlorosilane (~80%) and the anticipated product, alkylmethyldichlorosilane, was the minor component. With trichlorosilane the two products were obtained in roughly equal amounts.

Nickel(II) complexes with monodentate phosphines, or with nitrogen-containing ligands, were ineffective.

The hydrogen–halogen interchange observed with these catalysts possibly occurs as a result of disproportionation of the silicon hydride prior to hydrosilation. Catalytic disproportionations of silicon hydrides catalyzed by complexes of rhodium and iridium have been observed previously (25).

IV. RHODIUM COMPLEXES AS HYDROSILATION CATALYSTS

The complexes $[(CO)_2RhCl]_2$ (**8**) and $[(C_2H_4)_2RhCl]_2$ showed the highest activity; $RhCl(CO)(PR_3)_2$ [R = Et (**9**) or Ph (**10**)] intermediate activity; and $RhCl(PPh_3)_3$ (**11**) the lowest activity (24). The sensitivity of this reaction to oxygen throws some doubt on the validity of such reactivity sequences.

Contrary to the experience with platinum, several cases of oxidative additions of silicon hydrides to rhodium complexes to yield stable, characterizable adducts have been reported. Charentenay et al. (24) prepared adducts according to Eq. 9.

$$RhCl(PPh_3)_3 + R_3SiH \rightleftarrows \underset{\mathbf{(12)}}{RhCl(PPh_3)_2H(SiR_3)} + PPh_3 \qquad (9)$$

The chemistry of compound **12a** (R = Cl) yields an interesting insight into some subtle features of coordination catalysis (32). Although the original preparation of **12a** contained solvent molecules (24), it has subsequently been obtained free of solvent (30) and preliminary crystallographic data indicate the structure **13**.

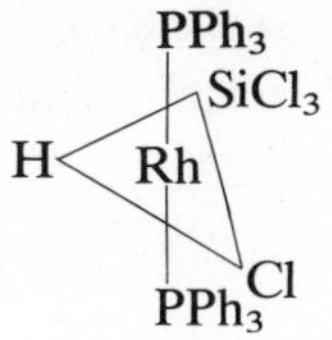

PPh$_3$
SiCl$_3$
H Rh
Cl
PPh$_3$

(**13**)

Neither **13** nor its precursor in Eq. 9 is a very effective catalyst for hydrosilation. The reason for its lack of catalytic activity may be a reflection of the demonstrable inability of **13** to coordinate further ligands. Rather than relieving the apparent coordination unsaturation of **13**, strongly binding ligands such as CO, or phosphines, induce reductive elimination of silicon hydride, that is, the back reaction of Eq. 9. The same is true for olefins, and it is interesting to speculate as to the critical factors influencing the competition between reductive elimination of the silicon hydride and the successful coordination and insertion of the olefin molecule. A likely reason for the failure of **13** to coordinate a sixth ligand is the unduly large steric requirements of the phosphine and the $SiCl_3$ ligands. It has been well established that the oxidative addition–reductive elimination equilibria for silicon hydrides with group VIII complexes are only weakly *exo*-free energetic and consequently may be either promoted, or completely discouraged by relatively minor changes in the metal complex, or the silicon hydride (5,24,33). It is not unreasonable that steric effects could thus completely inhibit oxidative addition. The exceptional activity of tetracarbonyl-μ-dichlorodirhodium(I) and tetrakis-(ethylene)-μ-dichlorodirhodium(I) as hydrosilation catalysts may be attributed to the absence of unduly bulky ligands.

A rather puzzling experimental fact is the observation that hydridocarbonyltris(triphenylphosphine)rhodium(I) (**14**) is a quite effective catalyst for hydrosilation yet does not undergo observable oxidative addition of silicon hydrides (11,31). On the other hand, the iridium analog (**15**) undergoes ready oxidative addition of silicon hydrides, yet is completely ineffective as a hydrosilation catalyst (22). All the foregoing observations point to the successful coordination of olefin into the hydrosilated metal complex as critical to the effective catalytic functioning of the complex. Presumably complex **13** is totally incapable of coordinating an olefin molecule without reductive elimination of silicon hydride, whereas complex **15**, and its silicon hydride adducts, are too substitution-inert to allow rapid exchange of phosphine for olefin ligand. On the other hand, complex **14** shows a higher substitutional lability than **15**, while its reduced tendency to oxidatively add silicon hydride is more than compensated for by the ability of the oxidative adduct to adopt six-coordination as a result of the smaller size of a hydride relative to a chloride ligand.

The oxidative addition of silicon hydrides to **14** and **15** requires predissociation of a phosphine ligand (34), thus

$$MH(CO)(PR_3)_3 \rightleftarrows MH(CO)(PR_3)_2 + PR_3 \quad (M = Rh \text{ or } Ir) \qquad (10)$$

$$MH(CO)(PR_3)_2 + R_3SiH \rightleftarrows \underset{\textbf{(16)}}{MH_2(CO)(PR_3)_2(R_3Si)} \qquad (11)$$

Consequently the formation of **16** or its lack of formation is not an indication of the absolute stability of **16**, but is rather an indication of the relative effectiveness of the phosphine and silicon hydride in competing for the four-coordinate intermediate. An example of this is the fact that the trimethylsilyl analog of **16** (M = Ir) cannot be isolated pure when prepared by reactions 10 and 11 because phosphine causes the complex to revert to starting material (22). An alternative synthesis (Eqs. 11*a* and 11*b* where L′ = CO, L = PPh_3, M = Ir) in which no excess phosphine is present gives a complex so stable that no reductive elimination of silicon hydride is observed below 100–120°C (35).

$$MClL'L_2 + R_3SiH \rightleftharpoons MHCl(SiR_3)L'L_2 \rightarrow [MHL'L_2] + R_3SiCl \quad (11a)$$

$$[MHL'L_2] + R_3SiH \rightleftharpoons MH_2(SiR_3)L'L_2 \quad (11b)$$

Reactions 11*a* and 11*b* also suggest a route whereby catalytically active rhodium hydride complexes could be formed in hydrosilation reactions catalyzed by rhodium halo-complexes (32).

It is evident from the foregoing that the presence of phosphine-type ligands may considerably reduce the activity of metal complex hydrosilation catalysts. This effect has in fact proved useful in preparing platinum complex catalysts with "latent" activity" (36). Although phosphines kill the room temperature activity of the usual platinum hydrosilation catalysts, they show restored activity at higher temperatures.

The rhodium complex **16** is exceptional for its low isomerization activity. Hydrosilation of excess 1-pentene with phenyldimethylsilane in the presence of this catalyst left 80% of the excess olefin unisomerized after all of the silane had reacted (31).

A useful heterogeneous coordination catalyst for hydrosilation was reported by Čapka et al. (58). This catalyst, and several modifications, was prepared by reaction of $RhCl_3 \cdot nH_2O$ with a diphenylphosphinated, cross-linked, polystyrene resin. The catalyst showed a high activity for hydrosilation of olefins, could be easily filtered from the products, and re-used many times even after storage in air.

V. COBALT CARBONYL AS A HYDROSILATION CATALYST

Following the recognition of the metal-complex-catalyzed hydrosilation of olefins as a special case of the more general class of hydrometallation coordination catalysts, $Co_2(CO)_8$ became an obvious choice as a potential catalyst for hydrosilation. The considerably detailed understanding of the function of cobalt carbonyls in hydroformylation and hydrogenation reactions also provided an exceptional opportunity for unraveling the details of their function in hydrosilation.

Both $Co_2(CO)_8$ and $HCo(CO)_4$ show high catalytic activity for the addition of silicon hydrides to olefins (37,38). Useful rates of reaction are obtained under ambient conditions with catalyst concentrations of about

0.001 *M*. Use of forcing conditions does not offer much advantage with these catalysts, since the reactive intermediates are quite susceptible to irreversible thermal deactivation.

The cobalt carbonyl-catalyzed reaction, like that catalyzed by platinum or rhodium complexes, is effected through a closed loop sequence consisting of numerous steps. Some of these steps have been unequivocally identified, but others have been hypothesized as analogous to steps known to occur in other cobalt carbonyl-catalyzed reactions.

The first step in the $Co_2(CO)_8$-catalyzed reaction is the formation of $HCo(CO)_4$ and silylcobalt carbonyl according to Eq. 12.

$$Co_2(CO)_8 + R_3SiH \rightarrow HCo(CO)_4 + R_3SiCo(CO)_4 \qquad (12)$$

Since the silyl cobalt carbonyls are inactive as hydrosilation catalysts, it may be reasonably assumed that the next step in the catalytic cycle involves the $HCo(CO)_4$. By analogy with the proposed mechanism for the hydroformylation reaction, the hydrocarbonyl is expected to react with olefin to yield alkylcobalt carbonyl, thus

$$HCo(CO)_n + R'CH{=}CH_2 \rightleftharpoons R'CH_2CH_2Co(CO)_n \qquad (13)$$

where n is generally believed to be 3. The reversibility of Eq. 13 accounts for the occurrence of olefin isomerization during hydrosilation in the same manner as that normally catalyzed by $HCo(CO)_4$. The final step of the catalytic cycle is the attack of silicon hydride on alkylcobalt carbonyl to regenerate $HCo(CO)_4$ and alkylsilane.

$$RCH_2CH_2Co(CO)_n + R_3SiH \rightarrow HCo(CO)_n + R_3SiCH_2CH_2R' \qquad (14)$$

The alternative possibility, shown in Eq. 15, was rejected because no hydrosilation product could be isolated from a mixture of silylcobalt carbonyl, $HCo(CO)_4$, and olefin.

$$R'CH_2CH_2Co(CO)_n + R_3SiCo(CO)_4 \rightarrow R_3SiCH_2CH_2R' + Co_2(CO)_8 \qquad (15)$$

Inclusion of an equivalent of silicon hydride in the latter reaction produced some hydrosilation product derived from the silylcobalt carbonyl. This result was subsequently shown to be a consequence of the exchange reaction:

$$R_3SiCo(CO)_4 + R''_3SiH \rightleftharpoons R''_3SiCo(CO)_4 + R_3SiH \qquad (16)$$

This reaction is also responsible for the stereospecific Si–H/Si–D exchange reactions reported by Sommer et al. (15).

Reactions 12, 13, 14, and 16 are not single-step processes and the details of how they occur are still obscure. The most appealing hypothesis at the present time is that the d^8 Co(I) complexes participate in oxidative addition reactions similar to those of Rh(I) and Ir(I) (38*a*). A reversible oxidative addition provides a natural explanation for reaction 16, thus in addition to explaining exchange, reaction 17 is also consistent with the observed retention of configuration at silicon, a feature common to other reactions involving oxidative addition of silicon hydrides.

$$R''_3Si\text{–}Co(CO)_3(CO) + R''_3SiH \rightleftharpoons (R_3Si)(H)(SiR''_3)Co(CO)_3 + CO$$

$$\rightleftharpoons R''_3Si\text{–}Co(CO)_4 + R_3SiH \qquad (17)$$

A side reaction that destroys active catalyst is the reaction of silicon hydride with $HCo(CO)_4$ to produce silylcobalt carbonyl and H_2. It is likely that this reaction occurs by way of oxidative addition:

$$HCo(CO)_4 + R_3SiH \rightleftharpoons \underset{(\mathbf{17})}{(H)_2(R_3Si)Co(CO)_3} + CO \rightleftharpoons R_3SiCo(CO)_4 + H_2 \qquad (18)$$

The rate of H_2 production by reaction 18, relative to the direct decomposition of $HCo(CO)_4$, is greater for siloxanes than for alkyl and arylsilanes. This fact is in accord with the known greater rate of reductive elimination of the latter relative to the former (11,33).

It is likely that reaction 14 also proceeds by oxidative addition of silicon hydride, followed by reductive elimination of alkylsilane. An alternative oxidative addition route for hydrosilation would be the formation of **17** by oxidative addition of silicon hydride to $HCo(CO)_4$ followed by olefin insertion and reductive elimination of alkylsilane. Either of these mechanisms is in accord with the presently available facts and, as in the case of hydrogenation catalyzed by **11** (39,40), it is possible that both are operative.

A study of the infrared spectra of reaction mixtures from cobalt carbonyl-catalyzed hydrosilations reveals that most of the cobalt is present in the form of silyl-cobalt carbonyl and an unidentified transient species with a unique strong absorption in the 1950–2000 cm^{-1} region (37). This species is undoubtedly the key catalytic intermediate and its concentration parallels the rate of the hydrosilation reaction. The unidentified band may be a Co–H band, or a carbonyl frequency of unusually low energy from a species where some of the carbonyl groups have been replaced by poorer π-acceptor ligands.

VI. IRON PENTACARBONYL AS A HYDROSILATION CATALYST

The first metal carbonyl reported to show activity as a hydrosilation catalyst was $Fe(CO)_5$ (41). This catalyst requires much more forcing conditions than cobalt carbonyl (100–140°C) and the reaction is far from clean. Reactions of ethylene give reasonable yields (60–80%) of hydrosilation product but substantial amounts of di-silylethylenes (20–40%) are also formed. With substituted ethylenes the catalyzed reaction proceeds according to reaction 19

$$X_3SiH + RCH{=}CH_2 \xrightarrow{Fe(CO)_5} X_3SiCH_2CH_2R \qquad (19a)$$

$$X_3SiH + RCH{=}CH_2 \xrightarrow{Fe(CO)_5} X_3SiCH{=}CHR + CH_3CH_2R \qquad (19b)$$

where $X = Cl$, OC_2H_5, or C_2H_5.

Reaction 19*a* is favored by a high ratio of silicon hydride to olefin. For example, with a 3:1 molar excess of triethylsilane to ethylene, the product is almost exclusively tetraethylsilane, but a 5:1 molar excess of ethylene gave almost exclusively vinyltriethsilane in high yield (>90% based on silane). The presence of chloride (or alkoxy) substituents at silicon, or of any substitution of ethylene favors the formation of vinylsilanes.

Besides reactions 19, terminal olefins undergo double-bond isomerization; a not unexpected result since $Fe(CO)_5$ catalyzes olefin isomerization in the absence of silanes. Unlike other hydrosilation catalysts, $Fe(CO)_5$ seems completely incapable of hydrosilating internal olefins.

Little detailed work on the mechanism of iron carbonyl catalyzed hydrosilation has been reported. The more forcing conditions necessary with $Fe(CO)_5$ are not unexpected in view of its greater inertia towards substitution than the cobalt carbonyls. A similar analogy exists in the case of catalytic hydrogenation for the carbonyls of iron (42) and cobalt (43).

Most of the empirical rules of coordination catalysis suggest that a coordinatively saturated species such as $Fe(CO)_5$ will require two substitution processes to become involved in a catalytic cycle. One of these substitutions will replace a carbonyl ligand by olefin substrate and the other will allow oxidative addition of the silicon hydride by loss of a second carbonyl ligand. The sequence in which these steps occur is not known, but either possibility is likely. Direct oxidative addition of silicon hydrides to $Fe(CO)_5$ has been achieved under ultraviolet irradiation (44):

$$Fe(CO)_5 + R_3SiH \xrightarrow{\text{uv}} \underset{\mathbf{(18)}}{R_3Si(H)Fe(CO)_4} + CO \tag{20}$$

Under purely thermal conditions, adducts of the type $(R_3Si)_2Fe(CO)_4$ are obtained. Adducts such as **18** should be more labile to olefin substitution than $Fe(CO)_5$ as a result of the high *trans* effects of H and R_3Si.

In the $Fe(CO)_5$-catalyzed hydrogenation of dienes the experimental data supports the hypothesis that substitution of CO by olefin is the initial step (42). However, the much greater experimental concentrations and greater reactivities of silicon hydrides compared to hydrogen increase the likelihood that oxidative addition will compete effectively with olefin substitution in hydrosilation. The most fascinating aspect of the iron carbonyl catalyzed reaction is the formation of substitution products as illustrated in reaction 19*b*. This type of reaction is quite commonly encountered in the coordination catalytic chemistry of olefins, particularly in palladium-catalyzed reactions of the general type (45) shown in Eq. 21.

$$CH_2{=}CH_2 + HX + Pd^{++} \rightarrow CH_2{=}CHX + [PdH]^+ + H^+ \tag{21}$$

Under normal circumstances the "hydride" produced in reaction 21 is scavenged by an inorganic oxidant to make the reaction catalytic.

As far as we know there is no authenticated case of reaction 21 occurring when $X = \cdot SiR_3$, but it is most probable that the $Fe(CO)_5$-catalyzed vinylation of silicon hydrides is of the same type. In the latter case the hydride is scavenged by an organic molecule, namely an olefin. For such a mechanism to operate, insertion of olefin into an Fe–Si bond must be a highly favored reaction, but there appears to be no previously discovered precedent for such a reaction (on the contrary, no insertion of olefins, or CO, could be induced into a Co–Si bond (37)). Given the assumption that insertion into an Fe–Si bond can occur, the following mechanism accounts for the experimental observations:

$$Fe(CO)_5 \rightleftharpoons Fe(CO)_4 + CO \tag{22}$$

$$Fe(CO)_4 + R_3SiH \rightleftharpoons (R_3Si)HFe(CO)_4 \tag{23}$$

$$(R_3Si)HFe(CO)_4 + CH_2{=}CH_2 \rightleftharpoons (R_3Si)HFe(CO)_3(C_2H_4) + CO \quad (24)$$

$$(R_3Si)HFe(CO)_3(C_2H_4) \rightarrow HFe(CO)_3(C_2H_4SiR_3) \quad (25)$$

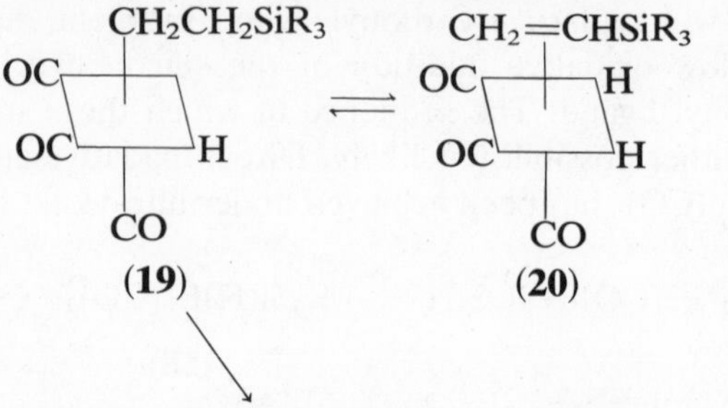

$CH_2CH_2SiR_3$ / OC, OC, H, CO (**19**) $\rightleftharpoons$ $CH_2{=}CHSiR_3$ / OC, OC, H, H, CO (**20**) (26*a*)

19 $\searrow$

$$\text{“}Fe(CO)_3\text{”} + CH_3CH_2SiR_3 \quad (26b)$$

(**21**)

Hydrosilation may result either from path 26*b*, or from an alternative path involving insertion of olefin into Fe–H. The various observed reactions: hydrosilation, vinylation, isomerization, hydrogenation, and disilylation, all occurring simultaneously in this system, do not necessarily occur within the coordination sphere of the same type of complex. A species such as **20** may be responsible for the hydrogenation of olefin accompanying vinylation. On the other hand, hydrogenation may be effected by a different type of catalytic species. Species **20** has merit as a common site for vinylation/hydrogenation since it readily explains the effect of SiH/olefin ratio on the hydrosilation/vinylation ratio. In an olefin deficient system **20** will be long-lived and chances of the readdition of hydrogen to permit the occurrence of path 26*b* will be high. On the other hand, in an olefin-rich system substitution of vinylsilane by reactant olefin will be favored, followed by addition of hydrogen to reactant olefin.

The fact that introduction of electronegative groups at silicon favors the vinylation over hydrosilation is also reasonable, since reactions such as 21 usually require that either HX or X^- be a good nucleophile. The trialkylsilanes have a bond polarity in the direction $R_3Si^{\delta+}—H^{\delta-}$, while trichlorosilane has a bond moment close to zero. It thus seems more likely that a nucleophilic species R_3Si^- may result if some of the R groups are chloride.

In addition to the major products depicted in reaction 19, small amounts of disubstituted ethanes such as 1,2-bis(trialkylsilyl)ethanes are also produced. These products probably result from iron complexes containing two silyl ligands bonded to a single iron. Such compounds are known to form in thermal reactions of silicon hydrides with iron carbonyls (44).

The same types of products as those obtained from $Fe(CO)_5$-catalyzed hydrosilation are also produced in the presence of $Fe_2(CO)_9$. The latter gives rates of the same order as $Fe(CO)_5$ at temperatures about 45°C lower (41).

VII. METAL-COMPLEX-CATALYZED SILYLATION OF HYDRIDES

Many transition metals and their salts catalyze the condensation of silicon hydrides with protonic molecules. In particular, the condensations with alcohols, amines, and carboxylic acids have been widely studied (15):

$$ROH + R_3SiH \rightarrow ROSiR_3 + H_2 \qquad (27)$$

$$RNH_2 + R_3SiH \rightarrow RNHSiR_3 + H_2 \qquad (28)$$

$$RCOOH + R_3SiH \rightarrow RCOOSiR_3 + H_2 \qquad (29)$$

Sommer et al. showed that these reactions all proceed with inversion of configuration at silicon, as opposed to the hydrosilation of olefins and SiH/SiD exchange which both occurred with retention of configuration at silicon. Uncatalyzed nucleophilic substitution at silicon normally proceeds with retention of configuration.

All of the catalysts used by Sommer et al. were apparently heterogeneous and they proposed a tentative mechanism involving backside attack on a silicon bonded to a surface metal atom. There is no doubt, however, that very effective homogeneous catalysts exist for these reactions and a surface is not a prerequisite. Cobalt carbonyl is an excellent homogeneous catalyst and the mechanism of its function is now reasonably understood. Also, solutions of $RhCl_3 \cdot 3H_2O$ in organic protonic solvents remain homogeneous in the presence of excess silicon hydride and vigorously catalyze hydrogen evolution (11).

Reactions of silylcobalt carbonyls with protonic molecules have been studied (46) and proceed according to Eq. 30.

$$R_3SiCo(CO)_4 + XH \rightarrow R_3SiX + HCo(CO)_4 \qquad (30)$$

Since $HCo(CO)_4$ was known to react with silicon hydrides to regenerate the silylcobalt carbonyl (Eq. 18), a feasible catalytic cycle could be achieved through combination of Eqs. 18 and 30. This objective was realized experimentally for triethylsilane with a variety of alcohols and carboxylic acids (47). However, no catalysis was observed with phenol, or with diethylamine. The lack of reaction with amines is not surprising since even reaction 30 goes only in very poor yield (ca. 20%) and the formation of ammonium tetracarbonylcobaltate would prevent reaction of the $HCo(CO)_4$ with silicon hydride.

The reasons for the lack of reactivity of phenol are less certain. It has been suggested that a change in mechanism may occur in going from basic to acidic reagents; phenol, because of its intermediate acidity, manifests a slow reactivity by either mechanism. The more basic reagents probably displace tetracarbonylcobaltate by backside attack on silicon.

$$R'OH + R_3SiCo(CO)_4 \longrightarrow \left[\begin{matrix} R' \\ \quad \diagdown \\ \quad\quad O{-}SiR_3 \\ \quad \diagup \\ H \end{matrix}\right]^+ \left[Co(CO)_4 \right]^- \quad (\mathbf{21}) \longrightarrow R'OSiR_3 + HCo(CO)_4 \tag{31}$$

Thus the reaction of methanol with an optically active silylcobalt carbonyl has been shown to result in inversion of configuration at silicon (15). Numerous examples of analogs of **21** with aprotic bases such as pyridine (48), DMF (49), and trimethylamine (50) are known. With more acidic reagents a proton-induced electrophilic displacement of silicon from cobalt may be more favored.

That a considerable positive charge density exists on the silicon in silyl-transition complexes is suggested by the ease with which they attack cyclic ethers. Tetrahydrofuran is readily polymerized by $Co_2(CO)_8$ in the presence of excess triethylsilane (47). Since polymerization of THF is normally achieved only with acid catalysts, the sequence depicted in reaction 32 appears to be a reasonable rationalization.

$$R_3SiCoL + \text{(THF)} \longrightarrow \left[R_3Si{-}\overset{+}{O}(C_4H_8) \right]^+ \left[CoL \right]^- \xrightarrow{THF} \left[(C_4H_8)\overset{+}{O}{-}(CH_2)_4OSiR_3 \right] \left[CoL \right]^- \tag{32}$$

Silylhydridobis(bis-1,2-diphenylphosphinoethane)iridium(III) salts also attack THF, but hydride transfer to metal seems to occur rather than polymerization(33).

$$\left[(R_3Si)H(diphos)_2Ir(III)\right]^+ + \text{THF} \longrightarrow \left[H_2(diphos)_2Ir(III)\right]^+ + CH_2{=}CHCH_2CH_2OSiR_3 \quad (33)$$

A similar reaction occurs with $(R_3Si)H_2(CO)(Ph_3P)_2Ir(III)$. The ease with which attack on THF occurs increases with alkyl (as opposed to alkoxy, or chloro) substitution on the silicon.

VIII. REMARKS ON FREE RADICALS IN HYDROSILATION

The hydrosilation of olefins may also be catalyzed by peroxides and by ultraviolet light. These catalyses are clearly free radical and it is tempting to suggest that the metal-catalyzed reactions are similar. This point of view has a number of proponents (1). We have tried, on the other hand, to explain these catalyses in terms of the known chemistry of the platinum metals, emphasizing the decreased activation energy of reactions occurring within the coordination sphere. Although metal complexes catalyze many free radical reactions, these catalyses involve metal complexes which readily undergo a one-electron change in valence. The metal complexes that we have discussed as hydrosilation catalysts, on the other hand, have stable valence states separated by two electrons. Changes in valence thus occur chiefly by reactions such as "oxidative addition" and "reductive elimination." These reactions, together with the well-documented olefin insertion reaction, make up the basis of our suggested mechanism. Similar mechanisms stressing "coordination catalysis" had been proposed earlier for reactions such as olefin hydrogenation and olefin hydroformylation, and it is not surprising that interpretation of the metal-catalyzed hydrosilation reaction in terms of a "coordination catalysis" received a ready acceptance. It is perhaps therefore appropriate that we should point out that this interpretation is at best only a first approximation to the truth and that there is still much to be understood.

One of the most telling arguments against a free radical mechanism is the lack of effect of free radical inhibitors (ref. 1, p. 232). Further, metal-catalyzed hydrosilations occur with retention of configuration at silicon and this is paralleled by retention of configuration at silicon in the stoichiometric reactions of silicon hydrides with transition metal complexes. These two facts, however, can still be reconciled with the participation of free radicals. Thus, free radical inhibitors will have no significant effect if the steps in the radical chain are very much faster than reaction of the radicals with the inhibitor. Under these conditions the lifetime of the radical will be very small and, if it is small enough, racemization will be incomplete. These possibilities should be given serious consideration in view of the discovery that a large number of reactions of main group organometallics proceeds via radical paths in which the radical steps are too fast for many conventional inhibitors to have an effect (51). The autoxidation of a variety of metal alkyls for example (eq. 34) proceeds via the propagation reactions 35 and 36.

$$MR_n + O_2 \rightarrow ROOMR_{(n-1)} \tag{34}$$

$$(M = Mg, Zn, Cd, B, Al)$$

$$R\cdot + O_2 \rightarrow RO_2\cdot \tag{35}$$

$$RO_2\cdot + MR_n \rightarrow ROOMR(_{(n-1)} + R\cdot \tag{36}$$

The rate constant for reaction 36 when $M = B$ is approximately 10^6 faster than for the corresponding reaction of hydrogen, reaction 37, which occurs in the autoxidation of hydrocarbons (52).

$$RO_2\cdot + RH \rightarrow RO_2H + R\cdot \tag{37}$$

Reactions such as these could only be inhibited by the most reactive free radical inhibitors such as galvinoxyl or 2,6-di-*t*-butyl-4-methoxyphenol.

A demonstration that a reaction involving silyl free radicals can occur with retention of configuration at silicon has already been mentioned (16). It seems likely that reactions that involve the homolytic fission of a transition metal to silicon bond will also be found to involve retention of configuration at silicon.

We have often pointed out the close similarity between homogeneous catalytic hydrogenation and homogeneous catalytic hydrosilation. The general acceptance of a "coordination catalysis" mechanism for the former has assisted the acceptance of a similar mechanism in the latter case. However, it should be pointed out that, in one case at least, catalytic hydrogenation is believed to involve hydrogen radicals. Thus one proposed mechanism for hydrogenation of olefins with pentacyanocobaltate(II) involves the homolytic fission of a transition metal hydride (53). The hydrogen atom is taken up by an olefin as in reaction 38.

$$[HCo(CN)_5]^{3-} + {>}C{=}C{<} \rightleftharpoons H{-}C{-}C\cdot + [Co(CN)_5]^{3-} \tag{38}$$

The two species formed in reaction 38 can combine to form an organocobalt complex and must do so faster than the rate of rotation about the carbon–carbon bond, since use of a metal deuteride has established a stereospecific *cis* addition (54). This is, of course, a somewhat special case since the stable valence states of cobalt are here 2 and 3 which facilitates a homolytic fission of the transition metal hydride. Further, the metal hydride is substitution inert to the coordination of olefin. Pentacyanocobaltate(II) can therefore be regarded as a catalyst that acts partly as a "coordination catalyst" since hydrogen enters the coordination sphere, and partly as a free radical initiator. It is conceivable that other catalysts would be intermediate in nature by virtue of other combinations of properties. Similar reasoning applied to hydrosilation catalysis would lead us to expect a variety of possible catalysts with mechanisms varying all the way from pure "coordination catalysis" to conventional free radical. The latter type of catalyst would be more likely found among the more substitution inert complexes such as $Fe(CO)_5$ and those with stable valence states only one electron apart such as cupric/cuprous.

In view of considerations such as the above, it would seem profitable to examine unusual catalysts such as $Fe(CO)_5$ and the oxygen activated catalysts more closely.

Later evidence has shown that, under special circumstances, both oxidative addition (59) and insertion into Pt–H (60) bonds may be catalyzed by free radical initiators.

IX. ADDENDUM

This chapter was originally completed in 1973. Since then a large number of reports on the use of hydrosilation as a synthetic procedure have appeared. In spite of this, there has been remarkably little advance in the fundamental understanding of the mechanism, or mechanisms, of the reactions. Continuing in the spirit of the original review, we do not intend to present an exhaustive review of the more recent literature on hydrosilation; instead we shall draw attention to several areas of current research that add significantly to the completeness of our understanding of mechanism, or present opportunities for the synthesis of new types of product. (This addendum covers the literature up to mid–1975.)

Some important support for a nonradical mechanism for metal-ion-catalyzed hydrosilation resulted from studies of intramolecular hydrosilation of 4-pentenylsilanes (61,62). Such reactions gave cyclic products in which five-rather than six-membered rings predominated. This observation makes a free radical mechanism unlikely since silyl radicals with a double bond in the δ-position are known to give mainly six-membered rings. A five-membered ring would naturally result from extrusion of a catalytic metal ion from a six-membered ring intermediate, produced by successive oxidative addition of Si–H and intramolecular insertion of the pendant olefinic group into the metal hydride.

The failure of palladium compounds to catalyze hydrosilation of simple olefins has been difficult to understand in view of the close similarity of the chemistries of Pd and Pt, and the efficiency of Pd in many other homogeneous catalyses. The demonstration by Tsuji et al. (63) that many palladium compounds, and even metallic palladium, become active catalysts in the presence of phosphine (or phosphite) ligands has removed this anomaly. The experimental evidence strongly suggests that the active species are Pd(0) complexes which undergo olefin coordination and oxidative addition of Si–H analogous to the mechanism outlined in Fig. 1. Since a combination of *tert*-phosphine and silicon hydride was shown to convert finely divided Pd metal to the bis(phosphine)palladium(0) complex, it would seem that the role of the ligand is to prevent loss of homogeneous catalyst as metal, a problem that does not generally arise with platinum-based catalysts.

The facile addition of silicon hydrides to Pt(0) complexes has been further demonstrated by Eaborn et al. (64). Although these observations, together with those on Pd(0) described above, amplify the possibility that the catalytic cycle in Fig. 1 should include Pt(0) $\rightleftharpoons$ Pt(II) reactions, there is also new evidence that the Pt(II) $\rightleftharpoons$ Pt(IV) reactions are equally plausible. The extensive work of Glockling and co-workers (65) on reactions of silylplatinum(II) complexes with various group IV hydrides clearly demonstrated the availability of facile oxidative-addition–reductive elimination cycles between Pt(II) and Pt(IV) for group IV ligand exchange. Eaborn et al. (66) have also shown that the products of reaction between bis(dimethylphenylphosphine)dimethylplatinum(II) and silicon hydrides conform to those expected from a sequence of oxidative-addition–reductive elimination reactions.

The use of nickel complexes as catalysts has been described. Kiso et al. (67) found bis(cyclopentadienyl)nickel to be quite active for hydrosilation of α-olefins (67). The same study showed dialkyl(bipyridyl)nickel(II) to be a poor catalyst and bis(trichlorosilyl)(bipyridyl)nickel(II) to be inactive. An example of selectivity was reported by Čapka et al. (68) who found that $Ni(CO)_4$ catalyzed addition of trichlorosilane or methyldichlorosilane to styrene to give 1-phenyl-2-trichlorosilylethane or methyldichlorosilylethane, the addition occurring in the opposite sense to that found with platinum catalysts. A similar specificity for terminal trichlorosilylation of styrene was observed with a $[C_5H_5(CO)Ni]_2$ catalyst (69).

The number of carbonyl compounds active as hydrosilation catalysts has been extended by the discovery that $Cr(CO)_6$ is photocatalytically active for hydrosilation of 1,3-dienes (70). The reaction, once initiated, continues in the absence of irradiation, but higher rates were achieved by continued irradiation. The catalyst was specific for terminal silylation and 1,4-addition while showing high selectivity for substrate geometry. For example, *trans*-1,3-pentadiene was hydrosilated to the exclusion of *cis*-1,3-pentadiene when a mixture of the two isomers was reacted. The exhibition of catalytic activity by $Cr(CO)_6$ under photolysis conditions is not surprising since coordinatively unsaturated Cr(O) species were shown to be active hydrogenation catalysts (71). Although few mechanistic details of these reactions have been explored, they probably proceed by olefin coordination and oxidative addition steps [Cr(O) $\leftrightarrow$ Cr(II); $d^6 \leftrightarrow d^4$] analogous to those proposed for other catalyst systems. The catalytic intermediate, $Cr(CO)_3$, resulting from photolysis of $Cr(CO)_6$ may also be generated thermally from $Cr(CO)_3(CH_3CN)_3$. The use of this catalyst for stereospecific hydrogenations suggests its use for similar hydrosilations (72).

The synthetic scope of the olefin hydrosilation reaction as a route to alkylsilicon compounds has been extended by the interesting observation that chloroplatinic acid catalyzed hydrosilation of internal olefins by SiH_2Cl_2 gives internally silylated products in high yield (73). The effects of double bond shift seem negligible in the case of SiH_2Cl_2, unlike the case of *sym*-tetramethyldisiloxane where, although a significant amount of internally silylated products were observed with 2- and 3-hexene, the major product was the terminal silyl derivative (74). Most silanes give only terminal silylation products with acyclic olefins. The factors that govern the relative rates of Si–C reductive elimination and double-bond migration are too poorly understood either to explain this pattern of behavior, or to predict which classes of silicon hydride will favor internal silylation over double-bond migration and terminal silylation.

The sterochemistry of the hydrosilation of acetylenes was shown to be quite different under the influence of a $(Ph_3P)_3RhCl$ catalyst. Contrary to the case of chloroplatinic acid-catalyzed addition, where ~ 100% *trans* product is observed, the rhodium catalyst gives high yields of *cis* products (75). Since it is difficult to explain the appearance of *cis* products by a counter-thermodynamic isomerization, it is tempting to speculate that this is an example of a reaction going by a free radical route, since the free radical initiated hydrosilation of acetylenes was shown to proceed by *trans* addition to give predominantly *cis* products (76).

The possibility that hydrosilation of acetylenes may involve insertion of an acetylene into a metal–silicon bond was raised by the work of Kiso, Tamao, and Kumada (77). The following stoichiometric reactions were shown to occur:

$$\mathrm{Ni(bipy)_2 + 2HSiCl_3 \rightarrow (bipy)Ni(SiCl_3)_2 + H_2} \quad (39)$$

$$\mathrm{(bipy)Ni(SiCl_3)_2 + RC{\equiv}CR' \longrightarrow (bipy)Ni \leftarrow \left\| \begin{matrix} Cl_3Si & & R \\ R' & & SiCl_3 \end{matrix} \right.} \quad (40)$$

The stereochemistry of the olefinic product was found to depend on the nature of R and R′. Whether these reactions can be combined to give a catalytic di-silylation of acetylenes remains to be seen. A closely related di-silylation of an acetylene has been achieved by what appears to be an oxidative addition of an Si–Si bond to Ni(O) thus (78):

$$\text{(cyclic } t\text{-Bu–C=CH bridged by } SiF_2\text{–}SiF_2) + \mathrm{Ni(CO)_4} \longrightarrow \text{(ring: } t\text{-Bu–C=CH–}SiF_2\text{–}\mathrm{Ni(CO)_2}\text{–}SiF_2\text{)} \quad (41)$$

$$\text{(ring: } t\text{-Bu–C=CH–}SiF_2\text{–}\mathrm{Ni(CO)_2}\text{–}SiF_2\text{)} + \mathrm{(CH_3)_3CC{\equiv}CH} \longrightarrow \text{(di-}t\text{-Bu-1,4-disilacyclohexadiene, } SiF_2\text{)} + \text{(isomer)} \quad (42)$$

The most active, and perhaps most interesting, new developments in hydrosilation involve additions to unsaturated systems other than C═C and C≡C, and particularly to prochiral substrates. The hydrosilation of ketones and aldehydes has attracted a great deal of attention (79,80). To date the most effective catalysts appear to be the phosphinerhodium(I) complexes of the type $(R_3P)_3RhX$ which are also particularly suitable for modification with chiral ligands. Kagan et al. (81) studied the reaction with a catalyst prepared *in situ* by reaction of $[RhCl(C_2H_4)_2]_2$ with (+)-2,3-0-isopropylidene-2,3-dihydroxy-1,4-bis(diphenylphosphino(butane), diop. With a prochiral ketone, such as acetophenone and a monohydrosilane, the asymmetric induction in the alkoxy group was quite small (0–10% optical yield). A dramatic improvement in optical yield (20–60%) was obtained with dihydrosilanes such as diphenylsilane or methylphenylsilane. Corriu and Moureau (82) confirmed the observations of Kagan et al. (81) and showed further that there was also substantial asymmetric induction at silicon when the reaction was carried out with a prochiral dihydrosilane, for example, 1-$NpPhSiH_2$. The highest optical yield of 46% was obtained for the reaction of pentan-3-one with 1-$NpPhSiH_2$. When both ketone and silane were prochiral, asymmetric induction at both Si and C was observed. The optical purities of the two fragments of the products were in general similar (~30–50%) but definitely different, as would be expected since the steric constraints on oxidative addition and migration of the silane fragment are quite different from those on coordination and insertion of the ketone.

Still further improvement of the optical yields (up to 80%) of asymmetric induction at Si and C was achieved by using a chiral ketone and a prochiral dihydrosilane with the (diop)RhCl catalyst (83). The efficacy of using a chiral silicon hydride with a prochiral ketone has not yet been reported. The application of stereospecific displacement of alkoxide by hydridic reagents raises the possibility of recycling chiral hydrosilane in such reactions and rendering possible a greater than 1:1 turnover of ketone per unit of silane.

The (diop)RhCl catalyst is also effective for the hydrosilation of imines. Although the intermediate silylamines were not isolated, the reaction appears to go essentially to completion and the optically active amine products were obtained by hydrolysis in high chemical yield (84). The highest reported optical yield (65%) was obtained by reacting diphenylsilane with $C_6H_5(CH_3)C{=}NCH_2C_6H_5$ at 2°C. The optical yield was very dependent on temperature and the difference in activation parameters for formation of (*S*)- and (*R*)- isomers were reported to be $\Delta\Delta H^{\ddagger}_{S-R} = -3$ kcal/mole and $\Delta\Delta^{\ddagger} = -8.3$ cal/mole/degree.

Ojima et al. (85) reported the hydrosilation of isocyanates, under the influence of palladium catalysts, as a route for the synthesis of formamides. With 1-naphthyl isocyanate, triethysilane yielded the N-silylated product, whereas the C-silylated product was obtained with butyl and phenyl isocyanates. In all cases the silylated product underwent facile methanolysis to the substituted formamide.

REFERENCES

1. C. Eaborn and R. W. Bott give an excellent review of the synthesis and reactions of organosilicon compounds in A. G. MacDiarmid, Ed., *The Organometallic Compounds of the Group IV Elements*, Vol. I, Part 1, Ch. 2, Marcel Dekker, New York, 1968.
2. B. A. Bluestein, U.S. Pat. 2,971,970 (1961); *Chem. Abstr.*, **55,** 12356 (1961); *J. Am. Chem. Soc.*, **83,** 1000 (1961).
3. R. N. Meals, *Pure Appl. Chem.*, **13,** 141 (1966).
4. J. L. Speier, J. A. Webster, and G. H. Barnes, *J. Am. Chem. Soc.*, **79,** 974 (1957).
5. A. J. Chalk and J. F. Harrod, *J. Am. Chem. Soc.*, **87,** 16 (1965).
6. E. K. Pierpoint, Brit. Pat. 923,710 (1963); *cf.* Belg. Pat. 609,997 (1962); *Chem. Abstr.*, **57,** 16657 (1962).

7*a*. J. L. Speier and J. C. Saam, *J. Am. Chem. Soc.*, **80,** 4104 (1958).

7*b*. *Ibid.*, **83,** 1351 (1961).

7*c*. J. W. Ryan and J. L. Speier, *ibid.*, **86,** 895 (1964).
7*d*. H. M. Bank, J. C. Saam, and J. L. Speier, *J. Org. Chem.*, **29,** 792 (1964).
7*e*. A. G. Smith, J. W. Ryan, and J. L. Speier, *ibid.*, **27,** 2183 (1962).
7*f*. V. A. Ponomarenko,V. G. Cherkaev, A. D. Petrov, and N. A. Zadorozhnyi, *Izvest. Akad. Nauk S.S.S.R., Otdel. Khim. Nauk,* **1958,** 247; *Chem. Abstr.*, **52,** 12751 (1958).

8. T. G. Selin and R. West, *J. Am. Chem. Soc.*, **84,** 1863 (1962).
9. J. Chatt, C. Eaborn, and P. N. Kapoor, *J. Organometal. Chem.*, **13,** 21 (1968).
10. F. Glockling and K. A. Hooton, *J. Chem. Soc. A*, **1968,** 1006.
11. J. F. Harrod, unpublished results.
12. J. K. Whittle and G. Urry, *Inorg. Chem.*, **7,** 560 (1968).
13. R. A. Benkeser, M. L. Burrows, L. E. Nelson, and J. V. Swisher, *J. Am. Chem. Soc.*, **83,** 4385 (1961); R. A. Benkeser, R. F. Cunico, S. Dunny, P. R. Jones, and P. G. Nerleker, *J. Org. Chem.*, **32,** 2634 (1967).
14. H. G. Kimila and C. R. Warner, *J. Org. Chem.*, **29,** 2845 (1964).
15. L. H. Sommer, J. E. Lyons, and H. Fujimoto, *J. Am. Chem. Soc.*, **91,** 7051 (1969); L. H. Sommer, K. W. Michael, and H. Fujimoto, *ibid.*, **89,** 1519 (1967); L. H. Sommer and J. D. Citron, *J. Org. Chem.*, **32,** 2470 (1967).
16. A. G. Brook and J. H. Duff, *J. Am. Chem. Soc.*, **91,** 2119 (1969); H. Sakurai, M. Murakami, and M. Kumada, *J. Am. Chem. Soc.*, **91,** 519 (1969); P. J. Krusic and J. K. Kochi, *J. Am. Chem. Soc.*, **91,** 3938 (1969).
17. J. F. Harrod and A. J. Chalk, *J. Am. Chem. Soc.*, **88,** 3491 (1966).
18. R. Cramer and R. V. Lindsey, *J. Am. Chem. Soc.*, **88,** 3534 (1966).
19. M. Orchin in D. D. Eley, H. Pines, and P. D. Weisz, Eds., *Advances in Catalysis,* Vol. XVI, Academic Press, New York, 1966, p. 1.
20. R. Heck, *J. Am. Chem. Soc.*, **90,** 5518 (1968).
21. H. Alper and P. C. Leport, *J. Am. Chem. Soc.*, **91,** 7554 (1969).
22. J. F. Harrod, D. F. R. Gilson, and R. Charles, *Can. J. Chem.*, **47,** 2205 (1969).
23. J. Halpern and J. B. Milne, *Actes Congr. intern. catalyse,* 2^{e}, Paris 1960, **1,** 445, 1961; *Chem. Abstr.*, **55,** 21946 (1961).
24. F. de Charentenay, J. A. Osborn, and G. Wilkinson, *J. Chem. Soc. A*, **1968,** 787.
25. A. J. Chalk, unpublished results.
26. H. van Bekkum, F. van Rantwijk, and T. van der Putte, *Tetrahedron Lett.*, **1969,** 1.
27. R. L. Augustine and J. van Peppen, *Chem. Commun.*, **1970,** 495, 497, and 571.
28. D. D. Lehman, D. F. Shriver, and I. Wharf, *Chem. Commun.*, **1970,** 1486.
29. S. Takahashi, T. Shibano, and N. Nagihara, *Chem. Commun.*, **1969,** 161.
30. R. N. Haszeldine, R. V. Parish, and D. J. Parry, *J. Chem. Soc. A*, **1969,** 683.
31. A. J. Chalk, *J. Organometal. Chem.*, **21,** 207 (1970).
32. A. J. Chalk, *Trans. N. Y. Acad. Sci.*, **32,** 481 (1970).
33. J. F. Harrod and C. A. Smith, *J. Am. Chem. Soc.*, **92,** 2699 (1970).
34. J. F. Harrod and C. A. Smith, *Can. J. Chem.*, **48,** 870 (1970).
35. A. J. Chalk, *Chem. Commun.*, **1969,** 1207.
36. A. J. Chalk, U.S. Pat. 3,188,300, 1965; *Chem. Abstr.*, **63,** 7043 (1965).
37. A. J. Chalk and J. F. Harrod, *J. Am. Chem. Soc.*, **89,** 1640 (1967).
38. Y. L. Baay and A. G. MacDiarmid, *Inorg. Chem.*, **8,** 986 (1969).
39. J. P. Candlin, A. R. Oldham, *Faraday Soc. Disc.*, **46,** 60 (1968).
40. J. A. Osborn, F. H. Jardine, J. F. Young, and G. Wilkinson, *J. Chem. Soc. A*, **1966,** 1711.
41. R. K. Friedlina, E. C. Chukovskaya, J. Tsao, and A. N. Nesmeyanov, *Dokl. Akad. Nauk SSSR,* **132,** 374 (1960); A. N. Nesmeyanov, R. K. Friedlina, E. C. Chukovskaya, R. G. Petrova, and A. B. Belyavksy, *Tetrahedron,* **17,** 61 (1961).

42. E. N. Frankel, E. A. Emken, H. M. Peters, V. L. Davison, and R. O. Butterfield, *J. Org. Chem.*, **29,** 3292 (1964).
43. E. N. Frankel, E. P. Jones, V. L. Davison, E. Emken, and H. J. Dutton, *J. Am. Oil Chemists Soc.*, **42,** 130 (1965).
44. W. Jetz and W. A. G. Graham, *J. Am. Chem. Soc.*, **91,** 3375 (1969); W. A. G. Graham and W. Jetz, *Inorg. Chem.*, **10,** 4 (1971); *idem*, p. 1159.
45. C. W. Bird, *Transition Metal Intermediates in Organic Synthesis*, Academic Press, New York, 1967, Ch. 4.
46. B. J. Aylett and J. M. Campbell, *J. Chem. Soc. A*, **1969,** 1910; Y. L. Baay and A. G. MacDiarmid, *Inorg. Chem.*, **8,** 986 (1969).
47. A. J. Chalk, *Chem. Commun.*, **1970,** 847.
48. B. J. Aylett and J. M. Campbell, *J. Chem. Soc. A*, **1969,** 1920.
49. J. M. Burlitch, *J. Am. Chem. Soc.*, **91,** 4562 (1969).
50. J. F. Bald, Jr., A. D. Barry, R. E. Highsmith, A. G. MacDiarmid, and M. A. Nasta, *Abstr. Fourth Intern. Conf. Organometal. Chem., Bristol*, A5 (1969).
51. A. G. Davies, *Angew. Chem. Intern. Ed. Engl.*, **9,** 741 (1970), and references therein; A. G. Davies, K. J. Ingold, B. P. Roberts, and R. Tudor, *J. Chem. Soc. B*, **1971,** 698.
52. K. W. Ingold, *Chem. Commun.*, **1969,** 911.
53. J. Halpern and L-Y. Wong, *J. Am. Chem. Soc.*, **90,** 6665 (1968), and references therein.
54. L. M. Jackson, J. A. Hamilton, and J. M. Lawlor, *J. Am. Chem. Soc.*, **90,** 1914 (1968).
55. K. Yamamoto, T. Hayashi, and M. Kumada, *J. Organometal. Chem.*, **28,** C37 (1971).
56. K. Yamamoto, T. Hayashi, and M. Kumada, *J. Am. Chem. Soc.*, **93,** 5301 (1971).
57. M. Kumada, Y. Kiso, and M. Umeno, *Chem. Commun.*, **1970,** 611.
58. M. Čapka, P. Svoboda, M. Černý, and J. Hetflejs, *Tetrahedron Lett.*, **1971,** 4787.
59. J. S. Bradley, D. E. Connor, D. Dolphin, J. A. Labinger, and J. A. Osborn, *J. Am. Chem. Soc.*, **94,** 4043 (1972).
60. H. C. Clarke, private communication.
61. H. Sakurai, T. Hirose, and A. Hosomi, *J. Organometal. Chem.*, **86,** 197 (1975).
62. J. V. Swisher and H. H. Chen, *J. Organometal. Chem.*, **69,** 83 (1974).
63. J. Tsuji, M. Hara, and K. Ohan, *Tetrahedron*, **30,** 2143 (1974).
64. C. Eaborn, B. Ratcliff, and A. Pidcock, *J. Organometal. Chem.*, **65,** 181 (1974).
65. F. Glockling and J. J. I. Pollock, *J. Chem. Soc., Dalton Trans.*, **1975,** 497.
66. C. Eaborn, A. Pidcock, and B. Ratcliff, *J. Organometal. Chem.*, **66,** 23 (1974).
67. Y. Kiso, K. Tamao, and M. Kumada, *J. Organometal. Chem.*, **76,** 95 (1974).
68. M. Čapka, P. Svoboda, and J. Hetflejs, Collect. *Czech. Chem. Commun.*, **38,** 3830 (1973).
69. P. Svoboda, P. Sedlmayer, and J. Hetflejs, *Ibid.*, **38,** 1235 (1973).
70. M. S. Wrighton and M. A. Schroeder, *J. Am. Chem. Soc.*, **96,** 6235 (1974).
71. J. Nasielski, P. Kirch, and L. Willputte-Steinert, *J. Organometal. Chem.*, **27,** C13 (1971).
72. M. A. Schroeder and M. S. Wrighton, *J. Organometal. Chem.*, **74,** C29 (1974).
73. R. A. Benkeser and W. C. Muench, *J. Am. Chem. Soc.*, **95,** 285 (1974).
74. H. M. Bank, J. C. Saam, and J. L. Speier, *J. Org. Chem.*, **29,** 792 (1964).
75. I. Ojima, M. Kumagai, and Y. Nagai, *J. Organometal. Chem.*, **66,** C14 (1974).
76. R. A. Benkeser, M. L. Burrows, L. E. Nelson, and J. V. Swisher, *J. Am. Chem. Soc.*, **83,** 4385 (1961).
77. Y. Kiso, K. Tamao, and M. Kumada, *J. Organometal. Chem.*, **76,** 95 (1974); *idem*, p. 105.

78. Chi-Wen Cheng and Chao-Shivan Liu, *Chem. Commun.*, **1974,** 1014.
79. I. Ojima, M. Nihonyagi, and Y. Nagai, *Chem. Commun.*, **1973,** 38.
80. C. Eaborn, H. Pidcock, and K. Odell, *J. Organometal. Chem.*, **63,** 93 (1973).
81. W. Dumont, J-C. Poulin, T-P. Dang, and H. B. Kagan, *J. Am. Chem. Soc.*, **95,** 8295 (1973).
82. R. J. P. Corriu and J. E. E. Moreau, *J. Organometal. Chem.*, **64,** C51 (1974).
83. R. J. P. Corriu and J. E. E. Moreau, *J. Organometal. Chem.*, **91,** C27 (1975).
84. N. Langlois, T-P. Dang, and H. B. Kagan, *Tetrahedron Lett.*, **1973,** 4865.
85. I. Ojima, S. Inaba, and Y. Nagai, *Tetrahedron Lett.*, **1973,** 4363.

Catalysis of Symmetry-Restricted Reactions by Transition Metal Compounds

JACK HALPERN, *Department of Chemistry, The University of Chicago, Chicago, Illinois* 60637

I. INTRODUCTION

The concept of orbital symmetry conservation encompassed by the Woodward-Hoffmann rules and related views has been applied with considerable success in recent years to interpretation of certain problems of chemical reactivity, especially in organic chemistry (1,2). According to these rules, the slowness of certain reactions may be understood in terms of nonconservation of the symmetries of the stable, filled orbitals of the system during the concerted transformation of reactants to products. Familiar examples of such "thermally forbidden" processes include various four-center reactions such as the hydrogenation of olefins, the fusion of two olefins to cyclobutane (*suprafacial* 2+2 cycloaddition) and the reverse cycloreversion, as well as certain other processes such as disrotatory cyclobutene–butadiene transformations.

The observed reactivity for a given process is, of course, governed by the rates of the nonconcerted pathways, as well as of the concerted pathways, through which the process can occur. In the case of the cyclobutane–diolefin transformation (**1** → **3**) the reactivities associated with both pathways are low; that for the concerted pathway (i.e., via **2a**) because of nonconservation of orbital symmetry, and those for the nonconcerted pathways (i.e., via **2b** or **2c**) because of the considerable endothermicity of the uncompensated dissociation of a carbon–carbon bond (ca. 50 kcal/mole) even in a strained ring system such as cyclobutane.

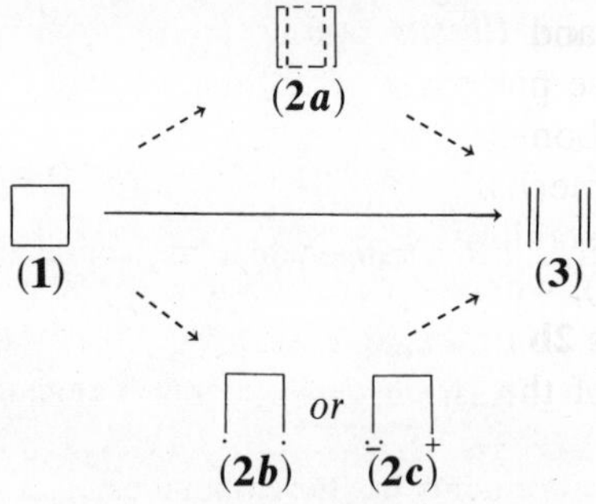

Many instances have been reported of the catalysis of such "symmetry-restricted" reactions by transition metal compounds. Besides such long familiar processes as the homogeneous catalytic hydrogenation of olefins, known examples of this type of catalysis now encompass many reactions involving valence isomerization and other carbon–carbon bond rearrangement processes—both intramolecular and intermolecular. Thus, it has been found that rhodium(I) complexes catalyze various cyclobutane–diolefin transformations such as the valence isomerizations of quadricyclane to norbornadiene (3,4), of hexamethylprismane to hexamethyldewarbenzene (5), of cubane to *syn*-tricyclooctadiene, and of the latter to cyclooctatraene (6). Silver(I) also catalyzes a number of symmetry-restricted reactions such as the isomerization of cubane to "cuneane" (7), the analogous rearrangements of the homocubyl and 1,1′-bishomocubyl systems (8–15), as well as the transformation of tricyclooctadiene to cyclooctatetraene (the latter interpreted as a sequence of disrotatory cyclobutene–butadiene conversions) (16,17). Rhodium(I) and silver(I) both catalyze bicyclobutane–butadiene rearrangements (18–35). Other examples of this type of catalysis include nickel(0)-catalyzed 2+2 olefin cycloadditions (36–39), as well as the metathesis (or disproportionation)

of olefins catalyzed by compounds of tungsten and molybdenum (40–45). While this review will be concerned primarily with the mechanistic aspects of such carbon–carbon bond rearrangement processes, many of the same themes also extend to other metal-catalyzed reactions such as hydrogenation and other olefin addition reactions.

II. GENERAL MECHANISTIC CONSIDERATIONS

Two alternative (although not necessarily mutually exclusive) interpretations were proposed for the above catalytic effects, namely (*i*) removal of the symmetry constraints on the otherwise-"forbidden" concerted pathways by interaction of the reacting system with the *d* orbitals of the catalyst (17,46,47), and (*ii*) the opening up of, otherwise prohibitively endothermic, stepwise pathways involving the initial breaking (or formation) of only one carbon–carbon bond through an "oxidative addition" or heterolytic splitting mechanism (48,49). The first of these interpretations corresponds to the stabilization of the configuration **2a** by interaction with the catalyst (M), whereas the second involves the stabilization of a configuration such as **2b** or **2c**, that is, to intermediate σ-bonded organometallic complexes of the type **4a**, **4b** and **4c**, respectively.

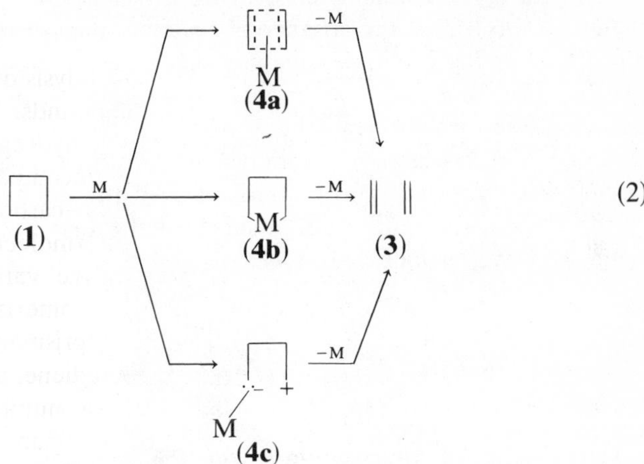

(2)

The terms "concerted" and "stepwise," as used in the present context, refer to the overall process of going from organic reactants to products. Thus, a "stepwise" mechanism, in accord with the usual meaning of the designation, implies a mechanism involving a discrete organometallic intermediate such as **4b** or **4c**. It should be recognized, however, that the distinction between "concerted" and "stepwise" may also be applied to each component step of a stepwise sequence, for example the oxidative addition reaction, **1** → **4b** and the subsequent metallocycle–product rearrangement, **4b** → **3**. (The latter step may be loosely identified as a "β-carbon elimination," by analogy with the well-known β-hydride elimination through which transition metal alkyls frequently undergo decomposition to yield olefins.) At least in the case of square planar d^8 complexes such as those of rhodium(I), there is reason to believe that each of these steps is a concerted process and that each is "symmetry-allowed" according to the usual orbital symmetry conservation selection rules. The catalysis of reaction 1 by rhodium(I) according to the mechanistic scheme **1** → **4b** → **3** may thus be ascribed to the opening up by the catalyst of a stepwise reaction path involving a sequence of two steps, each of which is concerted and symmetry allowed.

III. RHODIUM(I)-CATALYZED REARRANGEMENTS OF CUBANE AND RELATED COMPOUNDS

The rearrangement of cubane to *syn*-tricyclooctadiene (**5** → **7**) has proved to be a convenient probe for the examination of the mechanisms of catalysis of cyclobutane–diolefin transformations (6). The advantages of this system are (*i*) the absence of reactive sites other than cyclobutane rings (such as the cyclopropane rings in quadricyclane), which eliminates possible contributions from mechanisms other than those intrinsically involving the cyclobutane rings, and (*ii*) the high symmetry of the molecule and availability of equivalent substitution sites, which permit mechanistic information to be derived from the influence of substituents on the rates of rearrangement of substituted cubanes and the distribution of the resulting isomeric products.

$$(\mathbf{5}) \xrightarrow[k_3]{\mathrm{Rh^I}} (\mathbf{6})\ \mathrm{Rh} \xrightarrow{-\mathrm{Rh^I}} (\mathbf{7}) \qquad (3)$$

The results of investigations on the catalysis of reaction 3 by rhodium(I) complexes of the type, $[\mathrm{Rh^I_2(diene)_2Cl_2}]$ (where diene = norbornadiene,1,5-cyclooctadiene, etc.) provide convincing support for the stepwise mechanism depicted by Eq. 3, in which the rate-determining step is the oxidative addition to Rh(I) of one of the strained C–C bonds of the substrate. The following evidence supports this conclusion (6):

(*i*) The reaction exhibits the catalytic second-order rate-law corresponding to Eq. 4. The rate is not influenced by addition of excess diene indicating that the mechanism does not involve a pre-equilibrium coordination of the substrate to the catalyst by displacement of a diene ligand but, instead, proceeds through a rate-determining reaction of cubane with the intact $[Rh_2(diene)_2Cl_2]$ molecule.

$$\text{Rate} = k_3[Rh_2(\text{diene})_2Cl_2][\text{cubane}] \tag{4}$$

(*ii*) Oxidative addition of cubane to rhodium(I) is directly demonstrated by the reaction of cubane with $Rh_2(CO)_4Cl_2$ to yield a stable isolable acylrhodium adduct (**8**), in accord with Eq. 5. This reaction presumably involves oxidative addition of the C–C bonds of cubane, followed by the familiar migratory insertion of a coordinated CO ligand. Similarity of the patterns of relative rate constants of reactions 3 and 5, for a series of different substituted cubanes ($k_3/k_5 \approx 4 \times 10^2$) reinforces the conclusion that both reactions proceed through similar rate-determining steps and, hence, that the rate-determining step of the catalytic reaction 3 is the proposed oxidative addition step. Treatment of the oxidative adduct **8** with a ligand such as PPh_3 results in "extrusion" of the cycloketone **9**. This constitutes a synthetically useful route for the "insertion" of CO into strained C–C bonds (50).

4 [cubane] $+ 2Rh_2(CO)_4Cl_2 \xrightarrow{k_5}$ [cubane–Rh(CO)(Cl)–C(=O)]$_4$ (5)

(**8**)

8 $+ 8PPh_3 \longrightarrow$ [cubane ketone, C=O] $+ 4RhCl(CO)(PPh_3)_2$ (6)

(**9**)

(*iii*) The distribution of products from reactions 5 and 6 of the substituted cubane **10** ($R = COOCH_3$) with $Rh_2(CO)_4Cl_2$, (34% **12**, 66% **15**), parallels that from the $Rh_2(diene)_2Cl_2$-catalyzed isomerization of **10** according to reaction 3, (30% **13**, 70% **16**). In each case the product distribution is that predicted to arise from approximately statistical competition between the two sterically favored modes of oxidative addition of

10 leading to the intermediates **11** and **14** in the expected ratio 1:2.

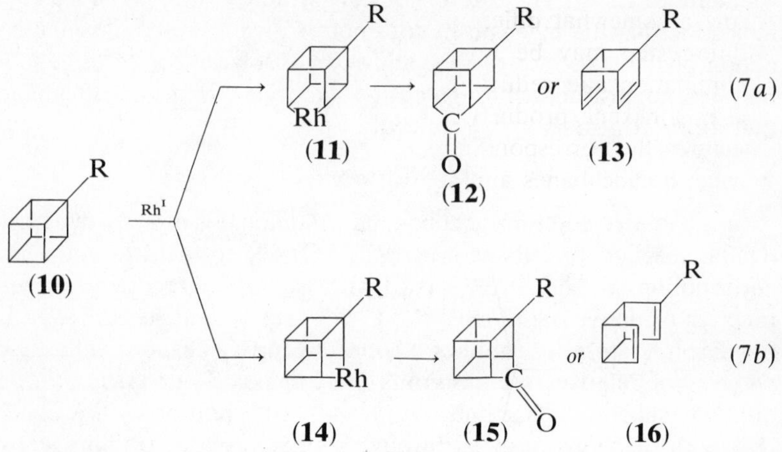

The rhodium(I)-catalyzed rearrangements of homocubane (**17**) and substituted homocubanes, according to Eq. 8, were found to exhibit similar characteristics and have been similarly interpreted.

(**17**) $\xrightarrow{Rh^I}$ (**18**) $\xrightarrow{-Rh^I}$ (**19**) (8)

Convincing evidence (including the isolation of stable acyl-rhodium adducts from the reaction of quadricyclane with $Rh_2(CO)_4Cl_2$) was also obtained for an analogous nonconcerted oxidative addition mechanism for the rhodium(I)-catalyzed valence isomerization of quadricyclane to norbornadiene according to Eq. 9 (4). It seems highly likely that other rhodium(I)-catalyzed "symmetry-restricted" reactions, such as the transformation of hexamethylprismane to hexamethyldewarbenzene (5), and the valence isomerization of *exo*-tricyclo[3.2.1.$0^{2,4}$]oct-6-ene (as also concluded by Katz and Cerefice) (51), also proceed through such stepwise mechanisms.

(**20**) $\xrightarrow{Rh^I}$ (**21**) $\xrightarrow{-Rh^I}$ (**22**) (9)

The rhodium(I)-catalyzed rearrangements of noncage fused-cyclobutane compounds such as tricyclooctane (**23**) or bicyclohexane (**24**) follow a somewhat different course. In such cases the initial oxidative addition step may be followed by the sequence of rhodium hydride β-elimination and reductive elimination steps (reactions 10*b* *or* 11) to give cyclohexene products (51). Because of the constraints of the cage structures the corresponding rearrangement pathways are precluded for cubane, homocubanes and quadricyclane.

(**23**) Rh^I → [Rh intermediate] → $-$Rh^I (<20%) (10*a*)

→ [H–Rh intermediate] → $-$Rh^I (>80%) (≡) (10*b*)

(**24**) Rh^I → [Rh intermediate] → [H–Rh intermediate] → $-$Rh^I (≡) (11)

In such cases the reactions with $Rh_2(CO)_4Cl_2$ may be correspondingly modified from those of the cage compounds such as cubane and quadricyclane (exemplified by Eqs. 5 and 6) to yield, instead of ketones, the corresponding aldehydes in accord with the scheme of Eq. 12 (52).

(**24**) $Rh_2(CO)_4Cl_2$ → [Rh–C(=O) intermediate] → [H–Rh–C(=O) intermediate] → $-$Rh^I [O=C–H product] (≡ O=C(H)–cyclohexene)

(12)

Analogous stepwise mechanisms involving metallocycle intermediates have also been invoked for 2+2 cycloaddition reactions, for example the cyclodimerization of norbornadiene (Eq. 13) (53,54). The formation of such a metallocycle (**26**) through the reaction of iridium(I) with norbornadiene was actually demonstrated (54).

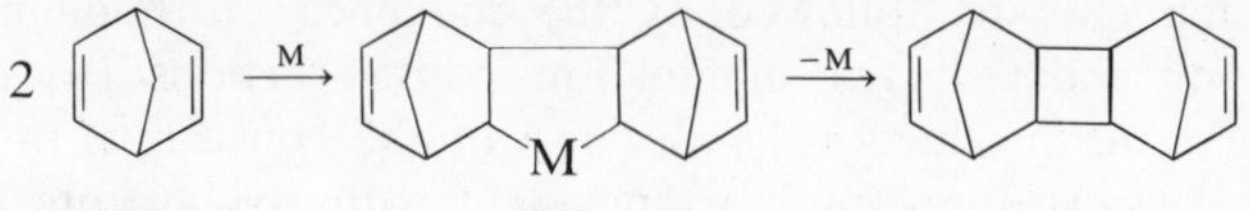

$$2\ (\mathbf{25}) \xrightarrow{M} (\mathbf{26}) \xrightarrow{-M} \qquad (13)$$

IV. NICKEL(0)-CATALYZED REACTIONS

In view of the well known tendency of nickel(0) complexes ($3d^{10}$ configuration) to undergo oxidative addition to form nickel(II) adducts, such complexes might be expected to exhibit catalytic patterns similar to those of rhodium(I). Such parallels are reflected in the catalysis by bis(1,5-cyclooctadiene)nickel(0) (in the presence of methyl acrylate) of the rearrangements of the 1,1′-bishomocubane **27** (where $R = COOCH_3$) to the corresponding dienes, according to Eq. 4, a process that is interpreted in terms of the intermediate nickel(II) oxidative adducts, **28** and **29** (56).

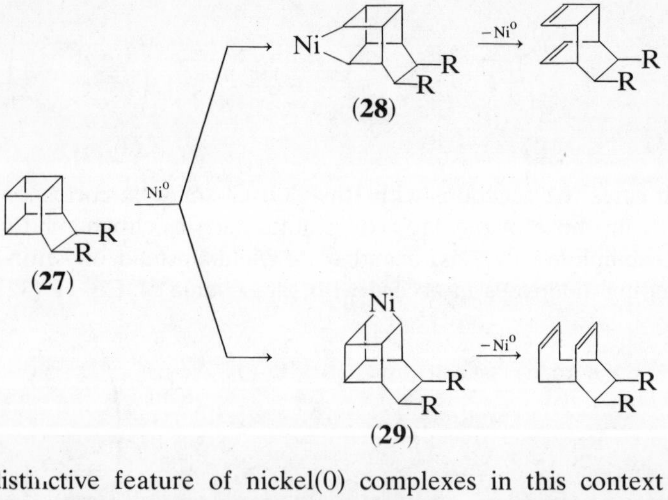

$$(\mathbf{27}) \xrightarrow{Ni^0} (\mathbf{28}) \xrightarrow{-Ni^0} \qquad (14a)$$

$$(\mathbf{27}) \xrightarrow{Ni^0} (\mathbf{29}) \xrightarrow{-Ni^0} \qquad (14b)$$

A distinctive feature of nickel(0) complexes in this context is their ability to catalyze intermolecular processes involving carbon–carbon bond formation. Thus, bis(acrylonitrile)nickel(0), catalyzes not only the isomerization of quadricyclane to norbornadiene, but also the cycloaddition of acrylonitrile to quadricyclane to form the adduct **31**. This result may be interpreted in terms of the scheme of reaction 15 according to which the two products arise from alternative reaction paths involving the common intermediate nickel(II) oxidative adduct, **30** (57).

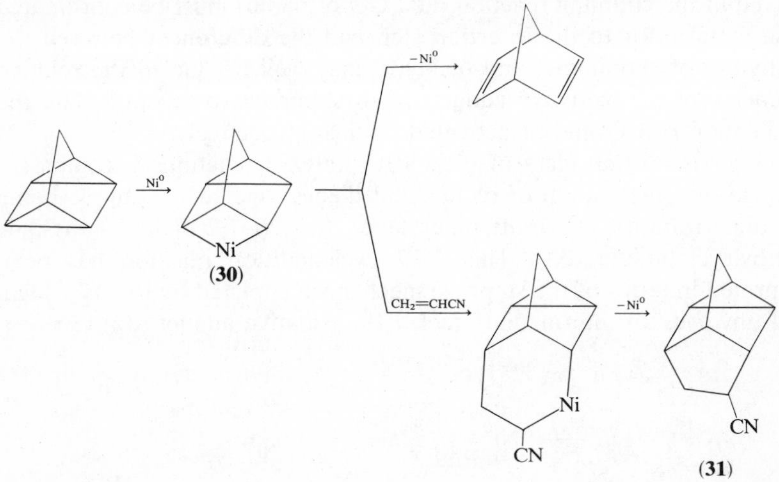

(15*a*)

(15*b*)

The ability of nickel(0) complexes to catalyze such cycloaddition reactions of strained carbon–carbon bonds with activated (electron-deficient) olefins appears to be fairly widespread, another example (also interpreted in terms of an initial oxidative addition step) being that depicted in reaction 16, where CHZ=CHZ is dimethyl maleate (58).

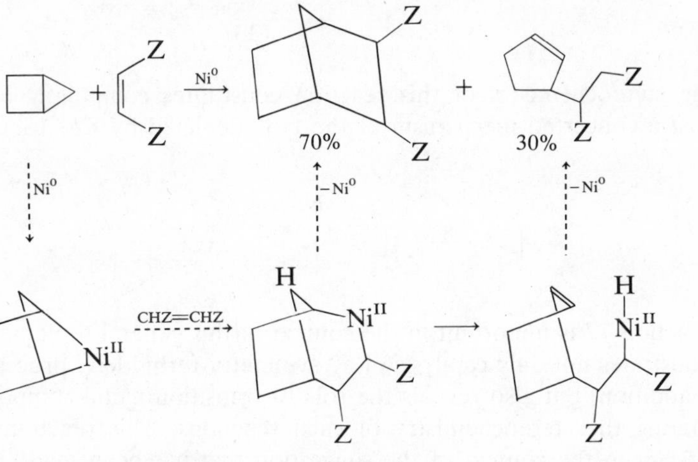

(16)

The catalysis of such intramolecular carbon–carbon bond formation by nickel(0) complexes parallels the rhodium(I) promoted carbonylations of strained hydrocarbons depicted by Eqs. 5, 6, and 12. However, the ability of rhodium(I) to catalyze such coupling reactions does not extend to olefins. In each case it seems likely that the unsaturated molecule involved in the coupling reaction (i.e., CO or olefin) must be coordinated to the metal prior to the insertion step and the differences between the selectivities of rhodium(I) and nickel(0) may well be due to the relative tendencies of the oxidative adducts of these metals to accommodate the coordination of CO and of activated olefins, respectively.

Another important class of nickel(0)-catalyzed reactions encompasses the cyclo-oligomerization of olefins and dienes, one such example being the dimerization of *trans*-piperylene to *cis*-1,2-divinyl-*trans*-3,4-dimethylcyclobutane (**33**). This 2+2 cycloaddition reaction has been interpreted in terms of the stepwise mechanism depicted by Eq. 17 which, again, involves an intermediate nickel(II) oxidative adduct (**32**) (36–39).

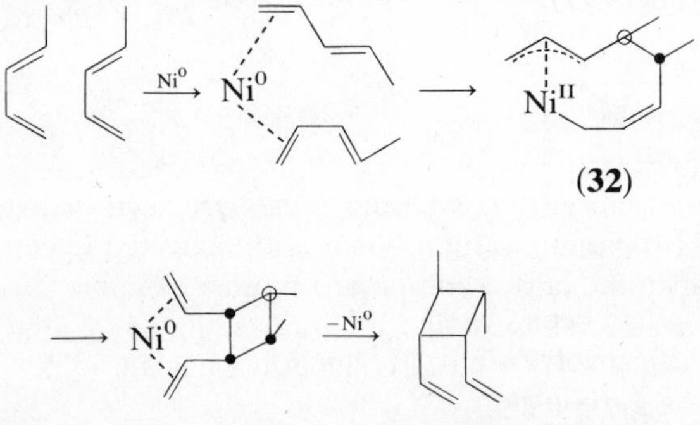

Ni^0 Ni^0 Ni^{II}

(**32**)

(17)

Ni^0 $-Ni^0$

(**33**)

The stereochemistry of this reaction constitutes compelling evidence against a concerted mechanism of the type depicted by Eq. 18.

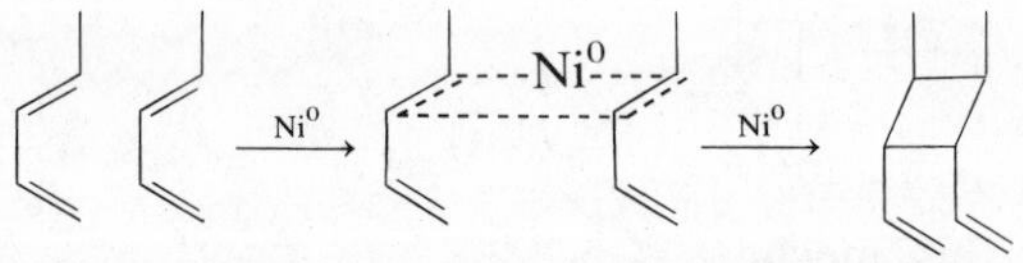

Ni^0 Ni^0 Ni^0 (18)

Reaction 17 is important in the context of this general topic because it demonstrates not only catalysis of a "symmetry-forbidden" process (2+2 cycloaddition) but also reveals the role of transition metal compounds in regulating the stereochemistry of such reactions. This reaction is also significant in the context of the suggestion that has been made that the preference of reactions such as 3 and 9 for stepwise pathways involving oxidative addition steps, rather than for concerted catalytic mechanisms of the type depicted by Eq. 18, is due to driving forces derived from the considerable relief of strain associated with such reactions (47). It had been suggested that preference for concerted mechanisms of the latter type would be exhibited by reactions in which such driving forces were absent. However, the preference of reaction 17 for a stepwise catalytic mechanism, despite the absence of any contribution to relief of strain, clearly does not support this view.

V. SILVER(I)-CATALYZED REARRANGEMENTS OF CUBANE AND RELATED MOLECULES

Silver(I) has been found to catalyze a quite different course of rearrangement of cubane from that of rhodium(I), leading to cuneane (**34**) in accord with Eq. 19 (7).

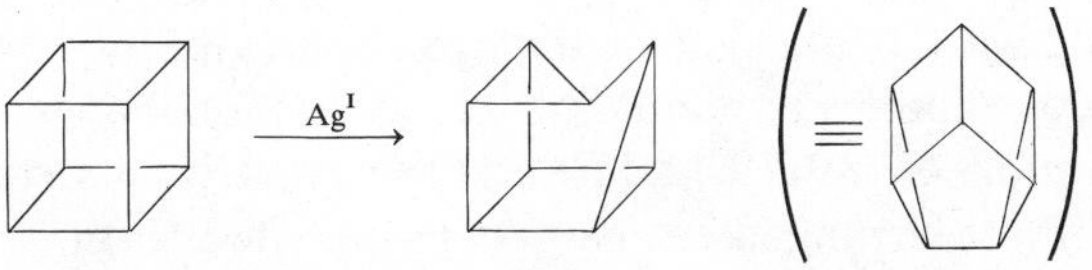

$$\xrightarrow{Ag^{I}} \quad (\equiv) \qquad (19)$$

(**34**)

This reaction had been interpreted in terms of the mechanism depicted by Eq. 20 which involves initial heterolytic (electrophilic) cleavage of a strained carbon–carbon bond by silver(I) accompanied by conventional cyclobutyl-cyclopropylcarbinyl rearrangement of the resulting carbonium ion (48,59).

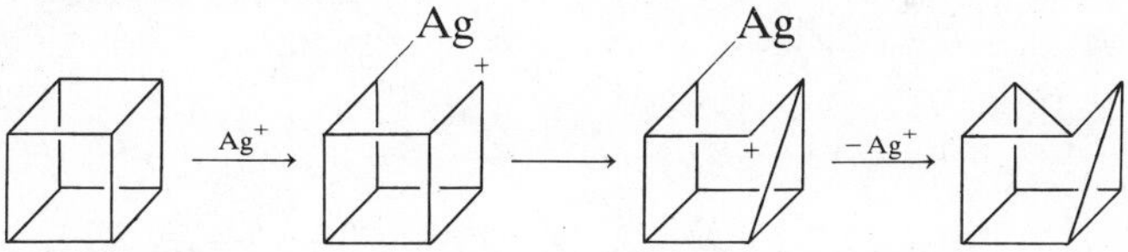

$$\xrightarrow{Ag^{+}} \ \text{Ag} \ + \ \longrightarrow \ \text{Ag} \ + \ \xrightarrow{-Ag^{+}} \qquad (20)$$

Support for this mechanism is provided by studies on substituted cubanes (**35**), which reveal both a large reduction rate by electron-withdrawing substituents (7,60) as well as a pattern of product distribution that provides a direct test of the mechanism. It is anticipated, in accord with the scheme of Eq. 21, that substituents which destabilize carbonium ions (e.g., $R = COOCH_3$) should favor the formation of isomer **37a**, whereas stabilizing substituents (e.g., $R = CH_2OAc$) should favor isomer **37b**. The experimental findings are in accord with these predictions (7).

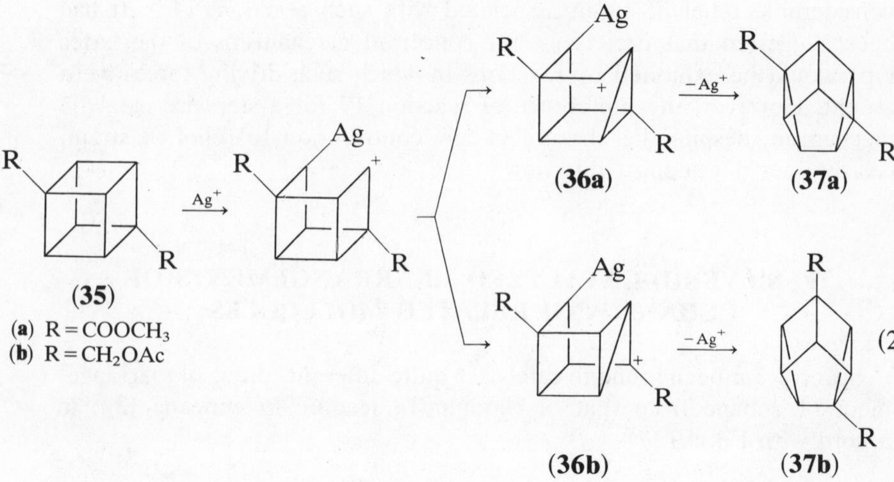

(35) →(Ag⁺) → (36a) →(−Ag⁺) (37a) (21a)

(36b) →(−Ag⁺) (37b) (21b)

(35)
(**a**) R = $COOCH_3$
(**b**) R = CH_2OAc

(**36a**) (**37a**)

(**36b**) (**37b**)

Analogous silver(I)-catalyzed rearrangements were observed for other cyclobutane-containing compounds such as homocubanes (**38**) (10,13), 1,8-bis-homocubanes (**39**) (9,11,12), and the *syn*-tricyclooctane (**40**) (61). In each case the rearrangement can be interpreted through mechanisms similar to that proposed for cubane, that is, involving heterolytic splitting of a strained carbon–carbon bond by electrophilic attack of silver(I) and cyclobutyl–cyclopropylcarbinyl rearrangement of the resulting carbonium ion. Such mechanisms are depicted in Eqs. 22–24. In the case of 4-methylhomocubane, kinetic evidence was found for the initial pre-equilibrium formation of a silver(I)–homocubane complex (13).

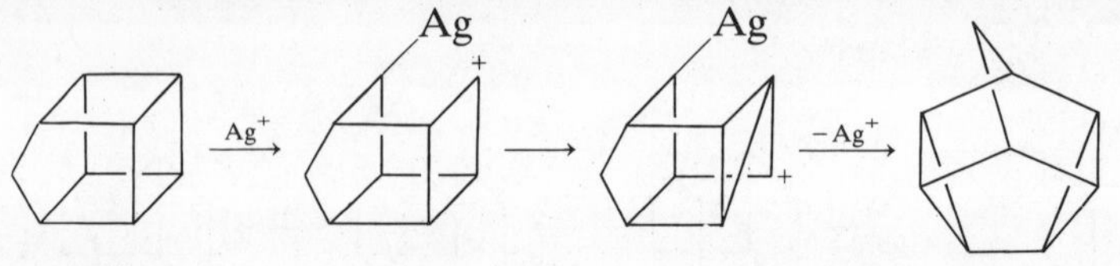

(**38**) (22)

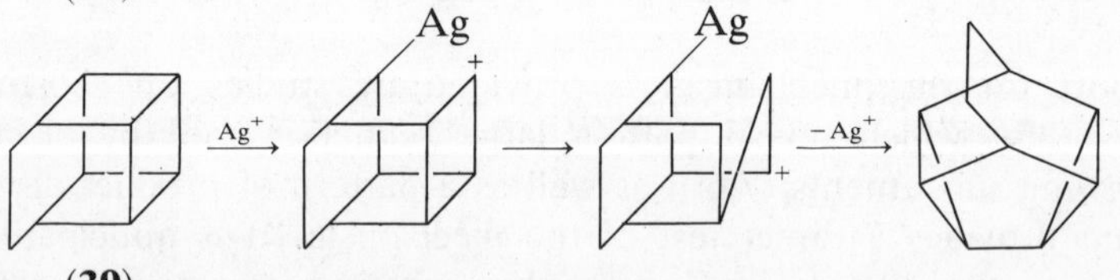

(**39**) (23)

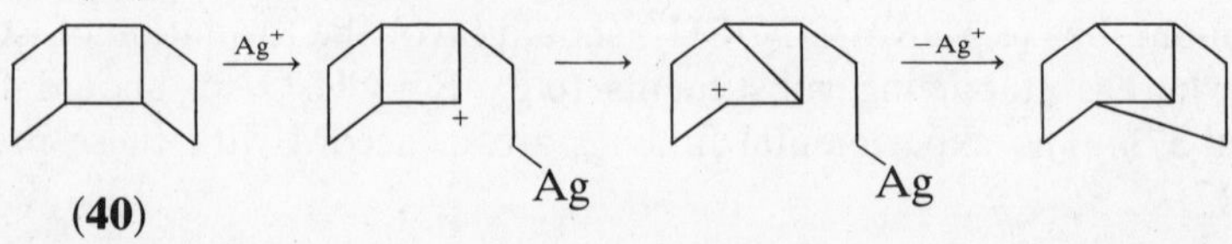

(**40**) (24)

Attention is directed to noteworthy parallels between the different modes of reaction of rhodium(I) and silver(I) with strained carbon–carbon bonds, reflected in the above mechanisms, and the characteristic reactions of corresponding metal complexes with molecular hydrogen. Thus, Rh(I) and other d^8 complexes typically dissociate H_2 through oxidative addition (Eq. 25) (48), whereas Ag(I) has been shown to dissociate H_2 heterolytically by electrophilic attack (Eq. 26) (62).

$$L_4Rh^I + H_2 \rightarrow L_4Rh\begin{matrix} \diagup H \\ \diagdown H \end{matrix} \tag{25}$$

$$Ag^+ + H_2 \rightarrow AgH + H^+ \tag{26}$$

In this context, it is not unexpected that complexes of palladium(II) which is isoelectronic ($4d^8$) with rhodium(I) but which is considerably more electrophilic than the latter, can catalyze both types of rearrangement of cubane and related compounds, as depicted in Eq. 27 (63,64). A variety of factors, especially the nature of the ligands coordinated to palladium(II), influence the preference among these alternative modes of reaction.

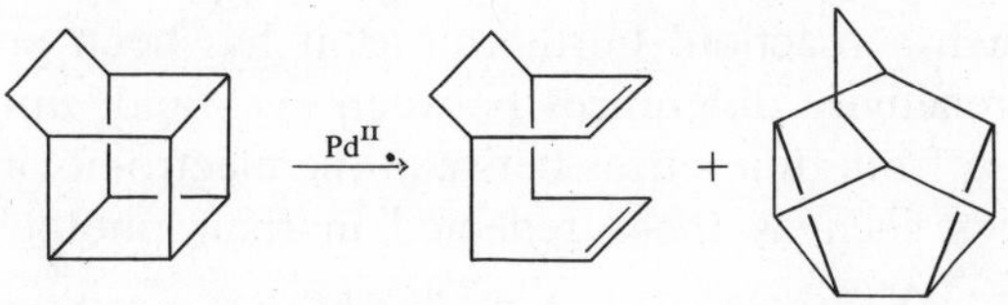

$$\xrightarrow{Pd^{II}} \quad + \tag{27}$$

In some cases, the two alternative modes of reaction of metal ions with H_2 or with strained C–C bonds, that is, oxidative addition *versus* electrophilic heterolytic cleavage *may* have a common origin. Thus, as depicted in Eq. 28, oxidative addition of hydrogen or C–C to a sufficiently electrophilic metal atom may be followed by heterolytic cleavage of a M–H or M–C bond of the resulting oxidative adduct, the overall reaction being equivalent to electrophilic displacement of a proton or carbonium ion. This mode of reaction seems plausible for Pd(II) or Pt(II) but not for Ag(I) or Hg(II) since Pd(IV) and Pt(IV) are considered accessible intermediate oxidation states, whereas Ag(III) and Hg(IV) are not. It is more likely that heterolytic splitting by Ag(I) or Hg(II) proceeds through direct electrophilic displacement without intermediate oxidative addition.

$$M^{I} + H_2 \longrightarrow \left[\begin{array}{c} H \\ \ \ \ \ \ \ \ M^{III} \\ H \end{array} \right] \longrightarrow H\text{—}M^{I} + H^{+} \qquad (28a)$$

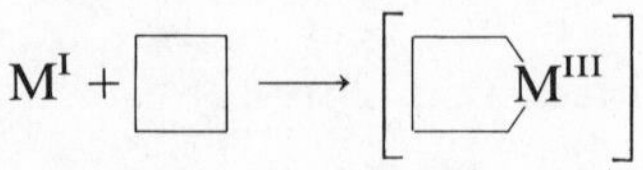

$$M^{I} + \square \longrightarrow \left[\square M^{III} \right] \longrightarrow \overset{+}{\square} M^{I} \qquad (28b)$$

Whereas, our proposed stepwise (i.e., oxidative addition) mechanism for the rhodium(I)-catalyzed rearrangements of cubane and related compounds (e.g., Eq. 3) has been generally accepted, there is still substantive disagreement about the general validity of our proposed stepwise carbonium ion mechanism (e.g., Eqs. 20 and 24) for the corresponding silver(I)-catalyzed rearrangements. An alternative view is that such rearrangements may proceed through concerted mechanisms involving "multicenter" interaction of Ag^{+} with a cyclobutane face of the substrates, instead of localized reaction with one C–C bond (61). One of the observations that has been cited in support of the latter view is that silver(I)-catalyzed rearrangements of *anti*-tricyclooctane (where such multicenter interaction might be expected to be sterically unfavorable), is much slower than the corresponding rearrangement of the *syn* isomer depicted in Eq. 24 (61). It should, however, be noted that the same rearrangement for the *anti* isomer is precluded by virtue of the inaccessibility of the highly strained *trans*-fused cyclopropane product that would result from such a reaction. Furthermore, it has been pointed out (12) that marked reactivity differences between *syn*- and *anti*-tricyclooctane could well arise from differences between the electronic properties of the two compounds such as those reflected in their photoelectron spectra (65).

An alternative mechanism for the silver(I)-catalyzed rearrangement of the *syn*-tricyclooctane (**40**) was proposed (66) which is a variant of that depicted by Eq. 24 and which also involves intermediate carbonium ions. This variant is depicted by Eq. 29:

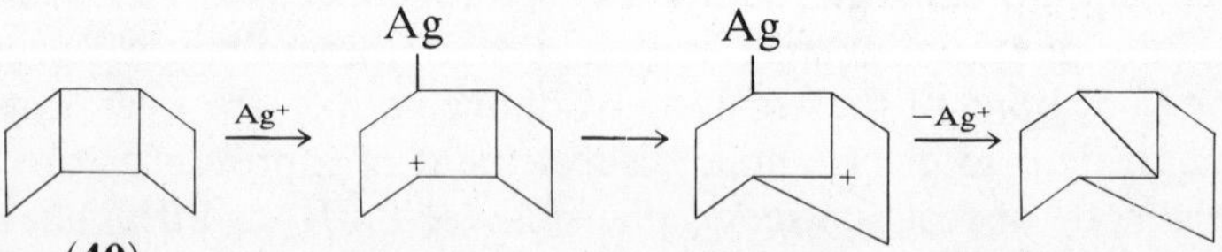

Ag Ag

$\xrightarrow{Ag^{+}}$ + $\longrightarrow$ + $\xrightarrow{-Ag^{+}}$ (29)

(**40**)

This interpretation also readily accommodates the observation that the *anti*-isomer does not undergo a corresponding silver(I)-catalyzed rearrangement.

VI. CATALYZED REARRANGEMENTS OF BICYCLOBUTANES

Bicyclo(1.1.0)butane and its substituted derivatives (e.g., **41**) although characterized by high strain energies (ca. 65 kcal/mole), possess remarkable thermal stabilities. Rearrangement to butadienes at elevated temperatures has been shown to proceed through a concerted $[\sigma 2_s + \sigma 2_a]$ mechanism, the corresponding $[\sigma 2_a + \sigma 2_a]$ being thermally disallowed on orbital symmetry grounds (67). Complexes of several metals including silver(I), rhodium(I), and palladium(II) catalyze a variety of reactions of such bicyclobutanes and, under certain conditions, reverse the stereochemistry of the rearrangement to butadiene so that the overall change corresponds formally to the thermally forbidden $[\sigma 2_a + \sigma 2_a]$ process. These relationships are exemplified by Eq. 30 (21).

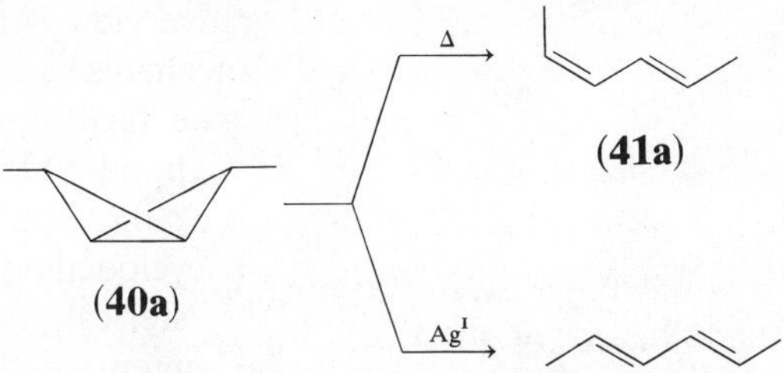

Δ → **(41a)** (30*a*)

(40a)

Ag^I → **(41b)** (30*b*)

The patterns of catalysis of the rearrangements of bicyclo(1.1.0)butanes are complex and reflect the influences of substrate substituents, the nature of the metal, the nature of the ligands, the solvent and other reaction conditions (18–35). Although extensively studied, the nature and origin of these influences and the mechanistic features of the various catalytic processes that characterize this system are still far from fully elucidated.

The silver(I)-catalyzed rearrangements of bicyclobutanes to butadienes encompass two quite different reactions (i.e., **42** → **43** and **42** → **44**) which may be distinguished through appropriate isotopic or substituent labeling of the skeletal carbon atoms. The first process corresponds to cleavage of the C_1–C_2 and C_3–C_4 bonds, and the second to cleavage of the C_1–C_2 and C_1–C_3 bonds (with an accompanying C_4 to C_1 H-shift).

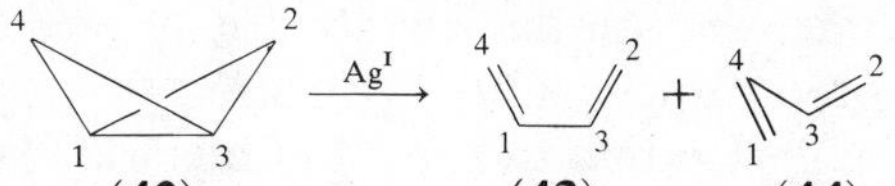

4 2 1 3 **(42)** $\xrightarrow{Ag^I}$ 4 2 1 3 **(43)** + 4 2 1 3 **(44)** (31)

The mechanism depicted by Eq. 32 accommodates many of the observations relating to this system (23). The initial step in this mechanism, which resembles those of the other silver(I)-catalyzed reactions discussed above, involves the heterolytic splitting of one of the strained C–C bonds through electrophilic attack with formation of an intermediate organometallic carbonium ion (**45**). The latter may rearrange to **43** or, through a further organometallic intermediate, the silver–carbene complex or argentocarbonium ion (**46**), to **44**.

(**42**) $\xrightarrow{Ag^I}$ (**45**) → (**46**) $\xrightarrow[\text{H-Shift}]{-Ag^I}$ (**44**) (32*b*)

(**45**) → $\xrightarrow{-Ag^I}$ (**43**) (32*a*)

The evidence that has been cited (23) in support of the above mechanism includes (*i*) the influence of alkyl substituents on the product distribution which corresponds to a preference for the reaction path involving the more substituted carbonium ion (i.e., C_2, C_4-substitution favoring **43** and C_1, C_3-substitution favoring **44**, and (*ii*) the formation of methyl ethers such as **47** when the reaction is carried out in methanol. It has been proposed that such ethers are formed through solvolytic interception of intermediate carbonium ions (Eq. 33), but this interpretation is not unequivocal and a variety of observations suggest that the catalytic mechanisms in protic solvents such as methanol may well be different from those in nonprotic solvents (68–71).

$\xrightarrow[-Ag^I]{\text{MeOH}}$ (**47**) (33)

The influence of methanol on the silver(I)-catalyzed rearrangement of quadricyclane to norbornadiene (**20** → **22**) also raises questions of interpretation. It was reported that treatment of quadricyclane with $AgBF_4$ in methanol–dioxane solvent mixtures gives rise to a mixture of norbornadiene and the ethers, methoxynortricyclane (**49**), and dimethoxynortricyclane (**50**), and that formation of the ethers is accompanied by precipitation of metallic silver (72,73). It was suggested (73) that the two products arise through independent reaction paths—the catalytic rearrangement to quadricyclane being a concerted process while the solvolytic products arise through intermediate carbonium ions, thus:

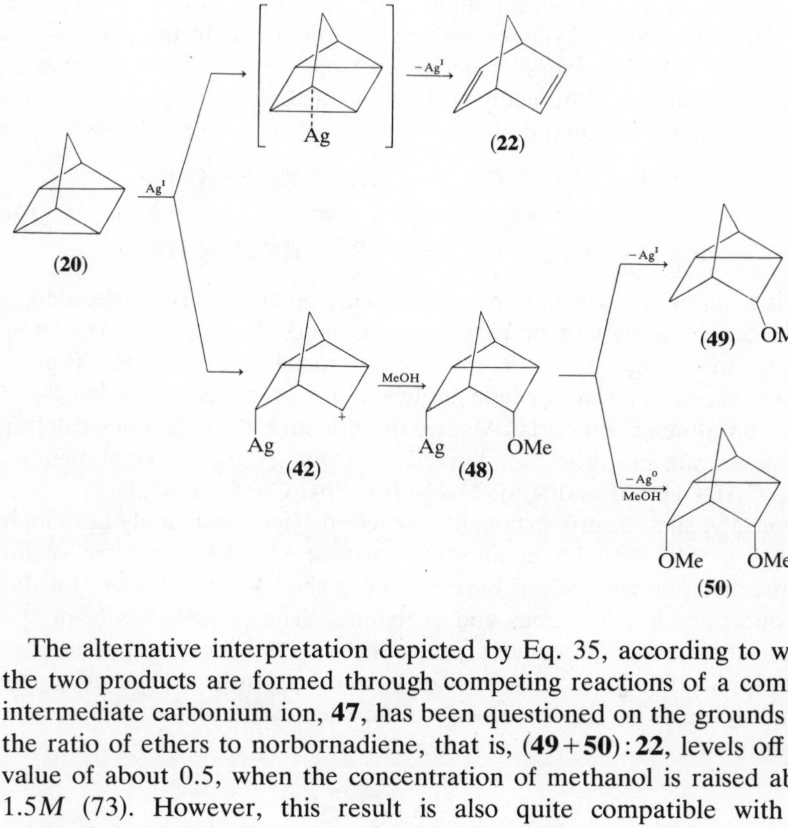

(20) $\xrightarrow{Ag^I}$ [Ag complex] $\xrightarrow{-Ag^I}$ **(22)** (34*a*)

(20) $\xrightarrow{Ag^I}$ **(42)** $\xrightarrow{MeOH}$ **(48)** $\xrightarrow{-Ag^I}$ **(49)** OMe (34*b*)

(48) $\xrightarrow[MeOH]{-Ag^0}$ **(50)** OMe OMe (34*c*)

The alternative interpretation depicted by Eq. 35, according to which the two products are formed through competing reactions of a common intermediate carbonium ion, **47**, has been questioned on the grounds that the ratio of ethers to norbornadiene, that is, (**49**+**50**):**22**, levels off at a value of about 0.5, when the concentration of methanol is raised above 1.5*M* (73). However, this result is also quite compatible with the mechanistic scheme of Eq. 35, since the expected preferential solvation of Ag^+ by methanol could well lead to such "saturation" behavior even at relatively low methanol concentrations.

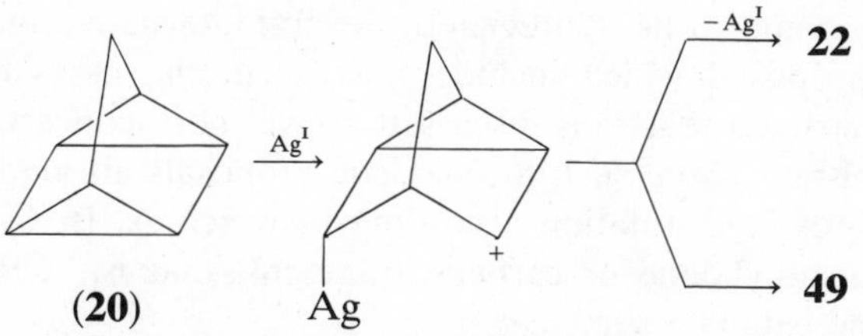

(20) $\xrightarrow{Ag^I}$ [Ag, +] $\xrightarrow{-Ag^I}$ **22** (35*a*)

[Ag, +] $\longrightarrow$ **49** (35*b*)

VII. OLEFIN METATHESIS

Another important class of reactions encompassed by the theme of "symmetry-restricted" processes is the *metathesis* of olefins. First recognized in 1964 (74), such reactions (for which the designations olefin *disproportionation* and *dismutation* have also been used) involve the redistribution of the alkylidene moieties of olefins as in the example of Eq. 36 (40–45). The overall reaction corresponds to a $[2_s+2_s]$ process and is, accordingly, "thermally forbidden" according to the criteria of orbital symmetry conservation.

$$\begin{matrix} R_1CH{=}CHR_2 & & R_1CH{=}CHR_3 & & R_1CH{=}CHR_4 \\ + & \rightleftarrows & + & \rightleftarrows & + \\ R_3CH{=}CHR_4 & & R_2CH{=}CHR_4 & & R_2CH{=}CHR_3 \end{matrix} \tag{36}$$

This reaction, for which there are virtually no close earlier precedents, is induced by a variety of heterogeneous and homogeneous catalysts, notably involving group VI transition metal components. Typical heterogeneous catalysts include molybdenum or tungsten carbonyls or oxides on alumina supports. Among the effective homogeneous catalyst systems are those derived from the combinations $WCl_6 : C_2H_5OH : C_2H_5AlCl_2$ or $Mo(NO)_2(PPh_3)_2Cl_2 : C_2H_5AlCl_2$.

Typically, such disproportionation reactions yield, ultimately (although not necessarily initially) equilibrium mixtures of the accessible olefin constituents. Similar catalysts have also been shown to be effective for the disproportionation of dienes and acetylenes. The process has been applied commercially to convert propylene to ethylene and butenes. A novel application(75,76) involves the synthesis of large carbocyclic compounds through successive fusions of smaller cyclic olefins of the type shown in Eq. 37.

$$2\,\|\,(CH_2)_n \rightleftharpoons (H_2C)_n \overset{=}{\underset{=}{}} (CH_2)_n \tag{37}$$

The mechanistic features of this unprecedented class of reactions have only recently come to be appreciated. Several alternative interpretations have been proposed which include, just as in the cases of the other symmetry-restricted reactions discussed above, both concerted and stepwise mechanisms. Three such mechanistic proposals are depicted below, using an abbreviated notation according to which A, B, C and D each represents an alkylidene or carbene fragment, that is, $:CR_1R_2$, and M represents the catalytic site.

A. Concerted Four-Center Mechanism

$$\begin{array}{c} A \\ \| \\ B \end{array} + \begin{array}{c} C \\ \| \\ D \end{array} \xrightarrow{M} \left[\begin{array}{c} A\text{—}C \\ |\quad\ \ | \\ B\text{—}D \\ M \end{array}\right] \ or \ \left[\begin{array}{ccc} A & & C \\ & M & \\ B & & D \end{array}\right] \xrightarrow{-M} \begin{array}{c} A{=}C \\ + \\ B{=}D \end{array} \quad (38)$$

(51) **(52)**

The basis for the widespread mechanistic proposals of this type appears to be principally "topological." Thus, the identities of the original olefin combinations are lost in the symmetrical four-center intermediate resulting in scrambling of the component alkylidene fragments. Objections to such mechanisms include (*i*) the absence of reasonable precedents for species resembling the proposed intermediates **51** and **52**, (*ii*) the expectation that species such as **52** are energetically inaccessible, and (*iii*) evidence that free cyclobutanes are neither significant by-products nor intermediates in olefin metathesis.

B. Formation of Metallocycles through Oxidative Addition

$$\begin{array}{c} A \\ \| \\ B \end{array} + \begin{array}{c} C \\ \| \\ D \end{array} \xrightarrow{M} \left[\begin{array}{c} A\text{—}C \\ |\quad\ \ | \\ B\quad D \\ M \end{array}\right] \rightleftharpoons \left[\begin{array}{c} A\text{—}C \\ |\qquad M \\ B\text{—}D \end{array}\right] \xrightarrow{-M} \begin{array}{c} A{=}C \\ + \\ B{=}D \end{array}$$

(53a) **(53b)**

$$\left[\begin{array}{c} A \\ B \qquad C \\ M \\ \| \\ D \end{array}\right] \quad (39)$$

(54)

This type of mechanism derives plausibility from the many known examples of formation of five-membered metallocycles through oxidative addition, starting either with cyclobutanes or with olefins, as depicted by Eqs. 3 and 13. In addition, some direct support is provided by the results of the "model experiment" depicted in Eq. 40, in which a synthetic procedure expected to generate the metallocycle **55**, yielded a mixture of isotopically labeled ethylenes which seems to confirm not only that the proposed metallocycle decomposes into olefinic products but, also, that it undergoes the 1,2-metal shift (i.e., **53a** → **53b** necessary to scramble the alkylidene moieties and effect olefin metathesis (77). One possible mechanism of such a 1,2-shift is that of carbene elimination-insertion reflected in the sequence **53a** → **54** → **53b** of Eq. 39 (78).

$$WCl_6 + \text{(D,D,Li,Li-substituted butane)} \longrightarrow [\text{W metallacycle}] \longrightarrow \begin{cases} 88\%\ C_2H_3D \\ 6\%\ C_2H_4 \\ 6\%\ C_2H_2D_2 \end{cases} \quad (40)$$

C. One-Carbene Chain Mechanism

$$\text{Initiation: } M \xrightarrow{A{=}B} M{=}B, \text{ etc.} \quad (41)$$

Propagation:

$$\begin{matrix} C \\ \| \\ D \end{matrix} \xrightarrow{M=B} \left[\begin{matrix} C{=}D \\ \downarrow \\ M{=}B \end{matrix}\right] \longrightarrow \left[M \begin{matrix} C \\ \diamond \\ B \end{matrix} D \right] \longrightarrow \left[\begin{matrix} B{=}D \\ \downarrow \\ M{=}C \end{matrix}\right] \xrightarrow{-M=C} B{=}D \quad (42a)$$

$$\begin{matrix} A \\ \| \\ B \end{matrix} \xrightarrow{M=C} \left[\begin{matrix} B{=}A \\ \downarrow \\ M{=}C \end{matrix}\right] \longrightarrow \left[M \begin{matrix} B \\ \diamond \\ C \end{matrix} A \right] \longrightarrow \left[\begin{matrix} A{=}C \\ \downarrow \\ M{=}B \end{matrix}\right] \xrightarrow{-M=B} A{=}C \quad (42b)$$

This mechanism differs from the first two in one important respect, namely in predicting the *initial* products of metathesis of a given pair of olefins A=B and C=D, to be a statistical mixture of all the possible alkylidene combinations, whereas the other two mechanisms predict specific "pairwise" exchange. This distinction is recognizable in certain cases (e.g., in the metathesis of a mixture of cyclic and acyclic olefins) and provided the evidence on which the first proposal of mechanism **C** in 1970 was based (79,80). More recent confirmation of the predictions of mechanism **C** is provided by the results of studies on dienes such as 1,7-octadiene which undergo intramolecular metathesis of the type depicted in Eq. 43. The pairwise exchange mechanisms **A** and **B** predict that

$$\text{(1,7-octadiene)} \xrightarrow{M} \text{(cyclohexene)} + \begin{matrix} CH_2 \\ \| \\ CH_2 \end{matrix} \quad (43)$$

initial ethylene products of metathesis of a 1:1 mixture of 1,7-octadiene and 1,1,8,8-d_4-1,7-octadiene should be a 1:1 mixture of C_2H_4 and C_2D_4. On the other hand, the one-carbon chain mechanism **C** predicts an initial 1:2:1 mixture of C_2H_4:C_2H_3D:C_2D_4. Recent experimental evidence confirms the latter result (81,82).

The one-carbene exchange mechanism also receives support from "model" experiments with stable carbene complexes which demonstrate the occurrence of stoichiometric carbene exchange such as that depicted by Eq. 44 (83,84).

$$(CO)_5W{=}C(Ph)(Ph) + CH_2{=}C(Ph)(OMe) \rightarrow (CO)_5W{=}C(Ph)(OMe) + CH_2{=}C(Ph)(Ph) \quad (44)$$

At this stage the evidence that olefin metathesis occurs through the stepwise one-carbon chain mechanism depicted by Eqs. 41 and 42 rather than through a concerted four-center mechanism (i.e., through Eq. 38) appears overwhelming. The mechanistic features that remain to be clarified concern the initiation step, that is, the mechanism of initial formation of the metal carbene complex. One possible such mechanism is direct C=C bond cleavage:

$$2M + C{=}C \rightarrow 2M{=}C \quad (45)$$

or

$$M + C{=}C \rightarrow M({=}C)({=}C) \quad (46)$$

This type of reaction has been observed (85) for activated, electron-rich, olefins of the type $(R_2N)_2C{=}C(NR_2)_2$, but seems less likely for simple alkenes. A more likely path of metal-carbene generation in the case of catalysts activated by metal alkyls (e.g., WCl_6–$Zn(CH_3)_2$) is that depicted by Eq. 47, which also accounts for the observed formation of CH_4 (86). There are several precedents for such a proposed step involving metal–carbene complex formation by α-hydrogen migration from an alkyl ligand (87,88).

$$WCl_6 \xrightarrow{Zn(CH_3)_2} \left[W(CH_3)(CH_3) \right] \longrightarrow \left[W({=}CH_2)(H)(CH_3) \right] \longrightarrow W{=}CH_2 + CH_4 \quad (47)$$

(**55**) (**56**)

For catalysts that do not involve metal alkyl components [e.g., $M(CO)_6$ on alumina] the initial metal–carbene complexes are presumably derived from the olefins themselves. One possible route for accomplishing this is that depicted by Eqs. 45 or 46. Other possible routes are provided by reaction 39, which directly encompasses the carbene complex **54**, as well as providing for the formation of metal carbene complexes from the intermediate metal alkyls **53a** and **53b** through α-hydrogen migration processes similar to Eq. 47. It is not unlikely that several such processes contribute to the initiation step of olefin metathesis.

In this connection, it seems appropriate to recognize that the driving forces for the α-migration reactions, such as those encompassed by Eqs. 39 and 47, through which the reactive metal–carbene complexes involved in olefin metathesis are probably formed, are derived from electron deficiency of the metal center. Thus, an α-migration step, such as **53a** → **54** or **55** → **56**, results in a two-electron increase in the metal valence shell. In line with this, catalytic activity for olefin metathesis is typically associated with the early transition metals and is inhibited by electron donors such as amines. A corollary of this is the expectation that metathesis catalysts will exhibit poor tolerance for functionalities and, indeed, this seems to be the case. Most reports of olefin metathesis involve hydrocarbon substrates and only a few examples of the metathesis of functionalized olefins have been documented (89).

VIII. CONCLUDING REMARKS

This account has dealt with the catalysis by metal complexes of a variety of organic reactions which have in common the feature of being "thermally forbidden" according to the selection rules of orbital symmetry conservation. The reactions discussed include valence isomerization, intramolecular rearrangements involving hydrogen migrations, as well as cycloadditions and other intermolecular four-center reactions.

In virtually every case the available evidence appears to favor stepwise mechanisms involving discrete organometallic intermediates, usually containing metal–carbon σ-bonds. Each of the component steps of such stepwise mechanisms (for example oxidative addition, migratory insertion, etc.) may be a concerted process and, thus, governed by orbital symmetry selection rules. However, in no case do the *overall* reactions appear to proceed by a concerted pathway involving multicenter interactions of the metal with the substrate, for example four-center interaction of the type depicted by the sequence **1** → **4a** → **3** of Eq. 2 or by Eqs. 18 or 37. The theoretical basis, upon which such concerted catalytic pathways were originally proposed, appears sound and such pathways cannot be ruled out as a general possibility. However, their occurrence remains to be conclusively demonstrated.

Although this discussion has focused on reactions involving the formation or breaking of carbon–carbon bonds, similar considerations govern the catalysis of other four-center reactions such as the hydrogenation of olefins. For such reactions also, concerted four-center mechanisms of the type depicted by Eq. 48a can be conceived. However, it has long been appreciated that catalytic hydrogenation reactions are not concerted processes but, instead, proceed through stepwise pathways involving σ-bonded metal hydrides and organometallic intermediates. This is exemplified by the generally accepted mechanism of rhodium(I)-catalyzed hydrogenation depicted in Eq. 48b (90).

$$\left.\begin{matrix} \mathrm{C{=}C} \\ + \\ \mathrm{H{-}H} \end{matrix}\right\} \xrightarrow{\mathrm{M}} \left[\begin{matrix} \mathrm{M} \\ \mathrm{C{-}C} \\ \mathrm{H{\cdots}H} \end{matrix}\right] \xrightarrow{-\mathrm{M}} \begin{matrix} \mathrm{C{-}C} \\ \mathrm{H\quad H} \end{matrix} \qquad (48a)$$

$$\left[\begin{matrix} \mathrm{C{=}C} \\ \downarrow \\ \mathrm{M} \\ \mathrm{H\quad H} \end{matrix}\right] \longrightarrow \left[\begin{matrix} \mathrm{C{-}C} \\ \mathrm{M\qquad H} \\ \mathrm{H} \end{matrix}\right] \xrightarrow{-\mathrm{M}} \begin{matrix} \mathrm{C{-}C} \\ \mathrm{H\quad H} \end{matrix} \qquad (48b)$$

The interpretation of the catalysis of symmetry-restricted reactions by metal complexes has been pervaded by controversies concerning the concerted *versus* stepwise character of the catalytic mechanisms. Many of these controversies have focused on substantive distinctions, for example the distinction between the alternative mechanism of olefin metathesis depicted by Eqs. 38, 39, and 41–42. In some cases, however, the distinctions appear to be of a semantic nature. It has been claimed, for example, that the facile rearrangement of the cyclobutene–iron carbonyl complex depicted by Eq. 49 proceeds through a *concerted disrotatory cyclobutene–butadiene transfromation* of the coordinated ligand, whereas the corresponding rearrangement of the *free* ligand is symmetry-forbidden (91).

$$\underset{\mathbf{(57)}}{(CO)_4Fe{-}[\text{tricyclic diene}]} \underset{CO}{\overset{-CO}{\rightleftharpoons}} \left[(CO)_3Fe{-}[\text{tricyclic diene}]\right] \longrightarrow (CO)_3Fe{-}[\text{bicyclic triene}] \qquad (49)$$

Whereas this interpretation of reaction 49 as a concerted rearrangement is undoubtedly correct, it would be misleading to infer, by analogy, that processes such as the rhodium(I)-catalyzed rearrangement of cubane to *syn*-tricyclooctadiene (**5** → **7**) also proceed through "concerted" mechanisms rather than through the stepwise mechanism depicted by Eq. 3. The successive steps of the latter mechanism correspond to the formation of the organometallic intermediate (**5** → **6**) and to its subsequent rearrangement (**6** → **7**). The preformed organometallic *reactant* (57) of reaction 49 corresponds formally to the *intermediate* **6** of reaction 3. Reaction 49 is thus the formal analog, not of the overall reaction 3, but only of the second step of the latter reaction (i.e., **6** → **7**). The designation "concerted" is considered appropriate for this step (and, correspondingly, for reaction **49** but not for the overall reaction 3) which must clearly be described as "stepwise."

Unfortunately, with exception of two later general treatments (92,93), most of the attempts that have been made at theoretical treatment of the catalysis of symmetry-restricted reactions by metal complexes (17,46,47) relate to concerted pathways, involving multicenter metal-substrate interactions, and are therefore not directly applicable to most of the processes with which this account has been concerned.

REFERENCES

1. R. B. Woodward and R. Hoffmann, *Angew. Chem. Intern. Ed.*, **8,** 781 (1969), and references therein.
2. M. J. S. Dewar, *Angew. Chem. Intern. Ed.*, **10,** 761 (1971) and references therein.
3. H. Hogeveen and H. C. Volger, *J. Am. Chem. Soc.*, **89,** 2486 (1967).
4. L. Cassar and J. Halpern, *Chem. Commun.*, **1970,** 1082.
5. H. Hogeveen and H. C. Volger, *Chem. Commun.*, **1967,** 1133.
6. L. Cassar, P. E. Eaton, and J. Halpern, *J. Am. Chem. Soc.*, **92,** 3515 (1970).
7. L. Cassar, P. E. Eaton, and J. Halpern, *J. Am. Chem. Soc.*, **92,** 6366 (1970).
8. W. G. Dauben, M. G. Buzzolini, C. H. Schallhorn, and D. L. Whalen, *Tetrahedron Lett.*, **1970,** 787.
9. W. G. Dauben, C. H. Schallhorn, and D. L. Whalen, *J. Am. Chem. Soc.*, **93,** 1446 (1971).
10. L. A. Paquette and J. C. Stowell, *J. Am. Chem. Soc.*, **92,** 2584 (1970); *ibid.*, **93,** 2459 (1971).
11. L. A. Paquette and R. S. Beckley, *J. Am. Chem. Soc.*, **97,** 1084 (1975).
12. L. A. Paquette, R. S. Beckley, and W. B. Farnham, *J. Am. Chem. Soc.*, **97,** 1089 (1975).
13. L. A. Paquette, J. S. Ward, R. A. Boggs, and W. B. Farnham, *J. Am. Chem. Soc.*, **97,** 1101 (1975).
14. L. A. Paquette, R. S. Beckley, and T. McCreadle, *Tetrahedron Lett.*, **1971,** 775.
15. H. H. Westberg and H. Ona, *Chem. Commun.*, **1971,** 248.
16. W. Merk and R. Pettit, *J. Am. Chem. Soc.*, **89,** 4788 (1967).
17. R. Pettit, H. Sugahara, J. Wristers, and W. Merk, *Disc. Faraday Soc.*, **19,** 291 (1969), and references therein.

18. L. A. Paquette, G. R. Allen, Jr., and R. P. Henzel, *J. Am. Chem. Soc.*, **92,** 7002 (1970).
19. P. Gassman and F. J. Williams, *J. Am. Chem. Soc.*, **92,** 7631 (1970).
20. P. Gassman and T. J. Atkins, *J. Am. Chem. Soc.*, **93,** 1042, 4591 (1971); *ibid.*, **94,** 7748 (1972).
21. M. Sakai, H. Yamaguchi, H. H. Westberg, and S. Masamune, *J. Am. Chem. Soc.*, **93,** 1043 (1971).
22. M. Sakai and S. Masamune, *J. Am. Chem. Soc.*, **93,** 4610 (1971).
23. M. Sakai, H. H. Westberg, H. Yamaguchi, and S. Masamune, *J. Am. Chem. Soc.*, **93,** 4611 (1971).
24. L. A. Paquette and S. E. Wilson, *J. Am. Chem. Soc.*, **93,** 5934 (1971).
25. L. A. Paquette, *Acc. Chem. Res.*, **4,** 280 (1971).
26. L. A. Paquette, R. P. Henzel, and S. E. Wilson, *J. Am. Chem. Soc.*, **93,** 2335 (1971).
27. P. Gassman and T. Nikai, *J. Am. Chem. Soc.*, **93,** 5897 (1971); *ibid.*, **94,** 2877, 5497 (1972).
28. G. Dauben and A. J. Kielbania, Jr., *J. Am. Chem. Soc.*, **94,** 3669 (1972).
29. P. Gassman and F. J. Williams, *Chem. Commun.*, **1972,** 80; *J. Am. Chem. Soc.*, **94,** 7733 (1972).
30. P. Gassman, G. R. Meyer, and F. J. Williams, *J. Am. Chem. Soc.*, **94,** 7741 (1972); *Chem. Commun.*, **1971,** 842.
31. P. Gassman, T. J. Atkins, and J. T. Lumb, *J. Am. Chem. Soc.*, **94,** 7757 (1972).
32. L. A. Paquette, S. E. Wilson, R. P. Henzel, and G. R. Allen, Jr., *J. Am. Chem. Soc.*, **94,** 7761 (1972).
33. L. A. Paquette, S. E. Wilson, and R. P. Henzel, *J. Am. Chem. Soc.*, **93,** 1288 (1971); **94,** 7771, 7780 (1972).
34. G. Zon and L. A. Paquette, *J. Am. Chem. Soc.*, **95,** 4456 (1973); *ibid.* **96,** 215 (1974).
35. L. A. Paquette and G. Zon, *J. Am. Chem. Soc.*, **96,** 203, 224 (1974).
36. P. Heimbach, P. W. Jolly, and G. Wilke, *Advan. Organometal. Chem.*, **8,** 29 (1970).
37. P. Hembach and H. Hey, *Angew. Chem. Intern. Ed.*, **9,** 528 (1970).
38. P. Heimbach, *Angew. Chem. Intern. Ed.*, **12,** 975 (1973).
39. P. Heimbach, in R. Ugo, Ed., *Aspects of Homogeneous Catalysis*, Vol. II, Reidel, 1974, p. 79.
40. G. C. Bailey, *Catalysis Rev.*, **3,** 37 (1969).
41. R. L. Banks, *Top. Curr. Chem.*, **25,** 39 (1972).
42. W. B. Hughes, *Organometal. Chem. Syn.*, **1,** 341 (1972).
43. N. Calderon, *Acc. Chem. Res.*, **5,** 127 (1972).
44. J. C. Mol and J. A. Moulign, *Advan. Catal.*, **24,** 131 (1974).
45*a*. R. J. Haines and G. J. Leigh, *Chem. Soc. Rev.*, **4,** 155 (1975).
45*b*. N. Calderon, E. A. Ofstead and W. A. Judy, *Angew. Chem. Intern. Ed.*, **15,** 401 (1976).
46. F. D. Mango and J. H. Schachtschneider, *J. Am. Chem. Soc.*, **89,** 2484 (1976); *ibid.*, **93,** 1123 (1971), and references therein.
47. F. D. Mango, *Top. Curr. Chem.*, **45,** 39 (1974), and references therein.
48. J. Halpern, *Acc. Chem. Res.*, **3,** 386 (1970), and references therein.
49. J. Halpern, *Proc. 14th Intern. Comf. Coord. Chem.*, Toronto, 1972, p. 698.
50. J. Blum, C. Zlotogorski, and A. Zoran, *Tetrahedron Lett.*, **1975,** 1117.
51. T. J. Katz and S. Cerefice, *J. Am. Chem. Soc.*, **91,** 2405, 6519 (1969).
52. J. Halpern, J. Blum, and M. Sohn, unpublished results.
53. T. J. Katz and N. Acton, *Tetrahedron Lett.*, **1967,** 2601.
54. A. R. Fraser, P. H. Bird, S. A. Bezman, J. R. Shapley, R. White, and J. A. Osborn, *J. Am. Chem. Soc.*, **95,** 597 (1973).

55. R. Noyori, in Y. Ishii and M. Tsutsui, Eds., *Organotransition-Metal Chemistry*, Plenum Press, 1975, p. 231.
56. H. Takaya, M. Yamakawa, and R. Noyori *Chem. Lett.*, **1973,** 781.
57. R. Noyori, I. Umeda, H. Kawauchi, and H. Takaya, *J. Am. Chem. Soc.*, **97,** 812 (1975).
58. R. Noyori, T. Suzuki, and H. Takaya, *J. Am. Chem. Soc.*, **93,** 5896 (1971).
59. J. E. Byrd, L. Cassar, P. E. Eaton, and J. Halpern, *Chem. Commun.*, **1971,** 40.
60. G. F. Koser, *Chem. Commun.*, **1971,** 388.
61. J. Wristers, L. Brener, and R. Pettit, *J. Am. Chem. Soc.*, **92,** 7499 (1970).
62. A. H. Webster and J. Halpern, *J. Phys. Chem.*, **61,** 1239 (1957).
63. W. G. Dauben and A. J. Kielbania, Jr., *J. Am. Chem. Soc.*, **93,** 7345 (1971).
64. (*a*) L. A. Paquette, R. A. Boggs, W. B. Farnham, and R. S. Beckley, *J. Am. Chem. Soc.*, **97,** 1112 (1975).
(*b*) L. A. Paquette, R. A. Boggs, and J. S. Ward, *J. Am. Chem. Soc.*, **97,** 1118 (1975).
65. R. Gleiter, E. Heilbronner, M. Hekman, and H.-D. Martin, *Chem. Ber.*, **106,** 28 (1973).
66. P. E. Eaton, L. Cassar, R. A. Hudson and D. R. Hwang, *J. Org. Chem.*, in press.
67. G. L. Closs and P. E. Pfeffer, *J. Am. Chem. Soc.*, **90,** 2452 (1968).
68. L. A. Paquette, S. E. Wilson, G. Zon, and J. A. Schwartz, *J. Am. Chem. Soc.*, **94,** 9222 (1972).
69. P. G. Gassman and T. Nakai, *J. Am. Chem. Soc.*, **94,** 5497 (1972).
70. P. G. Gassman and R. R. Reitz, *J. Am. Chem. Soc.*, **95,** 3057 (1973).
71. W. G. Dauben, A. J. Kielbania, Jr., and K. N. Raymond, *J. Am. Chem. Soc.*, **95,** 7166 (1973).
72. G. F. Koser, P. R. Pappas, and S. M. Yu, *Tetrahedron Lett.*, **1973,** 4943.
73. R. Pettit, J. S. McKennis, W. Slegeir, W. H. Starnes, Jr., T. Devon, R. Case, J. C. Wagnon, L. Brener, and J. Wristers, *Ann. N. Y. Acad. Sci.*, **239,** 22 (1974).
74. R. L. Banks and G. C. Bailey, *Ind. Eng. Chem., Prod. Res. Develop.*, **3,** 170 (1964).
75. E. Wasserman, D. A. Ben-Efraim, and R. Wolovsky, *J. Am. Chem. Soc.*, **90,** 3286 (1968).
76. R. Wolovsky, *J. Am. Chem. Soc.*, **92,** 2132 (1970).
77. R. H. Grubbs and T. K. Brunck, *J. Am. Chem. Soc.*, **94,** 2538 (1972).
78. C. G. Biefeld, H. A. Eick, and R. H. Grubbs, *Inorg. Chem.*, **12,** 2166 (1973).
79. J.-L. Hérisson and Y. Chauvin, *Makromol. Chem.*, **141,** 161 (1970).
80. J.-P. Soufflet, D. Commerenc, and Y. Chauvin, *Compt. Rend. Acad. Sci. Paris*, 276C, 169 (1973).
81. T. J. Katz and J. McGinnis, *J. Am. Chem. Soc.*, **97,** 1592 (1975).
82. R. H. Grubbs, P. L. Burk, and O. D. Carr, *J. Am. Chem. Soc.*, **97,** 3265 (1975).
83. C. P. Casey and T. J. Burkhardt, *J. Am. Chem. Soc.*, **96,** 7808 (1974).
84. D. J. Cardin, M. J. Doyle, and M. F. Lappert, *Chem. Commun.*, **1972,** 927.
85. B. Cetinkaya, P. Dixneuf, and M. F. Lappert, *Chem. Commun.*, **1973,** 206.
86. E. L. Muetterties, *Inorg. Chem.*, **14,** 951 (1975).
87. R. R. Schrock, *J. Am. Chem. Soc.*, **96,** 6796 (1974).
88. N. J. Cooper and M. L. H. Green, *Chem. Commun.*, **1974,** 208, 761.
89. E. Verkuijlen and C. Boelhouwer, *Chem. Commun.*, **1974,** 793.
90. B. R. James, *Homogeneous Hydrogenation*, Wiley, 1973, and references therein.
91. W. Slegeir, R. Case, J. S. McKennis, and R. Pettit, *J. Am. Chem. Soc.*, **96,** 287 (1974).
92. R. Fukui and S. Inagaki, *J. Am. Chem. Soc.*, **97,** 4445 (1975).
93. R. Noyori, M. Yakamawa and H. Takaya, *J. Amer. Chem. Soc*, **98,** 1471 (1976).

Index